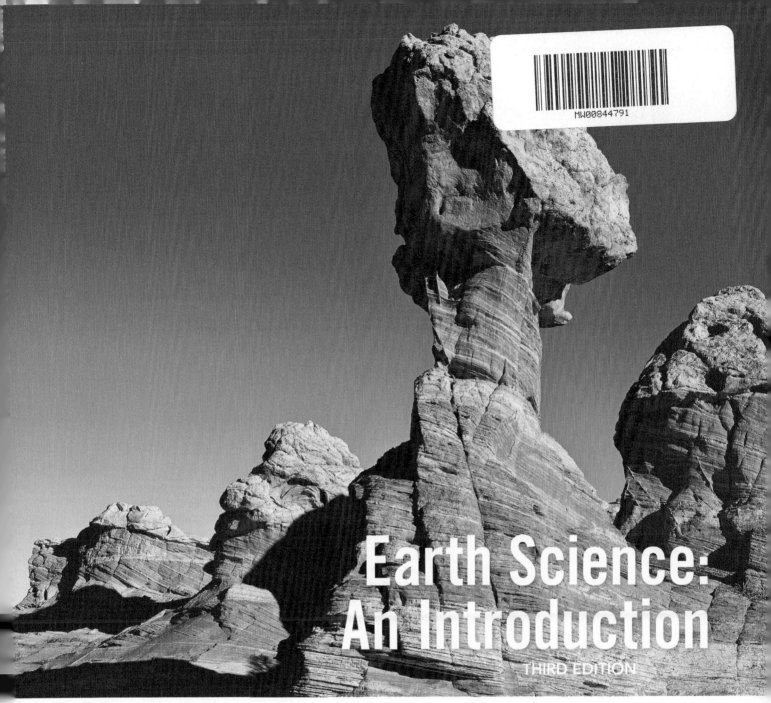

Earth Science: An Introduction

THIRD EDITION

Marc S. Hendrix
The University of Montana
Missoula, Montana

Graham R. Thompson
Professor Emeritus
The University of Montana
Missoula, Montana

Jonathan Turk

Australia • Brazil • Canada • Mexico • Singapore • United Kingdom • United States

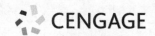

Earth Science: An Introduction,
Third Edition
Hendrix/Thompson/Turk

Product Director: Thais Alencar

Product Team Manager: Kelsey Churchman

Product Manager: Lauren Bakker

Production Manager: Julia White

Learning Designer: Linda Man

Content Manager: Nicole Evans

Product Assistant: Vanessa Desiato

Marketing Manager: Andrew Stock

Digital Delivery Lead: Susan Ordway

Art Director: Helen Bruno

Manufacturing Planner: Becky Cross

Intellectual Property Analyst:
    Christine Myaskovsky

Project Manager: Erika Mugavin

Production Service: Lumina Datamatics, Inc.

Cover image(s): James Hager/Collection Mix:
    Subjects/Getty Images

For product information and technology assistance, contact us at
Cengage Customer & Sales Support, 1-800-354-9706 or
support.cengage.com.

For permission to use material from this text or product,
submit all requests online at **www.cengage.com/permissions.**

Library of Congress Control Number: 2018968241

Soft-cover Edition ISBN: 978-0-357-11656-2

Loose-leaf Edition ISBN: 978-0-357-12006-4

Cengage
200 Pier 4 Boulevard
Boston, MA 02210
USA

Cengage is a leading provider of customized learning solutions with employees residing in nearly 40 different countries and sales in more than 125 countries around the world. Find your local representative at **www.cengage.com.**

To learn more about Cengage platforms and services, visit **www. cengage.com.** To register or access your online learning solution or purchase materials for your course, visit **www.cengagebrain.com.**

Printed at CLDPC, USA, 12-21

BRIEF CONTENTS

CONTENTS

4

GEOLOGIC TIME 60

5

GEOLOGIC RESOURCES 87

6

THE ACTIVE EARTH 119

EARTHQUAKES AND THE EARTH'S STRUCTURE   145

VOLCANOES AND PLUTONS   170

MOUNTAINS   199

WEATHERING, SOIL, AND EROSION   223

FRESHWATER   252

WATER RESOURCES   281

GLACIERS AND GLACIATIONS   309

THE ATMOSPHERE 410

ENERGY BALANCE IN THE ATMOSPHERE 430

MOISTURE, CLOUDS, AND WEATHER 449

20 CLIMATE 477

21 CLIMATE CHANGE 492

22 MOTIONS IN THE HEAVENS 515

23 PLANETS AND THEIR MOONS 530

24

STARS, SPACE, AND GALAXIES 557

APPENDICES

PREFACE

From our very beginnings, humans have depended on the Earth for survival. The food we eat, air we breathe, and water we drink all exist here and now, because Earth has undergone roughly 4.5 billion years' worth of evolution. In other words, Earth and the current environment is a product of continuous change that resulted—at this point in geologic time—in a planet capable of supporting humans.

If the entire history of Earth were compressed into a single 24-hour period, all of human civilization would have occurred within the very last second, as would have the appearance of Earth's earliest preserved human remains (Homo sapiens). From the standpoint of Earth history, humans are very recent newcomers. The evolution of humans was made possible because Earth itself had evolved to a point in which its temperature, atmospheric composition, surface water chemistry, and food sources were compatible with human survival needs. The continued survival of humans on Earth will depend on whether future changes to the solid Earth, its water, atmosphere, and biology remain within the range of human tolerance.

Today, not only is the Earth and its water, air, and ecology changing rapidly but the changes that are taking place are moving our planet in a direction that is less likely to favor the survival of humans as a species. Changes to our atmosphere, oceans, soils, and ecology are causing the most severe rate of species extinction in 4.5 billion years of Earth's history, and we as humans are living in the midst of this global species extinction event. Either we as humans will adapt to these changing conditions by implementing sustainable uses of Earth's resources or Earth no longer will be able to support human civilization as it exists currently.

Today, more rock and soil on Earth's continents is moved annually by humans than is moved by natural erosion. This fact alone has enormous downstream consequences for Earth's future, because the landscapes and underlying subsurface geology originated over geologic time and the rate of their human-caused change today far surpasses the time needed for them to redevelop. Moreover, as we will explore in this book, planet Earth consists of many interdependent, constantly changing natural systems. The dynamics of each system directly or indirectly affects the others, and perturbation of one system often has unforeseen consequences for another. For example, the shift to biofuels and their increased consumption, particularly in the United States, has supported the widespread development of palm oil plantations in southeastern Asia, a region naturally covered by tropical rainforests. Replacing the rain forests with palm oil plantations requires leveling and plowing of the region. Doing so causes a huge and rapid release of carbon dioxide as the organic-rich rain forest floor is churned up and quickly decays. Ironically, the increased use of palm oil as a fuel has accelerated the release of carbon dioxide into the atmosphere and not decreased it, as intended.

This book is intended to provide a foundational understanding of Earth and the physical, chemical, and biological processes that have shaped it and that continue to shape it today. The entire 4.5-billion-year history of Earth and the big changes in the solid Earth, oceans, atmosphere, and biota over geologic time are examined. Like all modern texts on Earth Science, this book addresses processes that take place over geologic time, including plate tectonics, planetary differentiation, and the evolution of species. In addition, however, this text also examines the current and ongoing changes in Earth's systems taking place over historic time frames, including ocean acidification, global climate change, and the ongoing mass extinction of species. Thus, this book is intended to provide a reference frame for the rapid and unpredictable environmental changes taking place today.

This book is the third edition of *Earth* (2011 and 2014). The first edition was written by Graham R. Thompson and Jon Turk. Both the second edition and this book were substantially revised by Marc S. Hendrix.

This book presents the planet Earth as an integrated system involving the solid Earth, water, the atmosphere, and living organisms and their co-evolution over geologic time at a level suitable for science or nonscience undergraduate students or advanced high-school students. Along with descriptions of Earth's internal and surficial geologic processes, this book examines Earth's oceans, atmosphere, and life forms and the evolution of each through geologic time. The first chapter introduces the four Earth Systems, and Chapters 2 and 3 focus on the materials of solid Earth—rocks and minerals. Geologic time, the evolution of life, and the processes of fossilization are covered in Chapters 4 and 5. Following are four chapters describing geological processes associated with Earth's interior. These chapters expand on plate tectonics, earthquakes, magmatic processes, and mountain building. Chapters 10, 13, and 14 describe surficial processes and how these have shaped Earth's natural landscapes.

Along with the focus on geologic time and Earth's internal and surficial processes, *Earth Science: An Introduction* focuses on Earth's fresh water (Chapter 11) and oceans

(Chapters 15 and 16). Following are three chapters dedicated to understanding Earth's atmosphere (Chapters 17 and 19). Chapters 20 and 21 address Earth's climate and climate change. The last three (Chapters 22–24) examine the moon, planet, and solar system.

Each major topic in this book is viewed through the lens of human impact, both in terms of how Earth's evolution has benefited human society and in terms of how human activity is altering Earth and its solid, liquid, gaseous, and living components. To these ends, one entire chapter is dedicated to Earth's mineral and energy resources (Chapter 5), and another is dedicated to Earth's water resources (Chapter 12). Both chapters are presented against the complicated global sociopolitical backdrop surrounding the extraction, transport, and use of mineral, energy, and water resources.

ACKNOWLEDGMENTS

This book results from over a decade of scientific work and writing by the three contributing authors, and it is not possible to recognize all of the people, both scientists and nonscientists, who have helped to make its publication possible. Included among those whose scientific interactions, insights, and discussions helped to shape the contents of this book (knowingly or not) are Steve Graham, Don Winston, James Sears, James Staub, Bill Woessner, Steve Sheriff, George Stanley, Andrew Wilcox, and Rebecca Bendick. Special thanks must go to product manager Lauren Bakker at Cengage Learning, who initiated this third edition, and to our learning designer Lauren Oliveira and content manager Nicole Evans, who not only kept us on task but also superbly edited and managed the content for this edition as well as provided a fresh perspective on this edition. Thanks to senior designer Helen Bruno for the fresh design. I would also like to recognize marketing manager Andrew Stock and market development manager Roxanne Wang.

I am deeply grateful for the support of my parents, Carol A and Sherman S. Hendrix, and my wife Brigette, without whom the writing of this book would not have been possible. I dedicate this book to our sons Gabriel and Michael Hendrix, whose natural curiosity and frequent requests for explanations as to the book's detailed contents both opened my eyes and kept me on my toes.

Marc S. Hendrix
Missoula, Montana

Central California coastline showing elements of the geosphere (sea cliffs), atmosphere (sky), hydrosphere (ocean), and biosphere (vegetation).

1

EARTH SYSTEMS

INTRODUCTION

Earth is sometimes called the water planet or the blue planet because azure seas cover more than two-thirds of its surface. Earth is the only planet or moon in the Solar System in which water falls from clouds as rain, runs across the land surface, and collects in extensive oceans. It is also the only body we know of that supports life.

THE EARTH'S FOUR SPHERES

Imagine walking along a sandy beach as a storm blows in from the sea. Wind whips the ocean into whitecaps, while large waves crash onto shore. Blowing sand stings your eyes as gulls overhead frantically beat their wings en route to finding shelter. In minutes, blowing spray has soaked your clothes. A hard rain begins as you hurry back to your vehicle. During this adventure, you have experienced the four major spheres of Earth. The beach sand underfoot is the surface of the **geosphere**, or the solid Earth. The rain and sea are parts of the **hydrosphere**, the watery part of our planet. The blowing wind belongs to the **atmosphere**. Finally, you, the gulls, the beach grasses, and all other forms of life in the sea, on land, and in the air are parts of the **biosphere**, the realm of organisms.

You can readily observe that the atmosphere is in motion, because clouds drift across the sky and wind blows against your face. In the biosphere, animals—and to a lesser extent plants—also move. Flowing streams, crashing waves, and falling rain are all familiar examples of motion in the hydrosphere. Although it is less apparent on a day-to-day

basis, the geosphere is also moving and dynamic. Vast masses of solid rock flow very slowly within the planet's interior. Continents move, while intervening ocean basins slowly open, then collapse. Mountains rise and then erode into sediment. Throughout this book, we will study many of these phenomena to learn which energy forces set matter in motion and how these motions affect the planet on which we live. ●**FIGURE 1.1** shows schematically all the possible interactions among the spheres.

●**FIGURE 1.2** shows that the geosphere is by far the largest of the four spheres. The Earth's radius is about 6,400 kilometers, roughly the same distance as Miami to Anchorage. Despite this great size, nearly all of our direct contact with Earth occurs at or very near its surface. The deepest well penetrates little more than 12 kilometers, less than two-tenths of 1 percent of the distance to Earth's center. The oceans make up most of the hydrosphere, and although it can extend as deep as 11 kilometers, the ocean floor averages only about 4 kilometers in depth. Most of Earth's atmosphere lies within 30 kilometers of the surface, and the biosphere is a thin shell about 15 kilometers thick.

The Geosphere

Our Solar System coalesced from a frigid cloud of dust and gas rotating slowly in space. The Sun formed as gravity pulled material toward the swirling center. At the same time, rotational forces spun material in the outer cloud into a thin disk. Eventually, small grains of matter within the disk stuck together to form fist-sized masses. These planetary "seeds" then accreted to form rocky clumps, which grew to form larger bodies, called *planetesimals*, 100 to 1,000 kilometers in diameter. Finally, the planetesimals consolidated to form the planets. This process was completed about 4.6 billion years ago.

As the Earth coalesced, gravity caused the rocky chunks and planetesimals to accelerate so that they slammed together at high speeds. Particles heat up when they collide, so the early Earth warmed as it formed. Later, asteroids, comets, and more planetesimals crashed into the surface, generating additional heat. At the same time, radioactive decay heated the Earth's interior. These three processes caused the early Earth to become so hot that much of the planet melted as it formed.

Within the molten Earth, the denser materials sunk toward the center, while the less dense materials floated toward the top, creating a layered structure. Today, the geosphere consists of three major layers: a dense metallic **core**, a less dense rocky **mantle**, and an even less dense surface **crust** (Figure 1.2).

The temperature of modern Earth increases with depth. At its center, Earth is 6,000°C—as hot as the Sun's surface. The core is composed mainly of iron and nickel. The outer core is molten metal. However, the inner core, although hotter yet, is solid because the great pressure compresses the metal to a solid state.

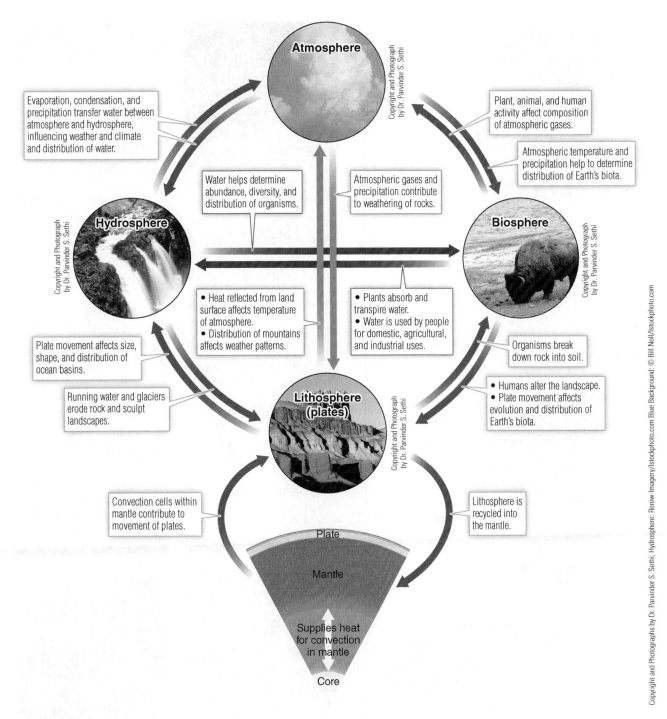

Atmosphere

Evaporation, condensation, and precipitation transfer water between atmosphere and hydrosphere, influencing weather and climate and distribution of water.

Plant, animal, and human activity affect composition of atmospheric gases.

Atmospheric temperature and precipitation help to determine distribution of Earth's biota.

Water helps determine abundance, diversity, and distribution of organisms.

Atmospheric gases and precipitation contribute to weathering of rocks.

Hydrosphere

Biosphere

• Heat reflected from land surface affects temperature of atmosphere.
• Distribution of mountains affects weather patterns.

• Plants absorb and transpire water.
• Water is used by people for domestic, agricultural, and industrial uses.

Plate movement affects size, shape, and distribution of ocean basins.

Organisms break down rock into soil.

• Humans alter the landscape.
• Plate movement affects evolution and distribution of Earth's biota.

Running water and glaciers erode rock and sculpt landscapes.

Lithosphere (plates)

Convection cells within mantle contribute to movement of plates.

Lithosphere is recycled into the mantle.

Plate

Mantle

Supplies heat for convection in mantle

Core

● **FIGURE 1.1** All of Earth's cycles and spheres are interconnected.

geosphere The solid Earth, consisting of the entire planet from the center of the core to the outer crust.

hydrosphere All of Earth's water, which circulates among oceans, continents, glaciers, and atmosphere.

atmosphere The gaseous layer above the Earth's surface, mostly nitrogen and oxygen, with smaller amounts of argon, carbon dioxide, and other gases. The atmosphere is held to Earth by gravity and thins rapidly with altitude.

biosphere The zone of Earth comprising all forms of life in the sea, on land, and in the air.

core The dense, metallic, innermost region of Earth's geosphere, consisting mainly of iron and nickel. The outer core is molten, but the inner core is solid.

mantle The rocky, mostly solid layer of Earth's geosphere lying beneath the crust and above the core. The mantle extends from the base of the crust to a depth of about 2,900 kilometers.

crust The outermost layer of Earth's geosphere, ranging from 4 to 75 kilometers thick and composed of relative low-density silicate rocks.

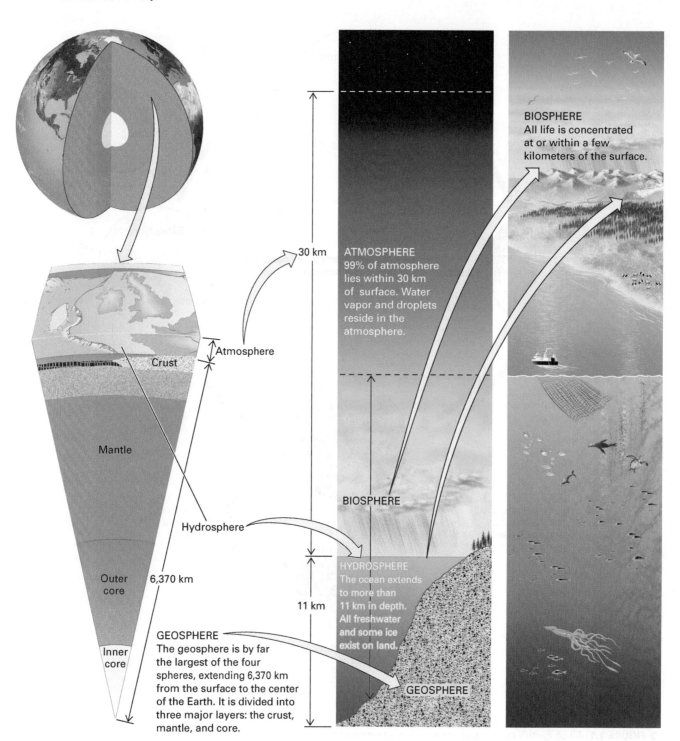

● **FIGURE 1.2** The geosphere is the largest component of Earth. It is surrounded by the hydrosphere, the biosphere, and the atmosphere.

The mantle surrounds the core and lies beneath the crust. The physical characteristics of the mantle vary with depth. From its upper surface to a depth of about 100 kilometers, the outermost mantle is relatively cool, strong, and hard. However, below a depth of 100 kilometers, rock making up the mantle is so hot that it is weak, soft, *plastic* (a solid that will deform permanently), and flows slowly—like cold honey. Even deeper in the mantle, pressure overwhelms temperature, and the rock becomes strong again.

The crust is the outermost layer of rock extending from the ground surface or bottom of the ocean to the top of the mantle. The crust ranges from as little as 4 kilometers thick beneath the oceans to as much as 75 kilometers thick beneath the continents. Even a casual observer sees that the crust includes many different rock types: some are soft, others hard, and they come in many colors, as you can see in ●**FIGURE 1.3**.

The relatively cool, hard, and strong rock of the uppermost mantle is similar to that of the crust. Together

(A)

Copyright and Photograph by Dr. Parvinder S. Sethi

Courtesy of Graham R. Thompson/Jonathan Turk

(B)

● **FIGURE 1.3** Earth's crust is made up of different kinds of rock (A) A grand view of the sandstones, siltstones, limestones, and mudstones of the Grand Canyon, as seen from the South Rim. Horizontal layers of sediment were deposited, then buried and hardened to form sedimentary rock. Hundreds of millions of years later, the layers of sedimentary rock were uplifted while the Colorado River eroded downward through them, exposing the view we see today. Note people for scale. (B) The granite of Baffin Island in the Canadian Arctic is gray, hard, and strong.

these layers make up the **lithosphere**, which averages about 100 kilometers thick.

According to the theory of plate tectonics, developed in the 1960s, the lithosphere is divided into seven major and eight smaller segments called **tectonic plates**. These tectonic plates float on the relatively hot, weak, plastic mantle rock beneath and move horizontally with respect to each other (●**FIGURE 1.4**). For example, North and South America are currently moving west relative to Eurasia and Africa about as fast as your fingernails grow. These continental movements are causing the Atlantic Ocean to grow larger and the Pacific Ocean to shrink. In a few hundred million years—almost incomprehensibly long on a human time scale but brief when compared with planetary history—Asia and North America may collide, completely collapsing the Pacific Ocean and crumpling the leading edges of the continents together into a giant mountain range. In later chapters, we will learn how the theory of plate tectonics explains earthquakes, volcanic eruptions, and the formation of mountain

ranges, as well as many other processes and events that have created our modern Earth and its environment.

Scientific awareness, much less scientific understanding, of plate tectonics did not occur until after World War II and it was not widely accepted among scientists until the late 1960s. Since then, advances in modern technology have accelerated, and so have the tools available for scientists to probe Earth's interior and study its dynamics. Similarly, accelerating computational capacity and new tools are constantly being developed and deployed to explore Earth's oceans and atmosphere. In aggregate, these new tools have generated immense volume of new data available for scientific exploration of Earth's systems. Electronic accessibility to peer-reviewed scientific publications describing the results and analysis of that scientific exploration broadens the global community of scientists and educated nonscientists alike. All in all, it is a marvelous time to study Earth Sciences. Earth is dynamic, and we are just beginning to appreciate how much so that is.

The Hydrosphere

The hydrosphere includes all of Earth's water, which circulates among oceans, continents, glaciers, and the atmosphere. ●**FIGURE 1.5** shows the proportion of water in each of these areas. Oceans cover 71 percent of Earth and contain 97.5 percent of its water. Ocean currents transport heat across vast distances, altering global climate.

---

lithosphere The cool, rigid, outer part of Earth, which includes the crust and the uppermost mantle, is about 100 kilometers thick and makes up Earth's tectonic plates.

tectonic plates The segments of Earth's outermost, cool, rigid shell, comprising the lithosphere. Tectonic plates float on the weak, plastic rock of the asthenosphere beneath.

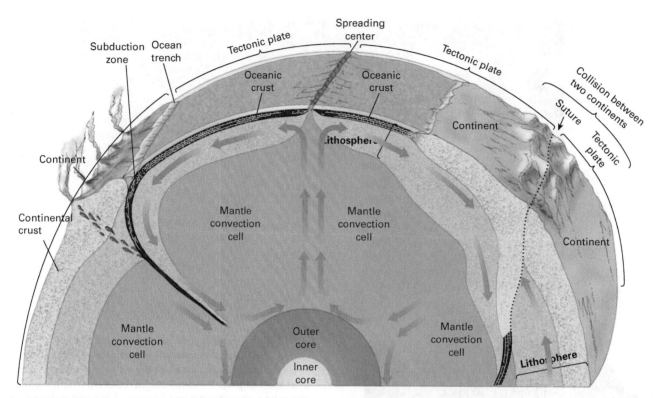

● **FIGURE 1.4** The lithosphere is composed of the crust and the uppermost mantle. It is a 100-kilometer-thick layer of strong rock that floats on the underlying plastic mantle. The lithosphere is broken into seven major segments, called tectonic plates, that glide horizontally over the plastic mantle at rates of a few centimeters per year. In the drawing, the thickness of the mantle and the lithosphere are exaggerated to show detail.

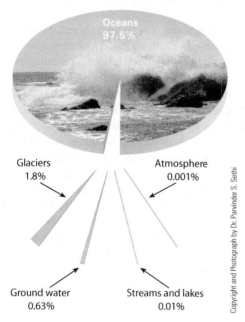

Copyright and Photograph by Dr. Parvinder S. Sethi

● **FIGURE 1.5** The oceans contain most of Earth's surface water. Most freshwater is frozen into glaciers. Most available freshwater is stored underground as groundwater.

About 1.8 percent of Earth's water is frozen in glaciers. Although glaciers cover about 10 percent of Earth's land surface today, they covered much greater portions of the globe as recently as 18,000 years ago. During this glacial period, nearly all of Canada, the British Isles, and Scandinavia were covered with ice that was locally up to 4 kilometers thick. In North America, the ice was so massive that it caused the underlying lithosphere to sag beneath its weight. Hudson Bay in Canada exists because the glacial ice melted faster than the lithosphere was able to bounce back. Today, parts of the Hudsons Bay region are undergoing over 10 cm of uplift per year as the lithosphere continues to recover from the weight of the ice. As shown on ●**FIGURE 1.6,** upward velocity rates of the ground surface decrease south of Hudson Bay to zero in the Great Lakes region. South of the Great Lakes, the ground surface is slowly subsiding downward. The boundary between upward-rebounding and slowly subsiding regions (green line in Figure 1.6) corresponds very roughly to the maximum advance of continental glaciers during the Last Glacial Maximum about 18,000 years ago.

Only about 0.64 percent of Earth's total water exists on the continents as a liquid. Although this is a small proportion, freshwater is essential to life on Earth. Lakes, rivers, and clear, sparkling streams are the most visible reservoirs of continental water, but they constitute only 0.01 percent of Earth's water. In contrast, **groundwater**—which occurs in pores, spaces within soil and rock of the upper few kilometers of the geosphere—is much more voluminous and accounts for 0.63 percent of Earth's water. Only a minuscule amount of water, 0.001 percent, exists in the atmosphere, but

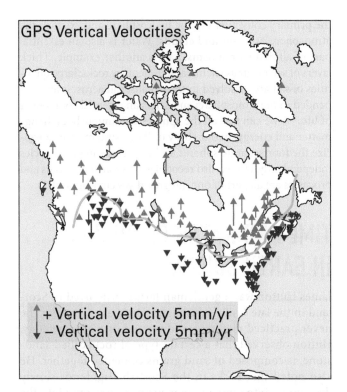

GPS Vertical Velocities

↑ + Vertical velocity 5mm/yr
↓ – Vertical velocity 5mm/yr

● **FIGURE 1.6** Map showing vertical velocities of the ground surface in North America. Each blue arrow shows downward movement of the ground surface, with the length of the arrow proportional to the velocity measurement as shown in the key. Each red arrow shows upward movement of the ground surface, with the length of each arrow similarly proportional to the velocity magnitude. The green line demarcates the boundary between upward and downward moving regions and is approximately coincident with the southernmost advance of continental ice sheets during the Last Glacial Maximum, approximately 18,000 years ago.

because it is so mobile, this atmospheric water profoundly affects both the weather and the climate of our planet.

The Atmosphere

The atmosphere is a mixture of gases: mostly nitrogen and oxygen, with smaller amounts of argon, carbon dioxide, and other gases. It is held to Earth by gravity and thins rapidly with altitude. Ninety-nine percent is concentrated in the first 30 kilometers, but traces of atmospheric gas occur as far as 10,000 kilometers above Earth's surface.

The atmosphere supports life, because animals need oxygen and plants need both carbon dioxide and oxygen. In

---

groundwater Subsurface water contained in the soil and bedrock of the upper few kilometers of the geosphere, comprising about 0.63 percent of all water in the hydrosphere.

system Any combination of interacting components that form a complex whole.

ecosystem A complex community of individual organisms interacting with each other and with their physical environment and functioning as an ecological unit in nature.

addition, the atmosphere supports life indirectly by regulating climate. Air serves as both a blanket and a filter, retaining heat at night and shielding us from direct solar radiation during the day. Wind transports heat from the equator toward the poles, cooling equatorial regions and warming temperate and polar zones.

The Biosphere

The biosphere is the zone that life inhabits. It includes the uppermost geosphere, the hydrosphere, and the lower parts of the atmosphere. Sea life concentrates near the surface, where sunlight is available. Plants also grow on Earth's surface, with roots penetrating a few meters into the soil. Animals live on the surface, fly a kilometer or two above it, or burrow a few meters underground. Large populations of bacteria live under glacial ice and in rock to depths of as great as 4 kilometers. Some organisms live on the ocean floor, including entire thriving ecosystems based off mineral-rich, super-heated water spewing from vents along the mid-ocean ridges. A few windblown microorganisms drift at heights of 10 kilometers or more. Despite these extremes, the biosphere is a very thin layer at Earth's surface.

Plants and animals are clearly affected by Earth's environment: Organisms breathe air, require water, and thrive in a relatively narrow temperature range. Terrestrial organisms ultimately depend on soil, which is part of the geosphere. But plants and animals also alter the atmosphere through respiration and contribute organic matter to the geosphere when they die.

EARTH SYSTEMS

A **system** is any assemblage or combination of interacting components that forms a complex whole. For example, the human body is a system composed of bones, nerves, muscles, and a variety of specialized organs. Each organ is discrete, yet all the organs interact to produce a living human. For example, blood nurtures the stomach, and the stomach helps provide energy to maintain the blood.

Systems are driven by the flow of matter and energy. Thus, a person ingests food, which contains both matter and chemical energy, and inhales oxygen. Waste products are released through urine, feces, sweat, and exhaled breath. Some energy is used for respiration and motion, and the remainder is released as heat or stored as fat.

A single system may be composed of many smaller ones. For instance, the human body contains hundreds of millions of bacteria, each of which is its own system. Many of these bacteria are essential to the functioning of human metabolic processes such as digestion.

In addition, humans are part of their local **ecosystem**, which is defined as a complex community of organisms and their environment functioning as an ecological unit in nature. Therefore, to understand the human body system, we must study smaller systems (e.g., bacteria) that exist within the body, while also exploring how humans interact with their larger ecosystems.

But we're not finished yet. Individual ecosystems interact with climate systems, ocean currents, and other Earth systems. Thus:

- The size of systems varies dramatically.
- Large systems contain numerous smaller systems.
- Systems interact with one another in complex ways.

As we have learned, Earth is composed of four major systems: geosphere, hydrosphere, atmosphere, and biosphere. Each of these large systems is subdivided into a great many interacting smaller ones. For example, a single volcanic eruption is part of a system. Energy from deep within Earth melts rock, forming magma. Some of this magma escapes during the eruption, along with volcanic gases that react chemically with surface materials. But this volcanic eruption is driven by the distribution and movement of heat within Earth's interior, which is also a system. Volcanic ash and certain gases spewed skyward during the eruption can affect local weather and cool Earth's climate, thereby becoming part of these systems. Heat from the eruption can also rapidly melt glaciers growing near the summit of the volcano, affecting the local hydrologic system. In this book, we will study systems of all sizes and illustrate many of the complex interactions among them.

Earth's surface systems—the atmosphere, hydrosphere, and biosphere—are ultimately powered by the Sun. Wind is powered by uneven solar heating of the atmosphere, ocean waves are driven by the wind, and ocean currents move in response to wind or differences in water temperature or density. Luckily for us, Earth receives a continuous influx of solar energy, and it will continue to receive this energy for another 5 billion years or so.

In contrast, Earth's interior is powered by the decay of radioactive elements and by residual heat from the primordial coalescence of the planet. We will discuss these sources of heat in later chapters.

Fundamental to our study of Earth systems are several energy and material cycles. A **cycle** is a sequential process or phenomenon that returns to its beginning and then repeats itself over and over. During the course of these cycles, matter and energy both are always conserved. They never simply disappear, although either may continuously change form. For example, water evaporates from the ocean into the atmosphere, falls to Earth as rain or snow, and eventually flows back to the oceans. Ultraviolet radiation that we cannot see enters Earth's atmosphere from the sun and heats up the surface which in turn emits heat. In this book, we will examine the rock cycle (Chapter 3), the hydrologic cycle or water cycle (Chapter 11), the carbon cycle (Chapter 17), and the nitrogen cycle (Chapter 17). We will also explore the critical role that energy and transformations of energy from one form to another play in Earth's oceans (Chapters 15 and 16) and atmospheric processes (Chapters 17–19).

Because matter exists in so many different chemical and physical forms, most materials occur in all four of Earth's major spheres—geosphere, hydrosphere, atmosphere, and biosphere. Water, for example, is chemically bound into clays and other minerals as a component of the geosphere. It is the primary constituent of the hydrosphere and exists in the atmosphere as vapor and clouds. Water is also an essential part of all living organisms. Salt is another example. Thick layers of salt occur as chemical sedimentary rocks; large quantities of salt are dissolved in the oceans; salt aerosols are suspended in the atmosphere; and salt is an essential component of life. As we can see, all the spheres continuously exchange matter and energy. In our study of Earth systems, we categorize the four separate spheres and numerous material cycles independently, but we also recognize that Earth materials and processes are all part of one integrated system (Figure 1.1).

TIME AND RATES OF CHANGE IN EARTH SCIENCE

James Hutton was a gentleman farmer who lived in Scotland in the late 1700s. Although trained as a physician, he never practiced medicine and instead turned to geology. Hutton observed that a certain type of rock, called sandstone, is composed of sand grains cemented together. He also noted that rocks in the Scottish Highlands slowly decompose into sand and that streams carry sand into the Lowlands. He inferred that sandstone is composed of sand grains that originated from the erosion of ancient cliffs and mountains.

Hutton tried to deduce how much time was required to form a thick bed of sandstone. He studied sand grains slowly breaking away from rock outcrops. He watched sand bouncing down streambeds. Finally, he traveled to beaches and river deltas where sand was accumulating. By estimating the time needed for thick layers of sand to accumulate on beaches, Hutton concluded that sandstone must be much older than human history.

Hutton had no way of measuring the magnitude of geologic time. However, modern geologists have learned that certain radioactive materials in rocks can be used as clocks to record the passage of time. Using these "clocks" and other clues embedded in Earth's crust, in the Moon, and in meteorites fallen from the Solar System, geologists estimate that Earth formed 4.6 billion years ago.

The primordial Earth was vastly different from our modern world. There was no crust as we know it today, there were no oceans, and the diffuse atmosphere was entirely different from the modern one. There were no living organisms.

No one knows exactly when or how the first living organisms evolved, but we know that life existed at least as early as 3.8 billion years ago, 800 million years after the planet formed. For the following 3.3 billion years, life evolved slowly, and although some multicellular organisms developed, most of the biosphere consisted of single-celled organisms. Organisms rapidly became more complex, abundant, and varied about 542 million years ago. The dinosaurs flourished between 225 million and 65 million years ago. *Homo sapiens* and our direct ancestors have been on Earth for 5 to 7 million years, or roughly only one-tenth of 1 percent of the planet's history.

In his book *Basin and Range*, John McPhee offers a metaphor for the magnitude of geologic time.[1] If the history of Earth were represented by the old English measure of a yard—the distance from the king's nose to the end of his outstretched hand—all of human history could be erased by a single stroke of a file on his middle fingernail. ●**FIGURE 1.7** summarizes Earth history in graphical form.

Geologists routinely talk about events that occurred millions or even billions of years ago. For example, about 1.7 billion years ago, the granite now forming Mount Rushmore cooled from a melt and crystallized. About a half billion years ago, the Appalachian Mountains began to form from tectonic crumpling of the eastern part of the North American plate. About 150 million years ago, a blanket of mud containing the remains of tiny plankton accumulated in deep water off the West Coast of North America. That sediment has since hardened to sedimentary rock that was subsequently added to the westernmost edge of North America by tectonic plate motions and today forms the bedrock on which the northern end of the Golden Gate Bridge rests. (●**FIGURE 1.8**).

There are two significant consequences of the vast span of geologic time:

1. Events that occur slowly become significant. If a continent moves a few centimeters a year, the movement makes no noticeable alteration of Earth systems over decades or centuries. But over hundreds of millions of years, the effects are significant.
2. Improbable events occur regularly. The chances are great that a large meteorite won't crash into Earth tomorrow, or next year, or during the next century. But during the past 500 million years, several catastrophic impacts have occurred, and they will probably occur sometime in the future.

James Hutton deduced that sandstone forms when rocks slowly decompose to sand, the sand is transported to lowland regions, and the grains cement together. This process occurs step by step—over many years. Hutton's conclusions led him to formulate the principles now known as **gradualism** and **uniformitarianism**. The principle of gradualism states that geologic change occurs over long periods of time, by a sequence of almost-imperceptible events.

---

cycle A sequential process or phenomenon that returns to its beginning and then repeats itself over and over.

gradualism A principle stating that geological change occurs as a consequence of slow or gradual accumulation of small events, such as the slow erosion of mountains by wind and rain. More recently, scientists studying biological evolution use the term to describe a theory of evolution that proposes that species change gradually in small increments.

uniformitarianism A principle stating that the geologic processes operating today also operated in the past.

---

1. John McPhee, *Basin and Range* (New York: Farrar, Straus & Giroux, 1982).

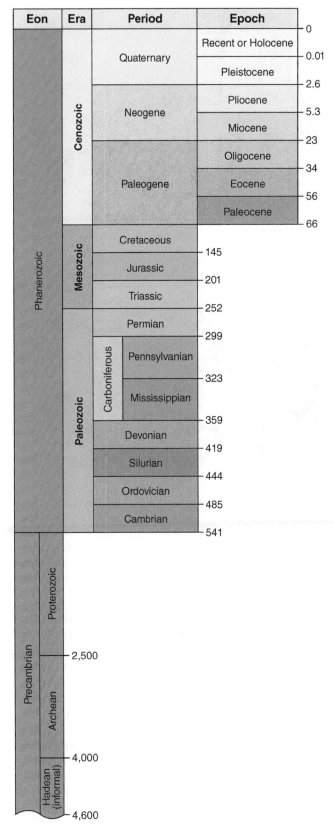

Eon	Era	Period	Epoch	
		Quaternary	Recent or Holocene	0
				0.01
			Pleistocene	
	Cenozoic			2.6
		Neogene	Pliocene	
				5.3
			Miocene	
				23
			Oligocene	
		Paleogene		34
			Eocene	
				56
			Paleocene	
				66
Phanerozoic		Cretaceous		
				145
	Mesozoic	Jurassic		
				201
		Triassic		
				252
		Permian		
				299
		Carboniferous — Pennsylvanian		
				323
	Paleozoic	Carboniferous — Mississippian		
				359
		Devonian		
				419
		Silurian		
				444
		Ordovician		
				485
		Cambrian		
				541
Precambrian	Proterozoic			
				2,500
	Archean			
				4,000
	Hadean (informal)			
				4,600

● **FIGURE 1.7** The geologic time scale. Note the great length of time before multicellular organisms became abundant about 541 million years ago.

(A)

(B)

● **FIGURE 1.8** (A) The north end of the Golden Gate Bridge, shown here, was built on rock that was deposited as mud in the deep ocean about 150 million years ago. (B) The mud hardened to form these layers of rock. The sharp folds in the rock layers were produced by tectonic forces related to the western boundary of the North American Plate.

Uniformitarianism stipulates that geologic processes operating today also operated in the past. Thus, scientists can explain events that occurred in the past by observing slow changes occurring today. Sometimes this idea is summarized in the statement "The present is the key to the past."

However, not all geologic change is gradual. William Whewell, a geologist in the 19th century, argued that geologic change was sometimes rapid. He wrote that the geologic past may have "consisted of epochs of paroxysmal and catastrophic action, interposed between periods of comparative tranquility"[2] (earthquakes and volcanoes are examples of catastrophic events). Whewell argued for the principle of **catastrophism**—that occasionally huge catastrophes have altered the course of Earth history.

2. Cited in Charles Lyell, *Principles of Geology*, vol. 2 (London: Murray, 1832), *Quarterly Reviews* 47: 103–123.

Today, geologists know that Hutton's gradualism and Whewell's catastrophism are both correct. Over the great expanses of geologic time, slow, relatively uniform processes alter Earth. In addition, infrequent catastrophic events radically modify the path of slow change.

Gradual Change in Earth History

A good example of a slow but continuous Earth process is the movement of tectonic plates. Although too slow to be observed directly without sensitive instruments, over geologic time, the movement of tectonic plates alters the shapes of ocean basins, forms lofty mountains and plateaus, generates earthquakes and volcanic eruptions, and affects our planet in many other ways.

Catastrophic Change in Earth History

Although the chance that you will experience a large earthquake this year is very small, it nonetheless exists. In fact, deadly earthquakes occur each year somewhere on the planet. The largest recorded earthquake in the United States occurred in Alaska on September 2, 1964. Although the Alaskan earthquake caused significant loss of property and is attributed to 139 human deaths, six other catastrophic earthquakes have caused a greater loss of life and property in the twenty-first century: the 2011 Tohoku Earthquake and tsunami in Japan, the 2004 Sumatra-Andaman Earthquake and tsunami in Indonesia, and large earthquakes that occurred in Haiti in 2010, Sichuan, China, in 2008, and Kashmir, Pakistan, in 2005. Though each horrific, even these catastrophes are minuscule compared with events that have occurred in the past.

When geologists study the 4.6 billion years of Earth history, they find abundant evidence of catastrophic events that are highly improbable in a human lifetime or even in human history. For example, clues preserved in rock and sediment indicate that giant meteorites have smashed into our planet, vaporizing portions of the crust, spreading dense dust clouds over the sky, and leaving large craters (such as the one in ●**FIGURE 1.9**). Geologists have suggested that some meteorite impacts (much larger than that shown in Figure 1.9) have almost instantaneously driven millions of species into extinction. As another example, when the continental glaciers receded at the end of the last ice age, huge floods occurred that were massively larger than any floods during human history (●**FIGURE 1.10**). In addition, periodically throughout Earth history, catastrophic volcanic eruptions have changed environmental conditions for life around the globe.

catastrophism The principle stating that infrequent catastrophic geologic events alter the course of Earth history, in contrast to the principle of gradualism.

● **FIGURE 1.9** Meteor Crater, Arizona.

● **FIGURE 1.10** These giant current ripples in Washington state were formed thousands of years ago by catastrophic floods released by the sudden draining of a large glacial lake upstream of this location. The floodwaters poured across eastern and central Washington and eventually reached the Columbia River, shown beyond the giant ripples.

THRESHOLD AND FEEDBACK EFFECTS

Earth systems often change in ways that are difficult to predict. Two mechanisms that contribute to the challenges and complexities facing earth science today are discussed next.

Threshold Effects

Consider a single rainstorm in Southern California. If the rain is gentle enough and it has not rained much recently, all of the rain falling will soak into the soil. If the rain lasts long enough or becomes heavy enough, a point will be reached where the soil becomes saturated and water begins to flow overland. Once runoff begins to flow across already saturated soil, a very different situation exists, especially on hill slopes where the overland flow runs together. There, even small additional increases in rainfall can cause the saturated soil to begin to move downslope as a landslide. Each year, landslides in Southern California are a serious threat, although widespread landsliding occurs only during especially wet years. The landslide shown in ●FIGURE 1.11 occurred in March 1995, following an unusually wet spring. Part of the landslide moved again in 2005, killing 10 people and causing millions of dollars in damage.

A **threshold effect** occurs when the environment initially changes slowly (or not at all) in response to a small perturbation, or deviation from the norm in a system; but after the threshold is crossed, an additional small perturbation causes rapid and dramatic change. In the example above, the initial rainfall simply soaked into the soil, but once it became saturated and additional rain began to flow across the surface, the saturated soil could no longer maintain its slope and the landslide began to move.

If we don't understand a system thoroughly, it is easy to be deluded into a false sense of complacency. Imagine, in the case just described, that we didn't know about landslide hazards. If we observed that normal amounts of rainfall simply soaked in or produced a muddy stream, we might conclude that a landslide was not possible. But once the threshold level of saturation was crossed, the landslide did begin to move.

Populations of plants, animals, and people also respond to thresholds. In 1944, the U.S. Coast Guard released 29 reindeer onto St. Matthew Island in the Bering Sea as an emergency food supply for military personnel stationed there during World War II. The military withdrew from St. Matthew Island at the end of the war, and because there were no natural predators on the island, the reindeer population increased dramatically (●FIGURE 1.12A). In 1957, 13 years after the release of the reindeer, wildlife biologist Dave Klein visited the island and counted 1,350 reindeer, most of which appeared well fed and healthy. To an untrained eye, the introduction was successful, because the reindeer were healthy and multiplying. But during that visit, Klein noticed some local overgrazing and trampling of lichens (the main food source for the reindeer), foreshadowing trouble ahead. He returned in 1963 and counted 6,000 reindeer, although this time the average size was down, as was the ratio of adults to yearlings. Klein also noticed obvious signs of overgrazing during that trip; the reindeer herd was eating more food than was growing

Mark Reid/U.S. Geological Survey

● **FIGURE 1.11** This landslide in Southern California formed during a threshold event in 1995 as heavy rains infiltrated the ground and saturated it to the point where it began to move downslope. Part of this landslide moved again in 2005, killing 10 people and destroying several homes.

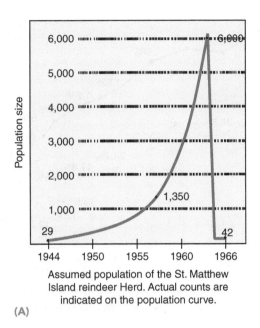

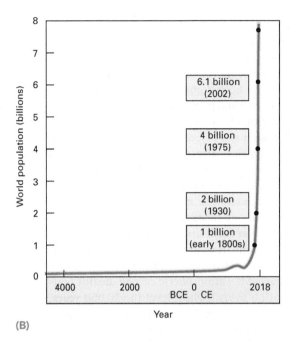

FIGURE 1.12 (A) Population curve for reindeer on St. Matthew Island. (B) Population curve for humans since the 1700s. For most of human history, there were fewer than a half-billion people on Earth. As of 2019, 7.7 billion people inhabited our planet.

and was rapidly heading toward a threshold. Everything was fine as long as the reindeer could move across the island to find new sources of food, but when the entire island became overgrazed, suddenly there was nothing to eat—and the reindeer starved. When Klein next visited St. Matthew Island, in 1966, he was greeted by hundreds of bleached skeletons and counted only 41 emaciated survivors, a population drop of over 99 percent in three years or less. By the 1980s, all of the reindeer had perished.

Feedback Mechanisms

A **feedback mechanism** occurs when a small perturbation in one component affects a different component of Earth's systems, which then amplifies the original effect and perturbs the system even more, leading to an even greater effect. For example, as Earth warms, the rate at which organic matter in soil decomposes also increases. The increased rate of decomposition releases more carbon dioxide into the atmosphere. The additional carbon dioxide gas enhances the greenhouse effect, causing Earth's climate to warm further and promoting even faster rates of decomposition. Although no one fully understands the eventual ramifications of this feedback loop for Earth's climate, Chapter 21 explores what scientists do know.

---

threshold effect A reaction whereby the environment initially changes slowly (or not at all) in response to a small perturbation, but after the threshold is crossed, an additional small perturbation causes rapid change.

feedback mechanism A reaction whereby a small initial perturbation in one component affects a different component of Earth's systems, which amplifies the original effect, which perturbs the system even more, which leads to an even greater effect, and so on.

HUMANS AND EARTH SYSTEMS

Ten thousand years ago, a mere eyeblink in Earth history, human population was low and technology was limited. Sparsely scattered bands hunted and gathered for their food, built simple dwellings, and warmed themselves with small fires. The impact of humans was not appreciably different from that of many other species on the planet. But with the invention of agriculture and the growth of cities, problems arose in some cultures. In his recent book *Collapse*, Jared Diamond documents several case histories where people settled in a region, built a society, and consumed resources faster than they were naturally replaced.[3] Then, like the reindeer on St. Matthew Island, populations crashed. For example, at the height of the Mayan civilization, between the 3rd and the 9th centuries, between 1 and 2 million people are estimated to have lived in a region about 800 miles square, resulting in a population density of between 1,800 and 2,600 people per square mile. (By comparison, according to the 2010 U.S Census, Los Angeles had a population density of 7,544.6 people per square mile.) Then, due to a variety of human and environmental factors, the Mayan population crashed and the cities were abandoned, as an estimated 90 to 95 percent of the people died. In a similar manner, the Anasazi civilizations in the American Southwest disappeared in the 13th century, and Viking colonies starved to death in Greenland. All three of these human tragedies were exacerbated by climate change and human-to-human interactions, such as war. Nevertheless, Diamond points out that after thresholds were reached, the populations didn't just decline back to a sustainable level; rather, they collapsed catastrophically.

As of June 2019, there were 7.7 billion people on Earth (**FIGURE 1.12B**). Farms, pasturelands, and cities covered

---

3. Jared Diamond, *Collapse: How Societies Choose to Fail or Succeed* (New York: Penguin Group, 2005).

40 percent of the land area of the planet. We diverted half of all the freshwater on the continents and controlled 50 percent of the total net terrestrial biological productivity. As the land and its waters have been domesticated and manipulated, other animal and plant species have perished. The species extinction rate today is 1,000 times more rapid than it was before the Industrial Revolution. In fact, species extinction hasn't been this rapid since a giant meteorite smashed into Earth's surface 65 million years ago, leading to the demise of the dinosaurs and many other species. We are presently living through the largest documented extinction on Earth since the age of the dinosaurs.

Emissions from our fires, factories, electricity-generating plants, and transportation machines have raised the atmospheric carbon dioxide concentration to its highest level in 420,000 years. Since the Industrial Revolution, Earth's mean annual temperature has increased by 0.8°C.

Roughly 20 percent of the human population lives in wealthy industrialized nations where food, shelter, clean drinking water, efficient sewage disposal, and advanced medical attention are all readily available. As of 2016, 767 million people worldwide survived on less than $1.90 per day (●**FIGURE 1.13**). One-eighth to one-half of the human population on this planet is malnourished, and 14 million infants or young children starve to death every year. If you were a member of a family in the Himalayan region of northern India—where you spent hours every day gathering scant firewood or cow dung for fuel, where meals consisted of an endless repetition of barley gruel and peas, where you brought your sheep and goats into the house during the winter so you could all huddle together for warmth—your experience of the Earth and its systems would be quite different from that of your life today.

Yet no matter where we live and regardless of whether we are rich or poor, our survival is interconnected with the flow of energy and materials among Earth systems. We affect Earth, and at the same time Earth affects us; it is the most intimate and permanent of marriages.

Predictions for the future are, by their very nature, uncertain. More than 200 years ago, in 1798, the Reverend Thomas Malthus argued that the human population was growing exponentially, while food production was increasing much more slowly. As a result, humankind was facing famine, accompanied by "misery and vice."[4] But technological advancements have led to an exponential increase in food production, and today food is plentiful in many regions. At the same time, however, serious shortages exist, especially in parts of Africa.

Many scientists are deeply concerned about our future: John Holdren, professor at the John F. Kennedy School of Government, argues that:

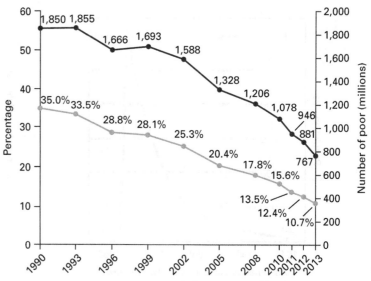

Source: http://www.Worldbank.Org/en/topic/poverty/overview

—— Percentage of global population who live on $1.90 or less per day
—— Number of people who live below $1.90 a day (right axis)

●**FIGURE 1.13** These two curves show that global poverty has fallen steadily since 1990 although remains at high levels. As of 2013, 767,000,000 people survived on $1.90 or less per day, or 10.7% of all humans.

[The problem is] not that we are running out of energy, food, or water but that we are running out of environment—that is, running out of the capacity of air, water, soil, and biota to absorb, without intolerable consequences for human well-being, the effects of energy extraction, transport transformation and use.[5]

Three broad trends are currently facing humanity:

1. The human population is large and continuing to increase—although the rate of increase is slowing. As of June, 2019, the U.S. Census Bureau estimated the world population at 7.6 billion. According to the United Nations, world population is projected to reach 9.8 billion in 2050, and 11.2 billion in 2100.
2. At the same time, extreme poverty is decreasing (Figure 1.13). In India and China, with one-third of the world's population, economic output and standards of living are rising rapidly. When people have money, they consume more resources.
3. As both population and per capita consumption rise, pollution and pressure on Earth's resources also continue to rise.

Professor Diamond, quoted earlier, argues that we are in a "horse race" with the environment.[6] One horse, representing increased human consumption and the resulting environmental degradation and resource depletion, is galloping

---

4. Thomas Malthus, *An Essay on the Principle of Population*, 1st ed. (Penguin Classics, 1798).

5. John Holdren, "Energy: Asking the Wrong Question," *Scientific American* 286 (January 2002): 65–67.

6. Jared Diamond, *Collapse: How Societies Choose to Fail or Succeed* (New York: Penguin Group, 2005).

toward an unsustainable society with a possible catastrophic collapse. The other horse, powered by improved technology, increased awareness, and declining rates of population growth, leads us toward a happy, healthy, sustainable society. Which horse will win?

This book will not provide answers—because there are none—but it will provide a basis for understanding Earth processes so that you can evaluate and appreciate issues concerning our stewardship of the planet. The table highlighted in green below suggests an environmental "to-do" list.

OUR ENVIRONMENTAL "TO-DO" LIST

1. Slow the increase in human population.
2. Reverse the buildup of greenhouse gases and the depletion of the ozone layer.
3. Slow the increase in human per capita resource consumption.
4. Conserve fossil fuels and firewood resources.
5. Conserve water and protect against causes of water shortages.
6. Prevent toxic chemicals from harming the environment.
7. Protect against soil erosion.
8. Halt the destruction of natural habitats such as forests, wetlands, prairies, and coral reefs.
9. Reduce our agricultural footprint so that some land area is left over to support the growth of natural forests.
10. Protect against the loss of ocean fisheries.
11. Protect endangered species so as not to lose biodiversity.
12. Protect indigenous species, and keep invasive species from destroying ecosystems.

Key Concepts Review

- Earth consists of four spheres: The geosphere is composed of a dense, hot, central core, surrounded by a large mantle that comprises most of Earth's volume, with a thin, rigid crust. The crust and mantle are composed of rock while the core is composed of iron and nickel. The hydrosphere is mostly ocean water. Most of Earth's freshwater is locked in glaciers. Most of the liquid freshwater lies in groundwater reservoirs; streams, lakes, and rivers account for only 0.01 percent of the planet's water. The atmosphere is a mixture of gases, mostly nitrogen and oxygen. Earth's atmosphere supports life and regulates climate. The biosphere is the thin zone that life inhabits.
- A system is composed of interrelated, interacting components that form a complex whole. Earth's four major systems—the four spheres—are subdivided into a great many interacting smaller ones. All the spheres continuously exchange matter and energy.
- Earth is about 4.6 billion years old; life formed at least 3.8 billion years ago; abundant multicellular life evolved

about 544 million years ago, and hominids have been on this planet for a mere 5 to 7 million years. The principle of gradualism states that geologic change occurs over a long period of time by a sequence of almost imperceptible events. In contrast, the principle of catastrophism postulates that geologic change occurs mainly during infrequent catastrophic events.
- A threshold effect occurs when the environment initially changes slowly (or not at all) in response to a small perturbation, but after the threshold is crossed, an additional small perturbation causes rapid and dramatic change. A feedback mechanism occurs when a small initial perturbation in one component affects a different component of Earth systems, amplifying the original effect, which perturbs the system even more and leads to an even greater effect.
- Due to our sheer numbers and technological prowess, humans have become a significant engine of change in Earth systems.

Important Terms

atmosphere (p. 2)

biosphere (p. 2)

catastrophism (p. 10)

core (p. 2)

crust (p. 2)

cycle (p. 8)

ecosystem (p. 7)

feedback mechanism (p. 13)

geosphere (p. 2)

gradualism (p. 9)

groundwater (p. 6)

hydrosphere (p. 2)

lithosphere (p. 5)

mantle (p. 2)

system (p. 7)

tectonic plates (p. 5)

threshold effect (p. 12)

uniformitarianism (p. 9)

Review Questions

1. List and briefly describe each of Earth's four spheres.

2. List the three major layers of Earth. Which is/are composed of rock, which is/are metallic? Which is the largest; which is the thinnest?

3. List six types of reservoirs that collectively contain most of Earth's water.

4. What is groundwater? Where in the hydrosphere is it located?

5. What two gases compose most of Earth's atmosphere?

6. How thick is Earth's atmosphere?

7. Briefly discuss the size and extent of the biosphere.

8. Define a system and explain why a systems approach is useful in Earth Science.

9. Briefly explain the statement: "Matter can be recycled, but energy cannot."

10. How old is Earth? When did life first evolve? How long have humans and their direct ancestors been on this planet?

11. Compare and contrast gradualism and catastrophism. Give an example of each type of geologic change.

12. Briefly explain threshold and feedback effects.

13. Briefly outline the magnitude of human impact on the planet.

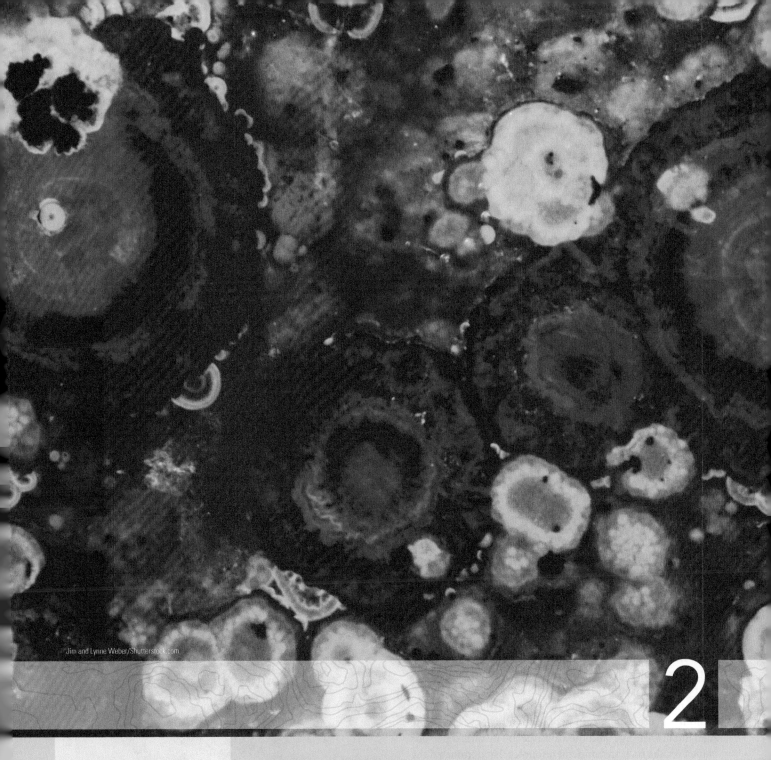

2

MINERALS

This specimen consists of blue azurite and green malachite. Both are hydrous copper carbonate minerals. The arrangement of minerals in this sample resulted from changes in water chemistry as the minerals slowly precipitated from solution.

LEARNING OBJECTIVES

LO1 Discuss the five properties of a mineral.

LO2 Explain why oil and coal are not minerals.

LO3 Define an element.

LO4 Distinguish between atom, cation, and anion.

LO5 Discuss the crystal structure of minerals and how it reflects the internal atomic structure.

LO6 Give examples of physical properties of minerals.

LO7 Describe the nine important mineral groups.

LO8 Categorize commercially important minerals.

LO9 Describe two harmful minerals and the effects on humans.

LO10 Discuss the effects mining and refining minerals can have on the environment.

INTRODUCTION

Minerals are the building blocks of rock, and therefore the geosphere is composed of minerals. Minerals make up soil and rock found at the Earth's surface, as well as the deeper rock of the crust and mantle. Any thorough study of Earth must include an understanding of minerals. But it is not sufficient to study minerals in isolation from the rest of the planet. Rather, we can learn more by observing the ways in which minerals interact with other Earth systems.

Most minerals in their natural settings are harmless to humans and other species. In addition, many resources that are essential to modern life are produced from rocks and minerals: iron, aluminum, gold, and all other metals; concrete, fertilizers, coal, and nuclear fuels. Both oil and natural gas are recovered from natural reservoirs in rock. The discovery of methods to extract and use these Earth materials has profoundly altered the course of human history. The Stone Age, the Bronze Age, and the Iron Age are historical periods named for the rock and mineral resources that dominated the development of civilization during those times. The Industrial Revolution occurred when humans discovered how to convert the energy stored in coal into useful work. The automobile age occurred because humans found vast oil reservoirs in shallow rocks of Earth's crust and built engines to burn the refined fuels. The computer age relies on the movement of electrons through wafer-thin slices of silicon, germanium, and other materials that are also crystallized from minerals.

However, some minerals are naturally harmful. For example, uranium-bearing minerals are radioactive, while sulfide-bearing minerals can dissolve and release harmful elements including lead and arsenic into the environment. Some common minerals become harmful to humans when they are crushed to dust during mining, quarrying, road building, and other activities. In this chapter, we will consider the nature of minerals and then return to a discussion of minerals as natural resources and as environmental hazards.

WHAT IS A MINERAL?

Pick up any rock and look at it carefully. You may see small, differently colored specks, such as those you see in the close-up photo of granite in ●**FIGURE 2.1.** Each speck is a mineral. In the photo, the white grains are the mineral feldspar, the black ones are biotite, and the glassy-gray ones are quartz. A rock is a mixture of minerals. Most rocks contain two to five abundant minerals plus minor amounts of several others. Few rocks are made of only one mineral.

Although we can define minerals as the building blocks of rocks, such a definition does not tell us much about the nature

(A)

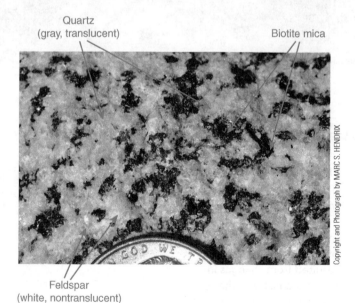

Quartz (gray, translucent)

Biotite mica

Feldspar (white, nontranslucent)

(B)

● **FIGURE 2.1** (A) The Bugaboo Mountains of British Columbia are made of granite. (B) Each of the differently colored grains in this close-up photo of granite is a different mineral. The white grains are feldspar, the black ones are biotite, and the glassy-gray ones are quartz. The top of the U.S. penny is for scale.

of minerals. More precisely, a **mineral** is a naturally occurring inorganic solid with a definite chemical composition and a crystalline structure. *Chemical composition* and *crystalline structure* are the two most important properties of a mineral: they distinguish any mineral from all others. Before discussing these two properties, however, let's briefly consider the other properties of minerals that this definition describes.

Naturally Occurring

A synthetic diamond can be identical to a natural one, but it is not a true mineral, according to our definition, because a mineral must form by natural processes. Like diamonds, most other gems that occur naturally can be manufactured by industrial processes. Natural gems are more highly valued than manufactured ones. For this reason, jewelers should always tell their customers whether a gem is natural or artificial, by prefacing the name of a manufactured gem with the term *synthetic*.

Inorganic

Organic substances are composed mostly of carbon that is chemically bonded to hydrogen. *Inorganic* compounds do not contain carbon–hydrogen bonds. Although organic compounds can be produced in laboratories and by industrial processes, plants and animals create most of Earth's organic material, because plant and animal tissue is organic. When these organisms die, they decompose to form other organic substances. Both coal and oil form by the decay of plants and animals. However, because of their organic properties and lack of a definite chemical composition, neither substance is a mineral. In addition, oil is a liquid, not a solid.

Some material that organisms produce is not organic. For example, limestone, one of the most common sedimentary rocks, commonly contains debris from the shells or skeletons of invertebrate organisms such as clams, snails, and corals. The shells and skeletal pieces, in turn, typically are made of the minerals calcite and aragonite. Although the organisms produce the calcite and aragonite, both are a mineral—an inorganic solid that formed naturally and has a definite chemical composition and crystalline structure.

Solid

All minerals are solids. Thus, ice is a mineral, but neither water nor water vapor is a mineral.

---

mineral   A naturally occurring inorganic solid with a definite chemical composition and a crystalline structure.

element   A substance that cannot be broken down into other substances by ordinary chemical means.

ion   An atom with an electrical charge, either positive or negative.

cation   A positively charged ion.

anion   An ion that has a negative charge.

THE CHEMICAL COMPOSITION OF MINERALS

Minerals are the fundamental building blocks of rocks, but what are minerals made of? Minerals and all other Earth materials are composed of chemical elements. An **element** is a fundamental component of matter that cannot be broken into simpler particles by ordinary chemical processes. Most common minerals consist of a small number of different chemical elements—usually two to five.

A total of 88 elements occur naturally in Earth's crust. However, eight elements—oxygen, silicon, aluminum, iron, calcium, sodium, potassium, and magnesium—make up more than 98 percent of the crust (**TABLE 2.1**). Each element is represented by a one- or two-letter symbol, such as O for oxygen and Si for silicon. In nature, most chemical elements have either a positive or negative electrical charge. For example, oxygen has a charge of negative 2 (–2) and silicon has a charge of plus 4 (+4). An atom with an electrical charge, whether it is positive or negative, is called an **ion**. A positively charged ion is a **cation**; an ion that has a negative charge is an **anion**. Ions with opposite charges are attracted to each other, like the positive end of a magnet attracts the negative end.

Recall that a mineral has a definite chemical composition. A substance with a definite chemical composition is made up of elements that are combined in definite proportions. Therefore, the composition of a mineral can be expressed as a chemical formula, written by combining the symbols of the individual elements with the number of each within the repeating molecular unit that forms the mineral crystal. A few minerals, such as gold and silver, consist of only a single element. Their chemical formulas, respectively, are Au (the symbol for gold) and Ag (the symbol for silver). Most minerals,

TABLE 2.1 The Eight Most Abundant Chemical Elements in the Earth's Crust*

Element	Symbol	Weight Percent	Atom Percent	Volume Percent[†]
Oxygen	O	46.60	62.55	93.80
Silicon	Si	27.72	21.22	0.90
Aluminum	Al	8.13	6.47	0.50
Iron	Fe	5.00	1.92	0.40
Calcium	Ca	3.63	1.94	1.00
Sodium	Na	2.83	2.64	1.30
Potassium	K	2.59	1.42	1.80
Magnesium	Mg	2.09	1.84	0.30
	Totals	98.59	100.00	100.00

*Abundances are given in percentages by weight, by number of atoms, and by volume.
[†]These numbers will vary somewhat as a function of the ionic radii chosen for the calculations.
Source: From Principles of Geochemistry by Brian Mason and Carleton B. Moore. © 1982 by John Wiley & Sons, Inc.

however, are made up of two to five essential elements. For example, the formula of quartz is SiO_2: it consists of one atom of silicon (Si) for every two of oxygen (O). Quartz from anywhere in the Universe has that exact composition. If it had a different composition, it would be some other mineral. The compositions of some minerals, such as quartz, do not vary by even a fraction of a percent. The compositions of other minerals vary slightly, but the variations are limited.

The 88 elements that occur naturally in Earth's crust can combine in many ways to form many different minerals. In fact, more than 3,500 minerals are known. However, the eight abundant elements commonly combine in only a few ways. As a result, only nine **rock-forming minerals** (or mineral "groups") make up most rocks of Earth's crust. We will describe the rock-forming minerals in section on "Mineral Classes and the Rock-Forming Minerals."

● **FIGURE 2.3** Photomicrograph of a thin slice (0.03 mm) of granite.

THE CRYSTALLINE NATURE OF MINERALS

Every mineral has a crystalline structure, and therefore every mineral is a crystal. A **crystal** is any solid element or compound whose atoms are arranged in a regular, periodically repeated three-dimensional pattern, like a wallpaper pattern that extends into three dimensions. The mineral halite (common table salt) has the composition $NaCl$: one sodium ion (Na^+) for every chlorine ion (Cl^-). ●**FIGURE 2.2A** is an exploded view of the ions in halite. ●**FIGURE 2.2B** is more realistic, showing the ions in contact. In both sketches, the sodium and chlorine ions alternate in orderly rows and columns intersecting at right angles. This orderly, repetitive arrangement of atoms is the **crystalline structure** of halite.

Most minerals initially form as tiny crystals that grow as layer after layer of atoms (or ions) are added to their surfaces. A halite crystal might grow, for example, as salty seawater evaporates from a tidal pool. At first, a tiny mineral grain might form, similar to the sketch of halite in ●**FIGURE 2.2**. This model shows a halite crystal containing 125 atoms; it would be only about one-millionth of a millimeter long on each side, far too small to see with the unaided eye. As evaporation continues, more and more sodium and chlorine ions would precipitate onto the corners, edges, and faces of the growing crystal.

A **crystal face** is a flat surface that develops if a crystal grows freely in an uncrowded environment, such as halite growing in a pool of evaporating seawater. When a mineral grows freely like this, it commonly forms a symmetrical crystal with perfectly flat faces that reflect light like a mirror. In nature, however, mineral crystals often impede the growth of adjacent crystals. For this reason, minerals rarely show perfect development of crystal faces. ●**FIGURE 2.3** is a photomicrograph (a photo taken through a microscope) of a thin slice of granite in which the crystals fit like pieces of a jigsaw puzzle. This interlocking texture developed because some crystals grew around others as molten magma cooled and solidified.

PHYSICAL PROPERTIES OF MINERALS

If chemical composition and crystal structure distinguish each mineral from all others, how does a geologist identify individual minerals in the field? For example, halite always consists of sodium and chlorine in a one-to-one ratio, with the atoms arranged in a cubic fashion. But if you pick up a crystal of halite, you cannot see the atoms. You could identify a sample of halite by measuring its chemical composition and crystal structure using laboratory procedures, but such analyses are expensive and time-consuming. Instead, geologists commonly identify minerals by visual recognition and then confirm the identification with simple tests.

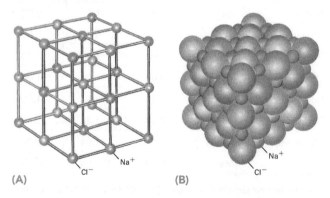

● **FIGURE 2.2** The orderly arrangement of sodium and chlorine ions in NaCl, or halite. The crystal model in (A) is exploded so that you can see into it; the ions are actually closely packed, as in (B).

Most minerals have distinctive appearances. Once you become familiar with common minerals, you will recognize them just as you recognize any familiar object. For example, an apple just looks like an apple, even though apples come in many colors and shapes. In the same way, quartz looks like quartz to a geologist. The color and shape of quartz may vary from sample to sample, but it still looks like quartz. Some minerals, however, look like others, so that their physical properties must be examined further to make a correct identification. For example, halite can look like calcite, quartz, and several other minerals, but halite tastes salty and scratches easily with a knife blade, two characteristics that distinguish it from the other minerals. Geologists commonly use properties such as crystal habit, cleavage, fracture, hardness, specific gravity, color, streak, and luster to identify minerals.

Crystal Habit

Crystal habit refers to the characteristic shape of an individual crystal and the manner in which aggregates of crystals grow. If a crystal grows freely, it develops a characteristic shape that the arrangement of its atoms controls. Some minerals occur in more than one habit. For example, ●**FIGURE 2.4A** shows quartz with a prismatic (pencil-shaped) habit, and ●**FIGURE 2.4B** shows a massive quartz in which numerous individual microscopic crystals intergrow to produce the habit.

Cleavage

Cleavage is the tendency of some minerals to break along flat surfaces. The surfaces are planes of weak bonds in the crystal. Some minerals, such as the micas, have one set of parallel cleavage planes (●**FIGURE 2.5**). Others have two, three, or even four different sets, as shown by the three photos in ●**FIGURE 2.6**. In some cases, the expression of the cleavage is excellent. For example, you can peel sheet after sheet from a

(A)

Sebastian Janicki/Shutterstock.com

(B)

ydl38885/Shutterstock.com

● **FIGURE 2.4** (A) Prismatic quartz grows as elongated crystals. (B) Massive rosy quartz shows no characteristic shape.

---

rock-forming minerals The nine minerals or mineral groups that are most abundant in the Earth's crust and that combine to make most rocks. They are olivine, pyroxene, amphibole, mica, the clay minerals, quartz, feldspar, calcite, and dolomite.

crystal A solid element or compound whose atoms are arranged in a regular, periodically repeated pattern.

crystalline structure The orderly, repetitive arrangement of atoms in a crystal.

crystal face The flat surface that develops if a crystal grows freely in an uncrowded environment. Under perfect conditions, the crystal that forms will be symmetrical.

crystal habit The characteristic shape of an individual crystal, and the manner in which aggregates of crystals grow.

cleavage The tendency of some minerals to break along flat surfaces, which are planes of weak bonds in the crystal. When a mineral has well-developed cleavage, sheet after sheet can be peeled from the crystal, like peeling layers from an onion.

woe/Shutterstock.com

● **FIGURE 2.5** Mica has a single, perfect cleavage plane.

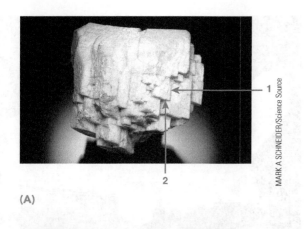

MARK A SCHNEIDER/Science Source

(A)

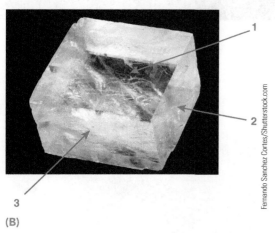

Fernando Sanchez Cortes/Shutterstock.com

(B)

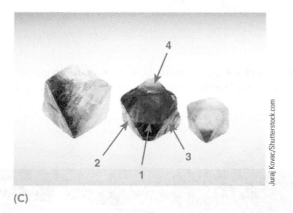

Juraj Kovac/Shutterstock.com

(C)

● **FIGURE 2.6** Some minerals have more than one cleavage plane. (A) The mineral feldspar has two well-developed cleavage planes that intersect at right angles. (B) Calcite has three sets of cleavage planes that do not intersect at right angles. (C) Fluorite has four cleavage planes. Each cleavage face has a parallel counterpart on the opposite side of the crystal.

mica crystal as if you were peeling layers from an onion. In other cases, cleavage is not well developed. Many minerals, such as quartz, have no cleavage at all because they have no planes of weak bonds. The number of cleavage planes, the expression—or quality—of cleavage, and the angles between cleavage planes all help in mineral identification.

A flat surface created by cleavage and a flat crystal face can appear to be very similar. However, a cleavage surface is repeated when a crystal is broken, whereas a crystal face is not.

Fracture

Fracture is the manner in which a mineral breaks other than along planes of cleavage. Many minerals fracture into characteristic shapes. For instance, a *conchoidal fracture* creates smooth, curved surfaces (●**FIGURE 2.7**). It is characteristic of quartz and olivine. Some minerals break into splintery or fibrous fragments. Most fracture into irregular shapes.

Hardness

Hardness is the resistance of a mineral to scratching and is one of the most commonly used properties for identifying a mineral. It is easily measured and is a fundamental property of each mineral because it is controlled by the bond strength between the atoms in the mineral. Geologists commonly gauge hardness by attempting to scratch a mineral with the tip of a pocketknife. If the metal scratches the mineral, the mineral is softer. If the knife tip cannot scratch the mineral, the mineral is harder.

To measure hardness more accurately, geologists use a scale based on 10 minerals, numbered 1 through 10 (**TABLE 2.2**). Each mineral is harder than those with lower numbers on the scale—so 10 (diamond) is the hardest and 1 (talc) is the softest. The scale is known as the **Mohs hardness scale**, after Friedrich Mohs, the Austrian mineralogist who developed it in the early 19th century.

As you can see from Table 2.2, the Mohs scale shows that a mineral scratched by quartz but not by orthoclase has a hardness between 6 and 7. Because the minerals of the Mohs scale are not always handy, it is useful to know the hardness values of common materials. A fingernail has a hardness of slightly more than 2, a pocketknife blade slightly more than 5, window glass about 5.5, and a steel file about 6.5. If you practice with a knife and the minerals of the Mohs scale, you can develop a "feel" for minerals with hardness of 5 and under by seeing how easily the blade scratches them.

Breck P. Kent

● **FIGURE 2.7** Some minerals without cleavage break along smoothly curved surfaces, called conchoidal fractures. This sample is smoky quartz.

TABLE 2.2 Minerals of the Mohs Hardness Scale

Minerals of Mohs Scale	Common Objects with Similar Hardness
1. Talc	
2. Gypsum	Fingernail
3. Calcite	Copper penny
4. Fluorite	
5. Apatite	Knife blade; window glass
6. Orthoclase	Steel file
7. Quartz	
8. Topaz	
9. Corundum	
10. Diamond	

When testing hardness, it is important to determine whether the mineral has actually been scratched by the object or the object has simply left a trail of its own powder on the surface of the mineral. To check, rub away the powder trail and feel the surface of the mineral with your fingernail for the groove of the scratch. Fresh, unweathered mineral surfaces must be used in hardness measurements, because weathering often produces a soft rind on minerals.

Specific Gravity

Specific gravity is the weight of a substance relative to that of an equal volume of water. If a mineral weighs 2.5 times as much as an equal volume of water, its specific gravity is 2.5. You can estimate a mineral's specific gravity simply by hefting a sample in your hand. If you practice with known minerals, you can develop a feel for specific gravity. Most common minerals have a specific gravity of about 2.7. Metals have a much higher specific gravity; for example, the specific gravity of lead is 11.3, silver is 10.5, and copper is 8.9. Gold has the highest specific gravity of all minerals at 19.

Color

Color is the most obvious property of a mineral, and it is often used in identification. But color can be unreliable, because small amounts of chemical impurities and imperfections in crystal structure can dramatically alter color. For example, corundum (Al_2O_3) is normally a cloudy, translucent, brown or blue mineral. The addition of a small amount of chromium can convert corundum to the beautiful, clear, red gem known as ruby. A small quantity of iron or titanium turns corundum into the striking blue gem called sapphire. Quartz also occurs in many colors, including white, clear, black, purple, and red, as a result of tiny amounts of impurities and minor defects in the perfect ordering of atoms.

Streak

Streak refers to the color of the fine powder of a mineral. It is observed by rubbing the mineral across a piece of unglazed porcelain known as a "streak plate." Many minerals leave a streak of powder on the plate with a diagnostic color. Streak is commonly more reliable for identification than the color of the mineral itself. For example, the mineral hematite (iron oxide) can occur both as a dull red mineral and in a shiny black form that closely resembles black mica, but both types will leave the same red powder on a streak plate.

Luster

Luster is the manner in which a mineral reflects light. A mineral with a metallic look, irrespective of color, has a metallic luster. Pyrite is a yellowish mineral with a metallic luster (●FIGURE 2.8). As a result, it looks like gold and is commonly called fool's gold. The luster of nonmetallic minerals is usually described by self-explanatory words such as *glassy, pearly, earthy,* and *resinous.*

---

fracture The manner in which minerals break, other than along planes of cleavage.

hardness The resistance of a mineral to scratching, controlled by the bond strength between its atoms.

Mohs hardness scale A scale, based on a series of 10 fairly common minerals and numbered 1 to 10 (from softest to hardest), used to measure and express the hardness of minerals.

specific gravity The weight of a substance relative to the weight of an equal volume of water.

streak The color of the fine powder of a mineral, usually obtained by rubbing the mineral on an unglazed porcelain streak plate.

luster The quality and intensity of light reflected from the surface of a mineral.

Nyura/Shutterstock.com

● **FIGURE 2.8** Pyrite, or fool's gold, has a metallic luster.

Other Properties

Properties such as reaction to acid, magnetism, radioactivity, fluorescence, and phosphorescence can be characteristic of specific minerals. Calcite and some other carbonate minerals dissolve rapidly in acid, releasing visible bubbles of carbon dioxide gas. The mineral magnetite displays an obvious, diagnostic attraction to magnets. Carnotite, a uranium-bearing mineral, emits radioactivity that can be detected with a scintillometer. Fluorescent materials emit visible light when they are exposed to ultraviolet light. Phosphorescent minerals continue to emit light after the external stimulus ceases.

MINERAL CLASSES AND THE ROCK-FORMING MINERALS

Geologists classify minerals according to their anions or anionic groups (**TABLE 2.3**). For example, **silicates** minerals all contain anionic groups made of silicon and oxygen bonded together to form $[SiO_4]^{-4}$; **carbonates** minerals all contain the anionic group $[CO_3]^{2-}$; and sulfides contain the anion **sulfide** (S^{1-}). The **native elements** are a small class of minerals, including pure gold and silver, that consist of only a single element.

Although more than 3,500 minerals are known to be in Earth's crust, only a small number—between 50 and 100—are

TABLE 2.3 Important Mineral Groups				
Group	Anion or Anionic Group	Chemical Formula	Mineral Name	Economic Use
Oxides	O^{2-}	Fe_2O_3	Hematite	Ore of iron
		Fe_3O_4	Magnetite	Ore of iron
		Al_2O_3	Corundum	Gemstone; abrasive
		$FeCr_2O_4$	Chromite	Ore of chromium
		H_2O	Ice	Solid form of water
Sulfides	S^{1-}	PbS	Galena	Ore of lead
		ZnS	Sphalerite	Ore of zinc
		FeS_2	Pyrite	Fool's gold
		$CuFeS_2$	Chalcopyrite	Ore of Copper
		Cu_5FeS_4	Bornite	Ore of copper
		HgS	Cinnabar	Ore of mercury
Sulfates	$[SO_4]^{3-}$	$CaSO_4 \cdot 2H_2O$	Gypsum	Plaster
		$CaSO_4$	Anhydrite	Plaster
		$BaSO_4$	Barite	Drilling mud
Native elements	N/A	Au	Gold	Electronics; jewelry
		Cu	Copper	Electronics
		C	Diamond	Gemstone; abrasive
		S	Sulfur	Sulfa drugs; chemicals
		C	Graphite	Pencil lead; dry lubricant
		Ag	Silver	Jewelry; photography
		Pt	Platinum	Catalyst
Halides	Cl^{1-}, F^{1-}, or Br^{1-}	$NaCl$	Halite	Common salt
		CaF_2	Fluorite	Steel making
		KCl	Sylvite	Fertilizer
Carbonates	$[CO_3]^{2-}$	$CaCO_3$	Calcite	Portland cement
		$CaMg(CO_3)_2$	Dolomite	Portland cement
		$CaCO_3$	Aragonite	Portland cement
Hydroxides	$[OH]^{1-}$	$FeO(OH) \cdot nH_2O$	Limonite	Ore of iron; pigments
		$Al(OH)_3 \cdot nH_2O$	Bauxite	Ore of aluminum
Phosphates	$[PO_4]^{3-}$	$Ca_5(PO_4)_3(F,Cl,OH)$	Apatite	Fertilizer
		$CuAl_6(PO_4)_4(OH)_8 \cdot 4H_2O$	Turquoise	Gemstone
Silicates	$[SiO_4]^{4-}$	The silicate mineral group makes up 92 percent of the Earth's crust. (Figure 2.10 summarizes the most common rock-forming silicate minerals.)		

common or valuable, and only nine rock-forming minerals make up most of the crust. These nine are important to geologists simply because they are the most common and abundant minerals. Seven of the rock-forming minerals are silicate minerals. The other two, calcite and dolomite, are carbonates.

Silicates

Silicates make up about 92 percent of Earth's crust. They are abundant for two reasons: First, silicon and oxygen are the two most plentiful elements in the crust. Second, silicon and oxygen combine readily.

To understand silicate minerals, remember that:

1. Every silicon atom in a silicate mineral surrounds itself with four oxygen atoms. The bonds between silicon and its four oxygen atoms are very strong.
2. The silicon atom and its four oxygen atoms form a pyramid-shaped structure, called the **silicate tetrahedron**, with silicon in the center and oxygen at the four corners, as illustrated in the two panels of ●**FIGURE 2.9**. The silicate tetrahedron is the fundamental building block of all silicate minerals.
3. Negatively-charged silicate tetrahedra combine with positively-charged elements (cations) in Earth's crust to form silicate minerals. The most common cations are aluminum, iron, calcium, potassium, sodium, and magnesium. Quartz is the only common silicate mineral that contains only silicon and oxygen.
4. Silicate tetrahedra commonly link together by sharing oxygen atoms to form chains, sheets, or three-dimensional networks as shown in ●**FIGURE 2.10**. In all silicate minerals except quartz, the chains, sheets, and networks of silicate tetrahedra formed by the sharing of oxygens are negatively-charged and must combine with positively-charged cations to form a neutral silicate mineral.

Rock-Forming Silicates

Silicate tetrahedra in silicate minerals link together in five different ways, forming five different silicate mineral structures. Each structure contains at least one of the rock-forming minerals. Figure 2.10 displays these structures as well as the most common silicate minerals in each group.

---

silicates Minerals whose crystal structure contains silicate tetrahedra.

carbonates Minerals containing the anionic group carbonate, CO_3^{2-}; an example is calcite ($CaCO_3$).

sulfides Minerals containing the anion sulfide, S^{1-}. An example is pyrite (FeS_2).

native elements Minerals that consist of only one element and thus the element occurs in the native state (not chemically bonded to other elements). Only about 20 elements occur in the native state as solids.

silicate tetrahedron The fundamental building block of all silicate minerals, a pyramid-shaped structure consisting of one silicon atom bonded to four oxygen atoms to form the anionic group silicate, $[SiO_4]^{4-}$.

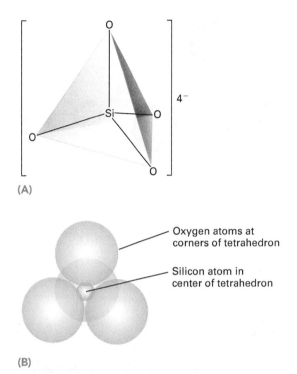

(A)

(B)

● **FIGURE 2.9** The silicate tetrahedron consists of one silicon atom surrounded by four oxygen atoms. It is the fundamental building block of all silicate minerals. (A) Schematic representation. (B) Proportionally accurate model.

●**FIGURE 2.11** shows that feldspar alone makes up about half of Earth's crust. Feldspars are divided into two main groups: plagioclase feldspar that contains calcium and alkali feldspar that contains potassium or sodium (called *alkali elements*) instead of calcium. Orthoclase is a common alkali feldspar and is abundant in granite and other rocks of the continents. Rocks of the oceanic crust contain about 50 percent plagioclase. Plagioclase and orthoclase often look alike and can be difficult to tell apart in hand samples.

Quartz comprises 12 percent of crustal rocks. It is widespread and abundant in most continental rocks but rare in oceanic crust and the mantle. Quartz is the only common silicate mineral that contains no cations other than silicon; it is pure SiO_2. It has a ratio of one 4^+ silicon for every two 2^- oxygens, so the positive and negative charges neutralize each other perfectly without the addition of other cations.

Pyroxene makes up another 11 percent of the crust. The basalt rock of ocean crust is made up almost entirely of two minerals, plagioclase feldspar and pyroxene.

Basalt commonly contains 1 or 2 percent olivine, but olivine is otherwise uncommon in Earth's crust. However, it makes up a large proportion of mantle rocks, and thus is an abundant and important mineral.

Amphibole, mica, and clay are the other rock-forming silicate minerals. Amphiboles and micas are common in granite and many other types of continental rocks but are rare in oceanic crust. Clay forms when rain and atmospheric gases chemically decompose other silicate minerals, and thus clay is common in soils. Clay minerals are especially common in mudstone, which is the most abundant of all sedimentary rocks and is described in Chapter 3.

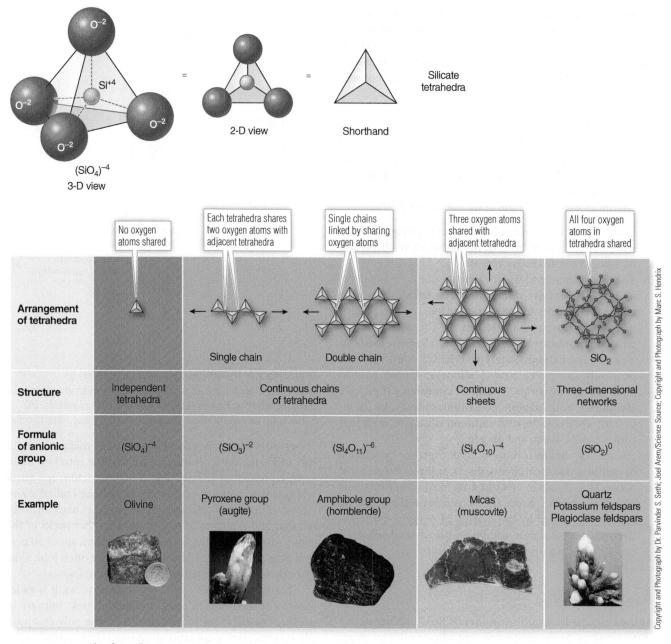

Silicate tetrahedra

2-D view

Shorthand

$(SiO_4)^{-4}$
3-D view

No oxygen atoms shared

Each tetrahedra shares two oxygen atoms with adjacent tetrahedra

Single chains linked by sharing oxygen atoms

Three oxygen atoms shared with adjacent tetrahedra

All four oxygen atoms in tetrahedra shared

Arrangement of tetrahedra		Single chain	Double chain		SiO_2
Structure	Independent tetrahedra	Continuous chains of tetrahedra		Continuous sheets	Three-dimensional networks
Formula of anionic group	$(SiO_4)^{-4}$	$(SiO_3)^{-2}$	$(Si_4O_{11})^{-6}$	$(Si_4O_{10})^{-4}$	$(SiO_2)^0$
Example	Olivine	Pyroxene group (augite)	Amphibole group (hornblende)	Micas (muscovite)	Quartz Potassium feldspars Plagioclase feldspars

Copyright and Photograph by Dr. Parvinder S. Sethi; Joel Arem/Science Source; Copyright and Photograph by Marc S. Hendrix

● **FIGURE 2.10** The five silicate mineral structures are based on sharing of oxygen atoms among silicate tetrahedra.

Modified from Klein, *Manual of Mineral Science, 22nd ed.*, John Wiley & Sons, Inc. 2002.

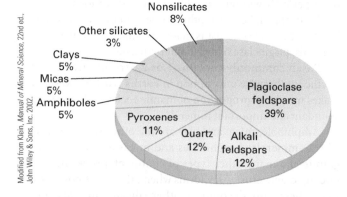

Nonsilicates 8%

Other silicates 3%

Clays 5%

Micas 5%

Amphiboles 5%

Pyroxenes 11%

Quartz 12%

Alkali feldspars 12%

Plagioclase feldspars 39%

● **FIGURE 2.11** More than 92 percent of Earth's crust is composed of silicate minerals. Feldspar alone makes up about 50 percent of the crust, and pyroxene and quartz constitute another 23 percent.

Carbonates

Carbonate minerals are much less common overall than silicates in Earth's crust, but many sedimentary rocks are formed from carbonate minerals, and these now cover large regions of every continent, as described in the following chapter. The shells and other hard parts of many marine organisms—including clams, snails, corals, and even certain types of plankton and green algae—are made of carbonate minerals. The hard pieces accumulate to form the abundant

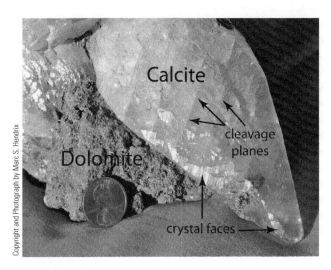

● **FIGURE 2.12** Calcite and dolomite are two common carbonate minerals. In this photograph, several large calcite crystals are growing from a finely crystalline dolomite matrix. Sunlight is reflecting off the faces of the calcite crystal, and one well-developed set of parallel cleavage planes is visible within the calcite crystal.

sedimentary rock called limestone (described in Chapter 3), which is comprised of calcite ($CaCO_3$) and is a rock type mined as a raw ingredient of cement. Another very common carbonate mineral is dolomite ($CaMg(CO_3)_2$), which forms a rock also called dolomite, or dolostone (●**FIGURE 2.12**).

COMMERCIALLY IMPORTANT MINERALS

Many minerals are commercially important even though they are not abundant. Our industrial society depends on metals such as iron, copper, lead, zinc, gold, and silver.

Ore minerals are minerals from which metals or other elements can be profitably recovered. A few, such as native gold and native silver, are composed of a single element. However, most metals are chemically bonded to other elements. Iron is commonly bonded to oxygen. Copper, lead, and zinc are commonly bonded to sulfur to form sulfide ore minerals, such as the galena in ●**FIGURE 2.13**.

Industrial minerals are not metal ores, fuels, or gems but have economic value nonetheless. Halite is mined for table salt, and gypsum is mined for plaster and sheetrock. Apatite and other phosphatic minerals are sources of the phosphate fertilizers crucial to modern agriculture. Calcite is a raw ingredient of cement. Native sulfur—used to manufacture sulfuric acid, insecticides, fertilizer, and rubber—is mined from the craters of dormant and active volcanoes, where it is deposited from gases emanating from the vents, like the one in ●**FIGURE 2.14**.

---

ore minerals Minerals from which metals or other elements can be profitably recovered.

industrial minerals Rocks or minerals that have economic value, exclusive of metal ores, fuels, and gems.

gem A mineral that is prized for its rarity and beauty rather than for industrial use.

● **FIGURE 2.13** Galena (lead sulfide) is the most important ore of lead.

● **FIGURE 2.14** Yellow native sulfur forming in a volcano vent in the Mutnovsky region, Kamchatka Peninsula, Russia.

A **gem** is a mineral that is prized primarily for its rarity and/or beauty, although some gems, such as diamonds, are also used industrially. Depending on its value, a gem can be either *precious* or *semiprecious*. Precious gems include diamond, emerald, ruby, and sapphire (●**FIGURE 2.15**). Several varieties of quartz—including amethyst, agate, jasper, and tiger's eye—are semiprecious gems. Garnet, olivine, topaz, turquoise, and many other minerals can also occur as attractive semiprecious gems (●**FIGURE 2.16**).

● **FIGURE 2.15** Sapphire is one of the most precious gems.

● **FIGURE 2.16** Topaz, a colorless crystal, is a popular semiprecious gem.

When you look at a lofty mountain or a steep cliff, you might not immediately think about the tiny mineral grains that form the rocks. Yet minerals are the building blocks of Earth. In addition, some minerals provide the basic resources for our industrial civilization.

As we will see throughout this book, Earth's surface and the human environment change as minerals react with the air and water, with other minerals, and with living organisms.

HARMFUL AND DANGEROUS ROCKS AND MINERALS

Most rocks and minerals in their natural states are harmless to humans and other organisms. That is not surprising, because all of us evolved among the minerals and rocks that make up Earth, and species would not survive if they were poisoned by their surroundings. A few rocks and minerals, however, are harmful and dangerous. Asbestos is one such mineral; it is a powerful *carcinogen*—a cancer-causing substance. Some rocks and minerals emit radon gas, a radioactive carcinogen. Other rocks and minerals contain toxic elements such as mercury, lead, arsenic, uranium, and sulfur.

In nature, most environmentally hazardous rocks and minerals are buried beneath the surface, where they are unavailable to plants and animals. Natural weathering and erosion expose them so slowly that they do little harm. Most minerals that contain toxic metals or other elements such as lead, mercury, and arsenic are also relatively insoluble. In most natural environments, they weather so slowly that they release the toxic materials in low concentrations. However, if pollution controls are inadequate, mining, milling, and smelting can concentrate and release hazardous natural materials at greatly accelerated rates, poisoning humans and other organisms.

Silicosis and Black Lung

Most common minerals such as feldspar and quartz are harmless in their natural states in solid rock. However, if these minerals are ground to dust, they can enter the lungs and cause serious and even fatal inflammation and scarring of the lung tissue, called silicosis. In advanced cases, the lungs become inflamed and may fill with fluid, causing severe breathing difficulty and low blood oxygen levels. On-the-job exposure to silica dust can occur in mining, stonecutting, quarrying, building and road construction, working with abrasives, sandblasting, and other occupations and hobbies. Intense exposure to silica may produce silicosis in a year or less, but it usually takes at least 10 or 15 years of exposure before symptoms develop. Silicosis has become less common since the Occupational Safety and Health Administration (OSHA) instituted regulations requiring the use of protective equipment that limits the amount of silica dust inhaled. Coal dust, inhaled by coal miners in large quantities before federal laws required dust suppression in coal mines, has similar effects and causes a disease called black lung.

Asbestos, Asbestosis, and Cancer

Asbestos is an industrial name for a group of minerals that crystallize as long, thin fibers. One type is a sheet silicate mineral called chrysotile, which has a crystal structure and composition similar to that of the micas and which forms tangled, curly fibers (●**FIGURE 2.17**). The other type of asbestos includes four similar varieties of amphibole that crystallize as straight, sharply pointed needles. Chrysotile has been the more valuable and commonly used type of asbestos, because the fibers are flexible and tough and can be woven into fabric. It was considered to be a less harmful form because some believed its curly fibers would be more easily expelled from the lungs. However, recent studies show that all forms of asbestos are carcinogenic.[1]

Asbestos is commercially valuable because it is flameproof, chemically inert, and extremely strong. For example,

● **FIGURE 2.17** Chrysotile asbestos is believed by some to be a less potent carcinogen but is still being actively removed from public buildings.

1. "Chrysotile Asbestos Fact Sheet," *Environmental Information Association*, 2009 (http://www.eia-usa.org/wp-content/post-files/chrysotile-fact-sheet1.pdf); "Scientific and Medical Facts about Chrysotile Asbestos Released by the Environmental Information Association and the ADAO," *Medical News Today*, April 29, 2009 (http://www.medicalnewstoday.com/articles/147938.php)

chrysotile fibers are eight times stronger than a steel wire of equivalent diameter. Asbestos has been used to manufacture brake linings, fireproof clothing, insulation, shingles, tile, pipe, and gaskets but now is allowed only in brake pads, shingles, and pipe.

In the early 1900s, asbestos miners and others who worked with asbestos learned that prolonged exposure to the fibers caused an irreversible respiratory disease called asbestosis. Later, in the 1950s and 1960s, studies showed that asbestos also causes lung cancer and other forms of cancer. One reason that so much time passed before scientists recognized the cancer-causing properties of asbestos is that the disease commonly does not develop until decades after exposure.

Although it is not clear how asbestos fibers cause cancer, it seems that the shapes of the crystals play an important role. Statistical studies also show that the sharp, pointed amphibole asbestos is a more potent carcinogen than the flexible chrysotile fibers, although both types cause cancer. In response to growing awareness of the health effects of asbestos, the Environmental Protection Agency (EPA) banned its use in building construction in 1978. However, the ban did not address the issue of what should be done with the asbestos already installed. In 1986, Congress passed a ruling called the Asbestos Hazard Emergency Response Act, requiring that all schools be inspected for asbestos. Public response has resulted in hasty programs to remove asbestos from schools and other buildings, at a cost of billions of dollars. But what is the real level of hazard?

Most asbestos in buildings is the less potent chrysotile, woven into cloth or glued into a tight matrix, and often the surface has been further stabilized by painting. Therefore, the fibers are not free to blow around. The levels of airborne asbestos in most buildings are no higher than those in outdoor air. Some scientists argue that asbestos insulation, despite being carcinogenic, poses no health danger if left alone; but when the material is removed, it is disturbed and asbestos dust escapes. Not only are workers endangered, but airborne asbestos can persist in a building for months after completion of an asbestos-removal project.

Radon and Cancer

Radon is one of a series of elements formed by the radioactive decay of uranium. Uranium occurs naturally in small concentrations in several minerals and in all types of rock, but it concentrates in two abundant rocks—granite and mudstone—that are described in the next chapter. It is also found in soil that formed from granite and mudstone, and in construction materials made from those rocks, such as aggregate or concrete. Radon is itself radioactive, and it decays quickly into other radioactive elements. Because of their radioactivity, radon and its decay products are carcinogenic, and because it is a gas, we inhale radon into our lungs.

Radon seeps from the ground into homes and other buildings, where it concentrates in indoor air and causes an estimated 5,000 to 20,000 cancer deaths per year among Americans. The risk of dying from radon-caused lung cancer in the United States is about 0.4 percent over a lifetime, much greater than the risk of dying from cancer caused by asbestos, pesticides, or other air pollutants and nearly as high as the risk of dying in an auto accident, from a fall, or in a fire at home.

Not all Americans are exposed to equal amounts of radon. Some homes contain very low concentrations of the gas; others have high concentrations. The variations in concentration are due to two factors: geology and home ventilation. Geology is important because some types of rocks, such as granite and mudstone, contain high concentrations of uranium and radon, while others contain relatively little.

As radon forms through the slow radioactive decay of uranium in bedrock, soil, or construction materials, it seeps into the basement of a home and circulates throughout the house. Radon concentrations are highest in poorly ventilated homes built on granite or mudstone, built on soil derived from these rocks, or constructed with concrete and concrete blocks containing these types of rocks. The highest home radon concentrations ever measured were found in houses built on the Reading Prong, a uranium-rich body of granite extending from Reading, Pennsylvania, through northern New Jersey and into New York. The air in one home in this area contained 700 times more radon than the EPA "action level"—the concentration at which the EPA recommends that corrective measures be taken to reduce the amount of radon in indoor air.

People should ask two questions in regard to radon hazards: "What is the radon concentration in my home?" and "If it is high, what can be done about it?" Because radon is radioactive, it can be measured with a simple detector available at most hardware stores or from local government agencies, for about $25. If the detector indicates excessive radon, several solutions are possible for lowering the radon level. One solution is to extend a ventilation duct either from the basement or from below the basement floor to the outside of the house. The ventilation duct prevents radon from accumulating in or below the basement by venting it directly to the outdoors. Another solution is to pump outside air into the house to keep indoor air at a slightly higher pressure than the outside air. This positive pressure prevents gas from seeping from soil or bedrock into the basement. Despite these solutions for lowering radon levels, it is impossible to avoid exposure to radon completely, because it is everywhere—in outdoor air as well as in homes and other buildings. But it is relatively easy and inexpensive to minimize exposure and thus avoid a significant cause of cancer.

Acid Mine Drainage and Heavy Metals Contamination

Sulfide ore minerals are combinations of metals such as lead, zinc, copper, cadmium, mercury, and other heavy metals with sulfur. Some contain arsenic or other toxic elements in

● **FIGURE 2.18** The Rio Tinto (translated *Red River*) in Spain owes its name to the color resulting from acid mine drainage. The Rio Tinto flows past a large open pit mining site where gold, silver, and copper have been extracted and processed since about 3,000 BCE. The mine is active today and contributes to the acidity of the stream, making it capable of dissolving and transporting heavy metals such as iron, lead, zinc, and cadmium. The yellow coating on rocks in the stream channel is produced by a coating of iron oxide.

DIGGING DEEPER

Here to Stay: Mercury

Mercury used to extract gold and silver from ore during the Montana frontier now reaches across the food chain.

One of the most challenging environmental problems associated with gold and silver smelting is the introduction of mercury into the environment. Mercury, symbolized Hg, is a very toxic metal that affects the brain, kidneys, and lungs and is extremely mobile once released into the environment. Mercury is the only native metal that is in the liquid state at room temperature. In this form, mercury exists as a native element with no charge (Hg^0). However, mercury also can exist as a positively charged ion (Hg^{2+}) capable of forming complexes with inorganic and organic compounds. The chemically reactive nature of mercury results in it being quickly dispersed into the environment once released.

Because it exists as a liquid metal at room temperatures, mercury can be used to recover tiny flecks of gold from crushed rock or sediment (● **FIGURE 2.19**). When the crushed rock or sediment is mixed with liquid mercury, the gold metal and mercury metal stick together. The mercury/gold mixture naturally separates from the rest of the nonmetallic rock matrix and then is heated in a crucible. The heat drives off the mercury as a vapor, while leaving the gold behind. The mercury vapor disperses quickly into the atmosphere, often by clinging to tiny dust particles that also are released by the smelting process. The mercury-laden dust and vapor enter the global atmospheric cycle and eventually combine with precipitation to settle on plants, soil, and lake and river surfaces, where it enters the base of the food chain.

Recent environmental investigations in western Montana have used the fish-eating bird of prey, the osprey, as a means of pinpointing specific river drainages in which mercury contamination exists as a legacy of historic silver and gold mining and smelting during the 19th and early 20th centuries. Osprey exclusively eat fish, and the fish feed on aquatic insects that in turn eat mercury-contaminated microorganisms in the streams. With each step in the food chain, the mercury is concentrated—or biomagnified—so that osprey, at the top of the food chain, can contain very high levels of the metal in their body tissue (● **FIGURE 2.20**). It is therefore not surprising that those stream drainages in western Montana that have hosted long histories of gold mining and smelting are inhabited by osprey with the highest mercury levels in their blood (● **FIGURE 2.21**).

Escape of toxic elements such as lead, mercury, cadmium, and arsenic into the environment from mining and smelting activities is a global problem. Although areas with historic mining are far likelier to have elevated mercury concentrations, the recent detection of mercury in the sediment of isolated

● **FIGURE 2.19** Nineteenth-century hard-rock miners in Montana. These miners commonly used mercury to extract gold and silver from the ore they recovered.

addition to the metals. These minerals are mined for their metals, which are essential to modern industrial societies.

However, mining and refining sulfide ore minerals can create serious air and water pollution problems. When these minerals are mined or refined without adequate pollution control, sulfur escapes into streams, groundwater, and the atmosphere, where it forms hydrogen sulfide and sulfuric acid. Waterborne sulfides poison aquatic organisms, and atmospheric sulfur compounds contribute to acid precipitation.

The sulfuric acid that forms from the weathering of sulfide ore minerals can cause other minerals from the rock in and around the mining site to be dissolved. Dissolution of these minerals releases into solution the metals they contain in their crystal structure. Once dissolved, the metals can quickly become part of the groundwater and surface water systems if not contained. Arsenic, lead, zinc, and copper are several of the metals that are common pollutants in areas affected by acid mine drainage (●FIGURE 2.18).

● **FIGURE 2.20** A scientist from The University of Montana uses a cherry picker to access an osprey nest to collect blood samples from the young birds for mercury analysis.

● **FIGURE 2.21** Mercury that occurs in the streams and lakes is concentrated in aquatic insects, fish, and osprey, a bird that eats fish exclusively. By analyzing the mercury levels in the blood of osprey, scientists can identify drainages affected by mercury pollution from historic mining activities.

alpine lakes in the western United States has caused many scientists to conclude that atmospheric delivery of this toxic metal from distant industrial sources, particularly in Asia, is occurring.

Because it is chemically reactive and capable of combining with negatively charged organic molecules, mercury tends to occur naturally in sediments rich in organic matter. Coal, a rock containing over 50 percent organic carbon, is used worldwide as a source of energy. Unfortunately, however, the increasingly widespread burning of coal since the advent of the Industrial Revolution has released much mercury into the atmosphere, and areas formerly considered pristine now contain detectable mercury from atmospheric fallout.

The magnitude of this problem is likely to grow in the future as the demand for world energy increases. Coal is plentiful (particularly in the United States and China) and is a relatively cheap source of energy. In 2018, an estimated 150,000 fully-loaded train hopper cars of coal passed from coal mines in Montana and Wyoming westward to the coast, where the coal was loaded on ships bound for Asia. Each train hauled between 100 and 125 fully loaded boxcars of coal. Ironically, much of the mercury contained in these coal trains returns to America with the jet stream and contributes to our

mercury problem (●FIGURE 2.22). Thus, although the consumption of large quantities of coal is providing much of the energy needed for global economic growth, the burning of the coal comes with serious environmental challenges, including the release of mercury, that must be considered by current and future generations.

Shovel and Texture Photos: Copyright Shutterstock.com

● **FIGURE 2.22** Several trains fully loaded with coal from Montana and Wyoming pass west daily to the port of Seattle, where the coal is loaded onto barges heading for Asia. Ironically, much of the mercury released by the burning of this coal in Asia returns to North America on the jet stream.

Key Concepts Review

- Minerals are the substances that make up rocks. A mineral is a naturally occurring, inorganic solid with a definite chemical composition and a crystalline structure.
- Each mineral consists of chemical elements bonded together in definite proportions, so that its chemical composition can be given as a chemical formula.
- Every mineral has a crystalline structure—an orderly, periodically repeated arrangement of its atoms—and therefore every mineral is a crystal. The shape of a crystal is determined by the shape and arrangement of its atoms. Every mineral is distinguished from others by its chemical composition and crystal structure.
- Most common minerals are easily recognized and identified visually. Identification is aided by observing a few physical properties, including crystal habit, cleavage, fracture, hardness, specific gravity, color, streak, and luster.

- Although more than 3,500 minerals are known in Earth's crust, only the nine rock-forming minerals make up most of Earth's crust. They are feldspar, quartz, pyroxene, amphibole, mica, clay, olivine, calcite, and dolomite. The first seven on this list are silicates; their structures and compositions are based on the silicate tetrahedron, in which a silicon atom is surrounded by four oxygen atoms to form a pyramid-shaped structure. Two carbonate minerals, calcite and dolomite, are also sufficiently abundant to be called rock-forming minerals.
- Ore minerals, industrial minerals, and gems are important for economic reasons.
- Most minerals and rocks are environmentally safe in their natural states, but some can release environmentally hazardous materials when they are mined, milled, or smelted.

Important Terms

anion (p. 19)

cation (p. 19)

carbonates (p. 24)

cleavage (p. 21)

crystal (p. 20)

crystal face (p. 20)

crystal habit (p. 21)

crystalline structure (p. 20)

element (p. 19)

fracture (p. 22)

gem (p. 27)

hardness (p. 22)

industrial minerals (p. 27)

luster (p. 23)

ion (p. 19)

mineral (p. 19)

Mohs hardness scale (p. 22)

native elements (p. 24)

ore minerals (p. 27)

rock-forming minerals (p. 20)

silicates (p. 24)

specific gravity (p. 23)

streak (p. 23)

sulfides (p. 24)

silicate tetrahedron (p. 25)

Review Questions

1. What properties distinguish minerals from other substances?

2. Explain why oil and coal are not minerals.

3. What does the chemical formula for quartz, SiO_2, tell you about its chemical composition? What does $KAlSi_3O_8$ tell you about orthoclase feldspar?

4. What is an atom? An ion? A cation? An anion? An anionic group? What role does each play in minerals?

5. Every mineral has a crystalline structure. What does this mean?

6. What are the factors that control the shape of a well-formed crystal?

7. What is a crystal face?

8. What conditions allow minerals to grow well-formed crystals? What conditions prevent their growth?

9. List and explain the physical properties of minerals that are most useful for identification.

10. Why do some minerals have cleavage whereas others do not? Why do some minerals have more than one direction of cleavage planes?

11. Why is color often an unreliable property for mineral identification?

12. List the rock-forming minerals. Why are they called rock-forming? Which are silicates? Why are so many of them silicates?

13. Draw a three-dimensional view of a single silicate tetrahedron. Draw the five arrangements of tetrahedra found in the rock-forming silicate minerals. How many oxygen ions are shared between adjacent tetrahedra in each of the five configurations?

14. Make a table with two columns. List the basic silicate structures in the left column. In the right column, list one or more examples of rock-forming minerals for each structure.

15. Explain how mining can release harmful or poisonous materials from rocks that were benign in their natural environment.

Bill45/ShutterStock.com

ROCKS

Boulders, cobbles, and pebbles along the shoreline of Swiftcurrent Lake in Glacier National Park, Montana. These rocks were eroded from the surrounding mountains by glaciers and streams, transported downslope by gravity, and partially rounded by wave energy on the lake shoreline.

INTRODUCTION

Earth is solid rock to a depth of 2,900 kilometers, where the mantle meets the liquid outer core. Even casual observation reveals that rocks are not all alike. The great peaks of the Sierra Nevada in California are hard, strong granite. The red cliffs of the Utah desert are soft sandstone. The top of Mount Everest is limestone, containing the fossil remains of small marine invertebrates.

The marine fossils atop Mount Everest tell us that this limestone formed in the sea. What forces lifted the rock to the highest surface elevation on Earth? Where did the vast amounts of sand in the Utah sandstone come from? How did the granite of the Sierra Nevada form?

All of these questions ask about the processes that formed the rocks and changed them throughout geologic history. In this chapter, we will study rocks, what they are made of, and how they form.

ROCKS AND THE ROCK CYCLE

A rock is a solid aggregate of one or more minerals. Based on how the rocks form, geologists define three main categories: igneous rocks, sedimentary rocks, and metamorphic rocks.

Earth's interior is hot and dynamic. The high temperature that exists within a few hundred kilometers of the surface can cause solid rock to melt, forming a molten liquid called **magma**. Because the liquid magma is less dense than the surrounding solid rock, the magma rises slowly toward Earth's surface. As it rises, the magma cools. **Igneous rock** forms when cooling magma solidifies.

All rocks may seem permanent and unchanging over a human lifetime, but this apparent permanence is an illusion created by our short observational time frame. Over geologic time, water and air attack rocks of all kinds at Earth's surface through the process called **weathering**, breaking them down into smaller particles. These particles—including gravel, sand, clay, and all other fragments weathered and eroded from rock—are transported away from the site of weathering by streams, glaciers, wind, and gravity. The particles eventually accumulate in loose, unconsolidated layers called **sediment**. Sand on a beach and mud on a lake bottom are examples of sediment. **Sedimentary rock** forms when sediment particles become cemented or compacted into solid rock.

Weathering processes also form dissolved ions such as sodium, calcium, and chloride. Weathering of solid rock causes some of the minerals in it to change or dissolve altogether, thereby liberating positively charged cations and negatively charged anions. These ions are transported away from the site of weathering by streams and groundwater. Many households contain water softeners to remove these dissolved ions so that they don't precipitate and clog the home's plumbing system.

Most of the dissolved ions formed by weathering eventually are carried to the sea. There, marine organisms such as clams, oysters, and corals extract dissolved calcium from seawater. They combine the calcium with carbon dioxide that dissolves in seawater from Earth's atmosphere, and use it to form their shells and other hard parts. After the organisms die, the remains of those shells accumulate. Precipitation of mineral cements between the shells and other particles results in the formation of sedimentary rock called limestone. Thus, limestone, one of the most common sedimentary rocks, forms by direct interactions among the geosphere, the hydrosphere, the biosphere, and the atmosphere.

A **metamorphic rock** forms when heat, pressure, or hot water alters any preexisting rock. For example, when rising magma intrudes the Earth's crust, it heats the rock around it. Although the heat may not be sufficient to melt rock surrounding the magma, it can bring about other changes. For example, the mineral crystals in the heated rock may grow or change their internal arrangement. Such alterations in the size, shape, and arrangement of mineral

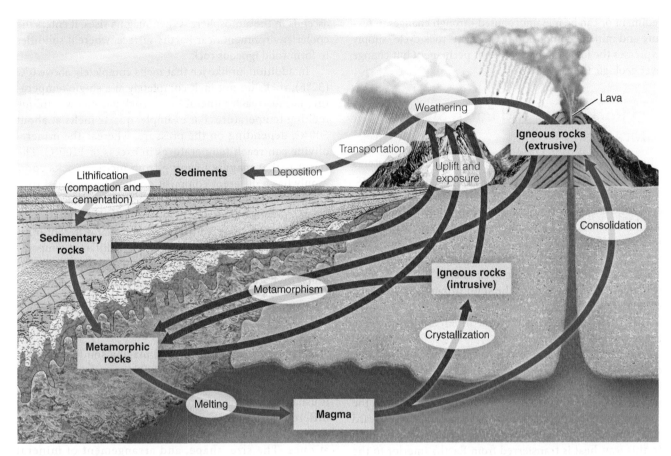

● **FIGURE 3.1** The rock cycle shows that rocks change over time. The arrows show the paths that rocks can follow as they change.

crystals within the rock are changes in the rock's **texture**. In addition, temperature and pressure within the Earth's crust may cause entirely different mineral types to grow that are stable at those changed temperatures and pressures. Thus,

---

magma Molten rock generated from melting of any rock in the subsurface; cools to form igneous rock.

Igneous rock Rock that forms when magma cools and crystallizes.

weathering The decomposition and disintegration of rocks and minerals at Earth's surface by chemical and physical processes.

sediment Solid rock or mineral fragments that are transported and deposited by wind, water, gravity, or ice; that are weathered by natural forces, precipitated by chemical reactions, or secreted by organisms; and that accumulate in loose, unconsolidated layers.

Sedimentary rock Rock formed when sediment becomes compacted and cemented through the process of lithification.

metamorphic rock A rock formed when igneous, sedimentary, or other metamorphic rocks recrystallize in response to elevated temperature, increased pressure, chemical change, and/or deformation.

texture The size, shape, and arrangement of mineral grains, or crystals, in a rock.

rock cycle The sequence of events in which rocks are formed, destroyed, altered, and reformed by geological processes.

the process of metamorphism can change both the texture and mineral types within a preexisting rock. Unlike igneous rocks, however, metamorphic rocks form without completely melting.

No rock is permanent over geologic time; instead, all rocks undergo processes that change them from one of the three rock types to another. This continuous process is called the **rock cycle** (●FIGURE 3.1). For example, as sediment accumulates and is buried, it typically cements together to form a sedimentary rock. If sedimentary rock at the bottom of the accumulation is buried deeply enough, the rising temperature and pressure will cause it to convert to a metamorphic rock as it undergoes changes in texture and mineral composition. With additional burial and temperature increase, the metamorphic rock can melt, forming magma. The magma will then rise in the crust, slowly cool, and solidify to become igneous rock. Millions of years later, movement of Earth's crust might raise the igneous rock to the surface, where it will weather to form sediment. Rain and streams will then wash the sediment into a new basin, renewing the cycle.

The rock cycle does not follow a set order and can take many different paths. For example, all three rock types can melt to form magma and, eventually, an igneous rock. Similarly, all three rock types can be weathered to form sediment or can be metamorphosed through changes in texture and mineral composition. The term "rock cycle" simply expresses the idea that rocks are not permanent but change over geologic time.

The Rock Cycle and Earth Systems Interactions

The rock cycle illustrates several types of Earth systems interactions—interactions among rocks, the atmosphere, the biosphere, and the hydrosphere. Rain and air, aided by acids and other chemicals secreted by plants, decompose solid rocks to form large amounts of clay and other sediment. During these processes, water and atmospheric gases react chemically and become incorporated into the clay. Thus, these processes transfer water and air from the atmosphere and hydrosphere to the solid minerals of the geosphere. More rain then washes the clay and other sediment into streams, which carry it to a sedimentary basin where the clay is deposited.

The rock cycle is also driven by Earth's internal heat. For example, when layers of sediment are deposited and buried by younger layers, the deeper layers become heated and metamorphosed by Earth's heat. The same heat may melt the rocks to produce magma. But then the magma rises upward and perhaps even erupts onto Earth's surface from a volcano. In this way, heat is transferred from Earth's interior to the atmosphere.

Throughout this chapter, we emphasize these and other interactions among the atmosphere, biosphere, hydrosphere, and rocks of the geosphere to illustrate the point that those systems continuously exchange both energy and material so that Earth functions as a single, integrated system.

IGNEOUS ROCKS

Magma: The Source of Igneous Rocks

If you drill a well deep into in the middle of one of Earth's continents, you would find that the temperature within the crust rises about 30°C for every kilometer of depth. In the mantle between depths of 100 and 350 kilometers, the temperature is so high that rocks in some areas melt to form magma. Depending on its chemical composition and the depth at which it forms, the temperature of magma varies from about 600°C to 1,400°C. As a comparison, an iron bar turns red hot at about 600°C and melts at slightly over 1,500°C. Unlike ice, which *decreases* in volume when it melts to form water (ice floats), rocks *increase* in volume when they melt. When a rock melts to form magma, it expands by about 10 percent, so it is less dense than the rock around it and therefore rises as it forms—much as a hot air balloon

ascends in the atmosphere. When magma rises, it enters the cooler environment near Earth's surface, where it solidifies to form solid igneous rock.

In addition, unlike ice that melts completely above 0°C (32°F), rocks do not melt completely at a single temperature because each mineral in the rock has its own unique melting temperature. For example, quartz melts at about 500°C, depending on the pressure, whereas the mineral olivine can reach temperatures in excess of 1100°C. The result is that magma forming from melting rocks commonly has a different composition than the original rock because typically only those minerals with the lowest melting temperature melt, leaving behind those minerals with higher melting temperatures. The fact that different minerals melt—and freeze or crystallize—at different temperatures results in much of the variety of igneous rock types found on Earth.

Types of Igneous Rocks

Some igneous rocks form when magma solidifies within Earth's crust; other igneous rocks are created when magma erupts onto the surface. When magma solidifies, it usually crystallizes to form minerals. As magma cools, different minerals crystallize at different points during the cooling history—the minerals do not all crystallize at once. The size, shape, and arrangement of mineral crystals in an igneous rock are referred to as the rock's texture. Although some igneous rocks consist of mineral crystals that are too small to be seen with the unaided eye, others are made up of thumb-sized, or even larger, crystals (**TABLE 3.1**).

EXTRUSIVE (VOLCANIC) ROCKS When magma rises all the way through the crust to erupt onto Earth's surface, it forms **extrusive igneous rock**, also called *volcanic rock*. **Lava** is fluid magma that flows from a crack or a volcano onto Earth's surface. The term also refers to the rock that forms when lava cools and becomes solid.

TABLE 3.1 Igneous Rock Textures Based on Crystal Size

Name of Texture	Crystal Size
Glassy	No mineral crystals (obsidian)
Very finely crystalline	Too fine to see with unaided eye
Finely crystalline	Up to 1 millimeter
Medium crystalline	1 to 5 millimeters
Coarsely crystalline	More than 5 millimeters
Porphyry	Relatively large crystals in a finely crystalline matrix

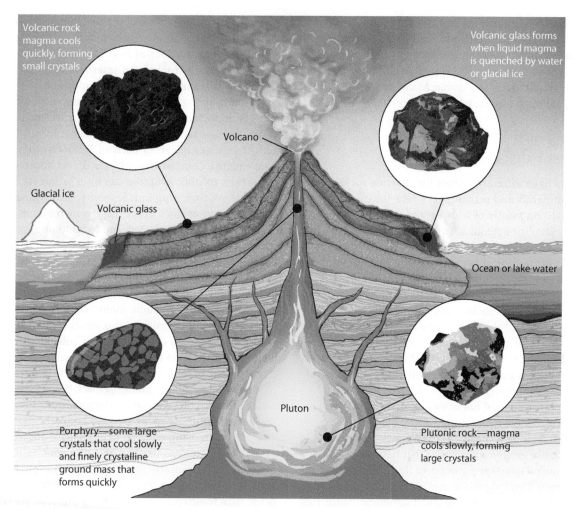

FIGURE 3.2 Differences in igneous rock textures resulting from variations in the length of time a magma has to cool before becoming solid igneous rock. Plutonic rock cools slowly as it remains deep underground, so its crystals are large. Volcanic rock commonly is finely crystalline because of the relatively short cooling time. Porphyry has large crystals, called phenocrysts, that occur within a more finely-crystalline ground mass. The phenocrysts form from minerals that crystallize slowly within the magma chamber prior to the eruption. The eruption brings the phenocrysts along with still-liquid magma to the surface. The ground mass forms from rapid cooling and crystallization of the liquid lava at the surface. Volcanic glass, or obsidian, forms from liquid magma that encounters water or ice upon eruption. The magma quenches so quickly that the atoms freeze in place and no crystals form.

After lava erupts onto the relatively cool Earth surface, it solidifies rapidly—in a period ranging from a few days to a few years. Crystals form but do not have much time to grow. As a result, many volcanic rocks consist of crystals too small to be seen with the unaided eye. Basalt is a common, very finely crystalline volcanic rock.

If erupting lava encounters glacial ice or cold seawater, it may solidify, or quench, within a few hours of erupting.

extrusive igneous rock Igneous rock formed from material that has erupted through the crust onto the surface of Earth; usually finely crystalline. Also called *volcanic rock*.

lava Fluid magma that flows onto Earth's surface from a volcano or fissure. Also, the rock formed by solidification of the same fluid magma.

intrusive igneous rock A rock formed when magma solidifies within Earth's crust without erupting to the surface; usually medium to coarsely crystalline. Also called *plutonic rock*.

Because the magma hardens so quickly, the atoms have no time to align themselves to form crystals. As a result, the atoms are frozen into a random chaotic pattern, as happens in glass. Volcanic glass is called obsidian (●**FIGURE 3.2**).

If magma rises slowly through the crust before erupting, some crystals may grow, while most of the magma remains molten. If this mixture of magma and crystals then erupts onto the surface, the magma solidifies quickly, forming porphyry, a rock with large crystals, called phenocrysts, embedded in a finely-crystalline matrix called ground mass (Figure 3.2).

INTRUSIVE (PLUTONIC) ROCKS When magma solidifies *within* the crust, without erupting to the surface, it is an **intrusive igneous rock**, also called plutonic rock, named after the Greek god of the underworld. The rock overlying the magma insulates it like a thick blanket, providing hundreds of thousands, or even millions, of years for the magma to crystallize. As a result, most plutonic rocks are medium-to-coarsely crystalline. Granite, the most abundant rock in continental crust, is a medium or coarsely crystalline plutonic rock. The crystals in granite are clearly visible. Many are a millimeter or so across, although some crystals may be much larger.

Naming and Identifying Igneous Rocks

Geologists have different names for igneous rocks based on their minerals and texture. Let's take for example two rocks consisting mostly of feldspar and quartz. Medium or coarsely crystalline igneous rock made up of these minerals is called granite. When igneous rock with these minerals is very finely crystalline, it is called rhyolite. You can see the difference in the panels of ●FIGURE 3.3. The same magma that solidifies slowly within the crust to form granite can also erupt onto Earth's surface to form rhyolite.

Like granite and rhyolite, most common igneous rocks are classified in pairs, with the two members of a pair containing the same minerals but with different crystal sizes. The crystal size depends mainly on whether the rock is volcanic (finely crystalline) or plutonic (coarsely crystalline).

The specific suite of minerals that make up an igneous rock determines the rock's composition. Igneous rock compositions fall along a spectrum from those containing the most silica, like granite and rhyolite that have so much silica that the mineral quartz is common, to those with so little silica that quartz is entirely absent and most of the minerals that do form are instead rich in iron and magnesium. Silica-rich igneous rocks are called *felsic*, whereas those with little

silica and abundant iron and magnesium are called *mafic*. Ultramafic rocks contain exceptionally low levels of silica and high concentrations of iron- and magnesium-bearing minerals. Igneous rocks with intermediate levels of silica simply are called intermediate in composition.

Understanding differences in the composition of igneous rocks is important because these differences relate directly to the geologic setting in which a magma forms, providing important clues to earlier parts of Earth's history. In addition, magma composition plays a direct control on how powerful a volcanic eruption can be. The largest, most catastrophic volcanic eruptions involve felsic magmas because the abundance of silica in the magma increases its viscosity, or stiffness, requiring the volcano to build up more internal pressure before erupting.

Once you learn to identify the rock-forming minerals in an igneous rock, it is easy to name a coarsely crystalline plutonic rock because the minerals are large enough to see. It is more difficult to name many volcanic rocks, because the minerals commonly are too small to identify. A field geologist often uses color to name a volcanic rock. Rhyolite is usually light in color because the silica-rich minerals characteristic of felsic rocks, such as quartz and feldspar, are light colored. Basalt is commonly black because of the abundance of dark-colored iron- and magnesium-bearing minerals. Many andesite rocks are gray or green (●FIGURE 3.4). In addition to using color to help identify different volcanic rocks, geologists use the heft of a rock to judge how dense it is. Ultramafic and mafic igneous rocks rich in iron- and magnesium-bearing minerals are notably more dense than felsic rocks in which these minerals are relatively uncommon.

(A)

(B)

(C)

(D)

● **FIGURE 3.3** (A) Granite sample showing coarse individual crystals of quartz, feldspar, and mica. (B) A porphyry showing coarse crystals, called "phenocrysts", within a more finely crystalline matrix. (C) Rhyolite made of up fine crystals. The pencil tip is pointing to a white feldspar crystal. (D) Obsidian— volcanic glass— that quenched before crystals could form, causing the atoms to freeze in place. Notice the curved fractures, called conchoidal fractures, characteristic of glass.

(A)

(B)

(C)

● **FIGURE 3.4** Compositional differences in igneous rocks. (A) Light-colored rhyolite, formed from silica-rich magmas; (B) Gray-colored andesite, intermediate-composition igneous rocks formed in volcanic arcs; (C) Dark-colored basalt, formed from mafic magmas rich in iron and magnesium and relatively poor in silica.

Common Igneous Rocks

Before proceeding with our discussion, you should become familiar with some terminology. Geologists commonly use the terms *basement rock, bedrock, parent rock,* and *country rock.* **Bedrock** is the solid rock that lies beneath soil or unconsolidated sediments. It can be igneous, metamorphic, or sedimentary. **Parent rock** is any original rock before it is changed by weathering, metamorphism, or other geological processes. The rock that already exists in an area and that is crosscut by younger intrusive igneous rocks or minerals deposits is called **country rock.** **Basement rock** is the igneous and metamorphic rock that lies beneath the thin layer of sediment and sedimentary rocks covering much of Earth's surface, thus forming the "basement" of the crust.

Granite and Rhyolite

As discussed above, granite contains mostly feldspar and quartz. Small amounts of black biotite or hornblende often give it a speckled appearance. Granite (and metamorphosed granite) is the most common rock in continental crust. It is found nearly everywhere as basement rock, beneath the relatively thin veneer of sedimentary rocks and soil that covers most of the continents. Granite is hard and resistant to

weathering; it forms steep, sheer cliffs in many of the world's great mountain ranges. Mountaineers prize granite cliffs for the steepness and strength of the rock.

As granitic magma (magma with the chemical composition of granite) rises through Earth's crust, some of it may erupt from a volcano to form rhyolite, while the remainder solidifies beneath the volcano, forming granite. Most obsidian forms from magma with a granitic (rhyolitic) composition. Today, an active granitic magma body that has erupted large volumes of rhyolite and produced abundant rhyolitic obsidian in the recent geologic past is the Yellowstone Volcano, where magma is located only a few kilometers below the surface. Over the past 2.1 million years, the Yellowstone Volcano has produced three catastrophic eruptions that have covered much of the continent with volcanic ash. In addition to these highly explosive eruptions, numerous flows of rhyolite have spread across the volcano, largely filling in its central crater. The longest of the flows are tens of kilometers in length, and several of the flows have an outer sheath of obsidian that formed by rapid quenching of the rhyolite by water or glacial ice (●**FIGURES 3.5** and **3.6**).

Basalt and Gabbro

Basalt consists of approximately equal amounts of plagioclase feldspar and pyroxene. It makes up most of the oceanic crust, as well as huge basalt plateaus on continents. Gabbro is the plutonic equivalent of basalt; it is mineralogically identical but consists of larger crystals. Gabbro is uncommon at Earth's surface, although it is abundant in deeper parts of the oceanic crust, where basaltic magma crystallizes slowly.

In North America, basalt is particularly abundant across much of the Pacific Northwest where a gigantic volume of basaltic magma erupted numerous times to form the Columbia River Plateau Basalts about 17 million years ago (●**FIGURE 3.7**). In addition, basalt is common across much

---

bedrock The solid rock that lies beneath soil or unconsolidated sediments; it can be igneous, metamorphic, or sedimentary.

parent rock Any original rock before it is changed by weathering, metamorphism, or other geological processes.

country rock The older rock already in an area, cut into by a younger igneous intrusion or mineral deposit.

basement rock The older igneous and metamorphic rock that lies beneath the thin layer of sedimentary rocks and soil covering much of Earth's surface; forms the "basement" of the crust.

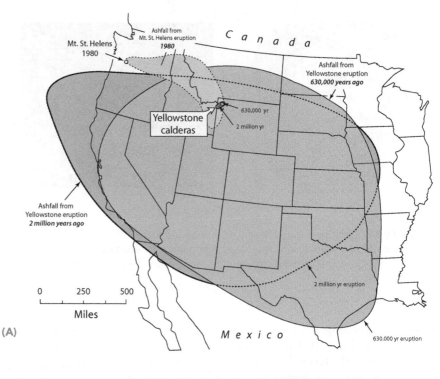

(A)

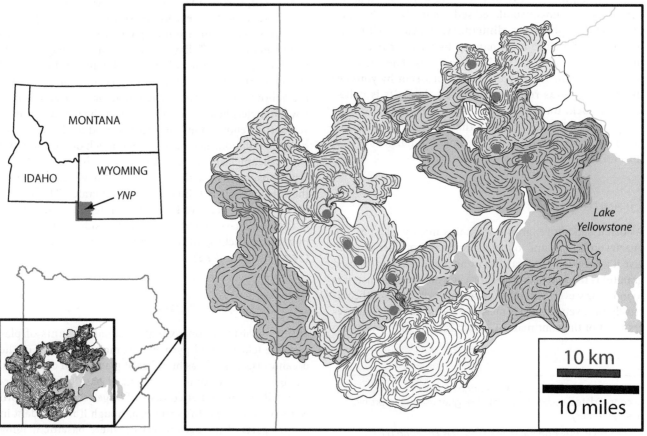

(B)

● **FIGURE 3.5** (A) The distribution of volcanic ash from the three largest eruptions to have occurred from the Yellowstone Volcano over the past 2.1 million years. (B) Much of Yellowstone National Park is covered by relatively young very large flows of rhyolite lava. This map shows the largest of the rhyolite flows, each with a different color. The red dots mark the location of known vents from which the lava erupted. The thin black lines on each flow are mapped "flow folds" formed by rumpling of the partially solidified surface of the flow. The longest of the rhyolite lava flows exceeds 20 kilometers in length.

(A)   (B)

● **FIGURE 3.6** (A) This cliff in Yellowstone is made mostly of volcanic glass formed at the leading edge of a flow of rhyolite lava that erupted about 500,000 years ago. (B) A close up of the volcanic glass. Notice the conchoidal fracture pattern that is characteristic of broken glass. The quarter is for scale.

● **FIGURE 3.7** Columbia River Plateau basalt exposed in eastern Washington State. Each layer was formed by a basaltic lava flow, with thousands of years separating each flow.

● **FIGURE 3.8** Palisades Sill, Hudson River Valley, made of basalt that was intruded when North America and Eurasia broke apart about 200 million years ago.

of the Great Basin of Utah and Nevada, where it is forming due to tectonic stretching of the crust and partial melting of the upper mantle. A particularly famous basalt locality is the Hudson River Valley in New York, where the Palisades Basalt formed about 175 million years ago as North America and Africa broke apart (●**FIGURE 3.8**).

Andesite and Diorite

Andesite is a volcanic rock intermediate in composition between basalt and granite. It is commonly gray or green and consists of plagioclase feldspar and dark minerals (usually biotite, amphibole, or pyroxene). It is named for the Andes Mountains, the volcanic chain on the western edge of South America, where andesite is abundant. Because it is volcanic, andesite is typically very fine grained. Diorite is the plutonic equivalent of andesite. It forms from the same magma as andesite and, consequently, often underlies andesitic mountain chains such as the Andes.

Andesite is most common in volcanic chains, like the Andes, that are called volcanic arcs because they typically form an arc-like pattern on geologic maps. Volcanic arcs are developed at plate boundaries where one tectonic plate descends beneath another and is slowly melted in the process. The Cascade Mountains of the Pacific Northwest are such a volcanic arc and are made mostly of andesite and diorite (●**FIGURE 3.9**). The Aleutian Island chain of Alaska is another volcanic arc where diorite and andesite are forming today.

Peridotite and Komatiite

Peridotite is an *ultramafic* (rich in magnesium and iron) igneous rock that makes up most of the upper mantle but is rare in Earth's crust. It is coarse grained and composed of olivine and small amounts of pyroxene, amphibole, or mica, but not feldspar. Komatiite is the finely crystalline, extrusive equivalent of peridotite. It is a rare rock type at the surface of today's Earth because modern volcanoes do not erupt

● **FIGURE 3.9** Overlooking the city of Seattle is Mount Rainier, an andesitic volcano that is part of the Cascade Volcanic Arc.

ultramafic magma—it's simply too dense to reach the surface and must pass through a much more felsic crust with minerals that melt at low temperatures before reaching the surface. However, during the early parts of Earth's history more than 4 billion years ago, the first crust to form from the molten planet was made of ultramafic rock. As a result, komatiites formed abundantly during Earth's early history, but today, they are preserved only in those few localities in which rocks of this age are preserved.

SEDIMENTARY ROCKS

Over geologic time, the atmosphere, biosphere, and hydrosphere weather rock, breaking it down to gravel, sand, silt, clay, and ions dissolved in water (●**FIGURE 3.10**). Weathering of rocks occurs by both chemical and physical processes. Glaciers, flowing water, gravity, and wind all erode the rock, transport the rock fragments downslope, and deposit them at lower elevations.

Dissolved ions resulting from weathering also are transported downslope, commonly all the way to the sea, where the ions are concentrated. Marine organisms such as clams, snails, certain kinds of green algae, and corals use some of these ions to form shells and other hard parts. After the organisms die, these hard parts accumulate to form limestone. Other ions can react chemically to produce a solid salt, through the process of **precipitation**. For example, the mineral halite (rock salt) precipitates from a lake on the floor of Death Valley.

Sedimentary rocks make up only about 5 percent of Earth's crust. However, because they form on Earth's surface, sedimentary rocks are widely spread in a thin veneer over underlying igneous and metamorphic rocks. As a result, they cover about 75 percent of continents.

Sedimentary rocks are broadly divided into four categories:

1. *Clastic sedimentary rock* is composed of particles of weathered rocks, such as sand grains and pebbles, called

(A)

(B)

● **FIGURE 3.10** The products of weathering. (A) Rounded gravel along the Flathead River, Montana. The gravel clasts are derived from the physical weathering and breakdown of bedrock, followed by transport and rounding of the clasts by the river. (B) The floor of Death Valley, where dissolved ions derived from weathered bedrock are transported into Death Valley and then concentrated through evaporation.

clasts, which have been transported, deposited, and lithified. (The generic term *clastic* refers to any rocks that are composed of fragments of older rocks.) This category includes conglomerate, sandstone, and mudstone (●**FIGURE 3.11**). Clastic sedimentary rock makes up about 85 percent of all sedimentary rock.

2. *Organic sedimentary rock* consists of the lithified remains of plants or animals. Coal is an organic sedimentary rock that contains such a high percentage of decomposed and compacted plant remains that the rock itself will burn (●**FIGURE 3.12**).

3. Chemical sedimentary rock forms by direct precipitation of minerals from solution. Rock salt, for example, forms when halite precipitates from evaporating seawater or saline lake water.

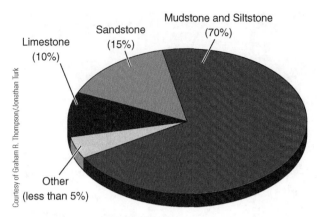

● **FIGURE 3.11** Earth's sedimentary rocks. Mudstone, siltstone, and sandstone are clastic rocks that make up about 85 percent of all sedimentary rocks. Limestone and other sedimentary rocks make up less than 15 percent.

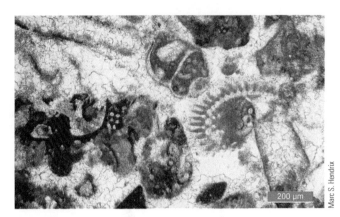

● **FIGURE 3.13** Bioclastic limestone from West Texas. This image was taken using a special microscope to examine a very thin slice of the rock. The scale bar (200 micrometers) is 0.2 millimeters long. Look for fragments (clasts) of fossils in the image.

● **FIGURE 3.12** Layers of coal and other sedimentary rocks in Utah.

4. *Bioclastic sedimentary rock* is composed of broken shell fragments and similar remains of living organisms. The fragments are clastic, but they have a biological origin. Many limestones are formed from broken shells and thus are bioclastic sedimentary rocks (●**FIGURE 3.13**).

Clastic Sedimentary Rocks

Clastic sediment is called gravel, sand, silt, or clay, in order of decreasing particle size. **TABLE 3.2** shows the main sediment particle sizes and their corresponding diameters. As

---

precipitation A chemical reaction that produces a solid salt, called a *precipitate*, from a liquid solution.

pore space The empty space between particles of rock, sediment, or soil.

compaction Increased packing together of sedimentary grains, usually resulting from the weight of overlying sediment; causes a decrease in porosity and contributes to lithification.

TABLE 3.2 Clastic Sediment Particle Types and Sizes

Sediment Particle Name	Sediment Particle Diameter (mm)
Boulder	>256
Cobble	64–256
Pebble	4–64
Granule	2–4
Sand	0.0625–2.0
Silt	0.0039–0.0625
Clay*	<0.0039

clastic particles ranging in size from boulders to coarse silt tumble downstream, their sharp edges are worn off and they become rounded. Finer silt and clay do not round effectively because they are so small and light that water and wind cushion them as they are buffeted along.

If you hold a pile of sand in your hand and dribble water onto it, the water will soak into the empty zone—called **pore space**—among the sand grains (●**FIGURE 3.14**). Commonly, sand and similar sediment have about 20 to 40 percent pore space. As we will see in Chapter 5, this pore space is very important because it is where economic quantities of water, petroleum, and natural gas typically occur.

As sediment accumulates and is buried, it is compressed by the weight of the overlying layers. The compression partially collapses the pore spaces between sediment grains, forcing water out (●**FIGURE 3.14B**). This process is called **compaction**.

As sediment is buried and compacted, groundwater slowly circulates through the remaining pore space. This water commonly contains dissolved ions that can precipitate in the pores, forming *cement* that binds the clastic grains firmly together into a hard rock (●**FIGURE 3.14C**). Calcite, quartz, and iron oxides are the most common cements in sedimentary rocks (●**FIGURE 3.15**).

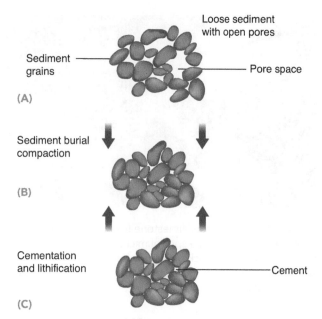

Loose sediment
with open pores

Sediment
grains

Pore space

(A)

Sediment burial
compaction

(B)

Cementation
and lithification

Cement

(C)

● **FIGURE 3.16** Conglomerate is lithified gravel. The individual clasts are clearly visible in this conglomerate from southwestern Montana. Lens cap for scale.

● **FIGURE 3.14** (A) Pore space is the open space between loose sediment grains. (B) Burial compaction squashes the grains together, reducing the pore space. (C) Cement fills some of the remaining pore space, lithifying the sediment. Some residual pore space commonly remains.

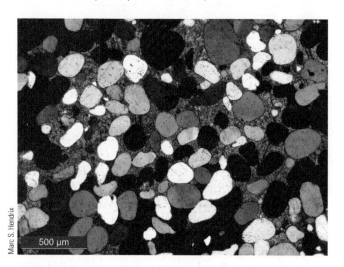

500 μm

● **FIGURE 3.15** Photomicrograph of sandstone from Missouri. Rounded grains of quartz sand are shown in various shades of gray. The original pore space in this rock has been completely filled by calcite cement, shown here in warm brown colors between the sand grains. The red bar in the lower left corner is 0.5 mm long.

The time required for **lithification** of sediment varies greatly. In some heavily irrigated areas of Southern California, calcite has precipitated from irrigation water to cement soils within only a few decades. In the northern Rocky Mountains, calcite has cemented some glacial deposits that are less than 20,000 years old. In contrast, sand and gravel deposited on the coastal plain of New Jersey as long ago as 95 million years can still be dug with a hand shovel.

Conglomerate is lithified gravel (●**FIGURE 3.16**). Each clast in a conglomerate is usually much larger than the individual mineral grains in the clast. In many conglomerates,

the clasts may be fist sized or even larger and therefore retain most of the characteristics of the parent rock. If enough is known about the geology of an area where conglomerate is found, it may be possible to identify exactly where the clasts originated. For example, a granite clast probably came from nearby granite bedrock.

The next time you walk along a gravelly stream bed, look carefully at the gravel. You will probably see sand or silt trapped among the larger clasts. In a similar way, most conglomerates contain fine sediment among the large clasts.

Sandstone consists of lithified sand grains (●**FIGURE 3.17**). The sand forms from the physical and chemical breakdown of preexisting rock. For example, weathering of granite typically produces sand-sized grains of quartz, feldspar, and other minerals. Streams, wind, glaciers, and gravity all can carry downslope sand formed by weathering. Sand grains carried in streams and by wind repeatedly undergo collisions as they bounce along. These impacts wear the sharp edges off of each grain, causing it to become rounded. Eventually, the sand grains accumulate to form a deposit of sand. Following burial, the deposit compacts and lithifies to form sandstone. Many sandstones consist predominantly of rounded quartz grains, because this hard mineral resists physical and chemical breakdown and therefore is concentrated by long periods of weathering and transport.

Mudstone is a clastic sedimentary rock that consists mostly of tiny clay minerals and lesser amounts of silt and organic particles (●**FIGURE 3.18**). Mudstone can vary widely in color, ranging from red to green to black. When weathered, some mudstone splits easily along very fine layering, called fissility. Such fissile mudstone usually is dark-colored due to the presence of abundant organic matter. As it is buried, this organic matter can convert to oil and gas, which is slowly expelled from the mudstone and rises toward the surface through pore spaces in the rock. In Chapter 5, we will learn how recent technological developments that permit the direct drilling and hydraulic fracturing ("fracking") of organic-rich black mudrock has resulted in entirely new sources of oil and gas and has significantly changed the global energy economy.

(A)   (B)   (C)

● **FIGURE 3.23** Common forms of limestone include the following. (A) Bioclastic limestone, made mostly of reworked shelly debris. (B) Reef limestone in this case showing several massive corals. (C) Oolite, limestone made of sand-sized ooids.

calcareous red algae in addition to calcareous microfossils called foraminifera and several kinds of calcareous green algae. Coral colonies can be branching or massive, and when broken down, they result in chunks of the original colony. Individual foraminifera are sand-sized and usually the calcareous skeleton, called a test, remains whole after the individual's death. Some calcareous green algae will break down into individual segments, whereas other types will break down into microscopic needles of aragonite. The diversity and abundance of marine organisms with skeletal hard parts made of calcite and aragonite is remarkable and has provided paleontologists with much information about changes to the marine ecosystem through geologic time.

Perhaps no paleo-environment has undergone a better documented set of changes in ecology through time than that of the carbonate reef. A reef is a wave-resistant framework of carbonate rock formed from the intergrowth of carbonate-forming organisms. In today's shallow marine reefs, corals are common as, for example, in the Great Barrier Reef of Australia (●**FIGURE 3.23B**). In contrast, about 100 million years ago, a type of clam was the dominant reef-building organism, and about 350 million years ago, a class of organisms called echinoderms was the main reef-building organism. Some reef-building organisms of the geologic past have no modern surviving counterparts, including stromatoporoids and archeocyathids—the main reef-builders of the Devonian and Cambrian Periods.

Many limestone-forming environments occur in shallow water that is close to being saturated with calcium carbonate, so in addition to broken-down skeletal debris from organisms with calcareous hard parts, limestones commonly contain sand-sized grains called ooids that form from the precipitation of aragonite and calcite on the surface of a sand- or silt-sized grain (●**FIGURE 3.23C**). Cross-sections of ooids show a well-developed concentric internal structure resulting from the layer-by-layer

● **FIGURE 3.24** The White Cliffs of Dover on the southern coast of England is made of chalk formed from the remains of calcareous plankton and nannoplankton.

precipitation of the carbonate minerals. Peloids—sand-sized internally structureless grains of carbonate mud—are another very common type of grain found in carbonate environments. Unlike ooids, which require agitation to roll the grain around, peloids form under more quiet water conditions.

Although reefs and reef-building organisms produce large pieces of bioskeletal debris—ranging in size from silt to boulders—many limestones are much finer-grained and the individual fossils are too small to see without a microscope. For example, **chalk** is a very fine-grained, soft, white bioclastic limestone made of the skeletons of planktonic nannofossils called coccolithophorids that float near the surface of the oceans and photosynthesize (●**FIGURE 3.24**). When they die, their remains sink to the bottom and accumulate to form chalk. Most fine-grained limestones contain an abundance of carbonate mud, or micrite, that are formed in the depositional environment either through the breakdown of larger particles, such as some types of calcareous green algae, or through the slow breakdown of calcareous debris to mud-sized particles by a type of microscopic algae. Some carbonate mud is formed through direct precipitation of tiny aragonite needles directly from seawater, forming isolated areas of white, cloudy water called "whitings." All of these components—skeletal debris, sand-sized carbonate grains that form in the environment (ooids and peloids), and carbonate mud —are common constituents of limestones.

---

carbonate rocks Bioclastic sedimentary rocks composed of the carbonate minerals (minerals based on the CO_3^{-2} anion).

chalk A very fine-grained limestone made up of the remains of tiny marine microorganisms.

Carbonate Rocks and Global Climate

Carbon dioxide is a greenhouse gas that traps heat in the atmosphere. Limestone is a hard, solid rock. Although it may seem counterintuitive, limestone rock is formed largely from carbon dioxide gas. Atmospheric carbon dioxide dissolves in seawater and forms CO_3 which then combines with dissolved calcium to form limestone rock, so formation of limestone removes carbon dioxide from both the sea and the air.

In Chapter 1, we mentioned briefly that Earth's outer shell is broken into several segments called *tectonic plates*. The plates float on the weak, plastic mantle below and glide across Earth, moving continents and oceans. You will learn in Chapter 6 that after a tectonic plate moves thousands of kilometers across Earth's surface, it eventually sinks deep into the mantle. In some instances, a sinking plate may carry limestone into the mantle, removing large amounts of carbon from Earth's surface and sequestering it in the deep mantle. In other cases, limestone on a sinking tectonic plate may become heated to the point that carbon is recycled back into the atmosphere as carbon dioxide is emitted during volcanic eruptions. In these ways, the greenhouse gas is cycled between surface limestone, deep-mantle rocks, and the atmosphere. Thus, interactions among limestone, the carbon dioxide in the atmosphere, and the carbon dioxide dissolved in the oceans are important determinants of global climate.

Physical Sedimentary Structures

Nearly all sedimentary rocks contain **sedimentary structures**—physical features that developed during or shortly after deposition of the sediment. Most sedimentary structures form through the interaction of moving water and loose grains of sediment. For example, when a stream moves sand along its bed, the size of the particles that can be moved will be limited by the strength of the flow. Only those particles small enough to be moved by the flow will do so; larger particles will be left behind. In this way, sediment is sorted by size as the water interacts with it. Environments with sustained high levels of wave energy, such as beaches, therefore typically produce sediments that are well-sorted, meaning that little size variation among particles exists. Generally, moderate to high levels of wind energy typical of many desert environments similarly produces well-sorted sand.

In addition to sorting the sediment by moving the grains, beach waves and desert winds will organize the sediment grains and produce sedimentary structures. Both sand dunes in a desert and sediment ripples in a stream are examples of sedimentary structures. Many sedimentary structures, like desert sand dunes, form little by little, as the wind current sorts and moves the sand into piles. As they are formed, many structures develop an internal architecture in addition to their external morphology. This internal architecture can provide important clues about how the sediment was transported and deposited to geologists examining ancient sedimentary rocks in cross-section (e.g., in an outcrop or road cut). Because sedimentary rocks form at Earth's surface,

● **FIGURE 3.25** Sedimentary beds from an ancient lake deposit in central Utah.

● **FIGURE 3.26** Cross-bedded sandstone in Zion National Park, southern Utah.

sedimentary structures and other features of sedimentary rocks also contain clues about environmental conditions at Earth's surface when the rocks were formed.

The most obvious and common sedimentary structure is **bedding**, or stratification—layering that develops as sediment is deposited (●**FIGURE 3.25**). Bedding forms because sediment accumulates layer by layer, with short pauses in sediment deposition represented by the surfaces separating individual beds.

Many beds of sandstone contain **cross-bedding**, in which sets of small beds are inclined to the main sedimentary layering (●**FIGURES 3.26** and **3.27**). Cross-bedding

---

sedimentary structures Any feature of sedimentary rock formed by physical processes during or shortly after deposition; examples include stratification, cross-bedding, ripple marks, and tool marks.

bedding Layering that develops as sediments are deposited; also called *stratification*.

cross-bedding A sedimentary structure in which wind or water deposits sets of beds that are inclined to the main sedimentary layering.

(A)

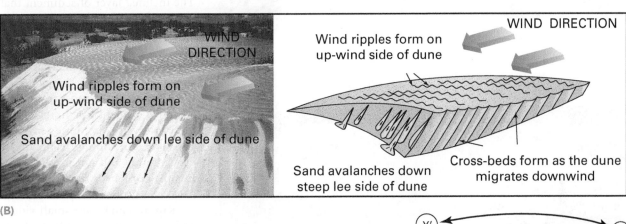

(B)

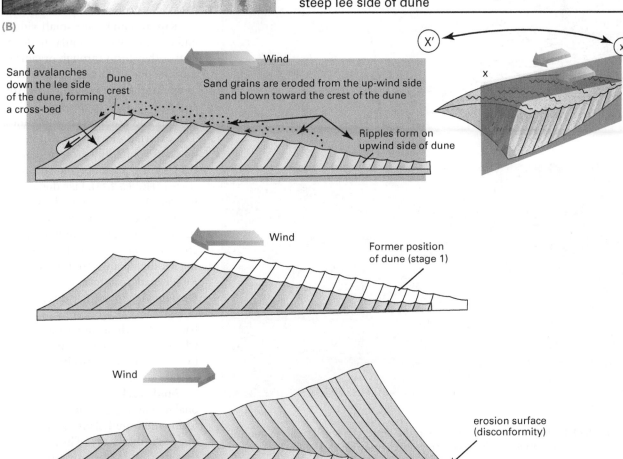

(C)

Doug McLean/Shutterstock.com

● **FIGURE 3.28** Ancient ripple marks preserved in sandstone.

Marc S. Hendrix

(A)

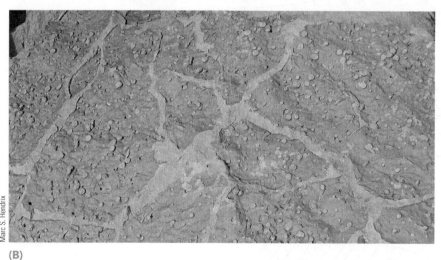

Marc S. Hendrix

(B)

● **FIGURE 3.29** (A) Mud cracks forming along the edge of a small pond in central Montana. (B) Ancient mud cracks in ancient sedimentary rock of Utah. The small round features are preserved burrows from insects, such as beetles or ants, that lived in the sediment.

forms in many environments where wind or water transports and deposits sediment. The wind and water currents organize the sand into semiparallel ridges called dunes. Continuous flow of water and wind causes sand grains to be eroded off the more gently sloping side of the dune that faces upstream, and these sand grains travel in the current to the brink of the dune. There, the sediment avalanches down the steep side of the dune that faces downstream. The inclined layer of sediment that results from the avalanche forms a cross-bed. As the dune develops through time, such avalanches form sets of cross-beds—an internal structure recognizable to the geologist (●**FIGURE 3.27B**). In most cases, only the cross-beds of a sediment dune are preserved—the dune's original external morphology typically is eroded away and not preserved so it is the cross-beds that provide the clue to the geologist.

Ripple marks are small ridges and troughs that are similar to dunes in that they form through organized movement and sorting of loose sediment grains by a moving flow. You commonly see ripples in shallow streams and around lake shorelines. The external morphology of ripple marks is commonly preserved along the bedding planes of ancient sandstone (●**FIGURE 3.28**). Like dunes, ripples also have characteristic internal structures called "ripple cross-lamination" that are recognizable in cross-sections of ancient rock where the bedding planes are not exposed. Ripple marks can form either by currents moving in a single direction, as in a stream, or by currents moving back and forth, as in a shallow lake where sand on the bottom is pushed forward and backward by waves on the lake surface.

Mud cracks are irregular polygonal downward-tapering cracks that form when mud shrinks as it dries out (●**FIGURE 3.29**). They indicate that the mud dried up after being deposited in water. Mud cracks are

Much more common are graded beds, formed when a flow carrying sediment slows down and the size of the sediment that is deposited decreases through time (●FIGURE 3.30). That is, the graded bed has the coarsest clasts at its base and the finest clasts at its top. Graded beds can be produced by a flood that wanes through time. Graded beds also are very common in deep marine environments where turbulent mixtures of sediment and water called turbidity currents move downslope, coming to rest eventually as the current slows down and eventually stops. Such graded beds are called turbidites.

Fossils and **biogenic structures** include remains and traces of plants or animals preserved in rock—any evidence of past life. Fossils include remains of shells, bones, or teeth; whole bodies preserved in mudstone, amber, or ice; and a variety of tracks, trails, and burrows that collectively are called "trace fossils" (●FIGURE 3.31). Fossils are discussed further in Chapter 4.

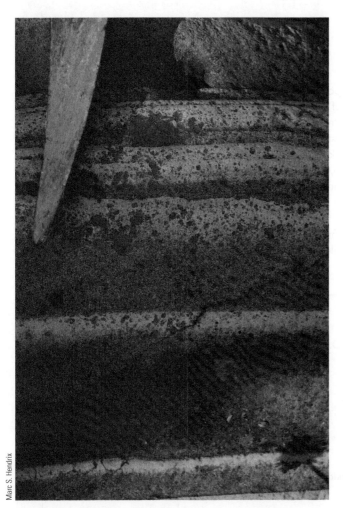

Marc S. Hendrix

● **FIGURE 3.30** Graded beds of sandstone, southern Tanzania. Each bed fines upward from coarse sand at the base, shown by the darker color, to fine sand at the top, shown by the lighter color. Tip of geology hammer for scale.

common in river floodplains, where mud deposited by a flood eventually dries out when the floodwaters recede. In rare cases, raindrop impressions are preserved on the bedding surfaces of mud that has dried out, although raindrop impressions are very delicate sedimentary structure that are rarely preserved.

---

ripple marks Small, semiparallel ridges and troughs formed mostly in sand by wind, water currents, or waves; often preserved when the sediment is lithified.

mud cracks Irregular polygonal downward-tapering fractures that develop when mud dries; may be preserved when the mud is lithified.

fossils The imprint, remains, or any other trace of a plant or animal preserved in rock.

biogenic structures Any physical trace left in the sedimentary record by a fossil organism; includes tracks, trails, burrows, and root casts.

Marc S. Hendrix

● **FIGURE 3.31** These burrows in dolomite from western Wyoming were made by ancient shrimp that burrowed into the loose sediment of the seafloor in order to build a dwelling that would help the shrimp evade predators. Collectively, the tracks, trails, burrows, and borings left behind by ancient organisms are called trace fossils.

Gravel Mining and Sustainability

Building the infrastructure for human society requires a tremendous volume of gravel. In 2016 alone, the United States Geological Survey (USGS) estimated that roughly 6,300 active sand and gravel operations were operating in the United States and that all 50 states were involved. The total value of construction sand and gravel that year was $8.9 billion, more than the estimated $8.5 billion value of the total amount of gold mined in the United States the same year. Gravel and sand are used primarily as concrete aggregates, in road building, as construction fill, as railroad ballast, and as roofing granules, among other uses.

Although we don't typically think of sand and gravel as being a finite resource, both are expensive to quarry and transport, and significant additional costs are incurred if it is necessary to sort the sediment into specific size classes. In addition, gravel in particular is a relatively uncommon sediment type that simply does not occur everywhere. There is very little gravel in the Gulf Coast region, for example, whereas it is common in Washington State.

Gravel is produced from the rapid weathering of bedrock in regions where mountainous topography exists. In mountainous regions in which glaciers recently formed or valleys into which melt water flowed and accumulated, gravel is especially common. As we will learn in Chapter 13, glaciers are very efficient at eroding bedrock and producing lots of sediment in the process, including gravel. Because glaciers and glacial lakes formed in many of the northern states, these states have more mineable gravel deposits.

In rapidly growing urban areas, such as the Wasatch Front in Utah, where Salt Lake City is located, the demand for gravel is especially high. Infrastructure construction in the Wasatch Front has relied largely on the extensive gravel deposits associated with the shorelines of glacial Lake Bonneville—a large freshwater interior lake that formed most recently about 18,000 years ago. Lake Bonneville stretched for over 20,000 square miles across much of northern Utah and Nevada and reached up to 1,200 feet depth. Today's Great Salt Lake is a remnant of the formerly much larger glacial Lake Bonneville (●FIGURE 3.32).

The mountains that surrounded glacial Lake Bonneville and the wind-driven waves that formed on the lake surface resulted in the formation of wide gravel beaches and large gravel bars as the gravel that washed into the lake was rolled around, rounded, sorted, and ultimately deposited. The wave energy of the lake not only sorted the gravel but piled it up into various bars, spits, and beaches along

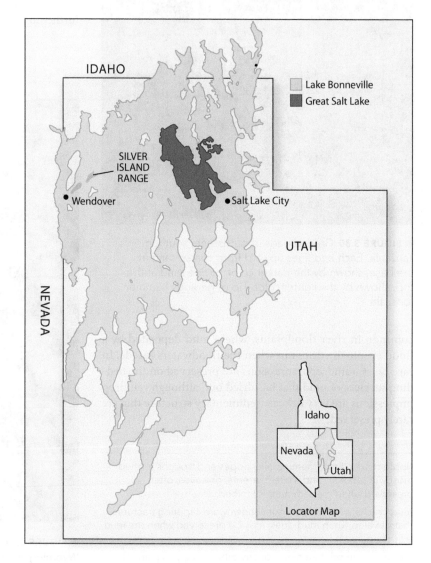

● FIGURE 3.32 Map of Lake Bonneville, a large lake that covered much of western Utah, eastern Nevada, and southern Idaho during the last glaciation about 18,000 years ago. The modern Great Salt Lake, also shown, is a remnant of the much larger Lake Bonneville.

the lake shorelines. These features were left behind as Lake Bonneville shrunk and remain today as evidence of the once extensive lake that covered much of the region.

In 1877, the shoreline gravels associated with Lake Bonneville were first described by Grove Karl (G.K.) Gilbert (●FIGURE 3.33), a geologist with the Powell Expedition. G.K. Gilbert not only mapped, sketched, and described the many landforms associated with the ancient lake shorelines but he was the first to recognize that they represented the direct evidence of an ancient lake that was many times larger than the present-day Great Salt Lake. When the U.S. Geological Survey was created in 1879, Gilbert was appointed to the position of "Senior Geologist," and in 1890, he published the survey's first monograph, a 438-page treatise simply entitled "Lake Bonneville." The shoreline deposits first described by G.K. Gilbert in the nineteenth century since have become famous examples of shoreline deposits, familiar to generations of geology students.

Today, the high demand for gravel in the rapidly growing Wasatch Front has resulted in extensive mining of the Bonneville shoreline deposits first described by G.K. Gilbert (●FIGURE 3.34). Unfortunately, the removal of these deposits through gravel mining destroys the information they preserve about the evolution and natural history of Lake Bonneville. In addition, mining of the gravel removes altogether an iconic set of landforms tied directly to the region's recent geologic past (●FIGURE 3.35). Some of the gravel shorelines are already mined out, and many are in danger of disappearing in the near future as rapid construction continues in the Wasatch Valley and the surrounding region. Once these gravels and the landforms containing them are gone, we will have lost a unique set of natural resources and much information about the glacial lake history they represent.

usgs.gov copyright

(A)

AMERICAN PHILOSOPHICAL SOCIETY/Science Source

(B)

● **FIGURE 3.33** (A) G.K. Gilbert, famous geologist who first described the iconic landforms associated with the former shorelines of Lake Bonneville in parts of Utah. Many consider Gilbert to be the greatest of all American geologists. (B) G.K. Gilbert in the field.

Plate XVI by H.H. Nichol, artist, and G.K. Gilbert, geologist, in Gilbert 1884

● **FIGURE 3.34** Sketch of Lake Bonneville shoreline by G.K. Gilbert. Gilbert made this sketch in 1877 when he first described the landforms associated with the former shorelines.

Utah Geological Survey, State of Utah

● **FIGURE 3.35** Photo of Lake Bonneville shoreline gravel deposit with mining operation. Continued mining of the gravels will remove the landforms and the natural history they represent.

METAMORPHIC ROCKS

A potter forms a delicate vase from moist clay. She places the soft piece in a kiln and slowly heats it to 1,000°C. As temperature rises, the clay minerals decompose. Atoms from the clay then recombine to form new minerals that make the vase strong and hard. The breakdown of the clay minerals, growth of new minerals, and hardening of the vase all occur without melting the solid materials.

Metamorphism (from the Greek words for "changing form") is the process by which rising temperature and pressure, or changing chemical conditions, transform rocks and minerals. Metamorphism occurs in solid rock, like the transformations in the vase as the potter fires it in her kiln. Small amounts of water and other fluids speed up the metamorphic mineral reactions, but the rock remains solid as it changes. Metamorphism can change any type of parent rock: sedimentary, igneous, or even another metamorphic rock.

A mineral that does not decompose or change in other ways, no matter how much time passes, is a "stable" mineral. A stable mineral can become unstable when environmental conditions change. Three types of environmental change affect mineral stability and cause metamorphism: rising temperature, rising pressure, and changing chemical composition (usually caused by an influx of hot water). For example, when the potter put the clay in her kiln and raised the temperature, the clay minerals decomposed because they became unstable at the higher temperature. The atoms from the clay then recombined to form new minerals that were stable at the higher temperature.

Similarly, if hot water seeping through bedrock carries new chemicals to a rock, those chemicals may react with the rock's original minerals to form different minerals that are stable in the new chemical environment. Thus, metamorphism occurs because each mineral is stable only within a certain range of temperature, pressure, and chemical environment. If temperature or pressure rises above that range, or if chemicals are added or removed from the rock, the rock's original minerals may decompose and their components recombine to form new minerals that are stable under the new conditions.

Metamorphic Grade

The **metamorphic grade** of a rock is the intensity of metamorphism that formed it. Temperature is the most important factor in metamorphism, and therefore, grade closely reflects the temperature of metamorphism. Because temperature increases with depth in Earth, a general relationship exists between depth and metamorphic grade (●**FIGURE 3.36**). Low-grade metamorphism occurs at shallow depths, less than 10 kilometers beneath the surface, where temperature is no higher than 300°C to 400°C. Medium-grade conditions,

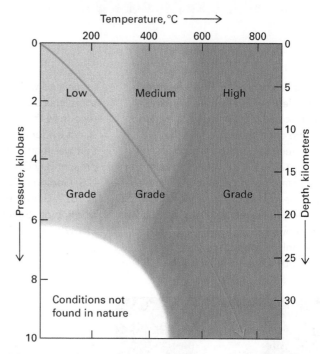

● **FIGURE 3.36** Metamorphic grade increases with depth because temperature and pressure rise with depth. The blue arrow traces the path of increasing temperature and pressure with depth in crust typically found in a stable continental interior.

where temperatures are between 400°C and 600°C, exist at depths between about 10 and 40 kilometers. High-grade conditions are found deep within the continental crust and in the upper mantle, 40 to 55 kilometers below Earth's surface. The temperature there is 600°C to 800°C, close to the melting point of rock. High-grade conditions can develop at shallower depths, however, in rocks adjacent to hot magma. For example, today, metamorphic rocks are forming beneath Yellowstone Park, where hot magma lies close to Earth's surface.

Metamorphic Changes

Metamorphism commonly alters both the texture and mineral content of a rock.

Textural Changes

As a rock undergoes metamorphism, some mineral grains grow larger and others shrink. The shapes of the grains may also change. For example, the fossiliferous limestone shown in ●**FIGURE 3.23A** is a sedimentary rock, but if it is subject to high temperature and pressure, the small calcite crystals that make up both the fossils and the cement between them will recrystallize, with some crystals growing at the expense of others. In the process, the fossils will be slowly destroyed and the rock will transform into marble, a metamorphic rock composed of calcite but with a texture consisting instead of large, interlocking crystals.

Micas are common metamorphic minerals that form when many different parent rocks undergo metamorphism. Micas are shaped like pie plates. When metamorphism occurs without deformation—without causing the rocks to change

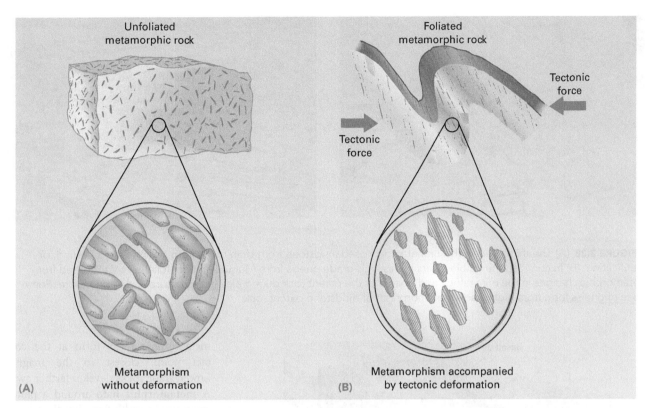

● **FIGURE 3.37** (A) When metamorphism occurs without deformation, platy micas grow with random orientation. (B) When deformation accompanies metamorphism, platy micas orient perpendicular to the force squeezing the rocks, forming foliated metamorphic rocks. The wavy lines represent original sedimentary layers that have been folded by the deformation.

shape—the micas grow with random orientations, like pie plates flying through the air (●FIGURE 3.37A). However, when tectonic force squeezes rocks as they are heated during metamorphism, the rock deforms into folds. When rocks are folded as mica crystals are growing, the micas grow with their flat surfaces perpendicular to the direction of the squeezing. This parallel alignment of micas (and other minerals) produces the metamorphic layering called **foliation** (●FIGURE 3.37B).

The foliation layers can range in thickness from a fraction of a millimeter to a meter or more. Metamorphic rocks commonly will break parallel to foliation planes to form **slaty cleavage**.

Although metamorphic foliation and the slaty cleavage that results can resemble sedimentary bedding (stratification), the two types of layering are different in origin. Foliation results

from the alignment and segregation of metamorphic minerals during metamorphism; it forms at right angles to the direction of maximum compressive stress acting on the rocks. Stratification develops because sediments are deposited layer by layer.

Mineralogical Changes

Sometimes, when a parent rock contains only one mineral, metamorphism transforms the rock into one composed of the same mineral but with a coarser texture. Limestone converting to marble is one example of this generalization: both rocks consist of the mineral calcite, but their respective textures are very different. Another example is the metamorphism of quartz sandstone to quartzite, a rock composed of recrystallized quartz grains.

In contrast, metamorphism of a parent rock containing several minerals usually forms a rock with new and different minerals *and* a new texture. For example, a typical sedimentary mudstone contains large amounts of clay as well as quartz feldspar and several other minerals (●FIGURE 3.38A). When heated, some of those minerals decompose, and their atoms recombine to form new minerals such as mica, garnet, and a different kind of feldspar. ●FIGURE 3.38B shows a rock called gneiss that was formed when metamorphism altered both the texture and the minerals of mudstone. If migrating fluids alter the chemical composition of a rock, new minerals invariably form.

---

metamorphism The process by which rocks change texture and mineral content in response to variations in temperature, pressure, chemical conditions, and/or deformation.

metamorphic grade The intensity of metamorphism that formed a rock; the maximum temperature and pressure attained during metamorphism.

foliation The layering in metamorphic rocks resulting from regional dynamothermal metamorphism.

slaty cleavage A metamorphic foliation producing a parallel fracture pattern that cuts across the original sedimentary bedding.

(A)

(B)

● **FIGURE 3.38** (A) The thin layers in this unmetamorphosed mudstone from Utah result from sedimentary bedding, or stratification. (B) In contrast, the visible layers in this high-grade gneiss from Montana reflect foliation that resulted from metamorphic changes in the mineralogy and texture of the parent rock during elevated pressure and temperature. Some types of gneiss form from high-grade metamorphism of mudstone parent rock.

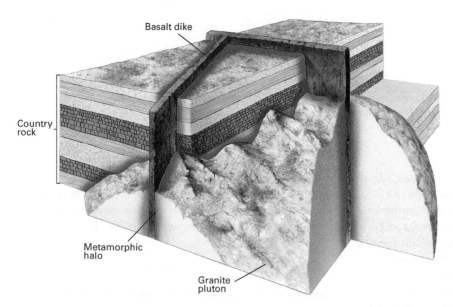

● **FIGURE 3.39** A halo of contact metamorphism, shown in red, surrounds a pluton. The later intrusion of the basalt dike (a sheet-like, intrusive rock) metamorphosed both the pluton and the existing sedimentary rock, creating a second metamorphic halo.

Types of Metamorphism and Metamorphic Rocks

Recall that rising temperature, rising pressure, and changing chemical environment cause metamorphism. In addition, deformation resulting from the movement of Earth's crust causes foliation to develop and thus strongly affects the texture of a metamorphic rock. Four different geologic processes create these changes.

CONTACT METAMORPHISM Contact metamorphism occurs where hot magma intrudes cooler rock of any type—sedimentary, metamorphic, or igneous. The highest grade

metamorphic rocks form at the contact point, closest to the magma. Lower-grade rocks develop farther away. A **metamorphic halo** around a pluton can range in width from less than a meter to hundreds of meters, depending on the size and temperature of the intrusion and the effects of water or other fluids (●**FIGURE 3.39**). Contact metamorphism commonly occurs without deformation. As a result, the metamorphic minerals grow with random orientations—like the pie plates flying through the air—and the rocks develop no foliation. The rock **hornfels** is a common unfoliated metamorphic rock that forms by contact metamorphism of mudstone.

BURIAL METAMORPHISM Burial metamorphism results from the burial of rocks in a sedimentary basin. Younger sediment may bury the oldest layers to depths greater than 10 kilometers in a large basin. Over time, temperature and pressure increase within the deeper layers until burial metamorphism begins.

Burial metamorphism is occurring today in the sediments underlying many large deltas, including the Mississippi River delta, the Amazon Basin on the east coast of South America, and the Niger River delta on the west coast of Africa. Like contact metamorphism, burial metamorphism occurs without deformation. Consequently, metamorphic minerals grow with random orientations and, like contact metamorphic rocks, burial metamorphic rocks are unfoliated.

REGIONAL DYNAMOTHERMAL METAMORPHISM Regional dynamothermal metamorphism occurs where

METAMORPHISM OF MUDSTONE

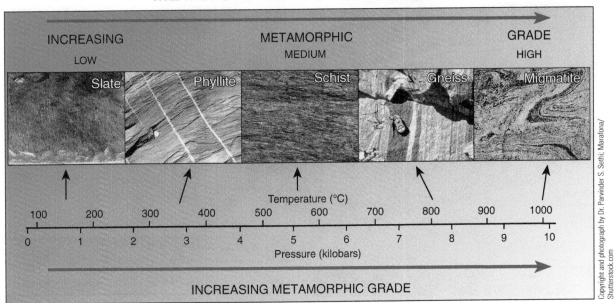

● FIGURE 3.40 As mudstone is metamorphosed, it undergoes changes in texture and mineral content. Low-grade metamorphism changes mudstone to slate, a dull, finely crystalline rock that is harder than mudstone. With increased metamorphism, slate transforms to phyllite, a rock characterized by shiny, foliated surfaces such as those seen here. Schist forms at a higher metamorphic grade and has crystals big enough to see with the unaided eye. Gneiss, a high-grade metamorphic rock, is coarsely crystalline and characterized by segregated layers of lighter-colored and darker-colored minerals. Migmatite forms by partial melting of a gneiss and typically shows well-developed folds formed when the rock behaved like a plastic.

major crustal movements build mountains and deform rocks. The term *dynamothermal* simply indicates that the rocks are being deformed and heated at the same time. It is the most common and widespread type of metamorphism and affects broad regions of Earth's crust.

Magma rises and heats large portions of the crust in places where tectonic plates converge (see Chapter 6). The high temperatures cause new metamorphic minerals to form throughout the region. The rising magma also deforms the hot, plastic country rock as it forces its way upward. At the same time, the movements of the crust squeeze and deform rocks. As a result of all these processes acting together, regionally metamorphosed rocks are strongly foliated and are typically associated with mountains and igneous rocks (**●FIGURE 3.40**). Regional dynamothermal metamorphism produces zones of foliated metamorphic rocks tens to hundreds of kilometers across. For example, large portions of the Appalachian Mountains in the eastern United States have undergone dynamothermal metamorphism and exhibit well-developed foliation.

Among the best examples of the changes in texture and mineral content that accompany metamorphism are those that occur in mudstone with increasing metamorphic grade. The most abundant type of sedimentary rock, mudstone consists mainly of clay minerals, quartz, and feldspar. The mineral grains are too small to be seen with the unaided eye. As regional metamorphism begins, clay minerals break down and mica and chlorite replace them. These new, platy minerals grow perpendicular to the direction of squeezing. As a result, the rock develops slaty cleavage and is called slate. With rising temperature and continued deformation, the micas and chlorite grow larger, and wavy or wrinkled surfaces replace the flat slaty cleavage, creating phyllite—which has a glossy appearance compared to slate, which is more dull. In both slate and phyllite, the mica and chlorite are too small to be seen with the unaided eye.

As the temperature continues to rise, the mica and chlorite grow large enough to be seen without a microscope and foliation becomes very well developed. Rock of this type is called schist. Schist forms approximately at the transition from low to intermediate metamorphic grades.

At high metamorphic grades, light- and dark-colored minerals often separate into bands that are thicker than the layers of schist, to form a rock called gneiss (pronounced "nice"). At the highest metamorphic grade, the rock begins to melt, though only partially, forming small veins of granitic magma. When metamorphism wanes and the rock cools, the magma veins solidify to form migmatite, a mixture of igneous and metamorphic rock.

metamorphic halo The zone surrounding an intrusive igneous body in which the country rock has been metamorphosed by heat and hydrothermal fluids from the cooling magma.

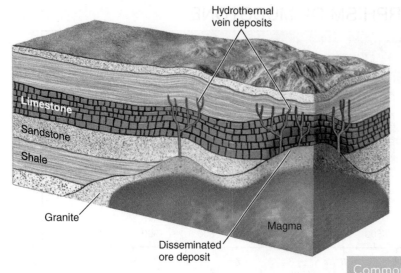

Common Metamorphic Rocks	
Unfoliated	Foliated
Marble	Slate
hornfels	Phyllite
quartzite	Schist
	Gneiss
	Migmatite

Ason Patrick Ross/Shutterstock.com

Common Igneous Rocks			
Felsic	Intermediate	Mafic	Ultramafic
Obsidian	Andesite	Basalt	Peridotite
Granite	Diorite	Gabbro	
Rhyolite			

Common Sedimentary Rocks			
Clastic	Bioclastic	Organic	Chemical
Conglomerate	Limestone	Chert	Evaporite
Sandstone	Some dolomite (dolostone)	Coal	Some dolomite (dolostone)
Siltstone	Coquina		
Mudstone	Chalk		

● **FIGURE 3.41** Hydrothermal ore deposits form when hot water dissolves metals from country rock and deposits them in fractures and surrounding country rock.

HYDROTHERMAL METAMORPHISM Water is a chemically active fluid; it attacks and dissolves rocks and minerals. If the water is hot, it attacks minerals even more rapidly. Hydrothermal metamorphism (also called *hydrothermal alteration* or *metasomatism*) occurs when hot water and ions dissolved in the hot water react with a rock to change its chemical composition and minerals.

Most rocks and magma contain very low concentrations of metals such as gold, silver, copper, lead, and zinc. For example, gold makes up 0.0000002 percent of average crustal rock, while copper makes up 0.0058 percent and lead comprises 0.0001 percent. Although the metals are present in very low concentrations, hydrothermal solutions sweep slowly through large volumes of country rock, dissolving and accumulating the metals as they go. The solutions then deposit the dissolved metals where they encounter changes in temperature, pressure, or chemical environment (●**FIGURE 3.41**). In this way, hydrothermal solutions scavenge and concentrate metals from average crustal rocks and then deposit them locally to form ore. Hydrothermal ore deposits are discussed further in Chapter 5.

Key Concepts Review

- Geologists divide rocks into three groups, depending upon how the rocks formed. Igneous rocks solidify from magma. Sedimentary rocks form from clay, sand, gravel, and other sediment that collects at Earth's surface. Metamorphic rocks form when any rock is altered by temperature, pressure, or an influx of hot water. The rock cycle summarizes processes by which rocks continuously recycle in the outer layers of Earth, forming new rocks from old ones. Rock cycle processes exchange energy and materials with the atmosphere, the hydrosphere, and the biosphere.

- Extrusive (or volcanic) igneous rocks are fine-grained rocks that solidify from magma that has erupted onto Earth's surface. Granite and basalt are the two most common igneous rocks. Intrusive (or plutonic) igneous rocks are medium- to coarse-grained rocks that solidify within Earth's crust.
- Sediment forms by the weathering of rocks and minerals. It includes all solid particles such as rock and mineral fragments, organic remains, and precipitated minerals. It is transported by streams, glaciers, wind, and gravity; is deposited in layers; and eventually

is lithified to form sedimentary rock. Mudstone, sandstone, and limestone are the most common kinds of sedimentary rock.

- When a rock is heated, when pressure increases, or when hot water alters its chemistry, both the minerals and textures of the rock change in a process called metamorphism. Contact metamorphism affects rocks heated by a nearby igneous intrusion. Burial metamorphism alters rocks as they are buried deeply within Earth's crust. In regions where tectonic plates converge, high temperature, deformation from rising magma, and plate movement all combine to cause regional dynamothermal metamorphism. Hydrothermal metamorphism is caused by hot solutions soaking through rocks and is often associated with emplacement of ore deposits. Slate, schist, gneiss, and marble are common metamorphic rocks.

Important Terms

basement rock (p. 39)

bedding (p. 48)

bedrock (p. 39)

biogenic structures (p. 51)

carbonate rocks (p. 46)

chalk (p. 47)

compaction (p. 43)

conglomerate (p. 44)

country rock (p. 39)

cross-bedding (p. 48)

extrusive igneous rock (p. 36)

foliation (p. 55)

fossils (p. 51)

igneous rock (p. 34)

intrusive igneous rock (p. 37)

lava (p. 36)

lithification (p. 44)

magma (p. 34)

metamorphic grade (p. 54)

metamorphic halo (p. 56)

metamorphic rock (p. 34)

metamorphism (p. 54)

mud cracks (p. 50)

mudstone (p. 44)

parent rock (p. 39)

peat (p. 45)

pore space (p. 43)

precipitation (p. 42)

ripple marks (p. 50)

rock cycle (p. 35)

sediment (p. 34)

sedimentary rock (p. 34)

sedimentary structures (p. 48)

slaty cleavage (p. 55)

texture (p. 35)

weathering (p. 34)

Review Questions

1. Explain what the rock cycle tells us about Earth processes.
2. Describe ways in which rock cycle processes exchange energy and materials with the atmosphere, the hydrosphere, and the biosphere.
3. What are the three main kinds of rock in Earth's crust?
4. How do the three main types of rock differ from each other?
5. What is magma? Where does it originate?
6. Describe how igneous rocks are classified and named.
7. What are the two most common kinds of igneous rock?
8. Describe and explain the differences between plutonic and volcanic rock.
9. What is sediment? How does it form?
10. How do sedimentary grains become rounded?
11. Where in your own area would you look for rounded sediment?
12. Describe how sediment becomes lithified.
13. What are the differences among mudstone, sandstone, and limestone?
14. Explain why almost all sedimentary rocks are layered, or bedded.
15. How does cross-bedding form?
16. What is metamorphism? What factors cause metamorphism?
17. What kinds of changes occur in a rock as it is metamorphosed?
18. What is metamorphic foliation? How does it differ from sedimentary bedding?
19. How do contact metamorphism and regional metamorphism differ, and how are they similar?

4

GEOLOGIC TIME
Earth's History in the Rocks

View of the Grand Canyon, looking North. Over one mile of vertical topographic relief exists between the tree-covered canyon rim on the skyline and the Colorado River in the foreground. The unlayered rock exposed directly above the river is Precambrian gneiss and schist. This rock, which formed deep within the crust, was uplifted and eroded prior to Cambrian time. After more than a billion years of erosion, the Cambrian Tapeats sandstone was deposited. The Tapeats is the prominent cliff-forming layer sitting directly over the Precambrian metamorphic rocks. The red- and tan-colored layers in the distance are younger Paleozoic sedimentary rocks.

EARTH ROCKS, EARTH HISTORY, AND MASS EXTINCTIONS

In many locations across Europe and later in North America, 19th-century geologists found thick sequences of sedimentary rock layers containing abundant fossil organisms preserved in them. The geologists also found layers of rock containing few or no fossils lying above the fossil-rich rocks. Even higher in the sequence of rock layers, they found fossils again, but of organisms very different from those in the older rocks. Most surprising, many of the most abundant fossilized organisms in lower layers never appeared again in the younger rocks. Those organisms simply had disappeared from the surface of Earth forever, as though suddenly extinguished, never to return.

Today, we know through this fossil record that more than once, a sudden, catastrophic event has abruptly decimated life on Earth, causing a **mass extinction** in which many types of organisms simply were wiped out. Following each extinction event, new life-forms slowly emerged and these new life-forms recolonized the planet.

The idea that life on Earth underwent catastrophic extinctions due to geologic events that affected the entire planet was not accepted by many geologists of the 20th century. Instead, they suggested that perhaps the rocks containing evidence of more gradual changes in Earth's life-forms simply had been destroyed by erosion or were never deposited. According to this reasoning, evidence for the gradual decline of species that became extinct and the slow emergence of new species simply was missing from the rock record or had not been found. Following the conventional wisdom of the time, scientists concluded that the extinctions of old life-forms and the emergence of new ones occurred slowly, as a result of gradually changing conditions.

As travel became easier in the 20th century and more of Earth's rock record could be reached and studied by scientists, many searched for fossiliferous rocks of the appropriate ages to fill in the gaps and provide evidence for gradual extinctions—but these rocks were never found. Instead, the scientists found an abundance of geologic evidence indicating that the near-instantaneous mass extinctions occurred at least five times in the geologic past, each time decimating life on Earth as it existed at that time (●FIGURE 4.1).

The most dramatic extinction occurred about 252 million years ago, at the end of the Permian Period. At that time, 90 percent of all species in the oceans suddenly died out. On land, about two-thirds of reptile and amphibian species and 30 percent of insect species also vanished. In his

LEARNING OBJECTIVES

LO1 Describe three possible causes geologists have proposed for mass extinctions.

LO2 Explain how carbon is transferred between the atmosphere and oceans.

LO3 Describe how large transfers in carbon between the atmosphere and the oceans can cause major changes in global temperature.

LO4 Give examples of how the formation and/or breakup of a supercontinent can alter Earth's carbon budget.

LO5 Express the difference between relative and absolute age.

LO6 Describe how each of the four relative age dating principles are used by geologists.

LO7 Describe how to recognize an unconformity in the field.

LO8 Describe three different types of unconformities.

LO9 Explain how geologists use index fossils and key beds to correlate sedimentary rocks layers.

LO10 Graph the percent abundances of parent and daughter isotopes over the course of four half-lives.

LO11 Order the following geologic time subdivisions from shortest to longest: age, eon, epoch, era, period.

LO12 Explain the differences in how rocks are correlated on the basis of lithostratigraphy, biostratigraphy, and sequence stratigraphy.

book describing this extinction, author Douglas Erwin exclaims that it was "the closest life has come to complete extermination since its origin."[1]

The death of most life-forms at the end of the Permian Period left huge ecological voids in the biosphere. Ocean ecosystems changed as new organisms emerged in an environment relatively free of predators and competition. On land, terrestrial animals including dinosaurs slowly appeared and proliferated. Plants too underwent major changes that led to the **evolution** of the first flowering plants. About 65 million years ago, at the end of the Cretaceous Period, another catastrophic extinction wiped out up to 50 percent of Earth's genera, including the dinosaurs. Small mammals survived this disaster, facing a new world free of the efficient predators that had hunted them. Grasses also evolved after the Cretaceous extinction event, leading to the establishment of widespread grasslands ecosystems. Eventually, humans evolved.

What kinds of catastrophic events caused these sudden extinctions? Why would species that had survived for tens of millions of years or more disappear so rapidly? Here we

---

1. Douglas Erwin, *The Great Paleozoic Crisis* (New York: Columbia University Press, 1993), 187.

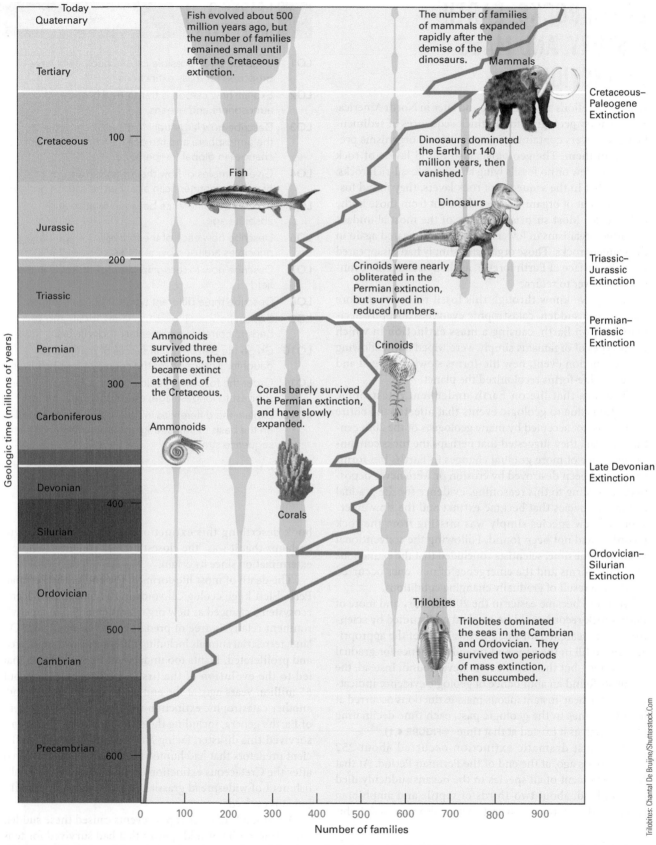

Today
Quaternary

Tertiary

Cretaceous 100

Jurassic

200

Triassic

Permian

300

Carboniferous

Devonian

400

Silurian

Ordovician

500

Cambrian

Precambrian 600

Geologic time (millions of years)

Fish evolved about 500 million years ago, but the number of families remained small until after the Cretaceous extinction.

The number of families of mammals expanded rapidly after the demise of the dinosaurs. Mammals

Dinosaurs dominated the Earth for 140 million years, then vanished. Dinosaurs

Crinoids were nearly obliterated in the Permian extinction, but survived in reduced numbers. Crinoids

Fish

Ammonoids survived three extinctions, then became extinct at the end of the Cretaceous. Ammonoids

Corals barely survived the Permian extinction, and have slowly expanded. Corals

Trilobites dominated the seas in the Cambrian and Ordovician. They survived two periods of mass extinction, then succumbed. Trilobites

Cretaceous–Paleogene Extinction

Triassic–Jurassic Extinction

Permian–Triassic Extinction

Late Devonian Extinction

Ordovician–Silurian Extinction

0 100 200 300 400 500 600 700 800 900
Number of families

Trilobites: Chantal De Bruijne/Shutterstock.Com

● **FIGURE 4.1** The red line shows that the number of families of organisms, indicated by the thickness of the blue-shaded areas, has varied throughout Earth history. Sudden leftward shifts of the red line indicate mass extinctions. The numbers in the left column indicate millions of years ago.

consider three hypotheses for mass extinctions. Each of these hypotheses involves large-scale interactions among Earth's four major systems. The geosphere, atmosphere, hydrosphere, and biosphere all were involved in each type of mass extinction. The trigger that initiated the radical changes among the four Earth systems and caused the mass extinction is the main difference between each hypothesis.

Extraterrestrial Impacts

In the late 1970s, the father-and-son team of Walter and Luis Alvarez were studying sedimentary rocks deposited when the dinosaurs and so many other life-forms were becoming extinct. In one layer, they found abundant dinosaur fossils. Just above it, they found very few fossils of any kind and no dinosaur fossils. Between these two rock layers they found a thin, sooty layer of clay. They brought samples of the clay to the laboratory and found that it contained high concentrations of the element iridium. This discovery was surprising because iridium is rare in Earth rocks. Where did it come from?

Although rare in Earth's crust, iridium is more abundant in meteorites. In a landmark paper published in 1980, Walter and Luis Alvarez suggested that 65 million years ago, a meteorite roughly 10 kilometers in diameter hit the Earth with explosive energy many thousands of times greater than that of today's entire global nuclear arsenal.[2] The collision vaporized both the meteorite and Earth's crust at the point of impact, forming a plume of hot dust and gas that rose into the upper atmosphere and circled the globe. This thick, dark cloud blotted out the Sun, significantly reducing the amount of solar energy that reached Earth's surface for up to several years. Particularly hard hit were microscopic algae and other tiny photosynthesizing organisms in the oceans that were unable to survive under the reduced levels of solar energy reaching the surface. The widespread die-off of photosynthesizing organisms in the oceans greatly reduced the base of the food chain and contributed to the extinction of larger marine organisms such as ammonites and plesiosaurs, a kind of swimming dinosaur. The reduced solar energy reaching the surface also resulted in less vegetation, ultimately eliminating all dinosaurs and many reptiles. Analysis of the fossil record has suggested that nothing with a body mass exceeding 25 kilograms (about the size of a large dog) survived.

The now-famous 1980 paper by the Alvarez team created a considerable stir among geologists, in part because the paper cited compelling evidence for the impact but did not identify the actual impact site. At about the same time, geophysicist Glen Penfield was analyzing magnetic surveys of the Yucatan Peninsula for the Mexican oil company Pemex, when he discovered the presence of a large, symmetrical semicircle in the Gulf of Mexico on the northwest side of the peninsula. Obtaining data from an older gravity survey

of the peninsula, Penfield noticed that a second semicircle was present onshore and that the two semicircles together formed a large, circular structure about 70 kilometers across. Although Penfield and his coworkers suspected that the circular structure might be the "missing" impact site, they were not able to secure any rock samples from the area to examine for evidence of an impact, and Pemex did not allow them to publish the geophysical data behind their discovery. Instead, they published a short abstract and presented a brief talk in 1981, announcing the discovery of a large igneous body in the Yucatan Peninsula. It was not until 1991—14 years after the Alvarez paper—that a team of scientists (among them Penfield) had amassed sufficient data, including rock samples from old Pemex oil wells in the area, to publish the first paper[3] describing the site as an impact structure. Since then, many analyses of the impact site and surrounding areas have suggested that it occurred 65 million years ago, coinciding with the Cretaceous extinction event.

Although much evidence in support of the Yucatan site as a major impact structure has been published since 1991, its relation to the mass extinction at the end of the Cretaceous Period is still the subject of vigorous debate. Careful analyses of fossils preserved in sediment cores recovered from the crater have indicated that the impact actually occurred about 300,000 years prior to the extinction event. In addition, other possible large impact sites of Late Cretaceous age have been described in the North Sea, the Ukraine, and India; at the latter location, an impact crater more than 500 kilometers in diameter—the biggest on Earth—has been reported (●FIGURE 4.2).

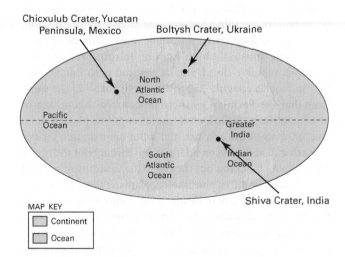

● **FIGURE 4.2** Strong evidence suggests that several large meteorites struck Earth about 65 million years ago. Many scientists think the impact site in the Yucatan Peninsula of Mexico contributed to the extinction event at the end of the Cretaceous Period, although impact sites in the Ukraine and India likely also played a role.

---

2. Luis W. Alvarez, Walter Alvarez, Frank Asaro, and Helen V. Michel, "Extraterrestrial Cause for the Cretaceous-Tertiary Extinction," *Science* 208, no. 4448 (1980): 1095–1108.

3. Alan R. Hildebrand, Glen T. Penfield, David A. Kring, Mark Pilkington, Antonio Z. Camargo, Stein B. Jacobsen, and William V. Boynton, "Chicxulub Crater: A Possible Cretaceous/Tertiary Boundary Impact Crater on the Yucatan Peninsula, Mexico," *Geology* 19 (1991): 867–871.

Although the Yucatan impact site was hailed for many years as the "smoking gun" that caused the Cretaceous terminal extinction event, these more recent findings cast doubt on that simple explanation. Some scientists have suggested that widespread volcanism in India, combined with severe greenhouse conditions at the end of the Cretaceous, caused significant environmental stress that reduced the population sizes of many species (or eliminated them entirely) prior to the Yucatan impact. These scientists have suggested that the impact itself did not catastrophically kill a healthy, thriving biosphere, but rather pushed an already very-stressed biosphere over the threshold that led to a major extinction. Although the extraterrestrial impact hypothesis still is very popular overall, reports of other major impact sites of Late Cretaceous age and uncertainty about the role that factors such as volcanism may have played will keep scientific interest fixed for the foreseeable future on the cause(s) of the Cretaceous extinction event.

Volcanic Eruptions

A volcanic eruption ejects gas and fine ash into the atmosphere. The volcanic ash serves like an umbrella that reflects sunlight back into space and thus causes climate to cool. One volcanic gas, sulfur dioxide, forms small particles called *aerosols* that also reflect sunlight and cool Earth. Carbon dioxide is another gas emitted by volcanoes. It is a greenhouse gas that can cause warming of Earth's atmosphere. Thus, volcanoes erupt materials capable of causing both atmospheric cooling and warming. In many eruptions, the cooling overwhelms the warming effects. This cooling can occur rapidly—within a few weeks or months of the eruption.

Scientists have noted that some mass extinctions coincided with unusually high rates of volcanic activity. For example, massive flood basalts erupted onto Earth's surface in Siberia roughly 252 million years ago—at the same time that the Permian extinction occurred. According to one hypothesis, explosive volcanic eruptions blasted massive amounts of volcanic ash and sulfur aerosols into the air. The ash and aerosols spread like a pall throughout the upper atmosphere, blocking out the Sun. Earth's atmospheric and oceanic temperatures plummeted, plants withered, and animals starved or froze to death. If this hypothesis is correct, the Permian extinction (and perhaps others) resulted from a classic Earth systems interaction, in which a volcanic eruption occurring within the geosphere altered the composition of the atmosphere, with profound consequences for atmospheric and marine temperatures, and for life.

Supercontinents and Earth's Carbon Dioxide Budget

In modern oceans, cold polar seawater sinks to the seafloor and flows as a deep-ocean current toward the equator. This current transports oxygen to the deep-ocean basins and forces the deep water to rise near the equator, where it is warmed. The current thus mixes both heat and gases throughout the oceans and the atmosphere.

Surface marine organisms that photosynthesize absorb atmospheric carbon dioxide and use it to produce organic tissue. These organisms then die and settle to the seafloor, thereby removing carbon from the atmosphere and transporting it to the seafloor. Some of the carbon converts back to carbon dioxide as organisms living in the deeper ocean water consume the settling organic matter, and some of this carbon dioxide is returned to the atmosphere as modern ocean currents mix the deep and shallow seawater.

Ocean currents are partially controlled by the positions of the continents. Recall from Chapter 1 that the continents slowly drift around the globe. Several times in Earth history, all the continents have joined together to form one giant supercontinent. One such supercontinent, called Pangea, assembled during Permian time (●FIGURE 4.3). The growth of Pangea also caused a single global ocean to be formed.

Computer models and the rock record both suggest that the assembly of all continents into a single landmass and the development of a single global ocean largely prevented mixing of the surface water and the ocean depths during Permian time.[4] The lack of vertical mixing caused the chemistry of the ocean water to change drastically. As ocean water stagnated, two major changes in water chemistry began to take place in some parts of the Permian Ocean. First, Permian seawater lost most or all of its dissolved oxygen as organic matter that washed in from the continents or from dead planktons from the surface waters sank and were decomposed by oxidizing bacteria. Oxidizing bacteria require oxygen to decompose organic matter, so eventually the available oxygen was used up, resulting in a condition called anoxia—no measurable dissolved oxygen. Once the oxygen was gone and the oxidizing bacteria with it, sulfate-reducing bacteria took over and continued to decompose the available organic

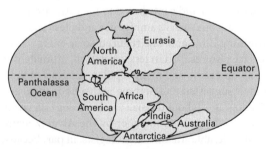

● **FIGURE 4.3** Computer models of circulation in the ocean suggest that during Permian time, assembly of all continents into the supercontinent Pangea prevented vertical mixing of ocean water. The lack of circulation caused carbon dioxide removed from the atmosphere to accumulate in the deep ocean as calcium carbonate, causing Earth's climate to cool.

4. A. H. Knoll, R. K. Bambach, D. E. Canfield, and J. P. Grotzinger, "Comparative Earth History and Late Permian Mass Extinction," *Science* 273 (July 26, 1996): 452–457.

matter. Sulfate-reducing bacteria convert sulfate dissolved in the seawater to sulfide in order to obtain energy for growth. This conversion produces the by-product hydrogen sulfide, a colorless, poisonous gas. In some parts of the Permian Ocean, not only was most or all of the dissolved oxygen removed, but the water became so full of dissolved hydrogen sulfide that it began to come out of solution and form tiny bubbles, a condition known as euxinia.

The anoxic, euxinic conditions that developed in some parts of the Permian Ocean were deadly to nearly all species of metacellular marine life living near or on the seafloor.

Organisms living near the surface were less affected because oxygen from the atmosphere could still mix with the water from wave energy to keep the water chemistry at the surface from becoming anoxic. Some surface planktons were able to thrive and in so doing continued to remove carbon dioxide from the atmosphere and supply organic carbon to the bottom waters which remained anoxic and euxinic (●FIGURE 4.4A). Removal of carbon dioxide from the atmosphere caused a decrease in greenhouse warming and resulted in global cooling. Some scientists have suggested that this global cooling led directly to the formation of continental ice sheets.

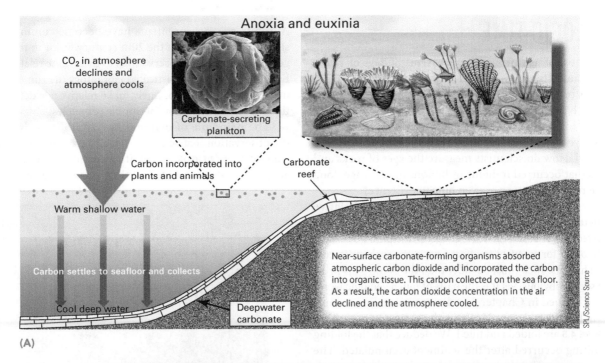

(A)

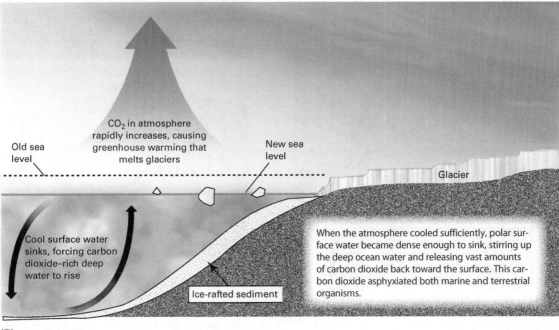

(B)

● FIGURE 4.4 (A) Global cooling in the Late Permian Period. (B) The Permian extinction.

When water cools, it becomes denser. The Late Permian global cooling eventually reached a threshold at which polar surface water became denser than the deep water. At this point, the cool, dense surface water sank, displacing the carbon dioxide–rich deep water and causing it to rise toward the sea surface, where massive amounts of carbon dioxide were released into the atmosphere (●FIGURE 4.4B). According to this hypothesis, the high levels of atmospheric carbon dioxide asphyxiated much life both in the seas and on the continents and caused the Permian extinction event. Later, new organisms evolved when the carbon dioxide levels fell.

GEOLOGIC TIME

While most of us think of time in terms of days or years, Earth scientists commonly refer to events that happened millions or billions of years ago. In Chapter 1 you learned that Earth is approximately 4.6 billion years old. Yet humans and our humanlike ancestors have existed for only 5 to 7 million years, and recorded history is only a few thousand years old. How do scientists measure the ages of rocks and events that occurred millions or billions of years ago, long before recorded history or even human existence?

Scientists measure geologic time in two ways. **Relative age** determination refers only to the order in which events occurred and is based on a few simple principles. For example, in order for an event to affect a rock, the rock must exist first. Thus, the rock must be older than the event. This principle seems obvious, yet it is the basis of much geologic work. As you learned in Chapter 3, sediment normally accumulates in horizontal layers. However, the sedimentary rocks in ●FIGURE 4.5 are folded and tilted. We deduce that the folding and tilting occurred after the sediment accumulated. The

order in which rocks and geologic features formed can usually be interpreted by such observation and logic.

Absolute age is age in numeric years. For example, the dinosaurs became extinct about 66 million years ago. The Permian extinction occurred 252 million years ago. The last catastrophic eruption of the Yellowstone Volcano occurred 630,000 years ago. Absolute age tells us both the order in which events occurred and the amount of time that has passed since they occurred.

RELATIVE GEOLOGIC TIME

Absolute age measurements have become common only since the second half of the 20th century. Prior to that time, geologists used field observations to determine relative ages. Even with today's sophisticated laboratory instrumentation, most field geologists routinely make relative age determinations based on the following simple principles:

The **principle of original horizontality** is based on the observation that sediment usually accumulates in horizontal layers (●FIGURE 4.6A). If sedimentary rocks lie at an angle, as in ●FIGURE 4.6B, or if they are folded as you saw in

(A)

(B)

● **FIGURE 4.6** (A) The principle of original horizontality tells us that most sedimentary rocks were originally deposited in horizontal layers, like these in western North Dakota. (B) When we see tilted sedimentary rocks such as these in northwestern Wyoming, we infer that they were tilted *after* the layers were deposited.

● **FIGURE 4.5** The limestone forming this cliff in Montana was deposited about 320 million years ago as horizontal layers of sediment in a shallow sea. About 200 million years later, tectonic forces uplifted and folded the layers.

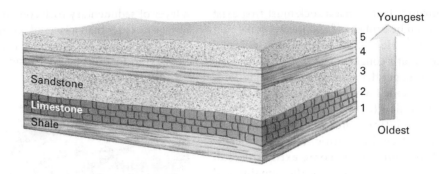

● **FIGURE 4.7** The principle of superposition. In a sequence of sedimentary beds, the oldest bed (labeled *1*) is the lowest, and the youngest (labeled *5*) is on top.

Figure 4.5, we can infer that tectonic forces tilted or folded them after they formed.

The **principle of superposition** states that in an undeformed sequence of layered sedimentary rocks, the oldest layers are at the bottom and the youngest layers are at the top. For example, in ●**FIGURE 4.7**, the sedimentary layers become progressively younger in the order 1, 2, 3, 4, and 5. Two notable exceptions to this principle occur if tectonic forces have turned the layers of sedimentary rock upside down or have caused an older layer to be shoved up and over a younger layer.

The **principle of crosscutting relationships** is based on the obvious fact that a rock must first exist before anything can happen to it. ●**FIGURE 4.8** shows an igneous intrusion—called a *dike*—cutting across conglomerate. Clearly, the dike must be younger than the conglomerate. ●**FIGURE 4.9** shows sedimentary rocks intruded by three granite dikes. Dike 2 cuts dike 1, and dike 3 cuts dike 2, so 2 is younger than 1, and 3 is the youngest. The sedimentary rocks must be older than all the dikes.

● **FIGURE 4.8** An igneous dike cutting across conglomerate in northwestern Wyoming. By the principle of crosscutting relationships, the dike must be younger than the conglomerate.

relative age An approach to determining the timing of geologic events based on the order in which they occurred, rather than on the absolute number of years ago in which they occurred.

absolute age The age of a geologic event based on the absolute number of years ago it occurred.

principle of original horizontality The principle that most sediment is deposited as nearly horizontal beds, and therefore most sedimentary rocks started out with nearly horizontal layering.

principle of superposition The principle that in undisturbed layers of sediment or sedimentary rock, the age becomes progressively younger from bottom to top; younger layers always accumulate on top of older layers.

principle of crosscutting relationships The obvious principle that a rock or feature must first exist before anything can happen to it; thus, if a dike of basalt cuts across a layer of sandstone, the basalt is younger.

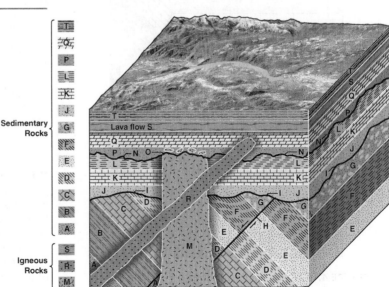

● **FIGURE 4.9** Relative age dating principles determine the order of events corresponding to the formation of each rock unit, fault, and unconformity in this block diagram. A is oldest and T is youngest.

A similarly obvious fact is that a rock must first exist before pieces of it can be broken off and incorporated into another geologic unit. Hence, the **principle of included fragments** requires that a conglomerate that contains clasts of granite must be younger than the age of the granite. Similarly, pieces of sandstone incorporated into a basalt flow must be older than the flow (●**FIGURE 4.10**).

The theory of evolution states that life-forms have changed throughout geologic time. Species emerge, persist for a while, and then become extinct. Recall from Chapter 3 that fossils consist of the remains and other traces of prehistoric life preserved in sedimentary rocks. Fossils are useful in determining relative ages of rocks, because different animals and plants lived at different times in Earth history. For example, trilobites (●**FIGURE 4.11**) lived from about 521 million to about 252 million years ago, and the first dinosaurs appeared about 240 million years ago.

The principle of superposition tells us that sedimentary rock layers become younger from bottom to top. If the rocks formed over a long time, different fossils appear and then vanish from bottom to top, in the same order in which the organisms evolved and then became extinct. Rocks containing dinosaur bones must be younger than those containing trilobite remains.

The **principle of faunal succession** states that species succeeded one another through time in a definite and recognizable order and that the relative ages of sedimentary rocks can therefore be recognized from their fossils. For example,

a layer of sedimentary rock containing trilobite fossils will always be older than a layer of sedimentary rock with dinosaur bones, because trilobites evolved and went extinct before the dinosaurs evolved (●**FIGURE 4.12**).

● **FIGURE 4.11** Trilobites were marine creatures that thrived in Earth's ancient seas before the time of the dinosaurs.

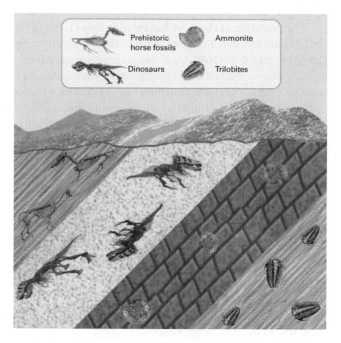

● **FIGURE 4.12** Layers of sedimentary rock can be relatively dated based on the fossils they contain. According to the Principle of Faunal Succession, layers containing older fossils were deposited before sedimentary layers containing younger fossils. Therefore, in tilted beds such as those shown above, the sedimentary layers containing trilobites were deposited before those containing ammonites. Both of these fossils appeared before dinosaurs. The layer containing fossil bones from pre-historic horses is the youngest. Trilobites were extinct before dinosaurs evolved, and horses did not appear until after the extinction of the dinosaurs.

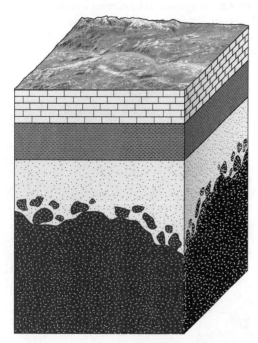

● **FIGURE 4.10** A rock first has to exist before pieces of it can be broken off and incorporated into another rock. In the example shown, pieces of the basalt have been eroded and deposited in sand that later lithified to sandstone. By the Principle of Included Fragments, the basalt is older.

Interpreting Geologic History from Fossils

Paleontologists study fossils to understand the history of life and evolution. The oldest-known fossils are traces of bacteria-like organisms that lived about 3.5 billion years ago. A much younger fossil consists of the frozen and mummified remains of a Bronze Age man, recently found in a glacier near the Italian-Austrian border.

Fossils also allow geologists to interpret ancient environments, because most organisms thrive in specific environments and do not survive in others. For example, all modern species of coral live in marine environments. Corals do not survive in fresh or even brackish (slightly salty) water. Yet today, fossil corals are abundant in the rocks of many mountain environments, including the Canadian Rockies, the European Alps, and the Tian Shan (translated "Heavenly Mountains") of western China. Each of these mountain ranges are hundreds of kilometers or more away from the nearest ocean. The fossil corals tell us that the rocks exposed in these mountains were once submerged beneath marine waters. Today, geologists understand that tectonic processes raised the marine rocks and their fossils to their present altitudes.

Although some individual fossils can be used to recognize ancient depositional environments, like corals representing a normal marine environment, geologists more often use the entire fossil assemblage in a rock formation to interpret its depositional environment. For example, if the collection of fossils in a rock sequence includes numerous different species of body fossils and a wide variety of trace fossils, geologists might infer that the environment was widely suitable for inhabitation. Shallow marine environments in tropical or temperate waters are such environments—the rich ecosystem supports a diversity of life-forms. On the other hand, if only one or two fossil types occur in a sedimentary rock sequence and the trace fossil diversity also is low, geologists might infer that the environment of deposition was stressful and only a few types of specialized organisms could survive. Deserts represent such a hostile environment of deposition.

UNCONFORMITIES AND CORRELATION

The walls of the Grand Canyon are composed of sedimentary rocks lying on older igneous and metamorphic rocks (●FIGURE 4.13). Their ages range from about 200 million years to nearly 2 billion years. The principle of superposition tells us that the deepest sedimentary rocks are the oldest, and the

rocks become younger as we climb up the canyon walls. However, no principle assures us that the rocks formed continuously from 2 billion to 200 million years ago. The rock record may be incomplete. Suppose that no sediment was deposited for a period of time, or that erosion removed some sedimentary layers before younger layers accumulated. In either case, a gap would exist in the rock record. We know that in an undeformed layered sequence of sedimentary rock, any one layer is younger than the layer below it, but without more information we do not know how much younger.

MINDTAP From Cengage

Animation: Types of Unconformities

MINDTAP From Cengage

Animation: History of Unconformities

P.Burghardt/Shutterstock.com

● **FIGURE 4.13** The walls of the Grand Canyon are composed of sedimentary rocks lying on older igneous and metamorphic rocks. This photo shows the dark-colored PreCambrian schist of 1.7 billion year age in the deepest parts of the canyon. Nonconformably overlying the dark schist is horizontally layered reddish-brown Tapeats Sandstone, which is about 525 million years old. Overlying the Tapeats are younger layers of sedimentary rock, including the Redwall Limestone deposited about 340 million years ago. The base of the Redwall Limestone is highlighted by the dashed white line.

---

principle of included fragments The principle that a rock unit must first exist before pieces of it can be broken off and incorporated into another rock unit.

principle of faunal succession The principle that species succeeded one another through time in a definite order, so the relative ages of fossiliferous sedimentary rocks can be determined by their fossil content.

Unconformities

Layers of sedimentary rocks are **conformable** if they were deposited without detectable interruption. An **unconformity** is a bounding surface that exists between two rock units and represents an interruption in deposition, usually of long duration. Some rock may have been eroded during the time frame represented by the unconformity, but eventually deposition of sediment resumed and buried the unconformity. Thus, an unconformity is a physical feature (a bounding surface) that represents a long interval of time for which no geologic record exists in that place. The missing record may involve billions of years.

Several types of unconformities exist. In a **disconformity**, the sedimentary layers above and below the unconformity are parallel (●**FIGURE 4.14**). Some disconformities are produced by downward or lateral (sideways) erosion through previously deposited layers of sediment through the process of reworking. For example, river channels and deep sea submarine channels both can erode downward and sideways through time, resulting in a disconformity that curves upward at the margins of the former channel (●**FIGURE 4.15**).

Although some disconformities, like the submarine channel example shown in Figure 4.15, are relatively easy to recognize in a good exposure of sedimentary rocks, many disconformities are much more subtle and difficult to recognize without detailed investigation of the rock texture, composition, sedimentary structures, and fossil content. For example, layers of sediment that accumulate slowly on parts of the seafloor can all consist of similar-looking mud. In such mud-rich sediment sequences, disconformities that may represent tens or hundreds of thousands of years can be very difficult to recognize without detailed study of the mud itself. Another example of a subtle disconformity is shown in ●**FIGURE 4.16**. In this case, the disconformity was formed by a break in deposition on a shallow marine sand bar (●**FIGURE 4.17**). In some cases, disconformities can be identified by determining the ages of rocks, using methods based on fossils and absolute dating, described below.

In an **angular unconformity**, tectonic activity tilted older sedimentary rock layers and erosion planed them off before the younger sediment accumulated (●**FIGURES 4.18** and **4.19**). Although some disconformities involve the erosional truncation of underlying beds as shown in Figure 4.15, the erosion of the underlying beds was produced by lateral (sideways) erosion in a former channel. The rocks under the disconformity were not tectonically tilted prior to their partial removal by erosion.

Disconformities are **architectural elements** that contribute significantly to the three-dimensional shape of sediment layers in which they occur. In the case of a great cathedral or modern skyscraper, the architectural elements include the

Marc S. Hendrix

● **FIGURE 4.14** These layers of sandstone in Tanzania were deposited in the deep lake as pulses of sediment delivered to the lake floor by large, possibly seasonal floods. Each flood deposit starts out as fine sand but becomes more coarse-grained as the flood event builds and more sediment is delivered to the lake. Separating the three individual flood deposits are disconformities, marked by the red arrows. Each disconformity was formed by a long pause between successive floods during which no deposition of sediment took place on the lake floor. The "missing time" between flood events is represented by each disconformity.

Marc S. Hendrix

● **FIGURE 4.15** Example of a disconformity in West Texas. In this example, a submarine channel, represented by the light brown sandstone on the right, eroded downward and laterally into previously deposited beds of fine sandstone and mudstone, shown on the right. Notice that the beds on the left are cut off by the sandstone, indicating that they were partially removed through the erosion that formed the disconformity. A rock hammer (circled) is sitting on the disconformity and provides scale.

● **FIGURE 4.16** The horizontal bedding plane behind the yellow field book is a disconformity formed on a river-mouth bar such as that illustrated in Figure 4.17. Small holes in the rock (red arrows) below the disconformity were formed by burrowing shrimp during the break in deposition. Eventually, deposition resumed and buried the disconformity.

Copyright and Photograph by Marc S. hendrix

← Disconformity

Shrimp burrow

arrangement, size, and shape of a building's support structure, framing, window placement, embellishments, and so on. Similarly, sedimentary architectural elements include the thickness and lateral dimensions of sediment beds, the 3-D shape of disconformities within a sediment sequence, and the type and geometry of sedimentary structures present within the beds. The architectural elements exposed in ancient sequences of sedimentary rocks provide very useful information for geologists concerned with interpreting the depositional environments represented by the sedimentary deposits.

A **nonconformity** is an unconformity in which sedimentary rocks lie on igneous or metamorphic rocks. The nonconformity shown in Figure 4.13 represents a time gap of about 1 billion years. Perhaps the best exposed and widespread example of a nonconformity is the unconformity that separates metamorphic and igneous rocks that make up the basement rocks of Earth's continents and the overlying sedimentary rocks that have been deposited on top of it. Because it is so widespread and represents such a significant break in the geologic record of the continents, this nonconformity commonly is called "The Great Unconformity."

conformable A term describing sedimentary layers that were deposited continuously, without detectable interruption.

unconformity A physical surface between layers of sediment or sedimentary rock that represents an interruption in the deposition of sediment, including an interruption that involves partial erosion of the older layer. Also the physical surface between eroded igneous and overlying sedimentary layers deposited following erosion. Types of unconformities include disconformity, angular unconformity, and nonconformity.

disconformity A type of unconformity in which the sedimentary layers above and below the unconformity are parallel.

angular unconformity An unconformity in which younger sediment or sedimentary rocks rest on the eroded surface of tilted or folded older rocks.

architectural element A top-down hierarchical classification for describing the 3-D spatial arrangement of sedimentary deposits at scales ranging from an entire sedimentary basin to individual grains. The overall geometry of each sedimentary lithology, all surfaces of erosion, and all sedimentary structures are examples of architectural elements.

nonconformity A type of unconformity in which layered sedimentary rocks lie on an erosion surface cut into igneous or metamorphic rocks.

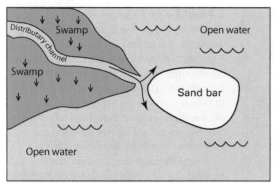

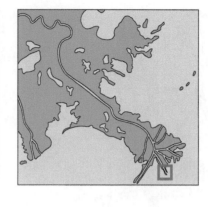

Bar forms beyond mouth of river channel; sediment accumulates, building upward to near sea-level.

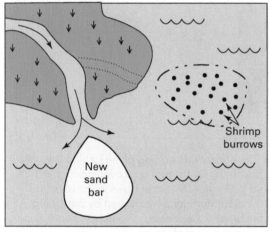

River mouth shifts to a different location, forming a new bar; original bar is abandoned, sediment stops accumulating, and upper surface is colonized by burrowing shrimp.

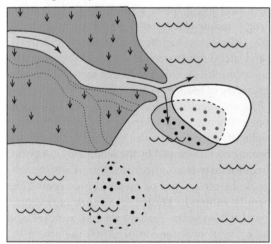

River mouth shifts again to a location near the original bar; new bar forms and partly overlaps original bar, covering disconformity.

● **FIGURE 4.17** A disconformity is created when there is a pause in the deposition of sediment, sometimes accompanied by minor erosion. For example, when a river channel shifts its position, the sandbar formed at its mouth no longer receives sediment and begins to subside. Commonly, organisms will burrow into this stable surface before the river channel shifts back and deposition resumes on the bar.

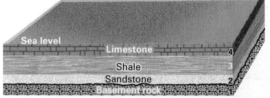

Sediment is deposited in horizontal layers and lithifies to form rock.

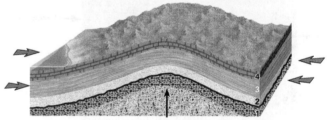

Rocks are compressed, uplifted, and tilted.

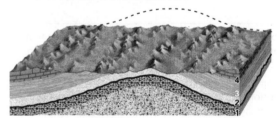

Erosion exposes folded rocks.

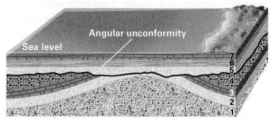

Rocks sink below sea level and new sediment is deposited.

● **FIGURE 4.18** An angular unconformity develops when older sedimentary rocks are folded and partially eroded before younger sediment accumulates.

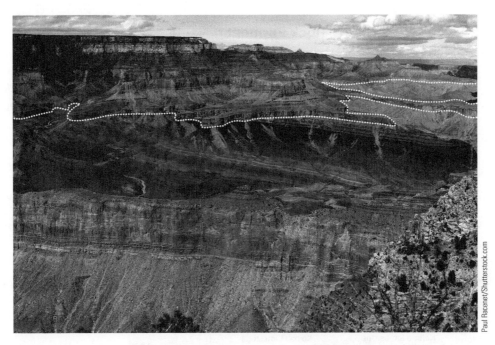

● **FIGURE 4.19** In this classic angular unconformity in the Grand Canyon, flat-lying Cambrian sedimentary rocks rest atop tilted and partially eroded Precambrian sedimentary rocks. The angular unconformity is highlighted by the dashed white line.

Correlation

Ideally, geologists would like to develop a continuous history for each region of Earth by interpreting rocks that formed in that place throughout geologic time. However, there is no single place on Earth where rocks were continuously formed and preserved. Consequently, in any one place, the **rock record**—that is, the rocks that currently exist on Earth and contain the record of its history—is full of unconformities, or historical gaps. To assemble as complete and continuous a record as possible, geologists gather geological evidence from many localities and match rocks of the same age from different localities in a process called **correlation** (●FIGURE 4.20).

The correlation of rock of the same age over great distances is at the heart of reconstructing Earth's ancient systems and recognizing when and how these changes affected the surface of the Earth. Correlating rock sequences from one region on earth to another is complicated by the great variety of environments that existed at any single time in Earth's geologic past, not to mention the massive changes that Earth's systems have undergone over the past 4.6 billion years, since the planet first formed.

In order to understand what even a small part of Earth's surface looked like at any point during the geologic past, geologists must be able to correlate rocks that formed in different environments that existed at the same point in time. Today, for example, the east coast region of North America includes the Appalachian Mountains, a very gently-sloping mostly flat coastal plain, the shoreline environment, and the shallow marine settings offshore ranging from the ankle-deep swash zone to the outer part of the continental shelf over 100 kilometers beyond the shoreline. Beyond the continental shelf are environments of the deeper ocean. Each of these environments is very different, yet all exist today—at this point in geologic time. Therefore, the primary aim of correlation is the recognition of time-equivalence between different sequences of physical sedimentary rock.

If you follow a single sedimentary bed from one place to another, then it is clearly the same layer in both places. It is natural to suppose that because the layer can be physically traced from one place to another that the layer therefore must be the same age everywhere. This supposition is almost never the case, however, and has misled geologists for many decades. Although a layer may be physically traced, the result does not reflect a set of environments that changed in space, but rather a set of environments that change in space and through time. For example, many rivers will migrate laterally over many decades or hundreds of years as one bank erodes and the opposite bank builds out. The result is a layer of sandstone that was not deposited all at once, but instead grew laterally as the river slowly migrated (●FIGURE 4.21).

---

rock record The rocks that currently exist on Earth and contain the record of its history.

correlation The process of establishing the age relationship of rocks or geologic features from different locations on Earth; can be done by comparing sedimentary characteristics of the layers or the types of fossils found in them. There are two types of correlation: *time correlation* (age equivalence) and *lithologic correlation* (physical continuity of the rock unit).

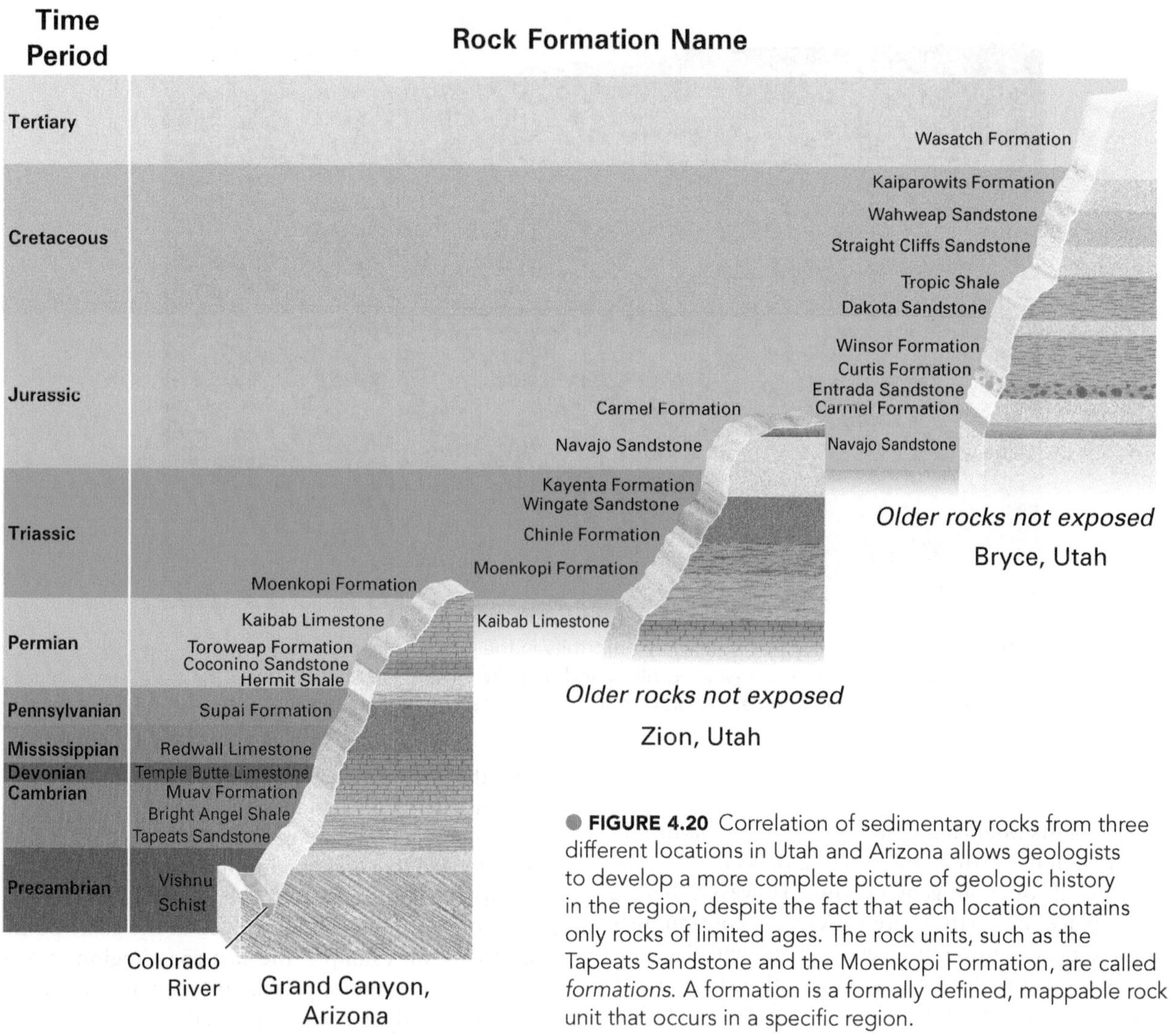

Time Period

Rock Formation Name

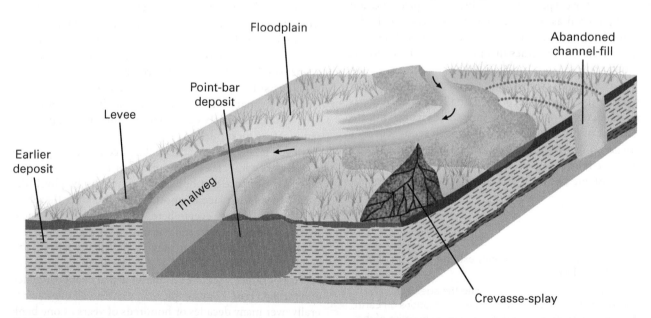

Tertiary

Cretaceous

Jurassic

Triassic

Permian

Pennsylvanian

Mississippian

Devonian

Cambrian

Precambrian

Wasatch Formation

Kaiparowits Formation
Wahweap Sandstone
Straight Cliffs Sandstone

Tropic Shale
Dakota Sandstone

Winsor Formation
Curtis Formation
Entrada Sandstone
Carmel Formation

Navajo Sandstone

Older rocks not exposed
Bryce, Utah

Carmel Formation

Navajo Sandstone

Kayenta Formation
Wingate Sandstone

Chinle Formation

Moenkopi Formation

Kaibab Limestone

Older rocks not exposed
Zion, Utah

Moenkopi Formation

Kaibab Limestone
Toroweap Formation
Coconino Sandstone
Hermit Shale

Supai Formation

Redwall Limestone
Temple Butte Limestone
Muav Formation
Bright Angel Shale
Tapeats Sandstone

Vishnu Schist

Colorado River

Grand Canyon, Arizona

● **FIGURE 4.20** Correlation of sedimentary rocks from three different locations in Utah and Arizona allows geologists to develop a more complete picture of geologic history in the region, despite the fact that each location contains only rocks of limited ages. The rock units, such as the Tapeats Sandstone and the Moenkopi Formation, are called *formations*. A formation is a formally defined, mappable rock unit that occurs in a specific region.

Floodplain

Abandoned channel-fill

Point-bar deposit

Levee

Earlier deposit

Thalweg

Crevasse-splay

● **FIGURE 4.21** Single layers of sedimentary rock may not be commonly deposited all at the same time. In the example above, a river channel migrates laterally through time as one bank erodes while the opposite builds out. As it migrates over hundreds or thousands of years, the channel will leave behind a layer of sandstone that is one physically continuous layer but that does not have one single depositional age.

Source: The classical point-bar model for deposition in a meandering stream channel (after J.R.L. Allen, 1964, 1970b).

In a larger scale example, suppose you are attempting to trace a layer of sandstone deposited by a beach. If the relative sea level rose and slowly inundated a continental interior, the beach would also migrate inland. However, those beach deposits closer to the edge of the continent would be older than the beach deposits further inland. Thus, although the entire layer would be physically continuous, it would not be the same age everywhere and therefore would not correlate in time (Figure 4.21). Over a few hundred meters, the age difference within the sandstone layer may be unimportant, but over hundreds of kilometers it could vary by millions of years.

So there are two kinds of correlation: time correlation refers to age equivalence, and lithologic correlation refers to the continuity of a rock unit. The two are not always the same, because some rock units, such as the sandstone, were deposited at different times in different places (●FIGURE 4.22). To construct a record of Earth history and a geologic time scale, Earth scientists use several types of evidence to correlate rocks of the same age.

Index fossils are a very useful tool for correlation, because they can precisely indicate the ages of sedimentary rocks. The best index fossils are abundantly preserved in rocks, are geographically widespread, existed as a species or genus for only a relatively short time, and are easily identified in the field. Floating or swimming tiny marine organisms make some of the best index fossils because they spread rapidly and widely throughout the seas and they evolved quickly. The shorter the time that a species existed, the more precisely the index fossil reflects the age of a rock.

In many cases, the presence of a single type of index fossil is sufficient to establish the age of a rock. More commonly, an assemblage of several fossils is used to date and correlate rocks (●FIGURE 4.23). For example, the exact position of the Cretaceous-Tertiary boundary in the sediment core from the Yucatan meteor impact site, described above, was determined by using tiny floating marine organisms as index fossils.

Another correlation tool is the **key bed**—a thin, widespread, easily recognized sedimentary layer that was deposited rapidly and simultaneously over a wide area. Many volcanic eruptions eject great volumes of fine, glassy volcanic ash into the atmosphere. Wind carries the ash over great distances before it settles. Some historic ash clouds

have encircled the globe. When the ash settles, it forms a depositional layer, and the glassy fragments within the ash layer typically crystallize to form clay. Because such volcanic eruptions occur at a precise point in time, the ash deposit is the same age everywhere. Thus, the thin, sooty, iridium-rich clay layer deposited roughly 65 million years ago from debris of a giant meteorite impact (described at the beginning of this chapter) is a classic example of a key bed. A more recent key bed is the thin layer of white volcanic ash shown in ●FIGURE 4.24 from western Montana. The layer of ash in the photo came from a massive eruption that occurred 11,200 years ago from Glacier Peak, an active volcano northeast of Seattle, Washington.

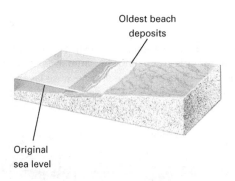

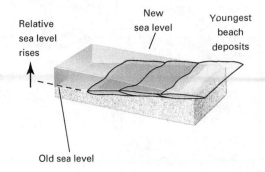

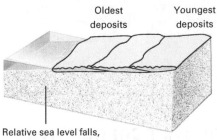

● FIGURE 4.22 As relative sea level rises slowly through time, a beach migrates inland, forming a single sand layer that is of different ages at different localities.

---

index fossils A fossil that dates the layers where it is found because it came from an organism that is abundantly preserved in rocks, was widespread geographically, and existed as a species or genus for only a relatively short time.

key bed A thin, widespread, easily recognized sedimentary layer that can be used for correlation because it was deposited rapidly and simultaneously over a wide area.

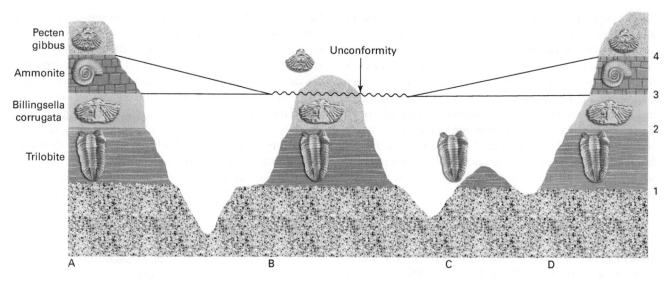

Pecten gibbus

Ammonite

Billingsella corrugata

Trilobite

Unconformity

A   B   C   D

4
3
2
1

● **FIGURE 4.23** Index fossils demonstrate age equivalency of sedimentary rocks from widely separated localities. Sedimentary beds containing the same index fossils are interpreted to be of the same age. In this figure, the fossils show that at locality B, layer 3 is missing because layer 4 lies directly on top of layer 2. Either layer 3 was deposited and then eroded away before layer 4 was deposited at locality B, or layer 3 was never deposited there. At locality C, all layers above layer 1 are missing, either because of erosion or because they were never deposited.

Marc S. Hendrix

● **FIGURE 4.24** Example of a key bed. The white layer above the pen is volcanic ash from an eruption of Glacier Peak, near Seattle, about 11,200 years ago. This photo was taken in western Montana over 500 kilometers east of the volcano.

ABSOLUTE GEOLOGIC TIME

Geologists have a challenging task—they are attempting to measure the absolute age of events that occurred before history was recorded. Measuring absolute age relies on (1) a process that occurs at a constant rate, such as Earth rotating once every 24 hours, and (2) some way to keep a cumulative record of that process, such as marking a calendar each time the Sun rises. Measurement of time with a calendar, a clock, an hourglass, or any other device depends on these two factors.

Fortunately, scientists have found a natural process that occurs at a constant rate and accumulates its own record: it is the radioactive decay of elements that are present in many rocks. Thus, many rocks have built-in calendars.

To begin to understand radioactivity, we need a basic understanding of the atom. An atom consists of a small, dense nucleus surrounded by a cloud of electrons. A nucleus consists of positively charged *protons* and neutral particles called *neutrons*. All atoms of any given element have the same number of protons in the nucleus. However, the number of neutrons may vary. **Isotopes** are atoms of the same element with different numbers of neutrons. For example, all isotopes of carbon contain 6 protons, but only carbon-12 also has 6 neutrons. Carbon-13 has 7 neutrons, and carbon-14 has 8 neutrons.

Many isotopes are *stable*, meaning that they do not change with time. Carbon-12 is one such isotope; if you study a sample of carbon-12 for a million years, all of the atoms would remain unchanged. However, other isotopes are unstable or radioactive. Given time, their nuclei spontaneously decay, forming a different isotope and commonly a different element. For example, carbon-14, the one unstable isotope of carbon, spontaneously decays to form nitrogen-14.

A radioactive isotope such as carbon-14 is known as a *parent isotope*. An isotope created by radioactivity, such as nitrogen-14, is called a *daughter isotope*.

Many common elements, including carbon and potassium, consist of a mixture of radioactive and nonradioactive isotopes. With time, the radioactive isotopes decay, but the nonradioactive ones do not. Some elements, such as uranium, consist only of radioactive isotopes. The amount of uranium on Earth slowly decreases as it decays to other elements, such as lead.

Radioactivity and Half-Life

If you watch a single atom of carbon-14, when will it decompose? This question cannot be answered, because any particular carbon-14 atom may or may not decompose at any time. Each atom has a certain probability of decaying at any time. Averaged over time, half of the atoms in any sample of carbon-14 will decompose in 5,730 years. The **half-life** is the time it takes for half of the atoms in a radioactive sample to decompose. The half-life of carbon-14 is 5,730 years. Therefore, if 1 gram of carbon-14 were placed in a container, 0.5 gram would remain after 5,730 years, 0.25 gram after 11,460 years, and so on. Each radioactive isotope has its own half-life; some half-lives are fractions of a second, and others are measured in billions of years.

isotopes Atoms of the same element that have different numbers of neutrons.

half-life The time it takes for half of the atoms of a radioactive isotope in a sample to decay.

radiometric dating The process of measuring the absolute age of rocks, minerals, and fossils by measuring the concentrations of radioactive isotopes and their decay products.

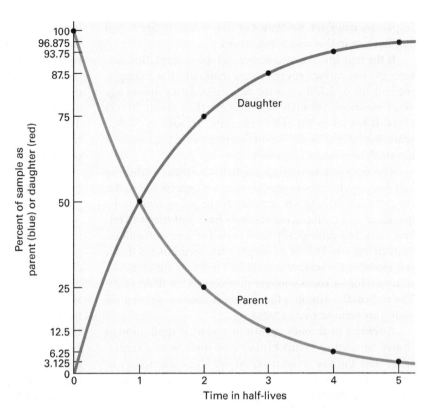

● **FIGURE 4.25** As a radioactive parent isotope decays to its daughter isotope, the proportion of parent decreases (blue line) and the amount of daughter increases (red line). The half-life is the amount of time required for half of the parent to decay to daughter. At time zero, when the radiometric calendar starts, a sample is 100 percent parent. At the end of one half-life, 50 percent of the parent has converted to daughter. At the end of two half-lives, 25 percent of the sample is parent and 75 percent is daughter. Thus, by measuring the proportions of parent and daughter in a rock, the rock's age in half-lives can be obtained. Because the half-lives of all radioactive isotopes are well known, it is simple to convert age in half-lives to age in years.

Radiometric Dating

Two properties of radioactivity are essential to the calendars in rocks. First, the half-life of a radioactive isotope is constant. It can be measured in the laboratory and is unaffected by geologic or chemical processes, so radioactive decay occurs at a known, constant rate. Second, as a parent isotope decays, its daughter accumulates in the rock. The longer the rock exists, the more the daughter isotope accumulates. **Radiometric dating** is the process of determining the absolute ages of rocks, minerals, and fossils by measuring the relative amounts of parent and daughter isotopes.

●**FIGURE 4.25** shows the relationships between age and relative amounts of parent and daughter isotopes. At the end of one half-life, 50 percent of the parent atoms have decayed to daughter. At the end of two half-lives, the mixture is 25 percent parent and 75 percent daughter. To determine the age of a rock, a geologist measures the proportions of parent and daughter isotopes in a sample and compares the ratio to a similar graph. Consider a hypothetical parent–daughter pair having a half-life of 1 million years. If we determine that a rock contains a mixture of 25 percent parent isotope and

75 percent daughter, ●**FIGURE 4.19** shows that the age is two half-lives, or in this case 2 million years.

If the half-life of a radioactive isotope is short, that isotope gives accurate ages for young materials. For example, the half-life of 5,730 years for carbon-14 dating allows age determinations for materials younger than about 50,000 years. It is useless for older materials, because by 50,000 years, the amount of carbon-14 that remains in the sample is too small to measure accurately.

The opposite limitation exists for isotopes with long half-lives. Such isotopes provide accurate ages for old rocks, but not enough daughter accumulates in young rocks to be measured. For example, rubidium-87 has a half-life of 47 billion years. In a geologically short period of time—10 million years or less—so little of its daughter has accumulated that it is impossible to measure accurately. Therefore, rubidium-87 is not useful for rocks younger than about 10 million years. The radioactive isotopes that are most commonly used for dating are summarized in **TABLE 4.1**.

Advances in absolute age dating and their application to slowly unraveling Earth's history have undergone a renaissance over the past quarter century or so, as technology in general has advanced. Today, for example, it is possible to determine with **accuracy** and **precision** the age of individual grains of sand collected from an ancient sandstone. By isolating datable mineral grains, most commonly the mineral zircon, from a sandstone sample, geologists can measure the age of each grain, forming an age spectrum—a histogram of ages from all the individual grains measured. The age spectrum tell geologists the ages of rock that was eroded to form the sandstone. Along with information about the composition of the grains—that is, what types of minerals and rock fragments are represented—age spectra provide a direct link to the provenance (from French "provenir," meaning "to come from") of the sample. In addition, by applying the principle of included fragments to ages determined from individual grains of sand from an ancient sandstone, geologists are able to determine the maximum depositional age of the layer of sandstone from which the grains came. The grains must first have existed before they were eroded, transported, and deposited in the layer of sandstone that was collected by the geologist for analysis.

Sequence Stratigraphy—Interpreting the Rock Record through Time

Geology is both an analytic and historic science. That is, geology uses the principles established by physicists and chemists to analyze the rock record systematically, but unlike the timeless principles upon which analytic science is based, such as Newton's Law of Gravitation, the evolution of Earth has taken place along a one-way timeline. Time has a direction, and insofar as understanding the ancient history of the Earth is concerned, geology events are bound by time. Earth history relies on a one-way record of physical remains left behind through the vast passage of geologic of time. This so-called rock record is very fragmented, because erosion has removed many older rock bodies that had formed during earlier phases of Earth's history. Nevertheless, the rock record was produced through time that travels in one direction.

Ultimately, understanding and accurately reconstructing Earth's history relies on the ability of humans to place the physical evidence that is left behind (i.e., the "rock record") into a temporal framework. In many geologic applications, relative age-dating is sufficient to reconstruct the geologic events that transpired. However, the precision and accuracy

TABLE 4.1	The Most Commonly Used Isotopes in Radiometric Age Dating			
Isotope		**Half-Life of Parent (Years)**	**Effective Dating Range (Years)**	**Minerals and Other Materials That Can Be Dated with This Isotope**
(Parent)	**(Daughter)**			
Carbon-14	Nitrogen-14	5,730 ± 30	100–50,000	Anything that was once alive: wood, other plant matter, bone, flesh, or shells; also, carbon in carbon dioxide dissolved in groundwater, deep layers of the ocean, or glacial ice
Potassium-40	Argon-40 Calcium-40	1.3 billion	50,000–4.6 billion	Muscovite Biotite Hornblende Whole volcanic rock
Uranium-238 Uranium-235 Thorium-232	Lead-206 Lead-207 Lead-208	4.5 billion 710 million 14 billion	10 million–4.6 billion	Zircon Uraninite and pitchblende
Rubidium-87	Strontium-87	47 billion	10 million–4.6 billion	Muscovite Biotite Potassium feldspar Whole metamorphic or igneous rock

with which geologists can determine the absolute (numerical) age of geological events is among the biggest hurdles in reconstructing Earth history.

For over 150 years, geologists used fossils to determine the age of sedimentary rocks. One of the earliest scientists to recognize that fossils in the rock record were related to time was the Frenchman George Cuvier. By reassembling the bones of fossil vertebrate organisms that had been unearthed from ancient rock or sediment, Cuvier was able to demonstrate convincingly to his contemporaries that many types of large vertebrates no longer exist today and therefore must have gone extinct. Cuvier assembled and described the first plesiosaur from a set of bones collected in England. He recognized that different types of teeth found in Ohio and transported back to Europe belonged to an extinct form of elephant-like creature, known as the Mammoth. By the end of his career, Cuvier had reassembled the skeletons from and described dozens of now-extinct organisms.

Cuvier's work influenced a now-famous scientist of the 1800s—Charles Darwin. In 1831, Darwin left his home in England for a sailing journey that took him around the world. His main goal was to describe the modern plants, animals, and environments that he encountered along the way and in this sense was very much a sort of "natural scientist-explorer." When he arrived in the Galapagos Islands, west of modern-day Ecuador, Darwin discovered that each island had its own unique species of finch, each with a particular set of adaptations not found anywhere else. Darwin reasoned that because the islands were physically separated by enough space, it prevented finches from island-hopping and that, through time, finch species on each island evolved independently through the process of **natural selection**. Natural selection—Darwin's most famous contribution to science—posits that as Earth environments change, individual species will change with it so as to be better adapted to the new environmental conditions. Those individuals with the specific trait needed to adapt to the new conditions would be stronger than those individuals without it and therefore more likely to have offspring. In this manner, the favorable trait would be preferentially passed down to future generations. Those species which developed the key trait are *naturally selected* to survive; those species which did not, go extinct.

Through the early works of Cuvier, Darwin, and others, scientists of the 1800s began to realize that many fossil organisms evolved through time and went extinct. Dinosaurs

evolved and now are extinct, for example. Moreover, dinosaur remains are only found in rock layers of a specific age. By utilizing observations such as this, and by making detailed catalogs of the specific fossils that occurred across different sedimentary layers at a single locality, geologists quickly recognized that the fossils themselves provided an indication of the age of the rock.

Through much of the late 1800s and early 1900s, thousands of paleontologists spent their careers studying, describing, and cataloging the detailed assemblage of fossils in rock sequences around the world. Not only were the bones of fossil vertebrates studied, but invertebrate fossils were widely collected and studied, as were tiny fossils called microfossils such as foraminifera or radiolaria—two types of single-celled protists with tiny shells that commonly are fossilized. As a result of the vast body of scientific literature that resulted from the many decades of detailed paleontologic study of Earth's fossil record, geologists developed the ability to recognize the age of a rock unit by the fossils it contains. In fact, for well over a century, fossils were used to subdivide and correlate rock units across the globe.

By the mid-20th century, however, it became clear through detailed work that fossil occurrence relied not only on time but also on environment. As environments changed through time in one place, some fossil organisms simply migrated to another place. As a result, the age of many fossils was not the same everywhere—the age could vary laterally as organisms moved around. This realization contradicted the long-held assumption that fossil occurrence is strictly dependent on time.

Instead of correlating beds containing the same fossils, a technique called biostratigraphy, or correlating rocks of similar lithology, called lithostratigraphy, today geologists correlate geologic deposits based on their age—or chronology—of deposition; that is, their chronostratigraphy. The key challenge in this regard is figuring out how to determine which rocks are time-correlative, if neither fossils nor simple tracing of rock layers actually quite work for this purpose. The answer to this challenge developed in North America, as geologists studied the sedimentary layers across the continent and began to use modern geophysical tools to provide direct images of the large-scale architectural elements of sedimentary rocks in the subsurface.

Early workers of North American stratigraphy began to notice that across the continent, unconformities occurred during certain periods of time and that no rock of these periods was preserved anywhere. The unconformities appeared to represent long periods of erosion or non-deposition, either one of which produces an unconformity. The same workers also noticed that for other long periods of time, sediments containing marine fossils were deposited most or all the way across the continent, unlike today.

One of the first geologists to describe these widespread unconformities and the bodies of rock in between was Larry Sloss, a Professor of Geology at Northwestern University in Chicago from 1947 to his death in 1996. By carefully cataloging the type and ages of sedimentary rocks at different locations across North America, Sloss determined that the widespread unconformities subdivided the rock record into meaningful packages, or **sequences**. To clearly differentiate

---

accuracy the degree to which a measurement reflects the measure's 'true' value; the degree of 'correctness' of a quantity

precision the degree to which multiple measurements are close to each other, irrespective of their accuracy

natural selection the process by which the hereditary traits of individuals within a population change so as to provide a competitive advantage. Those individuals with the competitive advantage tend to survive (i.e., are naturally selected) and produce more offspring with the same key trait. Over many generations the entire population evolves to possess the trait.

sequence a succession of layered sediments or sedimentary rocks that are genetically related and deposited during one full cycle of change in available accommodation space, such as one full cycle of sea level change.

these unconformity-bound sequences from the European fossil-based geologic timescale, Sloss named each sequence after a tribe of North American natives in whose traditional homeland sedimentary rocks of the appropriate age were widespread. For example, he named the Sauk Sequence after a tribe in Wisconsin where rocks belonging to the sequence are widely exposed (●FIGURE 4.26). ●FIGURE 4.27 shows the Sauk Sequence and lower Tippecanoe Sequence as mapped by Sloss along four separate lines of cross-sections in North America.

After publishing his initial ideas about North American stratigraphic sequences, Sloss began compiling the records of Phanerozoic sedimentary rocks across Eurasia in order to see whether the ages of the major sequences corresponded to those from North America. In many cases, similar sequences could be identified across Eurasia and North America, although the exact timing of the beginning and end of each sequence and the size of each sequence—how much of the continent each sequence covered—varied somewhat.

In the late 1960s, three of Sloss's graduate students were studying a set of sedimentary rocks in the mid-continent in which the different layers passed back and forth between those containing marine fossils and those containing land-dwelling fossils. Along with Sloss, the three students—Peter Vail, Robert Mitchum, and John Sangree—began to develop the idea that the rock layers were deposited as sea level fluctuated globally enough so that sometimes the continent was covered by marine water and sediment containing marine fossils was deposited, and other times the region was a terrestrial landscape with rivers and forests. They reasoned that a series of glacial intervals, commonly called "Ice Ages" caused global sea level to fluctuate during the time the sediments were deposited. When ice built up during a glacial interval, sea level dropped and the North American continent was exposed. When ice melted between glacial intervals, sea level rose high enough to inundate the continents, creating marine environments with marine organisms.

After their graduate work, Vail, Mitchum, and Sangree went to work at the research lab at Exxon and there they had access to a huge volume of geologic information from the subsurface. They quickly adopted their techniques for

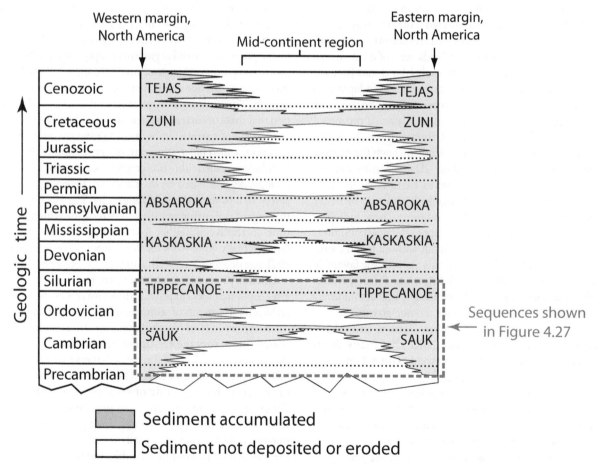

● **FIGURE 4.26** Stratigraphic sequences of North America, as first defined by Larry Sloss in 1963. To underscore the idea that these sequences were defined from the North American rock record, as opposed to the European rock record which to that point had been the basis for definition of the geologic time scale, Sloss named each sequence after a North American native tribe in whose traditional homeland rocks of the named sequence occurs. Sloss recognized that the unconformities separating each sequence (shown in white) also occur widely across North America and that each unconformity represents more missing time (longer period of erosion) in the middle of the continent compared to the margins. The marine character of the rocks suggested to Sloss that each sequence was deposited during a significant rise in sea level that largely inundated the continental surface.

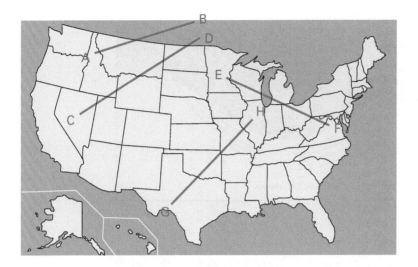

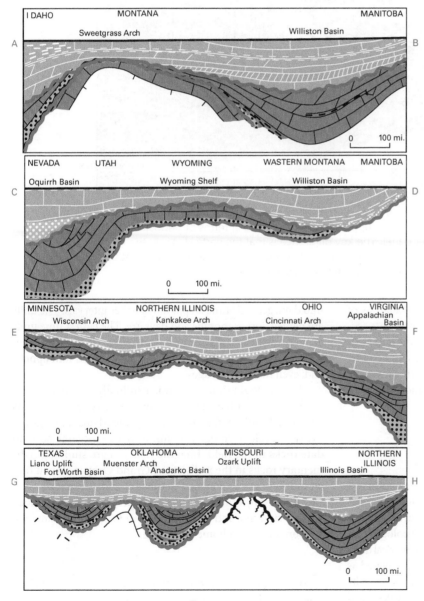

separating rock units into unconformity-bounded sequences to the subsurface data they studied at Exxon. They recognized similar successions of rock sequences and unconformities on the margins of different continents, implying that global sea level did indeed rise and fall synchronously and in so doing controlled the depositional environments (and their fossils) left behind in the rock record. Over time, the concepts and techniques developed by the Exxon team became known as **sequence stratigraphy**, which is the framework most widely used today to correlate rock units and understand how the depositional environments those rock units reflect changed both through space and time.

To understand how sequence stratigraphy works, consider the position of the shoreline. If sea level somehow rises, the shoreline will move landward. If sea level falls, the shoreline will move seaward. As the shoreline moves, all the depositional environments associated with the nearshore region also will migrate. For example, terrestrial environments will transform through time into marine environments as sea level inundates a particular location. Conversely, many formerly marine and nearshore environments become exposed and eroded as the shoreline recedes toward the ocean and river processes become important.

Sequence stratigraphy, then, is the organization of sedimentary rock bodies into depositional units that are bounded by unconformities and that contain a set of key stratigraphic surfaces, each of which represents a single point in time. For example, if sea level rises to a maximum position, then begins to fall again, it will leave behind a recognizable change that is expressed as a single bedding surface developed at that point in time. By tracing this bedding surface from locality to locality, geologists are able to use it as a timeline for subdividing the sequence of rock above and below the surface. By recognizing bedding surfaces that correspond to different positions of the shoreline through time, geologists are able

● **FIGURE 4.27** These four cross sections were published by Larry Sloss in 1963 and show the Sauk Sequence and part of the overlying Tippecanoe Sequence. The squiggly red lines are sequence boundaries - unconformities that separate different sequences. The small inset map shows the location of each cross section.

Source: modified from Sloss, L.L., 1963, Sequences in the cratonic interior of North America: Geological Society of America Bulletin, v. 74, p. 93–114.

sequence stratigraphy the study of sedimentary rock relationships using a chronostratigraphic framework in which genetically-related stratigraphic units that are separated by surfaces of erosion or non-deposition are recognized based on their lithology, fossil content, sedimentary structures, stratal geometry and other physical properties.

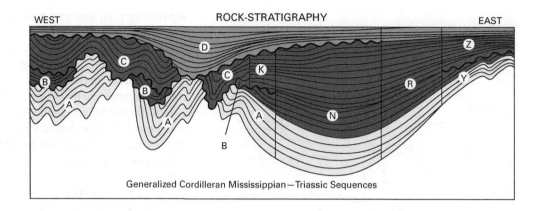

ROCK-STRATIGRAPHY

WEST   EAST

Generalized Cordilleran Mississippian—Triassic Sequences

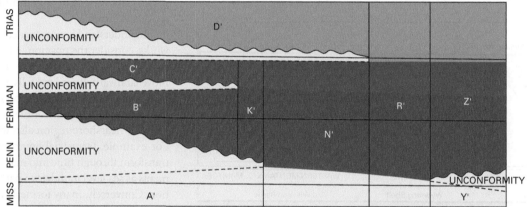

TIME-STRATIGRAPHY

● **FIGURE 4.28** Stratigraphy viewed in terms of a physical cross section showing major unconformities and bedding surfaces in the upper panel and viewed in terms of deposition through time (time stratigraphy) in the lower panel. The key difference is that the vertical axis in the upper panel is depth below the ground surface, whereas in the lower panel it is time with oldest at the bottom and youngest at the top. In both panels that horizontal axis is the same.

to erect a chronostratigraphic framework for the analysis of the sedimentary rock record. ●**FIGURE 4.28** shows a cross section of sedimentary rocks as they might occur over hundreds of kilometers, with different sequences of rock separated by unconformities. The lower panel in Figure 4.28 shows the same set of rocks recast into a chronostratigraphic framework through sequence stratigraphy.

THE GEOLOGIC TIME SCALE

As mentioned earlier, no single locality exists on Earth where a complete sequence of rocks formed continuously throughout geologic time. However, geologists have analyzed and correlated rocks from many localities around the world to create the **geologic time scale**, a formal subdivision of Earth time based on numerical dates (●**FIGURE 4.29**). The geologic time scale is frequently revised as more rock units are dated radiometrically and more fossil assemblages are described around the world.

Just as a year is subdivided into months, months into weeks, and weeks

into days, geologic time is split into smaller intervals. The units are named, just as months and days are. The largest time units are **eons**, which are divided into **eras**. Eras are subdivided, in turn, into **periods**, which are further subdivided into **epochs** and, even further, into **ages**.

The geologic time scale was originally constructed on the basis of relative age determinations. When geologists developed radiometric dating, they added absolute ages to the time scale. Geologists commonly use the time scale to date rocks in the field. Imagine that you are studying sedimentary rocks in the walls of the Grand Canyon. If you find an index fossil or a key bed that has already been radiometrically dated by other scientists, you know the age of the rock you are studying and you do not need to send the sample to a laboratory for radiometric dating.

The Earliest Eons of Geologic Time: Precambrian Time

The earliest eons—the Hadean, Archean, and Proterozoic Eons—often are not subdivided at all, even though together

MINDTAP
From Cengage

Animation:
Continental accretion

● **FIGURE 4.29** The geologic time scale is a record of geologic time divided into eons, eras, periods, and epochs. Time is given in millions of years (for example, 1,000 stands for 1,000 million, which is 1 billion) and is not drawn to scale here. Little is known about events that occurred during the early part of Earth's history. Therefore, the first 4 billion years are given relatively little space on this chart, while the more recent Phanerozoic Eon, which spans only 541 million years, receives proportionally more space.

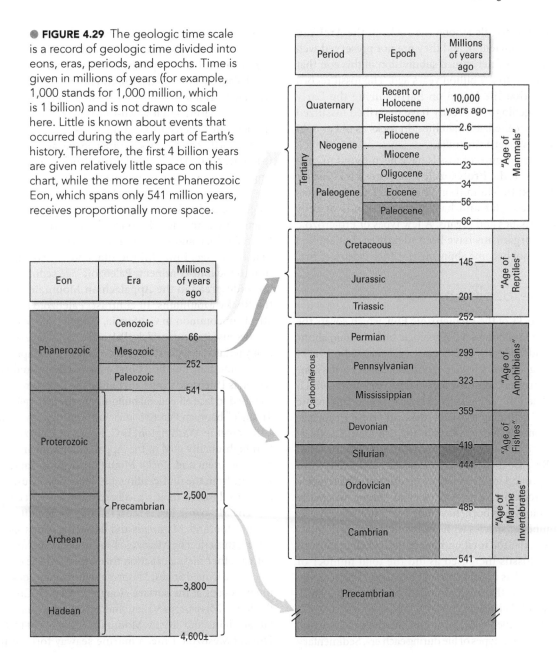

geologic time scale A chronological arrangement of geologic time subdivided into eons, eras, periods, epochs, and ages.

eons The largest unit of geologic time. The most recent eon, the Phanerozoic Eon, is further subdivided into eras.

eras A geologic time unit. Eons are divided into eras and, in turn, eras are subdivided into periods.

periods A geologic time unit longer than an epoch and shorter than an era.

epochs A geologic time unit longer than an age and shorter than a period.

ages The shortest period of geologic time. Epochs are divided into ages.

Precambrian A term referring to all of geologic time before the Paleozoic Era, encompassing approximately the first 4 billion years, or roughly 90 percent, of Earth's history. Also refers to all rocks formed during that time.

they constitute a time interval of 4 billion years—nearly 90 percent of Earth's history. These early eons are commonly referred to with the informal term **Precambrian**, because they preceded the Cambrian Period, when fossil remains became very abundant.

The **Hadean Eon** (Greek for "beneath the Earth") is the earliest time in Earth history and ranges from the planet's origin 4.6 billion years ago to 3.8 billion years ago. Only a few Earth rocks are known that formed during the Hadean Eon, and no fossils of Hadean age are known, making it impossible to subdivide the Hadean Eon based on fossils.

Hadean Eon The earliest time in Earth's history, ranging from 4.6 billion years ago to 3.8 billion years ago.

There are few fossils among the rocks of the **Archean Eon** (Greek for "ancient"), and they are not preserved well enough to allow for finely tuned subdivision of this eon that spanned from 3.8 to 2.5 billion years ago. The fossil record does indicate that life began on Earth during the Early Archean, and geologists recently have described fossilized algal structures, called stromatolites, in rocks as old as 3.45 billion years in western Australia.

More diverse groups of fossils have been found in sedimentary rocks of the **Proterozoic Eon** (Greek for "earlier life"), 2.5 billion to 543 million years ago. The most complex are multicellular and have different kinds of cells arranged into tissues and organs. A few types of Proterozoic shell-bearing organisms have been identified, but shelled organisms did not become abundant until the Paleozoic Era.

The Phanerozoic Eon

Sedimentary rocks of the **Phanerozoic Eon**, which covers the most recent 543 million years of geologic time, contain abundant fossils. Additionally, sedimentary rock and sediments of Phanerozoic age are very widely exposed on all of Earth's continents and form much of the surficial geology upon which human civilization exists. In North America, Phanerozoic sediments and sedimentary rocks make up nearly all surface exposures from the Florida Keys to the San Juan Islands of Washington State, and from southern California to the tip of Cape Cod. Because they are widely exposed and literally form much of the ground immediately under our feet, Phanerozoic sedimentary rocks in North America and Europe, where the science of geology was born, have long been studied.

The number of species with shells and skeletons dramatically increase, but the total number of individual organisms preserved as fossils increased as did the total number of species preserved as fossils. In rocks of earliest Phanerozoic time and younger, the most abundant fossils are the hard, tough shells and skeletons.

Subdivision of the Phanerozoic Eon into three eras is based on the most common types of life during each era. Sedimentary rocks that formed during the **Paleozoic Era** (Greek for "old life") contain fossils of early life-forms, such as invertebrates, fishes, amphibians, reptiles, ferns, and cone-bearing trees. The Paleozoic Era ended abruptly about 252 million years ago when the terminal Permian mass extinction wiped out 90 percent of all marine species, two-thirds of reptile and amphibian species, and 30 percent of insect species.

In North America, rocks of Paleozoic age are widespread (●**FIGURE 4.30**). Almost all of the bedrock geology of the New England states is Paleozoic in age. In New York State, only the bedrock of the Adirondack Mountains is not Paleozoic. Paleozoic bedrock also occurs widely across the mid-continent in Oklahoma, Missouri, Illinois, Indiana, Ohio, and West Virginia, and surrounding states including Pennsylvania, Tennessee, Iowa, Minnesota, and Wisconsin. In addition to the mid-continent, Paleozoic-age sedimentary rocks are widespread in the Appalachian Mountains and within the Rocky Mountains. In contrast, Paleozoic rocks are relatively uncommon in California, Oregon, and Washington.

Because it occurs widely, Paleozoic rocks have been used in many industrial applications historically. For example, almost all of the limestone building stone quarried in the mid-continent region is Paleozoic in age, including the limestone from Indiana used to face all Federal buildings in the area called "Federal Triangle," between Pennsylvania and Constitution Avenues and 14th Street in Washington, D.C. Rocks of Paleozoic age contain high-quality coal in the Appalachians, oil and gas in the mid-continent and Rocky Mountain regions. Sand of Paleozoic age from the mid-continent is being heavily used today to develop oil wells, a topic we will explore in the next chapter.

The **Mesozoic Era** is most famous for the dinosaurs that roamed the land. Mammals and flowering plants also evolved during this era. The Mesozoic Era ended 66 million years ago with another mass extinction that wiped out the dinosaurs.

In the United States, Mesozoic sedimentary rocks and sediments occur at the surface along the Atlantic Coastal Plain, the lower Mississippi Valley, and across large portions of the Great Plains and Rocky Mountain regions (●**FIGURE 4.31**). During Mesozoic time, a marine seaway formed in North

● **FIGURE 4.30** Paleozoic rock in Brewster County, Texas.

Luigi Trevisi/Shutterstock.com

● **FIGURE 4.31** Mesozoic rock in Dunraven Bay, Wales.

America, and the seaway—called the "Cretaceous Interior Seaway"—stretched from the Gulf of Mexico to the Arctic Ocean. Sediments deposited in the seaway formed sedimentary rocks belonging to the youngest of Larry Sloss's formally named marine sequences. He called it the "Tejas Sequence" after the widespread exposures of Mesozoic rock in the Texas region.

During the **Cenozoic Era** (Greek for "recent life"), mammals and grasses became abundant. Humans have evolved and lived entirely in the Cenozoic Era. In the United States, almost all of the unconsolidated sediment that occurs at the surface is Cenozoic in age, including widespread glacial deposits across the northern states. Sedimentary rock and sediment of Cenozoic age covers the entire surface of Florida and Louisiana, and all of the geology of the Hawaiian Islands is Cenozoic in age. Cenozoic sediments also occur widely across other southeastern and Gulf Coast states, in a large swath across the Great Plains, within the broad valleys of the Rocky Mountain states and across far western states (●**FIGURE 4.32**).

The eras of Phanerozoic time are subdivided into periods, the time unit most commonly used by geologists. Some of the periods are named after special characteristics of the rocks formed during those periods. For example, the Cretaceous Period is named from the Latin word for chalk (*creta*), after chalk beds of this age in Africa, North America, and Europe.

Other periods are named for the geographic localities where rocks of that age were first described. For example, the Jurassic Period is named for the Jura Mountains of France and Switzerland. The Cambrian Period is named for Cambria, the Roman name for Wales, where rocks of this age were first studied.

In addition to the abundance of fossils, another reason that details of Phanerozoic time are better known than those of Precambrian time is that many of the older rocks have been deformed, metamorphosed, and eroded. It is simple probability that the older a rock is, the greater the chance that metamorphism or erosion has destroyed the rock or the features that record Earth's history.

Archean Eon A division of geologic time 3.8 to 2.5 billion years ago. The oldest-known rocks formed at the beginning of, or just prior to, the start of the Archean Eon.

Proterozoic Eon The portion of geologic time occurring during a period from 2.5 billion to 541 million years ago.

Phanerozoic Eon The most recent 541 million years of geologic time, including the present, represented by rocks that contain evident and abundant fossils.

Paleozoic Era The part of geologic time occurring during a period from 541 to 251 million years ago. During this era invertebrates, fishes, amphibians, reptiles, ferns, and cone-bearing trees were dominant.

Mesozoic Era The part of geologic time roughly from 251 to 65 million years ago. Dinosaurs rose to prominence and became extinct during this era.

Cenozoic Era The latest of the three Phanerozoic eras, 65 million years ago to the present.

Marc S. Hendrix

● **FIGURE 4.32** The sedimentary rocks exposed in Bryce Canyon National Park, Utah are Cenozoic in age.

Key Concepts Review

- At least six catastrophic mass extinctions have occurred in geologic history. The greatest, 248 million years ago, annihilated 90 percent of all marine species, two-thirds of reptile and amphibian species, and 30 percent of insect species.
- Scientists measure geologic time in two ways. Relative age measurement refers only to the order in which events occurred. Absolute age is measured in years.
- Determinations of relative time are based on geologic relationships among rocks and the evolution of life-forms through time. The criteria for relative dating are summarized in a few simple principles: the principle of original horizontality, the principle of superposition, the principle of crosscutting relationships, and the principle of faunal succession.
- Layers of sedimentary rock are conformable if they were deposited without major interruptions. An unconformity represents a major interruption of deposition and a significant time gap between formation of successive layers of rock. In a disconformity, layers of sedimentary rock immediately above and below the unconformity are parallel. An angular unconformity forms when lower layers of rock are tilted and partially eroded prior to deposition of the upper beds. In a nonconformity, sedimentary layers lie on top of an erosion surface developed on igneous or metamorphic rocks. Correlation is the process of establishing the age relationship of rocks from different locations on Earth by comparing characteristics of the layers or the fossils found in those layers. Index fossils and key beds are important tools in time correlation—the demonstration that sedimentary rocks found in different geographic localities formed at the same time.
- Absolute time is measured by radiometric dating, which relies on the fact that radioactive parent isotopes decay to form daughter isotopes at a fixed known rate as expressed by the half-life of the isotope. The cumulative effects of the radioactive decay process can be determined by the accumulation of daughter isotopes in rocks and minerals.
- Worldwide correlation of rocks of all ages has resulted in the geologic column, a composite record of rocks formed throughout the history of Earth. The major units of the geologic time scale are eons, eras, periods, and epochs. The Phanerozoic Eon, the most recent 543 million years of geologic time, is finely and accurately subdivided because sedimentary rocks deposited at this time are often well preserved and they contain abundant well-preserved fossils. In contrast, Precambrian rocks and time are only coarsely subdivided because fossils are scarce and poorly preserved and the rocks are often altered.

Important Terms

absolute age (p. 66)

accuracy (p. 78)

architectural elements (p. 70)

ages (p. 82)

Archean Eon (p. 84)

Cenozoic Era (p. 85)

conformable (p. 70)

correlation (p. 73)

disconformity (p. 70)

eons (p. 82)

epochs (p. 82)

eras (p. 82)

evolution (p. 61)

geologic time scale (p. 82)

Hadean Eon (p. 83)

half-life (p. 77)

index fossils (p. 75)

isotopes (p. 77)

key bed (p. 75)

mass extinction (p. 61)

Mesozoic Era (p. 84)

natural selection (p. 79)

nonconformity (p. 71)

Paleozoic Era (p. 84)

periods (p. 82)

Phanerozoic Eon (p. 84)

Precambrian (p. 83)

precision (p. 78)

principle of crosscutting relationships (p. 67)

principle of faunal succession (p. 68)

principle of included fragments (p. 68)

principle of original horizontality (p. 66)

principle of superposition (p. 67)

Proterozoic Eon (p. 84)

radiometric dating (p. 77)

relative age (p. 66)

rock record (p. 73)

sequences (p. 79)

sequence stratigraphy (p. 81)

unconformity (p. 70)

Review Questions

1. Describe the two ways of measuring geologic time. How do they differ?

2. Give an example of how the principle of original horizontality might be used to determine the order of events affecting a sequence of folded sedimentary rocks.

3. Explain a conformable relationship in sedimentary rocks.

4. Describe the main difference between lithostratigraphy and chronostratigraphy.

Iurii/Shutterstock.com

Offshore drilling platforms for oil and gas exploration below the seafloor.

GEOLOGIC RESOURCES

LEARNING OBJECTIVES

LO1   Compare and contrast mineral resources and mineral reserves.

LO2   Discuss four different types of earth processes that can lead to the development of economic mineral resources.

LO3   List at least five mineral resources upon which the United States is 100% dependent on imports.

LO4   Explain how crude oil forms.

LO5   Describe the difference between a petroleum source rock and a conventional petroleum reservoir.

LO6   Identify two major technological advances that have made the production of unconventional "tight rock" reservoirs economic.

LO7   List three reasons why the construction of new nuclear power plants in the United States has decreased over the past few decades.

LO8   Briefly describe five renewable energy sources.

LO9   Describe a conventional petroleum reservoir and trap system.

LO10   Explain the difference between conventional and unconventional petroleum reservoirs.

LO11   Explain why the United States is expected to change from a net importer to a net exporter of energy before 2050.

LO12   List at least four technical solutions that can save energy in buildings, industry, or transportation.

INTRODUCTION

Since humanlike creatures emerged 5 to 7 million years ago, our use of geologic resources has become increasingly sophisticated. Early hominids used sticks and rocks as simple weapons and tools. Later prehistoric people used flint and obsidian to make more-effective weapons and tools, and they used natural pigments to create elegant art on cave walls. About 8000 BCE (Before Common Era), people learned to shape and fire clay to make pottery. Archaeologists have found copper ornaments in Turkey dating from 6500 BCE; 1,500 years later, Mesopotamian farmers used copper farm implements. Today, geologic resources provide iron for steel, silicon for making computer chips, and gasoline that powers most cars.

We use two types of geologic resources: **mineral resources** and **energy resources**. Mineral resources include all useful rocks and minerals. As we will see in the sections that follow, many mineral resources are naturally concentrated by processes that involve interactions among rock of the geosphere, atmospheric gases, and water from the hydrosphere. Humans have mined and refined these resources further to create the industrial world that has altered our planet. The primary energy resources of the early 21st century are coal, petroleum, and natural gas—all formed from the decayed remains of prehistoric plants and animals that have been altered by Earth systems processes.

MINERAL RESOURCES

Mineral resources include both metal ore and nonmetallic minerals. Recall from Chapter 2 that *ore* is rock sufficiently enriched in one or more minerals to be mined profitably. Geologists usually use the term to refer to metallic mineral deposits, and it is commonly accompanied by the name of the metal—for example, *iron ore* or *silver ore*.

Nonmetallic mineral resources refer to the useful rocks or minerals that are not metals—such as salt, building stone, sand, and gravel. When we think about "striking it rich" from mining, we usually think of gold. In 2016—the most recent year for which figures are available—the USGS estimated that the value of all the gold mined in the United States that year was $8.5 billion. However, the USGS also estimated in 2016 that more revenue was generated from mining gravel, sand, and crushed rock in the United States ($8.9 billion). Sand and gravel are mined from stream and glacial deposits, sand dunes, and beaches, whereas crushed stone is quarried from nonweathered igneous, metamorphic, or sedimentary bedrock. These nonmetallic resources are mixed with portland cement—a material produced by heating a mixture of crushed limestone and clay—to make concrete. Reinforced with steel, concrete is used to build roads, bridges, and buildings. Thus, reinforced concrete is one of the basic building materials of the modern world. In addition, many buildings are faced with stone—usually granite or limestone. Marble, slate, sandstone, and other rocks used for building are also mined from quarries cut into bedrock (●**FIGURE 5.1**).

There are many important metals and other elements that are fundamental parts of our lives and of the industries that produce a range of products in daily use. Some of these metals are familiar to us, such as iron, lead, copper, aluminum, silver, and gold. Others are less well known, such as molybdenum (rifle barrels), tungsten (light bulb filaments), and borax (soaps, antiseptics).

All mineral resources are nonrenewable: we use them up at a much faster rate than natural processes create them, although many can be recycled. Gold, for example, has been extracted and used by humans for over 6,000 years and today gold mines exist all over the world and on every continent except Antarctica. The gold has accumulated in different types of deposits in different locations over geologic time. Most individual gold deposits take hundreds of thousands or millions of years to develop, and many were formed millions of years ago.

According to Thomson Reuters GFMS, a major source of investment information for metals, if all the world's gold that has been mined by humans could be combined into a

● **FIGURE 5.1** A quarryman in China splits a large granite block with a sledgehammer. After he splits the rock, the circular saws in the background will cut it into thin slabs for floors and walls.

single giant cube, it would have sides that are 20.7 m (68 ft) long, corresponding to 171,300 metric tons (one metric ton is 1000 kilograms or one megagram). Other estimates vary

---

mineral resources Economically valuable geological materials including both metal ore and nonmetallic minerals.

energy resources Geologic resources—including petroleum, coal, natural gas, and nuclear fuels—used for heat, light, work, and communication.

nonmetallic mineral resources Economically useful rocks or minerals that are not metals; examples include salt, building stone, sand, and gravel.

---

but are generally in the same ballpark. Although it might seem that most gold has already been found, the USGS estimates that roughly 52,000 more metric tons exist in gold deposits that are not yet discovered, suggesting that gold will be mined for many years to come.

Gold is easily recycled through melting and has been recycled for millennia. In fact, almost all of the gold mined throughout history is still being used today. Your grandfather's gold watch may include some gold mined by the Romans 2000 years ago. The recent increase in technology changes this long tradition of recycling gold, however, because in many applications the amount of gold used is so small so as to make recycling noneconomic and thus impractical. As a result, for the first time in human history, gold is being consumed.

Gold represents an extreme example because it is so easily remelted and recycled and has held its value for so long. In almost all other cases, mineral ore extracted from the Earth is consumed at higher rates than gold. For example, much of the iron that is extracted from the Earth and concentrated to make steel is not recycled but rather simply is consumed and ultimately ends up in a landfill (●**FIGURE 5.2**).

ORE AND ORE DEPOSITS

If you pick up any rock and send it to a laboratory for analysis, the report will probably show that the rock contains measurable amounts of iron, gold, silver, aluminum, and other valuable metals. However, the natural concentrations of these metals are so low in most rocks that the extraction cost would be much greater than the income gained by selling

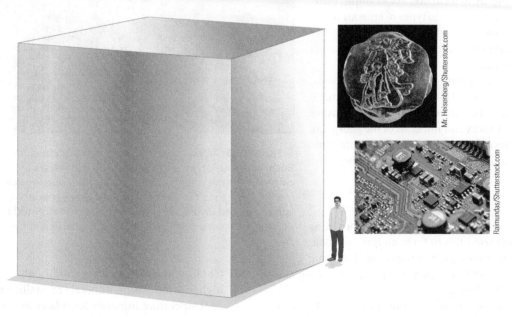

● **FIGURE 5.2** Gold as a nonrenewable resource. Despite the pursuits of alchemists over many centuries, gold has never been synthesized in the laboratory. Almost all of the gold ever mined is still in circulation, and if it could all be melted down and cast into one giant cube, it would measure only about 20 m on each side. Gold has been recycled since it was first used over 5,500 years ago as jewelry. Today, however, the small amount of gold used in some disposable technological devices has resulted in the deliberate consumption of gold for the first time in history.

TABLE 5.1 Comparison of Concentrations of Specific Elements in Earth's Crust with Concentrations Needed to Operate a Commercial Mine

Element	Natural Concentration in Crust (% by Weight)	Natural Concentration Required to Operate a Commercial Mine (% by Weight)	Enrichment Factor
Aluminum	8.0	24 to 32	3 to 4
Iron	5.8	40	6 to 7
Copper	0.0058	0.46 to 0.58	80 to 100
Nickel	0.0072	1.08	150
Zinc	0.0082	2.46	300
Uranium	0.00016	0.19	1,200
Lead	0.0001	0.20	2,000
Gold	0.0000002	0.0008	4,000
Mercury	0.000002	0.2	100,000

the refined metals. In certain locations, however, natural geologic processes have enriched metals many times above their normal concentrations. ●**TABLE 5.1** shows that the natural concentration of a metal in ore may exceed its average abundance in ordinary rock by a factor—called the *enrichment factor*—of more than 100,000.

Successful exploration for new ore deposits requires an understanding of the processes that concentrate metals to form ore. For example, platinum concentrates in certain types of igneous rocks. Therefore, if you were exploring for platinum, you would focus on those rocks rather than on sandstone or limestone.

With the exception of magmatic processes, which occur deep within the crust, the natural processes that concentrate ore minerals all involve interactions of rocks and minerals of the geosphere with water from the hydrosphere. The more common ore-forming processes are described below.

Magmatic Processes

Magmatic processes form mineral deposits as liquid magma solidifies to form an igneous rock. These processes create metal ores as well as some gems and nonmetallic mineral deposits including sulfur deposits and building stone.

Some large bodies of plutonic igneous rock, particularly those of *mafic* (high in magnesium and iron) composition, solidify in layers (●**FIGURE 5.3**). Each layer contains different minerals and is of a different chemical composition than adjacent layers. Some of the layers may contain rich ore deposits. The layering can develop by at least two processes:

● **FIGURE 5.3** An outcrop of layered mafic igneous rock from the Bushveld intrusion in South Africa. The dark layers are made of chromite crystals that settled to the bottom of the magma chamber more rapidly than the lower-density feldspar, making up the lighter-colored layers. The layering itself is interpreted to reflect multiple injections of magma into the magma chamber.

1. Cooling magma does not solidify all at once. Instead, higher-temperature minerals crystallize first, and lower-temperature minerals form later as the magma cools and the temperature drops. Most minerals are denser than magma. Consequently, early formed crystals may sink to the bottom of a magma chamber in a process called **crystal settling**. In some instances, ore minerals crystallize with other early formed minerals and accumulate in layers near the bottom of a pluton.

2. Some large bodies of mafic magma crystallize from the bottom upward. Thus, early formed ore minerals become concentrated near the base of the pluton.

The largest ore deposits found in layered mafic plutons are the rich chromium and platinum reserves of South Africa's Bushveld intrusion. The pluton is about 375 by 300 kilometers in area—roughly the size of the state of Maine—and about 7 kilometers thick. The Bushveld deposits contain more than 20 billion tons of chromium and more than 10 billion grams of platinum, the greatest reserves in any known deposit on Earth. The platinum alone is worth roughly $300 billion at 2018 prices.

Hydrothermal Processes

Hydrothermal processes—*hydro* for "water" and *thermal* for "heat," involving interactions between hot water or steam and rocks or minerals—are probably responsible for the formation of more ore deposits, and a larger total quantity of ore, than all other processes combined. To form a hydrothermal ore deposit, hot water dissolves metals from rock or magma. The metal-bearing solutions then seep through cracks or through permeable rock, where they precipitate to form an ore deposit.

---

magmatic processes Geologic processes that form ore deposits as liquid magma solidifies into igneous rock.

crystal settling A process in which the crystals that solidify first from a cooling magma settle to the bottom of the magma chamber because the minerals are more dense than magma; the ultimate result is a layered body of igneous rock, each layer containing different minerals.

hydrothermal processes Geologic processes in which hot water or steam dissolves metals and minerals from rocks or magma; the solutions then seep through cracks before cooling, to create ore deposits.

scavenging The process by which hydrothermal fluids sweep through large volumes of country rock and dissolve low concentrations of metals, concentrating them elsewhere as an ore deposit.

hydrothermal vein deposit A rich, sheetlike mineral deposit that forms when economically valuable minerals precipitate from hot water solutions along a fault or other fracture.

disseminated ore deposit A large, low-grade hydrothermal deposit in which metal-bearing minerals are widely scattered throughout a rock body; not as concentrated as a hydrothermal vein.

submarine hydrothermal ore deposits Ore deposits that form when hot seawater dissolves metals from seafloor rocks and then, as it rises through the upper layers of oceanic crust, cools and precipitates the metals.

black smokers A jet of black water spouting from a fracture or vent in the seafloor, commonly near a mid-oceanic ridge. The black color is caused by precipitation of fine-grained metal sulfide minerals as the hydrothermal solutions cool on contact with seawater.

Although water by itself is capable of dissolving minerals, most hydrothermal waters also contain dissolved salts. The presence of the salts greatly increases the water's ability to dissolve minerals. Therefore, hot, salty, hydrothermal water is a very powerful solvent, capable of dissolving and transporting metals.

Hydrothermal water comes from three sources—granitic magma, groundwater, and the oceans:

1. Granitic magma contains more dissolved water than solid granite rock. Thus, the magma gives off hydrothermal water as it solidifies. Because many ore metals do not fit neatly into the crystal structure of silicate minerals that form from a cooling granitic magma, these elements become concentrated in the hydrothermal waters.
2. Groundwater can seep into Earth's crust, where it is heated and forms a hydrothermal solution. The solution circulates through rock in the crust and dissolves ore metals, which later precipitate in concentrated form elsewhere. This scenario is common in volcanic areas where hot rock or magma heat groundwater at shallow depths.
3. In the oceans, hot, young basalt at a mid-oceanic ridge heats seawater as it seeps into cracks in the seafloor.

Refer again to Table 5.1, which shows that tiny amounts of all metals are found in average rocks of the Earth's crust. For example, gold makes up 0.0000002 percent of the crust, while copper makes up 0.0058 percent and lead 0.0001 percent. Although the metals are present in very low concentrations in country rock, hydrothermal solutions percolate through vast volumes of rock, dissolving or **scavenging** the metals and carrying them in solution. Where they encounter changes in temperature, pressure, or chemical environment, the solutions then can deposit the metals to form a local ore deposit (●**FIGURE 5.4**).

A hydrothermal vein deposit forms when dissolved metals precipitate in a fracture in rock. Ore veins range from less than a millimeter to several meters in width. A single vein can yield several million dollars worth of gold or silver. The same hydrothermal solutions may also soak into pores in country rock near the vein to create a large but much less concentrated **disseminated ore deposit**. Because they commonly form from the same solutions, rich ore veins and disseminated deposits are often found together. The history of many mining districts is one in which early miners dug shafts and tunnels to follow the rich veins. After the veins were exhausted, later miners used huge power shovels to extract low-grade ore from disseminated deposits surrounding the veins.

In volcanically active regions of the seafloor, near mid-ocean ridges and submarine volcanoes, seawater circulates through the hot, fractured oceanic crust. The hot seawater dissolves metals from the rocks and then, as it rises through the upper layers of oceanic crust, cools and precipitates the metals to form **submarine hydrothermal ore deposits**.

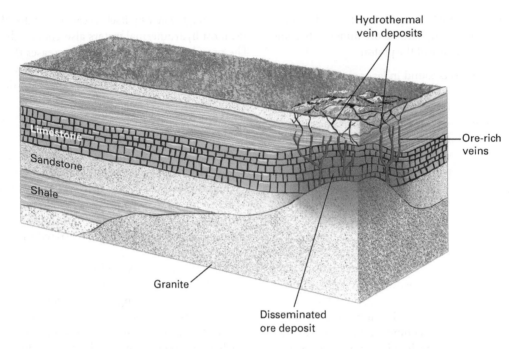

● **FIGURE 5.4** Hot water scavenges metals from crystallizing igneous rock and the country rock that surrounds it. The hydrothermal water then deposits metallic minerals in ore-rich veins that fill fractures in bedrock. It also deposits low-grade disseminated metal ore in large volumes of rock surrounding the veins.

● **FIGURE 5.5** A black smoker.

The metal-bearing solutions can be seen today as jets of black water, called **black smokers** (●FIGURE 5.5), spouting from fractures and vents in the Mid-Oceanic Ridge. The black color is caused by precipitation of fine-grained metal sulfide minerals as the solutions cool upon contact with seawater. The precipitating metals accumulate as chimneylike structures near the hot-water vent. Rich ore deposits form in such environments, but the cost to operate machinery in such great water depths is prohibitive.

Sedimentary Processes

PLACER DEPOSITS Gold is denser than any other mineral. Therefore, if you swirl a mixture of water, gold dust, and sand in a gold pan, the gold sinks to the bottom fastest. Differential settling also occurs in nature. Many streams carry silt, sand, and gravel with an uncommon small grain of gold. The gold settles fastest when the current slows down. Over years, currents agitate the sediment and the dense gold works its way into cracks and crevices in the streambed. Thus, grains of gold concentrate in gravel as well as in cracks and potholes eroded into the bedrock of the streambed, forming a **placer deposit** (●FIGURE 5.6). The prospectors who rushed to California in the Gold Rush of 1849, for example, mined placer deposits in conglomerate of Eocene age there.

PRECIPITATES Groundwater dissolves minerals as it seeps through soil and bedrock. In most environments, groundwater eventually flows into streams and then to the sea. Some of the dissolved ions, such as sodium and chloride, make seawater salty. In deserts, however, playa lakes develop with no outlet to the ocean. Water flows into the lakes but can escape only by evaporation. As the water evaporates, the dissolved salts concentrate until they precipitate to form evaporite deposits (see Chapter 3). The composition of the salt and specific salt minerals that form depend on the composition of dissolved ions transported to the basin, which in turn depend upon the bedrock in the region. Evaporite deposits in desert lakes include sodium chloride (table salt), borax, sodium sulfate, and sodium carbonate. These salts are used in the production of paper, soap, and medicines and for the tanning of leather.

Several times during the past 500 million years, shallow seas covered large regions of North America and all

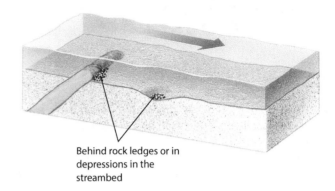

Behind rock ledges or in depressions in the streambed

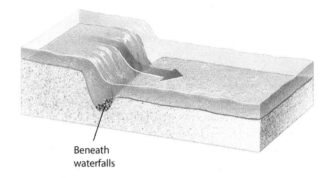

Beneath waterfalls

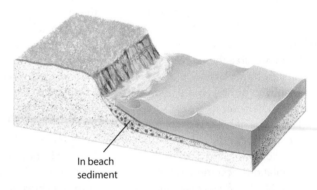

In beach sediment

● **FIGURE 5.6** Placer deposits form where water currents slow down and deposit high-density minerals.

other continents. At times, those seas were so weakly connected to the open oceans that water did not circulate freely between seas and the oceans. Consequently, evaporation concentrated the dissolved salts until they precipitated as marine evaporites. Periodically, new seawater from the open ocean would replenish the shallow seas, providing a new supply of salt. Thick marine evaporite beds, formed in this way, underlie much of North America. Table salt, gypsum (used to manufacture plaster and sheetrock), and potassium salts (used in fertilizer) are mined extensively from these deposits.

Most of the world's supply of iron is mined from sedimentary rocks called **banded iron formations**, which are

---

placer deposit A surface mineral deposit formed along stream beds, beneath waterfalls, or on beaches when water currents slow down and deposit high-density minerals.

banded iron formations Iron-rich sedimentary rocks composed of alternating iron-rich and silica-rich layers; source of most of the world's supply of iron.

Manganese Nodules

A rich source of strategically important metals rests on the deep-ocean floor.

Much of the Pacific Ocean floor is covered with golf ball– to bowling ball–sized **manganese nodules** (●FIGURES 5.7 and 5.8). A typical nodule contains 20 to 30 percent manganese, 6 percent iron, about 1 percent each of copper and nickel, and lesser percentages of other metals such as cobalt, zinc, and lead. At least 60 different elements have been reported from manganese nodules, several of which are metals with significant commercial and military applications. The metals are probably introduced into seawater by volcanic activity at mid-oceanic ridges, perhaps by black smokers. Certain specialized bacteria and algae on the seafloor are able to precipitate or **biomineralize** the metals, effectively forming a nodule seed that continues to grow as more metals are added to its outer layers.

Hundreds of billions of tons of manganese nodules lie on the seafloor, with the densest accumulations occurring in the Pacific Ocean. Most nodules occur at depths of around 5 kilometers, although some occur in water over 6 kilometers deep and some have been reported at less than 3 kilometers.

Since the 1970s, large-scale mining of the nodules has been discussed, although it has never been undertaken because the cost of recovering the nodules from these depths is prohibitive. However, the economics of mining nodules from the seafloor is changing because of recent increases in most metals prices and the strategic importance of the so-called rare earth elements that occur in manganese nodules. Rare earth elements are a suite of 17 different metals that are used in an ever-increasing array of high-tech equipment, including lasers, fiber optics, computer disk drives and memory chips, rechargeable

● **FIGURE 5.7** Manganese nodules cover large portions of the seafloor. These are from the central North Pacific Ocean at a depth of 5,000 meters.

● **FIGURE 5.8** Close-up of a manganese nodule from the South Pacific Ocean Penny for scale.

deposits composed of alternating iron-rich and silica-rich layers (●**FIGURE 5.9**). These iron-rich rocks precipitated from the seas between 2.6 and 1.9 billion years ago, as a result of rising atmospheric oxygen concentrations.

Weathering Processes

In environments with high rainfall, the abundant water dissolves and removes most of the soluble ions from soil and rock near Earth's surface. This process leaves the relatively insoluble ions in the soil to form **residual ore deposits**. Both aluminum and iron have very low solubilities in water. **Bauxite**, the principal source of aluminum, forms as a residual deposit, and in some instances iron also concentrates enough to become ore. Most bauxite deposits form in warm and rainy tropical or subtropical environments where

● **FIGURE 5.9** Banded iron formations from Michigan. The iron is concentrated as iron oxide in the metallic gray layers; the red layers are chert.

batteries, X-ray tubes, certain superconductors, and liquid crystal displays.

Because of their chemical properties, rare earth elements do not commonly concentrate in ore bodies, are expensive to refine, and create environmental problems when they are mined. Over the past 20 years, mining of rare earth elements has largely shifted to China because of lower labor costs and less restrictive environmental regulations. As a result, China now controls the vast majority of rare earth element production, with some estimates as high as 95 percent.

In 2010, China temporarily halted exports of rare earth elements to Japan. Although the embargo was short-lived, it set off political alarms around the world and got the attention of commodities investors because of the critical role rare earth elements play in modern technology. According to one report commissioned by an academic group from Columbia University for a prominent investment bank, "the risk of REE supply disruption is high. REEs mainly come from one source—China—and the supply chain is opaque and ill-understood, making it difficult to track how REEs move from the mine to the end-products." China now has over fifteen years of industrial mining and processing experience of REEs. REE extraction and concentration is very capital intensive and the extreme toxicity of REEs makes it socially costly to develop new production facilities in the United States. The Columbia report estimates that it would take 15 years to get a new REE extraction and processing facility built in the United States. In 2017, the U.S. Department of Energy announced the discovery of large REE deposits associated with coal-bearing Paleozoic rocks in Illinois, Kentucky, and Pennsylvania. They announced plans to begin bench-top experiments using mine drainage water as the feedstock for REE extraction. As of the writing of this book, the experiments were still underway and no plans to build a large-scale extraction and concentration facility have been announced. However, the critical need for REEs in modern society, coupled with the complete dependence on sources outside the United States, likely will continue to drive investigations and exploratory efforts like those involving the Paleozoic coal deposits forward well into the future.

Among the alternatives to diversifying the source of rare earth elements is large-scale mining of manganese nodules from the deep seafloor. As a result, detailed maps of the distribution of manganese nodules are now being rapidly developed, as are various systems for recovering nodules profitably from the deep seafloor. One can imagine robotic undersea video cameras locating the nodules and giant vacuums sucking them up and lifting them to a ship. But, because the seafloor is a difficult environment in which to operate complex machinery and the environmental consequences of such large-scale mining are not well understood, the question as to when the profitable harvest of manganese nodules will begin remains unanswered.

chemical weathering occurs rapidly. Thus, bauxite ores are common in Jamaica, Cuba, Guinea, Australia, and parts of the southeastern United States (●FIGURE 5.10).

manganese nodules A potato-shaped rock found on the ocean floor and rich in manganese and other metals precipitated from seawater through biomineralization.

biomineralize The process by which living organisms produce minerals.

residual ore deposits A mineral deposit formed from relatively insoluble ions left in the soil near Earth's surface after most of the soluble ions were dissolved and removed by abundant water.

bauxite A gray, yellow, or reddish-brown rock, composed of a mixture of aluminum oxides and hydroxides, that formed as a residual deposit; the principle source of aluminum.

mineral reserves A term to describe the known supply of ore in the ground; can be used on a local, national, or global scale.

MINERAL RESERVES VS. MINERAL RESOURCES

Mineral reserves are the known amount of ore in the ground that can be mined profitably. Reserves represent a working inventory of an economically extractable mineral commodity in a particular mine or on a national or global scale. Mineral resources, described at the beginning of this chapter, are all occurrences of a mineral commodity, including those only surmised to exist, that have present or anticipated future value.

Mining depletes mineral reserves by decreasing the amount of ore remaining in the ground, but reserves may also *increase* in two ways. First, geologists may discover new mineral deposits, thereby adding to the known amount of ore. Second, *subeconomic mineral deposits*—those in which the metal is not sufficiently concentrated to be mined at a profit—can become profitable if the price of that metal

Copyright and Photograph By Marc S. Hendrix

● **FIGURE 5.10**   This spheroidal texture is typical of bauxite, which is aluminum ore formed as a residual soil deposit by intense tropical weathering of aluminum-rich rocks. This bauxite is from northern Queensland in Australia. The pencil tip is pointing to concentric layering within a spheroid that has been broken.

cutting edge for over four millennia. Only when humans began to develop the ability to forge an even stronger metal—iron—did the Bronze Age of civilization end.

Today, copper wiring and plumbing are integral to the appliances, heating and cooling systems, and telecommunications links used every day in homes and businesses. Copper is used in the transmission of electricity and is an essential ingredient for alloys used to make motors, radiators, connectors, brakes, and bearings used in cars and trucks. According to the USGS, the amount of copper in an average modern car ranges from 20 kilograms (44 pounds) in small cars to 45 kilograms (99 pounds) in luxury and hybrid vehicles. Clearly, copper will continue to be a vital mineral resource to human society for long into the future.

increases or if improvements in mining or refining technology reduce extraction costs.

For example, in 1970, world copper resources, including all identified and undiscovered sources, were estimated at 1.6 billion metric tons. World reserves of copper—that portion of copper resources that could be profitably extracted—were estimated at only 280 million metric tons. Between 1970 and 2010, however, improved mining techniques and rising copper prices resulted in the production of about 400 million metric tons of copper, nearly 43 percent more than the 1970 global reserve estimate. Moreover, these two factors, improved technology and rising prices along with a continued strong demand have driven the exploration and production of new copper sources.

The strong and accelerating demand for copper exists because of the metal's excellent alloying properties which have made it exceptionally valuable when combined with other metals, such as zinc (to form brass), tin (to form bronze), or nickel (to form "cupronickel") which is especially resistant to corrosion by marine waters. Cupronickel also reduces the adhesion of marine life, such as barnacles, to boat hulls, thereby reducing drag and increasing fuel efficiency. Brass is more malleable and has better acoustic properties than pure copper or zinc; consequently, it is used in a variety of musical instruments, including trumpets, trombones, bells, and cymbals. Bronze (alloys of copper and tin) are especially fluid in the molten state and so are well-suited to casting because the metal flows into the nooks and crannies of the mold before solidifying. In fact bronze has so long been used by humans that the so-called Bronze Age corresponds to the dates between about 3500 BC and 1200 AD—when bronze represented society's technological

The Geopolitics of Metal Resources

Earth's mineral resources are unevenly distributed, and no single nation is self-sufficient in all minerals. Moreover, as technology evolves, the value of mineral and energy resources can change drastically. For example, salt was once used as currency. It drove the establishment of trade routes, caused armed conflicts between the Vatican and its subjects in Perugia (part of modern Italy) in 1540, and was at the source of a 12-year-long armed struggle along the Texas-Mexican border in the 1860s and 1870s. The discovery of large salt deposits and development of refrigeration since then has significantly reduced the strategic value of salt. Similarly, the discovery and refinement of large petroleum reserves has eliminated the strategic value of whale oil.

Historically, five nations—the United States, Russia, South Africa, Canada, and Australia—have supplied most of the mineral resources used by modern societies. However, today, China is the world's leading *producer* of many mineral resources, including aluminum, gold, iron, lead, phosphate rock (used mainly for fertilizer), tin, tungsten, and zinc. China is also the world's leading *exporter* of several mineral resources, including antimony, barite (used in drilling muds), graphite, tungsten, and rare earth metals critical to defense and other high-tech industries.

Although it is the lead producer and exporter of many mineral resources, China is also the world's largest *consumer* of many mineral commodities and has embarked on a strategic policy of purchasing the rights to many large mineral deposits around the world. Today, China's domestic supply and demand for various mineral commodities is high enough to directly affect the world mineral markets.

2017 U.S. NET IMPORT RELIANCE[1]

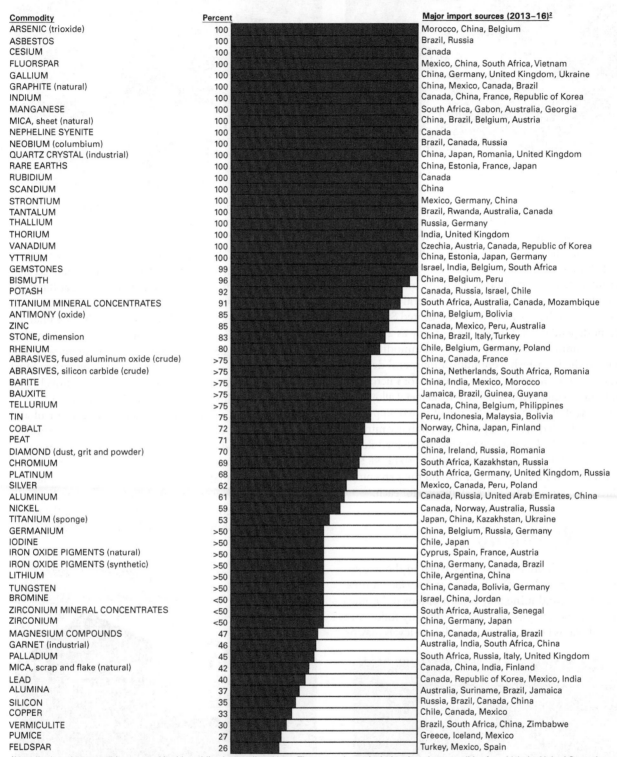

Commodity	Percent	Major import sources (2013–16)[2]
ARSENIC (trioxide)	100	Morocco, China, Belgium
ASBESTOS	100	Brazil, Russia
CESIUM	100	Canada
FLUORSPAR	100	Mexico, China, South Africa, Vietnam
GALLIUM	100	China, Germany, United Kingdom, Ukraine
GRAPHITE (natural)	100	China, Mexico, Canada, Brazil
INDIUM	100	Canada, China, France, Republic of Korea
MANGANESE	100	South Africa, Gabon, Australia, Georgia
MICA, sheet (natural)	100	China, Brazil, Belgium, Austria
NEPHELINE SYENITE	100	Canada
NEOBIUM (columbium)	100	Brazil, Canada, Russia
QUARTZ CRYSTAL (industrial)	100	China, Japan, Romania, United Kingdom
RARE EARTHS	100	China, Estonia, France, Japan
RUBIDIUM	100	Canada
SCANDIUM	100	China
STRONTIUM	100	Mexico, Germany, China
TANTALUM	100	Brazil, Rwanda, Australia, Canada
THALLIUM	100	Russia, Germany
THORIUM	100	India, United Kingdom
VANADIUM	100	Czechia, Austria, Canada, Republic of Korea
YTTRIUM	100	China, Estonia, Japan, Germany
GEMSTONES	99	Israel, India, Belgium, South Africa
BISMUTH	96	China, Belgium, Peru
POTASH	92	Canada, Russia, Israel, Chile
TITANIUM MINERAL CONCENTRATES	91	South Africa, Australia, Canada, Mozambique
ANTIMONY (oxide)	85	China, Belgium, Bolivia
ZINC	85	Canada, Mexico, Peru, Australia
STONE, dimension	83	China, Brazil, Italy, Turkey
RHENIUM	80	Chile, Belgium, Germany, Poland
ABRASIVES, fused aluminum oxide (crude)	>75	China, Canada, France
ABRASIVES, silicon carbide (crude)	>75	China, Netherlands, South Africa, Romania
BARITE	>75	China, India, Mexico, Morocco
BAUXITE	>75	Jamaica, Brazil, Guinea, Guyana
TELLURIUM	>75	Canada, China, Belgium, Philippines
TIN	75	Peru, Indonesia, Malaysia, Bolivia
COBALT	72	Norway, China, Japan, Finland
PEAT	71	Canada
DIAMOND (dust, grit and powder)	70	China, Ireland, Russia, Romania
CHROMIUM	69	South Africa, Kazakhstan, Russia
PLATINUM	68	South Africa, Germany, United Kingdom, Russia
SILVER	62	Mexico, Canada, Peru, Poland
ALUMINUM	61	Canada, Russia, United Arab Emirates, China
NICKEL	59	Canada, Norway, Australia, Russia
TITANIUM (sponge)	53	Japan, China, Kazakhstan, Ukraine
GERMANIUM	>50	China, Belgium, Russia, Germany
IODINE	>50	Chile, Japan
IRON OXIDE PIGMENTS (natural)	>50	Cyprus, Spain, France, Austria
IRON OXIDE PIGMENTS (synthetic)	>50	China, Germany, Canada, Brazil
LITHIUM	>50	Chile, Argentina, China
TUNGSTEN	>50	China, Canada, Bolivia, Germany
BROMINE	<50	Israel, China, Jordan
ZIRCONIUM MINERAL CONCENTRATES	<50	South Africa, Australia, Senegal
ZIRCONIUM	<50	China, Germany, Japan
MAGNESIUM COMPOUNDS	47	China, Canada, Australia, Brazil
GARNET (industrial)	46	Australia, India, South Africa, China
PALLADIUM	45	South Africa, Russia, Italy, United Kingdom
MICA, scrap and flake (natural)	42	Canada, China, India, Finland
LEAD	40	Canada, Republic of Korea, Mexico, India
ALUMINA	37	Australia, Suriname, Brazil, Jamaica
SILICON	35	Russia, Brazil, Canada, China
COPPER	33	Chile, Canada, Mexico
VERMICULITE	30	Brazil, South Africa, China, Zimbabwe
PUMICE	27	Greece, Iceland, Mexico
FELDSPAR	26	Turkey, Mexico, Spain

[1]Not all mineral commodities covered in this publication are listed here. Those not shown include mineral commodities for which the United States is a net exporter (abrasives, metallic; boron; clays; diatomite; gold; helium: iron and steel scrap; iron ore; kyanite; molybdenum; sand and gravel, industrial; selenium; soda ash; titanium dioxide pigment; wollastonite; and zeolites) or less than 25% import reliant (beryllium; cadmium; cement; diamond, industrial stones; gypsum; iron and steel; iron and steel slag; lime; magnesium metal; nitrogen (fixed)—ammonia; perlite; phosphate rock; sand and gravel, construction; salt; stone, crushed; sulfur; and talc). For some mineral commodities (hafnium and mercury), not enough information is available to calculate the exact percentage of import reliance.

● **FIGURE 5.11** List of mineral commodities and U.S. import dependence as of 2017.

Source: https://minerals.usgs.gov/minerals/pubs/mcs/

● **FIGURE 5.12** Machinery extracts coal from an underground coal mine.

Many other nations have few mineral resources. For example, Japan has almost no metal or fuel reserves; despite its modern economy and high productivity, it relies entirely on imports for metals and fuel.

Currently, the United States depends on dozens of other countries for the majority of its mineral consumption. According to the USGS, in 2017, the United States was 100 percent dependent on imports for as many as 21 different mineral commodities. Many of these commodities are important in the high-tech, communications, and defense industries. Examples include yttrium, essential for microwave communications equipment; cobalt, a critical metal alloy; and vanadium, essential for the manufacture of supercomputers. Because of the United States' dependence on many mineral and metal imports the U.S. government has maintained stockpiles of various strategic mineral commodities. ●**FIGURE 5.11** shows a list of 64 mineral commodities and the percentage upon which the United States relies on imports for each.

MINES AND MINING

Miners extract both ore and coal (described in the following section) from underground mines and surface mines. A large **underground mine** may consist of tens of kilometers of interconnected passages that commonly follow ore veins or coal seams (●**FIGURE 5.12**). The lowest levels may be

● **FIGURE 5.13** The Bingham Canyon, Utah, open-pit copper mine is the largest human-created excavation on Earth. It is over 4 kilometers in diameter and 1.2 kilometer deep.

● **FIGURE 5.14** Lignite (brown coal) being mined in Germany as a source of power.

several kilometers deep. In contrast, a **surface mine** is a hole excavated into Earth's surface. The largest human-created excavation on Earth is the open-pit copper mine at Bingham Canyon, Utah (●**FIGURE 5.13**). It is 4 kilometers in diameter and 1.2 kilometers deep and can be seen with the unaided eye from space. Since its beginning in 1873, the mine has produced about 14.5 million tons of copper, along with significant amounts of gold, silver, and molybdenum. Most modern coal mining is done by large power shovels that extract coal from huge surface mines (●**FIGURE 5.14**).

In the United States, the Surface Mining Control and Reclamation Act of 1977 requires mining companies to restore mined land so that it can be used for the same purposes for which it was used before mining began. In addition, a tax is levied to make it possible to reclaim land that was mined before the law was enacted. Enforcement and compliance of environmental laws waxes and wanes with the political climate in Washington. Yet environmental awareness has

● **FIGURE 5.15** This house tilted and broke in half as it sank into an abandoned underground coal mine.

increased dramatically over the past generation, and, overall, mining operations pollute much less today than they did 50 years ago. One of the big challenges for the future is to clean up old mines that were operated under lax or nonexistent environmental regulations of the past. In the United States, more than 6,000 unrestored coal and metal surface mines cover an area of about 90,000 square kilometers, almost as large as the state of Virginia. This figure does not include abandoned sand and gravel mines and rock quarries.

Although underground mines do not directly disturb the land surface, some abandoned mines collapse, and occasionally buildings have sunken into the resulting holes (●FIGURE 5.15). Over 800,000 hectares (2 million acres) of land in central Appalachia have settled into underground coal mine shafts.

Mining of both metal ores and coal also creates huge piles of *waste rock*—rock that must be removed to get at the ore or coal or that is left over after the refining of the ore. If the waste piles are not treated properly, rain erodes the loose rock and leaches toxic elements such as arsenic, sulfur, and heavy metals from the piles, choking the streams with sediment and polluting both stream water and groundwater.

ENERGY RESOURCES: COAL, PETROLEUM, AND NATURAL GAS

Coal, petroleum, and natural gas are called **fossil fuels** because they formed from the remains of plants and animals. Fossil fuels are not only nonrenewable but also unrecyclable. When a lump of coal or a liter of oil (petroleum) is burned, the energy dissipates and is, for all practical purposes, lost. Thus, our fossil fuel supply inexorably diminishes.

Coal

Coal (●FIGURE 5.16) is a sedimentary rock made mostly of organic carbon—enough that the rock will burn without refining. Coal-fired electric generating plants burn about 92 percent of the coal consumed in the United States and produce slightly less than 50 percent of the nation's electricity. Although it is easily mined and abundant in many parts of the world, coal emits air pollutants that can be removed only with expensive control devices. Mercury, in particular, is released into the atmosphere mainly through coal-fired power plants. Despite these drawbacks, coal is an abundant resource,

● FIGURE 5.16   Anthracite is a hard, compact variety of coal with the highest carbon count and lowest level of impurities of all coals.

with widespread availability projected to last beyond the 21st century.

In North America, large quantities of coal formed during the Carboniferous Period, between 360 and 286 million years ago, and later in Cretaceous and Paleocene times, when warm, humid swamps covered broad areas of low-lying land. When plants die in forests and grasslands, organisms consume some of the plant litter, and chemical reactions with oxygen and water decompose the remainder. As a result, little organic matter accumulates, except in the topsoil. In some swamps and bogs, however, plants grow

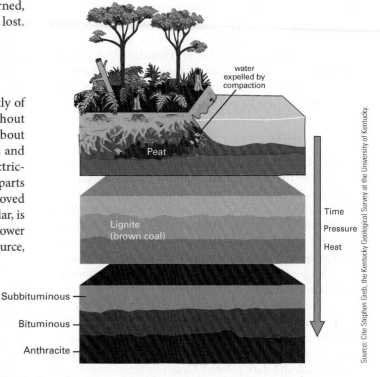

● FIGURE 5.17   Peat, lignite, and coal form as organic litter accumulates rapidly in a swamp and does not undergo complete decay. With subsequent burial, the organic litter compacts, expels water, and transforms to peat. With further burial and the addition of heat, peat will transform to lignite, then bituminous coal, and finally anthracite.

---

underground mine A mine consisting of subterranean passages that commonly follow ore veins or coal seams.

surface mine A hole excavated into Earth's surface for the purpose of recovering mineral or fuel resources.

fossil fuels Energy resources including petroleum, coal, and natural gas, which formed from the partially decayed remains of plants and animals; they are nonrenewable and unrecyclable.

TABLE 5.2 Heat Value (British Thermal Units per Mass of Fuel) for Different Grades of Coal, Compared to Liquid Petroleum and Natural Gas

Fuel Type	Heat Value (BTU/lb)
Peat	3,000 to 5,000
Lignite	7,000
Bituminous (soft coal)	12,000
Anthracite (hard coal)	14,000
Gasoline C10	112,114 to 116,090
Low sulfur diesel	128,488
Biodiesel	119,550
Propane	84,250
Liquefied Natural Gas (LNG)	21,240
Ethanol	76,330

and die so rapidly that newly fallen vegetation quickly buries older plant remains. The new layers prevent atmospheric oxygen from penetrating into the deeper layers, and decomposition stops before it is complete, leaving brown, partially decayed plant matter called *peat*.

Plant matter is composed mainly of carbon, hydrogen, and oxygen and contains large amounts of water. During burial, rising pressure expels the water and chemical reactions release most of the hydrogen and oxygen. As a result, the proportion of carbon increases until coal forms (●**FIGURE 5.17**). The grade of coal and the heat that can be recovered by burning coal can vary considerably depending on the carbon content. ●**TABLE 5.2** compares the amount of heat that can be produced by different grades of coal, compared to petroleum-based fuels, discussed next.

Petroleum

The word **petroleum** comes from the Latin for "rock oil" or "oil from the earth." Some natural oil seeps in Asia were used at least as long ago as Alexander the Great, and oil wells in China were hand dug with bamboo poles in 347 CE. In North America, the first commercial oil well was drilled in Titusville, Pennsylvania, in 1859, ushering in a new energy age. Crude oil, as it is called when pumped from the ground, is made up of thousands of different chemical compounds and ranges widely in consistency and color. Some petroleums are brown, waxy substances that are solid at room temperature but liquid at higher temperatures that exist within the Earth's crust. Some petroleums are yellowish or nearly clear liquids that resemble refined gasoline. Most are rather thick and dark colored. Once recovered from a well, crude oil is refined to produce propane, gasoline, heating oil, and other fuels (●**FIGURE 5.18**). Petroleum also is used to manufacture plastics, nylon, and other useful materials.

●**FIGURE 5.18** An oil refinery converts crude oil into useful products such as gasoline.

FORMATION OF PETROLEUM Petroleum forms from the accumulation of large quantities of organic matter in muddy sediment deposited in swamps, lakes, and marine waters. Most of the organic matter comes from algae, plant remains, and bacteria. Over millions of years, younger sediment buries this organic-rich mud to depths of a few kilometers, where rising temperature and pressure convert the mud to shale. At the same time, the elevated temperature and pressure cause the organic matter to convert to a solid organic substance called **kerogen**. At temperatures ranging from 50°C to about 160°C, the kerogen breaks down chemically, liberating small organic molecules. These organic molecules form petroleum.

The shale or other sedimentary rock in which oil originally forms is called the **petroleum source rock**. With time, some of the organic molecules in the source rock are expelled as liquid petroleum that seeps into the pore spaces in adjacent rock layers. Because petroleum is less dense than water in the pore spaces, the petroleum rises toward the surface through the network of pores in the rock. If it is not trapped along the way, the oil will migrate all the way to the surface, forming a natural oil seep. The La Brea Tar Pits in downtown Los Angeles is perhaps the most famous example of a natural oil seep (●**FIGURE 5.19**). Between 40,000 and 8,000 years ago, over 660 species of organisms became trapped in tar formed from the La Brea oil seeps, died, and were preserved.

In many circumstances, migrating petroleum will not reach the surface but rather will become trapped in a **conventional petroleum reservoir**. A conventional reservoir consists of oil-saturated porous rock that is like an oil-soaked sponge (●**FIGURE 5.20**). It is not an underground pool or lake

---

petroleum A complex liquid mixture of hydrocarbons, formed from decayed plant and animal matter, that can be extracted from sedimentary strata and refined to produce propane, gasoline, and other fuels. Also called *crude oil* or simply *oil*.

kerogen The waxy, solid organic material in oil shales that yields oil when the shale is heated; the precursor of liquid petroleum.

petroleum source rock The shale or other sedimentary rock from which oil or natural gas originates.

conventional petroleum reservoir A porous, permeable sedimentary rock that is saturated with trapped oil.

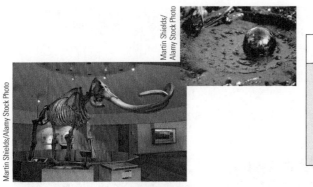

● **FIGURE 5.19** La Brea Tar Pits in the Los Angeles basin.

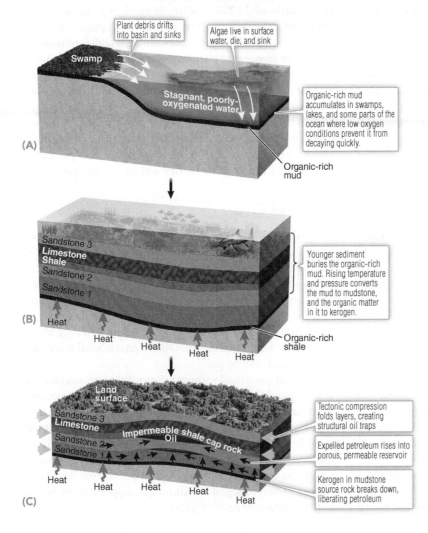

● **FIGURE 5.20** (A) Organic-rich mud accumulates in swamps, lakes, and some parts of the ocean where low oxygen conditions prevent it from decaying quickly. (B) Younger sediment buries the organic-rich mud. Rising temperature and pressure converts the mud to mudstone, and the organic matter in it to kerogen. (C) With continued heat, the kerogen breaks down, liberating petroleum that migrates out of the organic-rich source rock and into adjacent layers. Once there, the petroleum rises toward the surface until it is trapped. In this illustration, the oil is trapped where it encounters an impermeable cap rock in the crest of a dome-like fold in the rock layers.

● **FIGURE 5.21** A pump jack extracts oil from a conventional reservoir in Alberta, Canada. This pump jack is situated in a field of canola, a plant used to make biofuel (see "Biomass Energy").

● **FIGURE 5.22** An offshore oil-drilling platform extracts oil from below the seafloor.

of oil. Many conventional reservoirs form when the rising oil is trapped by an overlying layer of impermeable rock—that is, rock through which liquids do not pass quickly because the pore spaces are too small or are otherwise big enough but not interconnected, as with the isolated holes in Swiss cheese.

To extract petroleum from a conventional reservoir, an oil company drills a well into it. After the hole has been bored, the expensive drill rig is removed and replaced by a smaller rig that sets pipe in the borehole and perforates the pipe adjacent to oil-bearing layers so that the oil can flow from the rock into the pipe. Following this, a **pump jack** is installed to draw the petroleum up the pipe (●**FIGURE 5.21**). Fifty years ago, many conventional reservoirs lay near the surface and oil was easily pumped from shallow wells. But these reserves have been largely depleted, causing many modern oil companies to seek deeper reservoir targets, sometimes below the seafloor in water up to 3 kilometers deep.

Primary recovery techniques utilize the pressure in a conventional reservoir to push oil into the wellbore. As oil is removed, however, this pressure decreases to the point at which the remaining oil cannot be drawn into the wellbore. On average, more than half of the oil in a reservoir is too viscous to be pumped to the surface by conventional techniques and is left behind when the oil field has "gone dry." Additional oil can be extracted by **secondary and tertiary recovery techniques** involving the injection of water, detergent, pressurized gas, or other fluids into the reservoir. Secondary methods are employed first, and when those are exhausted, tertiary methods are used. In one simple secondary process, water is pumped into one well, called the "injection well." The pressurized water floods the reservoir, driving oil to nearby wells, where both the water and oil are extracted. At the surface, the water is separated from the oil and reused, while the oil is sent to the refinery. One tertiary process pumps detergent into the reservoir. The detergent dissolves the remaining oil and carries it to an adjacent well, where the petroleum is then recovered and the detergent recycled.

Because an oil well location occupies only a few hundred square meters of land, most cause relatively little environmental damage. However, oil companies are now extracting

petroleum from fragile environments such as the ocean floor and the Arctic tundra. To obtain oil from the seafloor, engineers build platforms on pilings driven into the ocean floor and mount drill rigs on these steel islands or use a drilling platform that floats but maintains its position through powerful stabilizing motors controlled by a precise GPS system (●**FIGURE 5.22**).

Despite great care, accidents occur during the drilling and extraction of oil. In 2010, the largest accidental marine oil spill in the history of the petroleum industry took place in the Gulf of Mexico when a blowout occurred on the seafloor below the *Deepwater Horizon* oil platform. The seafloor blowout caused an explosion on the drilling platform that killed 11 workers and injured 17 others. For three months, oil gushed uncontrollably from the seafloor before the well was finally capped and declared dead. By then, however, an estimated 4.4 million barrels of oil had been released into the environment. The oil spread throughout much of the Gulf of Mexico, poisoning marine life and disrupting marine and coastal ecosystems.

Natural Gas

Natural gas is an energy resource that forms in source rock or an oil reservoir when crude oil is heated above 100°C during burial and causes the organic molecules to break down to methane, CH_4, an organic molecule consisting of a single carbon atom bonded to four hydrogen atoms. Many conventional petroleum reservoirs contain a layer of oil-saturated rock, with a layer of gas-saturated rock above the heavier liquid petroleum. Other conventional reservoirs are saturated only by gas.

Natural gas is used without refining for home heating and cooking and for fueling large electricity generating plants. Because natural gas contains few impurities, it releases little sulfur or other pollutants when it burns, although, as with all fossil fuels, combustion of natural gas releases carbon dioxide, a greenhouse gas. This fuel produces fewer pollutants and is less expensive to produce than petroleum. At current consumption rates, global natural gas supplies are projected to last between 80 and 200 years.

UNCONVENTIONAL PETROLEUM AND GAS RESERVOIRS

Today, about 78 percent of energy used in the United States comes from petroleum, coal, and natural gas, which traditionally have been the cheapest fuels. However, the price of crude petroleum is subject to very large swings. For example, in inflation-adjusted July 2018 dollars, crude oil rose from a low of $17.48 in 1998 to a high of $161.23 in 2008 before crashing to $49.78 per barrel that same year. Less than 3 years later, in 2011, the price again had climbed to $127 per barrel before again crashing to $30.29 in early 2016. As of this writing (July 2018), the price of crude oil has climbed back to over $70 per barrel (●FIGURE 5.23).

In the past, many alternative forms of energy have been more expensive to develop than coal or oil and gas produced from conventional reservoirs. However, the cost of producing these alternative forms of energy has been decreasing, while the cost of producing traditional fuels has been increasing. As a result, a major restructuring of the global energy economy presently is underway. Many fuel sources that were uneconomical even a year ago are now economic, particularly given the development of new technologies.

Among the biggest of the new changes in the energy economy is the production of oil and gas from **unconventional reservoirs**. Unconventional reservoirs contain oil and/or gas, but until the past decade or so were not producible using conventional drilling, completion, and production technology. So the type of rock sought by the oil companies has changed from conventional reservoirs in which large pores are filled with oil that has migrated from its source rock to the reservoir where it is trapped. A conventional reservoir might be a coarse-grained sandstone with an abundance of large open pores that are filled with oil or gas.

Unconventional reservoirs typically have lower permeability than conventional reservoirs; that is, the pores that contain hydrocarbons in an unconventional reservoir are typically very small and not well connected. As a result, even though the rock is drilled and completed, it is not possible for the hydrocarbons to migrate through the network of tiny pores to the wellbore and so the hydrocarbons remain in the rock.

How small does a pore have to be for a molecule of oil to flow through it? The answer depends on the size and shape of the hydrocarbon molecule and also on the size, shape, and composition of the pore walls. The smallest hydrocarbon molecule that is produced from oil and gas wells is methane—CH_4, also known as dry gas. Methane is the primary type of gas that is delivered via pipeline to residential homes in the United States for heating and cooking. Methane molecules are 0.414 nanometers in diameter.

All other hydrocarbon molecules found in oil and gas are larger than methane, and some are thousands of times larger. Pores that are only a few nanometers across may allow for the passage of methane and restrict larger molecules typically found in liquid petroleum. However, pores that are slightly larger—micrometers in size—are big enough to permit the passage of most hydrocarbon molecules if the molecules don't stick to the pore walls. So most unconventional reservoirs are dominated by very small pore sizes, typically micropores or a mix of micropores and nanopores. Still, if a single well were to be drilled through this type of unconventional reservoir, the well would not be economically profitable because the hydrocarbon molecules will not be able to negotiate their way through the very elaborate network of natural micropores and nanopores. The reservoir would not release the liquid hydrocarbons.

Two major changes in technology, first developed in the U.S. oil and gas industry, have allowed the economic development of very low permeability, "tight rock" reservoirs that previously had not been possible: directional

pump jack The above-ground portion of a reciprocating piston pump in an oil well.

secondary and tertiary recovery techniques Methods of extracting oil or natural gas by artificially augmenting the reservoir energy or fluid composition, as by injection of water, pressurized gas, solvents, or other fluids.

natural gas A mixture of naturally occurring light hydrocarbons composed mainly of methane, CH_4, that is used for home heating and cooking and to fuel large electric generation plants.

unconventional reservoirs A sedimentary rock that is capable of producing oil with the application of special techniques, such as hydraulic fracturing.

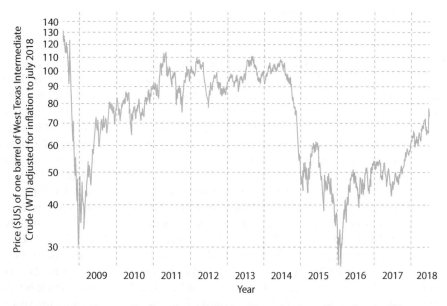

● **FIGURE 5.23** Spot price for a barrel of West Texas Intermediate Crude Oil over the past ten years.

Source: https://www.macrotrends.net/1369/crude-oil-price-history-chart

● **FIGURE 5.24** Directional drilling.

Source: https://rigzonenews.wordpress.com/ 2015/02/25/horizontal-drilling-howdo- they-get-it-to-go-sideways/
Source: http://youngpetro.org/2013/10 /14/how-to-get-entry-intodirectional- drilling-profession/

drilling and **hydraulic fracturing**. These technologies, explained below, significantly increased U.S. reserves of petroleum and natural gas because it made them accessible at a profitable level. According to the Energy Information Agency, in the next 10 years U.S. production of unconventional tight rock reservoirs will tip the country from a net importer of petroleum and natural gas to a net exporter of both.

The first major advance in technology was the ability to control the orientation of the borehole, called "directional drilling." For over 100 years since the first successful commercial oil well was drilled in 1859 in Titusville, Pennsylvania, nearly all oil wells were vertically drilled. In the 1970s, a new type of downhole drilling motor, called a mud motor, was developed that did not require the drill pipe to be rotated like a giant drill bit. Instead, drill pipe is "pushed" down the hole behind the mud motor. Drilling mud—an engineered slurry that is mixed on site—is pumped into the surface end of the drill string. The mud fills the entire interior of the

drill string and places an enormous hydraulic pressure on the end of the pipe where the mud motor and bit are placed. The hydraulic pressure causes the motor to spin the bit, and the drilling mud carries the small chips of rock, called cuttings, up the borehole along the outside of the drilling string and all the way to the surface. A well-site geologist samples and examines the cuttings at fixed intervals to determine what stratigraphic interval is represented (●**FIGURE 5.24**).

If a short section of drilling pipe containing a slight bend of one or two degrees is placed directly behind the mud motor, the borehole that is drilled will be slightly curved. As the well continues to be drilled, the small curve can be continued because a series of small gyroscopes at the leading edge of the drill string allows the inclination (angle from horizontal) and azimuth (compass direction) to be determined and signaled to the driller. The combined information about the total depth of the hole, its inclination, and azimuth is called a survey. By carefully monitoring the hydraulic pressure at the bottom of the hole, the rate of mud circulation, and real-time results from the survey at the bottom of the hole, the driller can control the drilling assembly and maintain the direction of the bend. A bend of one degree or less, if maintained as the well is drilled, will allow for a turn of 90 degrees over several hundred meters of drilling depth, so that the borehole changes from a vertical to a horizontal orientation or "lands" at the desired depth. Once the borehole is horizontal, drilling continues without the bent pipe behind the bit, and the borehole extends horizontally within the targeted geologic layer for one or two miles depending on the mineral lease.

The second major advance in technology that led to the recent development of unconventional tight rock hydrocarbon reservoirs is related to hydraulic fracturing, commonly called fracking. Hydraulic fracturing takes place once the borehole has been drilled, and steel casing (permanent pipe) is placed down the borehole all the way to its end. Once the casing is in place, the frac crew physically isolates a specific interval of the borehole to be hydraulically fractured, typically near or at the very end of the borehole and either perforates that interval or opens pre-engineered holes in the casing. This configuration allows the borehole to be filled with a dilute mixture of water and sand and connects the surface directly

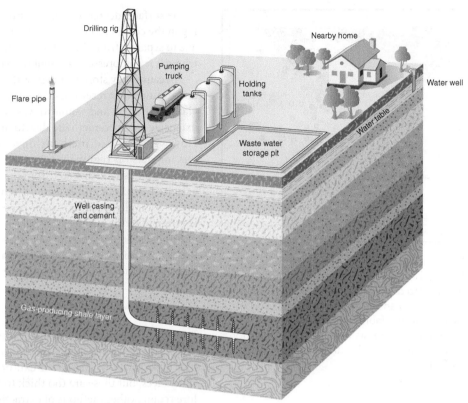

● FIGURE 5.25 Hydraulic fracturing.

and singularly to the interval of subsurface rock to be fractured. By using high-capacity pumps and large volumes of both water and sand, rock within the isolated interval is fractured, and the grains of sand that had been pumped down the hole are forced into the fractures, propping them open. Eventually, the fracturing stops. Typically, at this point, the pumping of water and sand is paused, a new section of pipe is isolated, and a new fracture set is initiated. Each separately fractured interval is called a "stage." Modern directionally drilled wells commonly involve 10 to 30 separate stages.

By directionally drilling wells that deviate from vertical to horizontal and then continue to drill horizontally within a single layer of sedimentary rock (so-called target landing zone) and then conducting multiple stages of hydraulic fracturing, the reservoir is transformed from one with very low permeability to one with a network of fractures propped open by sand. Moreover, the network of fractures is connected directly to the perforations in the well casing. If the hydrocarbons in the reservoir are under enough internal pressure from their slow conversion from kerogen to liquid petroleum and gas, this artificial permeability network provides the pathway into the wellbore and up the casing to the surface (●FIGURE 5.25).

Today, many organic-rich shale units in the United States, including the Marcellus Shale in New York and

Pennsylvania, the Eagleford and Barnett Shales in Texas, and the Bakken Shale in Montana and North Dakota, are developed by direct horizontal drilling and hydraulic fracturing. Moreover, according to the U.S. Geological Survey, roughly half of the industrial-grade sand that is quarried in the United States today is used for hydraulic fracturing process, providing some indication of the magnitude of this recent shift in exploration techniques.

Although directional drilling and hydraulic fracturing have revolutionized the oil and gas exploration industry, particularly in the United States, it has several negative consequences. The first is the large volumes of water that are used in the hydraulic fracturing process. According to the American Geosciences Institute, hydraulically fractured wells typically use over a million gallons of water, and most use several million. For reference, an Olympic-sized swimming pool holds 0.66 million gallons, so most hydraulically fractured wells use the equivalent of several Olympic-sized swimming pools.

A second consequence of pumping large volumes of water and sand down the borehole is that much of that water subsequently returns up the borehole during the earliest stages of production. Before much oil or gas can migrate through the newly formed network of fractures and into the wellbore, the frac fluid must get out of the way of hydrocarbons passing from the reservoir rock into the well. The fluid that returns to the surface in this initial "flow back" usually is very salty, contains chemicals left over from the drilling muds, and may contain significant quantities of dissolved or suspended hydrocarbons. The fluid is full enough of pollutants and produced in large enough volumes that the most

---

hydraulic fracturing The process of fracturing an unconventional reservoir—usually an organic-rich shale—by forcing large volumes of pressurized fluid into it.

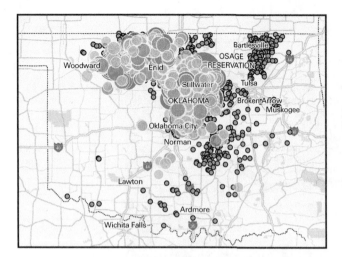

● **FIGURE 5.26** Earthquakes have become commonplace in central Oklahoma because disposal of waste fluids from hydraulic fracturing into deep injection wells is lubricating a set of northwest-trending ancient faults and increasing the pore pressures, causing the faults to slip and generate an earthquake. The blue dots are the location of the Injection wells. The orange circles are the locations of earthquakes during the year 2016 alone.

economic way to deal with it is to pump it down a disposal well. Disposal wells typically are very deep, and some reach all the way to crystalline basement rock. The large volume of frac fluid pumped down disposal wells is forced into the pores of the rocks at the bottom of the borehole. In some places, especially Oklahoma, the deepest parts of oil-producing sedimentary basins contain major faults that are historically inactive but still under significant tectonic stress. Fluid pumped into the pore systems near these faults inflates the pores and can lubricate the fault plane enough to cause it to slip, producing an earthquake. So-called induced seismicity is now common in some oil-producing basins and the potential for a large earthquake to result is a topic of concern (●**FIGURE 5.26**).

Coal Bed Methane

Although most commercial natural gas is produced from conventional reservoirs or through direct drilling and hydraulic fracturing of source rocks, about 7.5 percent of current U.S. gas production comes from coal seams. There, natural Earth heat and microbial activity slowly convert buried coal to **coal bed methane**, methane that is chemically bonded to coal. Coal bed methane reserves in the United States are estimated to be more than 700 trillion cubic feet (Tcf), roughly a 5-year supply at current rates of consumption.

Most coal beds have a high capacity to store water in small voids in the coal itself. As natural processes convert coal to methane, the gas dissolves in the groundwater within the coal. There, the gas is kept in solution by the pressure of overlying groundwater. Natural gas companies drill thousands of wells into the coal beds and pump the groundwater

to the surface, decreasing the pressure on the water remaining in the coal bed. The decreased pressure allows the methane to separate from the water. It is then piped to the surface, where it is compressed and sent to market.

Because they store so much water, coal beds are important groundwater reservoirs for farmers and ranchers, especially in the arid and semiarid western United States, where extensive coal bed methane development is now occurring. However, coal bed methane development has two serious impacts on regional agriculture and ecosystems. The extraction of so much water from the coal beds has depleted essential aquifers and lowered the water table over large areas. Secondly, coal bed water is commonly salty. After it is pumped to the surface, the salt water can poison streams and soils, rendering them useless for agriculture and wildlife. State and federal regulations on water extraction and disposal methods attempt to minimize these impacts.

Tar Sands

In some regions, large sand deposits called **tar sands** are permeated with heavy oil and **bitumen**, a sticky, tarlike hydrocarbon. Crude oil can be obtained from both substances, but these are too thick to be pumped and therefore require other methods of extraction.

The richest tar sands exist in Alberta (Canada), Utah, and Venezuela. In Alberta alone, tar sands contain an estimated 1 trillion barrels of petroleum, roughly 150 years of U.S. consumption at the 2010 rate. About 10 percent of this fuel is shallow enough to be surface mined. Tar sands are dug up and heated with steam to make the heavy oil and bitumen fluid enough to separate from the sand. The oil and bitumen are then treated chemically and heated to convert them to crude oil. At present, several companies mine the Alberta tar sands profitably. Once those reserves are depleted, a portion of the deeper deposits, comprising the remaining 90 percent of the reserve, can be extracted using subsurface techniques similar to those discussed for secondary and tertiary recoveries.

Despite its great size, much controversy has surrounded the development of the Alberta tar sands, because the refinement process uses water in very large volumes, which is then expensive to clean prior to being released back into the environment. This controversy has spilled over into the United States, where a major debate involving the construction and completion of the Keystone XL Pipeline has involved two U.S. presidents and a months-long protest at Cannon Ball, North Dakota, which ended in 2017. The Keystone XL Pipeline is being built as a short-cut to the Keystone Pipeline System, which is owned by TransCanada and connects the crude oil terminal in Hardisty, Alberta, Canada directly to the Bakken oil fields in eastern Montana and western North Dakota. From the Bakken oil fields, the Keystone XL pipeline will continue southeast to the oil pipeline hub at Steele City, Nebraska, from which the oil will continue southward to refineries on the Gulf Coast (●**FIGURE 5.27**).

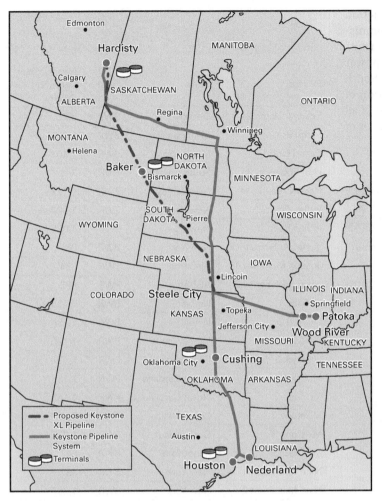

● **FIGURE 5.27** The Keystone Pipeline, including the recently controversial XL Keystone Pipeline, collects crude oil from western Canada, eastern Montana, western North Dakota, and delivers it to a major oil hub in Steele City, Nebraska.

ENERGY RESOURCES: NUCLEAR FUELS AND REACTORS

Nuclear fuels are radioactive isotopes that produce heat through nuclear reactions; the heat is used, in turn, to generate electricity in nuclear reactors. Uranium is the most commonly used nuclear fuel. These energy resources, like

---

coal bed methane Methane that is chemically bonded to coal. The methane can be recovered by removing the groundwater from a coal bed, which decreases the pressure and allows the methane to separate from the coal as a gas.

tar sands Sand deposits saturated with heavy oil and an oil-like substance called bitumen.

bitumen A thick, sticky, oil-like substance that permeates tar sands and can be converted to crude oil.

nuclear fuels Radioactive isotopes, such as those of uranium, used to generate electricity in nuclear reactors.

mineral resources, are nonrenewable, although uranium is abundant.

Every step in the mining, processing, and use of nuclear fuel produces radioactive wastes. The mine waste discarded during mining is radioactive. Enrichment of the ore produces additional radioactive waste. When a uranium nucleus undergoes fission in a reactor, it splits into two useless radioactive elements that must be discarded. After several months in a reactor, the concentration of useful uranium drops until the fuel is no longer viable. In some countries, spent fuel is reprocessed to recover useful uranium, but in the United States, this process is not economical, so the spent fuel is discarded as radioactive waste.

In the early 1970s, the nuclear industry in the United States was growing rapidly, and many energy experts predicted that nuclear energy would dominate the generation of the country's electric energy. These predictions have not been realized. Four factors have led to the decline of the nuclear power industry: (1) Construction of new reactors in the United States has become so costly, in part because of increased regulation, that electricity generated by nuclear power is more expensive than that generated by coal-fired power plants. (2) After major accidents at Three Mile Island in the United States (1979), Chernobyl in Ukraine (1986), and Fukushima Daiichi in Japan (2011), many people have become concerned about safety. (3) Serious concerns remain about the safe disposal of nuclear wastes. (4) The demand for electricity has risen less than expected during the past three decades.

After the 1979 Three Mile Island nuclear accident (●**FIGURE 5.28**), many plans for the construction of new nuclear power plants were cancelled or suspended, and

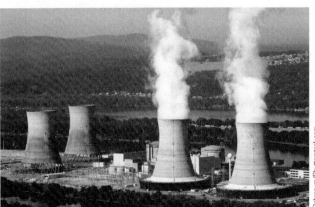

● **FIGURE 5.28** The nuclear power plant at Three Mile Island in central Pennsylvania. The two cooling towers and smaller dome-covered reactor (partly hidden from view) on the left were permanently shut down following the 1979 accident. The reactor and cooling towers on the right continue to operate.

for 34 years no permits were issued for the construction of new reactors by the U.S. Nuclear Regulatory Commission. Finally, in 2012, the Commission approved permits for four new reactors, two at an existing nuclear power plant in Georgia and two at an existing plant in South Carolina. Since then, the permitting and construction of new nuclear plants has continued. ●**TABLE 5.3** is a list of all nuclear power reactors operating in the United States as of April 2017.

TABLE 5.3 Nuclear Power Reactors Operating in the United States

Reactor	State	Type	Net Summer Capacity (MWe)
Arkansas Nuclear One 1	Arkansas	PWR	834
Arkansas Nuclear One 2	Arkansas	PWR	986
Beaver Valley 1	Pennsylvania	PWR	920
Beaver Valley 2	Pennsylvania	PWR	914
Braidwood 1	Illinois	PWR	1,178
Braidwood 2	Illinois	PWR	1,152
Browns Ferry 1	Alabama	BWR	1,101
Browns Ferry 2	Alabama	BWR	1,104
Browns Ferry 3	Alabama	BWR	1,105
Brunswick 1	North Carolina	BWR	938
Brunswick 2	North Carolina	BWR	932
Byron 1	Illinois	PWR	1,164
Byron 2	Illinois	PWR	1,136
Callaway	Missouri	PWR	1,190
Calvert Cliffs 1	Maryland	PWR	866
Calvert Cliffs 2	Maryland	PWR	861
Catawba 1	South Carolina	PWR	1,140
Catawba 2	South Carolina	PWR	1,150
Clinton	Illinois	BWR	1,065
Columbia 2	Washington	BWR	1,158
Comanche Peak 1	Texas	PWR	1,205
Comanche Peak 2	Texas	PWR	1,195
Cooper	Nebraska	BWR	764
Davis Besse	Ohio	PWR	894
Diablo Canyon 1	California	PWR	1,122
Diablo Canyon 2	California	PWR	1,118
Donald C. Cook 1	Michigan	PWR	1,009
Donald C. Cook 2	Michigan	PWR	1,060
Dresden 2	Illinois	BWR	902
Dresden 3	Illinois	BWR	895
Duane Arnold	Iowa	BWR	601
Edwin I. Hatch 1	Georgia	BWR	876

TABLE 5.3 Nuclear Power Reactors Operating in the United States   (*Continued*)

Reactor	State	Type	Net Summer Capacity (MWe)
Edwin I. Hatch 2	Georgia	BWR	883
Fermi 2	Michigan	BWR	1,124
Grand Gulf 1	Mississippi	BWR	1,401
H.B. Robinson 2	South Carolina	PWR	741
Hope Creek 1	New Jersey	BWR	1,172
Indian Point 2	New York	PWR	1,020
Indian Point 3	New York	PWR	1,035
James A. Fitzpatrick	New York	BWR	837
Joseph M. Farley 1	Alabama	PWR	874
Joseph M. Farley 2	Alabama	PWR	883
La Salle 1	Illinois	BWR	1,135
La Salle 2	Illinois	BWR	1,136
Limerick 1	Pennsylvania	BWR	1,120
Limerick 2	Pennsylvania	BWR	1,122
McGuire 1	North Carolina	PWR	1,158
McGuire 2	North Carolina	PWR	1,158
Millstone 2	Connecticut	PWR	868
Millstone 3	Connecticut	PWR	1,220
Monticello	Minnesota	BWR	647
Nine Mile Point 1	New York	BWR	637
Nine Mile Point 2	New York	BWR	1,287
North Anna 1	Virginia	PWR	948
North Anna 2	Virginia	PWR	944
Oconee 1	South Carolina	PWR	847
Oconee 2	South Carolina	PWR	848
Oconee 3	South Carolina	PWR	859
Oyster Creek 1	New Jersey	BWR	608
Palisades	Michigan	PWR	784
Palo Verde 1	Arizona	PWR	1,311
Palo Verde 2	Arizona	PWR	1,314
Palo Verde 3	Arizona	PWR	1,312
Peach Bottom 2	Pennsylvania	BWR	1,308
Peach Bottom 3	Pennsylvania	BWR	1,309
Perry 1	Ohio	BWR	1,240
Pilgrim 1	Massachusetts	BWR	682
Point Beach 1	Wisconsin	PWR	598
Point Beach 2	Wisconsin	PWR	598
Prairie Island 1	Minnesota	PWR	521
Prairie Island 2	Minnesota	PWR	519
Quad Cities 1	Illinois	BWR	908
Quad Cities 2	Illinois	BWR	911
R.E. Ginna	New York	PWR	582
River Bend 1	Louisiana	BWR	968

TABLE 5.3 Nuclear Power Reactors Operating in the United States (*Continued*)			
Reactor	State	Type	Net Summer Capacity (MWe)
Salem 1	New Jersey	PWR	1,170
Salem 2	New Jersey	PWR	1,158
Seabrook 1	New Hampshire	PWR	1,248
Sequoyah 1	Tennessee	PWR	1,152
Sequoyah 2	Tennessee	PWR	1,126
Shearon Harris 1	North Carolina	PWR	928
South Texas Project 1	Texas	PWR	1,280
South Texas Project 2	Texas	PWR	1,280
St. Lucie 1	Florida	PWR	981
St. Lucie 2	Florida	PWR	987
Surry 1	Virginia	PWR	838
Surry 2	Virginia	PWR	838
Susquehanna 1	Pennsylvania	BWR	1,260
Susquehanna 2	Pennsylvania	BWR	1,260
Three Mile Island 1	Pennsylvania	PWR	803
Turkey Point 3	Florida	PWR	802
Turkey Point 4	Florida	PWR	802
V.C. Summer	South Carolina	PWR	971
Vogtle 1	Georgia	PWR	1,150
Vogtle 2	Georgia	PWR	1,152
Waterford 3	Louisiana	PWR	1,160
Watts Bar 1	Tennessee	PWR	1,123
Watts Bar 2	Tennessee	PWR	1,122
Wolf Creek 1	Kansas	PWR	1,175
Total (99 units)			99,678

Source: US Energy Information Administration (EIA) data for January 2017, with Calvert Cliffs 2 uprated since then, Monticello and Peach Bottom 1&2 from PRIS. EIA webpage on U.S. Nuclear Generation and Generating Capacity shows total 99,316 MWe net summer capacity.

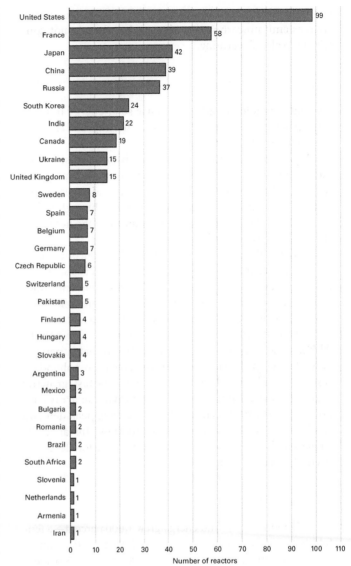

● **FIGURE 5.29** Bar graph showing the number of nuclear power plants currently operating in each of the world's countries.

Source: Data from https://www.statista.com/statistics/267158/number-of-nuclear-reactors-in-operation-by-country/

It is interesting to note that 83 of the 99 total number of operating nuclear plants in the United States in 2017 are located east of the Mississippi River or on the river itself. Only 16 of the 99 reactors are located west of the Mississippi. In addition to the simple need for more energy production in the more densely populated eastern part of the United States, the relatively arid west has easier and more abundant access to other alternative energy sources, especially solar, wind, and geothermal.

Outside the United States, nuclear power production has risen rapidly over the past 30 years, especially in China and Russia. ●**FIGURE 5.29** shows the total number of operating nuclear power plants worldwide as of 2018 broken down by country. Although the United States still has more nuclear reactors than any other country, others will soon catch up at the current rate of new construction outside the United States versus inside the United States. Between 2015 and 2018, for example, only one newly constructed reactor began operations within the United States (Watts Bar Two in Tennessee), compared to 28 new reactors for all other countries.

ENERGY RESOURCES: RENEWABLE ENERGY

Solar, wind, geothermal, hydroelectric, and wood and other biomass fuels are renewable—natural processes replenish them as we use them. Although the amount of energy produced today by renewable sources is small compared to that provided by fossil and nuclear fuels, renewable resources have the potential to supply all of our energy needs. As the prices of conventional fossil fuels have risen along with worldwide energy demand, some renewables have become economical. Except for biomass fuels, renewable energy sources emit no carbon dioxide and therefore do not contribute to global warming.

Solar Energy

Current technologies allow us to use solar energy in three ways: passive solar heating, active solar heating, and electricity production by solar cells.

A passive solar house is built to absorb and store the Sun's heat directly. In active solar heating systems, solar thermal collectors absorb the Sun's energy and use it to heat water. Pumps then circulate the hot water through radiators to heat a building, or the inhabitants use the hot water directly for washing and bathing.

A **solar cell**, or photovoltaic (PV) cell, produces electricity directly from sunlight. A modern solar cell is a *semiconductor*, a device that can conduct electrical current under some conditions but not others. Sunlight energizes electrons in the semiconductor, producing an electric current.

●**FIGURE 5.30** shows an installation of solar panels. Although solar power still accounts for less than 1 percent of world energy demand, solar energy is our most abundant resource, and PV cell production is the fastest-growing segment of the energy industry. Photovoltaic arrays are now competitive with electricity costs during peak demand times in many desert areas, especially those installed for single-family units. PVs are also cost-effective for electricity needs far from existing power lines.

Wind Energy

In the United States, wind energy grew 1,000 percent—a tenfold increase—in the decade between 2001 and 2011, and at the time of this writing (July 2018) accounts for 6.3 percent

● **FIGURE 5.30** A solar farm.

of the country's total electricity production according to the Energy Information Agency.

Wind energy production has grown rapidly because construction of wind generators is cheaper than building new fossil fuel–fired power plants. Wind energy is also clean and abundant. Today, gigantic wind farms generate electricity in Texas, California, and other states, and wind farms are now commonplace in many parts of Europe and elsewhere (●**FIGURE 5.31**). The main drawbacks to wind energy include its inconsistency, the conspicuous nature of the wind turbines (which some people view as unsightly), the death of birds and bats that collide with the turbine blades, and the noise generated by the blades. In addition, the best places to site a wind farm do not always have large nearby human populations, making it necessary to transmit the electricity generated across a set of power lines which may have to be built along with the wind farm. Economic and environmental costs are incurred by building new transmission lines, and the loss of electricity transmitted long distance is an economic consideration that limits the development of wind farms.

● **FIGURE 5.31** A wind farm.

● **FIGURE 5.32** A geothermal facility.

Geothermal Energy

Energy extracted from Earth's internal heat is called *geothermal energy* (●**FIGURE 5.32**). Geothermal plants typically collect underground steam from geysers, volcanoes, and hot springs and use the steam to spin turbines, which generate electricity. Naturally hot groundwater also can be pumped to the surface to generate electricity, or it can be used directly to heat homes and other buildings. Alternatively, cool surface water can be pumped deep into the ground, to be heated by subterranean rock, and then circulated to the surface for use. The United States is the largest producer of geothermal electricity in the world, with a production capacity of just over 3 gigawatts.

In the United States, most geothermal plants are located in the western states, because the region is more tectonically active and the geothermal gradient is higher. We will learn about tectonics in Chapter 6. The oldest, and also presently the largest, steam-driven geothermal plant in the United States is located at The Geysers, about 72 miles north of San Francisco. That plant alone is capable of generating 1.5 gigawatts of electricity, enough for about 375,000 average U.S. households.

Relative to other renewables such as wind and solar, geothermal energy can be used 24 hours a day, 7 days a week, so it has a larger **capacity factor**—a measure of the amount of the actual output of energy to the total possible output over some period of time. Because the wind does not blow all the time and the Sun does not shine all the time, these energy sources have a lower capacity factor.

Hydroelectric Energy

If a river is dammed, the energy of water dropping downward through the dam can be harnessed to turn turbines

that produce electricity. Hydroelectric generators supply roughly 19 percent of the world's electricity. In the United States, about 7 percent of our electricity comes from hydroelectric power.

Although hydroelectricity is renewable, it is entirely dependent on adequate runoff, and recent climate change has caused many of the large reservoirs in the western United States to be drawn down significantly. Inadequate runoff lowers the capacity factor of hydroelectric generating facilities. In addition, the construction of dams and formation of reservoirs destroys wildlife habitats, agricultural land, towns, and migratory fish populations. For example, the dams on the Columbia River and its tributaries are largely responsible for the demise of salmon populations in the Pacific Northwest. Undammed wild rivers and their canyons are prized for their aesthetic and recreational value. Large dams also are expensive to build, and few suitable sites remain.

For these reasons, the United States is unlikely to increase its production of hydroelectric energy. In fact, many historic dams—including some with hydroelectric power–generating capabilities—are being removed.

Biomass Energy

Biomass (plant-based) fuels provide many sources of energy. The burning of wood as a source of heat is familiar to all of us. Today, wood and agricultural products also are burned for the generation of steam and electricity at the industrial level. Additionally, biomass from oil-rich plants such as canola or corn are converted to liquid form to use as transportation fuels, such as ethanol. Much research currently is being directed toward the production of liquid fuels and other chemicals from biomass.

Biomass energy can be produced domestically in most countries, thereby creating local jobs and reducing foreign oil imports. However, production of biofuels is not always a net energy gain; in some cases, more energy is used in the production and processing of these fuels than can be extracted from them. In addition, burning of biomass produces carbon dioxide, a greenhouse gas, and releases particulates and other pollutants into the atmosphere.

Transmission of Electricity— The Hidden Costs

The building of new renewable energy plants, such as wind or solar farms, geothermal, hydroelectric, and biofuels plants all are based to some degree on the availability of the energy resource itself. Wind farms need wind, solar farms need sun, geothermal plants need geothermal energy, and most biofuels are generated in low-population agricultural regions. These places of energy resource availability usually don't correspond to a large local population of humans, so it is usually necessary to transmit the electricity generated in each of these cases to the cities and their associated industrial zones.

---

solar cell A device that produces electricity directly from sunlight; also sometimes called a *photovoltaic (PV) cell.*

capacity factor A measure of the actual to total potential output of an energy source over a period of time.

Installing new long-distance transmission lines is very expensive economically and very disruptive environmentally. In addition, the transmission of electricity itself requires energy, because the transmission lines lose energy through heat loss as the electrons pass through it, a phenomenon called the Joule Effect. The Joule Effect results in loss of between 8 and 15 percent of electricity during transmission across long distances. This loss is in many cases enough to render what might otherwise be a good site for a wind or solar farm or a new geothermal plant uneconomically feasible.

The Future of Renewable Energy Resources

According to the 2018 Annual Energy Outlook published by the US Energy Information Agency, the mix of energy fuels both produced and consumed by the United States will change between 2017 and 2050. On an absolute basis, the consumption of natural gas as an energy source will grow

the most during this period, but the development of non-hydroelectric renewable energy resources is projected to be the fastest-growing segment of the energy economy in percentage terms (●FIGURE 5.33). Much of this growth in non-hydroelectric renewables will offset the continued decline in the production of nuclear energy, but the percentage of overall energy produced in the United States by non-hydroelectric renewables will remain small compared to that produced by natural gas.

In addition, the agency forecasts that strong petroleum production coupled with relatively flat demand will allow the United States to become a net energy exporter within the next decade. Nearly all of the energy exported by the United States is projected to be in the form of petroleum and natural gas, and recovery of hydrocarbons mostly from tight oil reservoirs. The southwestern United States (parts of Texas and New Mexico) is expected to lead the growth in U.S. tight oil supplies between 2017 and 2050 (●FIGURE 5.34). In contrast, the eastern United States along with the southwest

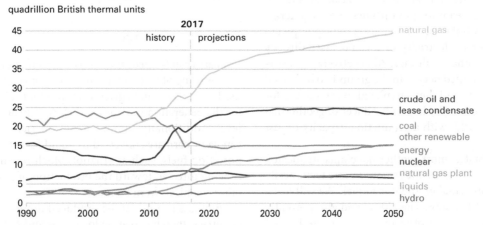

● **FIGURE 5.33** Historical and projected energy production.

Source: U.S. Energy Information Administration

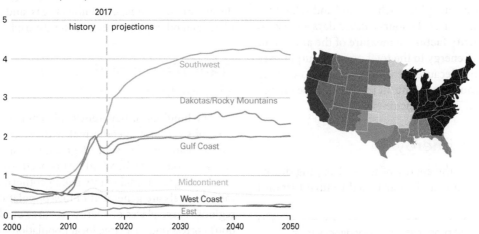

● **FIGURE 5.34** Historical and projected crude oil production from lower 48 onshore wells, organized by region. Crude oil from the Dakotas and Southwest (west Texas and east New Mexico) is projected to dominate.

Source: U.S. Energy Information Administration

World energy consumption by energy source
quadrillion Btu

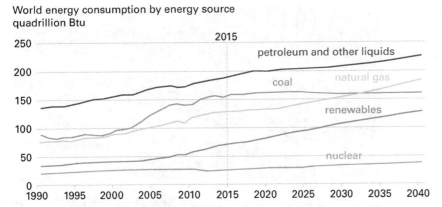

● **FIGURE 5.35** Historical and projected energy production for all countries.

Source: U.S. Energy Information Administration

is expected to lead the growth in natural gas production because tight gas reservoirs there are relatively shallow and widespread, an abundance of pipelines already are built, and there is a large population base to consume the gas and a coast to export it.

The global picture of energy production and consumption over the next several of decades is projected to be somewhat different than that of the United States. According to the International Energy Outlook published in 2017, between 2015 and 2040, world energy consumption is projected to increase by 28 percent, with more than half of the increase attributed to growth in Asia, especially China and India,

where strong economic growth drives increasing energy demand. Global energy consumption is expected to increase for all major fuel types except coal (●**FIGURE 5.35**) with renewables growing at the fastest rate. Global consumption from natural gas, petroleum, and nuclear all are projected to grow through 2050, despite the slowdown of nuclear reactor construction in the United States. The strong future growth projected for renewables through 2050 is forecast to lead to an increase in the global production of electricity from 21 percent in 2015 to 29 percent in 2050. Even with the projected decline in energy produced from coal, global energy production is projected to be dominated by petroleum and natural gas.

CONSERVATION AS AN ALTERNATIVE ENERGY RESOURCE

The single quickest and most effective way to decrease energy consumption and to prolong the availability of fossil fuels is to conserve energy (see ●**FIGURE 5.36**). Policies to

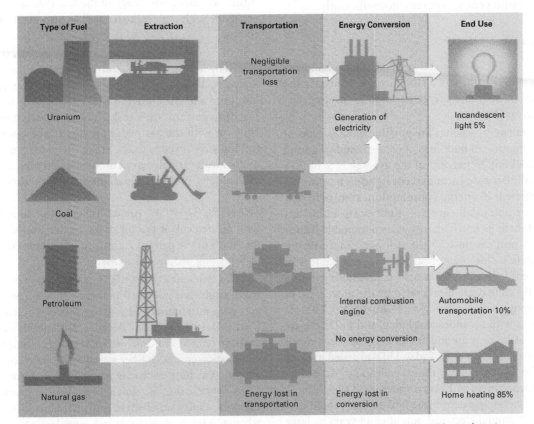

● **FIGURE 5.36** The end-use efficiencies of common energy-consuming systems. Home heating represents the only energy-consumption system that is even remotely efficient—and even there, 15 percent is wasted. Energy to produce incandescent lighting is 95 percent wasted; automobile transportation energy is 90 percent wasted.

improve energy efficiency are more cost-effective than building new power plants. Such policies help to reduce air pollution and dependence on oil imports while saving money for consumers and industry.

Energy conservation has helped to produce dramatic results in the United States, where expenditures for energy as a percentage of gross domestic product (GDP) fell from about 13.5 percent of GDP in 1980 to about 6 percent of GDP in 2000. Higher global energy prices since 2000 have caused expenditures for energy to rise again, to over 9 percent of GDP, despite the fact that the amount of energy used per person in the United States has fallen by about 12 percent over the same time period. Clearly, conservation is a critical component of keeping the total expenditure for energy in the country from continuing to rise. Some energy experts have suggested that if people in industrialized nations use more efficient equipment and develop more efficient habits, these nations could conserve as much as half of the energy they consume.

Energy use in the United States falls under three categories: buildings, industry, and transportation. Two types of conservation strategies can be applied in each of those categories. Technical solutions involve switching to more efficient implements. Social solutions involve decisions to use existing energy systems more efficiently.

Technical Solutions

BUILDINGS In 2017, residential and commercial buildings consumed about 39 percent of all the energy produced in the United States. Most of that energy was used for heating, air-conditioning, and lighting.

Significant energy savings are possible in all aspects of energy consumption in buildings. As one example, lighting accounts for about 20 percent of the average U.S. home's electricity bill. Because incandescent lighting is about 95 percent inefficient in energy consumption, savings in that area alone are potentially great. A fluorescent bulb consumes one-fourth as much energy as a comparable incandescent bulb and can last 10 times longer. In addition, new solid-state technology promises further advances in energy-efficient lighting. For example, light-emitting diodes (LEDs) are lights that are illuminated by the movement of electrons through a semiconductor (●**FIGURE 5.37**). There is no filament as with incandescent lights, and LEDs release almost no heat, so they are much more efficient, last much longer, and use far less energy. Today, LEDs are used in clock radios, jumbo TVs, and many other applications. According to the U.S. Department of Energy, widespread switching to LED lighting technologies over the next 15 years could save the equivalent annual electrical output of 44 years by large electrical power plants.

Pokpak Stock/Shutterstock.com

● **FIGURE 5.37** A bank of light-emitting diodes (LEDs).

INDUSTRY Industry consumes about 31 percent of the energy used in the United States (●**FIGURE 5.38**). In general, conservation practices are cost-effective, and many companies are taking advantage of the fact that saving energy is profitable, although industry still wastes great amounts of energy.

For example, about two-thirds of the electricity consumed by industry drives electric motors for machinery and tools. Most motors are inefficient because they run only at full speed and are slowed by brakes to operate at the proper speeds to perform their tasks. This approach is like driving your car with the gas pedal pressed to the floor and controlling your speed with the brakes. Replacing older electric motors with variable-speed motors would save vast amounts of electricity, but such replacement has been slow.

TRANSPORTATION About 28 percent of all energy, more than 50 percent of the oil usage in the United States, and one-third of the nation's carbon emissions are consumed for transporting people and goods.

The efficiency of standard gasoline-powered auto and truck engines is about 10 percent (Figure 5.36). Thus, we can save much energy by using more-efficient cars and trucks. Over the past few decades, automobile manufacturers have offered vehicles with increasingly efficient internal-combustion engines. Other avenues that auto companies are exploring include electricity and hydrogen power.

A *hybrid car* uses a small, fuel-efficient gasoline engine combined with an electric motor that assists the engine when accelerating. Hybrids consume less gas and produce less pollution per mile than conventional gasoline engines.

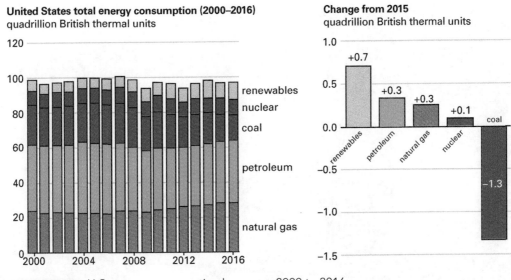

United States total energy consumption (2000–2016)
quadrillion British thermal units

Change from 2015
quadrillion British thermal units

● **FIGURE 5.38** U.S. energy consumption by source 2000 to 2016.

Source: https://www.eia.gov/todayinenergy/detail.php?id=30652

Current models of hybrid cars achieve fuel efficiencies ranging from 51/48 (highway/traffic) to 31/27 miles per gallon, depending on make and model, and they produce as much as 90 percent fewer harmful emissions than a comparable gasoline engine. Using hybrids and other energy-efficient vehicles, American motorists could achieve a 50 percent or greater increase in fuel economy, the equivalent of about one-third of current oil imports.

Another solution is the electric car. Battery-only and gasoline–electric hybrid cars have seen a recent increase in popularity and availability. If that trend continues, perhaps we can further reduce the cost of and dependence on petroleum. Challenges associated with battery-powered vehicles include the need to recharge the battery and draw energy off the electrical grid to do so, and the availability of key mineral resources needed to build the batteries, especially the element lithium.

According to the 2018 Mineral Markets Summary, published by the USGS, "Lithium supply security has become a top priority for technology companies in the United States and Asia. Strategic alliances and joint ventures among technology companies and exploration companies continued to be established to ensure a reliable, diversified supply of lithium for battery suppliers and vehicle manufacturers." In 2017, the most recent full year for which data are available, the United States imported more than 50 percent of its lithium, mostly from Argentina, Chile, and China which along with Australia are the world's top four producers of the element. Indeed, in less than two years between February 2016 and January 2018 alone, the world lithium price went from a record low to a record high, jumping nearly 150 percent in the process.

Lithium is mainly produced from a type of granite with extremely coarse crystals and rare minerals with rare elements. Pegmatite is an igneous rock that crystallizes at the final stages of cooling of a body of magma to a pluton.

The last bit of magma to cool contains an abundance of otherwise rare elements, like lithium and uranium, that do not normally fit into the crystals of common rock-forming minerals that make up plutonic rocks formed during magmatic cooling. Minerals crystallizing from these late-stage magmatic fluids do incorporate these unusual elements, including lithium, into their crystal structures. Pegmatite, the rock containing these minerals, is one major source of lithium. The other source is from deep underground brines that must be pumped and processed for the lithium to be recovered.

Social Solutions

Social solutions involve altering human behavior to conserve energy. Energy-conserving actions can be used in buildings, in industry, and in transportation. Some result in inconvenience to individuals. For example, if you choose to carpool rather than drive your own car, you save fuel but inconvenience yourself by coordinating your schedule with your carpool companions. High-mileage cars are on the market, but they will make an impact only if people make the social decision to use them. People argue that this social decision comes at a cost because light vehicles make the driver and passengers more vulnerable in case of an accident; but studies have shown that lighter, more agile vehicles, with better turning capacity and more effective braking, are actually safer than heavier SUVs.

An example of a modest change in lifestyle that could have a major positive impact on the balance between production of mineral and energy resources and their consumption and the environmental impacts of both is the role that plastics play in our society. From plastic straws and cup lids to children's toys, modern fabrics, car parts, and even satellite dishes, our society consumes plastics at a voracious rate and the consequences are only recently being brought to light.

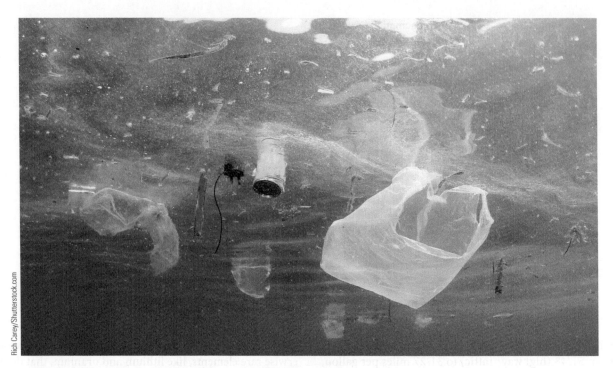

● **FIGURE 5.39** Photo of floating plastic garbage in the ocean.

All plastics are made from hydrocarbons. Plastics are not grown, nor are they mined; ultimately all plastics are synthesized from crude oil through chemical engineering. Not many people realize that their outdoor fabrics mostly are derived from crude oil, let alone their drinking straw.

The gigantic volume of plastics that are consumed globally has had a significant negative impact on the global environment because much of the consumed plastic ends up as waste that must be disposed, and a large amount finds its way into the environment as litter. Disposing of plastic through incineration generates a variety of toxic chemicals that are released into the atmosphere. Plastic that ends up in the environment as litter has accumulated in famous giant collections at certain spots in the middle of the ocean where ocean currents create a sort of slow vortex that draws in floating garbage (●**FIGURE 5.39**). In addition, when exposed to the environment, plastics break down to microplastics—tiny pieces of plastic—that are commonly mistaken for food by small vertebrates and invertebrates. Microplastics are having a large negative impact on the ocean ecosystem because it can fill the digestive gut of small organisms that mistake the floating bits of plastic as food and eat it. Eventually, the organism dies because its plastic-filled gut causes it to starve.

Today, awareness of the negative consequences of using so much disposable plastics in our society is increasing. For example, a recent campaign to never again use a plastic straw has been launched with the clever slogan "Don't suck." It is through awareness campaigns like this, along with individuals following through with necessary changes in behavior, that humans will be able to reverse some of the negative consequences of our "throw-away" society.

ENERGY FOR THE 21ST CENTURY

Is it possible to alter global energy production from a fossil fuel economy to an economy of renewable energy resources? In 2001, the UN's Intergovernmental Panel on Climate Change (IPCC) concluded that significant reduction of fossil fuel use is possible with renewable "technologies that exist in operation or pilot-plant stage today... without any drastic technological breakthroughs."[1] In other words, they suggested that if we vigorously develop all the renewable energy resources listed in section "Energy Resources: Nuclear Fuels and Reactors," the global economic system could absorb a drastic decline in petroleum production without massive disruptions. A year later, 18 prominent energy experts published a rebuttal in *Science*, proposing the exact opposite conclusion. Using almost the same phrases, with the simple addition of the word *not*, they argued that "energy resources that can produce 100 to 300 percent of present world power consumption without fossil fuels and greenhouse emissions do not exist operationally or as pilot plants."[2] Their basic counterargument is that global energy consumption is huge and renewable sources have low power densities.

1. Bert Metz, Ogunlade Davidson, Rob Swart, and Jiahua Pan, eds., *Climate Change 2001: Mitigation* (New York: Cambridge University Press, 2001), 8.

2. Martin I. Hoffert, Ken Caldeira, Gregory Benford, David R. Criswell, Christopher Green, Howard Herzog, et al., "Advanced Technology Paths to Global Climate Stability: Energy for a Greenhouse Planet," *Science* 298 (November 1, 2002): 981–987.

Thus, we do not have the available land, nor could we quickly build the required infrastructure to replace fossil fuels. In 2011, the IPCC again argued that close to 80 percent of world energy supplies could be met by the continued growth of renewable energy, provided they are backed by policies that enable their development. Four years later in 2015, the IPCC aggregated the risks posed by continued increases in global temperature into five "reasons for concerns." The IPCC went on to say "Measures exist to achieve the substantial emissions reductions required to limit the likely warming by 2 degrees" and that it would require 40–70 percent reduction in greenhouse gas emissions by 2050 and near zero emissions by 2100.

When experts disagree, it is difficult for laypersons to evaluate the merits of the contradictory arguments. But whoever is right, it is clear that if global energy demand significantly exceeds supply, the world will fall into unprecedented economic chaos. Commerce will slip into unimaginable depression. Food supplies will diminish, and food distribution will become expensive. Poor people, who are already on the edge of malnourishment, will starve. On the other hand, if we continue to base our future energy economy on hydrocarbons and ignore or fail to adequately mitigate the environmental consequences of global climate and environmental change that come with it, we risk pushing Earth's systems past the point at which the ongoing mass extinction will ultimately involve us. We can only hope that human ingenuity will combine with economic and political commitment to develop alternative energy resources before these catastrophes become reality.

Key Concepts Review

- Useful rocks and minerals are called mineral resources; they include both nonmetallic mineral resources and metals. All mineral resources are nonrenewable. Ore is rock sufficiently enriched in one or more minerals to be mined profitably; geologists usually use the term to refer to metallic mineral deposits.
- Four types of geologic processes concentrate elements to form ore. (1) Magmatic processes form ore as magma solidifies. (2) Hydrothermal processes transport and precipitate metals from hot water. (3) Sedimentary processes form placer deposits, evaporite deposits, and banded iron formations. (4) Weathering removes easily dissolved elements from rocks and minerals, leaving behind residual ore deposits such as bauxite.
- Mineral reserves are the known amount of ore in the ground.
- Metal ores and coal are extracted from underground mines and surface mines.
- One important energy resource is fossil fuels: coal, oil, and natural gas. Fossil fuels are nonrenewable and unrecyclable. Plant matter decays to form peat. Peat converts to coal when it is buried and subjected to elevated temperature and pressure. Petroleum forms from the remains of organisms that settle to the ocean floor or lake bed and are incorporated into source rock. The organic matter converts to liquid oil when it is buried and heated. The petroleum then migrates to a reservoir, where an oil trap retains it. Natural gas forms in source rock subjected to high temperature, and consequently many oil fields contain a mixture of oil with natural gas floating above the heavier liquid petroleum.
- Secondary and tertiary recovery can extract additional supplies of petroleum from old wells, tar sands, and oil shale.
- Nuclear power is expensive, and questions about the safety and disposal of nuclear wastes have diminished its future in the United States. Nuclear fuels, like mineral resources, are nonrenewable, although uranium is abundant. Inexpensive uranium ore will be available for a century or more.
- Solar, wind, geothermal, hydroelectric, and biomass fuels are renewable sources of energy.
- The single quickest and most effective way to decrease energy consumption and to prolong the availability of fossil fuels is to conserve energy.
- Alternative energy resources currently supply a small fraction of our energy needs but have the potential to provide abundant renewable energy.

Important Terms

banded iron formations (p. 93)

bauxite (p. 94)

biomineralize (p. 94)

bitumen (p. 106)

black smokers (p. 91)

capacity factor (p. 111)

coal bed methane (p. 106)

conventional petroleum reservoir (p. 100)

crystal settling (p. 90)

disseminated ore deposit (p. 91)

energy resources (p. 88)

fossil fuels (p. 99)

hydraulic fracturing (p. 104)

hydrothermal processes (p. 91)

hydrothermal vein deposit (p. 91)

kerogen (p. 100)

magmatic processes (p. 90)

manganese nodules (p. 94)

mineral reserves (p. 95)

mineral resources (p. 88)

natural gas (p. 102)

nonmetallic mineral resources (p. 88)

nuclear fuels (p. 107)

petroleum source rock (p. 100)

petroleum (p. 100)

placer deposit (p. 93)

pump jack (p. 102)

residual ore deposits (p. 94)

scavenging (p. 91)

secondary and tertiary recovery techniques (p. 102)

solar cell (p. 110)

submarine hydrothermal ore deposits (p. 91)

surface mine (p. 98)

tar sands (p. 106)

unconventional reservoirs (p. 103)

underground mine (p. 98)

Review Questions

1. Describe the two major categories of geologic resources.

2. Describe the differences between nonrenewable and renewable resources. List one example of each.

3. Discuss the formation of hydrothermal ore deposits.

4. Describe how fossil fuels such as coal, oil, and natural gas form.

5. What is ore? What are mineral reserves? Describe three factors that can change estimates of mineral reserves.

NASA

6

THE ACTIVE EARTH
Plate Tectonics

This photograph of the Sinai Peninsula was shot in 1991 by astronauts aboard the U.S. Space Shuttle *Columbia* from an altitude of about 283 kilometers. The northern end of the Red Sea splits into the Gulf of Suez (left) and the Gulf of Aqaba (right). All three water bodies have formed as a result of tectonic extension (pulling apart). Ongoing extension in the Red Sea and the Gulf of Aqaba is causing the African and Arabian Plates to separate. That plate boundary projects beyond the northern tip of the Gulf of Aqaba through the Dead Sea, seen in the distance. The large water body at the top of the photo is the Mediterranean Sea.

119

LEARNING OBJECTIVES

LO1   Describe four lines of evidence Alfred Wegener cited in support of continental drift.

LO2   On a cross-section, identify the layers of the Earth.

LO3   Describe the characteristics of the layers of the Earth.

LO4   Describe what role that the lithosphere and asthenosphere play in plate tectonic movements.

LO5   State the role that Harry Hess played in the discovery of plate tectonics.

LO6   Compare and contrast oceanic versus continental lithosphere in terms of thickness, density, and lithology.

LO7   Explain how new oceanic lithosphere is formed.

LO8   Describe how continents are formed.

LO9   Sketch a basic cross-section of a divergent, convergent, and transform plate boundary.

LO10   Explain the difference between an active and a passive continental margin.

LO11   Describe two mechanisms that have been proposed as driving horizontal plate movements.

LO12   State the relationship between a mantle plume and a hot spot, and name at least one active hot spot.

LO13   Describe the concept of isostatic adjustment using everyday items.

LO14   Describe an example of isostatic adjustment in geology.

LO15   Describe an example of a relationship between tectonics and climate.

LO16   Explain what is meant by a "supercontinent," and name the most recently formed supercontinent.

THE ORIGINS OF CONTINENTS AND OCEANS

About 5 billion years ago, a ball of dust and gas, one among billions in the universe, collapsed into a slowly spinning disc. The inner portion of the disc condensed to form our Sun, while the outer parts coalesced to form planets orbiting the Sun. Our Earth is one of those planets.

Earth began to form as particles of dust and gas were drawn together by gravity and began to collide. These collisions caused the coalescing particles to become hotter and hotter. Frozen crystals of carbon dioxide, methane, and ammonia melted as the spinning mass—early Earth—slowly heated up. Eventually, ice melted. The young planet grew hotter as asteroids, comets, and other space debris crashed into its surface. Additional heat was released by the decay of radioactive isotopes within its interior. By about 4.6 billion years ago, the planet became hot enough that it melted. Then, as the bombardment slowed down and radioactive isotopes decayed and became less abundant, much of Earth's heat radiated into space and our planet began to cool and solidify.

Today, although Earth's surface has cooled to temperatures that support living organisms, the interior remains hot, both from heat left over from the early melting event and from continued decay of radioactive isotopes. Consequently, Earth becomes hotter with depth. At its center, the Earth is close to 7,000°C—similar to the temperature of the Sun's surface. This internal heat causes earthquakes, volcanic eruptions, mountain building, and continual movements of the continents and ocean basins. These effects, in turn, profoundly affect our environment—Earth's atmosphere, hydrosphere, and biosphere. Earth's internal heat engine and its effects are described in the theory of **plate tectonics**, a simple theory that provides a unifying framework for understanding the way Earth works and how Earth systems interact to create our environment. The term *tectonics* is taken from the Greek *tektonikos*, meaning "construction."

Like most great scientific revolutions, the development of plate tectonics theory developed incrementally over many years, building on earlier observations, hypotheses, and theories. The story illustrates how a scientific theory evolves through the accumulation of evidence and how scientists rely on the work and discoveries of earlier scientists.

ALFRED WEGENER AND THE ORIGIN OF AN IDEA: THE CONTINENTAL DRIFT HYPOTHESIS

Although the theory of plate tectonics was not developed until the 1960s, it was foreshadowed early in the 20th century by a young German scientist named Alfred Wegener, who noticed that the African and South American coastlines on opposite sides of the Atlantic Ocean seemed to fit as if they were adjacent pieces of a jigsaw puzzle (●FIGURE 6.1). He realized that the apparent fit suggested that the continents had once been joined together and had later separated by thousands of kilometers to form the Atlantic Ocean.

Wegener was not the first to make this observation, but he was the first scientist to pursue it with additional research. Studying world maps and making paper cutouts of each continent that he could move around, Wegener realized that not only did the continents on both sides of the Atlantic fit together, but other continents, when positioned correctly, fit like pieces of the same jigsaw puzzle (●FIGURE 6.2). On his map, all the continents joined

---

plate tectonics A theory of global tectonics stating that the lithosphere is segmented into several plates that move about relative to one another by floating on and sliding over the plastic asthenosphere. Seismic and tectonic activity occur mainly at the plate boundaries.

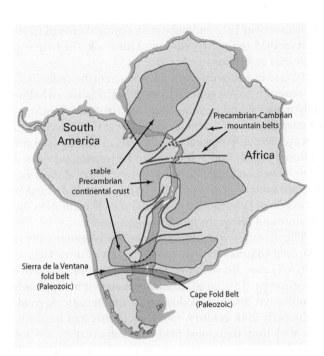

● **FIGURE 6.1** The African and South American coastlines appear to fit together like adjacent pieces of a jigsaw puzzle on Wegener's reconstruction map. Several of the distinctive rock types correlated between the two continents are shown. These include areas of Precambrian stable crust (green), Precambrian and Cambrian mountain belts (blue lines), and the Cape and Sierra de la Ventana Fold Belts of Paleozoic age (orange). The darker, brown regions are the continental shelves, which are the actual edges of the continents.

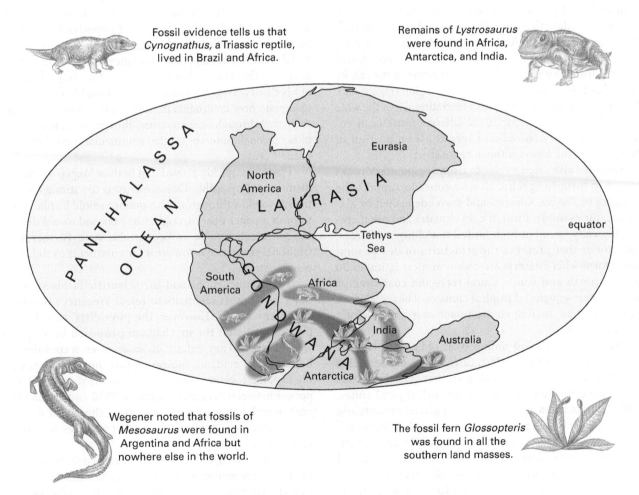

Fossil evidence tells us that *Cynognathus*, a Triassic reptile, lived in Brazil and Africa.

Remains of *Lystrosaurus* were found in Africa, Antarctica, and India.

Wegener noted that fossils of *Mesosaurus* were found in Argentina and Africa but nowhere else in the world.

The fossil fern *Glossopteris* was found in all the southern land masses.

● **FIGURE 6.2** Geographic distributions of plant and animal fossils on Wegener's map indicate that a single supercontinent, called Pangea, existed between about 300 and 200 million years ago.

together formed one supercontinent that he called **Pangea**, from the Greek root words for "all lands." The northern part of Pangea is commonly called **Laurasia** and the southern part **Gondwana**.

Wegener understood that the fit of the continents alone did not prove that a supercontinent had existed. He began seeking additional evidence in 1910 and continued work on the project until his death in 1930.

One line of evidence Wegener found in support of his hypothesis is the occurrence of uncommon rock types or distinctive sequences of rocks that are identical on one side of the Atlantic Ocean and the other. When he plotted the distinctive rocks on a map of Pangea, those presently on the east side of the Atlantic were continuous with their counterparts on the west side (Figure 6.1). For example, the deformed rocks of the Cape Fold Belt of South Africa are similar to rocks found in the Sierra de la Ventana Fold Belt of Argentina. Plotted on a map of Pangea, the two sequences of rocks appear as a single, continuous belt.

Using fossil evidence to support the existence of Pangea, Wegener compiled information regarding locations of certain fossil plant and animal species that could neither swim well nor fly so were unlikely to survive long oceanic crossings. Today, these fossils are found in Antarctica, Africa, Australia, South America, and India, all of which are separated by wide oceans. However, when Wegener plotted the same fossil localities on his reconstruction of Pangea, he found that they all occurred in the same region (Figure 6.2). Wegener deduced that rather than migrating across the wide oceans that presently separate the different fossil locations, each species had evolved and spread over a portion of Pangea *before* the supercontinent broke apart.

Wegener also cited evidence from sedimentary rocks known to form in specific climate zones to support the existence of Pangea. Glaciers and gravel deposited by glacial ice, for example, form in cold climates and are therefore typically found at high latitudes and high altitudes. Sandstones that preserve the structures of desert sand dunes form where deserts are common, near latitudes 30 degrees north and south. Coral reefs and coal swamps thrive in near-equatorial tropical climates. Thus, each rock type reflects an ancient environment characteristic of a specific latitude.

Wegener plotted 250-million-year-old glacial deposits on a map showing the modern distribution of continents (●**FIGURE 6.3A**). Notice on this map that glacial deposits would have formed in tropical and subtropical zones. ●**FIGURES 6.3B** and **6.3C** show the same glacial deposits and other geological indicators of climate plotted on Wegener's Pangea map. In Wegener's reconstruction of Pangea, the glaciers cluster neatly about the South Pole, coral reefs and coal both occur in equatorial position, and desert environments formed around 30 degrees north, similar to the modern distribution of these paleoclimate indicators.

Wegener's concept of a single supercontinent that broke apart to form the modern continents is called the theory of **continental drift**. Wegener first presented the framework

of his theory in 1912 and published a more thorough treatment in 1915 in the first edition of his book *The Origin of Continents and Oceans*.

In one of the great examples of scientific feudalism, the reaction to Wegener's hypothesis was overwhelmingly negative and in some cases exceptionally scathing. Thomas Chamberlin, geology professor at the University of Chicago, wrote, "Wegener's hypothesis in general is of the footloose type . . . and is less bound by restrictions or tied down by awkward, ugly facts than most of its rival theories." Stanford geology professor Bailey Willis remarked, "further discussion of [Wegener's hypothesis] merely incumbers the literature and befogs the mind."

The strongly negative reaction to Wegener's ideas regarding continental drift was the result of three factors. First, Wegener did not provide alternatives for his theory of continental drift, so some scientists viewed his approach as subjective, biased, and downright unscientific. Second, in the early 20th century, many geologists had begun to turn away from traditional field-based observations toward more-detailed, quantitative laboratory measurement of rock properties. As a result, many scientists considered Wegener's field-based evidence to be old-fashioned, unsophisticated, and vague. Third, and perhaps most importantly, Wegener had concentrated on developing evidence that continents had drifted and not on exactly *how* they could move. Perhaps as an afterthought to what he considered the important part of his theory, Wegener suggested two possible mechanisms to explain how continents moved: (1) that continents plow their way through oceanic crust, shoving it aside as a ship plows through water; or (2) that continental crust slides over oceanic crust.

Physicists quickly proved that both of Wegener's mechanisms were impossible. Oceanic crust is too strong for continents to plow through it. The attempt would be like trying to push a paper boat through heavy tar and would deform the continents into an unrecognizable state. Furthermore, frictional resistance is too great for continents to slide over oceanic crust.

These conclusions and the scientific fashions of the day caused most scientists to reject Wegener's theory of continental drift. However, the physicists' calculations proved only that the mechanism proposed by Wegener was incorrect. They did not disprove, or even consider, the huge mass of evidence indicating that the continents were once joined together. Nevertheless, in the roughly 30-year period between Wegener's death in 1930 and about 1960, continental drift was largely forgotten, although a few persistent geologists continued to report evidence in support of the idea.

Much of the hypothesis of continental drift is similar to modern plate tectonics theory. Modern evidence indicates that the continents were together much as Wegener had portrayed them in his map of Pangea. Today, most geologists recognize the importance of Wegener's contributions as being foundational to the development of plate tectonic theory as it exists today.

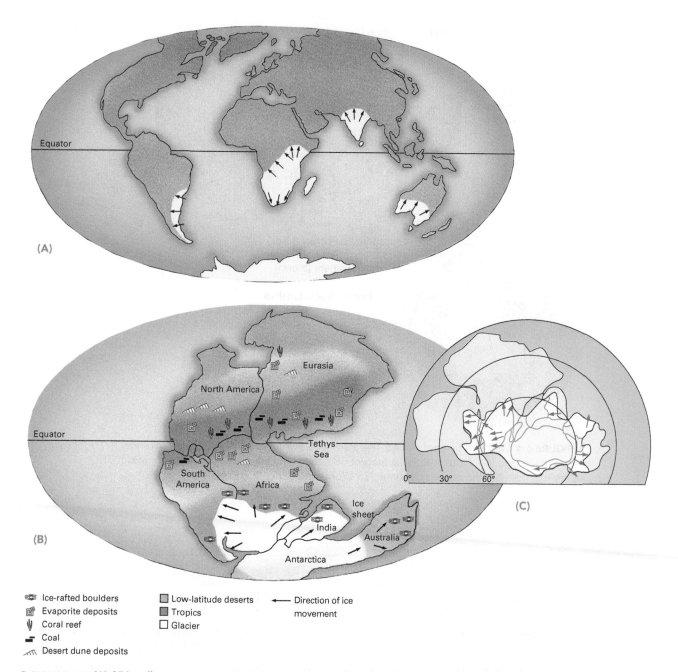

Ice-rafted boulders
Evaporite deposits
Coral reef
Coal
Desert dune deposits

Low-latitude deserts
Tropics
Glacier

← Direction of ice movement

● **FIGURE 6.3** (A) 250-million-year-old glacial deposits are displayed in white on a map showing the modern distribution of continents. The black arrows show directions of glacial movement, indicated by glacial features described in Chapter 13. Notice that many of the arrows are pointing from the shoreline toward the continental interior, a situation that is difficult to explain because it would require the glaciers to move upslope over long distances. (B) 300-million-year-old glacial deposits and other climate-sensitive sedimentary rocks plotted on Wegener's map of Pangea. (C) A view of Gondwana from the South Pole, showing the direction of ice movement 300 million years ago.

Pangea The supercontinent that existed when all Earth's continents were joined together, about 300 million to 200 million years ago, first identified and named by Alfred Wegener.

Laurasia The northern part of Pangea, consisting of what is now North America and Eurasia.

Gondwana The southern part of Pangea, consisting of what is now South America, Africa, Antarctica, India, and Australia.

continental drift The theory proposed by Alfred Wegener that Earth's continents were once joined together and later split and drifted apart. The continental drift theory has been replaced by the more complete plate tectonics theory.

THE EARTH'S LAYERS

The energy released by an earthquake travels through Earth as waves. After Wegener died and his theory was mostly forgotten, geologists discovered that both the speed and the direction of these waves change abruptly at certain depths, as the waves pass through Earth. They soon realized that these changes reveal that Earth is a layered planet. ●**FIGURE 6.4** and **TABLE 6.1** describe the layers. It is necessary to understand Earth's layers to consider the theory of plate tectonics.

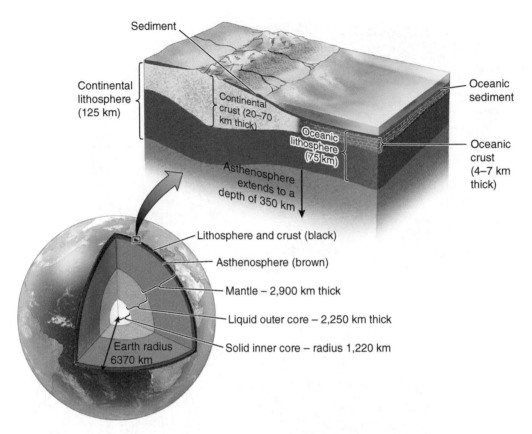

● **FIGURE 6.4** Earth is a layered planet. The insert is drawn on an expanded scale to show near-surface layering. Note that the average thickness of the lithosphere varies from about 75 kilometers beneath the oceans to about 125 kilometers beneath continents.

TABLE 6.1 The Layers of the Earth

Layer		Composition	Depth	Properties
Crust	Oceanic crust	Basalt	Extends from surface to between 4 and 7 km	Cool, hard, and strong
	Continental crust	Granite	Extends from surface to between 20 and 70 km	Cool, hard, and strong
Lithosphere	The crust and the uppermost portion of the mantle	Varies; the crust and the mantle lithosphere have different compositions	Extends from surface to between 75 and 125 km	Cool, hard, and strong
Mantle (excluding the uppermost portion, which is part of the lithosphere)	Asthenosphere	Plastic, ultramafic rock, mainly peridotite, throughout entire mantle; mineralogy varies with depth	Extends from base of lithosphere to 350 km	Hot, weak, and 1% or 2% melted
	Remainder of upper mantle		Extends from 350 to 660 km	Hot, under great pressure, and mechanically strong
	Lower mantle		Extends from 660 to 2,900 km	High pressure forms minerals different from those of the upper mantle
Core	Outer core	Iron and nickel	Extends from 2,900 to 5,150 km	Liquid
	Inner core	Iron and nickel	Extends from 5,150 km to the center of Earth	Solid

The Crust

The crust is the outermost and thinnest layer. Because it is cool relative to the layers below, the crust consists of hard, strong rock (Figure 6.4). Crust beneath the oceans differs from that of continents.

Oceanic crust is between 4 and 7 kilometers thick and is composed mostly of dark, dense basalt. In contrast, the average thickness of continental crust is about 20 to 40 kilometers, although under some mountain ranges it can be as much as 70 kilometers thick. Continents are composed primarily of granite, which is lighter colored and less dense than basalt.

The Mantle

The mantle lies directly below the crust. It is almost 2,900 kilometers thick and makes up 80 percent of Earth's volume. The mantle is composed mainly of peridotite, a rock that is denser than the basalt and granite of the crust.

Although the chemical composition may not vary much throughout the entire mantle, temperature and pressure increase with depth. ●FIGURE 6.5A shows that the temperature at the top of the mantle is near 1,000°C and that near the mantle/core boundary is about 3,300°C. These changes cause the strength of mantle rock to vary with depth. The differences in strength create layering. ●FIGURE 6.5B shows that internal pressure also increases with depth.

At this point, it is important to understand the effects of temperature and pressure on rocks. Most people understand that increasing temperature will eventually melt a rock. Less obvious, however, is the fact that high pressure *inhibits* melting, because rock expands by about 10 percent when it melts. High pressure makes it more difficult for a rock to expand and therefore impedes melting. If the combined effects of temperature and pressure are close to—but just below—a rock's melting point, the rock remains solid but loses strength, so it becomes weak and plastic. In such a weakened state, rock can flow slowly, similar to the way honey spills from a jar. At that point, if temperature rises or pressure decreases, the rock will begin to melt.

Because both temperature and pressure increase with depth in Earth, their combined effects change the physical properties of rocks with increasing depth. These changes create two distinctly different layers in the upper mantle. The strength of the rocks is very different between the two layers, although the composition of mantle rock in each layer is similar.

The Lithosphere

●FIGURE 6.5 shows that the uppermost mantle is cool and its pressure is low, conditions similar to those in the crust. Both factors combine to produce hard, strong rock similar to that of the crust. Recall from Chapter 1 that the outer part of Earth,

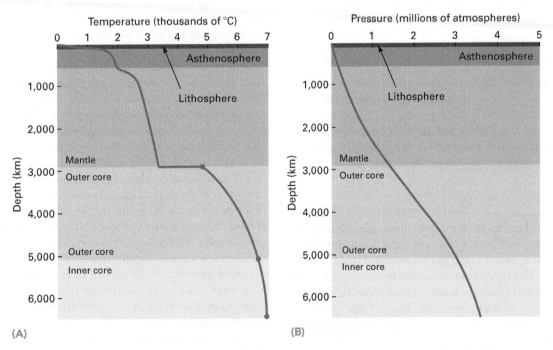

(A)   (B)

● **FIGURE 6.5** (A) Earth's internal temperature increases with depth. At the center of Earth, the temperature is close to 7,000°C, about as hot as the Sun's surface. For reference, the temperature of an oven baking a chocolate cake is about 175°C and a heated steel bar turns cherry red at 746°C. (B) In contrast with temperature, Earth's internal pressure increases almost linearly with depth.

including both the crust and the uppermost mantle, make up the *lithosphere*. The average thickness of the lithosphere is about 100 kilometers but ranges from about 75 kilometers beneath ocean basins to about 125 kilometers under the continents (Figure 6.4). The lithosphere, then, consists mostly of the cold, strong uppermost mantle; the crust is just a thin layer of buoyant rock forming the top of the lithosphere.

The Asthenosphere

At a depth varying from 75 to 125 kilometers beneath Earth's surface, the temperature and pressure conditions are close to the melting point of mantle rock. As a result, at this depth the mantle abruptly loses strength relative to the overlying rock and becomes weak and plastic (Figure 6.5). About 1 to 2 percent of the rock melts, although the rest remains solid. This weak, plastic, and partly molten character extends to a depth of about 350 kilometers, where increasing pressure overwhelms temperature and the rock becomes stronger again. This layer of weak mantle rock extending from about 100 to 350 kilometers deep is the **asthenosphere** (from the Greek for "weak layer"). The average temperature in the asthenosphere is about 1,800°C, although the temperature increases with depth as it does in other Earth layers. Pressure in the asthenosphere rises from about 35 kilobars[1] near the top to about 120 kilobars at the base.

If you apply force to a plastic solid, it deforms slowly, much like the spilled honey. The soft, plastic rock of the asthenosphere behaves in this way, relative to the strong, hard lithosphere that lies on top of it. The lithosphere is not rigidly supported by the rock beneath it, but instead floats on the soft, plastic rock of the asthenosphere. This concept of a floating lithosphere is important to our understanding of plate tectonics and Earth's internal processes.

The Mantle below the Asthenosphere

At the base of the asthenosphere, the increasing pressure overcomes the effect of rising temperature, and the strength of the mantle increases again (Figure 6.5). Although the mantle below 350 kilometers is stronger than the asthenosphere, it is not as strong as the lithosphere, but rather is plastic and capable of flowing slowly over geologic time.

The Core

As you learned in Chapter 1, the *core* is the innermost of Earth's layers. It is a sphere with a radius of about 3,470 kilometers, about the same size as Mars, and is composed largely of iron and nickel. The outer core is molten because of the high temperature and relatively lower pressure there. In contrast, the temperature of the inner core is close to 7,000°C, roughly similar to the temperature of the Sun's surface, and the pressure

is 3.5 million times that of Earth's atmosphere at sea level. This extreme pressure compresses the inner core to a solid, despite the fact that it is even hotter than the molten outer core.

THE SEAFLOOR SPREADING HYPOTHESIS

Following the death of Alfred Wegener in 1930, few geologists thought much about continental drift. Most regarded the arrangement of the continents and ocean basins as fixed, and this set of assumptions prevailed through the 1940s and 1950s.

The huge naval effort that was involved during WWII set in motion a series of new discoveries about the ocean floor that would ultimately disprove the "fixist" assumptions about the configuration of the continents and oceans. On the day after Pearl Harbor was bombed, a young Princeton Geology Instructor named Harry Hess volunteered for active duty in the U.S. Navy. Hess was already in the U.S. Naval Reserves, where he conducted research on the ocean floor using Navy submarines to carry out his experiments. Ten years before the war broke out, Hess had earned his Ph.D. in geology at Princeton and had subsequently joined the faculty there. As a young professor, Hess had participated in submarine-based deep-sea research with pre-eminent Dutch geophysicist Felix Vening Meinesz, who had discovered a series of unexplained systematic variations or "anomalies" in the strength of the Earth's gravity field along the long string of volcanic islands in the East and West Indies. Back at Princeton, Hess decided the best way to carry out similar work of his own was to join the Navy Reserves and carry out his work using U.S. submarines.

Once the war began and Hess became an active-duty Naval officer, he first worked to estimate the daily positions of German submarines and was later transferred to sea duty where he took part in several assault landings in the Pacific and eventually captained his own assault transport ship. While Hess was on active sea duty as a naval officer, he conducted a remarkable set of measurements using a special deep-sea fathometer that he had installed on his ship. Hess ordered that the fathometer be kept on all the time and that it print out a continuous chart of the ocean depth during his frequent criss-crossings of the Pacific. Over the course of the war, Hess thus recorded some 250,000 miles of soundings of the ocean floor.

Hess's wartime depth transects across the Pacific revealed that the floor of the ocean was covered by numerous flat-topped mountains that rose from the flat deep-ocean floor to much shallower depths of 2,000 or 3,000 feet. Hess recognized the features as ancient volcanoes that eventually became inactive and sank beneath the sea surface. As they sank, erosion by wave energy planed off the tops of the volcanoes. Hess called the features guyots, named after Arnold Guyot, the Swiss geologist who founded Princeton's Geology Department (●**FIGURE 6.7**).

MINDTAP
From Cengage

Animation:
Earth's magnetic
field and
paleomagnetism

---

1. 1 kilobar pressure is approximately equal to 987 atmospheres, or 14,504 pounds per square inch. An *atmosphere* (abbreviated *atm*) is a unit of pressure approximately equal to the air pressure at sea level.

One of the key results to come of this naval research effort to map the ocean floors after WWII was the observation that the guyots that stood closest to the ocean trenches were systematically older and also stood at a slight angle to guyots that stood further from the ocean trenches and closer to the mid-ocean ridges. Hess deduced that the guyots were transported from the mid-ocean ridges to the ocean trenches by the moving ocean floor. Guyots that were closest to the trench were tilted on the steeper slope of the ocean floor there. Hess calculated that the guyots were traveling about 5 inches per year.

In addition to the guyots, mapping the ocean floor during and after WWII revealed the presence of the largest mountain chain on Earth, now called the **Mid-Oceanic Ridge system** (●**FIGURE 6.8**). One branch of this huge submarine mountain range, called the **Mid-Atlantic Ridge**, lies directly in the middle of the Atlantic Ocean, halfway between North and South America to the west, and Europe and Africa to the east.

In 1962, Hess published his paper "History of Ocean Basins" in which he described his ideas about the movements of the ocean floor.

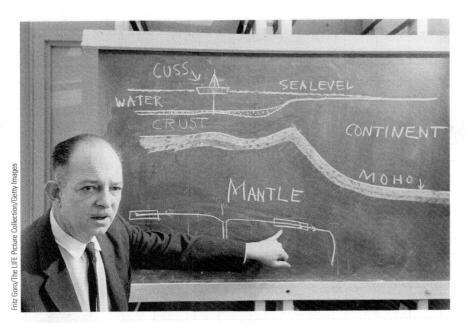

● **FIGURE 6.6** Harry Hess studied the ocean floor for over 40 years as a faculty member and as a reservist and officer in the U.S. Navy where he conducted research aboard submarines. During WWII, he installed a deep-sea fathometer on his ship and ordered it to be kept on at all times. He continued his research in the U.S. Navy after the war, and by then the military importance of having accurate maps of the seafloor was obvious. As a result of his prior experience, Hess was provided with so many resources to undertake this effort that later a close colleague remarked that at that time, "Hess had the entire U.S. Navy working for him as a data gathering agency." Hess was one of the first to recognize that new ocean crust is formed at spreading centers and old ocean crust is destroyed through subduction and he was also one of the first to recognize the connection between convecting mantle movements and seafloor spreading.

After WWII ended, Hess remained in the Navy Reserve where he eventually made the rank of Rear Admiral. Excited by his discovery of guyots in the Pacific, he continued to direct his naval research efforts at mapping the sea floor. Years later, the chair of Princeton's Geology Department while Hess was there was interviewed, and he remarked that following the war "Hess had the whole U.S. Navy working for him as a data collection agency."

The paper was so unusual that Hess described it as "an essay in geopoetry." At about the same time, a government geologist Robert Deitz had independently developed the same basic idea and Deitz coined the term "sea-floor spreading."

Both Hess and Dietz proposed convection—the circular movement of a heated substance, like hot air rising in a room—as the driving force behind the moving ocean floor. Building on an idea his Dutch colleague Meinesz had described many years earlier, Hess suggested that the earth's mantle was under such extreme heat and pressure that it could move like slowly deforming plastic. In his 1962 paper, Hess also made the following key observations:

1. Rising limbs of convecting mantle are present under mid-ocean ridges.
2. The upwelling mantle accounts for high heat flow observed under the mid-ocean ridges.
3. The entire ocean is "virtually swept clean" as it is replaced by new ocean floor every 300–400 million years. This constant replacement of the ocean floor accounts for the relatively thin veneer of sediments in the deep ocean and also accounts for the absence of oceanic crust older than Cretaceous age.

asthenosphere The portion of the upper mantle just beneath the lithosphere, extending from a depth of about 100 kilometers to about 350 kilometers below the surface of Earth and consisting of weak, plastic rock where magma may form.

Mid-Oceanic Ridge system The undersea mountain chain that forms at the boundary between divergent tectonic plates within oceanic crust. It circles the planet like the seam on a baseball, forming Earth's longest mountain chain.

Mid-Atlantic Ridge The portion of the Mid-Oceanic Ridge system that lies in the middle of the Atlantic Ocean, halfway between North America and South America to the west, and Europe and Africa to the east.

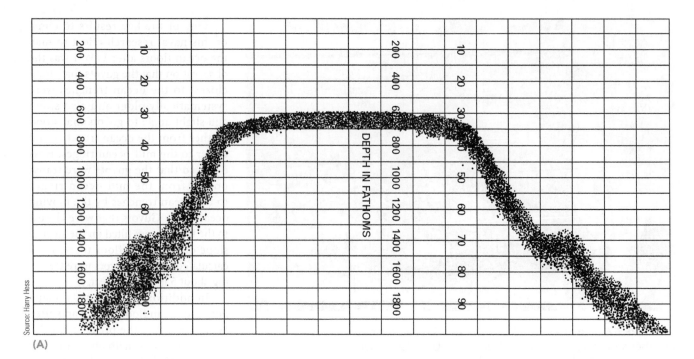

(A)

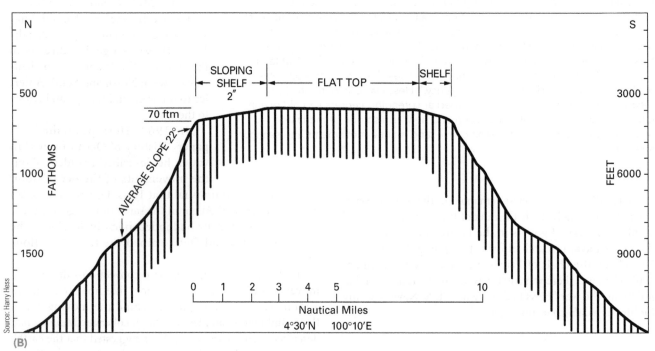

(B)

● **FIGURE 6.7** This image shows a trace of the Pacific Ocean floor obtained by Harry Hess aboard the ship he commanded during WWII. The trace shows a flat-topped mountain, called a guyot. Hess recognized these features as extinct volcanoes whose tops had been planed off by wave energy. Over the course of the war, Hess recorded over 250,000 miles of soundings of the Pacific Ocean floor—data that led Hess eventually to conclude that the ocean floors moved laterally great distances as a result of sea-floor spreading.

4. The continents are carried passively along by the convecting mantle. They do not plow through oceanic crust.

5. The ocean floor and its thin cover of sediments, as well as all volcanic seamounts and guyots also ride down into the "jaw crusher" of the descending limb of a mantle convection cell.

The 1962 paper by Hess paved the way for much additional research on the ocean basins during the rest of that decade. Among the most intriguing and ultimately revealing observations involved magnetism of the sea floor. As you learned in Chapter 3, oceanic crust is composed mostly of basalt, an igneous rock rich in iron. As basaltic lava

● **FIGURE 6.8** A color-coded image of the seafloor. The Mid-Oceanic Ridge system is a submarine mountain chain that encircles the globe like the seam on a baseball. On this image, the most visible parts of the Mid-Oceanic Ridge system are those segments that bisect the Atlantic Ocean and that extend to the southwest of the Gulf of California. Other parts of the Mid-Oceanic Ridge system are visible in the Indian Ocean east of Africa.

cools and forms solid rock, the iron-rich mineral crystals in the basalt operate like weak magnets. The magnetic fields of these minerals align parallel to the Earth's magnetic field. Thus, the basalt preserves a record of the orientation and strength of Earth's magnetic field at the time the rock cools.

By towing devices called magnetometers behind their research vessels, oceanographers were able to detect and record magnetic patterns in the basalt forming the deep-ocean floors. ●**FIGURE 6.9** shows the magnetic orientations of seafloor rocks near a part of the Mid-Atlantic Ridge southwest of Iceland. In this figure, purple stripes represent basalt with a magnetic orientation parallel to Earth's current magnetic field, called **normal magnetic polarity**. The intervening blue stripes represent rocks with magnetic orientations that are exactly opposite to the current magnetic field, called **reversed magnetic polarity**. Notice that the stripes form a symmetric pattern of normal and reversed polarity about the axis of the ridge, and that the central stripe is green, indicating that basalt at the ridge axis has a magnetic orientation parallel to that of Earth's magnetic field today.

Why do the seafloor rocks have alternating normal and reversed polarity, and why is the pattern symmetrically distributed across the Mid-Oceanic Ridge? In the mid-1960s, three scientists—a Cambridge graduate student named Frederick Vine; his professor, Drummond Matthews; and Lawrence Morley, a Canadian working independently of the other two—proposed an explanation for these odd magnetic

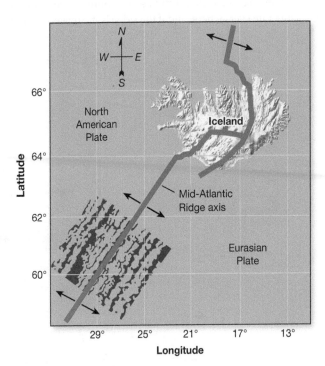

● **FIGURE 6.9** The Mid-Atlantic Ridge, shown in red, runs through Iceland. Magnetic orientation of seafloor rocks near the ridge is shown in the lower-left portion of the map. The purple stripes represent seafloor rocks with normal magnetic polarity, and the blue stripes represent rocks with reversed polarity. The stripes form a symmetrical pattern of alternating normal and reversed polarity on each side of the ridge.

---

normal magnetic polarity A magnetic orientation the same as that of Earth's current magnetic field.

reversed magnetic polarity Magnetic orientations in rock that are opposite to the current orientation of Earth's magnetic field.

patterns on the seafloor. They knew that other scientists had been studying the magnetism preserved in layers of basalt forming the Hawaiian Islands and discovered that Earth's magnetic field has reversed its polarity on the average of every 500,000 years during the past 65 million years. The data from Hawaii indicated that when a **magnetic reversal** of Earth's field occurs, the north magnetic pole becomes the south magnetic pole, and vice versa.

Vine, Matthews, and Morley suggested that the symmetrical magnetic stripes they observed in the seafloor were produced by the continuous spreading of newly formed oceanic crust away from the Mid-Oceanic Ridges, like two conveyor belts moving outward, away from each other (●**FIGURE 6.10**). They recognized that the seafloor and oceanic crust become older with increasing distance from the ridge axis, exactly as Hess had predicted in his 1962 paper. New basalt lava rises through cracks that form at the ridge axis as the two sides of the seafloor separate. As the lava cools and solidifies, the basalt records the strength and orientation of Earth's field. Because Earth's field periodically reverses, the magnetism preserved in the basalt of the ocean floor acquires a striped pattern.

At the same time as these seafloor magnetic patterns were being detected and explained, oceanographers discovered that the layer of mud overlying the seafloor basalt in most parts of the oceans typically is thinnest at the Mid-Oceanic Ridge and becomes progressively thicker at greater distance from the ridge. They reasoned that if mud settles onto the seafloor at the same rate everywhere, and if the ridge is the newest part of the seafloor, the mud layer would be thinnest there. Because oceanic crust is progressively older with increased distance from the ridge axis, more time has elapsed for mud to accumulate, so the layer of mud becomes progressively thicker.

The oceanographers also found that fossils in the deepest layers of mud overlying basalt are very young at the ridge axis but become progressively older with increasing distance from the ridge. This discovery, too, indicated that the seafloor becomes older with increasing distance from the ridge axis as had been predicted by Robert Deitz when he coined the term "sea-floor spreading" as a general model for the origin of all oceanic crust. By the end of the 1960s, the **seafloor spreading** hypothesis became the basis for development of the much broader theory of plate tectonics.

THE THEORY OF PLATE TECTONICS

Like many great unifying scientific ideas, the plate tectonics theory is simple. Briefly, it states that the lithosphere is a shell of hard, strong rock about 100 kilometers thick that floats on the hot, plastic asthenosphere (●**FIGURE 6.11**). As you

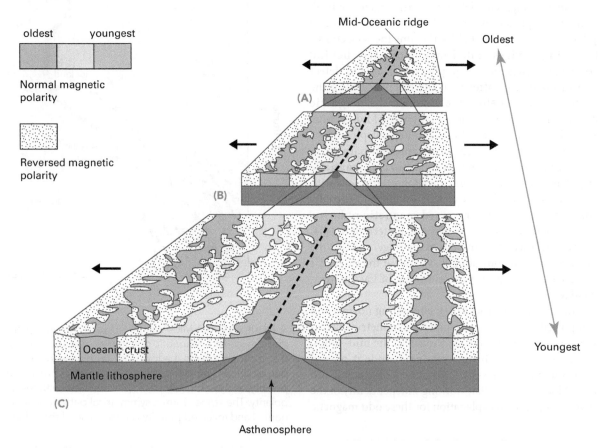

● **FIGURE 6.10** As new oceanic crust cools at the Mid-Oceanic Ridge, it acquires the magnetic orientation of Earth's field. Alternating stripes of normal (colored) and reversed (black stippled pattern) polarity record reversals in Earth's magnetic field that occurred as the crust spread outward from the ridge. The three frames show the evolution of the spreading center through time from oldest (A) to youngest (C).

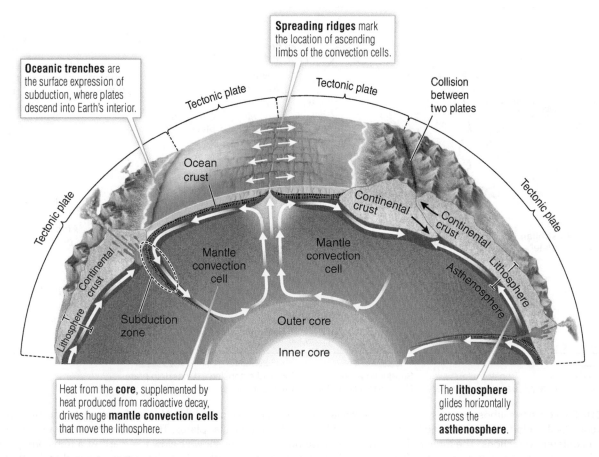

Spreading ridges mark the location of ascending limbs of the convection cells.

Oceanic trenches are the surface expression of subduction, where plates descend into Earth's interior.

Tectonic plate

Tectonic plate

Collision between two plates

Tectonic plate

Tectonic plate

Ocean crust

Continental crust

Continental crust

Tectonic plate

Mantle convection cell

Mantle convection cell

Asthenosphere

Lithosphere

Continental crust

Lithosphere

Subduction zone

Outer core

Inner core

Heat from the **core**, supplemented by heat produced from radioactive decay, drives huge **mantle convection cells** that move the lithosphere.

The **lithosphere** glides horizontally across the **asthenosphere**.

● **FIGURE 6.11** A cutaway view of Earth shows that the lithosphere glides horizontally across the asthenosphere. The top of the lithosphere includes the crust that forms continents and ocean basins, so both move horizontally at rates of a few centimeters each year. The mantle and lithosphere circulate in elliptical cells. In this illustration, the circulating cells involve the entire mantle, although some geologists have suggested that mantle convection may involve two layers, with relatively shallow convection above 660-kilometer depth and deeper convection below this depth. The thickness of the lithosphere is exaggerated here for clarity.

learned in Chapter 1, the lithosphere is broken into seven large (and several smaller) segments called *tectonic plates* (●**FIGURE 6.12**). They are also called *lithospheric plates* or, simply, *plates*—the terms are interchangeable. The tectonic plates slide slowly over the asthenosphere at rates ranging from less than 1 to about 16 centimeters per year, about as

magnetic reversal A change in Earth's magnetic field in which the north magnetic pole becomes the south magnetic pole and vice versa; has occurred on average every 500,000 years over the past 65 million years.

seafloor spreading The hypothesis that segments of oceanic crust are separating at the Mid-Oceanic Ridge.

plate boundary A fracture or boundary that separates two tectonic plates.

divergent boundary A plate boundary where tectonic plates move apart from each other and new lithosphere is continuously forming; also called a *spreading center* or a *rift zone*.

convergent boundary A plate boundary where two tectonic plates move toward each other and collide.

transform boundary A plate boundary where two tectonic plates slide horizontally past one another.

fast as your fingernails grow. Continents and ocean basins make up the upper parts of the lithospheric plates, so as the plates slide over the asthenosphere, the continents and oceans move with them.

A **plate boundary** is a fracture that separates one plate from another. Neighboring plates can move relative to one another at these boundaries in three ways, shown by the insets in Figure 6.12. At a **divergent boundary**, two plates move apart from each other; at a **convergent boundary**, two plates move toward each other; and at a **transform boundary**, they slide horizontally past each other. **TABLE 6.2** summarizes characteristics and examples of each type of plate boundary.

The great forces generated at plate boundaries build mountain ranges, cause earthquakes, and produce many of Earth's volcanoes. In contrast, the interior portions of plates usually are tectonically quiet because they are further from the zones where two plates interact.

Divergent Plate Boundaries

At a divergent plate boundary (also called a *spreading center* or a *rift zone*), two plates spread apart from one another,

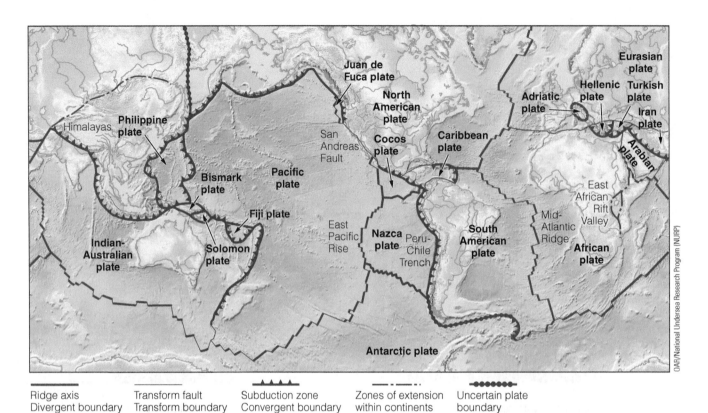

Ridge axis | Transform fault | Subduction zone | Zones of extension | Uncertain plate
Divergent boundary | Transform boundary | Convergent boundary | within continents | boundary

● **FIGURE 6.12** Earth's lithosphere is broken into seven large tectonic plates, called the African, Eurasian, Indian-Australian, Antarctic, Pacific, North American, and South American Plates and many smaller plates. The arrows show how the plates move in different directions. The three different types of plate boundaries are shown below the map: At a transform plate boundary, rocks on opposite sides of the fracture slide horizontally past each other. Two plates move toward each other at a convergent boundary. Two plates move apart at a divergent boundary.

TABLE 6.2 Characteristics and Examples of Plate Boundaries

Type of Boundary	Types of Plates Involved	Topography	Geologic Events	Modern Examples
Divergent	Ocean–ocean	Mid-Oceanic Ridge	Seafloor spreading, shallow earthquakes, rising magma, volcanoes	Mid-Atlantic Ridge
	Continent–continent	Rift valley	Continents torn apart, earthquakes, rising magma, volcanoes	East African Rift
Convergent	Ocean–ocean	Island arcs and ocean trenches	Subduction, deep earthquakes, rising magma, volcanoes, rock deformation	Western Aleutians
	Ocean–continent	Mountains and ocean trenches	Subduction, deep earthquakes, rising magma, volcanoes, rock deformation	Andes
	Continent–continent	Mountains	Deep earthquakes, rock deformation	Himalayas
Transform	Ocean–ocean	Major offset of Mid-Oceanic Ridge axis	Earthquakes	Offset of East Pacific Rise
	Ocean–continent	Linear, deformed mountain ranges	Earthquakes, rock deformation	Northern portion of San Andreas Fault
	Continent–continent	Linear, deformed mountain ranges	Earthquakes, rock deformation	Southern San Andreas Fault

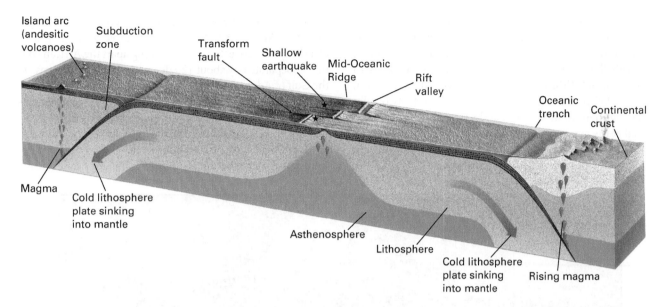

● FIGURE 6.13 Lithospheric plates move away from a spreading center by gliding over the weak, plastic asthenosphere. In the center of the drawing, new lithosphere forms at a spreading center. The lithosphere beneath the spreading center is only 10 or 15 kilometers thick, but it becomes thicker as the new lithosphere moves away from the spreading center and cools. At the sides of the drawing, old lithosphere sinks into the mantle at subduction zones.

as shown at the center of ●**FIGURE 6.13**. The underlying asthenosphere rises upward to fill the gap between the separating plates. As it rises, the decrease in pressure causes the hot asthenosphere to melt and form magma. As this magma continues to rise, it cools to form new crust. Most of this activity occurs at divergent plate boundaries within the ocean basins, but it also can occur between two continental plates that are rifting apart, as in East Africa.

Both the lower lithosphere (the part beneath the crust) and the asthenosphere are parts of the mantle and have similar chemical compositions. The main differences between the two layers are in temperature, pressure, and mechanical strength. The cool lithosphere is strong and hard, but the hot asthenosphere is weak and plastic. As the asthenosphere rises closer to Earth's surface between two separating plates, it cools and gains mechanical strength, and therefore *transforms into new lithosphere*. In this way, new lithosphere continuously forms at a divergent boundary.

At a divergent boundary, the rising asthenosphere is hot, weak, and plastic. Only the upper 10 to 15 kilometers cools enough to gain the strength and hardness of lithosphere rock. As a result, the lithosphere rock, including the crust and the upper few kilometers of mantle rock, can be as little as 10 or 15 kilometers thick at a spreading center. But as the lithosphere spreads, it cools from the top downward and thickens (Figure 6.13).

As it spreads outward and cools, the new lithosphere also thickens because the boundary between cool rock and hot rock migrates downward. Consequently, the thickness of the lithosphere increases as it moves away from

the spreading center. Think of ice freezing on a pond. On a cold day, water under the ice freezes and the ice becomes thicker. In a similar fashion, the cooling lithosphere thickens to about 75 kilometers beneath ocean basins and to about 125 kilometers beneath continents.

The Mid-Oceanic Ridge: Rifting in the Oceans

New lithosphere at an oceanic spreading center is hotter than older lithospheric rock farther away from the divergent boundary, and so the new lithosphere has lower density. Therefore, it floats to a higher level, forming the undersea mountain chain called the Mid-Oceanic Ridge system (Figure 6.8). But as lithosphere migrates away from a spreading center, it cools and becomes denser. As a result, the lithosphere sinks into the soft, plastic asthenosphere (Figure 6.13), causing the depth of the seafloor to increase with distance from the Mid-Oceanic Ridge.

Splitting Continents: Rifting in Continental Crust

A divergent plate boundary can split apart continental crust in a process called **continental rifting**. A rift valley develops in a continental rift zone because continental crust stretches and fractures. Basaltic magma from the base of the crust typically wells up through the newly formed fractures, and some of it reaches the surface to form basaltic volcanism. As the basalt magma works its way through the granitic continental crust, it partially melts it to form additional magma that is rhyolitic in composition. Typically, both basalt and rhyolite compositions erupt from evolving continental rifts.

---

continental rifting The process by which a continent is pulled apart at a divergent plate boundary.

As the continental crust continues to stretch and fracture, and as more basalt is intruded into it or erupted onto it, the surface of the rift begins to sink. As the process continues, eventually the continental crust separates into two sides with a zone of much thinner crust made of basalt in between. Continued rifting along the same zone causes the two sides of the continent to move apart and new oceanic crust to be formed at the active rifting zone located in the middle of the new ocean basin. With time, a narrow oceanic gulf like the Red Sea can spread to become a major ocean.

Today, one of the best places to study continental rifting is in eastern Africa (●FIGURE 6.14) where three converging arms of a rift system collectively are breaking apart the African continent. Separated by about 120 degrees, the three rift arms are the Gulf of Aden, the Red Sea, and the East Africa Rift System. ●FIGURE 6.15 shows the relative counter-clockwise movements of the Arabian Peninsula away from northeastern Africa as the Red Sea and Gulf of Aden continue to separate. The third arm of the rift—the East African Rift System—projects southward into East Africa where it is fracturing the crust and causing

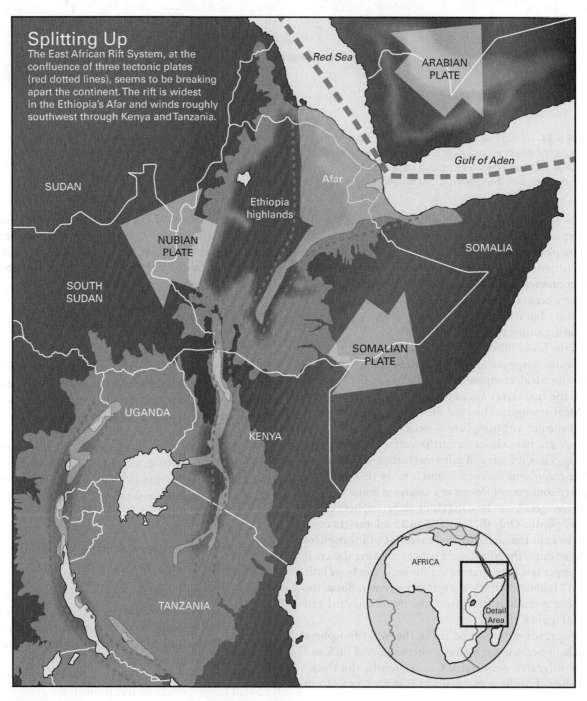

Splitting Up
The East African Rift System, at the confluence of three tectonic plates (red dotted lines), seems to be breaking apart the continent. The rift is widest in the Ethiopia's Afar and winds roughly southwest through Kenya and Tanzania.

● **FIGURE 6.14**   Today, a three-way set of rift arms exists in the Afar Triangle of East Africa. Two of the arms—the Red Sea and The Gulf of Aden—are actively forming new oceanic crust as the Arabian Plate moves away from the Nubian and Somali Plates. The third arm—the East Africa Rift System—is splitting apart the Somalian Plate from the Nubian Plate and rifting African continental lithosphere in the process.

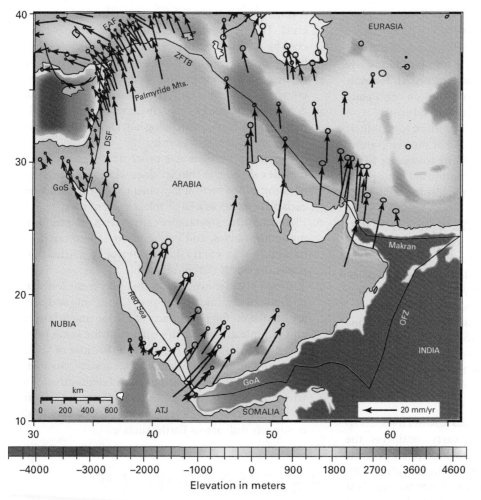

● **FIGURE 6.15** This map of the Arabian Peninsula and surrounding region is color-coded by elevation in meters as shown in the scale at the bottom. Also shown are arrows showing the speed and direction of plate motions at the location of each circle at the end of each arrow. The length of each arrow is proportional to the plate velocity at that spot. Note the 20 mm/yr reference arrow in the lower right. This set of measurements holds the African continent fixed, so all of the measurements shown on this map indicate that the Arabian Peninsula and Eurasia further north are rifting away from Africa along the Red Sea and Gulf of Aden rift zones.

Elevation in meters

the development of deep lakes in the rift valley adjacent to steep mountains on the rift margins. If this rifting continues, eastern Africa will separate from the main portion of the continent, forming at first a narrow gulf like the Red Sea and Gulf of Aden before widening to form a new ocean between the diverging portions of separated continental lithosphere.

Convergent Plate Boundaries

At a convergent plate boundary, two lithospheric plates move toward each other. Not all lithospheric plates are made of equally dense rock. Where two plates of different densities converge, the denser plate sinks into the mantle beneath the less dense one. This sinking process is called **subduction** and is shown on both the right and left sides of Figure 6.11. A **subduction zone** is a long, narrow belt where a lithospheric plate is sinking into the mantle. On a worldwide scale, the rate at which old lithosphere sinks into the mantle at subduction zones is equal to the rate at which

---

subduction The process in which two lithospheric plates of different densities converge and the denser one sinks into the mantle beneath the other.

subduction zone A long, narrow region at a convergent boundary where a lithospheric plate is sinking into the mantle during subduction; also referred to as *subduction boundary*.

new lithosphere forms at spreading centers. In this way, Earth maintains a global balance between the creation of new lithosphere and the destruction of old lithosphere.

Plate convergence can occur (1) between a plate carrying oceanic crust and another carrying continental crust, (2) between two plates carrying oceanic crust, and (3) between two plates carrying continental crust.

CONVERGENCE OF OCEANIC CRUST WITH CONTINENTAL CRUST Recall that oceanic crust is generally denser than continental crust. In fact, the entire lithosphere beneath the oceans is denser than continental lithosphere. When an oceanic plate converges with a continental plate, subduction occurs and the denser oceanic plate plunges into the mantle beneath the edge of the continent. As a result, many subduction zones are located at continental margins. Today, oceanic plates are subducting beneath the western edge of South America; along the coasts of Oregon, Washington, and British Columbia; and at several other continental margins shown in Figure 6.12. When the descending plate reaches the asthenosphere, large quantities of magma are generated by processes explained in Chapter 8. Magma rises through the lithosphere of the overriding plate; much of the magma reaches the surface, where it erupts from a chain of volcanoes that form parallel to the subduction zone. Chapter 9 describes how the Andes—a chain of volcanic mountains—

formed as a result of the subduction of a Pacific oceanic plate beneath the west coast of South America.

The oldest seafloor rocks on Earth are only about 200 million years old, because oceanic crust continuously is being destroyed where it subducts and is melted. In contrast, rocks as old as 4.03 billion years are found on continents because subduction consumes little continental crust.

Continental crust generally is too thick to subduct. In addition, continental crust is made of mostly granite, a lower-density rock than basalt. Relative to oceanic lithosphere, this lower density provides continental lithosphere with buoyancy and further inhibits its ability to subduct.

CONVERGENCE OF TWO PLATES CARRYING OCEANIC CRUST Recall that newly formed oceanic lithosphere is hot, thin, and of relatively low density, but as it spreads away from the Mid-Oceanic Ridge, it becomes older, cooler, thicker, and denser. Thus, the density of oceanic lithosphere increases with its age. When two oceanic plates converge, the older, denser one subducts into the mantle. Oceanic subduction zones are common in the southwestern Pacific Ocean (Figure 6.12) and also formed the Aleutian Islands. The effects of subduction zones on the geology of the seafloor are described in Chapter 15.

CONVERGENCE OF TWO PLATES CARRYING CONTINENTS If two converging plates carry continents, the relatively low density of the continental lithosphere prevents either plate from subducting deeply into the mantle. Continental lithosphere does not normally sink into the mantle at a subduction zone for the same reasons that a log does not sink into a lake: both are of lower density than the material beneath them. Rather, when two plates with continental lithosphere do collide, they crumple against each other and form a huge mountain chain. The Himalayas, the Alps, and the Appalachians all formed as a result of continental collisions. The formation of the Himalayas is described in Chapter 9.

Transform Plate Boundaries

A transform plate boundary forms where two plates slide horizontally past one another as they move in opposite directions (Figure 6.10). This type of boundary can occur in both oceans and continents and can result in frequent earthquakes. California's San Andreas Fault is a transform boundary between the North American Plate and the Pacific Plate.

THE ANATOMY OF A TECTONIC PLATE

The nature of a tectonic plate can be summarized as follows:

1. A plate is a segment of the lithosphere; thus, it includes the uppermost mantle and the overlying crust.

2. A single plate can carry both oceanic crust and continental crust. The average thickness of a lithospheric plate covered by oceanic crust is 75 kilometers, whereas that of lithosphere covered by a continent is 125 kilometers. Lithosphere is thinnest at oceanic spreading centers and thickest where continent–continent collisions are taking place.

3. A plate is composed of hard, mechanically strong rock.

4. A plate floats on the underlying hot, plastic asthenosphere and slides horizontally over it.

5. A plate behaves like a slab of ice floating on a pond. It may flex slightly, as thin ice does when a skater goes by, allowing minor vertical movements. In general, however, each plate moves as a large, intact sheet of rock.

6. Plate margins are tectonically active. Earthquakes, mountain ranges, and volcanoes are common at plate boundaries. In contrast, the interior of a lithospheric plate normally is tectonically stable.

7. Tectonic plates move at rates that vary from less than 1 to about 16 centimeters per year. Continents and oceans are carried on the upper parts of the moving lithosphere and migrate across Earth's surface at the same rates at which the plates move. For example, because of seafloor spreading in the Mid-Atlantic Ridge system, Manhattan Island is now 9 meters farther from London than it was when the Declaration of Independence was written in 1776. Alfred Wegener was correct in saying that continents drift across Earth's surface.

WHY PLATES MOVE: THE EARTH AS A HEAT ENGINE

After geologists had developed the theory of plate tectonics, they began to ask, "*Why* do the great slabs of lithosphere move?" Research[2] has shown that subduction can continue slowly all the way to the core–mantle boundary, to a depth of 2,900 kilometers. At the same time, hot rock rises from the deep mantle toward the surface to replace the lithosphere lost to subduction.

The term **convection** refers to the circulating flow of fluid material in response to heating and cooling. The process of *mantle convection* continually stirs the entire mantle as rock that is hotter than its surroundings rises toward Earth's surface and old plates that are colder than their surroundings sink into the mantle. In this way, the entire mantle–lithosphere system slowly circulates in cells that carry rock

---

2. Summarized by Richard A. Kerr, "Deep-Sinking Slabs Stir the Mantle," *Science* 275 (January 31, 1997): 613–615. Also see Hans-Peter Bunge and Mark Richards, "The Origin of Large Scale Structure in Mantle Convection: Effects of Plate Motions and Viscosity Stratification," *Geophysical Research Letters* 23 (October 15, 1996): 2987–2990.

---

convection The upward and downward flow of fluid material in response to density changes produced by heating and cooling. Convection occurs slowly in Earth's mantle and much more quickly in the oceans and the atmosphere.

from as deep as the core–mantle boundary toward Earth's surface and then back into the deepest mantle (Figure 6.11).

A soup pot on a hot stove illustrates the process of convection. Rising temperature causes most materials, including soup (or rock), to expand. When soup at the bottom of the pot is heated by the underlying stove, it becomes warm and expands. It then rises because it is less dense than the soup at the top. When the hot soup reaches the top of the pot, it flows along the surface until it cools and sinks (●FIGURE 6.16).

Although the circulation of soup in a hot pot provides a good illustration of convection, the circulation of the mantle is considerably more complex than that. Not only is the mantle not shaped like a pot, but the circulation of the mantle is driven by three processes: heat from the core below, radioactive decay of unstable isotopes within the mantle, and the cooling of the upper surface that is in contact with the lithosphere.

Today, the specifics of mantle convection are not well understood, although there is general agreement that mantle convection and plate tectonics are part of the same system and that mantle convection is the main mechanism for transport of heat away from Earth's interior to the surface. What is less certain is the structure of the circulating mantle rock. For example, upwelling of hot mantle beneath mid-ocean ridges is quite shallow and not related to deep-mantle circulation. In this case, it appears to be the diverging motion of the lithospheric plates that causes upwelling of mantle from shallow depths, and not the other way around.

Two other processes, shown in ●FIGURE 6.17, may facilitate the movement of tectonic plates. Notice that the base of the lithosphere slopes downward from a spreading

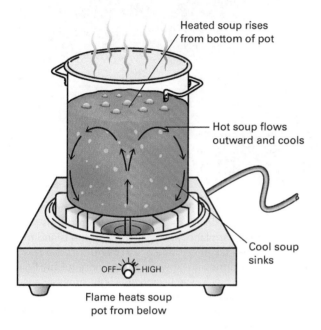

● FIGURE 6.16 Soup convects when it is heated from the bottom of the pot.

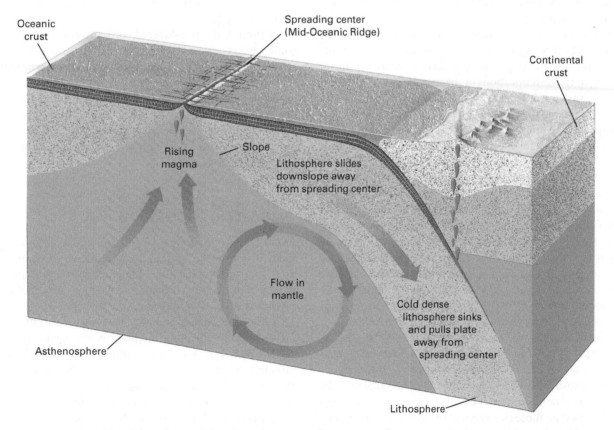

● FIGURE 6.17   New lithosphere glides downslope away from a spreading center. At the same time, the old, cool part of the plate sinks into the mantle at a subduction zone, pulling the rest of the plate along with it. (The steepness of the slope at the base of the lithosphere is exaggerated in this figure.)

center; the grade can be as steep as 8 percent, steeper than most paved highways. Calculations show that even if the slope were less steep, gravity would cause the lithosphere to slide away from a spreading center over the soft, plastic asthenosphere at a rate of a few centimeters per year. This downslope sliding of the lithosphere away from a spreading center is called "ridge push" and may contribute to the movement of plates.

As the lithosphere moves away from a spreading center and cools, it becomes denser. Eventually, old lithosphere may become denser than the asthenosphere below. Consequently, it can no longer float on the asthenosphere and sinks into the mantle in a subduction zone, pulling the trailing plate along with it in a phenomenon referred to as "slab pull." Both ridge push and slab pull are considered to contribute to the movement of a lithospheric plate as it slides over the asthenosphere.

Mantle Plumes and Hot Spots

In contrast to the huge, ridge-shaped mass of mantle that rises beneath a spreading center, a **mantle plume** is a relatively small rising column of plastic mantle rock that is hotter than surrounding rock. Some plumes appear to rise from great depths in the mantle, probably because small zones of rock near the core–mantle boundary become hotter and more buoyant than surrounding regions of the deep mantle. Others form as a result of heating in shallower portions of the mantle.

As pressure decreases in a rising plume, rock melts to form magma. The rising heat and magma produce a **hot spot** in the upper mantle which in turn heats the overlying lithosphere, forming a volcanic center. The Hawaiian Island chain is an example of a volcanic center over a hot spot. As demonstrated by the destructive lava flows on the Island of Hawaii in 2018, which destroyed dozens of homes, roadways, and farmland and even struck a boat of tourists with falling lava, the hot spot is very active and poses a constant hazard to its residents (●**FIGURE 6.18**). The volcanic center is in the middle of the Pacific tectonic plate because the plume originates deep in the mantle, far from any plate boundary and below the level of lateral plate motion.

Some researchers have suggested that the mantle consists of two primary layers, with each layer undergoing convection. The shallower layer, located above 660 kilometers in depth, behaves dynamically and is characterized by relatively rapid convection. Below 660 kilometers, convection is more sluggish. This two-layered mantle model explains why the chemical composition of basalts from mid-ocean ridges is different than those that erupt at hot spots such as Hawaii. Basalt erupting at mid-ocean ridges is part of the shallow convection system, where the mantle is well mixed. In contrast, the basalts erupting in Hawaii are more primitive and

● **FIGURE 6.18**   In May and June of 2018, Mount Kilauea, the most active volcano in the Hawaiian Islands began a series of eruptions that destroyed dozens of homes, farmland, and roadways on the Island of Hawaii. This aerial photo was taken on June 5, 2018 and shows a river of lava (light gray) in the lower right with slower-moving more rubbly flows on either side, shown in dark black. The lava river is light gray because its upper surface has begun to cool to form a thin skin of basalt. Below this skin, the red hot lava still flows and it emerges from the end of the flow where the large flames are located.

come from a plume of mantle welling up from the deeper convection system.

Although the two-layered convection model explains some observations, it is not consistent with others. For example, seismologists have been able to use variations in the velocity of earthquake waves passing through the mantle to identify zones of cooler temperature, attributed to mantle downwelling or old subducted oceanic lithosphere. These data suggest that the relatively cool remnants of some subducted slabs extend completely through the mantle to the core–mantle boundary.

SUPERCONTINENTS

Prior to 2 billion years ago, large continents as we know them today may not have existed. Instead, many—perhaps hundreds—of small masses of continental crust and island arcs similar to Japan, New Zealand, and the modern islands of the southwest Pacific Ocean probably dotted a global ocean basin. Then, between 2 billion and 1.8 billion years ago, tectonic plate movements brought these microcontinents together to form a single landmass called a **supercontinent**. After a few hundred million years, this supercontinent, called Nuna, developed rifts and broke into fragments. The fragments then separated, each riding away from the others on its own tectonic plate. About 1 billion years ago, the fragments of continental crust reassembled, forming a second supercontinent, called Rodinia. In turn, this continent fractured and the continental fragments reassembled into a third supercontinent about

300 million years ago, 70 million years before the appearance of dinosaurs. This third supercontinent is Alfred Wegener's Pangea, which began to break apart about 235 million years ago, in late Triassic time. The tectonic plates have continued their slow movement to create the mosaic of continents and ocean basins that shape the map of the world as we know it today and will continue to shape it into the future. One recent model of future plate motions suggests that the current configuration of continents will rearrange to form the next supercontinent, named Amasia, about 100 million years from now, near the present-day North Pole.

ISOSTASY: VERTICAL MOVEMENT OF THE LITHOSPHERE

If you have ever used a small boat, you may have noticed that the boat settles in the water as you get into it and rises as you step out. The lithosphere behaves in a similar manner. If a large mass is added to the lithosphere, the underlying asthenosphere flows laterally away from that region to make space for the settling lithosphere.

But how is mass added or subtracted from the lithosphere? One process that adds and removes mass is the growth and melting of large glaciers. When a glacier grows, the weight of ice forces the lithosphere downward and causes the asthenosphere to move laterally away from the depressed region. For example, the Hudson Bay region of Canada was depressed during the last glaciation by an ice sheet about 3,000 meters thick. Conversely,

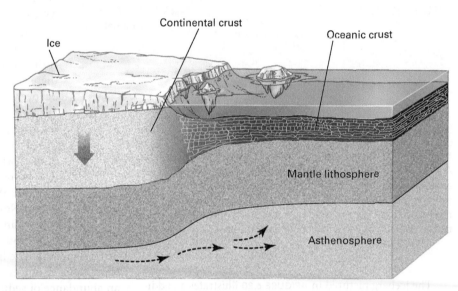

● **FIGURE 6.19**   Isostatic adjustment. The weight of an ice sheet causes continental lithosphere to sink in response to the added burden. Notice that thicker ice will depress the lithosphere to a greater degree.

when a glacier melts, the continent rises, or rebounds. Because the rate of rebound is slower than the rate at which the ice melts, the surface formerly below thick glacial ice can remain depressed for thousands of years after all the ice is gone. Thus, the Hudson Bay region of Canada remains below sea level, although it is rising at a rate of about one centimeter per year as the underlying asthenosphere slowly flows back into the region.

The Great Lakes, located near the former southern margin of the ice sheet, also are slowly rebounding following the melting of glacial ice. The rebound is faster on the northern side of the lakes because the ice was thicker there. Similarly, in Scandinavia, geologists have discovered ice-age beaches that are tens of meters above modern sea level. The beaches formed when glaciers had depressed the Scandinavian crust to sea level, but they now lie well above that elevation because the land rose as the ice melted.

The concept that the lithosphere is in floating equilibrium on the asthenosphere is called **isostasy**, and the vertical movement in response to a changing burden is called *isostatic adjustment* (●**FIGURE 6.19**).

---

mantle plume A relatively small rising column of mantle rock that is hotter than surrounding rock. As pressure decreases in a rising plume, the rock partially melts, forming magma.

hot spot The hot upper mantle rock located within a plume and associated with a volcanic center that forms on the overlying lithosphere.

supercontinent A continent, such as Alfred Wegener's Pangea, consisting of all or most of Earth's continental crust joined together to form a single, large landmass. At least three supercontinents are thought to have existed during the past 2 billion years, and each broke apart after a few hundred million years.

isostasy The concept that the lithosphere floats on the asthenosphere as an iceberg floats on water.

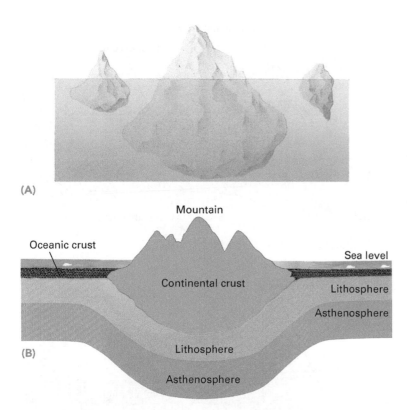

(A)

(B)

● **FIGURE 6.20** (A) Icebergs illustrate some of the effects of isostasy. A large iceberg has a deep root and also a high peak. (B) In an analogous manner, continental lithosphere extends more deeply into the asthenosphere beneath high mountains than it does under lower-elevation regions. In this cartoon, which is not to scale, the mountains were formed by the collision of two small continental fragments.

The iceberg pictured in ●**FIGURE 6.20** illustrates an additional effect of isostasy. A large iceberg has a high peak, but its base extends deeply below the water's surface. The lithosphere behaves in a similar manner. Continents rise high above sea level, and the lithosphere beneath a continent has a "root" that extends as much as 125 kilometers into the asthenosphere.

In contrast, most ocean crust lies approximately 5 kilometers below sea level, and oceanic lithosphere extends only about 75 kilometers into the asthenosphere. For similar reasons, high mountain ranges have deeper roots than do low plains, just as the bottom of a large iceberg is deeper than the base of a small one.

HOW PLATE TECTONICS AFFECT EARTH'S SURFACE

The movements of tectonic plates generate volcanic eruptions and earthquakes, which help shape Earth's surface. They also build mountain ranges and change the global distributions of continents and oceans. Tectonic activities strongly affect our environment in other ways by impacting global and regional climate, the atmosphere, the hydrosphere, and the biosphere.

Volcanoes

Most of Earth's volcanoes result from plate movements. At a divergent boundary (spreading center), hot asthenosphere oozes upward to fill the gap left between the two separating plates. Portions of the rising asthenosphere melt to form basaltic magma, which erupts onto Earth's surface. Thus, the Mid-Oceanic Ridge is in part a chain of submarine volcanoes. Volcanoes are common in continental rifts as well. For example, the longest continuously erupting lava lake in the world occurs at the volcano Erta Ale, located in Ethiopia within the northern portion of the East Africa Rift System (●**FIGURE 6.21**). Numerous other large volcanoes are scattered down the East Africa Rift System for over 1,000 kilometers.

Today, a large portion of western interior North America also is undergoing continental rifting with accompanying volcanism, although there the rifting and volcanics both are spread out over a larger region than in the comparatively narrow East African Rift System. The ongoing rifting in western North America has produced a geologic province called the "Basin and Range," which reaches from north of the Canadian border to south of the Mexican border and from the Wasatch Front of Utah to the Eastern Sierra Nevada of California (●**FIGURE 6.22**). The Basin and Range is so named because the extension has broken the crust into numerous tilted segments. The high-elevation parts of each segment form isolated ranges, and erosion of each range produces an abundance of sediment that fills in the lower-elevation basins (●**FIGURE 6.23**). Volcanics within the Basin and Range include mafic basalt, derived from the base of the crust, and felsic rhyolite, derived from partial melting of the granitic crust by the rising hot basaltic magma (●**FIGURES 6.24**).

In addition to continental or oceanic rift systems, huge quantities of magma also form in the descending lithosphere of a subduction zone. Some of the magma solidifies within the crust, and some erupts at the Earth's surface from a chain of volcanoes that forms parallel to the subduction zone. In North America, the Cascade Range in the Pacific Northwest and Canada is a group of volcanoes that collectively form such a chain. The Aleutian Islands of Alaska, the Andes Mountains in South America, and the volcanic islands of the Caribbean and the west Pacific Ocean all are examples of volcano chains formed in this manner (●**FIGURE 6.25**).

Earthquakes

Earthquakes are common at all three types of plate boundaries but are generally uncommon within the interior of a tectonic plate. As a result, changes in the land surface resulting from earthquakes are most common near plate boundaries. Quakes concentrate at plate boundaries simply because these are zones where one plate slips past another.

● **FIGURE 6.21** Erta Ale volcano in Ethiopia, the longest continuously erupting lava lake in the world.

● **FIGURE 6.23** An example of Basin and Range topography. This view across Death Valley was taken from the foothills of the Black Mountains (foreground) and shows the Panamint Range in the background.

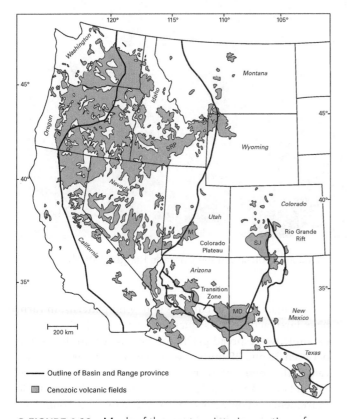

● **FIGURE 6.22** Much of the western interior portion of North America is undergoing active rifting, forming a region called the Basin and Range province. The tectonic stretching causes the crust to break into blocks that tilt. The high points form mountain ranges within the Basin and Range, and the low points called basins fill with sediment eroded from the nearby mountains.

(A)

(B)

● **FIGURE 6.24** Examples of geologically young volcanic rock from the North American Basin and Range extensional province. (A) The Owens River Gorge, a popular sport climbing destination, is cut through rhyolite of the Bishop Tuff. The massive eruption that produced this layer of volcanics took place about 767,000 years ago in southeastern California. (B) Devils Postpile, also in southeastern California, is basalt that erupted about 900,000 years ago.

The slippage of one plate relative to its neighbor is rarely smooth and continuous. Rather, stress builds up along the plate boundary and is stored in the rock as energy. Eventually, the stress level causes the rock to suddenly break, releasing the stored energy and causing rock on one side of the break to lurch violently past rock on the other. An earthquake is the vibration of the rock due to this sudden movement and rapid release of energy. During a single earthquake, the ground surface can be broken, forming a topographic feature called a **scarp**. Earthquakes also can cause major landslides and even change the course of rivers and streams. We will learn more about earthquakes in the next chapter.

scarp A break in the land surface caused by an earthquake.

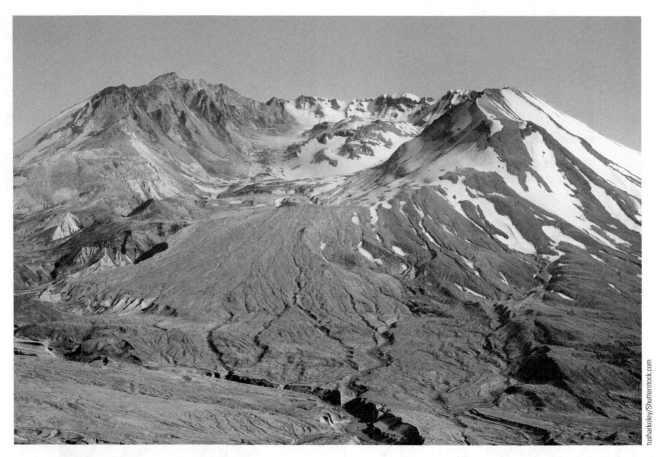

● **FIGURE 6.25**   Mount St. Helens, which last erupted in 1980, is an active volcano in the Cascade Range of Washington near a convergent plate boundary.

Mountain Building

Great chains of volcanic mountains form at rift zones because the new, hot lithosphere floats to a high level and large amounts of magma form in these zones. Along subduction zones, long, linear mountain chains form as magma from the descending oceanic lithosphere melts and some of it ascends to the surface. If two continents collide at a convergent plate boundary, the ground surface will rise for the same reason that a mound of bread dough thickens when you compress it from both sides. Such continent–continent collisions thrust great masses of rock upward, creating huge mountain chains such as the Himalayas, the Alps, and the Appalachians.

HOW PLATE TECTONICS AFFECT EARTH'S CLIMATE

The tectonic movements of the continents and the opening and closing of ocean basins through rifting and subduction profoundly alter Earth's oceanic and atmospheric systems. Ocean currents carry warm water from the equator toward the poles, and cool water from polar regions toward the equator, warming polar regions and cooling the tropics. Similarly, winds transport heat and moisture over the globe.

Changes in ocean currents and wind patterns, in turn, have far-reaching consequences for regional climate. For example, widespread glaciation across Antarctica began between 38 and 28 million years ago as seafloor spreading caused the continent to drift further away from the southern tips of South America, Africa, and Australia. These tectonic movements not only isolated Antarctica over the South Pole but also established the **Antarctic Circumpolar Current**, a strong west-to-east ocean current that continuously flows in a clockwise direction around the continent (●**FIGURE 6.26**). As it formed, this current prevented warmer water from the southern Atlantic, Pacific, and Indian Oceans from reaching Antarctica, effectively putting the continent into a deep freeze that continues today.

Tectonic movements also have altered the composition of the atmosphere and oceans, producing major changes in global climate. For example, some scientists have proposed that the Cenozoic uplift of the Tibetan Plateau in Asia—the largest such uplift in the world—caused a strengthening of the Asian monsoon that led to Earth's recent glaciations. According to this idea, the additional monsoonal rainfall and

---

Antarctic Circumpolar Current A strong west-to-east ocean current that circulates clockwise around Antarctica and prevents warmer water from getting close to the shores of the continent.

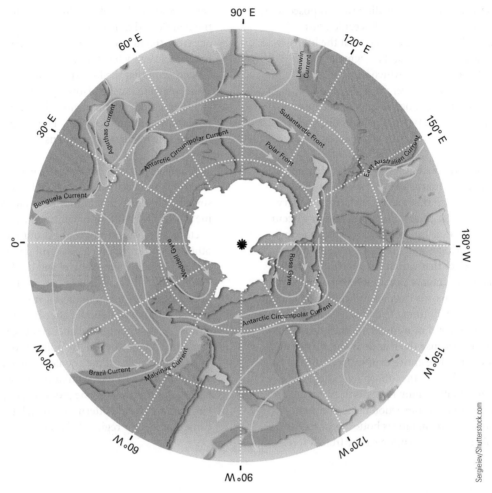

● FIGURE 6.26   The Antarctic Circumpolar Current is the strongest ocean current in the world and effectively isolates Antarctica from the rest of the planet, thereby keeping the continent in a deep freeze. The current was established between 38 and 28 million years ago when Antarctica rifted away from South America, Africa, and Australia and provides a good example of the close relationship between climate and tectonic changes.

Source: Department of the Environment and Energy Australian Antarctic Division

widespread exposure of new rock as the plateau was uplifted significantly increased the rate of chemical weathering of silicate minerals, including those containing calcium. Once in solution, the calcium ions combined with $[CO_3]^{2-}$ anionic groups to form calcium carbonate, $CaCO_3$. The $[CO_3]^{2-}$ was supplied by atmospheric CO_2 that dissolved in ocean water and changed to $[CO_3]^{2-}$. In this way, CO_2 (a greenhouse gas) was removed from the atmosphere, causing the Earth to cool and contributing to the development of major glaciations during late Cenozoic time.

With these changes in environments and climates come changes in the ecosystems supported by them; thus, tectonic processes also affect the biosphere. Streams and drainage patterns must respond to changes in slope direction and altered distributions of precipitation. Lakes may dry up, or new lakes form. Ultimately, plants and animals may die or migrate away, to be replaced by new species that are adapted to the new climatic conditions.

Key Concepts Review

- Alfred Wegener's hypothesis of continental drift foreshadowed the theory of plate tectonics, which provides a unifying framework for much of modern geology.
- Earth is a layered planet. The crust is its outermost layer and varies from 4 to 70 kilometers thick. The mantle extends from the base of the crust to a depth of 2,900

kilometers, where the core begins. The lithosphere is the cool, hard, strong outer 75 to 125 kilometers of Earth; it includes all of the crust and the uppermost mantle. The hot, plastic asthenosphere extends to 350 kilometers in depth. The core is mostly iron and nickel and consists of a liquid outer layer and a solid inner sphere.

- The hypothesis of seafloor spreading was proposed as a general model for the origin of all oceanic crust.
- Plate tectonics theory is the concept that the lithosphere floats on the asthenosphere and is segmented into seven major tectonic plates, which move relative to one another by gliding over the asthenosphere. Most of Earth's major geological activity occurs at plate boundaries. Three types of plate boundaries exist: (1) new lithosphere forms and spreads outward at a divergent boundary, or spreading center; (2) two lithospheric plates move toward each other at a convergent boundary; and (3) two plates slide horizontally past each other at a transform boundary.
- Volcanoes, earthquakes, and mountain building occur near plate boundaries. Interior parts of lithospheric plates are tectonically stable. Tectonic plates move horizontally at rates that vary from 1 to 16 centimeters per year. Plate movements carry continents across the globe, cause ocean basins to open and close, and affect climate and the distribution of plants and animals.
- Mantle convection and movement of lithospheric plates can occur because the mantle is hot, plastic, and capable of flowing. The entire mantle, from the top of the core to the base of the lithosphere, convects in huge cells. Horizontally moving tectonic plates are the uppermost portions of convection cells. Convection occurs because (1) the mantle is hottest near its base, (2) new lithosphere slides downslope away from a spreading center, and (3) the cold leading edge of a plate sinks into the mantle and drags the rest of the plate along.
- Supercontinents may assemble, split apart, and reassemble every few hundred million years.
- The concept that the lithosphere floats on the asthenosphere is called isostasy. When weight, such as a glacier, is added to or removed from Earth's surface, the lithosphere sinks or rises. This vertical movement in response to changing burdens is called isostatic adjustment.
- The movements of tectonic plates generate volcanic eruptions and earthquakes, which help shape Earth's surface. They also build mountain ranges and change the global distributions of continents and oceans. Tectonic activities strongly affect our environment in other ways—impacting global and regional climate, the atmosphere, the hydrosphere, and the biosphere.
- The tectonic movements of the continents and the opening and closing of ocean basins through rifting and subduction profoundly alter Earth's oceanic and atmospheric systems. Ocean currents carry warm water from the equator toward the poles, and cool water from polar regions toward the equator, warming polar regions and cooling the tropics. Similarly, winds transport heat and moisture over the globe. Changes in ocean currents and wind patterns, in turn, have far-reaching consequences for regional climate.

Important Terms

Antarctic Circumpolar Current (p. 142)

asthenosphere (p. 126)

continental drift (p. 122)

continental rifting (p. 133)

convection (p. 136)

convergent boundary (p. 131)

divergent boundary (p. 131)

Gondwana (p. 122)

hot spot (p. 138)

isostasy (p. 139)

Laurasia (p. 122)

magnetic reversal (p. 130)

mantle plume (p. 138)

Mid-Atlantic Ridge (p. 127)

Mid-Oceanic Ridge system (p. 127)

normal magnetic polarity (p. 129)

Pangea (p. 122)

plate boundary (p. 131)

plate tectonics (p. 120)

reversed magnetic polarity (p. 129)

scarp (p. 141)

seafloor spreading (p. 130)

subduction zone (p. 135)

subduction (p. 135)

supercontinent (p. 138)

transform boundary (p. 131)

Review Questions

1. Briefly describe Alfred Wegener's theory of continental drift. What evidence supported his ideas? How does Wegener's theory differ from the modern theory of plate tectonics?
2. Briefly describe the seafloor spreading hypothesis and the evidence used to develop the hypothesis. How does this idea differ from Wegener's theory and the theory of plate tectonics?
3. Draw a cross-sectional view of Earth. List all the major layers and the thickness of each.
4. Describe the physical properties of each of Earth's layers.
5. Describe and explain the important differences between the lithosphere and the asthenosphere.
6. What properties of the asthenosphere allow the lithospheric plates to glide over it?
7. Describe some important differences between the crust and the mantle.
8. Describe some important differences between oceanic crust and continental crust.

Kyodo News/Getty Images

The magnitude-9.0 earthquake that struck northern Japan on March 11, 2011, also triggered a huge tsunami. This photograph shows a giant whirlpool that was produced as the tsunami water drained away from the port of Oarai City, Ibaraki Prefecture. The boat is estimated to be over 20-feet long.

EARTHQUAKES AND THE EARTH'S STRUCTURE

7

INTRODUCTION

As we learned in Chapter 6, the Earth's tectonic plates slide slowly over the soft asthenosphere, about as fast as your fingernails grow. Friction often holds the two sides of a plate boundary locked together. Rocks stretch and compress across the boundary as forces build, but otherwise nothing big happens for long periods of time that can range from decades to multiple millennia. Then, suddenly, the two sides of the boundary snap free, releasing all at once the energy that had built up over time and causing an earthquake.

An earthquake is a classic example of a threshold effect. The interiors of plates move at a steady rate, but motion is constrained at plate boundaries, causing rocks there to slowly accumulate energy from ongoing plate movements. Once the threshold level of energy has accumulated, it is suddenly and cataclysmically released as an earthquake.

ANATOMY OF AN EARTHQUAKE

Rock appears rigid, but if you apply enough stress, rock will deform. When stress is applied to a rock, the rock can deform in one of three ways: (1) elastically, (2) by fracturing, or (3) plastically.

Under small amounts of stress, the rock deforms elastically. If the stress is removed, the rock returns to its original size and shape. A rubber band deforms elastically when you stretch it. The energy used to stretch the rubber band is stored in the elongated rubber. When the stress is removed, it springs back and releases the stored energy. In the same way, an elastically deformed rock will spring back to its original shape and release its stored elastic energy when the force is removed.

However, like a rubber band, every rock has a limit beyond which it cannot deform elastically. Under certain conditions, an elastically deformed rock may suddenly fracture. When large masses of rock in Earth's crust deform elastically and then fracture, vibrations are formed that travel through Earth and are felt as an earthquake.

Under other conditions, when its elastic limit is exceeded, a rock continues to deform, like the bending of a steel nail. This behavior is called **plastic deformation**. A rock that has deformed plastically keeps its new shape when the stress is released and, consequently, does not store the energy used to deform it. Therefore, earthquakes do not occur when rocks deform plastically. ●**FIGURE 7.1** shows an example of asphalt pavement that fractured during an earthquake—the

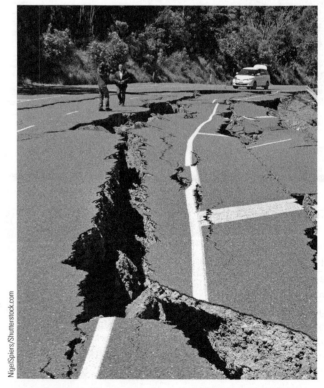

● **FIGURE 7.1** Ground shaking severely damaged this road in Christchurch, New Zealand, during an earthquake that occurred there on February 22, 2011.

NigelSpiers/Shutterstock.com

● **FIGURE 7.2** This rock deformed plastically when stressed.

elastic limit of the pavement was exceeded so it suddenly ruptured. In contrast, ●**FIGURE 7.2** shows a rock outcrop that was deformed plastically by forces deep within Earth's crust. Parallel rock layers were bent and twisted into an S-shaped fold over time while the rocks were hot and under pressure.

We learned in Chapter 6 that Earth's lithosphere is broken into seven large and several smaller tectonic plates. Although tectonic plates move at rates between 1 and 16 centimeters per year, friction prevents the plates from slipping past one another continuously. For time frames ranging from decades to millennia, the cool, strong rock near the plate boundary can stretch or compress elastically, while the edges remain locked and immobile. This elastic deformation strains the rock, causing potential energy to build, just like the stretched rubber band or a compressed spring. Then, when the strain reaches a critical threshold, rock suddenly fractures and movement occurs across the fracture, releasing all at once the energy that had built up previously. The ground rises and falls and undulates back and forth. Buildings topple, bridges fall, roadways break, and pipelines snap. An **earthquake** is a sudden motion or trembling of Earth caused by the abrupt release of energy that is stored in rocks.

When a rock fracture ruptures during an earthquake and rock on the two sides of the fracture move relative to one another, the fracture becomes a fault. Most movement of crustal rock occurs due to slippage along established faults because the friction binding the two sides of the fault is weaker than the rock itself (●**FIGURE 7.3**). As additional tectonic stress then builds and strain begins to accumulate in the rock again, it is more likely to move along the existing fault than to crack

somewhere else and create a new fault. However, new faults can form if the orientation and curvature of the existing faults no longer can most efficiently release stress that builds in the rock. Many faults develop as part of a larger group of faults that together release much of the strain that accumulates in a region. For example, thousands of separate normal faults exist today in the North American Basin and Range province, introduced in Chapter 6. Many of these faults form the geologic boundary separating each mountain range from its adjacent basin. The faults develop through time as the crust extends, tilting large blocks of it. As extension continues, the

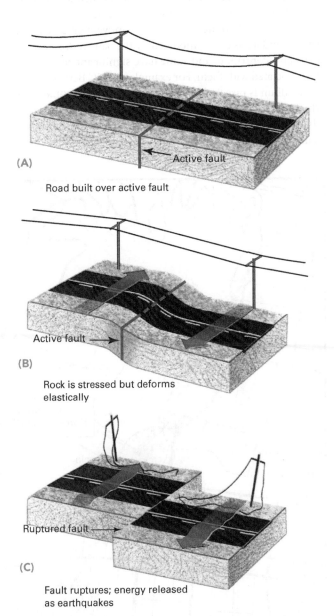

● **FIGURE 7.3** (A) A road is built across an old fault. (B) The rock stores elastic energy when it is stressed by a tectonic force. The stress causes deformation, called strain, to build. (C) At a critical point, the rock fractures, releasing the strain and returning the rock to its unstressed shape on either side of the fault.

---

plastic deformation Deformation that occurs without fracture after a rock's elastic limit is reached, and it continues to deform, like putty, while still solid. The rock keeps its new shape and does not store the energy used to deform it; thus, earthquakes do not occur when rocks deform plastically.

earthquake A sudden motion or trembling of Earth caused by the abrupt release of slowly accumulated elastic energy in rocks.

curved geometry of the first faults to develop makes them no longer the easiest way for the rocks to continue to accommodate the stretching—new faults develop with different orientations and the old faults no longer are active.

Although individual faults become active and inactive over geologic timescales, over historical timescales once a fault is recognized as active, it is presumed to remain so. Many historic earthquakes occur along established faults as strain is suddenly released. For example, the San Andreas Fault in Southern California lies along a tectonic plate boundary that has slipped many times in the past and will certainly again in the future (●**FIGURE 7.4**). The boundary includes numerous individual faults, some of which are active and some of which are not. Although much of the movement across the plate boundary is accommodated by the San Andreas, other faults also contribute and some have significant earthquake risk associated with them. For example, the UC Berkeley football stadium is bisected by the Hayward Fault (●**FIGURE 7.5**), which in places is slipping a rate of 5 millimeters per year. The Hayward Fault last generated an earthquake with estimated 7.0 magnitude in 1868 and is considered by the USGS

R. T. Wallace/USGS

● **FIGURE 7.4** California's San Andreas Fault, the source of many earthquakes, is the boundary between the Pacific Plate, on the left in this photo, and the North American Plate, on the right.

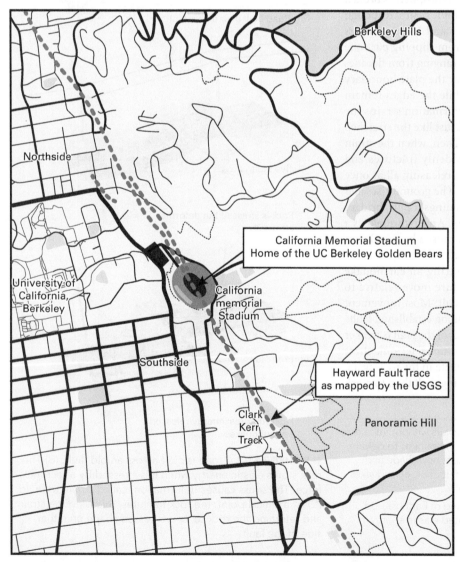

● **FIGURE 7.5** The Hayward Fault is an active fault in the Bay Area and has been mostly covered by urban infrastructure. The football stadium for the UC Berkeley Golden Bears is bisected by the fault, for example, as shown by this map of the fault trace by the USGS.

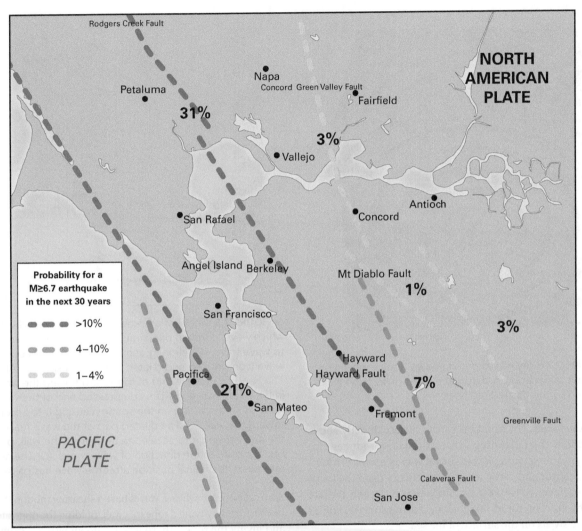

FIGURE 7.6 The San Francisco Bay Area straddles the boundary between the North American and Pacific Plates and is crisscrossed with individual faults that together accommodate the total slip across the plate boundary. The USGS has estimated the likelihood of each fault producing a magnitude 7.0 or greater earthquake within the next 30 years, shown above. The fault with the most risk in the Bay Area is the Hayward Fault (31%), with the San Andreas Fault second at 21%.

to be the most likely to generate a magnitude 7.0 or greater earthquake within the next 30 years (31% probability). For comparison, the San Andreas Fault, located across the San Francisco Bay Area from the Hayward Fault, is estimated to have only a 21% chance of generating a magnitude 7.0 earthquake over the same timeframe (●FIGURE 7.6).

During an earthquake, rock moves from a few centimeters to several meters. A single earthquake makes only small changes in the topography or geography of a region. But over tens of millions of years, the relentless motion of plates, accompanied by hundreds of thousands of earthquakes, changes the surface of the planet. In this way, mountains are raised, continents are split apart, and new ocean basins are formed.

---

seismic waves An elastic wave that travels through rock, produced by an earthquake or explosion.

seismology The study of earthquakes and the nature of Earth's interior based on evidence from seismic waves.

EARTHQUAKE WAVES

When buying a watermelon, one trick is to tap the melon gently with your knuckle. If you hear a sharp, clean sound, it is probably ripe; a dull thud indicates that it may be over-ripe and mushy. The watermelon illustrates two points that can be applied to Earth: (1) the energy of your tap travels through the melon, and (2) the nature of the melon's interior affects the quality of the sound.

A *wave* transmits energy from one place to another. A drumbeat travels through air as a sequence of waves, the Sun's heat travels to Earth as waves, and a tap travels through a watermelon in waves. Waves that travel through rock are called **seismic waves**. Earthquakes and explosions produce seismic waves. **Seismology** is the study of earthquakes and the nature of Earth's interior based on evidence from seismic waves.

The initial rupture point, where abrupt movement creates an earthquake, typically lies below the surface at a point

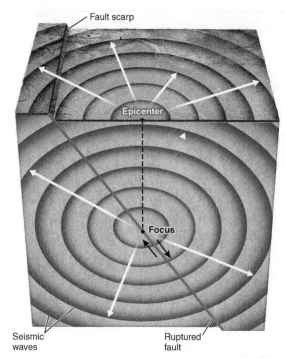

● **FIGURE 7.7** Body waves generated by abrupt movement on a fault radiate outward from the focus of an earthquake, as shown by the dashed red lines.

called the **focus**. The point on Earth's surface directly above the focus is the **epicenter**. An earthquake produces two main types of seismic waves: **Body waves** travel through Earth's interior and carry some of the energy from the focus to the surface (●**FIGURE 7.7**). **Surface waves** then radiate from the epicenter and travel along Earth's surface, somewhat like swells on the surface of the sea.

Body Waves

Body waves can be of two main types. **P waves** travel fast and are the first or "primary" seismic waves to reach an observer. They are compressional elastic waves that cause the rock to undergo alternating compression and expansion (●**FIGURE 7.8**). Consider a long spring such as the popular Slinky toy, with one end attached to a wall. If you stretch and then rapidly push the end of the Slinky, a compressional wave travels back and forth along its length. Each segment of the spring is compressed and then stretched as the wave passes by, so an ant sitting on the Slinky would move toward, then away, from the wall. Thus, the wave vibrates in the direction in which it propagates. P waves travel through air, liquid, and solid material. Next time you take a bath, immerse your head until your ears are under water and listen as you tap the sides of the tub. You are hearing P waves.

P waves travel at speeds between 4 and 7 kilometers per second in Earth's crust and about 8 kilometers per second in the uppermost mantle. For comparison, the speed of sound in air is only 0.34 kilometer per second, and the fastest jet fighters fly about 0.96 kilometer per second.

S waves, the other type of body waves, are slower than P waves and thus are the "secondary" waves to reach an observer.

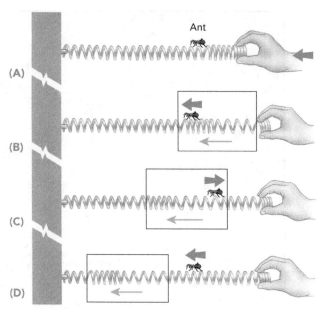

● **FIGURE 7.8** Model of a P wave (compressional wave) in which wave vibration direction (orange arrows) is parallel to wave travel direction (green arrows). Black box outlines wave (compressed and dilated portions of spring). (A) The hand pushes the end of the spring toward the wall, initiating the wave. (B) The compressed end of the wave reaches an ant sitting on the spring, causing it to accelerate toward the wall. (C) The dilated part of the wave reaches the ant, causing it to accelerate away from the wall, parallel but opposite to the direction of wave travel. (D) The ant returns to its original position after the wave has passed by.

Also called *shear waves*, S waves have a shearing motion that can be illustrated by tying a rope to a wall, holding the opposite end, and giving the rope a sharp up-and-down jerk (●**FIGURE 7.9**). Although the wave travels or propagates parallel to the rope, each segment of the rope moves at right angles to the rope length, so an ant sitting on the rope would move up and down as the wave passes by. A similar motion in an S wave produces shear stress in a rock—stress that acts in parallel but opposite directions—and gives the wave its name. S waves vibrate at right angles to the direction in which the wave is moving and travel between 3 and 4 kilometers per second in the crust.

Unlike P waves, S waves move only through solids. Because molecules in liquids and gases are only weakly bound to one another, they slip past each other and thus cannot transmit a shear wave.

Surface Waves

Surface waves travel more slowly than body waves. Surface waves undulate across the ground like the waves that ripple across the water when you throw a rock into a calm lake (although the actual wave mechanism is different). Two types of surface waves occur simultaneously: an up-and-down rolling motion called *Rayleigh waves* and a side-to-side vibration called *Love waves*, both named after the scientist who first described each wave type. During an earthquake, Earth's surface rolls like ocean waves and writhes from side to side like a snake (●**FIGURE 7.10**).

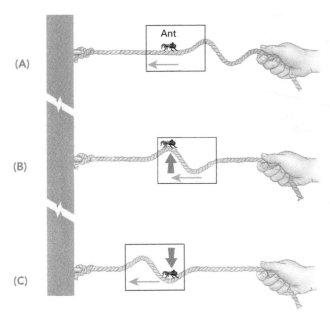

FIGURE 7.9 Model of an S wave (shear wave) in which wave vibration direction (orange arrows) is perpendicular to wave travel direction (green arrows). Blue box outlines wave. (A) Hand initiates shear wave by single up-and-down movement. (B) Front end of wave reaches ant, causing it to accelerate upward. (C) Back end of wave reaches ant, causing downward acceleration.

---

focus The initial rupture point of an earthquake, typically located below Earth's surface.

epicenter The point on Earth's surface directly above the initial rupture point (focus) of an earthquake.

body waves Seismic waves that travel through the interior of Earth, carrying energy from the earthquake's focus to the surface.

surface waves Seismic waves that radiate from the earthquake's epicenter and travel along the surface of Earth or along a boundary between layers within Earth.

P waves Body waves that travel faster than other seismic waves and are the first or "primary" waves to reach an observer; formed by alternating compression and expansion of rock parallel to the direction of wave travel.

S waves Seismic body waves that travel slower than P waves and are the "secondary" waves to reach an observer; sometimes called *shear waves* due to the shearing motion in rock caused as the waves vibrate perpendicular to the direction they travel.

seismograph An instrument that records seismic waves.

seismogram The physical or digital record of earthquake waves as measured by a seismograph.

Mercalli scale A scale of earthquake intensity that expresses the strength of an earthquake based on its destructive power and its effects on buildings and people; does not accurately measure the energy released by the quake.

Tōhoku earthquake An earthquake of moment-magnitude 9.0, the fourth largest ever recorded, that occurred on March 11, 2011, off the Japanese island of Honshu; produced a gigantic tsunami that contributed to the deaths of about 17,600 people and led directly to the third-worst nuclear accident in history.

Measurement of Seismic Waves

A **seismograph** is a device that records seismic waves. In early seismographs, a heavy weight was suspended from a spring. A pen attached to the weight was aimed at the zero mark on a piece of graph paper (●**FIGURE 7.11**). The graph paper was mounted on a rotating drum that was attached firmly to bedrock. During an earthquake, the graph paper jiggled up and down, but inertia kept the weight and its pen stationary. As a result, the paper moved up and down beneath the pen, while the rotating drum recorded this earthquake motion over time. This physical "written" record of Earth vibration is called a **seismogram** (●**FIGURE 7.12**). Modern seismographs use a coil fixed to a pendulum that is suspended in a magnetic field. When an earthquake causes the pendulum to move within the magnetic field, it creates an electric current that is amplified and recorded by a computer.

Today, the computer recording seismic wave information is part of the World Wide Web where it is instantly available to seismologists. The USGS hosts a large earthquake monitoring network, and information from thousands of deployed seismographs around the world is available as soon as it happens. Basic information provided on the monitoring network includes location, magnitude, and depth of each earthquake. For example, within the hour before this sentence was written, a total of 15 earthquakes were recorded globally; ten were on the island of Hawaii and probably related to the ongoing eruptions there.

Seismologists are able to use information from the worldwide network of seismographs to study the details of specific earthquakes, especially the largest ones. One such study is called finite fault modeling, which provide a description of the dynamics of the faulting event that caused the earthquake. Included in the finite fault model are estimates of the total slip distance that occurred on the fault during the earthquake, how much surface area on the fault was involved, and how long the slipping lasted. The USGS has finite fault model results for all of the largest earthquakes to occur over the past several years. ●**FIGURE 7.13** shows an example of the finite fault model for the devastating March 9, 2011 **Tōhoku earthquake** in Japan. In addition to information available directly through the USGS Earthquake Hazards Program, many excellent third-party web platforms for viewing and sharing maps and information about the most recent earthquakes are available.

Measurement of Earthquake Strength

Over the past century, geologists have devised several scales to express the strength of an earthquake. Before seismographs became widespread, geologists evaluated earthquakes on the **Mercalli scale**, which measures the intensity of an earthquake and is based on structural damage. On the Mercalli scale, an earthquake that destroyed many buildings was rated as more intense than one that destroyed only a few.

However, this system did not accurately measure the energy released by a quake because structural damage also depends on distance from the focus, the rock or soil beneath the structure, and the quality of construction. In 1935,

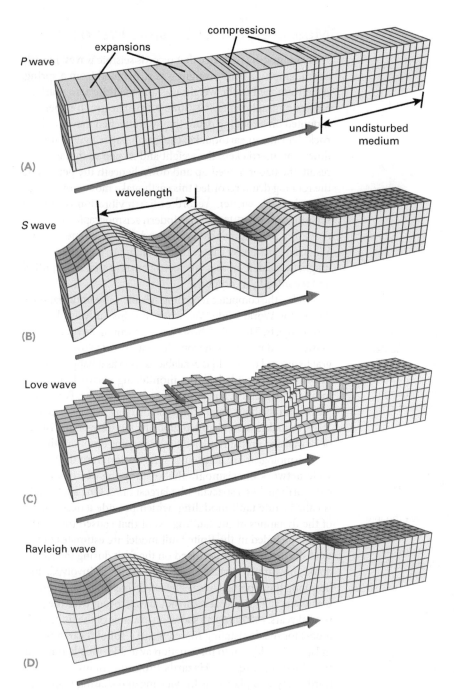

FIGURE 7.10 During an earthquake, both body waves and surface waves are generated. This diagram shows how each wave type affects the surface of the Earth. P waves compress and dilate the rocks in the direction of travel. S waves shear the rock up and down, perpendicular to the direction of travel. Love waves cause lateral (back and forth) shear that is restricted to the region near the ground surface. Rayleigh waves also compress and dilate the near surface while shearing it perpendicular to the direction of travel.

Charles Richter devised the **Richter scale** to express the amount of energy released during an earthquake. Richter magnitude is calculated from the height of the largest earthquake body wave recorded on a specific type of seismograph. The Richter scale is more quantitative than earlier intensity scales, but it is not a precise measure of earthquake energy. A sharp, quick jolt would register as a high peak on a Richter seismograph, but a very large earthquake can shake the ground for a long time without generating extremely high peaks. Thus, a great earthquake can release a huge amount of energy that is not reflected in the height of a single peak and that is not adequately expressed by Richter magnitude.

Modern equipment and methods enable seismologists to measure the amount of movement and the surface area of a fault that moved during a quake. The product of these two values and the shear strength of the faulted rock allow them to calculate the **moment magnitude**. Most seismologists now use moment magnitude rather than Richter magnitude, because it more closely reflects the total amount of energy released during an earthquake. An earthquake with a moment magnitude of 6.5 has an energy of about 10^{25}

Richter scale A scale of earthquake magnitude that expresses the amount of energy released; calculated from the amplitude of the largest body wave on a standardized seismograph, although not a precise measure of earthquake energy.

moment magnitude An earthquake scale in which the surface area of fault movement, the offset produced on the fault, and a measure of rock strength are multiplied together; moment magnitude scale closely reflects the total amount of energy released.

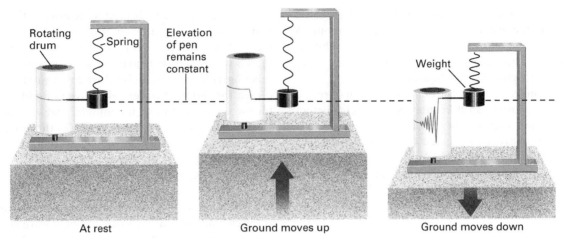

● FIGURE 7.11 A seismograph records ground motion during an earthquake. In early seismographs, the pen draws a straight line across the rotating drum when the ground is stationary. When the ground rises abruptly during an earthquake, it carries the drum up with it. But the spring stretches, so the weight and pen hardly move. Therefore, the pen marks a line lower on the drum. Conversely, when the ground sinks, the pen marks a line higher on the drum. During an earthquake, the pen traces a jagged line as the drum rises and falls.

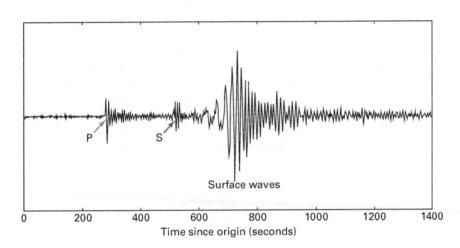

● FIGURE 7.12 This seismogram was recorded at a station in Belarus from a magnitude-5.0 earthquake in the North Atlantic. The arrows mark the first arrival of P waves (red) and S waves (blue).

● FIGURE 7.13 This figure shows finite fault modeling results for the March 9, 2011, 9.0-magnitude Tōhoku earthquake in Japan. The colors in the image correspond to the total slip distance for the fault that ruptured to cause the earthquake. As shown by the color key at the top of the figure, the warmest colors correspond to over 50 meters of slip during the earthquake. With this in mind, consider the horizontal and vertical scales represented by the cross section, which suggest that the cross-sectional area of the fault surface that moved measured tens of kilometers both laterally and vertically.

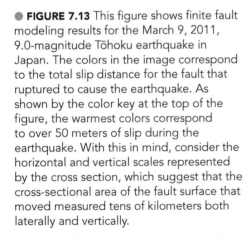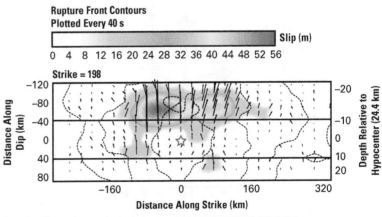

Source: https://earthquake.usgs.gov/earthquakes/eventpage/official20110311054624120_30#finite-fault

(1 followed by 26 zeros) ergs.[1] The atomic bomb that was dropped on the Japanese city of Hiroshima at the end of World War II released about that much energy. (Of course, atomic bombs also release intense heat and deadly radioactivity, while earthquakes only shake the ground.)

1. An erg is a standard unit of energy in scientific usage. One erg is a small amount of energy and is approximately equivalent to the energy released by a bug with a mass of 2 grams crawling at about 1 cm/sec; approximately 3×10^{12} ergs are needed to light a 100-watt lightbulb for 1 hour.

On both the moment magnitude and the Richter scales, the energy of the quake increases by a factor of about 30 for each whole number increment on the scale. Thus, a magnitude-6 earthquake releases roughly 30 times more energy than a magnitude-5 earthquake.

Moment magnitude depends on the strength of rocks, because a strong rock can store more elastic energy before it fractures than a weak rock can. The two largest earthquakes ever measured occurred in Chile in 1960 and Alaska in 1964. These quakes had moment magnitudes of 9.6 and 9.2—roughly 25,000 times and 7,500 times, respectively, greater than the energy released by the Hiroshima bomb.

Locating the Source of an Earthquake

If you have ever watched an electrical storm, you may have used a simple technique for estimating the distance between you and the place where the lightning strikes. After the flash of a lightning bolt, we count the seconds that pass before we hear the thunder. Although the electrical discharge produces thunder and lightning simultaneously, light travels much faster than sound. Light reaches us virtually instantaneously, whereas sound travels much more slowly through the atmosphere, at 340 meters per second. If the time interval between the lightning flash and the thunder is 1 second, then the lightning struck 340 meters away and was very close.

The same principle is used to determine the distance from a recording station to both the epicenter and the focus of an earthquake. Recall that P waves travel faster than S waves and that surface waves are slower yet. If a seismograph happens to be close to an earthquake epicenter, the different waves will arrive in rapid succession for the same reason that the thunder and lightning come close together when a storm is close. On the other hand, if a seismograph is located far from the epicenter, the S waves arrive after the P waves arrive, and the surface waves are even farther behind.

Geologists use a **travel-time curve** to calculate the distance between an earthquake epicenter and a seismograph. A travel-time curve is constructed using P-wave and S-wave arrival times from multiple stations at different locations for an earthquake with a known epicenter and occurrence time. The resulting graph can then be used to measure the distance between a recording station and an earthquake whose epicenter is unknown by plotting the time delay between the two wave types on the travel-time curve.

Travel-time curves were first constructed using data obtained from natural earthquakes. However, scientists do not always know precisely when and where an earthquake occurs. In the 1950s and 1960s, geologists studied seismic waves from atomic bomb tests to improve the travel-time curves, because they could verify both the location and timing of the explosions.

●**FIGURE 7.14** illustrates how the epicenter of an earthquake is located. Using seismograms from three or more

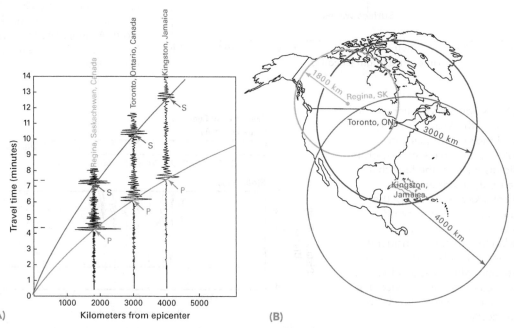

(A)   (B)

● **FIGURE 7.14** How earthquake locations are determined. (A) The distance from a seismic station to an earthquake is determined by using a travel-time curve. Seismograms from two stations in Canada and one in Jamaica are fit to the travel-time curve so that the first arrival of P waves (red arrows) and S waves (blue arrows) match the two curves on the graph. The distance to the earthquake from each seismic station is read off the x-axis of the graph. Note that the surface waves in these three seismograms have been removed for simplicity. (B) A circle centered on each of the three seismic stations is drawn on a map. The radius of each circle is the map distance to the earthquake from that station. The circles all intersect at the location of the earthquake epicenter, in this case near Las Vegas, Nevada.

seismic recording stations, a seismologist first plots the first arrival times for P and S waves on a travel-time curve. The corresponding value on the x-axis of the graph is the distance between the epicenter and each seismic station. The seismologist then plots a circle around each of the seismic station locations on a map. The radius of each circle is the linear distance to the epicenter from that station. The one point on the map where the three (or more) circles intersect is the location of the earthquake's epicenter.

---

travel-time curve A graph in which first arrival times of P and S earthquake waves are plotted against distance from epicenter; the separation distance between the resulting two curves determines the distance between the seismograph location and the epicenter.

San Andreas Fault zone A zone of strike-slip faults extending from Cape Mendicino in northern California to the northern Gulf of California in Mexico; fault zone forms the transform boundary between the Pacific Plate and the North American Plate and is the source of many earthquakes.

EARTHQUAKES AND TECTONIC PLATE BOUNDARIES

Before the theory of plate tectonics was developed, geologists recognized that earthquakes occur frequently in some regions and infrequently in others, but they did not understand why. Modern geologists know that most earthquakes occur along plate boundaries, where tectonic plates diverge, converge, or slip past one another (●FIGURE 7.15).

Earthquakes at a Transform Plate Boundary: The San Andreas Fault Zone

The populous region from San Francisco southward to San Diego straddles the **San Andreas Fault Zone**, which is a transform boundary between the Pacific Plate and the North American Plate. The fault zone, which ranges in width from <1 kilometer to several tens of kilometers, is made up of numerous individual faults, although almost all of the slip between the two plates occurs on only a handful of major

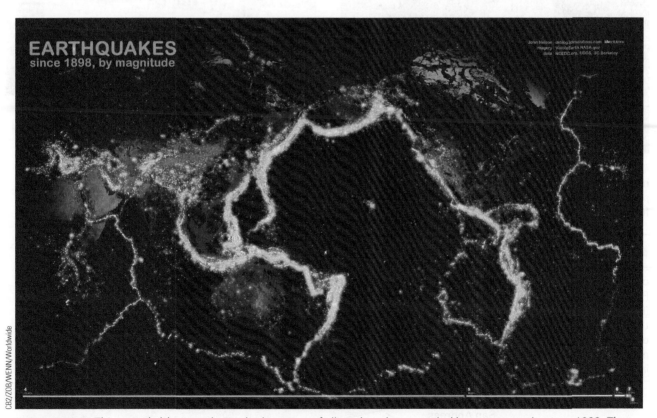

CBZ/ZOB/WENN/Worldwide

● FIGURE 7.15 This remarkable map shows the locations of all earthquakes recorded by seismographs since 1898. The earthquakes are shown as glowing dots, and the size of the dot corresponds to the magnitude of the earthquake. By far, the largest and most frequent earthquakes occur at convergent plate boundaries around the eastern, northern, and western margins of the Pacific Ocean. Frequent but smaller-magnitude earthquakes define the mid-oceanic divergent plate boundaries in the Atlantic, eastern and southern Pacific, and Indian Oceans. Also well defined are the East African Rift system and the Alpine-Himalayan mountain belt, in which a continent–continent collision is occurring.

faults. Of these, the San Andreas Fault is the biggest, accommodating up to 90% of total slip between the two plates in some places. The San Andreas Fault itself is vertical, but the rocks on opposite sides move horizontally. A fault of this type is called a **strike-slip fault** (●FIGURE 7.16). Plate motion stresses rock adjacent to the fault, and numerous smaller faults are generated as elastic energy accumulates and is released (●FIGURE 7.17). Thus, the San Andreas Fault zone comprises the San Andreas Fault itself and the smaller faults around it that distribute the strain across the broader zone. In the past few centuries, hundreds of thousands of earthquakes have occurred in this zone, including the infamous 1906 San Francisco earthquake (●FIGURE 7.18A and 7.18B).

(A)

(B)

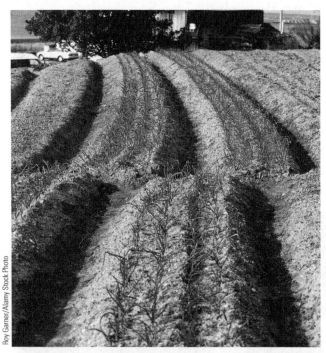

● **FIGURE 7.16** Rocks on opposite sides of a strike-slip fault move horizontally. In this photograph from Japan, a strike-slip fault that ruptured during an earthquake has displaced furrows in a field of spring onions.

● **FIGURE 7.18** (A) The 1906 earthquake and fire destroyed most of San Francisco. (B) Earthquake irony: This statue of the famous geologist Louis Agassiz fell about 9 meters (30 feet), landing head first and piercing a concrete sidewalk on the campus of Stanford University during the 1906 San Francisco earthquake.

● **FIGURE 7.17** An earthquake hazard map of the San Andreas Fault zone of California. (Percentages indicate the probability of an earthquake with a magnitude greater than 5 in the next 30 years.)

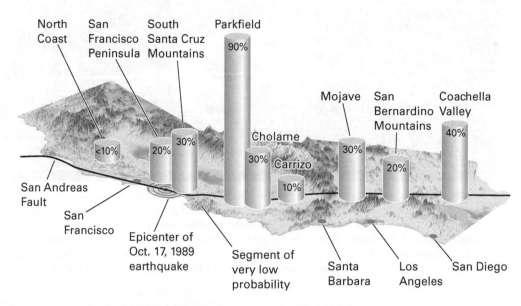

The Pacific and North American Plates move past one another in three different ways along different segments of the San Andreas Fault zone:

1. Along some portions of the fault zone, rocks slip past one another at a continuous, snail-like pace referred to as **fault creep**. This movement occurs without violent and destructive earthquakes because the rocks slip continuously and slowly.

2. In other segments of the fault, the plates pass one another in a series of small hops, causing numerous small, mostly nondamaging earthquakes.

3. Along the remaining portions of the fault, friction prevents slippage across the fault, although the Pacific and North American Plates continue to move relative to one another. In this case, rock across the locked fault segment deforms and stores elastic energy. Because the plates move past one another at an average of 3.5 centimeters per year, energy corresponding to 3.5 meters of elastic deformation accumulates over a period of 100 years. When the accumulated elastic energy exceeds the frictional resistance across the fault, the fault ruptures, producing a large, destructive earthquake.

Plates move slowly and silently every day. Rocks stretch, bend, and compress, unseen beneath our feet. Geologists understand the mechanisms of earthquakes and know for certain that more destructive quakes will occur along the San Andreas Fault. But no one knows when or where the "Big One" will strike.

Earthquakes at Convergent Plate Boundaries

Recall from Chapter 6 that in a subduction zone, a relatively cold, rigid lithospheric plate dives beneath another plate and slowly sinks into the mantle. In most places, the subducting plate slips past the plate above it with intermittent jerks, giving rise to numerous earthquakes. The earthquakes occur mostly along the upper part of the sinking plate, where it scrapes past the opposing plate. Additional earthquakes occur within the downward-moving slab as it is bent and extended upon being pulled into the mantle (●**FIGURE 7.19**). Collectively, these earthquakes form the **Benioff zone**, named after Hugo Benioff, the geologist who first recognized it. Many of the world's strongest earthquakes occur in subduction zones, including all five of the largest recorded earthquakes.

When two converging plates both carry continents, neither can sink deeply into the mantle. Instead, the crust buckles up to form high mountains, such as the Himalayas. Rocks fracture or slip during this buckling process, generating frequent earthquakes. The 2005 earthquake along the

---

strike-slip fault A vertical fault across which rocks on opposite sides move horizontally.

fault creep A continuous, slow movement of solid rock along a fault, resulting from a constant stress acting over a long time. Creeping faults do not usually have large earthquakes.

Benioff zone A three-dimensional zone of earthquake foci within and along the upper portion of a subducting plate; formed by release of strain as the subducting plate scrapes past the overriding plate.

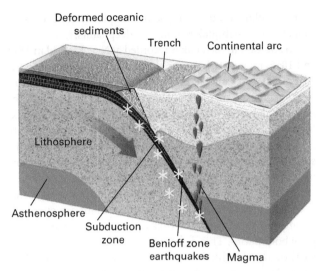

● **FIGURE 7.19** A descending lithospheric plate generates magma and earthquakes in a subduction zone. Earthquake foci, shown by stars, concentrate mostly along the upper portion of the subducting plate, called the Benioff zone.

India–Pakistan border was generated by the slow convergence of the subcontinent of India with Asia.

Earthquakes at Divergent Plate Boundaries

Earthquakes frequently shake the Mid-Oceanic Ridge system as a result of faults that form as the two plates separate. Blocks of oceanic crust drop downward along most mid-oceanic ridges, forming a rift valley in the center of the ridge. Only shallow earthquakes occur along the Mid-Oceanic Ridge, because the asthenosphere there rises to depths as shallow as 10 to 15 kilometers and is too hot and plastic to fracture.

Earthquakes in Plate Interiors

No major earthquakes have occurred in the central or eastern United States in the past 100 years, and no current lithospheric plate boundaries are known in these regions. However, the largest historical earthquake sequence in the contiguous 48 states occurred near New Madrid, Missouri. In 1811 and 1812, three shocks with estimated moment magnitudes between 7.3 and 7.8 altered the course of the Mississippi River and rang church bells 1,500 kilometers away in Washington, D.C. These earthquakes occurred within a zone of active extensional tectonics called the New Madrid Fault Zone and located in the lower Mississippi River Valley region.

Earthquakes in plate interiors are not as well understood as those at plate boundaries, but modern research is revealing some clues. Tectonic forces stretched North America in Precambrian time. As the continent pulled apart, rock fractured to create two huge fault zones that crisscross the continent like a giant *X*. Although the fault zones failed to develop into a divergent plate boundary, they remain weaknesses in the lithosphere. New Madrid lies at a major intersection of the faults—at the center of the *X*. As the North American Plate slides over the asthenosphere, it may pass over irregularities, or "bumps," in that plastic zone, causing slippage and earthquakes along the deep faults.

In addition to the quakes recorded at New Madrid in 1811 and 1812, major earthquakes also occurred there close to the years 1600, 1300, and 900. The evidence for these older earthquakes comes from a series of sand volcanoes produced by the quakes. The sand volcanoes formed when the ground shaking caused loose, saturated sand in the shallow subsurface to become pressurized and shoot violently upward through fissures in the ground. As it was ejected, the sand incorporated pieces of wood and pottery that were later dated.

In 2005, researchers reported in the journal *Nature* that the rocks adjacent to the New Madrid Fault are currently deforming about as fast as rocks adjacent to active tectonic plate boundaries. This deformation indicates that additional earthquakes are likely there in the future, and the USGS has assigned a high risk of future seismicity to the region as shown in ●**FIGURE 7.20**.[2]

2. R. Smalley Jr., M. A. Ellis, J. Paul, and R. B. Van Arsdale, "Space Geodetic Evidence for Rapid Strain Rates in the New Madrid Seismic Zone of Central USA," *Nature* 435 (June 23, 2005): 1088–1090.

EARTHQUAKE DAMAGE AND HAZARD MITIGATION

Violent Earth motion may toss a person to the ground and break the person's arm, but this motion, by itself, is seldom lethal. Most earthquake fatalities and injuries occur when falling structures crush people. Structural damage, injury, and death depend on the magnitude of the quake, its proximity to population centers, rock and soil types, topography, and the quality of construction in the region.

How Rock and Soil Influence Earthquake Damage

In many regions, bedrock lies at or near Earth's surface and buildings are anchored directly to the rock. Bedrock vibrates during an earthquake, and buildings may fail if the motion is violent enough. However, most bedrock returns to its original shape when the earthquake is

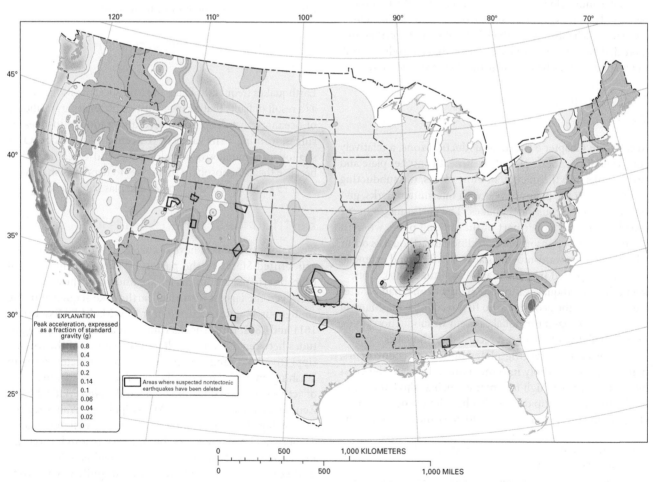

● **FIGURE 7.20** This map shows earthquake hazards in the United States as of 2014, the most recent long-term risk assessment was conducted by the USGS. The predictions are based on records of frequency and Mercalli magnitudes of historical earthquakes. Peak acceleration refers to how fast the ground surface would move up and down during an earthquake as a fraction of 1 g—the gravitational acceleration at the Earth's surface. Notice the polygon in central Oklahoma where earthquakes are suspected to be mostly related to deep injection of waste water as discussed in Chapter 5.

over, so if structures can withstand the shaking, they will survive. Thus, bedrock forms a desirable foundation in earthquake-prone areas.

In many places, structures are built on sand, silt, or clay. Sandy sediment and soil commonly settle during an earthquake. This displacement tilts buildings, breaks pipelines and roadways, and fractures dams. To avert structural failure in such soils, engineers drive steel or concrete pilings through the sand to the bedrock below. These pilings anchor and support the structures, even if the ground beneath them settles.

Mexico City provides one example of what can happen to clay-rich soils during an earthquake. The city is built on a high plateau ringed by even-higher mountains. European settlers built the modern city on water-soaked, clay-rich, sediment deposited in an ancient former lake. On September 19, 1985, an earthquake with a moment magnitude of 8.1 struck about 500 kilometers west of the city. Seismic waves shook the wet clay beneath the city and reflected back and forth between the bedrock sides and bottom of the basin, just as waves in a bowl of jelly bounce off the side and bottom of the bowl. The reflections amplified the waves, which destroyed more than 500 buildings and killed between 8,000 and 10,000 people. Meanwhile, there was comparatively little damage in Acapulco, a city much closer to the epicenter but built on bedrock.

If soil is saturated with water, the sudden shock of an earthquake can cause the grains to shift relative to one another. When this occurs, the pressure in the water filling the spaces between grains, called the pore water, increases. The pore pressure may rise sufficiently to temporarily suspend the grains in the water, so that they are no longer held together by frictional forces. In this case, the soil loses its shear strength—its resistance to shear—and behaves like a fluid in a process called **liquefaction**. A familiar example of liquefaction occurs when you tap your foot on saturated sand at the beach; the forces produced by your tapping cause compression waves to travel into the sand, liquefying it and producing a slurry of sand and water that flows over your toes. When soils liquefy on a hillside due to earthquake shaking, the resulting slurry flows downslope, carrying structures along with it. Landslides are common effects of earthquakes and are discussed further in Chapter 10.

---

liquefaction A geological process in which a soil or sediment loses its shear strength during an earthquake and becomes a fluid.

tsunami A large, destructive sea wave produced by an undersea earthquake or volcano; sometimes erroneously called a *tidal wave*, although it has nothing to do with astronomical tides.

Sumatra-Andaman earthquake An earthquake of moment-magnitude 9.2, the third largest ever recorded; occurred on December 26, 2004, off the Sumatran coast and produced a devastating tsunami that contributed to the deaths of an estimated 286,000 people.

Construction Design and Earthquake Damage

Most earthquake mortality occurs when buildings fall and crush the unfortunate inhabitants. Structure failure is caused by a variety of factors, including soil type and distance from the epicenter. But the tremendous differences in mortality between poor, less-developed countries and rich, developed countries are due to one simple factor: quality of construction. In addition to differences in the ability to withstand shaking associated with earthquakes, some building materials are susceptible to fires caused by rupturing of electrical or gas lines during the earthquake.

Some common framing materials used in buildings, such as wood and steel, bend and sway during an earthquake, but they resist failure. However, brick, stone, adobe (dried mud), and other masonry products are brittle and likely to fail during an earthquake. Although masonry can be reinforced with steel, in many regions of the world people cannot afford such structural reinforcement.

Thus, earthquake mortality is an example of the interface between natural systems and human ones. Forces deep within Earth, beyond human control, drive earthquake frequency and magnitude. But earthquake mortality is closely related to political and economic systems.

Tsunamis

When an earthquake occurs beneath the sea, part of the seafloor rises or falls. Water is displaced in response to the seafloor movement, forming a wave or series of waves. Sea waves produced by an earthquake are often called tidal waves, but they have nothing to do with tides. Therefore, geologists call them by their Japanese name, **tsunami**.

In the open sea, a tsunami is so flat that it is barely detectable. Typically, the crest may be only 1 to 3 meters high, and successive crests may be more than 100 to 150 kilometers apart. However, a tsunami may travel at 750 kilometers per hour. When the wave approaches the shallow water near shore, the base of the wave drags against the bottom and the water stacks up, increasing the height of the wave. The rising wall of water then flows inland. A tsunami can flood the land for as long as 5 to 10 minutes.

Today, the central part of the Indo-Australian Plate is subducting beneath the islands of Sumatra and Java. On December 26, 2004, a fault rupture spanning a distance of 1,500 kilometers—the longest fault rupture every recorded—occurred. Like a giant piece of paper being torn over a period of 10 minutes, the fault rupture propagated from south to north at a velocity of about 1.2 kilometers/second. The resulting magnitude-9.2 **Sumatra-Andaman earthquake** was the third-largest seismic event ever recorded and caused the seafloor to move westward about 6 meters and upward about 2 meters. This tremendous displacement of rock initiated a massive tsunami that radiated in all directions, killing an estimated 286,000 people along the Indian Ocean coastlines. Survivors reported that, moments prior to the deadly wave, coastal water retreated, exposing to the air wet mud in ocean bays. Then the wave raced inward, rearing upward as much as 24 meters, as high as a 10-story building, and inundating some areas 30 meters above sea level. Although mortality was highest in Sumatra, people died in coastal areas as far away as Port Elizabeth, South Africa, 8,000 kilometers from the epicenter.

An even larger tsunami occurred on March 11, 2011, when an earthquake with moment magnitude of 9.0 struck off the eastern coast of Japan. The 2011 Tōhoku earthquake was the largest ever reported from Japan and the fourth largest on Earth since the beginning of seismic instrumentation. The earthquake resulted from fault rupture where the Pacific Plate is subducting beneath the northern Japanese island of Honshu at a rate of about 8 to 9 centimeters per year. The rupture was only about 250 kilometers long, roughly half of that normally needed to produce an earthquake of this magnitude and only one-sixth the length of the rupture that caused the 2004 Sumatran-Andaman earthquake. However, the displacement across the ruptured fault in 2011 was estimated to be over 50 meters in some places—much more than many seismologists expected.

The 2011 Tōhoku earthquake lasted between 3 and 5 minutes and caused northern Japan to lurch about 4 meters closer to North America. This sudden change in the distribution of mass caused the Earth's rotational axis to shift between 10 and 25 centimeters and the Earth to rotate slightly faster, effectively shortening the length of a day by

● **FIGURE 7.21** A large ferry boat rests amid destroyed houses following the giant tsunami that struck the Miyagi Prefecture on March 11, 2011.

1.8 millionths of a second. Ultralow-frequency sound waves from the quake were even detected by satellites.

The Tōhoku earthquake produced a devastating tsunami that began to sweep ashore about an hour after the ground shaking and was estimated to be as much as 39 meters high in places (●**FIGURE 7.21**). The tsunami washed up to 10 kilometers inland, inundating over 550 square kilometers. Entire towns were overrun by the water, as were over 100 designated tsunami evacuation sites. The tsunami traveled all the way across the Pacific Ocean (●**FIGURE 7.22**).

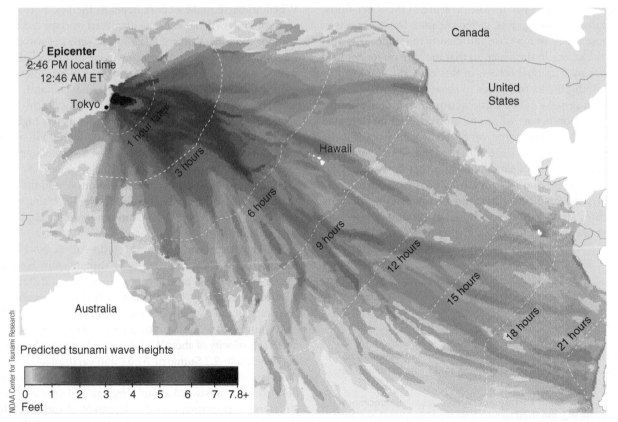

● **FIGURE 7.22** Map of the tsunami triggered by the 2011 Tōhoku earthquake in Japan. The tsunami traveled across the entire Pacific Ocean, taking over 21 hours to reach the coast of Chile. Note the thin strip of red along the west coasts of the Americas, indicating that the wave heights increased there due to the shallowing water depths.

Recent analysis of the fault rupture zone through oceanic drilling and analysis of data from the dense cluster of seismometers located in and around Japan have shown that the very large size of the tsunami resulted from the manner in which the fault broke (●FIGURE 7.23). The fault rupture began along the plate boundary about

West   Soft oceanic   Offshore   East
sediments   buoy

Strong crystalline rock   Pacific Plate (basalt)   Initial rupture propagates west, causing severe ground-shaking in Miyagi Prefecture

(A)

Fault rupture reverses direction, propagating eastward and upward. Rupture violently displaces seafloor and water upward.

(B)

Tsunami waves move oceanward and landward

(C)

Offshore buoy measures tsunami

Landward-moving tsunami waves are pushed upward by the shallowing bottom and bunched up by friction along the bottom

(D)

Tsunami rushes onshore, causing massive destruction

(E)

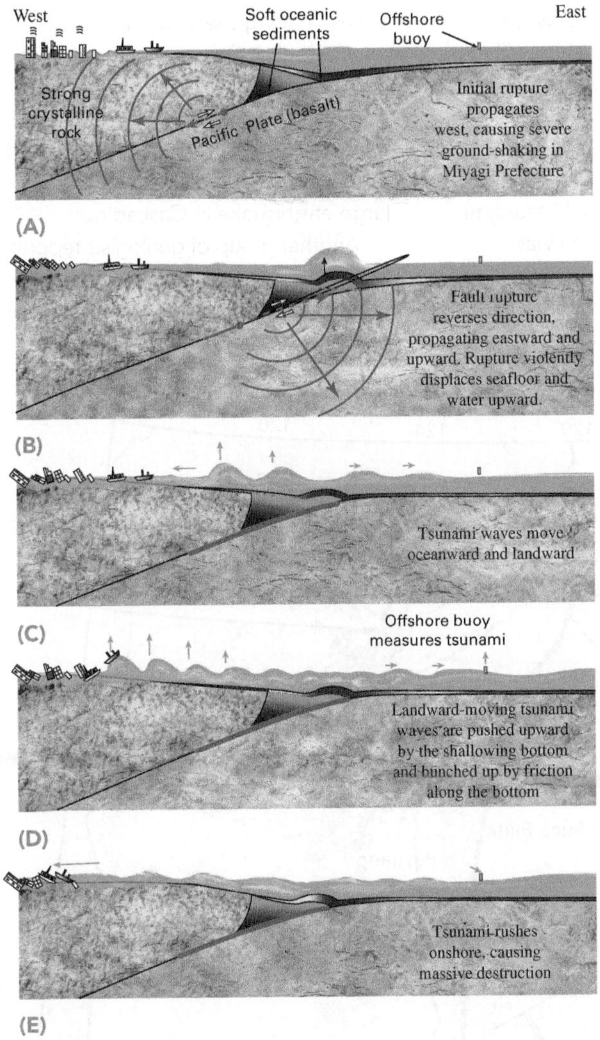

● **FIGURE 7.23** Evolution of the tsunami triggered by the 2011 Tōhoku earthquake in Japan. (A) The initial fault rupture started about 32 kilometers below the seafloor (red dot) and propagated to the west (dashed red line), causing severe ground shaking in the Miyagi Prefecture. (B) The fault rupture switched directions and propagated eastward and upward, violently punching upward both the soft sediment near the seafloor and the deep water overlying it. (C) The resulting large bulge of water spread out, moving out to sea and toward the shore. In the deep water of the open ocean, the tsunami traveled rapidly as a series of relatively small waves detected by offshore buoys. Tsunami waves traveling shoreward grew larger as the bottom shallowed. (D) As they approached the shoreline, the tsunami waves became even larger and bunched up due to friction along the bottom. (E) The tsunami rushed onshore, causing death and destruction.

32 kilometers below the seafloor. During the first 40 seconds, the fault plane broke westward and downward, through relatively strong rock. Then, to the surprise of many geologists and geophysicists, the rupture continued eastward and upward from its starting point for another 35 seconds. As this fault break grew shallower and began to involve relatively soft sediments near the seafloor, its size increased dramatically, violently deforming the sediments, punching the seafloor upward, and triggering the large tsunami.

The 2011 Tōhoku earthquake and tsunami killed over 15,000 people, with an additional 2,600 people missing and 6,000 injured. The earthquake also caused a dam to fail, widespread fires including one at an oil refinery, and the third-largest nuclear accident in history. The nuclear accident resulted when water from the tsunami flooded backup generators needed to power circulation pumps that cooled nuclear reactors at the Fukushima Daiichi nuclear power plant. The lack of cooling water caused the reactors to overheat and melt down, in turn causing several explosions that released radioactivity into the surrounding environment.

EARTHQUAKE PREDICTION

Long-Term Prediction

Long-term earthquake prediction recognizes that earthquakes have recurred many times along existing faults and will probably occur in these same regions again. For example, in the United States, although some faults exist in plate interiors (as in the New Madrid earthquake zone), the most active faults lie along the West Coast at plate boundaries (Figure 7.20) and within the actively extending Basin and Range Province.

Long-term prediction tells us *where* earthquakes are likely to occur. This information is useful because it allows engineers to establish strict building codes in earthquake-prone regions. However, long-term prediction gives only a vague idea of *when* the next earthquake will strike. Near New Madrid, earthquakes have occurred separated by 400, 300, and 200 years. The last one struck in 1812. Given an average 300-year interval, one might expect a quake in the year 2100, but such an analysis could be in error by hundreds of years.

Geophysicists studying earthquakes have shown that their distribution in time cannot be statistically distinguished from random. However, recent analyses of large earthquakes suggest that they appear to be grouped in time in a manner that is not random but rather suggests a synchroneity. The cause of the synchroneity is not known, but one interesting possibility posed by the authors of the study is that slight slowdown of the Earth's rotation causes internal stress to build up because the lithosphere slows down at a different speed than other parts of Earth's interior. As a

Earthquakes in Cascadia

Geologists use multiple lines of evidence to document a very large earthquake in the Pacific Northwest in year 1700

Convergent plate boundaries have produced virtually all of the 30 strongest earthquakes recorded since 1900, when seismographs began to become widely deployed. Conspicuously absent from the list of convergent boundaries that have produced Earth's strongest instrumented earthquakes is the *Cascadia subduction zone*, located off the coast of Northern California, Oregon, Washington, and southern British Columbia, where the Juan de Fuca Plate is subducting eastward beneath the North American Plate at a rate of 3 to 4 centimeters per year (●**FIGURE 7.24**). Although numerous small earthquakes have been recorded along the Cascadia subduction zone, no large-magnitude earthquakes have occurred there since the beginning of written records from the region around 160 years ago.

Although it has not produced any recent, large earthquakes, strong evidence suggests that the Cascadia subduction zone ruptured along its entire 900-kilometer length in January of 1700, producing a massive earthquake and tsunami. A similar earthquake today would devastate the region, both through the intense ground shaking and the likelihood that a large tsunami would strike only minutes later. Thus, geologists have carefully documented the physical and historical evidence left behind by the last "big one" to have resulted from this plate boundary, in order to better understand the seismic risk it poses.

Evidence supporting the occurrence of a very large earthquake along the Cascadia subduction zone includes physical geologic evidence, results from computer models,

historic documents from Japan, and oral legends from Native Americans living in the Pacific Northwest prior to occupation by white settlers.

In an ingenious study of written records from Japanese fishing villages, a group of Japanese geologists discovered that on January 25, 1700, a tsunami was observed at multiple villages along the Japanese coast. Unlike the devastating 2011 tsunami in Japan, the 1700 tsunami was not preceded by an earthquake, suggesting that the tsunami came from somewhere else. Historic records and

physical evidence from South America, Alaska, and Kamchatka for the year 1700 indicated that no earthquakes occurred along these subduction zones, leaving the geologists to conclude that Cascadia was the most likely source. Computer models used to simulate the tsunami wave recorded by the Japanese villagers also suggested that it probably resulted from a very large earthquake in Cascadia.

Another group of geologists reported the partial submergence of coastal forests along portions of the Washington coastline, likely as the result of an

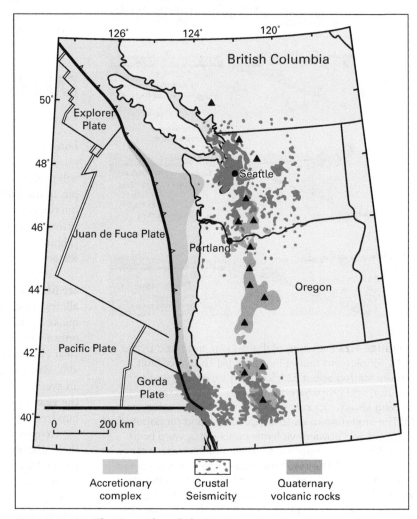

●**FIGURE 7.24** The Cascadia subduction zone.

earthquake that had dropped the coastal land surface roughly 300 years ago. Up to a meter of tidewater had submerged the trees, killing many and damaging others. By analyzing tree rings in the remnants of the killed trees and correlating their ring patterns with those from surviving trees situated on ground too high to have been inundated, the geologists were able to pin down the timing of submergence to the early winter of 1700, consistent with legends from Native Americans living in the region of a large wintertime earthquake that occurred at about this time. Other trees in the same regions that survived the inundations showed evidence of disturbance through tilting or submergence or both in the years immediately after 1700.

Armed with these results, another group of scientists examined the coastal marshes on the western side of Vancouver Island and discovered anomalously coarse layers of gravel and sand in sequences of peat and mud. The sheets of sand and gravel commonly contain marine fossils and become thinner and finer toward the land, consistent with deposition by a surge of marine water. In some cases, multiple layers of sand occur together, suggesting several surges from a group of waves. The scientists compared their results with regional maps of the tsunami produced by the 1964 Alaska earthquake and concluded that the layers of sand were tsunami deposits resulting from a large rupture along the Cascadia subduction zone (●FIGURE 7.25).

By documenting evidence from both sides of the Pacific that a large earthquake took place along the Cascadia subduction zone in January, 1700, geologists have concluded that only a rupture along the entire subduction zone could have produced an earthquake strong enough to account for the evidence of such a large tsunami. The moment magnitude of this earthquake is estimated to be 9.0, indicating that despite its relative seismic quiescence over the past 160 years, the Cascadia subduction zone is capable of earthquakes that indeed are among the largest on Earth.

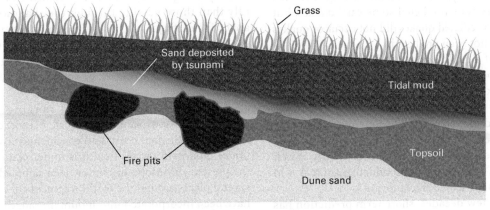

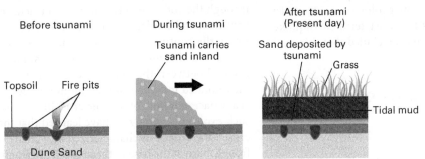

● **FIGURE 7.25** Sketch of field relations along a site on the Oregon coast, showing the layer of sediment interpreted to have been deposited by a tsunami resulting from the January 25, 1700 earthquake. This earthquake is interpreted to have had a magnitude of about 9.0 and resulted from rupture along the entire length of the Cascadia subduction zone. Since this earthquake, the Cascadia subduction zone has been relatively quiet.

result of differences in the rate of slowdown, every few decades Earth has a cluster of large earthquakes to relieve the stress that has accumulated.

Short-Term Prediction

Short-term predictions are forecasts that an earthquake may occur at a specific place and time. Short-term prediction depends on signals that immediately precede an earthquake.

Foreshocks are small earthquakes that precede a large quake by a few seconds to a few weeks. The cause of foreshocks can be explained by a simple analogy. If you try to break a stick by bending it slowly, you may hear a few small cracking sounds just before the final snap. Foreshocks are not a reliable tool for short-term prediction because they do not precede all earthquakes. Of those selected for study, foreshocks preceded only about half of the earthquakes. In addition, some swarms of small shocks thought to be foreshocks were not followed by a large quake.

Another approach to short-term earthquake prediction is to measure changes in the land surface near an active fault zone. Seismologists monitor unusual Earth movements with tiltmeters and laser surveying instruments because distortions of the crust may precede a major earthquake. This method has successfully predicted some earthquakes, but in other instances predicted quakes did not occur or quakes occurred that had not been predicted.

Other types of signals can be used in short-term prediction. When rock is deformed prior to an earthquake, microscopic cracks may develop as the rock approaches its rupture point. In some cases, the cracks release radon gas previously trapped in the rock and its minerals. The cracks may fill with water and cause the water levels in wells to fluctuate. Air-filled cracks do not conduct electricity as well as solid rock does, so the electrical conductivity of rock also decreases as the microscopic cracks form.

Over the past few decades, short-term prediction has not been reliable. Although some geologists continue to search for reliable indicators for short-term earthquake prediction, many geologists have concluded that this goal will remain elusive.

STUDYING THE EARTH'S INTERIOR

Earth's deepest borehole, located in northern Russia, extends to a depth of 12 kilometers or about one-third of the way through the crust. Despite the lack of deeper boreholes, scientists have learned a remarkable amount about Earth's structure by studying the behavior of seismic waves. Some of the principles necessary for understanding the behavior of seismic waves are as follows:

1. In a uniform, homogeneous medium, a wave radiates outward in concentric spheres and at constant velocity.

2. The velocity of a seismic wave depends on the nature of the material that it travels through. Thus, seismic waves travel at different velocities in different types of rock, varying with the rigidity and density of that rock.

3. When a wave passes from one material to another, it refracts (bends) and sometimes reflects (bounces back). Both refraction and reflection are easily observed in light waves. If you place a pencil in a glass half filled with water, the pencil appears bent. Of course the pencil does not bend; the light rays do. Light rays slow down when they pass from air to water, and as the velocity changes, the light waves *refract*. If you look in a mirror, the mirror *reflects* your image. In a similar manner, boundaries between Earth's layers refract and reflect seismic waves.

4. P waves are compressional waves and can travel through gases, liquids, and solids. S waves are shear waves and travel only through solids.

Discovery of the Crust–Mantle Boundary

In 1909, Andrija Mohorovičić discovered that seismic waves passing through the upper mantle travel more rapidly than those passing through the shallower crust. By analyzing earthquake arrival-time data from many different seismographs, he identified the boundary between the crust and the mantle. Today, this boundary is called the **Mohorovičić discontinuity**, or simply the *Moho*, in honor of its discoverer.

The Moho lies at a depth ranging from 4 to 70 kilometers. As explained previously, oceanic crust is thinner than continental crust, and continental crust is thicker under mountain ranges than it is under plains. Although the present-day Moho has never been sampled directly, in several places around the world fragments of oceanic crust have been incorporated into mountain ranges that formed through the collision between two continents or between a continent and a volcanic arc. Oceanic crust formerly separating the continents or the continent and arc was mostly consumed through subduction, but some can become trapped between the two converging masses and end up incorporated into the resulting mountain belt. An ophiolite is a surface exposure of oceanic crust in cross-section. From such exposures, geologists have learned that the base of the crust—the Moho—is characterized by a change in rock type from gabbro at the base of the crust to ultramafic rocks such as peridotite below the Moho in the uppermost mantle. This change in rock type affects the velocity of earthquake waves passing between the mantle and crust, as first noticed by Andrija Mohorovičić more than a century ago.

The Structure of the Mantle

The mantle is almost 2,900 kilometers thick and composes about 80 percent of Earth's volume. Much of our knowledge of the composition and structure of the mantle comes from seismic data. As explained earlier, seismic waves speed up abruptly at the crust–mantle boundary (●FIGURE 7.26). Seismic waves slow down again when they enter the plastic and partially melted asthenosphere at a depth between 75 and 125 kilometers. At the base of the asthenosphere 350 kilometers below the surface, seismic waves speed up again because increasing pressure overwhelms the temperature effect and the mantle becomes stronger and more rigid.

At a depth of about 660 kilometers, seismic wave velocities increase again because pressure is great enough there to produce denser minerals. The zone where the change occurs is called the **660-kilometer discontinuity**. The base of the mantle lies at a depth of 2,900 kilometers. Recent research has indicated that the base of the mantle, at the core–mantle boundary, may be so hot that despite the tremendous pressure, rock in this region is partially liquid.

Discovery of the Core

Using a global array of seismographs, seismologists can detect direct P and S waves up to 105 degrees from the focus of an earthquake. Between 105 and 140 degrees is a "shadow zone" where no direct P waves arrive at Earth's surface. This shadow zone is caused by a discontinuity, which is the mantle–core boundary. When P waves pass from the mantle into the core, they are refracted, or bent, as shown in ●FIGURE 7.27. The refraction deflects the P waves away from the shadow zone.

No S waves arrive beyond 105 degrees. Their absence in this region shows that they do not travel through the outer core. Recall that S waves are not transmitted through liquids, so the failure of S waves to pass through the outer core indicates that it is liquid.

Refraction patterns of P waves, shown in Figure 7.27, indicate that another boundary exists within the core. It is the boundary between the liquid outer core and the solid inner core. Although seismic waves tell us that the outer core is liquid and the inner core is solid, other evidence tells us that the core is composed of iron and nickel.

More detailed studies, conducted in the 1980s and 1990s, show that seismic waves travel at different speeds in different directions through the inner core—thus, the inner core is not homogeneous. One series of measurements suggests

---

foreshocks Small earthquakes that precede a large quake by a few seconds to a few weeks.

Mohorovičić discontinuity The boundary between the crust and the mantle, identified by a change in the velocity of seismic waves; also called the *Moho*.

660-kilometer discontinuity A boundary in the mantle, at a depth of about 660 kilometers, where seismic wave velocities increase because pressure is great enough that the minerals in the mantle recrystallize to form denser minerals.

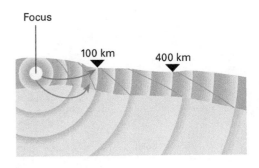

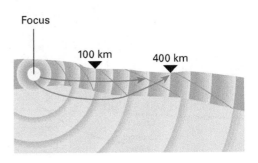

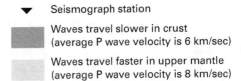

● **FIGURE 7.26** Velocities of P waves in the crust and the upper mantle. As a general rule, the velocity of P waves increases with depth. In some situations, waves that travel through the upper mantle, where P-wave velocities are high, reach a distant point on the surface before P-waves that have traveled a shorter distance through the crust, where velocities are lower.

that the inner core, lying within its liquid sheath, is rotating at a significantly faster rate than the mantle and crust. Other researchers have proposed that the solid inner core may convect just as the solid mantle does.

Density Measurements

The overall density of Earth is 5.5 grams per cubic centimeter (g/cm^3), but both crust and mantle have average densities less than this value. The density of the crust ranges from 2.5 g/cm^3 to 3.0 g/cm^3, and the density of the mantle varies from 3.3 g/cm^3 to 5.5 g/cm^3. Because the mantle and crust account for slightly more than 80 percent of Earth's volume, the core must be very dense to account for the average density of Earth. Calculations show that the density of the core must be 10 g/cm^3 to 13 g/cm^3, which is the density of many metals under high pressure.

Many meteorites are composed mainly of iron and nickel. Cosmologists think that meteorites formed at about the same time that the solar system did and that they reflect

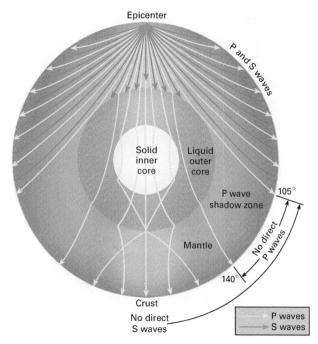

● **FIGURE 7.27** Cross section of Earth showing paths of seismic body waves. They bend gradually because of increasing pressure with depth. The body waves also bend sharply where they cross major layer boundaries in Earth's interior. Note that S waves do not travel through the liquid outer core and therefore are observed only within an arc of 105 degrees from the epicenter, creating an S-wave shadow elsewhere. P waves are refracted sharply at the core–mantle boundary, so there is a P-wave shadow from 105 to 140 degrees.

the composition of the primordial solar system. Because Earth coalesced from meteorites and similar objects, scientists conclude that iron and nickel must be abundant on Earth and that much of this iron and nickel exists as the metallic core.

EARTH'S MAGNETISM

Early navigators learned that no matter where they sailed, a needle-shaped magnet balanced on a point would align itself in an approximately north–south orientation. From this observation, they learned that Earth has a magnetic North Pole and a magnetic South Pole (●**FIGURE 7.28**).

Most likely, Earth's magnetic field is generated within the outer core. Metals are good conductors of electricity, and the metals in the outer core are liquid and very mobile. Two types of motion occur in the liquid outer core: (1) Because the outer core is much hotter at its base than at its surface, the liquid metal convects. (2) The rising and sinking currents of molten metal are then deflected by Earth's spin. These convecting, spinning liquid conductors are thought to generate Earth's magnetic field. The complex, three-dimensional paths of these constantly moving liquid conductors also cause the magnetic poles to move around by hundreds of miles over

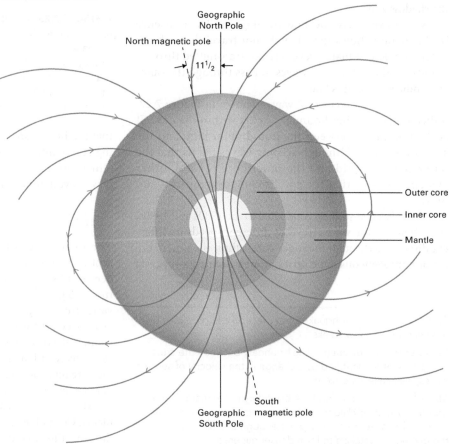

● **FIGURE 7.28** The magnetic field of Earth. Note that the magnetic North Pole currently is offset 11.5 degrees from the geographic North Pole.

historic time frames. The most recent survey of the magnetic North Pole, conducted in 2017, determined that the Pole is moving north-northwest at about 55 km per year. ●FIGURE 7.29 shows the movements of the magnetic North and South Poles over the past few centuries. In addition to causing the magnetic poles to move around, the constantly moving liquid of the outer core at times leads to polarity reversals in the magnetic field itself, so that the North Pole becomes the South Pole and vice versa. These polarity reversals occur over geologic time-frames and are recorded in basalt of the ocean floor as discussed in Chapter 6.

(A)

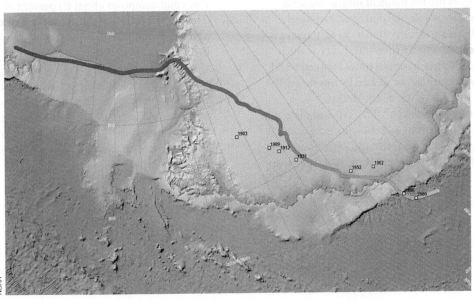

(B)

● **FIGURE 7.29** Maps showing changes in the location of the Magnetic North Pole (A) and South Pole (B) over the past few hundred years, according to the National Oceanic and Aeronautics Agency (NOAA). In both maps, the colored path shows the location of each magnetic pole between 1590 (blue end) and 2020 (yellow end), as calculated to computer modeling. The yellow squares in each map correspond to the observed location of each magnetic pole according to surveys conducted during the year next to each square. Note on both maps the magnetic pole is far from the geographic pole, where the lines of longitude shown in the thin gray lines converge.

Key Concepts Review

- An earthquake is a sudden motion or trembling of Earth caused by the abrupt release of slowly accumulated energy in rocks. Most earthquakes occur along tectonic plate boundaries. Earthquakes occur either when the elastic energy accumulated in rock exceeds the friction that holds rock along a fault, or when the elastic energy exceeds the strength of the rock and the rock breaks.

- Seismology is the study of earthquakes and the nature of Earth's interior based on evidence from seismic waves. An earthquake starts at the initial point of rupture, called the focus, typically lying below Earth's surface. The location on Earth's surface directly above the focus is the epicenter. Seismic waves include body waves, which travel through the interior of Earth, and surface waves, which travel on the surface. P waves are compressional body waves that cause alternate compression and expansion of the rock. They are the first or "primary" waves to reach an observer. S waves travel slower than P waves and are the "secondary" body waves to reach an observer. They consist of a shearing motion and travel through solids but not liquids. Surface waves travel more slowly than either type of body wave. Seismic waves are recorded on a seismograph; a record of earth vibration is called a seismogram. Early in the 20th century, geologists used the Mercalli scale to record the extent of earthquake damage. The Richter scale expresses earthquake magnitudes. Modern geologists use the moment magnitude scale to estimate the energy released during an earthquake. The distance from a seismic station to an earthquake is calculated by constructing a time-travel curve, which calibrates the difference in arrival times between S and P waves. The epicenter can be located by measuring the distance from three or more seismic stations.

- Earthquakes are common at all three types of plate boundaries. The San Andreas Fault zone is an example of a strike-slip fault along a transform plate boundary. Along some portions of the fault zone, rocks slip past one another at a continuous, snail-like pace called fault creep. In other regions, friction prevents slippage until elastic deformation builds and is eventually released in a large earthquake. Subduction zone earthquakes occur along the Benioff zone when the subducting plate slips suddenly. Earthquakes occur at divergent plate boundaries as blocks of lithosphere along the fault drop downward. Earthquakes occur in plate interiors along old faults.

- Earthquake damage is influenced by rock and soil type, construction design, and the likelihood of fires, landslides, and tsunamis.

- Long-term earthquake prediction is based on the observation that most earthquakes occur on preexisting faults at tectonic plate boundaries. Short-term prediction is based on occurrences of foreshocks, release of radon gas, and changes in the land surface, and the water table.

- Earth's internal structure and properties are known by studies of earthquake wave velocities and of the refraction and reflection of seismic waves as they pass through Earth. The boundary between the crust and the mantle, called the Mohorovičić discontinuity (or the Moho), lies at a depth ranging from 4 to 70 kilometers. Oceanic crust is thinner than continental crust, and continental crust is thicker under mountain ranges than it is under plains. The mantle is almost 2,900 kilometers thick and composes about 80 percent of Earth's volume. The upper portion of the mantle is part of the hard and rigid lithosphere. Beneath the lithosphere, the asthenosphere is plastic and partially melted. Beneath the asthenosphere, the mantle is stronger and less plastic; at the 660-kilometer discontinuity, the mineral composition of the mantle changes. Wave and density studies show that the outer core is liquid iron and nickel and the inner core is solid iron and nickel.

- Flowing metal in the outer core generates Earth's magnetic field.

Important Terms

660-kilometer discontinuity (p. 165)	Mercalli scale (p. 151)	seismogram (p. 151)
Benioff zone (p. 157)	Mohorovičić discontinuity (p. 164)	seismograph (p. 151)
body waves (p. 150)	moment magnitude (p. 152)	seismology (p. 149)
earthquake (p. 147)	P waves (p. 150)	strike-slip fault (p. 156)
epicenter (p. 150)	plastic deformation (p. 146)	Sumatra-Andaman earthquake (p. 159)
fault creep (p. 157)	Richter scale (p. 152)	surface waves (p. 150)
focus (p. 150)	S waves (p. 150)	Tōhoku earthquake (p. 151)
foreshocks (p. 164)	San Andreas Fault zone (p. 155)	travel-time curve (p. 154)
liquefaction (p. 159)	seismic waves (p. 149)	tsunami (p. 159)

Review Questions

1. Explain how energy is stored prior to, and then released during, an earthquake.
2. Give two mechanisms that can release accumulated elastic energy in rocks.
3. Why do most earthquakes occur at the boundaries between tectonic plates? Are there any exceptions?
4. Define focus and epicenter.
5. Discuss the similarities and differences among P waves, S waves, and surface waves.
6. Explain how a seismograph works. Sketch what an imaginary seismogram would look like before and during an earthquake.
7. Describe the similarities and differences between the Richter and moment magnitude scales. What is actually measured and what information is obtained?
8. Describe how the epicenter of an earthquake is located.
9. Discuss earthquake mechanisms at the three types of tectonic plate boundaries.
10. What is the Benioff zone? At what type of tectonic boundary does it occur?
11. Why do only shallow earthquakes occur along the Mid-Oceanic Ridge?
12. Discuss earthquake mechanisms at plate interiors.

Daisy Gilardini/Getty Images

8

VOLCANOES AND PLUTONS

Named Stromboli, this volcano located off the shore of southern Italy has been erupting almost continuously since 1932 and is among the most active on Earth. Stromboli's spectacular fountains of molten lava are often visible for long distances at night, earning it the name "The Lighthouse of the Mediterranean." The volcano is part of an island arc formed by subduction of the northern portion of the African Plate beneath the Eurasian Plate.

MAGMA

Animation:
Igneous rocks and
intrusive igneous
activity

In Chapter 3, we learned that rocks melt in certain environments to form magma. This process is one example of the constantly changing nature of rocks described by the rock cycle. Why do rocks melt, and in what environments does magma form?

Processes That Form Magma

Recall that the asthenosphere is the layer in the upper mantle that extends from a depth of about 100 kilometers to 350 kilometers. In that layer, the combined effects of temperature and pressure are such that 1 or 2 percent of the mantle rock is molten, as explained in Chapter 6. Although the majority of the asthenosphere is solid rock, it is so hot and so close to its melting point that large volumes of rock can melt with relatively small changes in temperature, pressure, or the volume of water present (●FIGURE 8.1).

INCREASING TEMPERATURE Everyone knows that a solid melts when it becomes hot enough. Butter melts in a frying pan and snow melts under the spring Sun. For similar reasons, an increase in temperature will melt a hot rock. Oddly, however, increasing temperature is the least important cause of magma formation in the asthenosphere.

DECREASING PRESSURE A mineral is a naturally occurring, inorganic solid composed of an ordered array of atoms bonded together to form a crystal. When a mineral crystal melts, the atoms become disordered and move freely, taking up more space than the solid mineral. Consequently, magma occupies about 10 percent more volume than the rock that melted to form it (●FIGURE 8.2). Keep in mind that this behavior is opposite to that which occurs when ice melts to form water. In that case, the water is more dense and occupies a smaller volume than the equivalent amount of ice. That's why ice cubes float. In contrast, cubes of solid igneous rock would sink if tossed into a pool of completely molten magma of the same composition.

If a rock is heated to its melting point on Earth's surface, it melts readily because there is little pressure to keep it from expanding. The temperature in the asthenosphere is hot enough to melt rock, but the high pressure prevents the rock from expanding, so it does not melt. This condition in which a rock is heated above the melting temperature

---

super-heating Heating of a substance above a phase-transition (gas to liquid or liquid to solid) without the transition occurring. For example, high pressure can keep solid rock from melting even though it is above its melting temperature.

pressure-release melting Melting caused by a decrease in pressure, expansion of rock volume, and melting. Usually occurs in the asthenosphere.

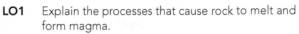
but does not melt is called **super-heating**. The increase in volume that would be required if the rock were to melt is not available because of the pressure, so the rock remains solid. However, if the pressure were to decrease, large volumes of super-heated asthenosphere rocks would melt.

Melting caused by decreasing pressure is called **pressure-release melting**. In the section "Environments of Magma Formation," we will see how certain tectonic processes decrease pressure on asthenosphere rocks, thereby melting them.

ADDITION OF WATER A rock containing small amounts of water generally melts at a lower temperature than an otherwise-identical dry rock. Consequently, the addition of water to rock that is near its melting temperature can cause the rock to melt. Certain tectonic processes, described shortly, add water to the hot rock of the asthenosphere to form magma.

Environments of Magma Formation

Magma forms abundantly in three tectonic environments: spreading centers, mantle plumes, and subduction zones. Let us consider each environment to see how rising temperature, decreasing pressure, and the addition of water can melt rock to create magma.

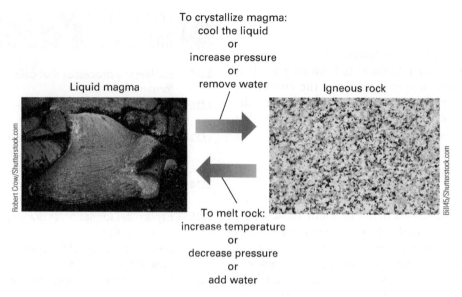

To crystallize magma:
cool the liquid
or
increase pressure
or
remove water

Liquid magma   Igneous rock

To melt rock:
increase temperature
or
decrease pressure
or
add water

● **FIGURE 8.1** The lower box shows that increasing temperature, the addition of water, and decreasing pressure all melt rock to form magma. The upper box shows that cooling, increasing pressure, and water loss all solidify magma to form igneous rock like both the gray-colored granite shown on the right.

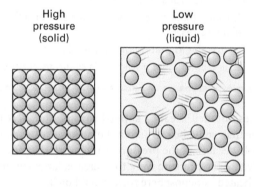

High
pressure
(solid)

Low
pressure
(liquid)

● **FIGURE 8.2** When most minerals melt, the volume of the resulting magma increases. As a result, high pressure favors the dense, orderly arrangement of a solid mineral and low pressure favors the random, less dense arrangement of molecules in liquid magma.

MAGMA PRODUCTION IN A SPREADING CENTER As lithospheric plates separate at a spreading center, hot, plastic asthenosphere wells upward to fill the gap (●**FIGURE 8.3**). As the asthenosphere rises, the surrounding pressure drops and pressure-release melting forms magma with a basaltic composition. Because the magma is of lower density than the surrounding rock, it rises buoyantly toward the surface.

Most of the world's spreading centers lie in the ocean basins, where they form the Mid-Oceanic Ridge system. The rising basaltic magma is injected into the spreading center where it solidifies to form new oceanic crust. Some of the magma erupts onto the seafloor. Once formed, the new oceanic crust then drifts away from the spreading center on both sides, riding atop the separating tectonic plates. Nearly all of Earth's oceanic crust is created in this way at the Mid-Oceanic Ridge system. In most places, the ridge lies beneath the sea. In a few places, such as Iceland, the ridge rises above sea level and basaltic magma erupts onto Earth's surface. Some spreading centers, such as the East African Rift or the North American Basin and Range Province, occur in continents, and here too basaltic magma erupts onto the surface in addition to magma with other compositions.

MAGMA PRODUCTION IN A MANTLE PLUME Recall from Chapter 6 that a mantle plume is a rising column of hot, plastic rock that originates within the mantle. The plume rises because it is hotter than the surrounding mantle and, consequently, is less dense and more buoyant. As a plume rises, pressure-release melting forms magma, which continues to rise toward Earth's surface (●**FIGURE 8.4**).

Because mantle plumes form below the lithosphere, they commonly occur within tectonic plates rather than at a boundary. For example, the Yellowstone Volcano—responsible for the volcanic activity, geysers, and hot springs in Yellowstone National Park—results from a shallow mantle plume that lies far from the nearest plate boundary. If a mantle plume rises beneath oceanic crust, volcanic eruptions build submarine volcanoes and volcanic islands. For example, the Hawaiian Islands are a chain of hot-spot volcanoes that formed over a long-lived mantle plume beneath the Pacific Ocean.

MAGMA PRODUCTION IN A SUBDUCTION ZONE In a subduction zone, the addition of water, decreasing pressure, and heat from friction all combine to form huge quantities of magma (●**FIGURE 8.5**). A subducting plate is covered by water-saturated oceanic sediments, and the upper portions of the underlying basalt also are saturated with water. As the

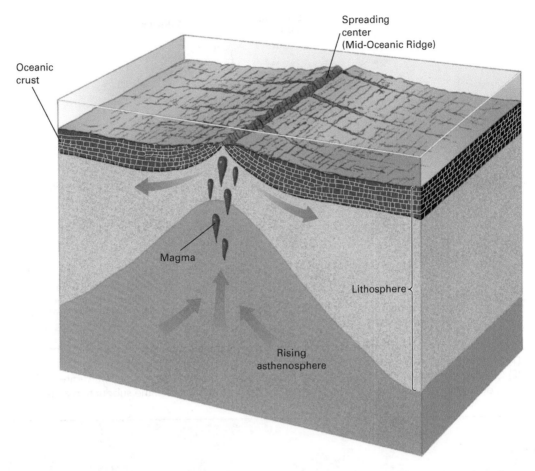

● **FIGURE 8.3** Pressure-release melting produces magma beneath a spreading center, where hot asthenosphere rises to fill the gap left by the two separating tectonic plates.

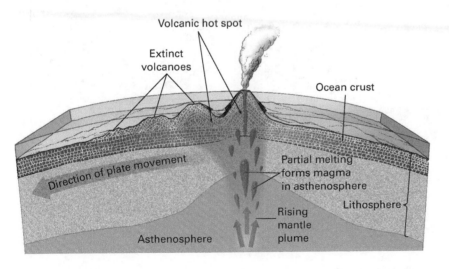

● **FIGURE 8.4** Pressure-release melting produces magma in a rising mantle plume. The magma rises to form a volcanic hot spot.

wet rock and sediments dive into the hot mantle, the heated water ascends into the hot asthenosphere directly above the sinking plate.

As the subducting plate descends, it drags plastic asthenosphere rock down with it, as shown by the elliptical arrows in Figure 8.5. Rock from deeper in the asthenosphere then flows upward to replace the sinking rock. Pressure decreases as this hot rock rises.

Friction generates heat in a subduction zone as the downgoing plate scrapes past the overriding plate. As Figure 8.5 shows, the addition of water, pressure release, and frictional heating combine to melt asthenosphere rocks in the zone where the subducting plate passes into the asthenosphere. The addition of water is probably the most important factor producing melting in a subduction zone, and frictional heating is the least important.

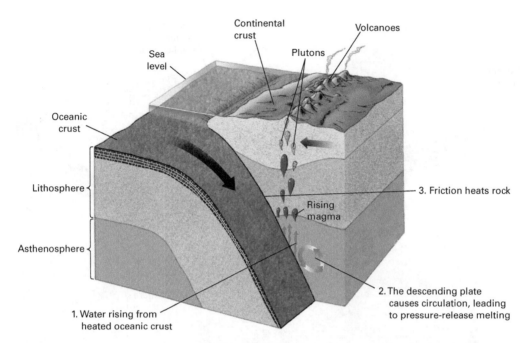

● **FIGURE 8.5** Three processes melt the asthenosphere to form magma at a subduction zone: (1) Geothermally heated water rises from wet oceanic crust on top of the subducting plate. (2) Circulation in the asthenosphere above the subducting plate causes local upward-directed flow, decreasing pressure on hot mantle rock. (3) Friction heats rocks in the subduction zone.

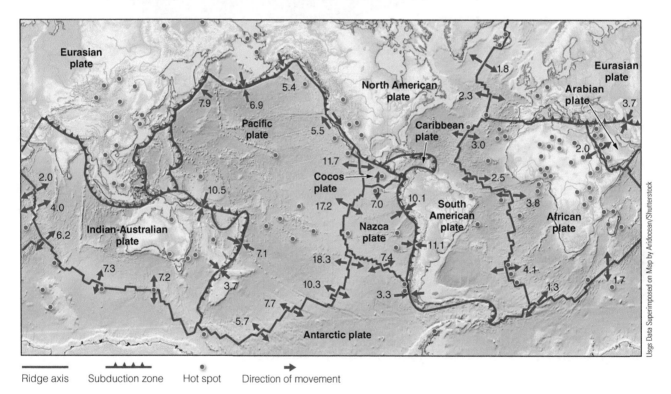

● **FIGURE 8.6** About 75 percent of Earth's active volcanoes lie in the Ring of Fire, a chain of subduction zones at convergent plate boundaries (heavy red lines with teeth) that encircles the Pacific Ocean. The red dots on this map are volcanic hot spots that occur within plates and not in association with a plate boundary.

The subduction process leads directly to the formation of large plutons and volcanoes. The volcanoes of the Pacific Northwest, the granite cliffs of Yosemite, and the Andes Mountains are all examples of volcanic and plutonic rocks formed through subduction. The Ring of Fire is a chain of active volcanoes that runs parallel to the subduction zones encircling the Pacific Ocean basin. About 75 percent of Earth's active volcanoes (exclusive of the submarine volcanoes at the Mid-Oceanic Ridge) lie in the Ring of Fire (●**FIGURE 8.6**).

BASALT AND GRANITE

Basalt and granite are the most common igneous rocks. Basalt makes up most of the oceanic crust, and granite is the most abundant rock in continental crust. Because of their abundance, it is interesting to consider how basalt and granite form.

Recall that basaltic magma forms by the melting of the asthenosphere. But the asthenosphere is peridotite, an ultramafic rock. Basalt and peridotite are quite different in composition: Peridotite contains about 40 percent silica (SiO_2), but basalt contains about 50 percent. Peridotite contains considerably more iron and magnesium than basalt. How, then, does peridotite melt to create basaltic magma? Why does the magma have a composition different from that of the rock that melted to produce it?

It is a general rule that a mixture of two or more minerals will begin to melt at a temperature lower than the melting point of any one of the minerals in its pure state. Remember that peridotite consists mainly of olivine and pyroxene, with small amounts of calcium feldspar. In one set of experiments designed to simulate the melting of mantle peridotite, pure olivine melted at 1,890°C, pure pyroxene melted at 1,390°C, and pure calcium feldspar melted at 1,550°C—but peridotite rock consisting of all three minerals began to melt at 1,270°C.

Furthermore, the composition of the first bit of melt is usually richer in silica than the rock that is undergoing this process of **partial melting**. Thus, when mantle peridotite begins to melt, the magma is of basaltic composition—about 20 percent richer in silica than peridotite. Because the new basaltic magma is less dense than the peridotite rock, the magma begins to rise toward Earth's surface. This process is called partial melting because only a small amount of the original peridotite melts to form basaltic magma, leaving silica-depleted peridotite in the asthenosphere.

Granite and Granitic Magma

Granite contains more silica than basalt and therefore melts at a lower temperature—typically between 700°C and 900°C. Thus, basaltic magma is hot enough to melt continental crust made of granite. Basaltic magma that forms beneath a continent and then rises into the continental crust will cause the crust to partially melt. Because the lower continental crust is hot, a small volume of basaltic magma can melt a large volume of lower continental crust to form granitic magma. Typically, the granitic magma rises a short distance and then solidifies within the crust to form granitic plutons. Most granitic plutons solidify at depths between about 5 and 20 kilometers. In continental rift zones where tectonic stretching has thinned the crust, some granitic magma reaches the surface where it erupts as rhyolite.

---

partial melting The process in which a silicate rock only partly melts as it is heated, forming magma that is more silica rich than the original rock.

Granite forms by this process in a subduction zone at a continental margin, a continental rift zone, and a mantle plume rising beneath a continent.

Andesite and Intermediate Magma

Igneous rocks of intermediate composition, such as andesite and diorite, form by processes similar to those that generate granitic magma. Their magmas contain less silica than granite, either because they form by the partial melting of continental lithosphere or asthenosphere with low silica content or because basaltic magma has mixed with granitic magma.

PARTIAL MELTING AND THE ORIGIN OF CONTINENTS

It is hypothesized that Earth melted shortly after its formation about 4.6 billion years ago. Magma at the surface then cooled to form the earliest crust. From the evidence of a few traces of very old crust combined with computer models of Earth's early formation, geologists surmise that the first crust was lava with the composition of peridotite.

Our explanation of the formation of granitic magma by the melting of granitic continental crust leaves us with two interesting questions: (1) When did granitic continents form? (2) If the early crust and mantle were composed of peridotite, how did granitic continental crust evolve at Earth's surface?

When Did Continents Form?

The 3.96-billion-year-old Acasta Gneiss in Canada's Northwest Territories is among Earth's oldest known rock. It is metamorphosed granitic rock, similar to modern continental crust, and this implies that at least some granitic crust had formed by early Precambrian time.

Other evidence of the formation of Earth's first granites comes from grains of a mineral called zircon found in a sandstone in western Australia. The zircon grains can be radiometrically dated, and the zircons from the Australian sandstone give radiometric dates of 4.4 billion years, although the sandstone was deposited more recently. Zircon commonly forms in granite. Geologists infer that the very old zircon initially formed in granite, which later weathered and released the zircon grains as sand. Eventually, the zircon grains became part of the younger sandstone. The presence of these zircon grains suggests that granitic rocks existed 4.4 billion years ago. Geologists have also found granitic rocks in Greenland and Labrador that are nearly as old as the Acasta Gneiss and the Australian zircon grains.

These dates tell us that some granitic continental crust probably existed by 4.4 billion years ago.

Partial Melting and the Origin of Granitic Continents

The differences in composition among the mantle, oceanic crust, and continental crust are reviewed in the following table:

Mantle (Peridotite)	Oceanic Crust (Basalt)	Continental Crust (Granite)
High magnesium and iron content; low silica	Less magnesium and iron and more silica compared with mantle rock	Low magnesium and iron; high silica

If the earliest crust had a composition similar to that of the mantle, we must describe how oceanic and continental crust evolved from mantle rocks. In section on "Basalt and Granite," we explained that partial melting produces magma that contains a higher proportion of silica than the rock from which the melt formed. Geologists infer that the earliest crust was peridotitic lava, formed from the melted mantle as Earth cooled from the surface downward after its early pervasive melting event. Later, partial melting of this primordial crust formed a basaltic crust that was richer in silica. Then, partial melting of the basalt probably formed intermediate rocks such as andesite, which underwent another partial melting to form the silica-rich granitic continents. The process of partial melting may explain how the silica-rich continents evolved in steps from the silica-poor mantle. But what tectonic processes caused the sequence of melting steps?

As stated earlier, in the modern Earth, magma forms in three geologic environments: spreading centers, subduction zones, and mantle plumes. Similar magma-forming environments may have existed in early Precambrian time, but geologists are uncertain which were most important. Some observations imply that Archean tectonics was similar to modern horizontal plate movements and that most magma formed at spreading centers and subduction zones. Other evidence indicates that horizontal plate motion was minor and that vertical mantle plumes dominated early Precambrian tectonics.

Tectonic Accretion

According to one hypothesis, heat-driven convection currents in a hot, active mantle initiated horizontal plate movement in the early Precambrian crust. The dense primordial crust dove into the mantle in subduction zones, where partial melting created basaltic magma. As a result, the crust gradually became basaltic.

At a later date, the earliest continental crust formed by partial melting of basaltic crust in a new generation of subduction zones. In Chapter 9, you will see that island arcs form today in the same manner. Therefore, the first continents probably consisted of small granitic or andesitic blobs—like island arcs—surrounded by basaltic crust

(●FIGURE 8.7A). Gradually, continued plate movement led to further subduction, and isolated islands coalesced to form microcontinents in a process called **tectonic accretion**. As the microcontinents grew through accretion, they were partially eroded, and the sediment that resulted accumulated and was buried and metamorphosed during the collisions (●FIGURE 8.7B).

Vertical Mantle Plume Tectonics

Several researchers have proposed, instead, that mantle plumes dominated early Precambrian tectonics. In this view, upwellings of mantle rock led to partial melting within the upper mantle. This magma then solidified to form basaltic crust. Continued partial melting eventually formed granitic continental crust.

As more information about the nature of early Earth accumulates through scientific study, some geologists have suggested that both vertical and horizontal mechanisms were important. According to one hypothesis, mantle plumes formed thick basaltic oceanic plateaus, which then oozed outward to initiate horizontal motion. This motion caused subduction and another melting episode that generated continental crust by partial melting of the basalt plateaus.

MAGMA BEHAVIOR

Once magma forms, it rises toward Earth's surface because it is less dense than surrounding rock. As it rises, two changes occur: (1) magma cools as it enters shallower and cooler levels of Earth, and (2) pressure drops because the weight of overlying rock decreases.

Recall from Figure 8.1 that cooling tends to solidify magma but decreasing pressure tends to keep it liquid. So, does magma solidify or remain liquid as it rises toward Earth's surface? The answer depends on the type of magma. Basaltic magma commonly remains liquid and rises to the surface to erupt from a volcano or flow onto the seafloor at the Mid-Oceanic Ridge. In contrast, granitic magma usually solidifies within the crust.

The contrasting behavior of granitic and basaltic magmas is a consequence of their different compositions. Granitic magma contains about 70 percent silica, whereas the silica content of basaltic magma is only about 50 percent. In addition, granitic magma generally contains up to 10 percent water, but basaltic magma contains only 1 to 2 percent water. These differences are summarized in the following table:

Typical Granitic Magma	Typical Basaltic Magma
70% silica; up to 10% water	50% silica; 1% to 2% water

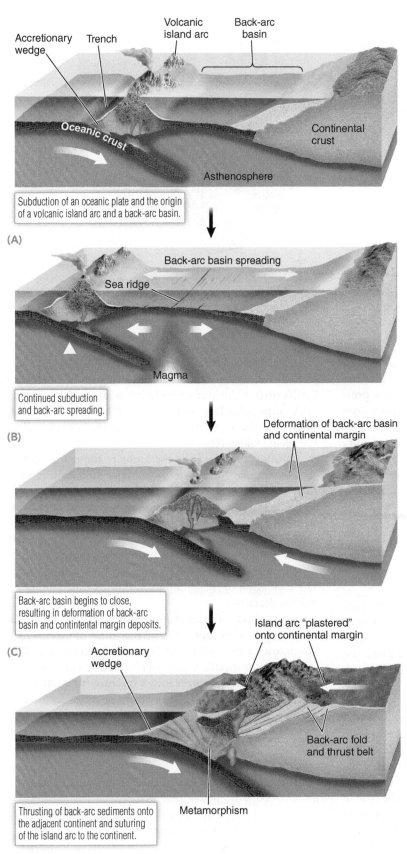

Accretionary wedge | Trench | Volcanic island arc | Back-arc basin

Oceanic crust

Asthenosphere

Continental crust

Subduction of an oceanic plate and the origin of a volcanic island arc and a back-arc basin.

(A)

Back-arc basin spreading

Sea ridge

Magma

Continued subduction and back-arc spreading.

(B)

Deformation of back-arc basin and continental margin

Back-arc basin begins to close, resulting in deformation of back-arc basin and contintental margin deposits.

(C)

Accretionary wedge

Island arc "plastered" onto continental margin

Back-arc fold and thrust belt

Thrusting of back-arc sediments onto the adjacent continent and suturing of the island arc to the continent.

Metamorphism

(D)

● **FIGURE 8.7** According to one model, early continents formed through the process of tectonic accretion. (A) A typical island arc showing the position of the trench and accretionary wedge. Note the presence of the back-arc basin behind the arc. (B) Back-arc basins originate from sea-floor spreading behind the arc. (C) Eventually, sea floor spreading ceases and the back-arc basins changes from extensional to compressional. (D) Complete collapse of the back-arc basin causes the island arc to collide with the continent and accrete to it.

Effects of Silica on Magma Behavior

In the silicate minerals, silicate tetrahedra link together to form the chains, sheets, and framework structures described in Chapter 2. Silicate tetrahedra link together in a similar manner in magma. They form long chains and similar structures if silica is abundant in the magma, but shorter chains if less silica is present.

Because of its higher silica content, granitic magma contains longer chains than does basaltic magma. The long chains become tangled, causing the magma to become stiff, or viscous. It rises slowly because of its viscosity and has ample time to solidify within the crust before reaching the surface. In contrast, basaltic magma, with its shorter silicate chains, is less viscous and flows more easily. Because of its fluidity, it rises rapidly to erupt at Earth's surface.

Effects of Water on Magma Behavior

A second difference between the two magmas is that granitic magma contains more water than basaltic magma does. Water lowers the temperature at which magma solidifies. If dry granitic magma solidifies at 700°C, the same magma with 10 percent water may not become solid until the temperature drops below 600°C.

tectonic accretion Tectonic accretion is the process of material such as island arcs and sediment being added to tectonic plates at subduction zones.

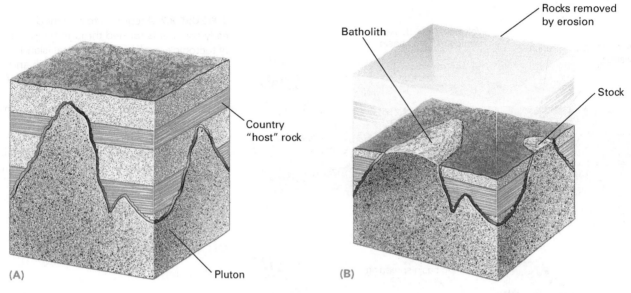

● **FIGURE 8.8** (A) A pluton is any large mass of intrusive igneous rock. (B) A batholith is a pluton with more than 100 square kilometers exposed at Earth's surface. A stock is similar to a batholith but has a smaller surface area.

Water tends to escape as steam from hot magma. But deep in the crust where granitic magma forms, high pressure prevents the water from escaping. As the magma rises, pressure decreases and water escapes. Because the magma loses water, its solidification temperature rises, causing it to crystallize. Water loss causes rising granitic magma to solidify within the crust. Because basaltic magmas have only 1 to 2 percent water to begin with, water loss is relatively unimportant. As a result, rising basaltic magma usually remains liquid all the way to Earth's surface, and basalt volcanoes are common.

PLUTONS

Recall from Chapter 3 that in most cases granitic magma solidifies within Earth's continental crust to form a large mass of igneous rock called a **pluton** (●**FIGURE 8.8A**). Many granite plutons measure tens of kilometers in diameter. How can such a large mass of viscous magma rise through solid rock?

If you place oil and water in a jar, screw the lid on, and shake the jar, oil droplets disperse throughout the water. When you set down the jar, the droplets coalesce to form larger blobs, which rise toward the surface, easily displacing the water as they ascend. Granitic magma rises in a similar way, except that the process is slower because it rises through solid rock. Granitic magma forms near the base of continental crust, where surrounding rock is hot and plastic. As the magma rises, it pushes aside the plastic country rock, which then slowly flows back to fill in behind the rising blobs of granitic magma.

After a pluton forms, tectonic forces may push that part of the crust upward, and erosion may expose parts of the

pluton at Earth's surface (●**FIGURE 8.8B**). A **batholith** is a pluton exposed across more than 100 square kilometers of Earth's surface. An average batholith is about 10 kilometers thick, although a large one may be 20 kilometers thick. A **stock** is similar to a batholith but is exposed over less than 100 square kilometers.

●**FIGURE 8.9** shows the locations of the major batholiths of western North America. Many mountain ranges, such as California's Sierra Nevada, contain large granite batholiths (●**FIGURE 8.10**). A batholith is commonly composed of numerous smaller plutons intruded sequentially over millions of years. For example, the Sierra Nevada batholith contains about 100 individual plutons, most of which were emplaced over a period of 50 million years. The formation of this complex batholith ended about 80 million years ago.

A large body of magma engulfs or pushes country rock aside as it rises. In contrast, a smaller mass of magma may flow into a fracture or between layers in country rock. A **dike** is a tabular, or sheetlike, body of plutonic igneous rock that forms when magma oozes into a fracture (●**FIGURE 8.11**). Dikes cut across sedimentary layers or other features in country rock and range from less than a centimeter to more than a kilometer thick (●**FIGURE 8.12**). A dike is commonly

---

pluton A body of intrusive igneous rock.

batholith A large pluton, exposed across more than 100 square kilometers of Earth's surface.

stock A pluton exposed over less than 100 square kilometers of Earth's surface; similar to a batholith, but smaller.

dike A sheetlike igneous rock, cutting through layers of country rock, that forms when magma is injected into a fracture.

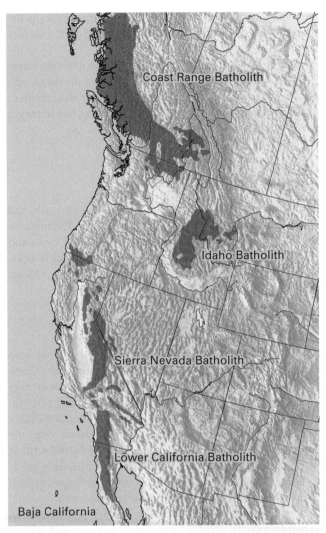

● **FIGURE 8.9** The large batholiths in western North America, shown here in brown, form high mountain ranges.

Coast Range Batholith

Idaho Batholith

Sierra Nevada Batholith

Lower California Batholith

Baja California

● **FIGURE 8.10** Most of California's Sierra Nevada is made of granite plutons, including these mountains in Yosemite National Park.

● **FIGURE 8.12** Here a dike of basalt (black) has intruded into older country rock. A hammer for scale is resting on the outcrop above the dike.

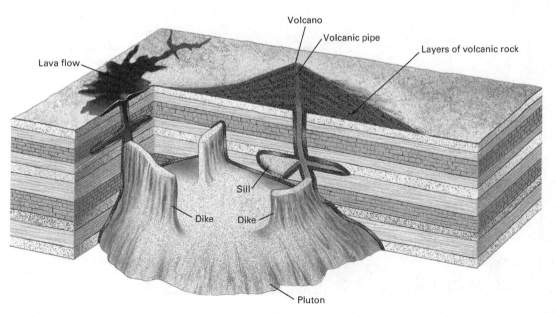

Volcano

Volcanic pipe

Layers of volcanic rock

Lava flow

Sill

Dike   Dike

Pluton

● **FIGURE 8.11** A large magma body may crystallize within the crust to form a pluton. Some of the magma may rise to the surface to form volcanoes and lava flows; some intrudes country rock to form dikes and sills.

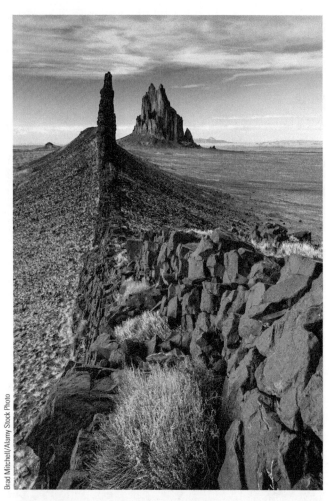

Brad Mitchell/Alamy Stock Photo

● **FIGURE 8.13** This large dike near Shiprock, New Mexico, has been left standing after the erosion of softer country rock made of sandstone.

more resistant to weathering than surrounding rock. As the country rock erodes, the dike is left standing on the surface (●**FIGURE 8.13**).

Magma that oozes between layers of country rock forms a sheetlike layer, called a **sill** (Figure 8.11). Like dikes, sills vary in thickness from less than a centimeter to more than a kilometer and may extend for tens of kilometers in length and width (●**FIGURE 8.14**).

VOLCANOES

A **volcano** is a hill or mountain formed from lava, ash, and rock fragments ejected through a volcanic vent. The material erupted from volcanoes creates a wide variety of rocks and landforms. Many islands, including the Hawaiian Islands, Iceland, and most islands of the southwestern Pacific Ocean, were built entirely through volcanic eruptions.

Lava and Pyroclastic Rocks

As you learned in Chapter 3, lava is magma that flows onto Earth's surface; the word also describes the rock that forms when the magma solidifies. Lava with low viscosity may continue to flow as it cools and stiffens, forming smooth, glassy-surfaced, wrinkled, or "ropy" ridges. This type of lava is called **pahoehoe** (pronounced "puh-HOY-hoy"), from the Hawaiian meaning "smooth" or "polished" (●**FIGURE 8.15**). If the viscosity of lava is higher, its surface may partially solidify as it flows. The solid crust breaks up as the deeper, molten lava continues to move, forming **aa** (pronounced "ah-ah") lava, with a jagged, rubbled, broken surface (●**FIGURE 8.16**). When lava cools, escaping gases such as water and carbon

James Steinberg/Science Source

Sill

● **FIGURE 8.14** This black basalt sill in Glacier National Park was injected between layers of older sedimentary country rock. The think white stripes above and below the black sill of basalt are zones where the country rock was altered (metamorphosed) by the hot basalt.

sill A sheetlike igneous rock, parallel to the grain or layering of country rock, that forms when magma is injected between layers.

volcano A hill or mountain formed from lava and rock fragments ejected from a volcanic vent.

pahoehoe Lava with a smooth, billowy, or ropy surface.

aa Lava that has a jagged, rubbly, broken surface.

vesicles Holes in lava rock that formed when the lava solidified before bubbles of gas or water could escape.

columnar joints Regularly spaced cracks that commonly develop in lava flows, grow downward starting from the surface, and typically form five- or six-sided columns.

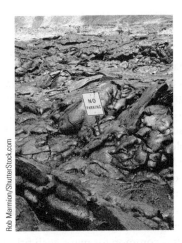

● **FIGURE 8.15** Pahoehoe lava from the 2018 volcanic eruptions on Hawaii partially buried this sign.

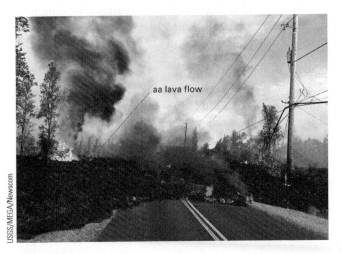

aa lava flow

● **FIGURE 8.16** An aa lava flow in 2018 on Hawaii.

● **FIGURE 8.17** Basaltic lava showing vesicles—gas bubbles preserved in the flow.

dioxide form bubbles in the lava. If the lava solidifies before the gas escapes, the bubbles are preserved as holes in the rock called **vesicles** (●**FIGURE 8.17**).

As hot lava cools and solidifies, it shrinks. The shrinkage pulls the rock apart, forming cracks that grow as the rock continues to cool. In Hawaii, geologists have observed this phenomenon while watching fresh lava cool: When a solid crust measuring only 0.5 centimeter thick had formed on the surface of the glowing liquid, five- or six-sided cracks developed. As the lava continued to cool and solidify, the cracks grew downward through the flow. Such cracks, called **columnar joints**, are regularly spaced and intersect to form five- or six-sided columns when viewed in cross section (●**FIGURE 8.18**).

(A)

(B)

● **FIGURE 8.18** (A) Columnar joints in basalt. (B) A view of columnar basalt columns from above. These basalt columns, from Devils Postpile National Monument in California, have been planed off and polished by a glacier.

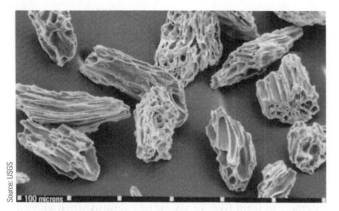

Source: USGS

● **FIGURE 8.19** Microscopic view of particles from a volcanic ash that erupted about 600,000 years ago from a volcano in northern California. The ash is made of pumice fragments, volcanic glass full of bubble holes. The bubble holes formed when liquid lava erupted and gas dissolved in the lava came out of solution and formed bubbles. The erupting magma was cooled so quickly that it formed pumice—volcanic glass with the bubbles trapped inside. The violence of the eruption fragmented the pumice and blew it into the atmosphere before it settled back to Earth far from the volcano.

If a volcano erupts explosively, it may eject a combination of hot gas, liquid magma, and solid rock fragments. A rock formed from this material is called a **pyroclastic rock** (from *pyro*, meaning "fire," and *clastic*, meaning "particles"). The smallest particles, called **volcanic ash**, consist of tiny fragments of glass that formed when liquid magma exploded into the air (●**FIGURE 8.19**). **Cinders** are volcanic fragments that vary in size from 4 to 32 millimeters.

Fissure Eruptions and Lava Plateaus

The gentlest, least explosive type of volcanic eruption occurs when magma is so fluid that it oozes from cracks in the land surface called **fissures** and flows over the land like water. Basaltic magma commonly erupts in this manner because of its low viscosity. Fissures and fissure eruptions vary greatly in scale. In some cases, lava pours from small cracks on the flank of a volcano. Fissure flows of this type are common on Hawaiian and Icelandic volcanoes.

In other cases, however, fissures extend for tens or hundreds of kilometers and pour thousands of cubic kilometers of basaltic lava onto Earth's surface. A fissure eruption of this type creates a **flood basalt**, which covers the landscape like a flood. It is common for many such fissure eruptions to occur in rapid succession and to create a **lava plateau**, covering thousands of square kilometers.

The Columbia River plateau in eastern Washington, northern Oregon, and western Idaho is a lava plateau containing 350,000 cubic kilometers of basalt (Figure 8.19). The lava is up to 3,000 meters thick and covers 200,000 square kilometers. The Columbia River basalts formed as a series of eruptions that began as early as 17.4 million years ago, peaked about 16 million years ago, and essentially ended about 12 million years ago. The individual flows are between 15 and 100 meters thick and some flowed for hundreds of kilometers. Much of the basalt erupted from a series of dikes where the states of Washington, Oregon, and Idaho all share a border. ●**FIGURE 8.20** shows a map of the Columbia River basalts and location of the Chief Joseph dike swarm—the source of much of the basalt.

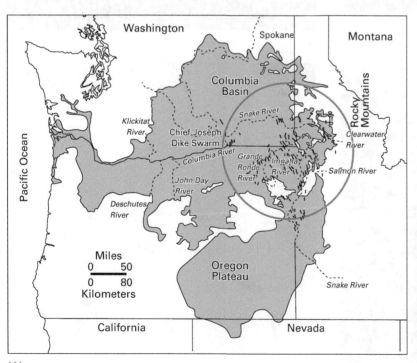

(A)

Zack Frank/ShutterStock.com

(B)

● **FIGURE 8.20** (A) The Columbia River basalt plateau, shown here in green, covers much of Washington, Oregon, and Idaho. The basalt erupted from vents and fissures from the St. Joseph's dike swarm in west-central Idaho and adjacent Washington and Oregon (red circle). (B) The Columbia River basalt.

TABLE 8.1 Characteristics of Different Types of Volcanic Features

Type of Volcanic Feature	Physical Form	Size	Type of Magma	Style of Activity	Examples
Basalt plateau	Flat to gentle slope	100,000 to 1,000,000 km² in area; 1 to 3 km thick	Basalt	Formed by gentle fissure eruptions	Columbia River plateau
Shield volcano	Slightly sloped, 6° to 12°	Up to 9,000 m high	Basalt	Gentle; some lava fountains	Hawaii
Cinder cone	Moderate slope	100 to 400 m high	Basalt or andesite	Ejections of pyroclastic material	Craters of the Moon, Idaho; Parícutin (Mexico)
Composite volcano	Alternate layers of flows and pyroclastics	100 to 3,500 m high	Variety of types of magmas and ash	Often violent	Vesuvius (Italy); Mount St. Helens; Aconcagua (Argentina)
Caldera	Circular depression, sometimes with steep walls	Less than 40 km in diameter	Rhyolite (Granite)	Formed by a violent cataclysmic explosion; potential for violent eruption remains	Yellowstone Volcano; San Juan Mountains

Volcano Types

Volcanoes differ widely in shape, structure, and size (Table 8.1). Lava and rock fragments commonly erupt from an opening called a **vent**, located in a **crater**, a bowl-like depression at the summit of the volcano that was itself created by volcanic activity (●FIGURE 8.21). As mentioned previously, lava or pyroclastic material may also erupt from a fissure on the flanks of the volcano.

pyroclastic rock Rock made up of liquid magma and solid rock fragments that were ejected explosively from a volcanic vent.

volcanic ash The smallest pyroclastic particles, less than 2 millimeters in diameter.

cinders Glassy, pyroclastic volcanic fragments 4 to 32 millimeters in size.

fissures Breaks, cracks, or fractures in rocks.

flood basalt Basaltic lava that erupts gently and in great volume from vents or fissures at Earth's surface, to cover large areas of land and form lava plateaus.

lava plateau A broad plateau covering thousands of square kilometers, formed by the accumulation of many individual lava flows that occur over a short period of geologic time.

vent An opening in a volcano, typically in the crater, through which lava and rock fragments erupt.

crater A bowl-like depression at the summit of a volcano, created by volcanic activity.

shield volcano A large, gently sloping volcanic mountain formed by successive flows of basaltic magma.

SHIELD VOLCANOES Fluid basaltic magma often builds a gently sloping mountain called a **shield volcano** (●FIGURE 8.22). The sides of a shield volcano generally slope away from the vent at angles between 6 and 12 degrees from horizontal. Although their slopes are gentle, shield volcanoes can be enormous. The height of Mauna Kea Volcano in Hawaii, measured from its true base on the seafloor to its top, is 10,200 meters (33,476 feet; ●FIGURE 8.23), making it the tallest mountain the world, exceeding the height of Mount Everest by over 1,200 meters (4,000 feet).

Although shield volcanoes, such as those of Hawaii and Iceland, erupt regularly, the eruptions are normally gentle

● **FIGURE 8.21** Hot gases rise from vents in the crater of Marum Volcano, in the South Pacific island nation of Vanuatu.

Courtesy of Science Graphics, Inc./Ward's Natural Science Establishment, Inc.

● **FIGURE 8.22** The Icelandic mountain Skjaldbreiður, meaning "broad shield," shows the typical low-angle slopes of a shield volcano.

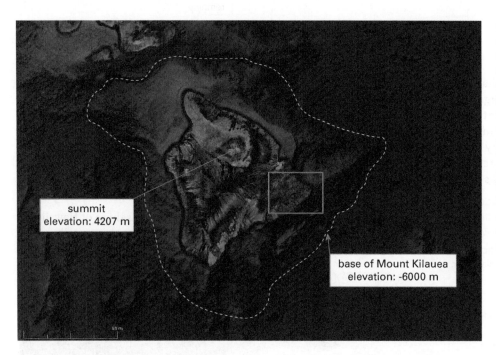

summit
elevation: 4207 m

base of Mount Kilauea
elevation: -6000 m

● **FIGURE 8.23** The island of Hawaii is the upper part of Mount Kilauea, the tallest mountain in the world with a height of 10,200 meters (33,476 feet) from its base to its top. Only the upper 40 percent of the mountain is exposed above sea level. This image shows the entire mountain from the base of the sea floor to the summit. The red box shows the outline of the map of the 2018 volcanic eruptions shown in Figure 8.24.

and rarely life threatening. Lava flows occasionally overrun homes and villages, but the flows advance slowly enough to give people time to evacuate.

Beginning in May 2018, however, Kilauea, the largest volcano on the Island of Hawaii, began a series of eruptions that took many volcanologists by surprise with its intensity and location (●**FIGURE 8.24**). The initial eruption began in Leilani Estates, a subdivision on the east side of the island, where on May 3, steaming ground cracks opened and began to spew lava. The governor issued a mandatory evacuation order for the neighborhood and by the next day, two homes had been destroyed by three actively erupting new vents, and high levels of toxic sulfur dioxide gas was present in the region. Lava

fountains over 300 feet tall were observed. By May 9, six days after the initial eruption, over a dozen eruptive fissures had opened in the Leilani Estates region, 1,700 people had been evacuated, and 27 homes had been destroyed.

Between May 12 and 22, four new fissures had begun to erupt several miles away in the lower Puna region of Hawaii. This new series of fissures produced so much lava that it overran everything in its path between the vents and the shoreline, including two highways. When the lava reached the shoreline and encountered seawater, it formed steam mixed with volcanic gases and particulates in a hazardous mixture called volcanic haze or laze. Within a few days, the lava began to build a small promontory into the ocean.

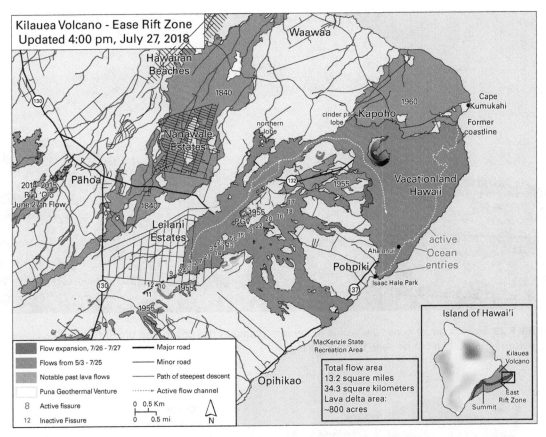

● **FIGURE 8.24** Map of the 2018 and notable earlier flows on the eastern tip of the Island of Hawaii as of July, 2018.

Twice during this period of effusive basaltic volcanism, the volcano's summit miles away erupted violently, sending clouds of ash 30,000 feet into the air.

By May 23, eruption at the new fissure sites in lower Puna sites weakened and shifted back up to the original series of fissures in the Leilani Estates region. Whereas the first set of eruptions involved mostly slow-moving aa lava flows, the second set involved much lower viscosity pahoehoe lava which poured across the land surface in fast-moving, spectacular streams that first pooled into a lava lake before spilling over in two places. One of the lava streams traveled northeast, crossed a road, and overran two geothermal wells in a geothermal plant, shutting it down. By June 3, the lava flow had reached the village of Kapoho—which it buried—and Vacationland Hawaii subdivision—which it obliterated. As of this writing (July 2018), over 700 homes have been destroyed by the 2018 Hawaiian eruptions, along with several sections of highway, a park, and a school.

CINDER CONES A **cinder cone** is a small volcano composed of pyroclastic fragments. A cinder cone forms when large amounts of gas accumulate in rising magma. When the gas pressure builds sufficiently, the entire mass erupts explosively, hurling cinders, ash, and molten magma into the air. The particles then fall back around the vent, to accumulate as a small mountain of pyroclastic debris. A cinder cone is usually active for only a short time, because once the gas escapes, the driving force behind the eruption is gone.

Cinder cones usually are symmetrical and can be steep (about 30°), especially near the vent, where ash and cinders pile up (●**FIGURE 8.25**). Most are less than 300 meters high, although a large one can be up to 700 meters high. A cinder cone erodes easily and quickly because the pyroclastic fragments are not cemented together.

COMPOSITE CONES A **composite cone**, sometimes called a **stratovolcano**, forms over a long period of time as a sequence of lava flows and pyroclastic eruptions. The hard lava covers the loose pyroclastic material and protects it from erosion (●**FIGURE 8.26A**).

---

cinder cone A small volcano, typically less than 300 meters high, made up of loose pyroclastic fragments blasted out of a central vent; usually active for only a short time.

composite cone A steep-sided volcano formed by an alternating series of lava flows and pyroclastic eruptions and marked by repeated eruption.

stratovolcano A steep-sided volcano formed by an alternating series of lava flows and pyroclastic deposits and marked by repeated eruption.

● **FIGURE 8.25** Several cinder cones are visible in the foreground of this photo; the broad form of the Hawaiian shield volcano Mauna Loa can be seen behind the fog.

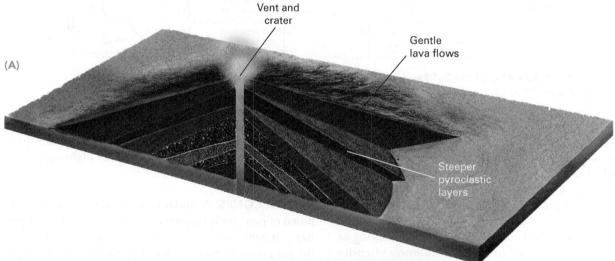

(A)

Vent and crater

Gentle lava flows

Steeper pyroclastic layers

(B)

● **FIGURE 8.26** (A) A composite cone consists of alternating layers of lava and loose pyroclastic material. (B) Mount Hood, in Oregon, is a composite cone.

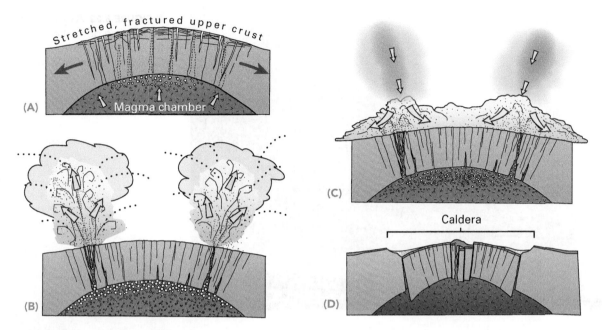

● **FIGURE 8.27** (A) When granitic magma rises to within a few kilometers of Earth's surface, it stretches and fractures overlying rock. Gas separates from the magma and rises to the upper part of the magma body. (B) The gas-rich magma explodes through fractures, rising as a vertical column of hot ash, rock fragments, and gas. (C) When the gas is used up, the column collapses and spreads outward as a high-speed pyroclastic flow. (D) Because so much material has erupted from the top of the magma chamber, the roof collapses to form a caldera.

Many of the highest mountains of the Andes and some of the most spectacular mountains of western North America are composite cones (●**FIGURE 8.26B**). Repeated eruptions are a trademark of a composite volcano. Mount St. Helens, in the state of Washington, erupted dozens of times in the 4,500 years preceding its most recent eruption in 1980. Mount Rainier, also in Washington, has been dormant in recent times but could become active again and threaten nearby populated regions.

VOLCANIC EXPLOSIONS: ASH-FLOW TUFFS AND CALDERAS

Although granitic magma usually solidifies within the crust, under certain conditions it rises to Earth's surface, where it erupts violently. Granitic magmas that rise to the surface contain only a few percent water, like basaltic magma. But decreasing pressure allows the small amount of dissolved water in granitic magmas to form a frothy, pressurized mixture of gas and liquid magma that may be as hot as 900°C. As the mixture rises to within a few kilometers of Earth's surface, it fractures overlying rocks and explodes skyward through the fractures, as shown in panels A and B of ●**FIGURE 8.27**. Think of a bottle of beer or soft drink. When the cap is on and the contents are under pressure, carbon dioxide gas is dissolved in the liquid. When you remove the cap, pressure decreases and bubbles rise to the surface. If conditions are favorable, the frothy mixture erupts through the bottleneck.

A large and violent eruption can blast a column of pyroclastic material 10 or 12 kilometers into the sky, and the column might be several kilometers in diameter. A cloud of fine ash may rise even higher—into the upper atmosphere. The force of material streaming out of the magma chamber can hold the column up for hours or even days.

Pyroclastic Flows

When much of the gas has escaped from the upper layers of magma, the eruption ends. The airborne column of ash, rock, and gas then falls back to Earth's surface, spreading over the land and funneling downstream valleys beyond (●**FIGURE 8.27C**). Such a flow is called a **pyroclastic flow** (●**FIGURE 8.28**).

When a pyroclastic flow stops, most of the gas escapes into the atmosphere, leaving behind a chaotic mixture

pyroclastic flow An extremely destructive incandescent mixture of volcanic ash, larger pyroclastic particles, minor lava, and hot gas that forms from collapse of an eruptive column and flows rapidly along Earth's surface.

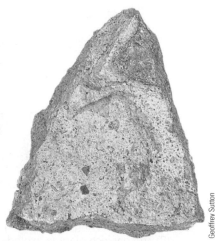

● **FIGURE 8.28** A pyroclastic flow descending down the slope of the Soufriere Hills Volcano on the Caribbean island of Montserrat in January 2010.

(B)

(A)

● **FIGURE 8.29** (A) Ash-flow tuff forms the spectacular cliffs of Smith Rock State Park near Bend, Oregon. (B) Ash-flow tuff forms when a pyroclastic flow comes to a stop. The fragments in the tuff are pieces of rock that were carried along with the volcanic ash and gas.

of volcanic ash and rock fragments called **ash-flow tuff** (Figures 8.27B and 8.27C). If the deposit is thick enough, the ash at the bottom of the pile begins to compact and may partially melt from the residual heat, causing the ash to fuse together and form a welded tuff. Typically, only the lower portion of a pyroclastic flow becomes welded; the upper portion usually remains a relatively porous accumulation of ash particles. Welded tuffs form several prized sport climbing areas in the western United States because of their strength (●**FIGURE 8.29**).

Calderas

After the gas-charged magma erupts, the roof of the magma chamber can collapse into the space that the magma had filled (●**FIGURE 8.27D**). Typically, the collapsing roof forms a circular depression, called a **caldera**. A large caldera may be 40

---

ash-flow tuff A volcanic rock formed when a pyroclastic flow solidifies.

caldera A large circular depression created by the collapse of the magma chamber after an explosive volcanic eruption.

The Destruction of Pompeii

A catastrophic series of eruptions from Italy's Mount Vesuvius buried several ancient Roman settlements in ash and remains a threat to Naples, Italy, today

In 79 CE, Mount Vesuvius erupted and destroyed the Roman cities of Pompeii, Herculaneum, and several neighboring villages near what is now Naples, Italy (●**FIGURE 8.30**). Prior to that eruption, the volcano had been inactive for about 700 years—so long that farmers had cultivated vineyards on the sides of the mountain all the way to the summit. During the eruption, a pyroclastic flow streamed down the flanks of the volcano, burying the cities and towns under 5 to 8 meters of hot ash. When archaeologists located and excavated Pompeii 17 centuries later, they found molds of inhabitants trapped by the pyroclastic flow as they attempted to flee or find shelter (●**FIGURE 8.30B**).

Some of the molds appear to preserve facial expressions of terror. After the 79 CE eruption, Mount Vesuvius returned to relative quiescence but became active again in 1631. It was frequently active from 1631 to 1944; in the 20th century, it erupted in 1906, 1929, and 1944.

Mount Vesuvius is a stratovolcano that was formed over 40,000 years ago by many eruptions that varied from gently flowing lava to the types of explosions that buried Pompeii. Geologists estimate that a total of about 50 cubic kilometers of magma has been erupted from Mount Vesuvius. Recent studies of seismic velocities beneath the volcano show that seismic waves suddenly slow from 6 to 2 kilometers per second at a depth of 8 kilometers, suggesting that molten magma still exists at that depth. One study mapped this low-velocity layer and concluded that it corresponded to a sill of magma that extended outward for at least 400 square kilometers below the volcano.

Because stratovolcanoes erupt frequently and remain active for long periods, and because a considerable body of magma underlies the volcano, geologists consider Mount Vesuvius a high risk for future eruptions. In 1841, the Vesuvius Observatory, the oldest volcanic observatory in the world, was built on the slopes of the mountain to monitor it and two other nearby volcanoes for signs of an imminent eruption. Today, about 3 million people live near Mount Vesuvius.

(A)

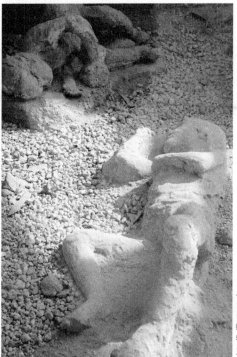

(B)

● **FIGURE 8.30** (A) The modern city of Naples has a population of about 3 million and sits at the foot of Mount Vesuvius. (B) Molds of two of the inhabitants of Pompeii who perished in the 79 CE eruption of Mount Vesuvius.

E. Auger, P. Gasparini, J. Virieux, and A. Zollo, Seismic evidence of an extended magmatic sill under Mt. Vesuvius, Science 294 (November 16, 2001): 1510–1512.

kilometers in diameter and have walls as much as a kilometer high. Some calderas fill up with volcanic debris; others maintain the circular depression and steep walls. For example, Figure 8.21 shows several generations of calderas at the summit of the Marum Volcano on the Southwest Pacific island of Vanuatu. Each caldera has a roughly circular outline and steep walls.

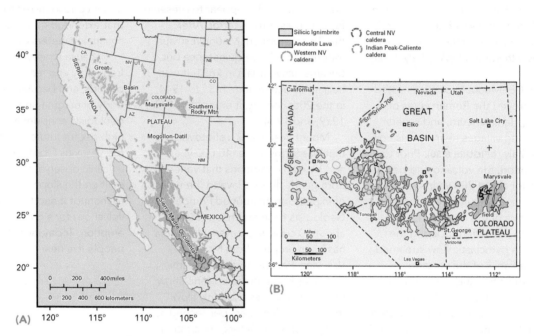

● **FIGURE 8.31** Map showing (A) the distribution of ash-flow tuffs across the North American southwest and (B) calderas formed between 36 and 31 million years ago in the central Basin and Range province.

DIGGING DEEPER

The Yellowstone Volcano

Centered in Yellowstone National Park, the Yellowstone Volcano has produced enormous, geologically recent eruptions and continues to be very active today

On March 1, 1872, President Ulysses S. Grant signed into law The Act of Dedication that officially founded Yellowstone National Park in Wyoming, Idaho, and Montana. That same year, Ferdinand V. Hayden, leader of the Geological and Geographical Survey of the Yellowstone region that had taken place the year before, dramatically described some of Yellowstone's geologic elements in his official report to the U.S. Congress. Hayden wrote:

From the summit of Mount Washburn, a bird's-eye view of the entire basin may be obtained, with the mountains surrounding it on every side without any apparent break in the rim. … It might be called one vast crater, made up of thousands of smaller volcanic vents and fissures out of which the fluid interior of the Earth, fragments of rock, and volcanic dust were poured in unlimited quantities.

Since Hayden's remarkable insightful words, geoscientists have indeed determined that the geology of Yellowstone National Park is dominated by the structures and rocks associated with a very large, active volcano that is

centered in the park and includes a caldera roughly 30 kilometers across (●**FIGURE 8.32**). The caldera formed about 640,000 years ago, following a gigantic eruption of the Yellowstone Volcano that far exceeded that of any eruption that has occurred anywhere on Earth during recorded human history. That eruption produced roughly 1,000 cubic kilometers of volcanic ash, enough to bury the entire state of Texas with a uniform blanket of ash nearly 1.5 meters (4.7 feet) thick! Prior to that, the Yellowstone Volcano produced two additional catastrophic, caldera-forming eruptions 1.3 million and 2.1 million years ago. The oldest of these eruptions was the largest

We usually think of volcanic landforms as mountain peaks, but the topographic depression of a caldera is an exception. ●**FIGURE 8.31** shows that calderas, ash-flow tuffs, and related rocks occur over a large part of western North America. The Yellowstone Volcano, Crater Lake in Oregon, and Long Valley in California are well-known examples.

RISK ASSESSMENT: PREDICTING VOLCANIC ERUPTIONS

TABLE 8.2 summarizes the major known volcanic disasters since the year 1500. The potential for such disasters in the future makes a volcanic eruption one of the greatest of all geologic hazards. It also makes risk assessment and prediction of volcanic eruptions an important part of modern science.

Approximately 1,300 active volcanoes are recognized globally, and nearly 6,000 eruptions have occurred in the past 10,000 years. These figures do not include the numerous submarine volcanoes of the Mid-Oceanic Ridge system. Many volcanoes have erupted recently, and we are certain that others will erupt soon. How can geologists predict an eruption and reduce the risk of a volcanic disaster?

Regional Prediction

Risk assessment for regional predictions is based both on the frequency of past eruptions and on potential violence. However, regional predictions based on the concentration of volcanoes in an area can only estimate *probabilities* and cannot be used to determine exactly when a particular volcano will erupt or the intensity of a particular eruption.

Short-Term Prediction

In contrast to regional predictions, short-term predictions attempt to forecast the specific time and place of an impending eruption. They are based on instruments that monitor an active volcano to detect signals that the volcano is about to erupt. The signals include changes in the shape of the mountain and surrounding land, earthquake swarms indicating movement of magma beneath the mountain, increased emissions of ash or gas, increasing temperatures or changing compositions of nearby hot springs, and any other signs that magma is approaching the surface.

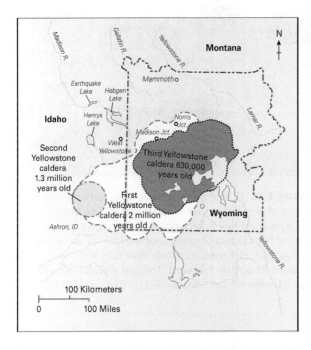

and produced 2,500 cubic kilometers of pyroclastic material, enough to cover all of Texas with more than 3.5 meters (11.5 feet) of ash!

The Yellowstone Volcano and the "supereruptions" that characterize it exist because a plume of hot rock from the upper mantle is slowly welling up below the Yellowstone region. As this hot mantle rock flows upward and decompresses, it partially melts, forming a pie-shaped reservoir of basaltic magma that accumulates at the base of the crust. Heat from this accumulating basalt partially melts the granitic continental crust above it, forming a second magma that is rhyolitic in composition. As more rhyolite magma forms, an interconnected network of rhyolite melt develops, and the magma within this network flows slowly toward the surface. At a depth of around 10 kilometers, however, the magma cools enough that most of it cannot continue upward, and it begins to accumulate there in a second magma chamber.

● **FIGURE 8.32** This map shows the locations of the three Yellowstone calderas that have formed over the past 2.1 million years.

(Continued)

The Yellowstone Volcano (Continued)

As rhyolite magma accumulates, pressure inside the magma chamber builds, causing a bulge of the ground surface that is roughly 10 kilometers across (Figure 8.27A). As the bulge grows, the rock around its perimeter cracks, forming a circular-shaped fracture, called a ring fracture, that typically starts at the surface and propagates downward toward the magma chamber with time. Eventually, one or more segments of the ring fracture intersect the magma chamber, initiating eruption of the volcano (Figure 8.27B).

Each of the three separate caldera-forming eruptions of the Yellowstone Volcano that have occurred over the past 2.1 million years produced a gigantic quantity of volcanic rock and an eruption that was far larger than anything that has occurred in recorded history. Each eruption began with highly explosive outbursts of volcanic ash, chunks of rock, and bits of magma from an opening at the surface called a vent. During the beginning of the eruption, volcanic material was likely hurled from the vent at supersonic speeds, producing a booming sound that would have been heard hundreds of kilometers away. As with large eruptions that have happened during recorded history, ash would have been blown over 15 kilometers into the atmosphere, blotting out the sunlight across much of the Yellowstone region.

As the eruption continued, tremendous quantities of ash, dust, hot gas, and rock were hurled skyward, producing a gigantic eruptive column over the vent. Eventually, the volcanic material forming the eruptive column began to fall back to Earth around the still-erupting vent. As this deluge of returning hot volcanic material encountered the ground surface, it spread out laterally, forming a pyroclastic flow (Figure 8.27C). Such

flows are capable of racing across the landscape at speeds of over 100 kilometers per hour and typically flow downslope into topographically low regions such as river valleys. Pyroclastic flows are so hot they are usually incandescent; typically nothing survives in their path.

Each caldera-forming eruption left behind a deposit of tuff (●FIGURE 8.33). At its base, the tuff that resulted consists of a loosely packed jumble of angular volcanic rock fragments and ash and is called an **air-fall tuff**. In contrast, the hot pyroclastic flows that developed as the eruption proceeded left behind an ash-flow tuff. Ash-flow tuffs accumulate so rapidly and contain enough heat that they partially melt after being deposited. As the eruption continues and more flows are added to the top of the deposit, their weight causes tuff near the base to compact and fuse together, forming a **welded tuff**.

The caldera itself forms after the main eruption, when the magma chamber is mostly depleted and the landscape is buried in newly deposited tuff. The weight of the tuff presses downward on the empty magma chamber, causing it to be downdropped along the ring fractures. Thus, the ring fractures become ring faults with a normal sense of movement, downdropping the caldera floor like a giant, circular-shaped keystone block and leaving behind a circular-shaped depression with steep crater walls.

Although the three main

caldera-forming eruptions of the Yellowstone Volcano were certainly the largest and most explosive eruptions to have come from the volcano, it has produced dozens of other very large eruptions that did not result in the formation of a caldera. In fact, the youngest of the three calderas, formed 640,000 years ago, is almost completely filled in by very large flows of rhyolite that have occurred more recently. The youngest of these flows occurred only 70,000 years ago and is over 20 kilometers long, up to 5 kilometers across, and locally over a hundred meters thick.

Today, numerous small earthquakes across the Yellowstone Volcano indicate that the volcano is still active, and precise leveling surveys of the ground surface elevation within the youngest caldera indicate that it is anything but stable. Much of the caldera floor rose about 90 centimeters (more than 3 feet) between 1923 and 1975 and continued to undergo uplift until 1984. After remaining stationary for

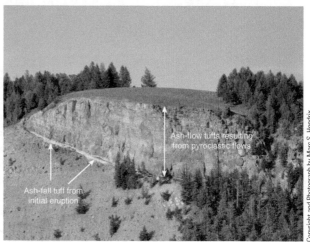

●**FIGURE 8.33** This photograph shows Member A of the Huckleberry Ridge Tuff in Yellowstone. The basal white layer is ash-fall tuff, deposited during the earliest phases of the caldera-forming eruption 2.1 million years ago. The main cliff consists of ash-flow tuffs from the pyroclastic flows that followed. The red discoloration of the rock below the tuff was caused by its residual heat.

about two years, the caldera floor then sunk about 18 centimeters (7 inches) over the next decade. More recent satellite-based measurements of the caldera floor show that it continues to undergo significant vertical movement.

Although the odds of a major, caldera-forming eruption occurring over the next few decades or century is extremely small, such an eruption would probably kill many thousands of people, entomb towns and cities beneath meters of ash, and change the courses of rivers and streams. It would probably raise a dust cloud in the upper atmosphere that would darken the Sun over the entire planet for months or years, cooling the atmosphere, altering global climate, and changing global ecosystems.

Because the Yellowstone Volcano is so geologically active, the U.S. Geological Survey has established the Yellowstone Volcanic Observatory to monitor the volcano for potential hazards to humans. Although the likelihood of a major eruption is very low, other hazards that have occurred during historic time and that are far more likely to occur in Yellowstone's near future include earthquakes, landslides, and hydrothermal explosions.

Hydrothermal explosions occur in regions in which the groundwater has been heated past the boiling point. The water does not boil because the weight of the groundwater above it produces a confining pressure. However, if that confining pressure is reduced, the superheated groundwater can suddenly flash to steam, fragmenting the overlying rock and propelling it upward along with water, steam, and mud. In 1989, an eyewitnessed hydrothermal explosion in Yellowstone ejected chunks of rock as large as a refrigerator and hurled debris up to 60 meters away. In 2006, a smaller hydrothermal explosion in Yellowstone was captured on camera (●**FIGURE 8.34**).

Much larger prehistoric hydrothermal explosions have left behind craters that range from a few hundred meters to over 2 kilometers across (●**FIGURE 8.35**).

Today, some geologists are concerned that global warming could reduce the amount of precipitation in the Yellowstone region and that this reduction in groundwater recharge could lower the regional water table. Such a lowering would reduce the confining pressure on superheated groundwater in some of Yellowstone's thermal areas and may lead to more frequent hydrothermal explosions.

Source: USGS

● **FIGURE 8.34** A small hydrothermal explosion photographed in Yellowstone Park in 2006.

---

air-fall tuff A tuff formed during an eruption by fallout of ash from the atmosphere.

welded tuff An ash-flow tuff that compacts from the weight of overlying tuff deposits and fuses together because of the residual heat from the pyroclastic flow.

Copyright and Photograph by Marc S. Hendrix

● **FIGURE 8.35** Indian Pond in Yellowstone Park is a crater left behind by a large hydrothermal explosion that occurred about 3,000 years ago.

TABLE 8.2 Some Notable Volcanic Disasters Involving 5,000 or More Fatalities, Since the Year 1500

Volcano	Country	Year	Primary Cause of Death and Number of Deaths				
			Pyroclastic Flow	Debris Flow	Lava Flow	Posteruption Starvation	Tsunami
Kelut	Indonesia	1586		10,000			
Vesuvius	Italy	1631	18,000				
Etna	Italy	1669			10,000		
Lakagigar	Iceland	1783				9,340	
Unzen	Japan	1792					15,190
Tambora	Indonesia	1815	12,000			80,000	
Krakatoa	Indonesia	1883					36,420
Pelée	Martinique	1902	29,000				
Santa Maria	Guatemala	1902	6,000				
Kelut	Indonesia	1919		5,110			
Nevado del Ruiz	Colombia	1985		>22,000			

In 1978, two U.S. Geological Survey (USGS) geologists, Dwight Crandall and Don Mullineaux, noted that Mount St. Helens had erupted more frequently and violently during the past 4,500 years than any other volcano in the contiguous 48 states. They predicted that the volcano would erupt again before the end of the 20th century.

In March 1980, about two months before the great May eruption, puffs of steam and volcanic ash rose from the crater of Mount St. Helens, and swarms of earthquakes occurred beneath the mountain. This activity convinced other USGS geologists that Crandall and Mullineaux's prediction was correct. In response, they installed networks of seismographs, tiltmeters, and surveying instruments on and around the mountain.

In the spring of 1980, the geologists warned government agencies and the public that Mount St. Helens showed signs of an impending eruption. The U.S. Forest Service and local law enforcement officers quickly evacuated the area surrounding the mountain, averting a much larger tragedy (●FIGURE 8.36).

U.S. Geological Survey

● **FIGURE 8.36** U.S. Geological Survey geologists accurately predicted the May 1980 eruption of Mount St. Helens.

VOLCANIC ERUPTIONS AND GLOBAL CLIMATE

A volcanic eruption can profoundly affect the atmosphere, the climate, and living organisms, thereby providing an excellent example of systems interactions. For instance, the 1991 eruptions of Mount Pinatubo in the Philippines produced the greatest ash and sulfur clouds in the latter half of the 20th century. Satellite measurements show that the total solar radiation reaching Earth's surface declined by 2 to 4 percent after the Pinatubo eruptions (●FIGURE 8.37). The following two years, 1992 and 1993, were a few tenths of a degree Celsius cooler than the temperatures of the previous decade. Temperatures rose again in 1994, after the ash and sulfur settled out. ●FIGURE 8.38 shows a compilation of five geologically recent eruptions and the temperature either observed or calculated from isotopic information preserved in ice cores.

Another example of a volcanic eruption that affected climate occurred in 1783 in Iceland, when the largely non-explosive eruption of the Laki crater occurred during June of that year. The eruption lasted nearly 9 months and produced a bluish haze of sulfur aerosols across Iceland that subsequently spread across Europe. This haze obscured the Sun, significantly reducing the solar energy reaching the surface.

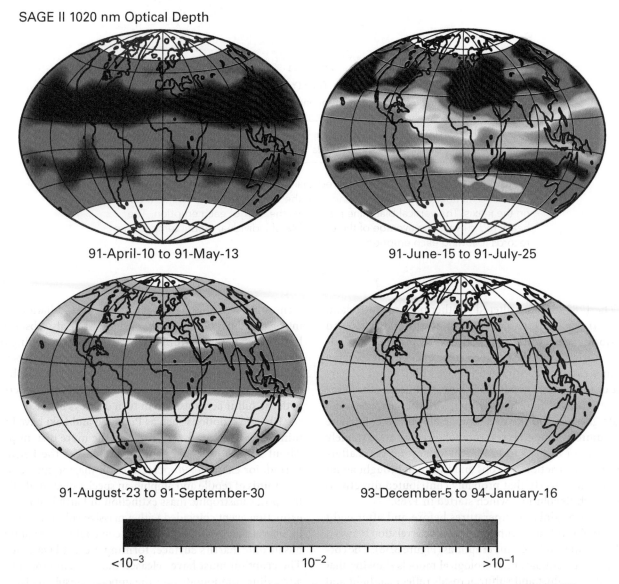

● **FIGURE 8.37** The images above were acquired by the Stratospheric Aerosol and Gas Experiment II (SAGE II) flying aboard NASA's Earth Radiation Budget Satellite (ERBS). The false-color images were taken during four different time spans, ranging from before the June 1991 eruption of Mount Pinatubo in the Philippines to two years after the eruption. The colors represent measurements of how much light is prevented from passing through the stratosphere by sulfur aerosols, which are small droplets of sulfuric acid that formed from sulfur dioxide erupted from Mount Pinatubo. Red colors represent the most blockage of light through the stratosphere by the sulfur aerosols. Notice how the volcanic plume gradually spreads across the entire globe and thus had a global-scale impact on climate.

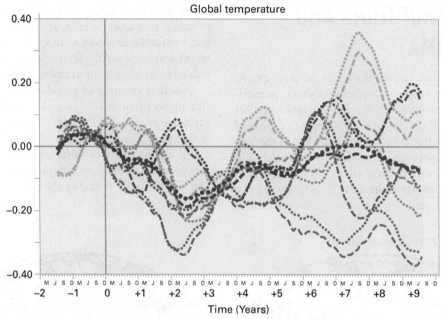

Global temperature

Impact of explosive volcanic eruptions on the main climate variability modes

| •••• HadCRU | – – – Krakatau | – – – Mt Agung | – – – Pinatubo |
| – – GISS | – – – Sta Maria | – – – El Chichón | – – – Ens. Mean |

● **FIGURE 8.38** This plot shows observed changes in global temperature on the y-axis and number of years before and after the start of a major volcanic eruption on the x-axis. The vertical black line represents the start of each eruption. The heavy red curve is the average of the ensemble of individual global temperature records following each eruption.

The 1783 eruption altered weather patterns in Iceland and Europe. In Iceland, violent thunderstorms and hailstorms killed cattle and destroyed crops. The crop failure resulting from the reduced solar energy and extreme weather events are estimated to have killed about 24 percent of the human population there. In Europe, the summer of 1783 was more like a winter, with the Sun remaining a pale ghost in the sky or a strange blood-red color. The cold summer temperatures were followed by an extremely harsh winter in 1783 to 1784, and for several years afterward the destruction of crops and livestock brought about famine and poverty that probably contributed directly to the French Revolution, which started in 1789.

A plot of global temperatures before and after eight recent major volcanic eruptions shows a correlation between global cooling and volcanic eruptions (Figure 8.38). The correlation substantiates meteorological models showing that high-altitude dust and sulfur aerosols reflect sunlight and cool the atmosphere. The primary impact of volcanic eruptions on the climate comes from the conversion of sulfur dioxide (SO_2) to sulfuric acid (H_2SO_4), which condenses to

form sulfur aerosols. These aerosols reflect radiation back into space, thus cooling Earth's surface, while also absorbing heat that radiates up from the Earth, thereby warming the stratosphere. The sulfur aerosols also change the types of chlorine and nitrogen molecules in the upper atmosphere. Chlorine monoxide (ClO) is produced through reaction with the sulfur aerosols, and in turn destroys ozone.

Historic eruptions such as Pinatubo and Laki have been minuscule compared with some in the more distant past. About 248 million years ago, at the end of the Permian Period, for example, 90 percent of all marine species and two-thirds of reptile and amphibian species died suddenly in the most catastrophic mass extinction in Earth history. This extinction event coincided with a massive volcanic eruption in Siberia that disgorged a million cubic kilometers of flood basalt onto Earth's surface, forming a great lava plateau. The eruption must have released massive amounts of ash and sulfur compounds into the upper atmosphere, leading to cooler global climates. Many geologists think that Earth cooled enough to cause, or at least contribute to, the Permian mass extinction.

Key Concepts Review

- Rocks of the asthenosphere partially melt to produce basaltic magma as a result of three processes: rising temperature, pressure-release melting, and addition of water. These processes occur beneath spreading centers, in mantle plumes, and in subduction zones to form both volcanoes and plutons.
- Basalt makes up most of the oceanic crust, and granite is the most abundant rock in continents. Basaltic magma forms by partial melting of mantle peridotite. Granitic magma forms when basaltic magma rises into and melts granitic rocks of the lower continental crust.
- Earth's earliest continents were probably formed by partial melting of the original peridotite crust to form basalt, and then by further partial melting of the basalt to form andesite and then granite.
- Basaltic magma usually erupts in a relatively gentle manner onto Earth's surface from a volcano. In contrast, granitic magma typically solidifies within Earth's crust. When granitic magma does erupt onto the surface, it often does so violently. These contrasts in behavior of the two types of magma are caused by differences in silica and water content.
- A pluton is any intrusive mass of igneous rock. A batholith is a pluton with more than 100 square kilometers of exposure at Earth's surface. A dike and a sill are both sheetlike plutons. Dikes cut across layering in country rock, and sills run parallel to layering.
- Magma may flow onto Earth's surface as lava or may erupt explosively as pyroclastic material. Fluid lava forms lava plateaus and shield volcanoes. A pyroclastic eruption may form a cinder cone. Alternating eruptions of fluid lava and pyroclastic material from the same vent create a composite cone.
- When granitic magma rises to Earth's surface, it may erupt explosively, forming ash-flow tuffs and calderas.
- Volcanic eruptions are common near a subduction zone, near a spreading center, and at a hot spot over a mantle plume, but are rare in other environments. Eruptions on a continent are often violent, whereas those in oceanic crust are gentle. Such observations form the basis of regional predictions of volcanic hazards. Short-term predictions are made on the basis of earthquakes caused by magma movements, swelling of a volcano, increased emissions of gas and ash from a vent, and other signs that magma is approaching the surface.
- Large volcanic episodes affect the atmosphere, climate, and living organisms.

Important Terms

aa (p. 180)	fissures (p. 182)	stock (p. 178)
air-fall tuff (p. 192)	flood basalt (p. 182)	stratovolcano (p. 185)
ash-flow tuff (p. 188)	lava plateau (p. 182)	super-heating (p. 171)
batholith (p. 178)	pahoehoe (p. 180)	tectonic accretion (p. 176)
caldera (p. 188)	partial melting (p. 175)	vent (p. 183)
cinder cone (p. 185)	pluton (p. 178)	vesicles (p. 181)
cinders (p. 182)	pressure-release melting (p. 171)	volcanic ash (p. 182)
columnar joints (p. 181)	pyroclastic flow (p. 187)	volcano (p. 180)
composite cone (p. 185)	pyroclastic rock (p. 182)	welded tuff (p. 192)
crater (p. 183)	shield volcano (p. 183)	
dike (p. 178)	sill (p. 180)	

Review Questions

1. Describe several ways in which volcanoes and volcanic eruptions can threaten human life and destroy property.
2. Describe three processes that generate magma in the asthenosphere.
3. Describe magma formation in a spreading center, a hot spot, and a subduction zone.
4. Describe the origin of granitic magma.

5. How much silica does average granitic magma contain? How much does basaltic magma contain?

6. How much water does average granitic magma contain? How much does basaltic magma contain?

7. Why does magma rise soon after it forms?

8. Explain why basaltic magma and granitic magma behave differently as they rise toward Earth's surface.

9. Many rocks and even entire mountain ranges at Earth's surface are composed of granite. Does this observation imply that granite forms at the surface?

10. Explain the difference between a dike and a sill.

11. How do a shield volcano, a cinder cone, and a composite cone differ from one another? How are they similar?

12. How does a composite cone form?

13. How does a caldera form?

14. Explain why additional eruptions of the Yellowstone Volcano seem likely. Describe what such an eruption might be like.

PhotoXite/Shutterstock.com

9

MOUNTAINS

Early morning view of the Teton Mountains, Wyoming, as seen from the northeast. The high peak is the Grand Teton; to its left and right are the Middle Teton and Mount Teewinot. The very abrupt, steep front of the mountain range results from offset along the Teton Fault, an active normal fault separating the sediment and rock of the flat valley in the foreground from the uplifted rock forming the rugged Teton Range. The fault runs along the base of the mountain front and is highlighted by the dotted red line. The two sunlit terraces on the valley floor are the tops of now-inactive glacial outwash fans that have been partially eroded by Snake River.

FOLDS AND FAULTS: GEOLOGIC STRUCTURES

How Rocks Respond to Tectonic Stress

Continents move at rates between 1 and 16 centimeters per year, about as fast as your fingernails grow. Although this movement is too slow to feel or sense in any ordinary way, try to imagine the immense forces involved as continent- or ocean-sized slabs of rock slowly grind past one another or collide head on in slow motion. In Chapters 7 and 8, we studied how tectonic activity generates earthquakes and volcanic eruptions. In this chapter, we will consider how rock crumples and buckles under tectonic forces to form mountains.

Stress is a force directed against an object. Stress is conventionally measured as force per unit area, or pressure. When a rock is stressed, the rock may deform elastically or plastically, or it may simply break by brittle fracture, as described in Chapter 7. Several factors control how a rock responds to stress:

1. *The nature of the rock.* Think of a quartz crystal, a gold nugget, and a rubber ball. If you strike quartz with a hammer, it shatters—that is, it fails by brittle fracture. In contrast, if you strike the gold nugget, it deforms in a plastic manner—it flattens and stays flat. If you hit the rubber ball, it deforms elastically and rebounds immediately, sending the hammer flying back at you. Initially, all rocks react to stress by deforming elastically by a slight amount. Near Earth's surface, where temperature is relatively low, different types of rocks behave differently to stress. Granite and quartzite tend to fracture in a brittle manner. Other rocks, such as mudstone, limestone, and marble, tend to deform in a more plastic manner.

2. *Temperature.* The higher the temperature, the greater the tendency of a rock to deform in a plastic manner. It is difficult to bend an iron bar at room temperature, but if the bar is heated to a red-hot temperature, it becomes plastic and bends easily.

3. *Pressure.* Both temperature and pressure increase during burial, and both of these factors promote plastic deformation. Thus, deeply buried rocks have a greater tendency to bend and flow under stress than do shallow rocks.

4. *Time.* Stress applied slowly, rather than suddenly, also favors plastic behavior. Over 100 years, marble park benches in New York City have sagged plastically under their own weight. In contrast, rapidly applied stress, such as the blow of a hammer, would cause the same marble to fracture.

Geologic Structures

Tectonic movement creates tremendous stress near plate boundaries; this stress deforms rocks. A **geologic structure** is any feature produced by rock deformation. Tectonic stress creates three types of structures: folds, faults, and joints.

FOLDS   A **fold** is a bend in rock. Some folded rocks display little or no fracturing, indicating that the rocks deformed in a plastic manner (●**FIGURE 9.1**). In other cases, folding occurs by a combination of plastic deformation and brittle fracture. Folds formed in this manner exhibit many tiny fractures.

● **FIGURE 9.1** This rock has undergone plastic deformation.

Figure 9.4 shows the basic parts of a fold. The sides of a fold are called the **limbs**. The **axial surface** of a fold is an imaginary surface that passes through all the points of maximum curvature on the fold. A fold arching upward is called an **anticline**, and one arching downward is a **syncline**. Even though an anticline is structurally a high point in a fold, anticlines do not necessarily form topographic ridges. Conversely, synclines do not always form valleys. For example, in ●**FIGURE 9.5**, a syncline lies beneath the peak and an anticline forms a saddle between two peaks, simply because some portions of the folded rock eroded faster than others.

The four common fold types shown in ●**FIGURE 9.6** are differentiated based on the orientation of their axial surface and the **facing direction** of the two fold limbs. In a symmetric fold the axial surface is vertical, and in an asymmetric fold it is tilted. Overturned folds also have a tilted axial surface, but one of the limbs faces down. In a recumbent fold, the axial surface is horizontal and one limb is completely upside down.

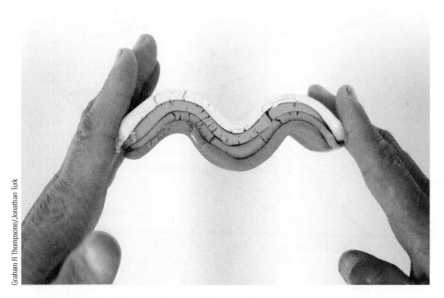

●**FIGURE 9.2** Clay is pliable and deforms into folds when compressed.

If you hold a sheet of clay between your hands and squeeze your hands together, the clay deforms into a sequence of folds (●**FIGURE 9.2**). This demonstration illustrates three characteristics of folds:

1. Folding usually results from compression. For example, the tightly folded limestone shown in ●**FIGURE 9.3** from southwest Montana formed by tectonic compression about 70 million years ago.
2. Folding always shortens the horizontal distances in rock. Notice in ●**FIGURE 9.4** that the distance between the two points A and A′ is shorter in the folded rock than it was before folding.
3. A fold usually occurs as part of a group of many similar folds.

---

stress The force exerted against an object, usually measured as force per unit area or pressure.

geologic structure Any feature produced by rock deformation, such as a fold or a fault; also refers to the combination of all such features in an area or region.

fold A bend in rock.

limbs The sides of a fold in rock.

axial surface An imaginary surface that connects all the points of maximum curvature in a fold.

anticline A fold in rock that arches upward; the oldest rocks are in the middle.

syncline A fold in rock that arches downward, and whose center contains the youngest rocks.

facing direction The direction, up or down, corresponding to the original stratigraphic up-direction of layered sedimentary rocks in which younger layers overlie older layers. In overturned and recumbent folds, one limb is facing down.

●**FIGURE 9.3** This limestone in southwest Montana was folded by tectonic compression about 70 million years ago.

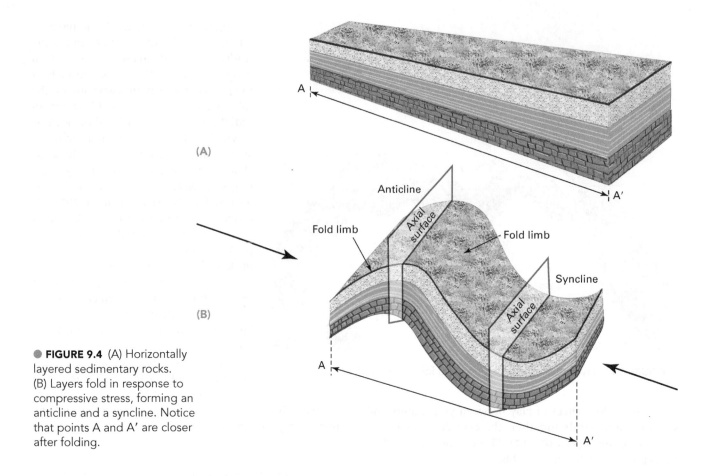

(A)

(B)

● **FIGURE 9.4** (A) Horizontally layered sedimentary rocks. (B) Layers fold in response to compressive stress, forming an anticline and a syncline. Notice that points A and A' are closer after folding.

● **FIGURE 9.5** On this peak in the Canadian Rockies, the syncline lies beneath the summit and the anticline forms the low point on the ridge.

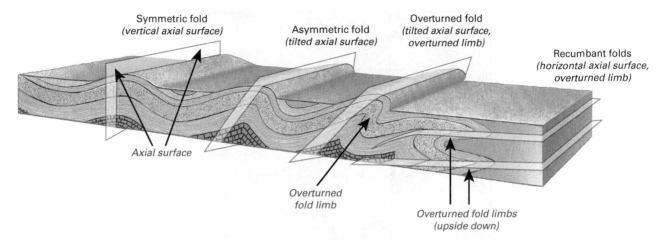

● FIGURE 9.6 Folds are classified on the basis of their axial surface orientations and the facing direction of the two limbs. Folds can be symmetrical (vertical axial surface) or asymmetrical (tilted axial surface). A fold with one tilted axial surface and one limb facing downward (i.e., upside down) is overturned. Recumbent folds also have one downward-facing limb but have a subhorizontal axial surface.

A circular or elliptical anticlinal structure is called a **dome**. A dome resembles an inverted cereal bowl. Sedimentary layering dips away from the top of a dome in all directions (**●FIGURE 9.7A**). A similarly shaped syncline is called a **basin** (**●FIGURE 9.7B**). Domes and basins can be small structures only a few kilometers or less in diameter. Alternatively, a large basin or dome can be hundreds of kilometers in diameter. These huge structures form by the sinking or rising of continental crust in response to vertical movements of the underlying mantle. An example of a large structural dome is the Black Hills in South Dakota (**●FIGURE 9.8**). Large structural basins include the Michigan basin (**●FIGURE 9.9**), which covers much of the state of Michigan, and the Williston basin (**●FIGURE 9.10**), which covers much of eastern Montana, northeastern Wyoming, the western Dakotas, and southern Alberta and Saskatchewan.

FAULTS A **fault** is a fracture along which rock on one side has moved relative to rock on the other side (**●FIGURE 9.11**). **Slip** is the distance that rocks on opposite sides of a fault have moved. Movement along a fault may be gradual, or the rock may move suddenly, generating an earthquake. Some faults occur as single fractures in rock; most, however, consist of numerous closely spaced fractures within a **fault zone** (**●FIGURE 9.12**). Fault zones can range in thickness from less

than 1 centimeter to over a kilometer. Over time, rock may slide hundreds of meters or many kilometers along a large fault zone.

Rock moves repeatedly along many faults and fault zones for two reasons: (1) Tectonic forces commonly persist in the same place over long periods of time (for example, at

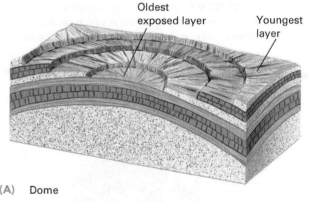

(A)   Dome

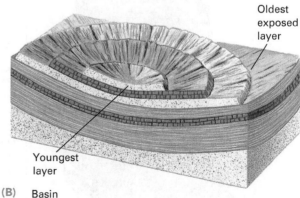

(B)   Basin

● FIGURE 9.7 (A) Sedimentary layering dips away from a dome in all directions. (B) Layering dips toward the center of a basin.

---

dome A circular or elliptical anticlinal structure resembling an inverted cereal bowl.

basin A bowl-shaped synclinal structure, commonly filled with sediment.

fault A fracture in rock along which one side has moved relative to the other side. Compare with *joint*.

slip The distance that rocks on opposite sides of a fault have moved relative to each other.

fault zone An area of numerous, closely spaced faults.

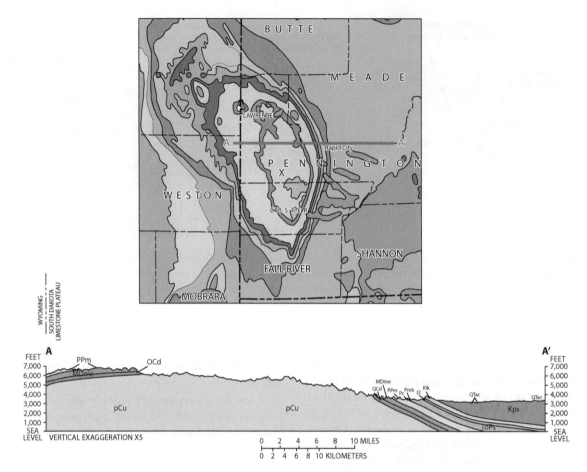

● **FIGURE 9.8** Geologic map and cross section of the Black Hills, South Dakota. The Black Hills are a classic example of a dome in which sedimentary rocks dip away from the center of the dome where the oldest rocks, in this case Precambrian basement rock, are exposed.

Source: https://d32ogoqmya1dw8.cloudfront.net/im ages/research_education/nativelands/pine ridge/geocs.gif

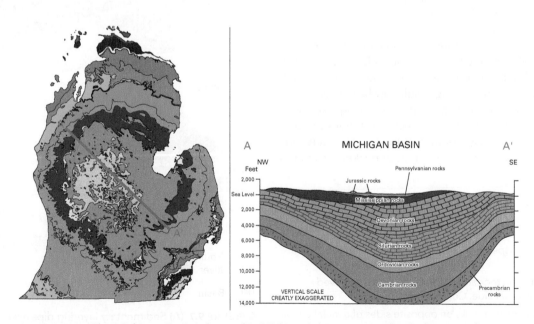

● **FIGURE 9.9** The Michigan Basin is located in Michigan's lower peninsula. Sedimentary layers on the margins of the basin dip toward its center where the youngest rocks are exposed.

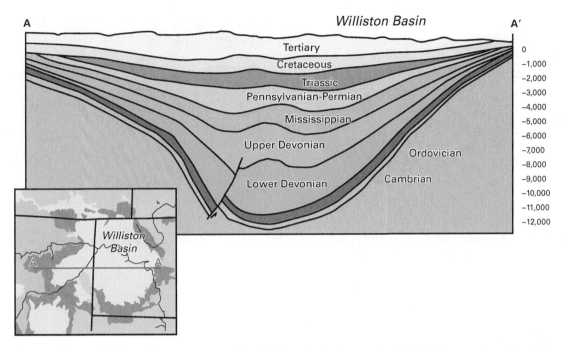

● FIGURE 9.10 Geological cross section of the Williston basin in western North Dakota, eastern Montana, and adjacent parts of Alberta and Saskatchewan.

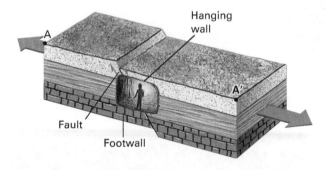

● FIGURE 9.11 A fault is a fracture along which rock on one side has moved relative to rock on the other side. The upper side of a fault is called the hanging wall, and the lower side is called the footwall.

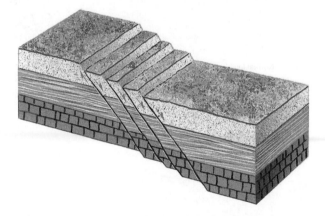

● FIGURE 9.12 Faults with a large slip commonly move along numerous closely spaced fractures, forming a fault zone.

a tectonic plate boundary). (2) Once a fault forms, it is easier for rock to move again along the same fracture than for a new fracture to develop nearby.

Hydrothermal solutions often precipitate rich ore veins in the fractures of a fault zone. Miners then dig shafts and tunnels along veins to get the ore. Many faults are not vertical, but dip at an angle. Therefore, many veins have an upper

---

hanging wall A term to describe the upper side of an inclined fault or vein (i.e., the rock hanging above one's head).

footwall A term to describe the lower side of an inclined fault or vein (i.e., the rock one would walk on).

normal fault A fault in which the hanging wall has moved downward relative to the footwall.

graben A wedge-shaped block of rock that has dropped downward between two normal faults, forming a valley.

and a lower side. Miners refer to the side that hangs over their heads as the **hanging wall** and the side they walk on as the **footwall** (Figure 9.11). These names are commonly used to distinguish between the two sides of a fault.

A **normal fault** forms where tectonic movement stretches Earth's crust, pulling it apart. As the crust stretches and fractures, the hanging wall moves down relative to the footwall. Notice that the horizontal distance between points on opposite sides of the fault, such as A and A′ in ●FIGURE 9.13, is greater after normal faulting occurs.

●FIGURE 9.14 shows a wedge or keystone-shaped block of rock called a **graben** that has dropped downward between pairs of normal faults. The word *graben* comes from the German word for "ditch or valley." (Think of a large block of

● **FIGURE 9.13** (A) A normal fault accommodates extension of the crust. (B) These layers of sandstone and shale have been offset about 30 centimeters by a small normal fault.

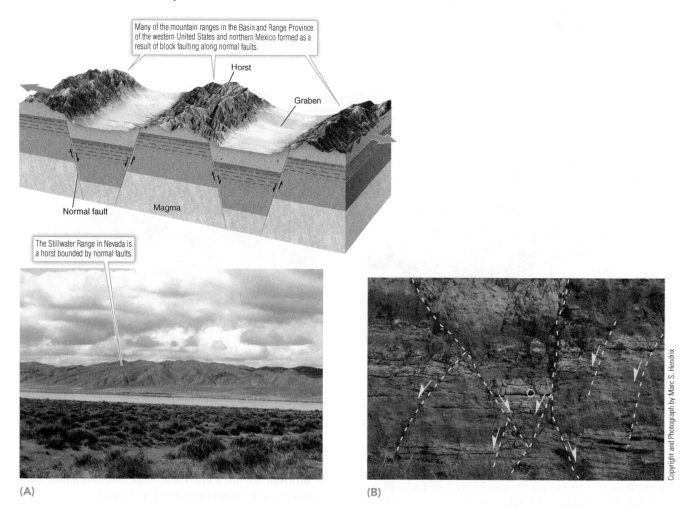

Many of the mountain ranges in the Basin and Range Province of the western United States and northern Mexico formed as a result of block faulting along normal faults.

Horst

Graben

Normal fault   Magma

The Stillwater Range in Nevada is a horst bounded by normal faults.

(A)   (B)

● **FIGURE 9.14** Horsts and grabens commonly form where tectonic forces stretch the crust over a broad area, generating a basin-and-range topography. (A) View across a down-dropped basin and a normal fault to the adjacent horst block. The normal fault is located at the foot of the mountain range. (B) A small graben and several normal faults in a roadcut along the Gulf of Corinth, central Greece. A camera lens (white circle) is on the graben for scale.

rock settling downward to form a valley.) If tectonic forces stretch the crust over a large area, many normal faults may develop, allowing numerous grabens to settle downward between the faults. The block of rock between two down-dropped grabens that appears to have moved upward relative to the grabens is called a **horst**.

Normal faults, grabens, and horsts are common where the crust is rifting at a spreading center, such as the Mid-Oceanic Ridge and the East African Rift zone, and where tectonic forces stretch a single plate, as in the Basin and Range Province of Utah, Nevada, and adjacent parts of western North America. They reflect an extension of Earth's crust.

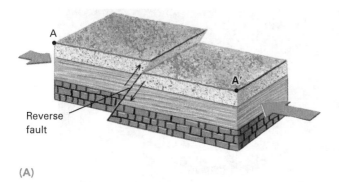

(A)

(B)

● **FIGURE 9.15** (A) A reverse fault accommodates crustal shortening when the crust is compressed. (B) This photograph shows several small reverse faults cutting across sandstone layers in Oregon.

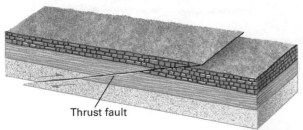

(A)

● **FIGURE 9.16** (A) A thrust fault is a low-angle, nearly horizontal compressional fault. (B) A classic, major thrust fault, called the Lewis Overthrust, in Montana. The Precambrian rock forms the hanging wall which has been pushed up and over younger Cretaceous rock which forms the footwall.

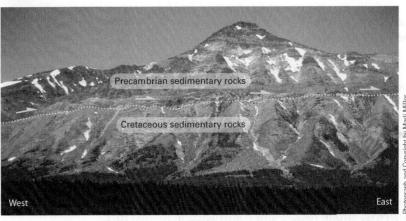

(B)

In contrast, horizontal compressive forces may fracture rock to produce a **reverse fault** (●FIGURE 9.15) in which the hanging wall moves upward relative to the footwall. In ●FIGURE 9.15A, the distance between points A and A′ is shortened by the faulting.

A **thrust fault** is a special type of reverse fault that is nearly horizontal (●FIGURE 9.16). In some thrust faults, the rocks of the hanging wall have moved many kilometers over

the footwall. For example, all of the rocks of Glacier National Park in northwestern Montana slid 50 to 100 kilometers eastward along one or more thrust faults to their present location. One of these faults is the Lewis Overthrust (●FIGURE 9.16B). These thrusts formed from about 180 to 55 million years ago as compressive tectonic forces built the mountains of western North America. During this time, widespread thrust faulting moved large slabs of rock, some even larger than that of Glacier Park, from west to east in a zone extending from Alaska to Mexico.

Recall from Chapter 7 that a *strike-slip fault* is one in which the fracture is vertical, or nearly so, and rocks on opposite sides of the fracture move horizontally past each other. A transform plate boundary is a strike-slip fault zone. As explained previously, the famous San Andreas Fault is the main fault within a zone of many strike-slip faults that

---

horst The block of rock between two grabens, which has moved relatively upward along normal faults as the grabens have settled downward.

reverse fault A fault in which the hanging wall has moved up relative to the footwall.

thrust fault A type of reverse fault that is nearly horizontal, with a dip of 45 degrees or less over most of its extent.

Copyright and Photograph by Marc S. Hendrix

● **FIGURE 9.17** Joints such as these in northern Queensland, Australia, are fractures along which the rock has not slipped.

collectively form the boundary between the Pacific Plate and the North American Plate.

JOINTS Unlike a fault, a **joint** is a fracture across which the rock on both sides does not slip. In Chapter 8, we discussed columnar joints that form in basalt as it cools. Tectonic forces also create joints (●**FIGURE 9.17**). Most rocks near Earth's surface are jointed, but joints become less abundant with depth because rocks become more plastic and less prone to fracturing at deeper levels in the crust.

Joints and faults are important in engineering, mining, and quarrying because they are planes of weakness in otherwise strong rock. Dams constructed in jointed rock often leak, not because the dams themselves have holes, but because water seeps into the joints and flows around the dam through the fractures. You can commonly observe seepage caused by such leaks in canyon walls downstream from a dam.

Folds, Faults, and Plate Boundaries

Each of the three types of plate boundaries—divergent, transform, and convergent—produces different tectonic stresses and therefore are typified by different kinds of folds and faults. Tectonic plates drift apart at a divergent boundary (the Mid-Oceanic Ridge system and continental rifts), stretching adjacent rock and producing normal faults and grabens. Relatively minor folding can occur in rock near normal faults, though typically the folding is not as intense as at convergent or transform settings.

At a transform boundary, friction often holds rock together as the plates undergo movement relative to each other. Small bends in the fault plane and interactions among different faults within the zone can cause rock near the plate boundary to be sheared and uplifted or sheared and downdropped, depending on the direction the fault bends relative to its offset direction. For example, the San Andreas Fault in California is a right-lateral strike-slip fault, meaning that each side of the fault moves to the right as viewed by a person standing on the opposite fault block. If in a birds-eye view, the fault bends to the left, then the right-lateral strike-slip motion will cause compression along the bend in addition to the horizontal shearing. A left-lateral strike-slip fault that bends to the right also will result in a zone where both horizontal shearing and compression occur. Such a fault bend is called a restraining bend (●**FIGURE 9.18**).

In southern California, the San Gabriel and San Bernadino Mountains (●**FIGURE 9.19**) exist because of a restraining bend in the San Andreas Fault. In fact, enough uplift has occurred along this restraining bend that these two coastal mountain ranges boast the only downhill ski areas in all of Southern California and the only ski areas anywhere in the California Coast Ranges. (All other California ski areas are in the Sierra Nevada.) In many cases, interactions between two faults can produce a similar result if the two faults are parallel but offset. Such a geometry is called

joint A fracture along which the rock on either side of the break does not move. Compare with *fault*.

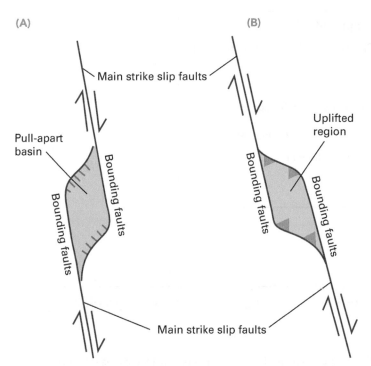

● FIGURE 9.18 (A) Releasing bends and releasing steps both involve extension in addition to horizontal shear. (B) Restraining bends and restraining steps both involve compression in addition to horizontal shear.

a restraining step, because the region between the two faults is compressed as it is sheared. At the same time, a right bend in a right-lateral fault (or a left bend in a left-lateral fault) causes extension in addition to lateral shearing. These types of bends are called releasing bends, and they cause the crust to be downdropped relative to the surrounding region. Similarly, a right step in a right-lateral fault is called a releasing step. The Santa Barbara basin and several other offshore basins between the California coastline and the Catalina Islands formed due to releasing bends and releasing steps in the San Andreas Fault system

Near a convergent plate boundary, compression commonly produces large regions of folds, reverse faults, and thrust faults. Folds and thrust faults are common in the mountains of western North America, the Appalachian Mountains, the Alps, and the Himalayas, all of which formed at convergent boundaries (**●FIGURE 9.20**).

Although plate convergence typically creates horizontal compression, in some instances crustal extension and normal faulting also occur at a convergent plate boundary. For example, the Himalayas and Tibetan Plateau formed by the collision between India and southern Asia. Yet, normal faults are common in both regions. We will describe how this occurs in the section on "The Himalayas: A Collision between Continents."

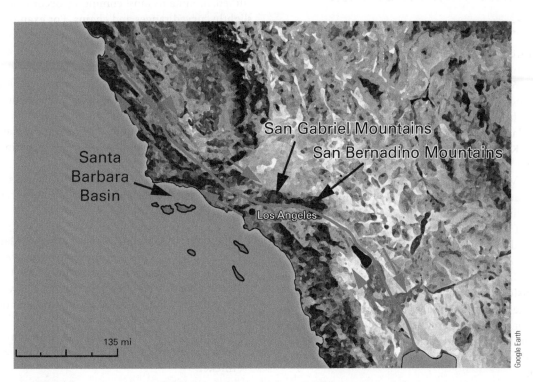

● FIGURE 9.19 In southern California, the San Andreas Fault has a major left step, causing compressional deformation that has resulted in uplift of the San Gabriel and San Bernadino Ranges near Los Angeles. Despite being located in southern California, both ranges are high enough to boast several downhill ski areas—the only ski areas in southern California.

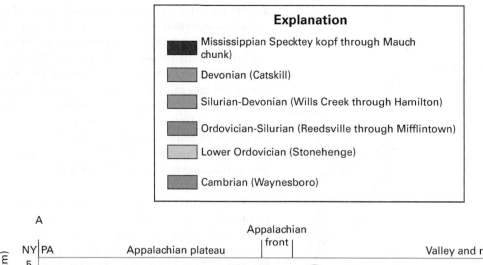

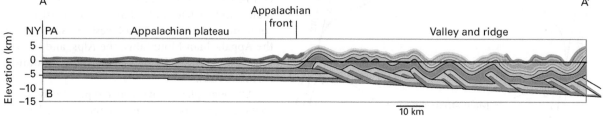

● **FIGURE 9.20** This cross section through a portion of the Appalachian Mountains in Pennsylvania shows numerous thrust faults and folds that were formed during continent–continent collision between North America and Africa.

MOUNTAINS AND MOUNTAIN RANGES

Tectonic forces have created mountains at each of the three types of tectonic plate boundaries. As you learned in Chapter 6, the world's largest mountain chain, the Mid-Oceanic Ridge system, formed at divergent plate boundaries beneath the ocean. Mountains also rise at divergent plate boundaries on land. Mount Kilimanjaro and Mount Kenya, two volcanic peaks near the equator, lie along the East African Rift. Other ranges, such as the San Gabriel Mountains of California, formed at transform plate boundaries. However, the volcanic island arcs of the southwestern Pacific Ocean, Alaska's Aleutian Islands, and the great continental mountain chains—including the Andes, Appalachians, Alps, Himalayas, and Rockies—all formed near convergent plate boundaries (●**FIGURE 9.21**). Folding and faulting of rocks, earthquakes, volcanic eruptions, intrusion of plutons, and metamorphism all occur at a convergent plate boundary. The term **orogeny** (from the Greek for "mountain producing") refers to the process of mountain building and includes all of these processes. The belt of rocks that is deformed in an orogeny is called an **orogen**.

Because plate boundaries are nearly linear or slightly curved, mountains most commonly occur as long, linear, or slightly curved ranges and chains. For example, the Andes extend in a narrow band along the west coast of South America, and the Appalachians form a gently curving uplift along the east coast of North America.

Recall that isostasy refers to a state of gravitational equilibrium between the lithosphere and asthenosphere such that the elevation at which a tectonic plate "floats" depends on its thickness and density. Where tectonic plates converge, the lithosphere becomes thicker, causing continental mountain ranges to rise isostatically.

Several processes can affect the thickness of the lithosphere in mountainous regions and, in turn, cause isostatic adjustment of the crust:

1. In a subduction zone, the descending slab generates magma, which rises to cool within the overlying lithosphere and forms plutons or erupts onto the surface to form volcanic peaks. Both the plutons and volcanic rocks thicken the lithosphere by adding volumes of new material to it.
2. Magmatic activity heats the lithosphere above a subduction zone, causing it to expand and become thicker.

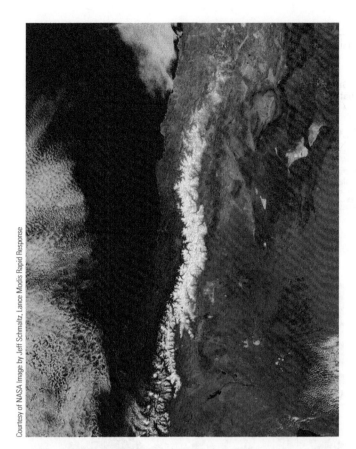

● **FIGURE 9.21** The central Andes Mountains as seen from satellite. Like many mountain ranges on Earth, the Andes is part of an orogen that formed parallel to the nearest plate boundary, in this case the subduction zone located offshore that separates the South American from Nazca Plates.

3. Compressive tectonic forces squeeze the crust horizontally and increase its thickness by folding and faulting.

4. In a region where two continents collide, such as the modern Himalayas, one continent may be forced beneath the other. This process, called **underthrusting**, can double the thickness of continental crust in the collision zone.

---

orogeny The process of mountain building; all tectonic processes associated with mountain building.

orogen The belt of rock that is deformed in an orogeny.

underthrusting The process by which one continent subducts beneath the other during a continent–continent collision.

island arc A gently curving chain of volcanic islands in the ocean formed by the convergence of two plates, each bearing ocean crust, and the resulting subduction of one plate beneath the other.

subduction complex The structurally complicated mass of rock consisting of deformed seafloor sediment and fragments of basaltic oceanic crust and upper mantle material that is scraped from the upper layers of the subducting slab in a subduction zone and added to the overriding plate.

5. In continental collision zones where crustal thickening is extreme, partial melting of the middle crust and the intense compression can cause lateral and upward extrusion of plastically deforming rock. This process has caused the southward extrusion of the Greater Himalayan Sequence, a large mass of high-grade metamorphic and igneous rocks.

As mountain peaks rise, streams, glaciers, and landslides erode them, carrying the sediment into adjacent valleys. Initially, when the mountains erode, they become lighter and rise isostatically, just as a canoe rises when you step out of it. Eventually, however, erosion wins over isostatic rebound. The Appalachians are an old range where erosion is now wearing away the remains of peaks that may once have been the size of the Himalayas.

Several of the processes that affect the isostatic position of a mountainous region are illustrated in ●**FIGURE 9.22A**, a cross section through the modern Himalayas and the Tibetan Plateau.

With this background, let's look at mountain building in three types of convergent plate boundaries: in the ocean, at a continental margin, and between continents.

ISLAND ARCS: SUBDUCTION WHERE TWO OCEANIC PLATES CONVERGE

An **island arc** is a volcanic mountain chain that forms where two plates carrying oceanic crust converge. The convergence causes the older, colder, and denser plate to sink into the mantle beneath the other plate, creating a subduction zone and an oceanic trench (●**FIGURE 9.23**). Magma forms in the subduction zone and rises to build submarine volcanoes. Typically, these volcanoes grow above sea level, creating an arc-shaped volcanic island chain that runs parallel to the trench and exists on the overriding plate.

A layer of sediment a half kilometer or more thick commonly covers the oldest basaltic crust of the deep seafloor. As the two plates converge, some of the sediment is scraped off of the subducting slab and plastered against the side of the overriding plate. Occasionally, slices of basalt from the oceanic crust, underwater mountains called seamounts, and even pieces of the upper mantle are scraped off and mixed in with the seafloor sediment. The process is like a bulldozer scraping soil from bedrock and occasionally knocking off a chunk of bedrock along with the soil. Rock and sediment added in this way to the over-riding plate forms a **subduction complex**, characterized by abundant folds, faults, and fractures that result from the scraping and compression (Figure 9.23).

Parts of the California Coast Ranges expose a large subduction complex that was built during the Mesozoic Era as the Farallon Plate subducted beneath western North

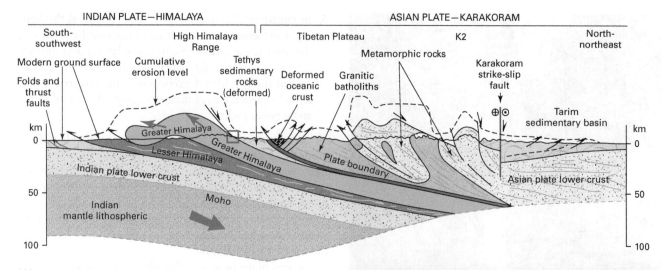

INDIAN PLATE—HIMALAYA   **ASIAN PLATE—KARAKORAM**

(A)

(B)

● **FIGURE 9.22** (A) Several factors affect the height of the Himalayas and the Tibetan Plateau mountainous regions. Note the vertical scale on the left. The regions have been uplifted by underthrusting of the Indian Plate under the Asian Plate. As the mountains grew upward, they underwent erosion. In this cross section, the volume of rock removed by erosion is represented by the area between the "cumulative erosion level" and the modern ground surface. The intense tectonic compression also has caused the southward extrusion of plastically deforming rock from the Greater Himalayan Sequence, like toothpaste being squeezed from a giant tube (small red arrows). East–west stretching of Tibet, probably due to the north–south compression, also has thinned the upper crust there. Collectively, these processes have caused the entire region to rise isostatically. Mount Everest is located in the small red box. (B) The Greater Himalayan Sequence is crystalline rock that has been squeezed southward from midcrustal depths. A major thrust fault forms the lower boundary of this rock mass, while a major normal fault forms the upper boundary. The normal fault puts deformed sedimentary rocks from the collapsed Tethys Sea over rocks of the Greater Himalayan Sequence. This photo of Mount Everest shows the position of the normal fault just below the mountain's summit.

America and the Sierra Nevada was an active continental arc. Although now completely subducted, the Farallon Plate transported oceanic crust eastward toward the convergent margin for tens of millions of years. The longevity of this ancient subduction zone caused the growth of a large subduction complex that incorporated many different elements of marine geology, including bits of oceanic crust, deep water sediments, and even bits of metamorphic rock from deeper in the subduction zone. Today, much of San Francisco, including the Golden Gate Bridge, is built on rocks belonging to the ancient subduction complex (●FIGURES 9.25 and 26).

Growth of the subduction complex occurs by underthrusting, in which sediment and rock are added to the bottom of the complex, forcing it to grow upward. In addition,

underthrusting thickens the crust, leading to isostatic uplift of the entire complex. Sediment eroded from the volcanic arc and transported towards the sea can be impounded by the rising subduction complex, forming a sedimentary basin called a **forearc basin** (●FIGURES 9.23 and 24).

Island arcs are abundant in the Pacific Ocean, where convergence of oceanic plates is common. The western Aleutian Islands and most of the island chains of the southwestern Pacific are island arcs (●FIGURE 9.27).

THE ANDES: SUBDUCTION AT A CONTINENTAL MARGIN

The Andes are the world's second-highest mountain chain, with 49 peaks above 6,000 meters, or nearly 20,000 feet (●FIGURE 9.28). The highest peak is Aconcagua, at 6,962 meters. The Andes rise from the Pacific coast of South America, starting nearly at sea level. Igneous rocks make up most of the Andes, although the chain also contains folded sedimentary rocks, especially in the eastern foothills.

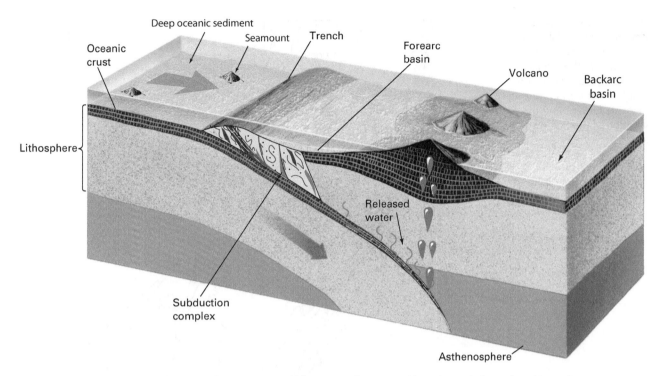

FIGURE 9.23 Formation of an island arc. Saturated oceanic sediment and basalt is subducted and introduces water into the overlying asthenosphere and base of the overriding plate, causing partial melting and formation of basaltic magma. The magma rises and erupts, building submarine volcanoes that eventually grow above sea level to form a volcanic mountain chain. A subduction complex forms at the top of the subduction zone and consists of highly deformed slices of oceanic sediment, along with fragments of oceanic crust and upper mantle scraped from the top of the subducting oceanic plate. A forearc basin forms when sediment accumulates between the arc and the crest of the subduction complex.

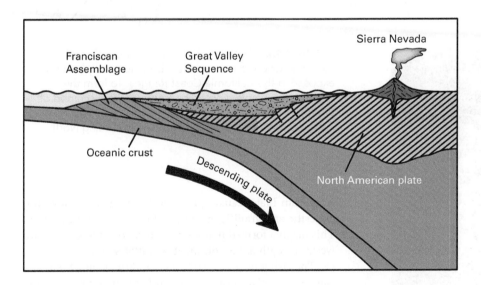

FIGURE 9.24 This cross section depicts the western margin of North America as it looked in California during Cretaceous time, before evolution of the San Andreas Fault. Oceanic lithosphere of the Farallon Plate subducted under North American continental lithosphere. The Sierra Nevada was an active continental arc, and much oceanic material from the downgoing Farallon Plate was scraped off to form the Franciscan Assemblage—a subduction complex that is presently exposed in much of California's coast ranges.

Subduction along the west coast of South America has a long geologic history that some geologists think extends as far back as the Paleozoic, when South America was part

forearc basin A sedimentary basin between the oceanic trench and the magmatic arc, either in an island arc or at an Andean margin.

of Gondwanaland. As the process of subduction proceeded, sediment and rock was scraped off of the downgoing plate and plastered onto the western edge of the continent, forming a subduction complex. Meanwhile, great volumes of basaltic magma were generated as water contained in the oceanic sediments and basalt of the downgoing slab were injected into the lithosphere of the overriding South

● **FIGURE 9.25** Simplified geologic map of California showing the widespread occurrence of the Franciscan Assemblage, a subduction complex that formed during Mesozoic time before the evolution of the San Andreas Fault.

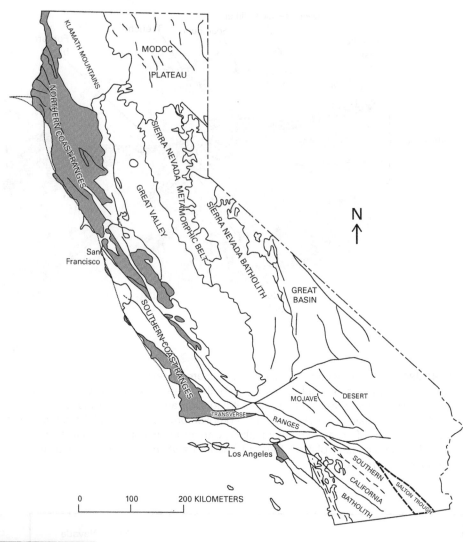

● **FIGURE 9.26** Outcrop photo of Franciscan Complex— the name given to the ancient subduction complex now exposed in the California Coast Ranges. To form the subduction complex, oceanic sediments were scraped off the top of the downgoing plate, complexly folded, and accreted to the western margin of the overriding North American Plate.

American Plate, causing it to partially melt and forming andesitic and granitic magma that rose toward the surface. This magma formed plutons and volcanoes along the entire western length of the continent (●**FIGURE 9.29**).

The rising magma also heated and thickened the crust, causing it to rise isostatically and form great peaks. These peaks are the Andes Mountains, which today form the backbone of South America. As subduction took place, much sediment was shed westward into a forearc basin formed between the arc and the subduction complex. With time, the subduction complex grew and the arc shifted to the east, causing the forearc basin to widen. Meanwhile, compression associated with the converging plates initiated a series of folds and thrust faults east of the Andes Mountains. As the thrust sheets stacked up, the weight of the

● **FIGURE 9.27** This satellite image was acquired on March 6, 2012, and shows an eruption of Pagan Island, part of an island arc associated with subduction along the Marianas Trench in the western Pacific.

● **FIGURE 9.28** The Cordillera Real mountain range in Bolivia's Andes rises over 6,000 meters.

stack caused the crust on the east side of the mountains to flex downward, producing a **foreland basin** that captured sediment eroded off of the thrust belt and arc behind it (●**FIGURE 9.29B**).

Thus, the Andes are a mountain chain consisting predominantly of igneous rocks formed by subduction at a continental margin. The chain also includes extensive sedimentary rocks on both sides of the mountains; those rocks formed from the sediment that eroded from the rising peaks. The Andes are a good general example of subduction at a continental margin, and this type of plate margin is called an **Andean margin**.

---

foreland basin A sedimentary basin formed by downward flexing of continental crust by the weight of thrust plates in a compressional orogen.

Andean margin A continental margin characterized by subduction of an oceanic lithospheric plate beneath a continental plate; also called a *continental arc*.

THE HIMALAYAS: A COLLISION BETWEEN CONTINENTS

The world's highest mountain chain, the Himalayas, separates China from India and includes the world's highest peaks (●**FIGURE 9.30**). If you were to stand on the southern edge of

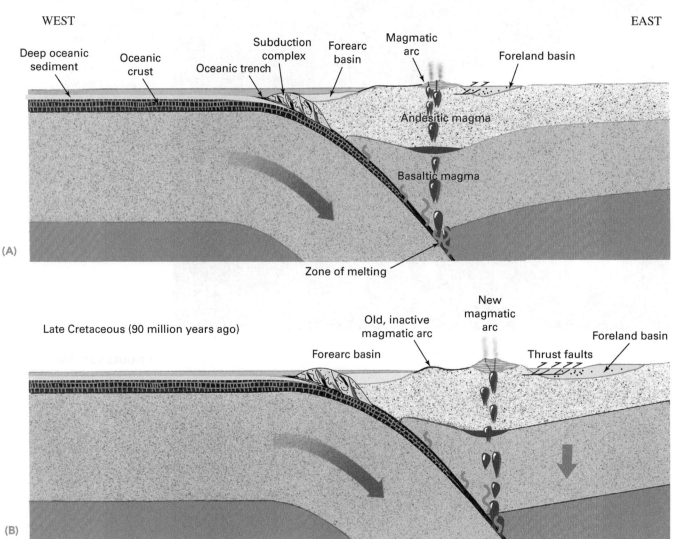

Early Cretaceous (140 million years ago)

WEST   EAST

(A)

Late Cretaceous (90 million years ago)

(B)

● **FIGURE 9.29** Development of the Andes, seen in cross section looking toward the north. (A) Subduction of oceanic lithosphere beneath the western side of the South American Plate has been ongoing for at least 140 million years. Basaltic magma produced by introduction of water from the downgoing slab rose to the overlying crust and partially melted it, forming andesitic magma. Eruption of this magma formed an arc on the overriding plate. Sediment eroded from the arc accumulated in a forearc basin between the arc and the subduction complex. Additional sediment accumulated on the east side of the arc in a foreland basin, formed where compression associated with the convergent margin caused thrust faulting, and the weight of the thrust sheets flexed the lithosphere downward. (B) As subduction continued, the size of the subduction complex and the width of the forearc basin grew, while arc magmatism migrated eastward. Continued compression in the foreland basin resulted in additional thrust faulting, downward flexure, and the accumulation of sediment there.

the Tibetan Plateau and look southward, you would see the high peaks of the Himalayas. Beyond this great mountain chain lie the rain forests and hot, dry plains of India. If you had been able to stand in the same place 200 million years ago and look southward, you would have seen only ocean. At that time, India was located south of the equator, separated from Tibet by thousands of kilometers of open ocean. The Himalayas had not yet begun to rise (●**FIGURE 9.31**).

Formation of an Andean-Type Margin

About 80 million years ago, a triangular piece of lithosphere that included present-day India had split off from a large mass of continental crust near the South Pole (●**FIGURES 9.31A** and **9.31B**). It began drifting northward toward Asia at a high speed—geologically speaking—perhaps as fast as

● **FIGURE 9.30** The Himalayas first began to rise as part of an Andean-style volcanic arc, but the modern range resulted from continent–continent collision between the Indian and Asian Plates. Pictured here is Ama Dablam, a 5,563-meter peak in eastern Nepal.

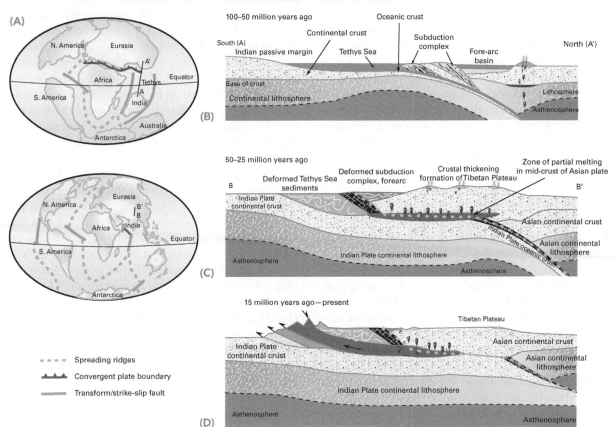

● **FIGURE 9.31** Evolution of the Himalayas and Tibetan Plateau. (A) The Indian subcontinent was part of Gondwana 200 million years ago and was separated from the southern margin of Asia by the Tethys Sea. (B) About 150 million years ago, Gondwanaland began to break apart through the process of tectonic rifting, and the Indian Plate was a major fragment resulting from this tectonic breakup. India began drifting northward by about 120 million years ago, as oceanic lithosphere forming its leading edge subducted beneath the southern continental margin of Asia. This process formed an oceanic trench, a subduction complex, and an Andean arc along Asia's southern margin. (In this reconstruction about 100 million years ago, the amount of oceanic crust between India and Asia is abbreviated to fit the diagram.) (C) By 50 million years ago, Indian continental lithosphere had begun to collide with and subduct under Asia. Great thicknesses of shallow marine oceanic sediments from the collapsed Tethys Sea were deformed by folding and thrust faulting in the collision zone. Compression from the continent–continent collision caused crustal thickening and melting of large volumes of magma in the middle crust. Collectively, these processes elevated the Tibetan Plateau. (D) By 15 million years ago, continued convergence of India with Asia led to the plastic extrusion of midcrustal-level (red) rocks upward and to the south, out of the collision zone. Today, these rocks form the Greater Himalayan Sequence.

20 centimeters per year. As this Indian plate started to move, its leading edge, consisting of oceanic crust, began to subduct beneath Asia's southern margin. This subduction formed an Andean-type continental margin along the southern edge of Tibet; volcanoes erupted and plutons with intermediate to granitic composition intruded the continent (Figure 9.31B).

Continent–Continent Collision

By about 50 million years ago, subduction had consumed all of the oceanic lithosphere between India and Asia, and the two continents began to collide (●**FIGURE 9.31C**). Plate convergence was accompanied by indentation of the Asian Plate, and the continental lithosphere of India began to subduct beneath the Asian continent. Many scientists have suggested that the low density of the downgoing Indian continental lithosphere caused it to subduct at a very shallow angle and underplate the Asian continent, while others have proposed that the Indian continental lithosphere was dense enough to subduct at a steep angle once oceanic sediments were scraped off of its upper surface. Recently, measurements of the velocity of seismic waves across the India-Asia collision show that the western part of the collision zone at the point of maximum indentation is characterized by flat subduction. In contrast, the eastern part of the convergent margin appears to be characterized by steeper subduction of the Indian continental lithosphere.

MINDTAP
From Cengage

Animation:
Continent-continent
collision

As India subducted beneath Asia, arc volcanism associated with previous subduction of oceanic crust ceased, and thick sequences of sediment that had accumulated on India's northern continental shelf were scraped off of the downgoing plate and pushed into great folds separated by major thrust faults.

The intense compression and the thickening of the lithosphere that resulted caused widespread melting of the middle crust. As the compression continued, this plastic rock in the core of the orogen was extruded southward and upward, like toothpaste from a giant tube. These southward-extruded rocks, called the Greater Himalayan Sequence, are bounded below by a major thrust fault and above by a major normal fault. A good exposure of this normal fault occurs just below the summit of Mount Everest (●**FIGURE 9.22B**).

The indentation of Asia by the Indian continent not only compressed Tibet in a north–south direction but it also acted like a giant wedge on the Asian continent. This wedge-like action pushed large blocks of lithosphere north of the collision zone laterally away from the comparatively rigid indenter (the Indian Plate) along huge strike-slip faults (●**FIGURE 9.32**). The continent–continent collision also caused structural reactivation of earlier-formed faults across much of central Asia, despite the fact that many of these faults are hundreds of

kilometers north of the India-Asia plate boundary. Deformation across this collective, complex network of faults by the ongoing collision has created the tall, linear mountain ranges and vast sedimentary basins of central Asia.

The Himalayas Today

Today, the Himalayas contain igneous, sedimentary, and metamorphic rocks. Many of the sedimentary rocks contain fossils of marine organisms that lived in the shallow sea of the Indian continental shelf. Volcanic rock and some plutonic rock formed when the range was an Andean margin. Large volumes of plutonic rocks also formed by midcrustal melting following the initiation of continent–continent collision. Rocks of all types were metamorphosed by the tremendous pressure and heat generated during subduction and continent–continent collision.

Although estimates of the convergence rate between India and Asia suggest that it slowed significantly when the continental lithosphere began to subduct, present-day estimates of the convergence rate range from about 3 to 5 centimeters per year. Roughly one-third of this convergence is thought to be taken up by compressional deformation in the Himalayas, while lateral extrusion of lithosphere north of the collision zone and crustal thickening across the region are thought to account for the other two-thirds of convergence.

The underthrusting of India beneath Tibet and the squashing of Tibet have greatly thickened the continental lithosphere under the Himalayas and the Tibetan Plateau to the north. Consequently, the region floats isostatically at high elevation (Figures 9.22A and 9.31D). Even the valleys lie at elevations of 3,000 to 4,000 meters, and the Tibetan Plateau has an average elevation of 4,000 to 5,000 meters. One reason why the Himalayas contain all of Earth's highest peaks is simply that the entire plateau lies at such a high elevation. From the valley floor to the summit, Mount Everest is actually smaller than Alaska's Denali (Mount McKinley), North America's highest peak. Mount Everest rises about 3,300 meters from base to summit, whereas Denali rises about 4,200 meters. The difference in elevation of the two peaks lies in the fact that the base of Mount Everest is at about 5,500 meters, but Denali's base is at 2,000 meters.

One controversial aspect of the India-Asia collision is explaining the widespread presence of seismically active, north–south-oriented normal faults, mostly in the Tibetan Plateau. Some geologists have suggested that these faults formed after growth of the plateau and resulted because the massive plateau was too weak to support its own weight, much like the collapse of a giant French soufflé. However, other geologists have demonstrated that movement on the normal faults was synchronous with the north–south compression that raised the plateau, so the normal faults

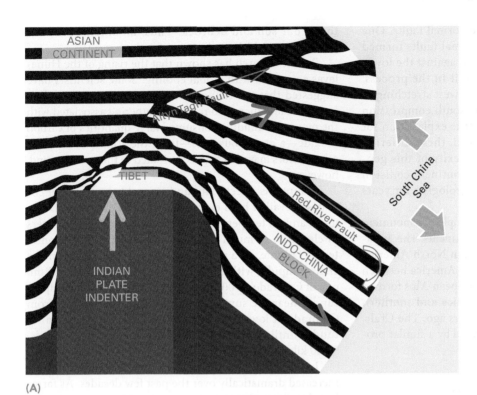

(A)

● **FIGURE 9.32** The continent–continent collision between India and Asia not only has created the highest mountains in the world, but it has resulted in large strike-slip faults that are accommodating the lateral movement of large parts of Asia out of the way of the Indian Plate. (A) This sketch shows the result of a famous experiment conducted in the 1970s in which a rigid block was slowly pushed horizontally into a block of clay with black and white stripes. As the experiment proceeded, the clay broke into several large pieces, each of which was separated by a strike-slip fault that accommodated the lateral extrusion of the pieces. (B) A simplified map of Asia showing the main fault associated with the India-Asia collisional orogen.

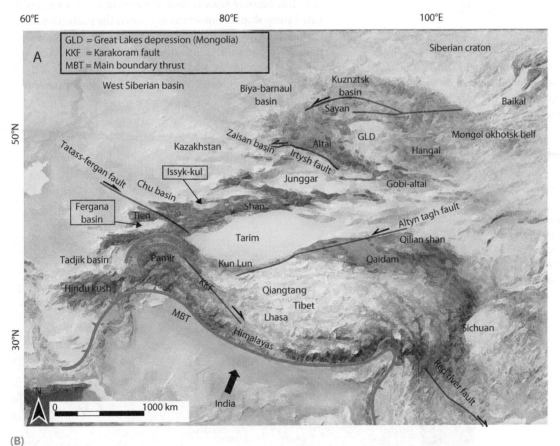

(B)

are not simply accommodating lateral spreading following uplift. Other geologists have pointed out that the normal faults terminate against major strike-slip faults that are accommodating the lateral extrusion of northern Tibet to the east—out of the way of the rigid Indian indenter plate. As northern Tibet is displaced to the east, the geologists reason, it stretches and produces the normal faults. One research group suggested that the normal faults formed as the subducting Indian Plate dragged against the lower surface of the Asian Plate, extending it in the process. Yet another group interpreted the east–west stretching as the natural consequence of the north–south compression that built the plateau. Whatever the true explanation for normal faulting of the Tibetan Plateau, these differing interpretations underscore the complexity of this geologic system and why this continent–continent collision will continue to attract the efforts of geologists for years to come.

The Himalayan chain is only one example of a mountain chain built by a collision between two continents. The Appalachian Mountains formed when eastern North America collided with Europe, Africa, and South America between 470 and 250 million years ago. The European Alps formed during collision between northern Africa and southern Europe beginning about 30 million years ago. The Urals, which separate Europe from Asia, formed by a similar process about 250 million years ago.

MOUNTAINS AND EARTH SYSTEMS

Air rises as it flows across a mountain range. Moisture condenses from rising air to produce rain or snow. Rain forms streams that race down steep hillsides, eroding gullies and canyons, while snow may accumulate to form glaciers that scour soil and bedrock. Thus, mountains promote precipitation, which erodes mountains.

Recent research has shown that the rising of the Himalayas coincided with global cooling. Some scientists have suggested that the rising mountains caused this cooling trend, because increased rates of weathering associated with the mountain building consumed large quantities of carbon dioxide—a greenhouse gas—from the atmosphere. Many other factors may also have contributed to the cooling trend, however.

Mountains affect plant and animal habitats. Cool climates prevail at high elevations, even at the equator. The mean annual temperature changes with each 1,000 meters of elevation as much as it does with each 1,000 kilometers of latitude. As a result, plant and animal communities change rapidly with elevation. For example, if you climb one of Colorado's 4,000-meter mountains, you reach alpine tundra near the summit. But if you traveled north from Colorado at low elevations, you would not reach similar tundra until you arrived at the Canadian Barren Lands, 3,000 kilometers away.

In many mountainous regions, human populations have increased dramatically over the past few decades. As farmland has become scarce, people have cut forests and cultivated steep slopes. Unfortunately, when the protective forest and native grasses are removed, soil erosion increases dramatically. Mud washes into streams and landslides carry soil downslope (●**FIGURE 9.33**). In this manner, human activity greatly increases erosion rates. We will explore these topics in the next chapter.

●**FIGURE 9.33** As the human population in mountainous regions increases, people have cut forests on steep hillsides to plant crops. This terraced hillside in Nepal collapsed during heavy monsoon rainfall.

kwanchai.c/Shutterstock.com

Key Concepts Review

- When stress is applied to rocks, the rocks can deform in an elastic or a plastic manner, or they may rupture by brittle fracture. The nature of the material, temperature, pressure, and rate at which stress is applied all affect rock behavior under stress. A geologic structure is any feature produced by deformation of rocks. Folds usually form when rocks are compressed.
- A fault is a fracture along which rock on one side has moved relative to rock on the other side. Normal faults are usually caused when rocks are pulled apart; reverse and thrust faults are caused by compression; and strike-slip faults form where blocks of crust slip horizontally past each other along vertical fractures. A joint is a fracture where the rock on either side has not moved.
- Mountains form when the crust thickens and rises isostatically. Their peak heights decrease when crustal rocks flow outward or are worn away by erosion.
- If two converging plates carry oceanic crust, a volcanic island arc forms.
- The Andes are a mountain chain consisting predominantly of igneous rocks formed at a type of continental margin called an Andean margin. An Andean margin is characterized by subduction of an oceanic lithospheric plate beneath a continental plate. Andean margins are dominated by granitic plutons and andesitic volcanoes. They also contain rocks of a subduction complex and sedimentary rocks deposited in a forearc basin.
- The Himalayas began with formation of an Andean-type continental margin along the edge of Tibet. Later, a continental plate carrying India collided with Tibet, producing compressional forces that created the mountain range. The geology of mountain ranges formed by continent–continent collisions, such as the Himalayas, is dominated by vast regions of folded and thrust-faulted sedimentary and metamorphic rocks and by earlier-formed plutonic and volcanic rocks.
- Mountains exert profound effects on climate, weather, and plant and animal habitats.

Important Terms

Andean margin (p. 215)
anticline (p. 201)
axial surface (p. 201)
basin (p. 203)
dome (p. 203)
facing direction (p. 201)
fault (p. 203)
fault zone (p. 203)
fold (p. 200)
footwall (p. 205)

forearc basin (p. 212)
foreland basin (p. 215)
geologic structure (p. 200)
graben (p. 205)
hanging wall (p. 205)
horst (p. 206)
island arc (p. 211)
joint (p. 208)
limbs (p. 201)
normal fault (p. 205)

orogen (p. 210)
orogeny (p. 210)
reverse fault (p. 207)
slip (p. 203)
stress (p. 200)
subduction complex (p. 211)
syncline (p. 201)
thrust fault (p. 207)
underthrusting (p. 211)

Review Questions

1. What is tectonic stress? Explain the main types of stress.
2. Explain the different ways in which rocks can respond to tectonic stress. What factors control the response of rocks to stress?
3. What is a geologic structure? What are the three main types of structures? What type(s) of rock behavior does each type of structure reflect?
4. At what type of tectonic plate boundary would you expect to find normal faults?

5. Explain why folds accommodate crustal shortening.

6. Draw a cross-sectional sketch of an anticline–syncline pair and label the limbs.

7. Draw a cross-sectional sketch of a normal fault. Label the hanging wall and the footwall. Use your sketch to explain how a normal fault accommodates crustal extension. Sketch a reverse fault and show how it accommodates crustal shortening.

8. Explain the similarities and differences between a fault and a joint.

9. In what tectonic environment would you expect to find a strike-slip fault, a normal fault, and a thrust fault?

10. What mountain chain has formed at a divergent plate boundary? What are the main differences between this chain and those developed at convergent boundaries? Explain the differences.

11. Explain why erosion initially causes a mountain range to rise and then eventually causes the peak heights to decrease.

12. Describe the similarities and differences between an island arc and the Andes. Why do the differences exist?

13. Describe the similarities and differences between the Andes and the Himalayan chain. Why do the differences exist?

14. Draw a cross-sectional sketch of an Andean-type plate margin to a depth of several hundred kilometers.

15. Draw a sequence of cross-sectional sketches showing the evolution of a Himalayan-type plate margin. Why does this type of boundary start out as an Andean-type margin?

Callahan Galleries/Moment/Getty Images

10

Weathering and erosion have removed a tremendous volume of rock from the Colorado Plateau, as seen here from Deadhorse Point State Park, Utah. The Colorado River runs through the canyon bottom.

WEATHERING, SOIL, AND EROSION

LO1 Explain the disappearance of early impact craters from the surface of the Earth.

LO2 Distinguish between weathering and erosion.

LO3 Classify types of weathering as either chemical, mechanical, or both.

LO4 Explain why most of Earth's sand is composed of quartz.

LO5 Discuss how the oxidation of minerals can lead to toxic sulfuric acid in streams and groundwater.

LO6 Describe the process of weathering by salt cracking.

LO7 Compare and contrast spheroidal weathering and abrasion.

LO8 Explain the effect of intensive agriculture on the ability of soil to absorb water.

LO9 Distinguish between the O, A, E, B, and C soil horizons.

LO10 Give examples of how climate controls soil formation.

LO11 Define the mass wasting terms flow, slide, and fall.

LO12 Distinguish between factors that effect the likelihood of slides, flows, and falls.

INTRODUCTION

As the Solar System formed, swarms of meteorites, comets, and asteroids crashed into the planets and their moons. These impacts formed huge craters—steep, circular depressions with a sharp berm and surrounding blanket of debris. The earliest impact craters on Earth's surface quickly disappeared because the planet was so hot that plastic and molten rock oozed inward to fill the craters. As the crust thickened, cooled, and hardened, additional bombardment formed new craters. Today, however, none of these ancient craters remain on Earth's surface. In contrast, the Moon is pockmarked with craters of all ages. Why have the craters vanished from Earth but been preserved on the Moon?

Over geologic time, tectonic processes such as mountain building, volcanic eruptions, and earthquakes continually change the surface of the Earth. In addition, Earth is of sufficient size to have retained its atmosphere and water. The atmosphere and hydrosphere weather and erode the surface rocks of the geosphere. The combination of tectonic activity and erosion has eliminated traces of early impact craters from Earth's surface. In contrast, the smaller Moon has lost most of its heat, so tectonic activity is nonexistent. In addition, the Moon's gravitational force is too weak to have retained an atmosphere or significant amounts of surface water. Thus, because neither wind nor flowing water is available to modify the Moon's surface, ancient impact craters have persisted there for billions of years.

WEATHERING AND EROSION

Recall from Chapter 3 that the process by which solid rock is decomposed to loose gravel, sand, clay, and soil is called weathering (●**FIGURE 10.1**).

Weathering involves little or no movement of the decomposed rocks and minerals. The weathered material simply accumulates where it forms. However, loose soil and other weathered material offer little resistance to rain or wind and are easily eroded. **Erosion** is the removal of rock mass from an area by rain, running water, wind, glaciers, or gravity. Agents of erosion may then carry the weathered material great distances and later deposit it as layers of sediment.

Weathering, erosion, transport, and deposition typically occur in an orderly sequence. For example, water freezes in a crack in granite and loosens a grain of quartz. A hard rain erodes the grain and washes it into a stream. The stream then transports the quartz to a beach at the seashore, where it is tumbled in the surf and becomes rounded (●**FIGURE 10.2**). During a large storm, the now rounded grain of quartz is swept offshore and deposited on the seafloor, where it may reside for hundreds of years or more. Later, another powerful storm again erodes—or **reworks**—the grain off the seafloor and suspends it in the turbulent water along with other sedimentary particles. Collectively, the sediment particles are carried into a submarine canyon and funneled into deep water as part of a **sediment-gravity flow**, a bottom-hugging fluid mixture of sediment and water that flows downslope. After several days of transport down the submarine canyon, the sediment-gravity flow comes to rest on the ocean floor in deep water, and the grain is finally deposited.

Weathering occurs by both mechanical and chemical processes. **Mechanical weathering** (also called

● **FIGURE 10.1** The base of this limestone sea cliff is being eroded by wave energy in the Gulf of Corinth, central Greece. In addition, freshwater is percolating downward through the limestone and partially dissolving it, forming holes and caves. Both of these processes are forms of weathering.

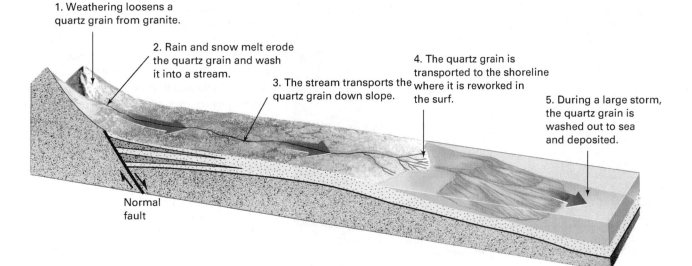

1. Weathering loosens a quartz grain from granite.

2. Rain and snow melt erode the quartz grain and wash it into a stream.

3. The stream transports the quartz grain down slope.

4. The quartz grain is transported to the shoreline where it is reworked in the surf.

5. During a large storm, the quartz grain is washed out to sea and deposited.

Normal fault

● **FIGURE 10.2** Weathering loosens a quartz grain from granite. Rain erodes the grain, washing it into a stream. The stream transports it to the shoreline, where the grain is reworked in the surf zone. Eventually, the grain is washed out to sea during a large storm and deposited.

physical weathering) reduces solid rock to small fragments but does not alter the chemical composition of rocks and minerals. Think of crushing a rock with a hammer: the fragments are no different from the parent rock except that they are smaller. In contrast, **chemical weathering** occurs when air and water chemically react with rock to alter its composition and mineral content. Chemical weathering is similar to the rusting of an old truck body; the final product differs both physically and chemically from the original material.

---

erosion The removal of weathered rocks that occurs when water, wind, ice, or gravity transports the material to a new location.

reworking The process by which sediment is deposited, then re-eroded and transported further.

sediment-gravity flow Sediment gravity flows occurs when gravity acts on the sediment particles and moves the fluid; this is in contrast to rivers where the fluid moves the particles.

mechanical weathering or physical weathering The disintegration of rock into smaller pieces by physical processes without altering the chemical composition of the rock.

chemical weathering The decomposition of rock when it chemically reacts with air, water, or other agents in the environment, altering its chemical composition and mineral content.

pressure-release fracturing A mechanical weathering process in which tectonic forces lift deeply buried rocks upward and then erosion removes overlying rock and sediment—the net result of which is to remove the pressure from overlying material, causing the rock to expand and fracture.

MECHANICAL WEATHERING

Five processes cause mechanical weathering: pressure-release fracturing, frost wedging, abrasion, organic activity, and thermal expansion and contraction. Two additional processes—salt cracking and hydrolysis-expansion—result from combinations of mechanical and chemical processes.

Pressure-Release Fracturing

Many igneous and metamorphic rocks form deep below Earth's surface. Imagine, for example, that a granitic pluton solidifies from magma at a depth of 15 kilometers. At that depth, the pressure from the weight of overlying rock is about 5,000 times that at Earth's surface. Over millions of years, tectonic forces may uplift the granite while erosional forces strip off the overlying layers of rock, ultimately exposing the granite at the Earth's surface. As the pressure on the rock diminishes, the rock expands, but because the rock is now cool and brittle, it fractures as it expands. This process is called **pressure-release fracturing**. Many igneous and metamorphic rocks that formed at depth but now lie at Earth's surface have fractured in this manner.

Frost Wedging

Although pressure-release fracturing occurs by processes involving only the geosphere, all other forms of weathering involve interactions among rocks of the geosphere and the hydrosphere, atmosphere, and biosphere.

● **FIGURE 10.3** Pressure-release fracturing contributed to the formation of Yosemite National Park's Half Dome, located in the Sierra Nevada of California. B) a closer view of large exfoliating slabs of granite near the summit.

Water expands by 7 to 8 percent when it freezes. If water accumulates in a crack and then freezes, the newly forming ice wedges the rock apart in a process called **frost wedging**. During spring and autumn in a temperate climate, water freezes at night and thaws during the day. Ice formation pushes rock apart but at the same time cements it together. During the day, when the ice melts, rock fragments come loose and tumble from a steep cliff. For this reason, experienced mountaineers try to travel in the early morning before ice melts.

Over time, a pile of loose, angular rock debris, called **talus**, will accumulate beneath many cliffs (●**FIGURE 10.4**). These rocks fall from the cliffs mainly as a result of frost wedging over many seasons. Typically, large open pore spaces exist between the rocks.

Abrasion

Rocks, grains of sand, and silt collide with one another when currents or waves carry them along a stream or beach. During these collisions, the sharp edges and corners of the particles are worn away and they become rounded. The mechanical wearing of rocks by friction and impact is called **abrasion**. Note that water itself is not abrasive—it is the collisions among rock, sand, and silt in the water that does the rounding.

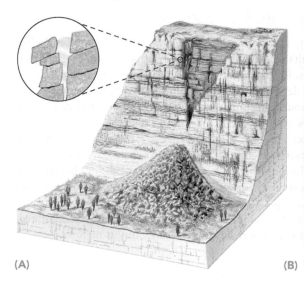

(A)

(B)

● **FIGURE 10.4** (A) Frost wedging dislodges rocks from cliffs and creates talus slopes consisting of loose, angular boulders with open pore spaces between. (B) Frost wedging produced this talus cone in Banff National Park, Canada. The angular blocks of rock forming the talus have resulted in so much porosity that the stream in the center of the photograph emerges from talus at the top of the photo, flows over bedrock for a short distance, then disappears back underground at the apex of the talus cone below.

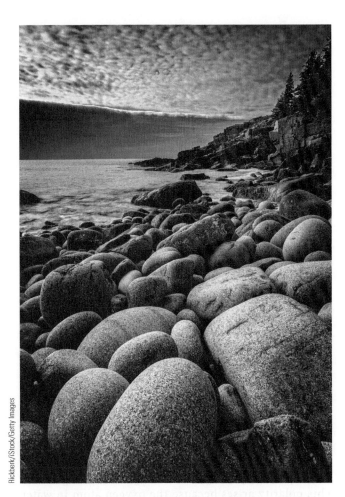

Rickberk/iStock/Getty Images

● **FIGURE 10.5** Wave energy along the shoreline of lakes and oceans can rapidly abraid and round rocks such as these.

Maxim Petrichuk/Shutterstock.com

(A)

Denis Burdin/Shutterstock.com

(B)

● **FIGURE 10.6** Wind-blown sand is a very powerful erosional agent. (A) Sandstorm in Altyn-Emel National Park, Kazakhstan. (B) Blowing sand sculpted and polished these unusual outcrops of weathered volcanic rock in the Libyan Desert.

Wind hurls sand and other small particles against rocks, sandblasting unusual shapes (●**FIGURE 10.6**). Because the wind does not normally carry sand particles high above the ground, they tend to weather the lower portion of the rock where some develop into balanced rock formations over the course of many years. In the past few years, the knocking over of balanced rocks has become a sport to a few people,

---

frost wedging A mechanical weathering process in which water freezes in a crack in rock, and the resulting expansion wedges the rock apart.

talus An accumulation of loose, angular rocks at the base of a cliff, created as rocks broke off the cliff as a result of frost wedging.

abrasion A mechanical weathering process that consists of the grinding and rounding of rock and mineral surfaces by friction and impact.

organic activity A mechanical weathering process in which a crack in a rock is expanded by tree or plant roots growing there.

thermal expansion and contraction A mechanical weathering process that fractures rock when temperature changes rapidly, causing the surface of the rock to heat or cool faster, and thereby to expand or contract faster, than the rock's interior.

depriving many of viewing these unique structures. Glaciers also are powerful agents of abrasion, as they drag rock clasts, sand, and silt across bedrock.

Organic Activity

If soil collects in a crack in bedrock, a seed may fall there and sprout. The roots work their way into the crack, expand, and may eventually widen the crack as they grow (●**FIGURE 10.7**). City dwellers often see the results of this type of **organic activity** when tree roots raise and crack concrete sidewalks.

Thermal Expansion and Contraction

Rocks at Earth's surface are exposed to daily and yearly cycles of heating and cooling. They expand when they are heated and contract when they cool. When temperature changes rapidly, the surface of a rock heats or cools faster than its interior, and as a result, the surface expands or contracts faster than the interior. The forces generated by this **thermal expansion and contraction** may fracture the rock.

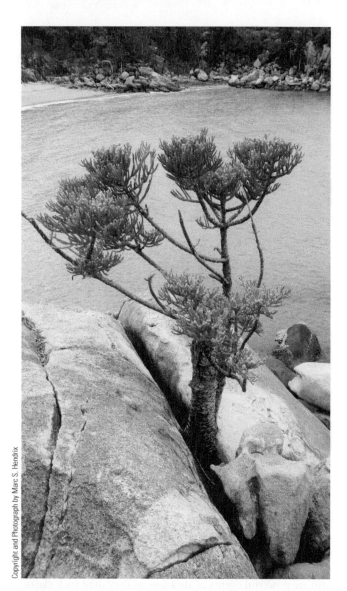

Copyright and Photograph by Marc S. Hendrix

● **FIGURE 10.7** This tree on the coast of Queensland, Australia, grew in a crack in granite. As it grew, the tree's roots widened the crack.

In mountains or deserts at midlatitudes, temperature may fluctuate from −5°C to +25°C during a spring day. This 30-degree difference is probably not sufficient to fracture rocks. In contrast to small daily or annual temperature changes, fire heats rock by hundreds of degrees. If you line a campfire with granite stones, the rocks commonly break as you cook your dinner. In a similar manner, forest fires or brush fires occur frequently in many ecosystems, producing cracked rock that is an important agent of mechanical weathering.

CHEMICAL WEATHERING

Rock is durable over a human lifetime. Over geologic time, however, air and water chemically attack rocks near Earth's surface. The most important processes of chemical weathering are dissolution, hydrolysis, and oxidation. Water, acids and bases, and oxygen in the atmosphere or in surface water or groundwater cause these processes to decompose rocks.

Dissolution

We are all familiar with the fact that some minerals dissolve readily in water and others do not. If you put a crystal of halite (rock salt, or table salt) in water, the crystal will rapidly dissolve to form a solution. The process is called **dissolution**. Halite dissolves so rapidly and completely in water that the mineral is rare in natural, moist environments. On the other hand, if you drop a crystal of quartz into pure water, only a tiny amount will dissolve and nearly all of the crystal will remain intact.

To understand how water dissolves a mineral, think of an atom on the surface of a crystal. It is held in place because it is attracted to the other atoms in the crystal by electrical forces called chemical bonds. At the same time, electrical attractions to the outside environment are pulling the atom away from the crystal. The result is like a tug-of-war. If the bonds between the atom and the crystal are stronger than the attraction of the atom to its outside environment, the crystal remains intact. If outside attractions are stronger, they pull the atom away from the crystal and the mineral dissolves.

Water (H_2O) is a polar molecule, meaning that it has both a positively charged and a negatively charged end. This polarity arises because the oxygen atom in water has a great affinity for electrons, which move away from the hydrogen atoms to concentrate around the oxygen and give that end of the molecule a negative charge. The two hydrogen atoms are then left with a slight positive charge because their electrons have moved closer to the oxygen atom. When a crystal of halite (NaCl) is dropped in a glass of water, the negatively charged ends of water molecules pull the positively charged sodium ions away from the halite, while the positively charged ends of water molecules remove the negatively charged chlorine ions (●**FIGURE 10.8**).

Rocks and minerals dissolve more rapidly when water is either acidic or basic. An acidic solution contains a high concentration of hydrogen ions (H+), whereas a basic solution contains a high concentration of hydroxyl ions (OH−). Acids and bases dissolve most minerals more effectively than pure water does because they provide more electrically charged hydrogen and hydroxyl ions to pull atoms out of crystals. For example, limestone is made of the mineral calcite ($CaCO_3$). Calcite barely dissolves in pure water but is quite soluble in acid. If you place a drop of strong acid on limestone, bubbles of carbon dioxide gas instantly form as the calcite dissolves.

Water found in nature is never pure. Atmospheric carbon dioxide dissolves in raindrops and reacts to form a weak acid called carbonic acid. This acidity can be

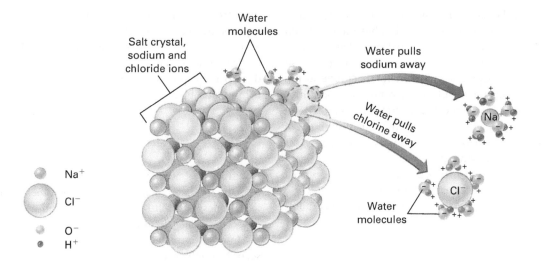

● **FIGURE 10.8** Halite (NaCl) dissolves in water (H₂O) because the attractions between the water molecules and the sodium and chloride ions are greater than the strength of the chemical bonds in the crystal. Note that although the oxygen atoms are labeled positive (+) and the hydrogen atoms are labeled negative (–), these are partial charges only.

intensified when rainwater percolates down through leaf litter on the ground. As a result, even the purest rainwater falling in the Arctic or on remote mountains is slightly acidic.

Air pollution can make rain even more acidic. When water vapor condenses to liquid, it requires a non-gaseous (liquid or solid) surface to do so. In the atmosphere, this surface is presented by tiny solid or liquid particles called **cloud condensation nuclei.** Cloud condensation nuclei are tiny particles of liquid or solid that typically are less than one micrometer (one millionth of a meter) in diameter. Air pollution creates condensation nuclei that are very acidic. For example, the burning of coal produces tiny sulfur-rich particles that serve as cloud condensation nuclei and that produce sulfuric acid when combined with water. More sulfur-rich particles are incorporated into the raindrops as they grow. By the time the raindrops are big enough to fall to Earth's surface, they have become acid rain.

Thus, water—especially if it is acidic or basic—dissolves ions from soil and bedrock and carries the dissolved material away. Groundwater also dissolves rock and can produce spectacular caverns in limestone (●**FIGURE 10.9**).

● **FIGURE 10.9** Caverns form when groundwater dissolves limestone. In contrast, stalactites, stalagmites, and columns—all shown in this photo from the Grotte de Demoiselles in southern France—form through re-precipitation of limestone from groundwater that is saturated with calcium carbonate.

Hydrolysis

In **hydrolysis,** water reacts with one mineral to form a new mineral that has water as part of its crystal structure. Most common minerals weather by hydrolysis. For example, feldspar, the most abundant mineral in Earth's crust, weathers by hydrolysis to form clay.

In contrast to feldspar, quartz resists chemical weathering because it dissolves extremely slowly and only to a small extent. It is also hard (H = 7 on the Mohs hardness scale; see Chapter 2) and has no cleavage, and so resists abrasion

---

dissolution A chemical weathering process in which mineral or rock dissolves, forming a solution.

cloud condensation nuclei Very small particles suspended in the atmosphere on which water vapor condenses.

hydrolysis A chemical weathering process in which a mineral reacts with water to form a new mineral that has water as part of its crystal structure.

during erosion. When granite weathers, the feldspar and other minerals decompose to form clay, but the unaltered quartz grains fall free from the rock. Hydrolysis has so deeply weathered some granites that quartz grains can be pried out with a fingernail at depths of several meters (●FIGURE 10.10). The rock looks like granite but has the consistency of sand.

Because quartz is so resistant to both mechanical and chemical weathering, it is a primary component in much of Earth's sand. After quartz is freed from its source rock, it is transported by water, wind, glacial ice, and gravity as sand-sized particles. Many of the quartz particles are transported to the shoreline where they are concentrated on beaches and deltas or are carried offshore by storm currents. Eventually, the sand lithifies to form sandstone.

Oxidation

Many elements react with atmospheric oxygen, O_2. Iron rusts when it reacts with water and oxygen. Rusting is an example of a more general weathering process called **oxidation**.[1] Iron is abundant in many minerals; if the iron in such a mineral oxidizes, the mineral decomposes.

Many valuable metals such as iron, copper, lead, and zinc occur as sulfide minerals in ore deposits. When these minerals oxidize during weathering, the sulfur reacts to form sulfuric acid, a strong acid. The sulfuric acid washes

into streams and groundwater, where it may harm aquatic organisms. In addition, the sulfuric acid can dissolve metals from the sulfide minerals, forming a metal-rich solution that can be toxic. These reactions are accelerated when ore is dug up and exposed by mining, and acid-mine drainage is a common problem at such sites.

Chemical and Mechanical Weathering Acting Together

Chemical and mechanical weathering can work together, often on the same rock and at the same time. After mechanical processes fracture a rock, water and air seep into the cracks to initiate chemical weathering.

In environments where groundwater is salty, saltwater seeps through pores and cracks in bedrock. When the water evaporates, the dissolved salts crystallize. The growing crystals exert tremendous forces that loosen mineral grains and widen cracks in a process called **salt cracking**. Thus, salt chemically precipitates in rock, and the growing salt crystals mechanically loosen mineral grains and break the rock apart.

Many sea cliffs show pits and depressions caused by salt cracking, because spray from the breaking waves brings the salt to the rock. Salt cracking is also common in deserts, where surface water and groundwater commonly contain dissolved salts (●FIGURE 10.11).

Granite typically fractures by **exfoliation**, a process in which plates or shells of weathered material split away like the layers of an onion (●FIGURE 10.12). The plates may be only 10 or 20 centimeters thick near the surface, but they thicken with depth. Because exfoliation fractures are usually absent below a depth of 50 to 100 meters, they seem to be a result of exposure of the granite at Earth's surface.

Although exfoliation is frequently explained as a form of pressure-release fracturing, many geologists have suggested

● **FIGURE 10.10** Chemical weathering in tropical northern Queensland, Australia, is breaking this coarsely crystalline granite (top) down to individual crystals of quartz and feldspar (bottom). The coin is 3 cm in diameter.

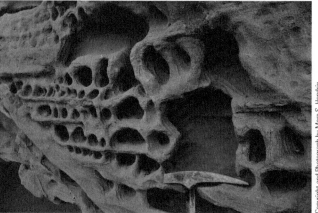

● **FIGURE 10.11** Crystallizing salt loosened sand grains to form these depressions in sandstone near Bridger, Montana. The rock hammer is for scale.

---

1. Oxidation is properly defined as the loss of electrons from a compound or element during a chemical reaction. In the weathering of common minerals, this usually occurs when the mineral reacts with molecular oxygen.

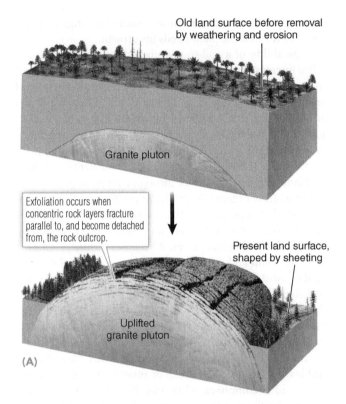

Old land surface before removal by weathering and erosion

Granite pluton

Exfoliation occurs when concentric rock layers fracture parallel to, and become detached from, the rock outcrop.

Present land surface, shaped by sheeting

Uplifted granite pluton

(A)

(B)

● **FIGURE 10.12** (A) Exfoliation occurs when concentric rock layers fracture and become detached from a granite outcrop. (B) Exfoliation has fractured this granite in Pinkham Notch, New Hampshire.

that hydrolysis-expansion may be the main cause of exfoliation. During hydrolysis, feldspars and other silicate minerals react with water to form clay. As a result of the addition of water, clays have a greater volume than the original minerals have. So a chemical reaction (hydrolysis) forms clay, and the mechanical expansion that occurs as the clay forms may cause exfoliation. This explanation is compatible with the observation that exfoliation concentrates near Earth's surface because water and chemical weathering are most abundant close to the surface.

Very commonly, the combined effects of mechanical and chemical weathering cause coarsely crystalline igneous rocks to form spheroidal shapes (●**FIGURE 10.13**). Igneous rocks exposed at the surface typically undergo pressure-

---

oxidation A chemical weathering process in which a mineral decomposes when it reacts with oxygen.

salt cracking A chemical weathering process in which salts that are dissolved in water in the pores of rock crystallize, exerting an outward pressure on pore walls and pushing the mineral grains apart.

exfoliation A weathering process resulting in fracture when concentric plates or shells split away from a main rock mass like the layers of an onion; frequently explained as a form of pressure-release fracturing, but many geologists consider it could result from hydrolysis-expansion.

spheroidal weathering The combined mechanical and chemical weathering of fractured crystalline bedrock into spheroidally shaped boulders; caused by the faster weathering rate of sharp bedrock corners (where at least three faces of rock can be attacked by weathering), over edges and the faster weathering of edges over faces.

● **FIGURE 10.13** Spheroidally weathered boulders of diabase, a coarsely crystalline igneous rock, in the Gettysburg National Military Park, Pennsylvania. During the Civil War battle that took place here in July of 1863, soldiers from both sides hid for protection behind some of these boulders.

release fracturing as a result of unroofing and jointing as a result of tectonic activity. The intersection of these different fracture planes initially produces rocks with flat faces and sharp corners and edges. As the rocks weather, however, their corners and edges are more exposed and susceptible to attack by hydrolysis and the mechanical expansion of the resulting clays. Through time, the corners and edges are slowly rounded, producing a spheroidally shaped boulder. It is important to understand that **spheroidal weathering** is not due to mechanical abrasion, but rather takes place as a rock is weathered in place, without moving.

SOIL

Bedrock breaks into smaller fragments as it weathers, and much of it decomposes to form sand, silt, and clay. Therefore, on most land surfaces, a thin layer of loose rock fragments, clay, silt, and sand overlies bedrock. This material is called **regolith**. **Soils** are the upper layers of regolith that contain organic matter and can support rooted plants.

Components of Soil

Soil is a mixture of mineral grains and rock fragments, organic material, water, and gas. The size and interconnectedness of pore spaces in soil and therefore the rate at which water and air can infiltrate the soil depends on the soil texture—the relative proportions of sand, silt, and clay. Soils rich in sand and silt contain pore spaces big enough to allow the infiltration of water and air. In contrast, clay-rich soils are so fine grained that pore spaces between the sediment particles are very small, inhibiting the transmission of air and water. Plants rooting in such soils often suffer from lack of oxygen. A **loam** is a soil with approximately equal parts of sand, silt, and clay. Such soils are well drained and may contain abundant organic matter, making them especially fertile and productive.

Clay minerals perform several important functions in soils. Clays can retain water by incorporating it into their crystal structure or by attracting water molecules to the charged flat surface of the platy clay crystals. The charged surfaces of clay minerals also can attract and hold other ions, such as calcium (Ca^{++}) or potassium ($K+$), commonly formed from the weathering of feldspars. By releasing H+

from organic acids, plant roots are able to exchange it for the Ca^{++} or $K+$ that the plant needs for growth.

The ability of a soil to absorb and exchange cations is called its **cation exchange capacity (CEC)** and is an important component of soil classification. In general, the more clay and organic matter in the soil, the higher its CEC. That proportion of a soil's CEC that is satisfied by cations Ca^{++}, Mg^{++}, $K+$, and $Na+$ (called the basic cations) is referred to as **percentage base saturation** and is also important in soil classification.

Organic matter in soils also contains nutrients necessary for plant growth. If you walk through a forest or prairie, you can find bits of organic **litter**—leaves, stems, and flowers that have not decomposed—on the soil surface. When this litter decomposes sufficiently that you can no longer determine the origin of individual pieces, it becomes **humus**. Humus is decay resistant and nutrient rich and is an essential component of most fertile soils. Humus-rich soils swell after a rain and shrink during dry spells. This alternate swelling and shrinking loosens the soil, allowing roots to grow into it easily. A rich layer of humus also insulates the soil from excessive heat and cold and reduces water loss from evaporation.

In intensive agriculture, farmers typically plow the soil and leave it exposed for weeks or months. Humus oxidizes in air and decomposes, while rain dissolves soil nutrients and carries them away. Farmers replace the lost nutrients with chemical fertilizers but rarely replenish the humus. As a result, much of the soil's ability to absorb and regulate water and nutrients is lost. Rainwater flowing over the plowed surface transports soil particles, excess fertilizer, and pesticide residues, polluting streams and groundwater.

Soil Horizons

Soils are subdivided into visibly, chemically, or physically distinct layers called **soil horizons** that form and change as the soil evolves. These horizons mainly are due to the **transformation** of soil constituents by the chemical, biological, or physical processes taking place in a soil over time and the downward **translocation** of sediment and ions within the soil profile. Five soil horizons can occur: O, A, E, B, and C (●**FIGURE 10.14**). Although some soils contain all five horizons as recognizable layers, many soils contain only a subset of the five horizons.

The uppermost layer is called the **O horizon**, named for its organic component. This layer consists mostly of organic litter and humus, with a small proportion of minerals. The next layer down, called the **A horizon**, is a mixture of humus, sand, silt, and clay. Together, the O and A horizons are called **topsoil**. A kilogram of average fertile topsoil contains about 30 percent organic matter, by weight, including approximately 2 trillion bacteria, 400 million fungi, 50 million algae, 30 million protozoa, and thousands of larger organisms such as insects, worms, nematodes, and mites.

Organic acids and carbon dioxide produced by decaying organic matter in the topsoil are transported downward into the **E horizon**. *E* stands for **eluviation**, referring to the process of **leaching** in which the acids produced in the topsoil dissolve ions such as calcium, silica, and iron from the

---

regolith The thin layer of loose, unconsolidated, weathered material that overlies bedrock. Some earth scientists and engineers use the terms *regolith* and *soil* interchangeably; soil scientists identify soil as only the upper layers of regolith.

soils The upper layers of regolith that support plant growth. Some earth scientists and engineers use the terms *soil* and *regolith* interchangeably.

loam The most fertile soil, a mixture especially rich in sand and silt with generous amounts of organic matter.

cation exchange capacity (CEC) Ability of a soil to release cations, typically by exchanging basic cations $K+$, $Na+$, Ca^{++}, or Mg^{++} for H+ with plant rootlets.

percentage base saturation The proportion of a soil's cation exchange capacity that is saturated by basic cations $K+$, $Na+$, Ca^{++}, or Mg^{++}.

litter Leaves, twigs, and other plant or animal materials that have fallen to the surface of the soil but have not decomposed.

humus The dark, organic component of soil consisting of litter that has decomposed enough so that the origin of the individual pieces cannot be determined.

soil horizon A layer of soil that is distinguishable from other layers because of differences in appearance and in physical and chemical properties.

transformation The change of soil constituents from one form to another, such as the hydrolysis of feldspar to clay.

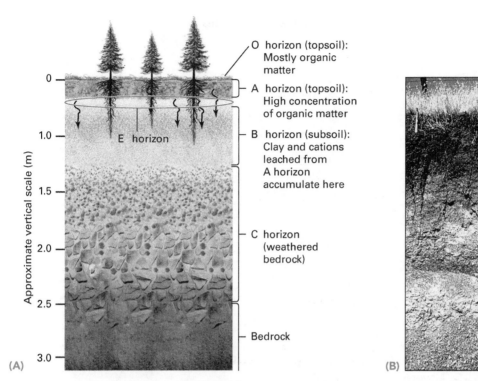

● FIGURE 10.14 (A) Five different soil horizons are commonly distinguished by color, texture, and chemistry. These include the O and A horizon, characterized by abundant organic matter; the E horizon from which organic acids percolating downward have leached cations and clays; the B horizon in which those cations and clays accumulate; and the C horizon, which consists mainly of weathered bedrock. (B) Many soils contain only a subset of these five horizons such as this mollisol (grassland soil) which is lacking the E horizon.

E horizon and translocate them downward. Clays are also translocated from the E horizon into deeper layers of soil. The loss of ions and clay from E horizons usually causes them to be sandy and gray (unpigmented). E horizons are

common in forested areas because forest litter is acidic and precipitation is abundant.

The clays and ions lost from the E horizon are translocated downward and added to the **B horizon**, also called the zone of accumulation or subsoil. The B horizon is a transitional zone between topsoil and the weathered parent rock below. Roots and other organic material can grow in the B horizon, but the total amount of organic matter usually is low. The lowest layer, called the **C horizon**, consists of partially weathered bedrock that grades into unweathered parent rock. This zone contains little organic matter.

translocation The vertical, usually downward, movement of physical or chemical soil constituents from one horizon to another.

O horizon The uppermost layer of soil, named for its organic component; the combined O and A horizons are called *topsoil*.

A horizon The layer of soil below the O horizon, composed of a mixture of humus, sand, silt, and clay; combines with the O horizon to form topsoil.

topsoil The fertile, dark-colored surface soil; the combined O and A soil horizons.

E horizon Soil horizon in which organic acids derived from overlying O and A horizons leach soluble cations and translocate them downward along with clays.

eluviation The removal and downward movement of dissolved ions and clays from the O, A, and E horizons by infiltrating water.

leaching The chemical dissolution of ions from the O and A soil horizons and their removal, usually downward into the B horizon where they accumulate.

B horizon The soil layer just below the A horizon, containing less organic matter and where ions and clays leached from the A and E horizon accumulate; also called *subsoil*.

C horizon The lowest soil layer, consisting of weathered bedrock.

soil order The highest hierarchical classification of soils by the National Resource Conservation Service. Twelve soil orders are recognized.

Soil Classification

Because of the close connection that soils have to humans living on Earth's surface, many different soil classification systems have been developed. Soils have been classified on the basis of their engineering properties as related to support for building foundations, their morphology, and their genesis. Many countries have established their own soil classification systems, although an international system also has been developed.

In the United States, the National Resources Conservation Service (NRCS) is responsible for defining and mapping soils across the country. **●FIGURE 10.15** is the current soils map for the United States as developed by the NRCS, and **●FIGURE 10.16** is the soils map for the world.

Dominant Soil Orders of the United States

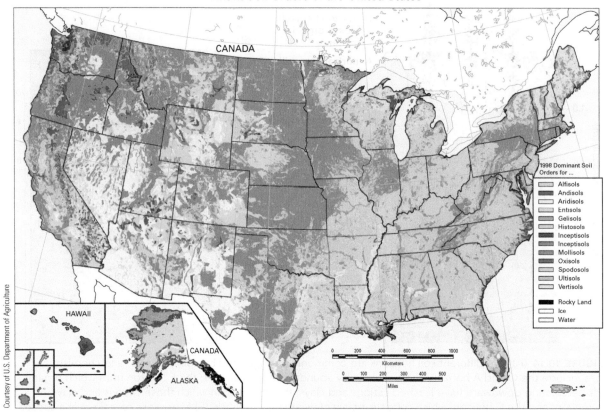

Courtesy of U.S. Department of Agriculture

1998 Dominant Soil Orders for …

- Alfisols
- Andisols
- Aridisols
- Entisols
- Gelisols
- Histosols
- Inceptisols
- Inceptisols
- Mollisols
- Oxisols
- Spodosols
- Ultisols
- Vertisols

- Rocky Land
- Ice
- Water

● **FIGURE 10.15** Map of soil orders across the United States.

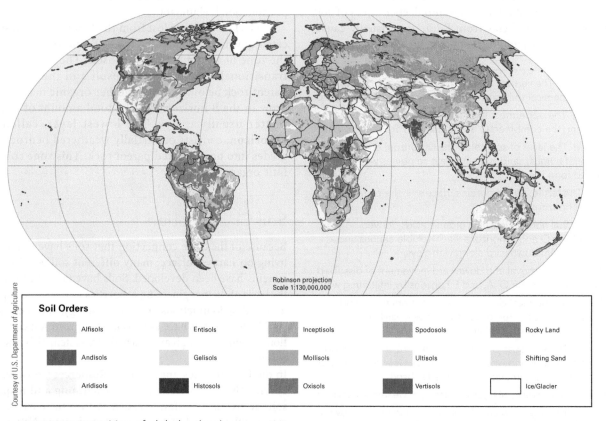

Courtesy of U.S. Department of Agriculture

Robinson projection
Scale 1:130,000,000

Soil Orders

Alfisols	Entisols	Inceptisols	Spodosols	Rocky Land
Andisols	Gelisols	Mollisols	Ultisols	Shifting Sand
Aridisols	Histosols	Oxisols	Vertisols	Ice/Glacier

● **FIGURE 10.16** Map of global soil orders.

The mapped soil elements in both Figures 10.15 and 10.16 are based on soil orders. A **soil order** is the highest hierarchical level of soil classification as defined by the NRCS and is based on the physical and chemical character, thickness, and color of the soil horizons. Each soil order is further subdivided into suborders, great groups, groups, families, and series. A **soil series** is the lowest category of soil classification and is generally given a name based on the locality where that soil was first mapped and described. Today, over 20,000 different soil series have been identified and mapped by the NRCS within the United States. Fifty of these soil series—one for each U.S. state—have been selected as a state soil because the soil series has special significance to that state. Twenty of these soil series have been officially legislated as state soils and share the same level of distinction as a state bird, flower, or motto.

●**FIGURE 10.17** shows typical examples of each of the twelve soil orders currently recognized by the NRCS, along with their associated landscapes and diagnostic characteristics. As explained below, each order is characteristic of a unique combination of the main soil-forming factors: parent rock, climate, biological activity, and topography.

Soil-Forming Factors

Why are some soils nutrient-rich and others nutrient-poor, some sandy and others loamy? Six factors control soil characteristics: parent rock, climate, rates of plant growth and decay, slope aspect and steepness, time, and transport of soil materials.

PARENT ROCK The texture and composition of soil depends partly on its parent rock. For example, when granite decomposes, the feldspar converts to clay and the rock releases quartz as sand grains. If the clay leaches from the E horizon into the B horizon, a sandy soil forms. In contrast, because basalt contains no quartz and is rich in finely crystalline feldspar, soil formed from basalt typically is rich in clay and contains little sand.

Both of the examples above involve the formation of soil directly on bedrock and are called **residual soils**. In contrast, **transported soils** do not develop from weathering of local bedrock; they develop from parent material (regolith) brought in from somewhere else. Despite the name, it is not the soil itself that is transported, but rather the regolith from which the soil is formed. Examples of transported soils include those formed on sediment deposited on a river floodplain during a flood or on deposits of windblown silt, called **loess**, which is derived from glacial erosion. Generally, soils formed on hillsides are residual soils and thin relative to valley bottoms, where thicker, transported soils typically form. Also, transported soils typically are more fertile than residual soils, because they consist of a wider variety of source materials and hence supply a greater variety of minerals and nutrients.

Andisol, one of the twelve soil orders, is linked specifically to its parent rock (●**FIGURE 10.17A**). Andisols contain high proportions of volcanic rock material, including volcanic glass, and primarily form as residual soils from the direct weathering of volcanic bedrock. In the continental United States, andisols are restricted to the large volcanic provinces of the Pacific Northwest. They are also common in Hawaii. Because volcanic rocks are commonly rich in nutrients, andisols are capable of supporting intense agricultural production.

Vertisol (●**FIGURE 10.17B**) is a soil order also classified mainly on the basis of its parent material. Characterized by an abundance of vertical cracks, vertisols form in clay-rich soil in which the clay expands and contracts with the addition and removal of water. Vertisols are common in regions where volcanic ash has weathered to clay and in the floodplains of some large river systems where much clay has been deposited.

CLIMATE Climate exerts a fundamental control over soil formation, mainly through average annual temperature and precipitation. Rates of chemical weathering reactions, plant growth, and plant decay are all higher in warmer climates than cold ones. Precipitation that infiltrates the soil can translocate soil particles and ions downward, leading to the development of soil horizons.

Although precipitation typically seeps downward through soil, several other factors related to climate can pull the water back upward. Plant roots suck soil water toward the surface. In arid climates, subsurface evaporation can cause upward movement of water. In addition, a process called **capillary action** draws water upward in the same way that water is drawn up into the holes of a sponge placed on a countertop spill of water. In a soil, capillary action is caused by the attraction of the water molecules to the soil particles and by the cohesiveness of the water itself.

In arid and semiarid regions, rainstorms typically are of short duration and little rain falls. Consequently, when the rain stops, capillary action, plant roots, and subsurface evaporation draw most of the water upward, where it escapes or is taken up by plants. As the water escapes, many of its

---

soil series The lowest hierarchical classification of soils by the National Resource Conservation Service. Over 20,000 soil series are recognized, including 50 designated "state soils."

residual soils A soil formed from the weathering of bedrock below.

transported soil A soil formed by the weathering of regolith that is transported from somewhere else and deposited.

loess An accumulation of windblown silt derived from glacial erosion.

andisol Young soil developed on volcanic parent material and containing abundant unweathered volcanic glass and other volcanic debris, resulting in a high cation exchange capacity and high fertility.

vertisol Clay-rich, dark-colored soil characterized by periodic desiccation and deep cracking. Nearly impermeable and sticky when wet.

capillary action The process by which water is pulled upward through the soil due to the natural attraction of water molecules to soil particles and the cohesion of water.

Soil order and key characteristics	Soil-forming environment	Example soil profile

(A) **Andisol**: young soil from volcanic parent material; contains >50% volcanic glass and other chemically-reactive volcanic debris, resulting in high CEC and high fertility. Capable of supporting intensive agriculture and supports very productive forests in Pacific Northwestern U.S.

Mount Rainier, Washington

(B) **Vertisol**: dark brown to black soil rich in expandable clay minerals. Characterized by deep vertical cracks when dry and commonly very sticky when wet. Organic debris can accumulate in open cracks, resulting in thick A horizon. Clay-rich character makes vertisols nearly impermeable when wet and favorable for rice production. Does not typically promote growth of trees.

Rice field during and after irrigation

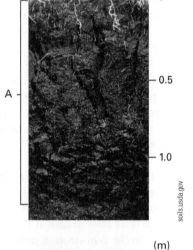

(C) **Aridisol**: desert soils characterized by low moisture content and little organic matter. Require sufficient time to develop. Subsoil is characterized by accumulation of mineral salts, especially calcium carbonate. Cementation of subsoil horizons can lead to formation of hardpan, and surface is prone to salinization if over-irrigated.

Sonoran Desert, southern Arizona

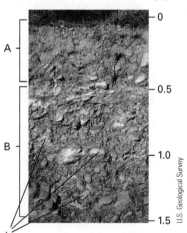

Abundant marble-sized calcium carbonate nodules

● **FIGURE 10.17** The main characteristics of each soil order, landscapes in which they occur, and example soil profile. (A) Andisol; (B) Vertisol; (C) Aridisol; (D) Alfisol; (E) Spodosol; (F) Mollisol; (G) Ultisol; (H) Oxisol; (I) Gelisol; (J) Histosol; (K) Entisol; and (L) Inceptisol.

Soil order and key characteristics	Soil-forming environment	Example soil profile

(D) **Alfisol**: form in semi-arid to humid environments, typically support hardwood forests; characterized by clay-rich subsoil with translocation of aluminum and iron to B horizon; also contain >35% base cations (Ca++, Mg++, K+); relatively high fertility and capable of supporting sustained agriculture.

Hardwood forest, eastern Wisconsin

(E) **Spodosol**: typically form in cool, moist environments and support coniferous forests. Texture is typically sand-dominated with little to no clay. The coarse-grained texture promotes downward transport of organic acids from A and E horizons to B horizon. Typically have a well-developed light-colored E horizon and red-brown B horizon resulting from downward translocation of aluminum and iron. Acidic soil, naturally infertile; requires addition of lime for cultivation.

Mixed coniferous-hardwood forest, Vermont

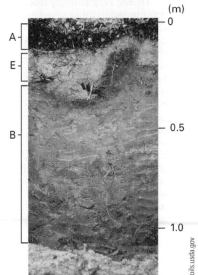

(F) **Mollisol**: mid-latitude grassland ecosystem soil characterized by thick, rich A horizon, high CEC, and high base saturation. High fertility of A horizon results from long-term addition of organic matter derived from plant roots. Downward translocation of base cations can produce B horizon rich in calcium carbonate or other salts. Mollisols are the most widespread soil order in the continental U.S. and are extensively cultivated.

Grassland prairie, Nebraska

● **FIGURE 10.17** (Continued)

Soil order and key characteristics	Soil-forming environment	Example soil profile

(G) **Ultisol:** Strongly leached soil formed in temperate-humid to tropical settings. Characterized by downward translocation and accumulation of clay in the B horizon and a red or yellow color resulting from the presence of iron oxides. Intense weathering has removed most of the base cations (Ca++, Mg++, K+) and base cation saturation is below 35%. Ultisols commonly form on older, stable geologic surfaces and have low fertility. The 'red clay soils' of the southeastern U.S. are ultisols.

Hardwood forest in southern Georgia

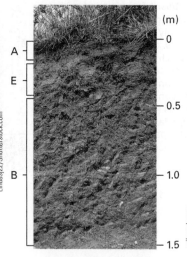

(H) **Oxisol:** Very strongly leached soil formed in tropical settings. Characterized by mixtures of quartz, clay, and minor organic matter. Oxisols form on very old geologic surfaces and typically lack well-developed horizons. Strong leaching has removed nearly all base cations, resulting in a very low CEC. They are extremely infertile soils with very low nutrient content. Nutrients that are present are contained within the standing vegetation, so if this is removed oxisols are very difficult to cultivate and erode easily.

Tropical rain forest in Vanuatu, south Pacific

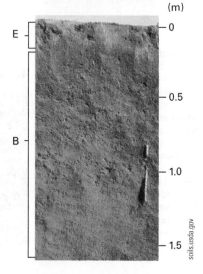

(I) **Gelisol:** Formed in high latitude or high elevation regions characterized by very cold temperatures and permafrost within 2 meters of the surface. Cold temperatures slow the decay of organic matter, and gelisols can be very organic-rich. Freezing and thawing activity has churned the soil, resulting in a general lack of well-developed horizons.

Tundra on Baffin Island, Canada

● **FIGURE 10.17** (*Continued*)

Soil order and key characteristics	Soil-forming environment	Example soil profile

(J) ***Histosol***: Characterized by thick, organic-rich accumulation in non-permafrost regions, typically bogs, marshes, or swamps. Abundance of standing water inhibits the decay of organic matter. Histosols are typically >20–30% organic carbon by weight, low density, and usually >40 cm thick. They are locally mined as peat.

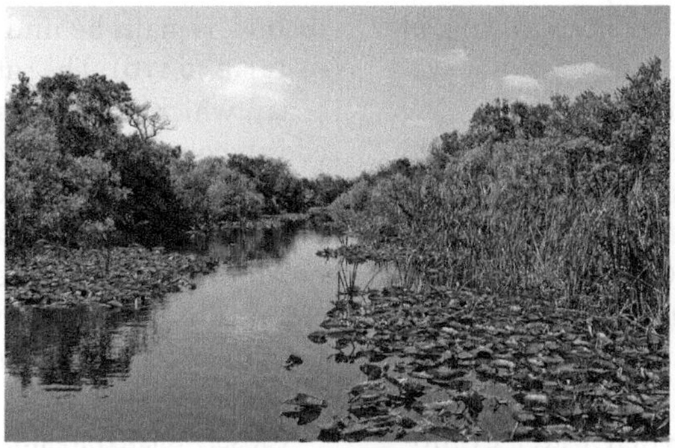

tonyz20/Shutterstock.com

Everglades National Park, Florida

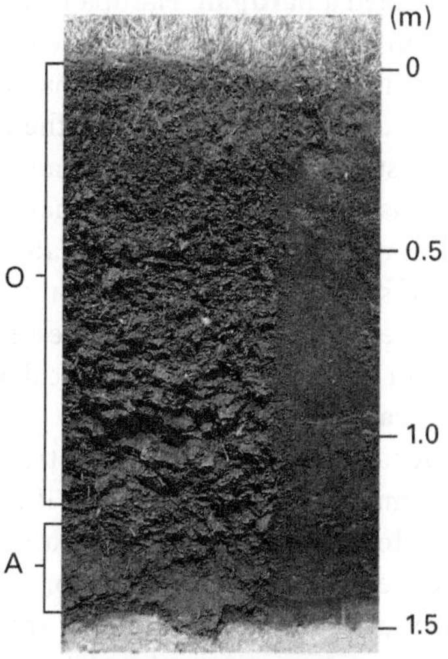

soils.usda.gov

(K) ***Entisol***: A very young soil with little or no evidence of horizon development except an A horizon. Typically developed on unconsolidated parent material. Many entisols are sandy and thin. Soils not classified under any other order are classified as entisols, so considerable diversity exists within this soil order. Globally, it is the most extensive soil order.

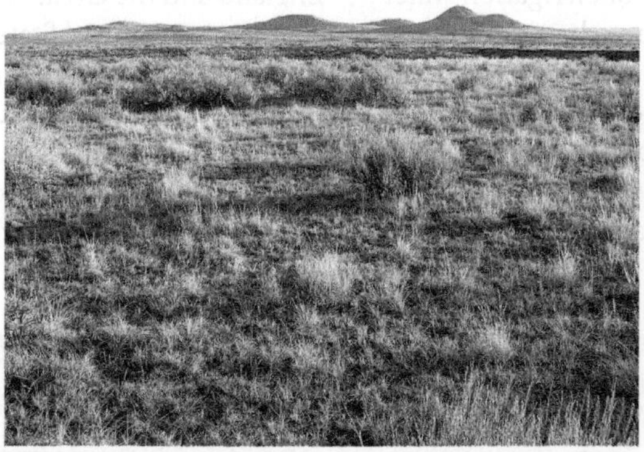

Copyright and Photograph by MARC S. HENDRIX

High plains near Broadus, southeastern Montana

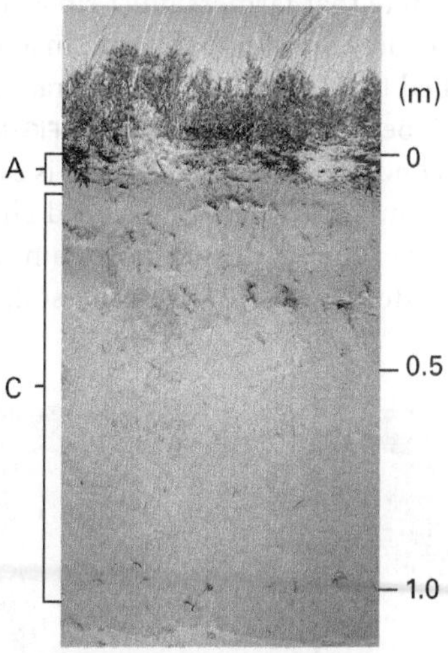

soils.usda.gov

(L) ***Inceptisol***: A young soil formed in sub-humid to humid environments; horizons are weakly developed but more pronounced than entisols. Inceptisols have lost some base cations but retain an abundance of unweathered minerals. They lack accumulated clays, aluminum, or organic matter from downward translocation and commonly form on moderate slopes, young surfaces, or resistant parent material. Variable fertility.

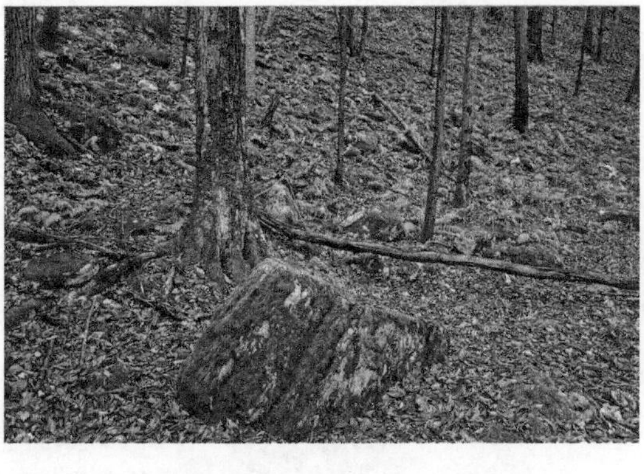

Andrew F. Kazmierski/Shutterstock.com

Pocono Mountains, Pennsylvania

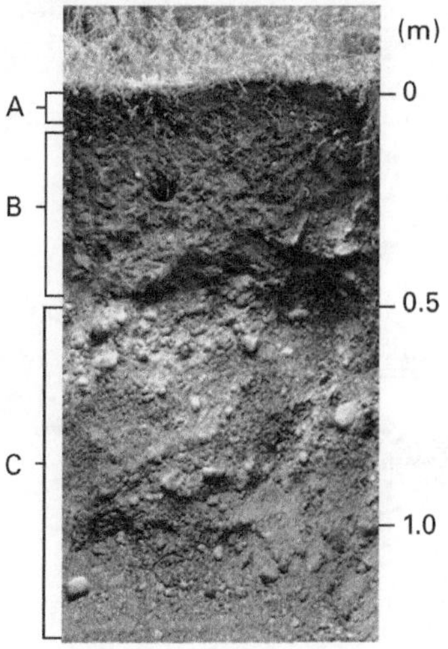

soils.usda.gov

FIGURE 10.17 *(Continued)*

dissolved ions precipitate in the B horizon, encrusting the soil with salts and in some cases forming an impervious layer called a **hardpan**. Hardpans can be very difficult to dig through and can be impenetrable to roots. Trees growing in soils with well-developed hardpans have shallow root systems and are easily uprooted by the wind.

Most arid and semiarid climates produce soils classified as **aridisols** (●FIGURE 10.17C), characterized by low concentration of organic matter, water deficiency, and accumulation of soluble salts in the B horizon. Commonly, upward movement of water in aridisols results in the precipitation of calcium carbonate nodules or a calcium carbonate hardpan called **calcrete**.

Because nutrients concentrate when water evaporates, many aridisols are fertile if irrigation water is available. However, salts from irrigation water can accelerate the development of calcrete. For example, in the Imperial Valley of Southern California, irrigation water contains high-enough concentrations of calcium carbonate to form a calcrete that farmers must rip apart with heavy machinery before continuing to farm. In addition, irrigation water applied to aridisols can concentrate so much salt in the soil that it becomes toxic to plants (●FIGURE 10.18). The evaporative concentration of salts in soil is called **salinization**.

In many semiarid to humid climates, rainstorms are of longer duration and more rain falls. As a result, water seeps downward through the soil, leaching soluble ions from both the A and E horizons and translocating these along with clays downward to the B horizon. The less-soluble elements—such as aluminum, iron, and some silicon—remain behind to form a soil type called **alfisol** (●FIGURE 10.17D). The subsoil in an alfisol is commonly rich in clay, which is mostly aluminum and silicon and has the reddish color of iron oxide; the prefix "alf" is derived from the chemical symbols for aluminum (Al) and iron (Fe). Alfisols typically form under a hardwood forest cover and contain relatively high percentages of base cations. For this reason, alfisols are relatively fertile and are widely used in agriculture. They are common in the Ohio River valley of the eastern United States.

Closely related to alfisols are **spodosols** (●FIGURE 10.17E), which are also typically formed in coniferous or mixed deciduous and coniferous forests. Spodosols form by intense translocation of base cations and clay downward, are relatively depleted of base cations, and typically have a well-developed E horizon. Spodosols are less fertile than alfisols and form mainly in the dense forests of northern New England and the Great Lakes region, as well as in parts of the southeastern United States.

Much of central North and South America and central Asia is characterized by a semiarid to semihumid climate, grassland ecosystem, and a soil type called **mollisol** (●FIGURE 10.17F). Mollisols are characterized by an A horizon that is nutrient enriched, has high organic content, and

● **FIGURE 10.18** Salts have poisoned this Wyoming soil. Saline water seeps into the depression and evaporates, to deposit white salt crystals on the ground and on the fence posts.

Courtesy of Graham R. Thompson/Jonathan Turk

is typically 60 to 80 centimeters thick. The high organic content and thick A horizon results from the long-term addition of organic matter from plant roots, which also produce a soft, granular texture. Fire and disruption by ants and earthworms also exert strong influences on mollisols. The fertility, thickness, and texture of the A horizon make mollisols among the most agriculturally productive soil type.

At the opposite extreme are soils formed in environments characterized by high temperatures and high rainfall. In such environments, so much water seeps through the forest litter that the organic acids formed leach away nearly all the soluble cations. Two soil orders, **ultisols** (●FIGURE 10.17G) and **oxisols** (●FIGURE 10.17H), are formed in such environments. Ultisols are commonly called "red clay soils" and are common in the southeastern United States, where heavy leaching has removed roughly two-thirds of the base cations from the soil. Ultisols can be used for agriculture, but

---

hardpan General term for a soil layer that is relatively impervious to water and impenetrable to plant roots. Commonly forms from precipitation of salts in a soil B horizon by either downward or upward translocation.

aridisol A soil formed in arid or semiarid environments and characterized by very low organic content, water deficiency, and precipitation of salts in the B horizon.

calcrete A hardpan that forms in the B soil horizon in arid and semiarid regions when calcium carbonate precipitates and cements the soil particles together.

salinization A process whereby salts accumulate in soil that is irrigated heavily, lowering soil fertility.

alfisol A soil formed in semiarid to humid climates, typically under hardwood cover. Characterized by accumulation of clay in the B horizon and relatively high fertility, making it productive for agriculture.

spodosol Sandy, acidic soil developed in moist, temperate environments, commonly in coniferous or mixed coniferous-deciduous forests. Leaching has translocated base cations downward, resulting in a well-developed E horizon.

mollisol Grassland soil characterized by rich A horizon, high cation exchange capacity, and B horizon rich in base cation salts; very fertile soil.

ultisol Strongly weathered soil formed in semihumid or humid environments. Intense weathering has removed most base cations, resulting in low fertility. Includes red clay soils of SE United States.

oxisol A soil formed in a hot, humid climate and characterized by intensive leaching of soluble cations from the A horizon, little ability to retain nutrients, and very poor fertility. Very insoluble iron and aluminum oxides are concentrated.

gelisol A high-latitude soil formed over permafrost that is no deeper than two meters. Characterized by an organic-rich A horizon that usually extends to the permafrost boundary.

histosol A very organic-rich soil, typically formed in a poorly drained area where stagnant water inhibits organic decay. Typically composed of thick O and A horizons. Can be mined as peat.

aspect The orientation of a slope with respect to the Sun; the direction toward which the slope faces.

---

typically must be fertilized. The name *oxisol* comes from "oxide" minerals, particularly the very insoluble iron oxides and aluminum oxides, which become concentrated. Oxisols are very infertile soils, are nearly devoid of soluble ions, and have little ability to retain nutrients. Thus, plants growing on oxisols must derive their nutrients almost entirely from decaying litter in the O horizon. Oxisols are typically colored yellow or red by iron oxide. Bauxite, the world's main type of aluminum ore (Chapter 5), is a highly aluminous oxisol. Oxisols are not common in the continental United States, but are found on Puerto Rico and Hawaii.

RATES OF PLANT GROWTH AND DECAY In the tropics, plants grow and decay rapidly and growing plants quickly absorb the nutrients released by decaying vegetation. Heavy rainfall and organic acids leach nutrients from the soil. Little humus accumulates and few nutrients are stored in the soil. Thus, even though the tropical rain forests support great populations of plants and animals, these depend on a rapid cycle of growth, death, and decay. Conversely, the Arctic tundra is so cold that organic matter in the soil decays very slowly, accumulating through time to form a significant reservoir of carbon. Such a soil, called a **gelisol** (from the Latin *gelare*, meaning "to freeze"), is characterized by permafrost within two meters of the surface and an A horizon that usually extends to the permafrost boundary (●FIGURE 10.17I). Gelisols may contain preserved organic matter, but they are unfertile because nutrients are typically highly leached above the permafrost zone.

Among the most fertile soils are the mollisols and alfisols of temperate prairies and forests. There, large amounts of plant litter drop to the ground in the autumn, but decay is slow during the winter. Plant growth during the growing season is not fast enough to extract all the nutrients from the soil. As a result, thick layers of nutrient-rich humus accumulate. Even more organic-rich are **histosols** (●FIGURE 10.17J), which are dominated by organic soil materials. Histosols are usually formed in bogs or swamps.

SLOPE ASPECT AND STEEPNESS **Aspect** refers to the orientation, or facing direction, of a slope with respect to the Sun. In the semiarid regions of the Northern Hemisphere, thick soils and dense forests cover the cool, shady north slopes of hills, but thin soils and grass dominate hot, dry, southern exposures (●FIGURE 10.19). The reason for this difference is that in the Northern Hemisphere more water evaporates from the hot, sunny, southern slopes. Therefore, fewer plants grow, weathering occurs slowly, and soil development is retarded. Plants grow more abundantly on the moister northern slopes, and more rapid weathering forms thicker soils.

TIME Chemical weathering occurs slowly in most environments, and time is therefore an important factor in determining the extent of weathering. Recall that most minerals weather to clay. In geologically young soils, weathering may be incomplete and the soils may contain many partly weathered

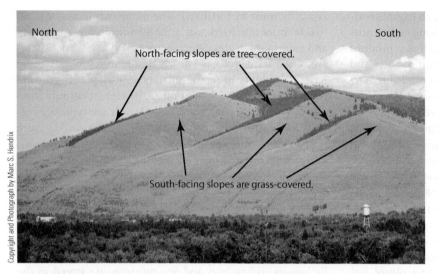

North

South

North-facing slopes are tree-covered.

South-facing slopes are grass-covered.

Copyright and Photograph by Marc S. Hendrix

● **FIGURE 10.19** In this photograph from the Missoula Valley of west-central Montana, thick forests cover the cool, shady slopes with a northern aspect, but grass and sparse trees dominate the hotter, drier southern-facing exposures.

mineral fragments. As a result, young soils are often sandy or gravelly. **Entisols** are soils that are so young that little or no evidence for soil horizons is present (●**FIGURE 10.17K**).

In addition to being sandy or gravelly, entisols are typically very thin. As weathering of an entisol continues, it commonly will evolve to an **inceptisol** (●**FIGURE 10.17L**), which is more mature but still lacks a B horizon in which downward translocated clays have accumulated. As weathering continues and clay and cations begin to translocate downward and accumulate, inceptisols will evolve to other soil orders containing a diagnostic B horizon.

MASS WASTING AND LANDSLIDES

Mass wasting is the downslope movement of earth material by gravity. The word **landslide** is a form of mass wasting that involves the creation of certain landforms.

All rock and sediment on the side of a hill or mountain is constantly being pulled downward by gravity, and any process that causes this soil to expand and shrink

DIGGING DEEPER

Soil Erosion and Human Activities

Cropland soils are rapidly eroding, posing a significant threat to food sustainability for future human generations

Weathering decomposes bedrock, and plants add organic material to the regolith to create soil at Earth's surface. However, soil does not accumulate and thicken throughout geologic time. If it did, Earth would be covered by a mantle of soil hundreds or thousands of meters thick and rocks would not exist at Earth's surface. Instead, all natural forms of erosion combine to remove soil about as fast as it forms.

Interactions with flowing water, wind, and glaciers all erode soil, and some weathered material simply slides downhill under the influence of gravity, as we will explore below. Once soil erodes, the particles of clay, sand, and gravel are carried downhill by the same agents that eroded them: streams, glaciers, wind, and gravity. On their journey, they may come to rest temporarily on a floodplain or lake bottom, but eventually most particles are eroded again, or reworked, and carried further downslope to the sea, where they accumulate in deltas or are swept offshore by marine currents. There,

the sediment is deposited, buried, and ultimately lithified to form sedimentary rocks.

It can take thousands of years for an eroded soil to reform. Thus, for all practical purposes, soil is a nonrenewable resource: once it is eroded, it is gone for generations.

Unfortunately, human activities have greatly accelerated soil erosion. One recent study concluded that humans move 10 times more sediment than the sum of all natural processes operating on the surface of the planet! The study also concluded that this massive

promotes the incremental **downhill creep** of soil particles. For example, when soil pore water freezes or when

---

entisol A very young soil typically lacking horizons and formed on unconsolidated parent material. All soils not classified with a different order are classified as entisols, so much diversity exists within this order.

inceptisol A young soil exhibiting weak horizons and developed in subhumid to humid environments. Typically retains abundant unweathered material.

mass wasting The downslope movement of earth material, primarily caused by gravity. (*See also landslide.*)

landslide A general term for mass wasting (the downslope movement of rock and regolith under the influence of gravity) and the landforms it creates.

downhill creep The gradual downhill movement, under the force of gravity, of soil and loose rock material on a slope. Facilitated by the freeze-thaw cycle, in which soil particles move orthogonal to the slope surface during freezing but directly downward during thawing.

flow Form of rapid mass wasting in which loose soil or sediment moves downslope as a slurry-like fluid, not as a consolidated mass; may occur slowly (less than 1 centimeter per year for some earthflows) or rapidly (several meters per second for some mudflows and debris flows).

---

expandable clay minerals in soil are wetted, the soil expands. As it expands, soil particles are pushed outward at a right angle to the slope surface. Later, when the soil shrinks, the pore water thaws, or the clay minerals dry out, gravity pulls the soil particles directly downward, not back toward the original slope surface. Over time, this process causes soil particles to slowly creep downslope (●**FIGURE 10.20**).

In contrast to downhill creep that takes place slowly, other forms of mass wasting can occur very quickly due to natural processes or human activities that destabilize a slope—a landslide. Rain, melting snow, or a leaking irrigation ditch all can add weight to and lubricate soil, causing it to move quickly downslope, especially on steep slopes.

TYPES OF RAPID MASS WASTING

Rapid forms of mass wasting fall broadly into three categories: flows, slides, and falls (●**FIGURE 10.21**). To understand these categories, think of a sand castle. Sand that is saturated with water flows down the face of the structure. During **flow**, loose, unconsolidated soil or sediment moves as a

displacement of sediment has resulted in a rate of cropland loss roughly 10 times faster than the rate at which cropland soils are formed.[2] Improper farming, livestock grazing, and logging all accelerate erosion. Plowing removes plant cover that protects soil. Logging removes forest cover, and the machinery breaks up the protective litter layer. Intensive grazing strips away protective plants, while hoof prints disrupt the soils. Rain, wind, and gravity then can erode the exposed soil easily and at a much faster rate than new soil is formed, leading to net soil loss. Soil erosion also contaminates waterways with silt as well as with herbicides and pesticides.

Soil can be difficult or expensive to renew once it has become degraded. Intense pressures on the world's soil resources have resulted in loss of productivity and biodiversity and increasing desertification of once-fertile lands. When farmers use proper conservation measures, soil can be preserved indefinitely or even improved. Some regions of Europe and China have supported continuous agriculture for centuries. However, in recent years, marginal lands on hillsides, in tropical rain forests, and along the edges of deserts have been brought under cultivation, greatly accelerating the global rate of soil loss. If left to continue, this global loss of soil resources will threaten the food sustainability for future human generations.

---

2. Wilkinson, B. E., "Humans as Geologic Agents: A Deep-Time Perspective," *Geology* 33, no. 3 (2005): 161–164.

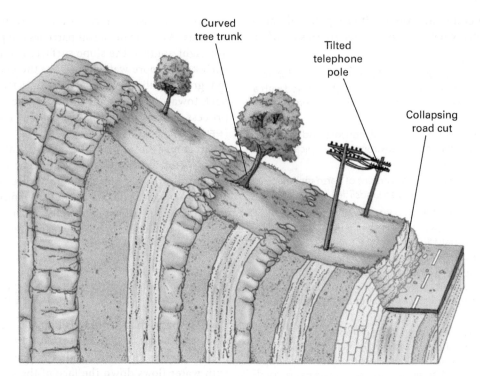

● **FIGURE 10.20** During creep, the land surface moves more rapidly than deeper layers, so objects embedded in rock or soil tilt downhill.

fluid. Some slopes flow slowly—at a speed of 1 centimeter or less per year. In contrast, mud with high water content can flow nearly as rapidly as pure water.

If you undermine the base of a sand castle, the wall may fracture and a segment of the wall may slide downward. Movement of a coherent block of material along a fracture is called a **slide**. Natural slides usually move over timeframes of a few seconds or minutes, although some slides take days or weeks to finish moving.

If you take a huge handful of sand out of the bottom of the castle, the whole tower topples. This rapid, free-falling motion of loose material is called **fall**. Fall is the most rapid type of mass wasting. In extreme cases, such as the face of a steep cliff, rock can fall at a speed dictated solely by the force of gravity and air resistance.

TABLE 10.1 outlines the characteristics of flow, slide, and fall. Also, refer back to Figure 10.21 as you read details of these three types of mass wasting, described next.

As the name implies, **creep** is the slow, downhill flow of rock or soil under the influence of gravity. A creeping slope typically moves at a rate of about 1 centimeter per year, although wet soil can creep more rapidly. During creep, the shallow soil or rock layers move more rapidly than deeper material moves. As a result, anything with roots or a foundation tilts downhill (see Figure 10.20).

Trees have a natural tendency to grow straight upward. As a result, when downhill creep of soil and shallow bedrock tilts a growing tree, the tree develops a J-shaped curve in its trunk, called pistol butt (●**FIGURE 10.22**). If you ever contemplate buying hillside land for a homesite, examine

the trees. If they have pistol-butt bases, the slope is probably creeping, and creeping soil can slowly tear a building foundation apart.

If heavy rain falls on unvegetated soil, the water can saturate the soil to form a slurry of mud and rocks called a debris flow, **earthflow**, or **mudflow** depending on the size and sorting of the particles. A slurry is a mixture of water and solid particles that flows as a liquid. Wet concrete is a familiar example of a slurry. It flows easily and is routinely poured or pumped from a truck.

The advancing front of a debris flow, earthflow, or mudflow often forms tongue-shaped lobes (●**FIGURE 10.23**).

---

slide Form of rapid mass wasting in which the rock or soil initially moves as a consolidated unit along a fracture surface.

fall Form of rapid mass wasting in which unconsolidated material falls freely or bounces down steep slopes or cliffs.

creep A form of mass wasting in which loose material moves very slowly downslope, usually at a rate of only about 1 centimeter per year and usually on land with vegetation. Trees on a creeping block tilt downhill and grow to have a trunk shaped like a pistol butt.

earthflow A viscous flow of fine-grained sediment or fine-grained sedimentary rock that is saturated with water and moves downslope as a result of gravity; usually slow moving, typically less than one to several meters per day.

mudflow A form of rapid mass wasting that involves the downslope movement, usually on unvegetated land, of fine-grained soil particles mixed with water; can be slow moving, as slow as 1 meter per year, or as fast as a speeding car.

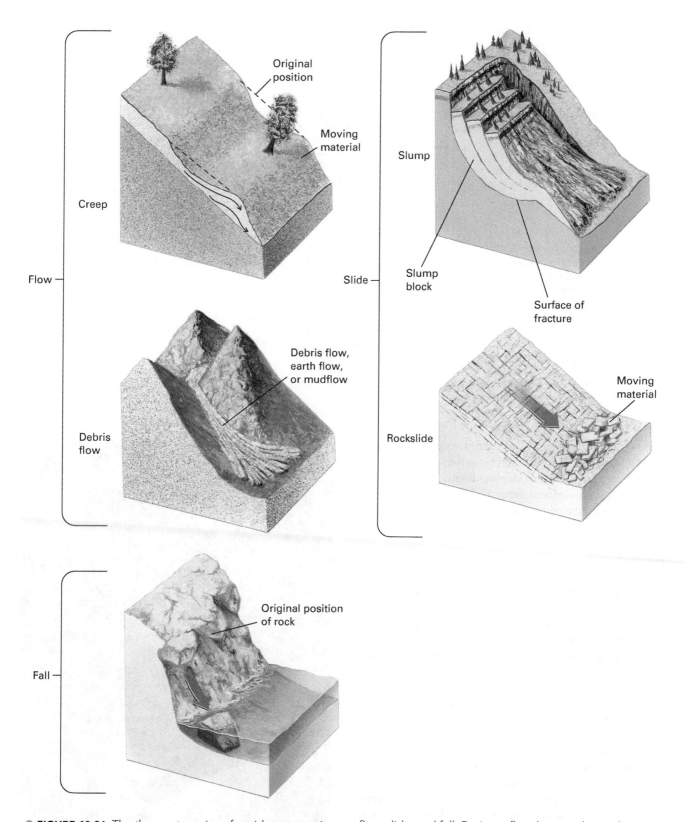

● **FIGURE 10.21** The three categories of rapid mass wasting are flow, slide, and fall. During a *flow*, loose soil or sediment moves as a fluid. During a *slide*, the entire slab of rock or soil moves downslope as a unit. In a *fall*, loose material bounces freely downslope.

TABLE 10.1 Categories of Mass Wasting

Type of Movement	Description	Subcategory	Description	Comments
Flow	Individual particles move downslope independently of one another, not as a consolidated mass. Typically occurs in loose, unconsolidated regolith.	Creep Debris flow Earthflow Mudflow	Rate of movement can range from visually imperceptible (<1 m/year) for creep to 100 km/hr or more for some mudflows. Flows typically are poorly sorted. Particle sizes range from large boulders in some debris flows to mostly silt and clay in mudflows.	Trees on creeping slopes are tilted downhill and develop pistol-butt shapes. Debris flows are common in arid regions with intermittent heavy rainfall, or can be triggered by a volcanic eruption. Earthflows usually develop in sloping, saturated fine-grained soils or saturated shale. Mudflows usually occur on unvegetated soil.
Slide	Material moves as consolidated blocks; can occur in regolith or bedrock.	Slump Rockslide	A block of Earth material slides downslope, usually with a backward rotation on an upward-concave, curved surface. A newly detached segment of bedrock breaks into fragments and tumbles downslope, usually with rapid movement.	Trees on slump blocks remain rooted and are tilted uphill.
Fall	Materials fall freely in air; typically occurs in bedrock.	—	—	Falls occur only on steep cliffs.

A slow-moving earthflow or mudflow travels at a rate of about 1 meter per year or slower, but others can move as fast as a car speeding along an interstate highway. A debris flow can pick up boulders and automobiles; it can also destroy houses, filling them with muddy sediment and even dislodging them from their foundations.

Slide

In some cases, a large block of rock or soil, or sometimes an entire mountainside, breaks away and slides downslope as a coherent mass or as a few intact blocks. Two types of slides occur: slumps and rockslides.

A **slump** occurs when blocks of material slide downhill over a gently curved fracture in rock or regolith. Trees remain rooted in the moving blocks. However, because the blocks rotate on the concave fracture, trees on the slumping blocks are tilted upslope (●FIGURE 10.24). Thus, you can distinguish slump from creep because a slump tilts trees uphill, whereas creep tilts them downhill. At the lower end of a large slump, the blocks often break apart and pile up to form a jumbled, hummocky topography.

It is useful to identify a slump, because it often recurs in the same place or appears on nearby slopes. Therefore, a slope that shows evidence of past slumping is not a good place to build a house.

During a **rockslide** (or rock avalanche), bedrock slides downslope over a fracture plane. Characteristically, the rock breaks up as it moves, and a turbulent mass of rubble tumbles down the hillside. In a large rockslide, the falling debris traps and compresses air beneath and among the tumbling

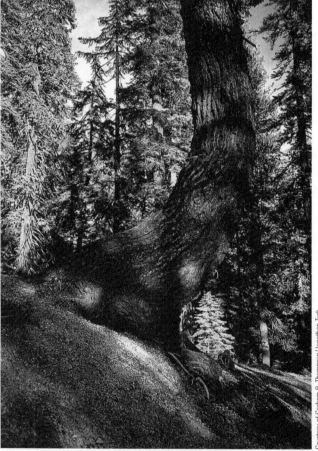

● **FIGURE 10.22** If a hillside creeps as a tree grows, the tree trunk develops pistol-butt shape.

● **Figure 10.23** Following the eruption of Mount Agung, a volcano on the Indonesian Island of Bali, the Yeh Sah River carried so much mud that it evolved into a debris flow. Here, people gather on the banks as the debris flow passes by.

Muhammad Fauzy Chaniago/Newzulu/Crowdspark/Alamy Stock Photo

● **FIGURE 10.24** Trees tilt back into the hillside on this slump along the Quesnel River in British Columbia. Slumping is facilitated by stream erosion at the base of the hill.

Graham R Thompsone/Jonathan Turk

blocks. The compressed air reduces friction and allows some rockslides to attain speeds of 500 kilometers per hour. The same mechanism allows a snow or ice avalanche to cover a great distance at high speed.

---

slump A type of slide in which blocks of material slide downslope as a consolidated unit over an upward-concave, curved fracture in rock or regolith; trees on the slumping blocks tilt uphill. The uphill portion of the slump usually consists of several tilted slide blocks, whereas the toe of the slump usually consists of rumpled, folded sediment.

rockslide A subcategory of slide mass wasting in which a segment of bedrock slides downslope along a fracture and the rock breaks into fragments and tumbles down the hillside; also called a *rock avalanche*.

Fall

If a rock dislodges from a steep cliff, it falls rapidly under the influence of gravity. Several processes commonly detach rocks from cliffs. Recall from our discussion of weathering that frost wedging can dislodge rocks from cliffs and cause rockfall. Rockfall also occurs when a stream or ocean waves undercut a cliff.

PREDICTING AND AVOIDING THE EFFECTS OF MASS WASTING

Every year, small rapid forms of mass wasting destroy homes and farmland. Occasionally, an enormous slide or flow buries a town or city, killing thousands of people. Rapid forms of mass wasting cause billions of dollars in damage every year. ●**FIGURE 10.25** shows three examples of mass wasting that have affected humans.

One of the most important tools in evaluating the risks associated with mass wasting, especially flows, slides, and falls, is understanding that these features commonly reoccur in the same area because the geologic conditions that cause mass wasting tend to be constant over a large area and for long periods of time. Thus, if a hillside has slumped, nearby hills may also be vulnerable to the same type of mass wasting. In addition, landslides and mudflows commonly follow the paths of previous slides and flows. If an old mudflow lies in a stream valley, future flows may follow the same valley.

Awareness and avoidance are the most effective defenses against mass wasting. Geologists evaluate landslide probability by combining data on soil and bedrock stability, slope angle, climate, and history of slope failure in the area. They include evaluations of the probability of a triggering event such as a volcanic eruption or earthquake. Building codes then regulate or prohibit construction in unstable areas. For example, according to the U.S. Uniform Building Code, a building cannot be constructed on a sandy slope steeper than 27 degrees, even though the angle of repose of sand is 30 to 35 degrees. So the law leaves a safety margin of 3 to 8 degrees. Architects can obtain permission to build on more-precipitous slopes if they anchor the foundation to stable rock.

Why Do Slides, Flows, and Falls Occur?

Imagine that you are a geological consultant on a construction project. The developers want to build a road at the base of a hill, and they wonder whether the road will be affected by slides, flows, or falls. What factors should you consider?

STEEPNESS OF THE SLOPE Obviously, the steepness of a slope is a factor in mass wasting. If frost wedging dislodges a rock from a steep cliff, the rock tumbles to the valley below. However, a similar rock is less likely to roll down a gentle slope.

(A)

(B)

(C)

● **FIGURE 10.25** Landslides cause billions of dollars in damage every year. (A) An earthquake of magnitude 6.7 struck Japan's northern island of Hokkaido on September 6, 2018, triggering these landslides which destroyed several houses. (B) This landslide on the island Gran Canaria, Spain, occurred after heavy rain in February, 2017. (C) On July 7, 2018, heavy rainfall triggered a very large landslide in western Iceland. The slide was initiated on a steep slope and ran for about 1.5 kilometers horizontally, blocking the Salmon River and causing the lake in the foreground to form.

TYPE OF ROCK AND ORIENTATION OF ROCK LAYERS If sedimentary rock layers dip in the same direction as a slope, a layer of weak rock can fail, causing layers over it to slide downslope. Imagine a hill underlain by shale, sandstone, and limestone oriented so that their bedding lies parallel to the slope, as shown in ●**FIGURE 10.26A**. If the base of the hill is undercut (●**FIGURE 10.26B**), the upper layers of sandstone and limestone may slide over the weak shale. In contrast, if the rock layers dip into the hillside, the slope may be stable even if it is undercut (●**FIGURES 10.26C** and **10.26D**).

Several processes can reduce the stability of a slope. Ocean waves or stream erosion can destabilize a slope, as can road building and excavation. Therefore, a geologist or engineer must consider not only a slope's stability before construction but also how the project might alter its stability.

THE NATURE OF UNCONSOLIDATED MATERIALS The **angle of repose** is the maximum slope or steepness at which loose sediment remains stable. If the slope becomes steeper than the angle of repose, the sediment slides. The angle of repose varies for different types of sediment. For example, rocks commonly tumble from a cliff, to collect at the base as angular blocks of talus. The angular blocks interlock and jam together. As a result, talus typically has a steep angle of repose, up to 45 degrees. In contrast, rounded sand grains do not interlock and therefore have a lower angle of repose—about 30 to 35 degrees (●**FIGURE 10.27**).

WATER AND VEGETATION To understand how water affects slope stability, think again of a sand castle. Even a novice sand castle builder knows that sand must be moistened to build steep walls and towers (●**FIGURE 10.28**); but too much water causes the walls to collapse. If only small amounts of water are present, the water collects only where one sand grain touches another. The surface tension and cohesion of the water binds the grains together. However, excess water fills the pores between the grains and exerts an outward pressure—called pore pressure—that pushes the grains apart. The excess water also lubricates the sand and adds weight to a slope. When some soils become water saturated, they flow downslope, just as the sand castle collapses. In addition, if groundwater collects on an impermeable layer of clay, shale, or even on permafrost, it may cause overlying rock or soil to become saturated and move easily.

Roots hold soil together and plants absorb water; therefore, a vegetated slope is more stable than a similar bare one. Many forested slopes that were stable for centuries slid when the trees were removed during logging, agriculture, or construction.

---

angle of repose The maximum slope or steepness at which loose material remains stable. If the slope becomes steeper than the angle of repose, the material slides.

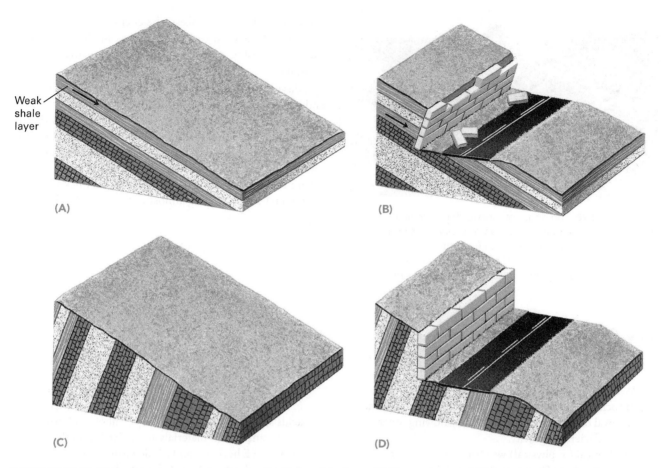

Weak
shale
layer

(A)

(B)

(C)

(D)

● **FIGURE 10.26** (A) Sedimentary rock layers dip parallel to this slope. (B) If a roadcut undermines the slope, the dipping rock provides a good sliding surface and the slope may fail. (C) Sedimentary rock layers dip at an angle to this slope. (D) The slope may remain stable even if it is undermined.

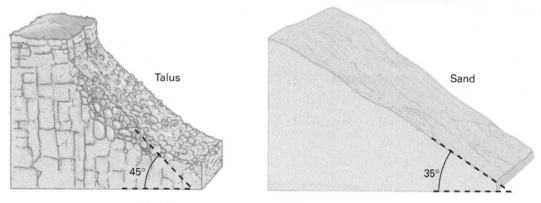

Talus

45°

Sand

35°

● **FIGURE 10.27** The angle of repose is the maximum slope at which a specific material can remain stable. Because the chunks of rock forming talus interlock and jam together, talus has a steeper angle of repose than sand, whose rounded grains do not interlock. See also Figure 10.4B.

Landslides are common in deserts and regions with intermittent rainfall. For example, Southern California has dry summers and occasional heavy winter rain. Vegetation is sparse because of summer drought and wildfires. When winter rains fall, bare hillsides often become saturated and slide. Landslides occur for similar reasons during infrequent but intense storms in deserts.

EARTHQUAKES AND VOLCANOES An earthquake may cause a landslide by shaking an unstable slope, causing it to move. Saturated soils are particularly prone to movement during earthquakes because seismic waves increase the pore pressure between soil particles, pushing them apart and causing the sediment to liquefy. If you have ever tapped with your foot on saturated sand at the beach,

Copyright and Photograph by Marc S. Hendrix

● **FIGURE 10.28** The angle of repose depends on both the type of material and its water content. Dry sand forms low mounds, but moistened sand can form the familiar steep-sided hills and towers of sand castles because the surface tension among the grains causes them to stick together.

you've seen the same process: the energy from the tapping of your foot travels through the sand as small compression waves, increasing its pore pressure and causing the sand to liquefy.

Mass wasting is common in earthquake-prone regions and in volcanically active areas. Steep volcanoes are prone to rock slides, particularly during earthquakes. In addition, a volcanic eruption may melt the snow and ice cap at the top of the volcano. As the meltwater flows downslope, it picks up loose sediment and ash and evolves into a debris flow (see Figure 10.23). Many volcanic regions contain thick sedimentary sequences consisting almost exclusively of debris flow deposits.

Key Concepts Review

- Weathering is the decomposition and disintegration of rocks and minerals at Earth's surface. Erosion is the removal of weathered rock or soil by flowing water, wind, glaciers, or gravity.
- Mechanical (or physical) weathering can occur by pressure-release fracturing, frost wedging, abrasion, organic activity, and thermal expansion and contraction.
- Chemical weathering is the alteration and breakdown of rock by changes in its chemistry and mineralogy. Dissolution, oxidation, and hydrolysis are among the most important chemical weathering processes. Chemical and physical weathering often operate together.
- Soil is the layer of weathered material overlying bedrock and consists of loose and weakly cemented sediment, humus, and ions in solution. Infiltration by

rain and snow melt causes downward translocation of silt and clay particles, soluble cations, and organic acids and, over time, the development of soil horizons. Most translocated materials are derived from the O, A, and E horizons and are deposited in the B horizon. The underlying C horizon is weathered bedrock.
- Mass wasting is the downslope movement of earth materials by gravity and includes incremental downhill creep of soil particles as well as rapid forms of mass wasting such as landslides.
- Rapid forms of mass wasting fall broadly into three categories: flows, slides, and falls.
- Earthquakes and volcanic eruptions trigger devastating mass wasting. Proper planning and engineering can avert much damage to human habitation.

Important Terms

A horizon (p. 232)	cation exchange capacity (CEC) (p. 232)	exfoliation (p. 230)
abrasion (p. 226)	chemical weathering (p. 225)	fall (p. 244)
alfisol (p. 240)	cloud condensation nuclei (p. 229)	flow (p. 243)
andisol (p. 235)	creep (p. 244)	frost wedging (p. 226)
angle of repose (p. 248)	dissolution (p. 228)	gelisol (p. 241)
aridisol (p. 240)	downhill creep (p. 243)	hardpan (p. 240)
aspect (p. 241)	E horizon (p. 232)	histosol (p. 241)
B horizon (p. 233)	earthflow (p. 244)	humus (p. 232)
C horizon (p. 233)	eluviation (p. 232)	hydrolysis (p. 229)
calcrete (p. 240)	entisol (p. 242)	inceptisol (p. 242)
capillary action (p. 235)	erosion (p. 224)	landslide (p. 242)

leaching (p. 232)
litter (p. 232)
loam (p. 232)
loess (p. 235)
mass wasting (p. 242)
mechanical weathering *or* physical
weathering (p. 225)
mollisol (p. 240)
mudflow (p. 244)
O horizon (p. 232)
organic activity (p. 227)
oxidation (p. 230)
oxisol (p. 241)

percentage base saturation (p. 232)
pressure-release fracturing (p. 225)
regolith (p. 232)
residual soils (p. 235)
reworking (p. 224)
rockslide (p. 246)
salinization (p. 240)
salt cracking (p. 230)
sediment-gravity flow (p. 225)
slide (p. 244)
slump (p. 246)
soil horizon (p. 232)
soil order (p. 233)

soil series (p. 235)
soils (p. 232)
spheroidal weathering (p. 231)
spodosol (p. 240)
talus (p. 226)
thermal expansion and
contraction (p. 227)
topsoil (p. 232)
transformation (p. 232)
translocation (p. 232)
transported soil (p. 235)
ultisol (p. 241)
vertisol (p. 235)

Review Questions

1. Explain the differences between mechanical weathering and chemical weathering.

2. What is a talus slope? What conditions favor the formation of talus slopes?

3. Characterize the four major horizons of a mature soil.

4. List and describe each of the factors that control slope stability.

Ikunl/Shutterstock.com

11

FRESHWATER

Streams, Lakes, Wetlands, and Groundwater

A portion of the Erawan Falls in Erawan National Park, located in Kanchanaburi Province, western Thailand. The Erawan Falls drop a total of 1,500 meters and comprise seven distinct tiers.

INTRODUCTION

Think of a pleasing, relaxing landscape, and you will probably envision a scene that involves water: a gentle surf lapping against white coral sand, with palm trees in the background; a gurgling brook flowing over moss-covered rocks, beneath a green forest canopy; or a sunrise burning through wispy fog on a mirror-smooth lake. Oceans cover two-thirds of Earth, and water also plays a seminal role in each of Earth's systems. We have already learned that water is an important agent in the rock cycle. Chapter 10 was also largely about water—as a corrosive chemical, a physical force when it freezes, and a medium that transports rock and sediment. This chapter explores the roles played by Earth's freshwater sources: streams, lakes, wetlands, and groundwater.

THE WATER CYCLE

The **hydrologic cycle**, or the water cycle, describes the continuous circulation of water among the four spheres: the hydrosphere (or watery part of the planet), the geosphere (the solid Earth), the biosphere (life-forms in the sea, on land, and in the air), and the atmosphere. About 1.3 billion cubic kilometers of water exist at Earth's surface. Of this huge volume, 97.5 percent is salty seawater, and another 1.8 percent is frozen into the great ice caps of Antarctica and Greenland. Although the hydrosphere contains a great volume of water, only 0.64 percent is fresh and available in streams and rivers, lakes, wetlands, and groundwater (●FIGURE 11.1).

Water evaporates from the continents and oceans to form water vapor in the atmosphere. This vapor eventually condenses and falls back to the surface as rain or snow. Most precipitation lands on the ocean, partly because it covers most of the planet. The precipitation that falls on the continents follows four paths, as illustrated in ●FIGURE 11.2:

1. Surface water flowing to the sea in streams and rivers is called **runoff**. This water may stop temporarily in a lake or wetland, but eventually it evaporates or flows to the oceans.
2. Much water seeps into the ground—the geosphere—to become part of a vast, subterranean reservoir of groundwater. Water infiltrating into the subsurface comes from direct precipitation and from surface water sources such as rivers and lakes. Although surface water is more conspicuous, 60 times more water is stored as groundwater than in all streams, lakes, and wetlands combined.

---

hydrologic cycle The continuous circulation of water among the hydrosphere, the atmosphere, the biosphere, and the geosphere; also called the *water cycle*.

runoff Surface water that flows to the oceans in streams and rivers.

Groundwater flows through pore spaces in sediment and bedrock in the subsurface but typically does so much more slowly than water flowing on the surface.

3. Most of the remainder of water that falls onto land evaporates back into the atmosphere. Water sucked upward

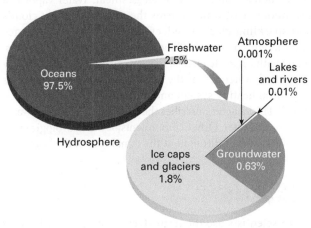

● FIGURE 11.1 Of Earth's water, 97.5 percent is in the oceans, and less than 1 percent is available as freshwater in streams and rivers, lakes, groundwater, and wetlands. Glacial ice presently accounts for 1.8% of Earth's freshwater.

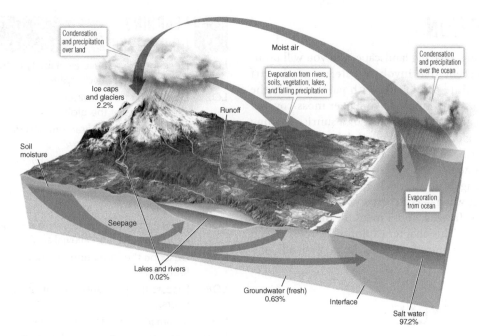

● **FIGURE 11.2** The hydrologic cycle shows that water circulates constantly among oceans, lakes, wetlands, and groundwater, Earth's atmosphere, its geosphere, and all organisms that live on the planet. Whole numbers and arrows shown in red indicate thousands of cubic kilometers of water transferred each year. The distribution of global water among Earth's major water storage reservoirs is shown as percentages.

through the root systems of plants also evaporates directly from plants as they breathe, in a process called **transpiration**.

4. A small amount of water is incorporated into the biosphere in the form of plant and animal tissue.

The hydrologic cycle refers to the movement of not only water but also energy from one part of the globe to another. Ocean currents transport huge quantities of heat from the equator toward the poles, thus cooling the equator and warming the higher latitudes. Evaporation is a cooling process, whereas condensation releases heat. Water vapor is a greenhouse gas, which warms the atmosphere, but clouds and sparkling glaciers reflect sunlight back out to space and thereby cool Earth. There are so many feedback cycles, many with opposite effects, that one scientist wrote, "Boldly simplified, water acts as the venetian blind of our planet, as its central heating system and as the fridge, all at the same time."[1] This chapter describes the movement of surface water and groundwater. Other components of the hydrologic cycle will be discussed throughout the remainder of the book.

STREAMS

Earth scientists use the term **stream** for all water flowing in a channel, regardless of the stream's size. The term *river* is commonly used for any large stream fed by smaller ones, called **tributaries**. In temperate climates, most streams run

year round, even during times of drought, because they are fed by groundwater that seeps into the streambed. These streams are called **perennial streams**. In contrast, most streams in arid environments are **intermittent streams** that flow only when water is provided by a precipitation event.

Stream Flow and Velocity

Three factors control stream velocity: gradient, discharge, and channel characteristics.

GRADIENT **Gradient** is the steepness, or vertical drop over a specific distance, of a stream. A tumbling mountain stream near the continental divide may drop 50 meters or more per kilometer, whereas a river in the upper Great Plains, like the Yellowstone River in eastern Montana, has much lower gradient of about 0.5 meters per kilometer. Considerably lower yet is the gradient of the lower Mississippi River, which has a gradient of 0.01 meters (1 cm) per kilometer (●**FIGURE 11.3**).

DISCHARGE **Discharge** is the volume of water flowing down a stream over a given period of time and usually is expressed as cubic meters per second (m³/sec) or cubic feet per second (cfs). At a given point in a stream, the velocity of its water increases when its discharge increases. Thus, a stream flows faster during flood, even though its gradient is unchanged.

When Huckleberry Finn floated down the lower Mississippi River near New Orleans on a raft, he probably drifted downstream at about 7 to 9 kilometers per hour (2 to 2.5 meters per second). In contrast, when Lewis and Clark

1. Heike Langenberg, Commentary introducing the special Insight section of "Climate and Water," *Nature* 419 (September 12, 2002): 187.

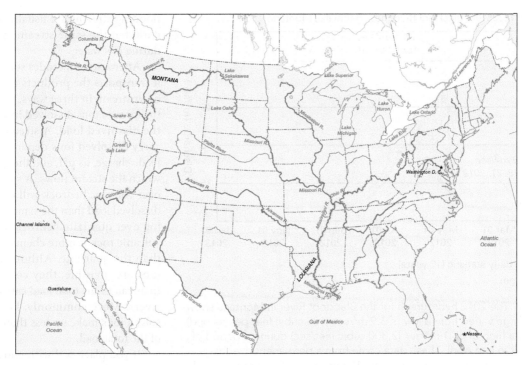

● **FIGURE 11.3** Map of major U.S. rivers highlighting the location of the Yellowstone River in Montana and the lower Mississippi River in Louisiana.

and the Corps of Discovery floated down the Yellowstone River, they drifted downstream at only 5 to 6 kilometers per hour (1.5 meters per second) even though the Yellowstone is about 50 times steeper than the lower Mississippi. The difference is discharge: the lower Mississippi has roughly 30 times as much water flowing through the channel than the Yellowstone River. The greater discharge of the lower Mississippi overcomes the steeper gradient of the Yellowstone. Thus, although it is counterintuitive, a large, lazy-appearing

river like the Mississippi is likely to flow more rapidly than a small, much steeper mountain stream—because of the larger river's greater discharge.

The largest river in the world is the Amazon, with an average discharge of 175,000 m^3/sec, enough water to fill 70 Olympic-sized swimming pools every second or enough to cover all of Manhattan with 18 centimeters (about 7 inches) of water every minute. In contrast, the Mississippi, the largest river in North America, has an average discharge of about 16,800 m^3/sec, enough to fill about 6.7 Olympic pools each second. In one minute of discharge, the Mississippi River would cover all of Manhattan with about 1.7 cm (5/8 inch) of water.

A stream's discharge can change dramatically from month to month or even during a single day. For example, the Blackfoot River, a mountain stream in western Montana, has a discharge of 30 to 35 m^3/sec during early summer, when mountain snow is melting rapidly. During the dry season in late summer, the discharge drops to about 5 m^3/sec (●**FIGURE 11.4**). Intermittent streams may be dry most of the time but become the site of a flash flood during a sudden thunderstorm.

CHANNEL CHARACTERISTICS Channel characteristics include the shape and roughness of a stream channel. The floor of the channel is called the **bed**, and the sides of the channel are the **banks**. Friction between flowing water and the stream channel slows current velocity. Consequently, water flows more slowly near the banks than near the center of a stream. If you paddle a canoe down a straight stream channel, you move faster when you stay away from the banks. The amount of friction depends on the roughness and shape

transpiration Direct evaporation of water into the atmosphere from the leaf surfaces of plants.

stream A moving body of water confined in a channel and flowing downslope; a *river* is a large stream fed by smaller ones.

tributaries A stream that feeds water into another stream or river.

perennial streams A stream that maintains some flow even during dry seasons.

intermittent streams A stream that does not maintain some flow all the time and may be dry for long periods.

gradient The steepness or vertical drop of a stream over a specific distance.

discharge The volume of water flowing downstream over a specified period of time, usually measured in units of cubic meters per second (m^3/sec) or cubic feet per second (cfs).

channel characteristics Features describing the shape and roughness of a stream channel.

bed The floor of a stream channel.

banks The rising slopes bordering the sides of a stream channel.

USGS gage 12335100; Blackfoot River near Helmville, Montana

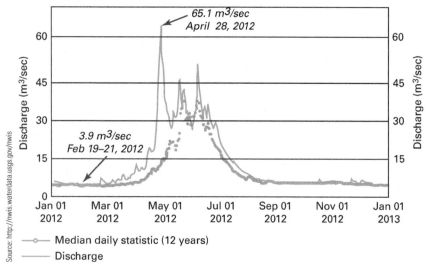

Source: http://nwis.waterdata.usgs.gov/nwis

— Median daily statistic (12 years)
— Discharge

● **FIGURE 11.4** The 2012 hydrograph for the Blackfoot River in Montana shows that the discharge varied from a low of 3.9 m³/sec (135 cubic feet per second) in February to a high of 65.1 m³/sec (2,300 cubic feet/sec) during a period of rapid snowmelt in late April. These data come from a permanent river gage that records discharge measurements every 15 minutes and is operated by the USGS.

of the channel. Boulders on the banks or in the streambed increase friction and slow a stream down, whereas the water flows more rapidly if the bed and banks are smooth. Likewise, a stream that is deep and narrow will flow faster than a broad, shallow stream with the same discharge, and a stream that is straight flows faster than a stream with lots of curves.

Stream Erosion and Sediment Transport

Streams shape Earth's surface by eroding soil and bedrock. The flowing water carries the eroded sediment downslope. A stream may deposit some of the sediment in the valley down which it flows, while it carries the remainder to a lake or to the sea, where the sediment accumulates.

Stream erosion and sediment transport depend on a stream's energy. A rapidly flowing stream has more energy to erode and transport sediment than a slow stream of the same size. The **competence** of a stream is a measure of the largest particle it can carry. A fast-flowing stream can transport cobbles and even boulders in addition to small particles. A slow stream carries only silt and clay.

The **capacity** of a stream is the total amount of sediment it can carry and is proportional to both current speed and discharge. Thus, a large, fast stream has a greater capacity than a small, slow one. Because the ability of a stream to carry sediment is proportional to both velocity and discharge, most sediment transport occurs during the few days each year when the stream is in flood. Relatively little sediment transport occurs during the remainder of the year. So

flooding streams are usually muddy and dark, but the same stream may be clear during low water.

After a stream erodes soil or bedrock, it transports the products of weathering downstream in three ways. One way is in the form of ions dissolved in water, called the **dissolved load**. A stream's ability to carry dissolved ions depends mostly on its discharge, its pH, and the geology over which the stream flows. A stream flowing over volcanic bedrock will have a higher dissolved load than the same stream flowing over quartzite bedrock, because the volcanic rock is more chemically reactive than the quartzite. Although dissolved ions are invisible, they comprise more than half of the total load carried by some rivers. More commonly, however, dissolved ions make up less than 20 percent of the total load.

If you place soil with equal amounts of sand, silt, and clay in a jar of water and shake it up, the sand grains settle quickly. But the smaller silt and clay particles remain suspended in the water as **suspended load**, giving it a cloudy appearance. Clay and silt are small enough that even the slight turbulence of a slow stream keeps them in suspension. A rapidly flowing stream can carry sand in suspension.

During a flood, when stream energy is highest, the rushing water can roll boulders and cobbles along the bottom as **bed load**. Sand also can move as part of the bed load by rolling or bouncing along the bottom while being carried downstream by the current between bounces.

Downcutting and Base Level

A stream can erode both downward into its bed and laterally against its banks. Downward erosion is called **downcutting** (●FIGURE 11.5). The **base level** of a stream is the deepest level to which it can erode. Most streams cannot erode below sea level, which is regarded as the ultimate base level.

In addition to ultimate base level, a stream may have a number of local, or temporary, base levels. For example, a stream stops flowing where it enters a lake. It then stops eroding its channel because it has reached a temporary base level (●FIGURE 11.6). A layer of rock that resists erosion may also establish a temporary local base level because it flattens the stream gradient, causing the stream to slow down and erosion to decrease. The top of a waterfall can be a temporary base level established by resistant rock. For example, Niagara Falls is formed by a resistant layer of dolomite overlying softer mudstone. The dolomite forms a temporary base level for the stream by resisting erosion, but the turbulent falling water undermines the soft mudstone at the base of the falls, causing the overhanging dolomite cap to collapse

● **FIGURE 11.5** The Blyde River in South Africa has eroded deeply into the bedrock of the region. Here at Bourke's Luck, the river has cut downward through sandstone to form these steep canyon walls and spectacular potholes.

● **FIGURE 11.6** A steep mountain stream flows from steep mountain drainages into the upper (right) end of a lake in the Canadian Rockies. The lake formed because the river had reached a temporary base level and could not erode any further downward. The lake drains out its lower (left) end.

● **FIGURE 11.7** Niagara Falls occurs at a temporary base level formed by a resistant layer of dolomite at the top of the falls. The softer mudstone underlying the dolomite is undercut at the base of the falls, causing them to migrate slowly upstream. The falls have eroded upstream 11 kilometers in the past 9,000 years and continue to do so today.

and the falls to migrate upstream. In this way, Niagara Falls has retreated 11 kilometers upstream since its formation about 9,000 years ago and is slowly eliminating the temporary base level formed by the dolomite (●**FIGURE 11.7**).

---

competence A measure of the largest particles that a stream can transport.

capacity The maximum quantity of sediment that a stream can transport at any one time.

dissolved load The total mass of ions dissolved in and carried by a stream at any one time; the ions are derived from chemical weathering.

suspended load The total mass of a stream's sediment load that is carried within the flow by turbulence and is free from contact with the streambed.

bed load The total mass of a stream's sediment load that is transported along the bottom or in intermittent contact with the bottom of the streambed.

downcutting Downward erosion by a stream into its bed, usually by cutting a V-shaped valley along a relatively straight path.

base level The deepest level to which a stream can erode its bed. The ultimate base level is usually sea level.

A stream like that in ●**FIGURE 11.8A** erodes rapidly in the steep places where its energy is high and deposits sediment in the low-gradient stretches where it flows more slowly. Over time, erosion and deposition smooth out the irregularities in the gradient. The resulting **graded stream** has a smooth, concave-upward profile (●**FIGURE 11.8B**). Once a stream becomes graded, there is no net erosion or deposition and the stream profile no longer changes. An idealized graded stream such as this one does not actually exist in nature, but many streams come close.

Sinuosity of a Stream Channel

A steep mountain stream usually downcuts rapidly compared with the rate of lateral erosion. As a result, it cuts a relatively straight channel with a steep-sided, V-shaped valley

(●**FIGURE 11.9**). The stream maintains its relatively straight path because it flows with enough energy to erode and transport any material that slumps into its channel.

In contrast, a low-gradient stream is less able to erode downward into its bed. Rather, much of the stream energy is directed against the banks, causing **lateral erosion**. Lateral erosion occurs where a low-gradient stream forms a series of bends, called **meanders** (●**FIGURES 11.10A** and **11.10B**). As a stream flows into a meander bend, the water pushes against the outside bank of the meander, much like your body leans against the inside of our car while going around a sharp curve at highway speed. As water is pushed against the outside bank,

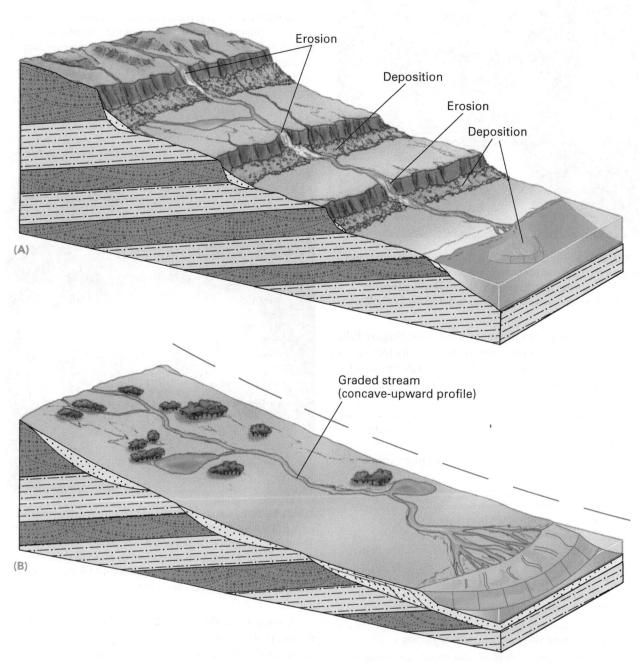

● **FIGURE 11.8** (A) An ungraded stream has many temporary base levels. (B) With time, the stream smooths out the irregularities, to develop an upward-concave graded profile that is illustrated here by the dashed red line.

it causes the bank to erode. The water also causes the deepest part of the channel to form adjacent to the outside bank, creating an asymmetric bottom profile from bank to bank.

The water piling up against the outside bank causes the flow around the inner bank to be both shallower and slower and promotes the deposition of sediment there, forming a **point bar** (Figure 11.10A). Through time, the point bar builds further and further out into the river as the outside bank of the meander is eroded back. In this manner, the meander slowly migrates laterally toward its outside bank. Occasionally, such

graded stream A stream with a smooth, concave-upward profile, in equilibrium with its sediment supply; it transports all the sediment supplied to it, with neither erosion nor deposition in the streambed.

lateral erosion The action of a low-gradient stream as it cuts into and erodes its outer bank while simultaneously depositing sediment onto its inner bank; results in slow lateral migration of the channel and, through time, formation of wide, flat alluvial valleys.

meanders A series of twisting curves or loops in the course of a stream.

point bar A deposit of sediment in the slower water on the inside of a meander.

● **FIGURE 11.9** A steep mountain stream eroded this V-shaped valley into soft mudstone in the Canadian Rockies.

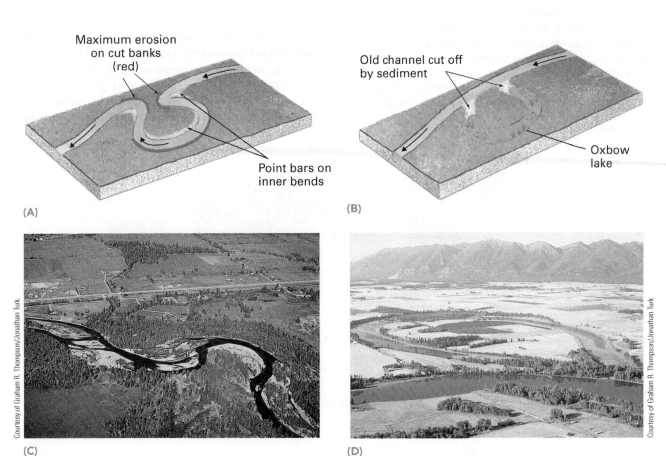

● **FIGURE 11.10** (A) A stream erodes the outsides of meanders (red) and deposits sand and gravel on the inside bends (yellow) to form point bars. (B) Meanders and point bars of the Bitterroot River in Montana. (C) Over time, a stream may erode through the neck of a meander to form an oxbow lake that is cut off from the river and rapidly colonized by aquatic plants. (D) This oxbow lake formed through the cutoff of a meander of the upper Flathead River in Montana.

lateral migration of a stream causes it to cut across the neck of a meander, forming a short-cut and isolating an old meander loop (**FIGURES 11.10C** and **11.10D**). Although it no longer has an active flow, the abandoned meander loop remains full of water and becomes an **oxbow lake**.

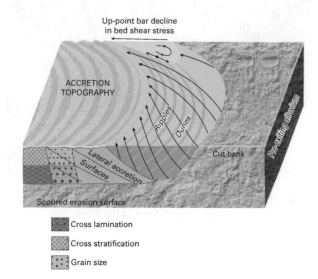

FIGURE 11.11 Dynamics of a river meander bend. As flow pushes against the outside bank of a meander bend, it causes erosion of the outer bank and formation of the deepest part of the channel close by. The slower flow along the inside bank causes sediment to be deposited there, forming a point bar. Through time, the channel migrates toward the outside bank as the point bar is extended.

Meandering streams typically form in regions of very low slope gradients. In addition, meandering streams typically carry a large suspended load that includes an abundance of clay. During floods, the mud is carried onto the adjacent floodplains where it settles out and is deposited (**FIGURE 11.11**). Subsequently, this clay-rich mud is exposed along the banks of the river, especially along the outside of meander bends. As anyone who has spent time making ceramics knows well, wet clay is sticky. This sticky nature allows river banks made of clay to become very steep or even overhanging without collapsing. The abundance of vegetation further helps stabilize the banks of meandering rivers. As a result, meandering streams typically flow in a single channel that changes slowly through the process of lateral migration described previously.

In contrast to meandering streams are **braided streams** that flow in many shallow, interconnecting channels (**FIGURE 11.12**). A braided stream forms where the supply of sediment exceeds the stream's capacity; that is, more sediment is available than the stream can carry. The excess sediment accumulates in the channel, forming a series of sediment bars. During low water, the stream flows simultaneously in multiple channels that pass around the bars. During floods, the bars may be completely submerged, and the higher discharge may cause the bars to migrate downstream a short distance, while changing shape in the process. As a result, braided streams often look different from year to year.

Unlike mud-dominated meandering streams, braided streams are dominated by their coarser-grained bed load. As a result, braided streams usually are found in regions with higher slope gradients where coarse-grained sediment

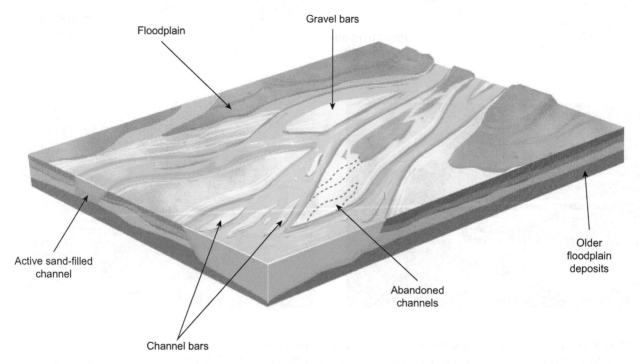

FIGURE 11.12 Braided streams form when too much sediment is available for the stream to carry. As a result, the stream piles the sediment up into bars and flows around them in multiple shallow, broad channels. Braided streams are dominated by bed load.

is generated through erosion. Braided streams also are common in both desert and glacial environments because both produce abundant coarse-grained sediment. A desert yields large amounts of sandy sediment because it has little or no vegetation to prevent erosion. Glaciers grind bedrock into sediment, which is carried by streams flowing from the melting ice. Unlike clay-rich banks of meandering streams, the banks of braided streams usually are made of coarser-grained sand and gravel particles that don't stick together, unlike clay-sized particles, but rather collapse at slopes steeper than the sediment's angle of repose. As a result, the relatively coarse-grained sediment forming the banks of braided rivers quickly is eroded from the bank into the channel, contributing to the high bedload characteristic of braided streams (●FIGURE 11.13).

● **FIGURE 11.13** Joe Creek in Canada's Yukon Territory is heavily braided because glaciers provide more sediment than the stream can carry.

---

oxbow lake A crescent-shaped lake created where a meander loop is cut off from a stream and the ends of the meander become plugged with sediment.

braided streams A stream that flows in many shallow, interconnecting channels that are usually separated by emergent sediment bars; formed because the stream's capacity has been exceeded by its sediment supply.

drainage basin The region that is drained by a single stream.

drainage divide A topographic high separating drainage basins.

Drainage Basins and Drainage Divides

Only a dozen or so major rivers flow into the sea along the coastlines of the United States (●FIGURE 11.14). Each is fed by a number of tributaries, which are in turn fed by smaller tributaries. The region drained by a single river is called a **drainage basin**. At the edge of each drainage basin is a **drainage divide**, an imaginary line on the ground

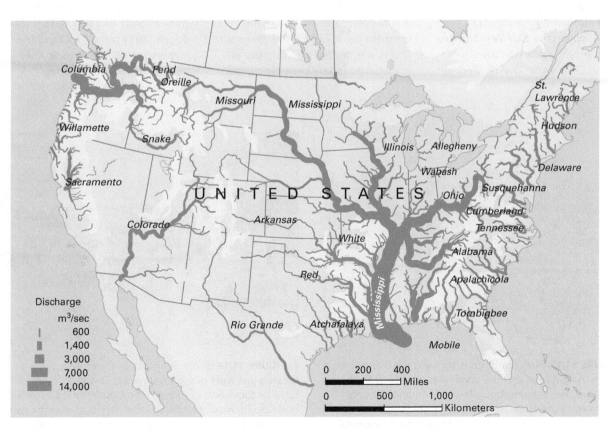

● **FIGURE 11.14** Most of the surface water in the United States flows to the sea from approximately a dozen major rivers. In this figure, the width of each river is proportional to its discharge.

The Richest Hill on Earth and Location of the Idaho Montana State Line

Although drainage divides are topographically obvious features in mountainous regions where they commonly occur along sharp ridge lines, in non-mountainous regions drainage divides can be topographically subtle features and difficult to locate exactly without the use of a map. Because they are not always obvious topographic features, drainage divides were historically not typically used to define the boundaries of U.S. states as they came into the union during the eighteenth and nineteen centuries; rather rivers filled this role.

A notable exception to this general rule is the boundary separating Montana from Idaho, which is located along the drainage divide separating the two sides of the Bitterroot Mountains. This drainage divide was chosen by the U.S. Congress and signed into law by President Lincoln in 1864, during the Civil War. Notably, the Bitterroot Divide was chosen over the Continental Divide which is located several hundred miles to the East. Had the Continental Divide been chosen, Idaho and Montana would have been roughly similar in size and Idaho would not have the narrow northern panhandle it does today.

Naturally, representatives from Idaho sought to make the Continental Divide the boundary between the two territories. However, before they could convince Congress to pass a bill defining the Continental Divide the official territorial boundary, representatives from Montana delivered thousands of dollars of gold bullion from Bannack, Montana—the capital of the soon-to-be formed Territory of Montana—to Washington, D.C. (●**FIGURE 11.15**). Whether this gold was delivered as a bribe to Congress or as a demonstration of the potential wealth the new Montana Territory could bring to the union is debated. However, before the Idahoans could object, the Bitterroot Divide was drafted by Congress as the legal boundary between the two territories, and it was signed into law by President Lincoln.

It is interesting to note that, had the Continental Divide been chosen for the state boundary, Idaho would have acquired the rich metal ore deposits of Butte, Montana. By 1864, gold placer deposits had been discovered at Butte, and since that time the ore deposits at Butte have been mined continuously, except for a three-year mining suspension in the mid-1980s. Although it wasn't until much later that Butte became known as the "Richest Hill on Earth," it is likely that as early as 1864, Montanans had a pretty good idea that Butte contained a big ore deposit and that they had decided to keep that deposit squarely within Montana's Territorial boundaries. As a result, following the establishment of the Bitterroot Divide as the official boundary, Montana went on to become the Treasure State and adopted the state slogan "oro y plata" (gold and silver) (●**FIGURE 11.16**). Idaho, meanwhile, became famous for potatoes.

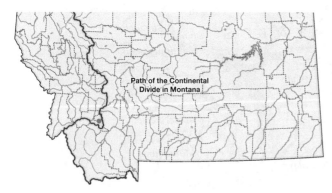

● **FIGURE 11.15** Map of Montana showing the major stream systems in light blue and the continental divide as the heavy brown line. The western boundary of Montana is shared with the state of Idaho. Notice that the southernmost portion of the Montana-Idaho line follows the continental divide, but the northern portion does not. If it had, the city of Butte with its rich precious metals deposits (red dot) would have been located in Idaho and not Montana.

Universal History Archive/Universal Images Group/Getty Images

● **FIGURE 11.16** The mining city of Butte, Montana, is located just west of the Continental Divide and would have become part of Idaho if the Continental Divide has been designated the official boundary between the two territories by the U.S. Congress in 1864. Instead, the territorial boundary was defined as the Bitterroot Divide, thus assuring that the rich metal ore deposits of Butte remained in Montana territory.

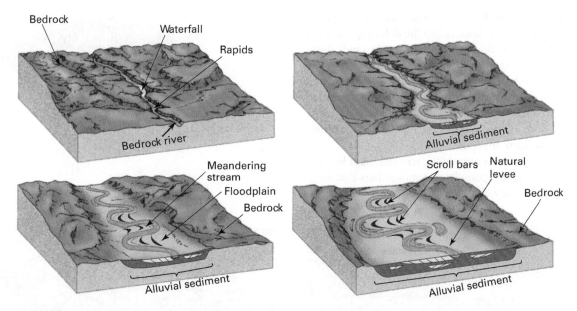

FIGURE 11.17 If tectonic activity does not uplift the land, over time streams erode the mountains away and widen the valleys into broad floodplains.

separating adjacent drainage basins. The most famous drainage divide in North America is the Continental Divide. Water runoff east of the Continental Divide drains into stream tributaries that ultimately lead to rivers flowing into the Atlantic Ocean, whereas runoff west of the Continental Divide eventually reaches the Pacific Ocean. Other important North American drainage divides separate the Columbia and Colorado River drainage basins west of the Continental Divide and the Rio Grande from the Mississippi River drainage basins east of the Continental Divide. Collectively these four drainage basins cover three-fourths of the lower 48 United States.

Stream Erosion and Mountains: How Landscapes Evolve

According to a model popular in the first half of the 20th century, streams erode Earth's surface and create landforms in an orderly sequence (**FIGURE 11.17**). At first, they cut steep, V-shaped valleys into mountains. Over time, the streams erode the mountains away and widen the valleys into broad floodplains. Eventually, the entire landscape becomes a large, featureless plain. However, if this were the only mechanism affecting Earth's surface during its 4.6-billion-year history, all landforms would have eroded to a flat plain. Why do mountains, valleys, and high plateaus still exist?

The model tells only half the story. Streams do continuously erode the landscape, wearing down mountains and

widening floodplains. But at the same time, tectonic activity may uplift the land and interrupt this simple, idealized sequence. In this way, Earth's hydrosphere and atmosphere work together with tectonic processes in the geosphere to create landforms. Geomorphology, from the Greek words for "earth forms," is a branch of geology that deals with the evolution of landforms through time.

STREAM DEPOSITION

When streams slow down, they deposit their bed load first. If water slows sufficiently, the suspended load may also settle to the bottom. Sediment deposited by moving water is called **alluvium**.

When alluvium is deposited, new landforms are created. For example, if a steep mountain stream flows onto a flat plain, its gradient and velocity decrease abruptly. As a result, the stream deposits most of its sediment in an accumulation called an **alluvial fan**, named because the shape of the deposit resembles a handheld fan (**FIGURE 11.18**). Alluvial fans are common in many arid and semiarid mountainous regions, where they commonly occur where a fault has caused a mountain range to be uplifted relative to the adjacent valley. In areas characterized by active faulting, it is common for the upper portion of an alluvial fan to be cut-off, or beheaded, from the lower portion of the fan as the fault continues to move through time. As the fault accumulates more offset through time, the lower portion of the fan slowly is buried as the valley fills with sediment, whereas the upper portion—the head—of the fan is uplifted as part of the mountain front where it is usually eroded (**FIGURE 11.19**).

A stream also slows abruptly where it enters the still water of a lake or an ocean. The sediment settles out to form a

---

alluvium Sediment deposited by moving water.

alluvial fan A fan-shaped accumulation of sediment created where a steep mountain stream rapidly slows down as it reaches a relatively flat plain.

delta, a nearly flat-topped landform with a steep frontal face. The upper surface of the delta, called the **delta top**, extends beyond the shoreline to the more steeply sloping, sandy **delta front**. Further offshore is the more muddy **prodelta**.

Both deltas and alluvial fans are commonly fan shaped and resemble the Greek letter *delta* (Δ). The fan shape is produced by **distributary channels**. Unlike tributary channels that merge in a downstream direction, distributary channels split in a downstream direction.

Distributary channels on a delta carry sediment beyond the shoreline, where it accumulates in mouth bars that form where the distributary channels end. (See ●**FIGURE 4.20**.)

●**FIGURE 11.18** This alluvial fan in Death Valley formed where a steep mountain stream deposited most of its sediment as it entered the flat valley. A road runs across the lower part of the fan.

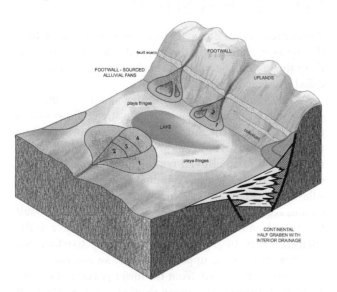

●**FIGURE 11.19** Many alluvial fans form in regions undergoing tectonic extension where normal faults separate mountain uplifts from subsiding basins. Alluvial fans that straddle the fault eventually are beheaded as the head of the fan is separated from the rest of the fan by fault movement. Eventually, the main body of the fan is buried in the subsiding basin, whereas the fan head is uplifted along the range front and eroded.

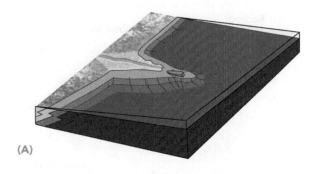

(A)

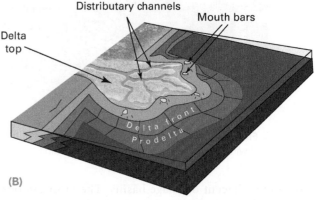

(B)

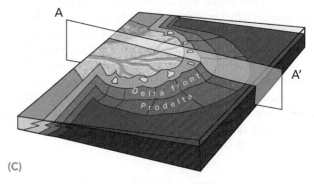

(C)

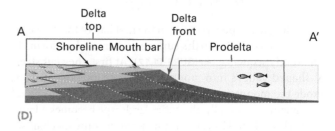

(D)

●**FIGURE 11.20** Deltas form and grow with time where a stream flows into a standing body of water and deposits its sediment. Much of the sediment in a delta is deposited underwater on the outer portion of the delta top, the more steeply sloping delta front, and the prodelta. (A) Delta formation is initiated as a stream begins to deliver sediment to a standing water body. A mouth bar is deposited on the delta top at the terminus of the stream channel. (B) Distributary channels (distributaries) develop as sediment is spread across the delta. (C) A mature delta with numerous distributary channels. (D) Cross section through a delta. The plane of cross section is shown in C.

As a mouth bar grows in size, flow from the distributary must diverge to get around it. At the same time, sediment is deposited on the bed of the distributary channel, causing it to become less efficient at transporting water and sediment. Ultimately these inefficiencies become so great that the distributary channel is abandoned and the water and sediment flows down a new channel established on a different part of the delta. This process of channel abandonment and reestablishment elsewhere is called **avulsion** and usually takes place when the river is in flood.

As sediment is added to alluvial fans and deltas, they build outward. Most alluvial fans are relatively small, covering areas ranging from less than a square kilometer to a few square kilometers (Figure 11.18). In contrast, a large delta may cover thousands of square kilometers (Figure 11.20). Although sediment is deposited on the delta top by flooding rivers, much sediment also is deposited beyond the mouths of the distributary channels, on the delta front and prodelta.

Even though deltas cover only a small fraction of Earth's total land surface, they are environments that involve each of Earth's four spheres. The delta is composed of solid sediment, but it is a watery zone where branched distributaries meet open water and much groundwater is contained between the individual sediment grains in the delta. Because floods regularly deliver nutrient-rich sediment to deltas, they typically are fertile, and delta ecosystems are rich in natural plant and animal life. The fertile soils and easy access to transportation systems make deltas desirable places for human habitation, but the elevation of delta land along the coast is barely above sea level and the river elevation, so it is certain to be flooded during atmospheric disturbances such as heavy rains and hurricanes.

---

delta A fan-shaped accumulation of sediment formed where a stream enters a lake or ocean; includes a nearly flat delta top that is partly onshore and partly offshore, a more steeply dipping delta front located offshore, and a muddy prodelta located further offshore at the base of the delta front.

delta top The upper surface of a delta, including the parts above and below water.

delta front The more steeply sloping, usually submerged, outer edge of a delta beyond the delta top.

prodelta The fine-grained, outermost edge of a delta, located offshore beyond the delta front.

distributary channels Channels that split from the main stream feeding a delta or alluvial fan and spread out across its surface, depositing sediment in the process.

avulsion The process by which a stream channel is abandoned and the water and sediment diverted down a new channel.

flood A relatively high stream discharge that overtops the stream banks, covering land that is not usually underwater.

floodplain That portion of a river valley adjacent to the channel; it is built upward by sediment deposited during floods and is covered by water during a flood.

FLOODS

When rainfall is heavy or snowmelt is rapid, more water flows down a stream than the channel can hold, creating a **flood**. During a flood, the stream overflows onto low-lying adjacent land called the **floodplain**. Massive floods occur somewhere in the world every year. The following three recent cases are good examples.

During the late spring and summer of 2008, heavy rain soaked the upper Midwest. Parts of eastern Iowa experienced 15 to 20 inches of rain in one month. As a result of the intense rainfall and abnormally heavy snowpack, the Mississippi River and nine Iowa rivers crested at or above record levels. In St. Louis, the Mississippi River crested 7 feet above normal. The Cedar River crested at over 32 feet, flooding 9.2 square miles of Cedar Rapids, Iowa, exceeding the highest recorded flood levels from 1929 and causing an estimated $1.5 billion damage in the city alone. The floods killed at least 24 people and injured hundreds more. Damages to the midwestern states exceed $6 billion.

In April of 2011, two major storm systems combined with the melting of the winter snowpack to swell the lower Mississippi River to record levels. The flood was big enough that the Army Corps of Engineers, responsible for managing runoff from the river, decided that it was necessary to open several "floodways," areas of historic floodplain that can be reconnected to the river during floods to lower the water level in the river and protect communities downstream. Although designated in the 1930s to serve as relief valves for future floods, the floodways are not empty lands, but rather contain numerous farms. On May 3, the Corps blasted a 3-kilometer hole in one of the Mississippi River levees in an effort to save the town of Cairo, Illinois. The breached levee caused the river level near Cairo to begin falling before the crest of the flood arrived, ultimately saving the town but flooding over 500 square kilometers of rural farmland in Missouri.

Eleven days later, the Corps began opening floodgates on the Morganza Spillway in Louisiana to prevent major flooding in Baton Rouge and New Orleans downstream. The opening of the floodgates diverted water into the upper reaches of the Atchafalaya River. Because floodgates had not been used since 1937, there was some risk that too much water would be diverted from the lower Mississippi River, causing it to abandon its current channel and reroute (avulse) down the Atchafalaya River. Fortunately, the river did not do this, but the 2011 Mississippi River floods caused an estimated $2 billion in damage and resulted in the evacuation of thousands of homes. Despite these significant damages, the management of the 2011 floods likely prevented a much worse scenario.

A much larger flood-related disaster in terms of human lives occurred in June of 2013 in northern India. Over a several day period, unusually heavy warm rain from the annual monsoon as well as several heavy cloudbursts fell in parts of the southern Himalaya Mountains. The rains melted large volumes of glacial ice in the mountains, sending huge quantities of water

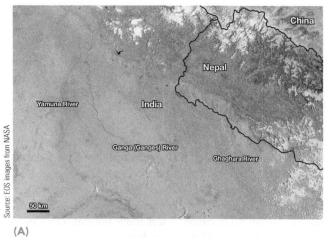

(A)

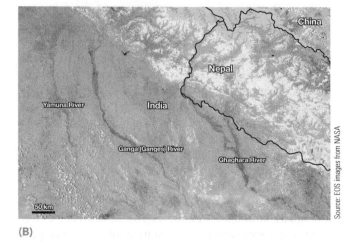

(B)

Source: EOS images from NASA

(C)

Total Rainfall (mm)

trace   ≤50   100   200   ≥300

● **FIGURE 11.21** The images in A and B were taken from the Earth Observation Satellite nine days apart. The first image shows a portion of northern India and neighboring Nepal on May 30, 2013, before the catastrophic floods that occurred over the following weeks. The second image shows the flood waters swelling all of the river systems leaving the southern Himalayan mountains as the runoff proceeded downstream. The map in C shows the total amount of rainfall that fell over six days across the region, resulting in the floods.

downstream toward the adjacent lowlands (●FIGURE 11.21). In addition, the rain saturated the ground and caused numerous landslides. The disaster that resulted was compounded by the proliferation of roads, hotels, shops, and multistory housing in landslide- and flood-prone regions and the construction of dams that blocked drainages and changed the distribution of water draining out of the mountain **tributary** system. The recently constructed dams and newly formed landslides blocked the flow of water downstream, causing villages to be inundated with water and mud. By June 28, 2013, over 1000 humans were reported dead and 70,000 were either stranded or missing.

Flood Control

Because deltas provide natural access to transportation waterways, contain level and fertile soil for agriculture and generally contain abundant wild game, deltas have long been desirable places for humans to live. However, flooding is a natural part of delta evolution, and all deltas undergo flood when sufficient discharge is supplied from the river that feeds it. As a result, humans have developed several strategies for controlling floods so that, despite their inevitable occurrence, the damage floods cause can be minimized. Flood control, while providing relief from flood waters for some, may exacerbate flood related damage for others.

ARTIFICIAL LEVEES AND CHANNELS An **artificial levee** is a wall built along the banks of a stream to prevent rising water from spilling out of the stream channel onto the floodplain. In the past 70 years, the U.S. Army Corps of Engineers has spent billions of dollars building 11,000 kilometers of levees along the banks of the Mississippi and its tributaries. Of course, levees can fail, resulting in significant flooding. In 2008, at least 20 levees along the Mississippi River failed, resulting in widespread flooding and damage across parts of Iowa and Indiana (●FIGURE 11.22).

Unfortunately, artificial levees create conditions that may increase both flood intensity and property damage. One factor is entirely human—the protection promised by levees encourages people to build in the floodplain. In the absence of a levee, people might decide to build on high ground that is safe from floods. But when levees are built, people are more likely to construct homes and businesses in harm's way.

Levees also can cause higher floods along nearby reaches of a river. By restricting the channel, artificial levees form a partial dam that raises the water level and increases the risk of flooding upstream (●FIGURE 11.23). In addition, because the stream cannot overflow levees during small floods, sediment that normally would be deposited on the floodplain is deposited within the channel, raising the level of the streambed

(●**FIGURES 11.24A** and **11.24B**). After several small floods, the entire stream may rise above its floodplain, contained only by the levees (●**FIGURE 11.24C**). This configuration creates the potential for a truly disastrous flood, because if the levee is breached during a large flood, the entire stream flows out of its channel and onto the floodplain. As a result of artificial levees and channel sedimentation, portions of the Yellow River in China now lie 10 meters above the adjacent floodplain. Thus, levees may solve flooding problems in the short term, but over a longer time frame they can cause even larger and more destructive floods.

Flood Control, the Mississippi River Delta, and Hurricane Katrina

In August 2005, Hurricane Katrina drove storm waters over the levees protecting New Orleans, flooding the city, chasing 1.3 million people from their homes, and causing approximately $200 billion in damages. This problem was exacerbated by decades of flood-control practices along the Mississippi.

In a natural system, a delta builds upward and extends seaward by sediment deposition, although the front of the delta is eroded by ocean waves and currents. In the past 50 years, deposition rates in many coastal deltas have decreased

● **FIGURE 11.22** Flooding in Iowa. When the levees along the Mississippi failed to hold the river, the results were catastrophic.

---

artificial levee A wall built along the banks of a stream to prevent rising floodwater from spilling out of the channel onto the floodplain.

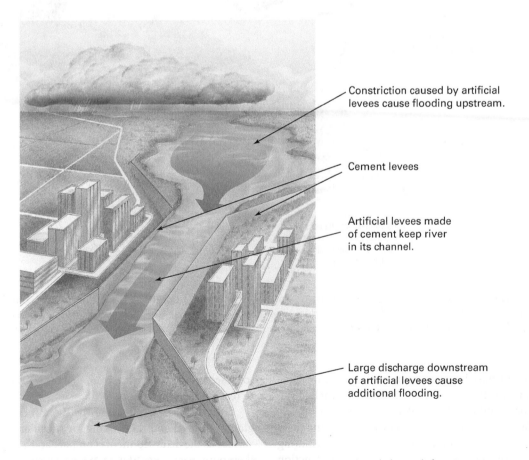

Constriction caused by artificial levees cause flooding upstream.

Cement levees

Artificial levees made of cement keep river in its channel.

Large discharge downstream of artificial levees cause additional flooding.

● **FIGURE 11.23** Artificial levees force a flooding river into a restricted channel, forming a partial dam that raises the flood level upstream from the restriction.

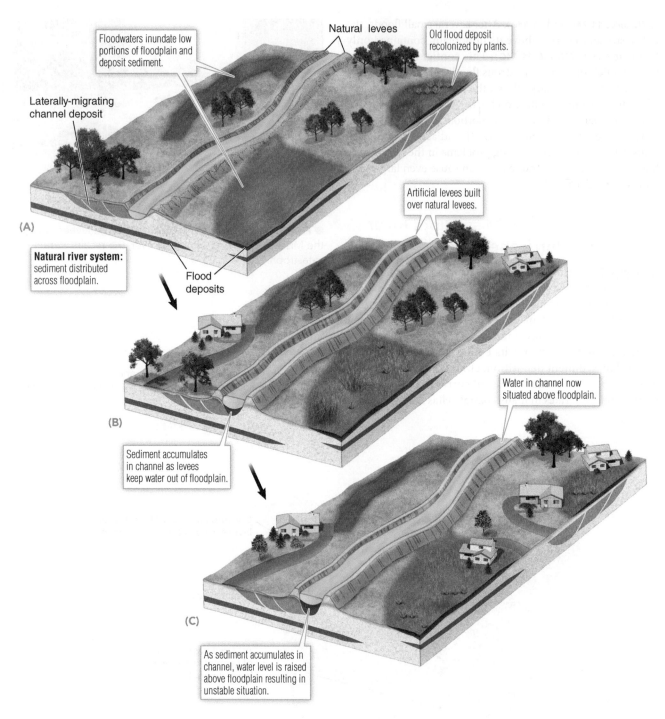

Floodwaters inundate low portions of floodplain and deposit sediment.

Natural levees

Old flood deposit recolonized by plants.

Laterally-migrating channel deposit

(A)

Natural river system: sediment distributed across floodplain.

Flood deposits

Artificial levees built over natural levees.

(B)

Sediment accumulates in channel as levees keep water out of floodplain.

Water in channel now situated above floodplain.

(C)

As sediment accumulates in channel, water level is raised above floodplain resulting in unstable situation.

● **FIGURE 11.24** (A) In the natural state, a flooding stream carries sediment from the stream channel onto the floodplain. (B) Artificial levees hold water in the river channel during what would otherwise be small floods, causing sediment to accumulate there. (C) Eventually, the channel rises above the floodplain, creating the potential for a disastrous flood.

significantly, in part because dams built upstream have trapped sediment that otherwise would have reached the delta. The construction of levees along distributary channels on the delta also has caused sediment to bypass the delta rather than be deposited on it during floods. Contributing to coastal erosion is the urban and agricultural destruction of natural plant communities that normally hold the soil

together and global sea level rise that causes the wave energy to reach further inland.

Other problems affecting the delta include the rising global sea level and the sinking, or subsidence, of the land surface as the delta sediments compact and expel water from their pore spaces (●**FIGURE 11.25**). As a result of all these processes, the Mississippi delta has shrunk. Between 1930 and

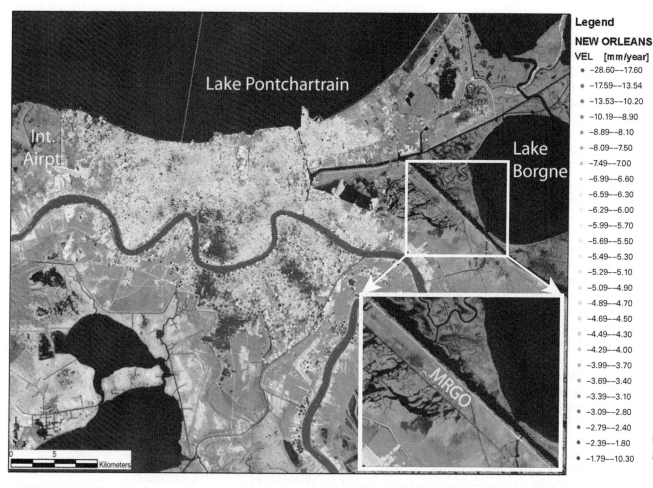

Legend

NEW ORLEANS

VEL   [mm/year]

- −28.60−−17.60
- −17.59−−13.54
- −13.53−−10.20
- −10.19−−8.90
- −8.89−−8.10
- −8.09−−7.50
- −7.49−−7.00
- −6.99−−6.60
- −6.59−−6.30
- −6.29−−6.00
- −5.99−−5.70
- −5.69−−5.50
- −5.49−−5.30
- −5.29−−5.10
- −5.09−−4.90
- −4.89−−4.70
- −4.69−−4.50
- −4.49−−4.30
- −4.29−−4.00
- −3.99−−3.70
- −3.69−−3.40
- −3.39−−3.10
- −3.09−−2.80
- −2.79−−2.40
- −2.39−−1.80
- −1.79−−10.30

Dixon T.H. et al., New Orleans Subsidence: Rates and Spacial Variation Measured by Permanent Scatterer Interferometry, Nature, 441, 587–586, 2006. Used with Permission.

● **FIGURE 11.25** The city of New Orleans and the Mississippi River appear in the center of this map showing the subsidence rates—the rates at which land in the area is sinking—in millimeters per year. MRGO on the map is the Mississippi River–Gulf Outlet channel connecting the Port of New Orleans to the Gulf of Mexico.

2012, nearly 5,000 square kilometers of the delta sank below sea level. One recent study estimated that by the year 2100, an additional 10,000 to 13,500 square kilometers will be submerged[2]—an area roughly the size of Connecticut.

The flooding of New Orleans in 2005 is an example of both a complex systems interaction and a threshold event. While Hurricane Katrina triggered the flood, human interference and development within a complex natural river/delta system set the stage. Over the previous decades, the global sea level rose while the delta subsided and slowly eroded. Natural buffers were removed, with no immediate ill effects. Then a catastrophic storm pushed waters inland. The levees failed, causing a huge human disaster.

Although it is easy to point to the hurricane as the cause of the 2005 disaster, the other factors that combined to make this such a devastating event were known well in advance. In 1995, the senior authors of this textbook wrote: "If current rates of erosion continue, the sea will … flood New Orleans … causing severe economic losses."[3] Tragically, this dire prediction was accurate.

FLOODPLAIN MANAGEMENT As we have shown, in many cases attempts at controlling floods either do not work or they shift the problem to a different time or place. An alternative approach to flood control is to abandon some flood-control projects and let the river spill out onto its floodplain. Of course, the question is what land should be allowed to flood. Every farmer and homeowner living on the floodplain wants to maintain the levees that protect it. Currently, federal and state governments are establishing wildlife reserves in some floodplains. Because no development is allowed in these reserves, they will be permitted to flood during the next high water. However, a complete river management plan involves complex political and economic considerations.

2. Michael D. Blum and Harry H. Roberts, "Drowning of the Mississippi Delta Due to Insufficient Sediment Supply and Global Sea-Level Rise," *Nature Geoscience* 2 (June 28, 2009): 488–491.

3. Jonathan Turk and Graham Thompson, *Environmental Geoscience* (Philadelphia: Saunders College Publishing, 1995), 428.

LAKES

Lakes and lakeshores are attractive places to live and play. Clean, sparkling water, abundant wildlife, beautiful scenery, aquatic recreation, and fresh breezes all come to mind when we think of going to the lake. Despite their great value, however, lakes are fragile and ephemeral.

The Life Cycle of a Lake

A **lake** is a large inland body of standing water that occupies a depression in the land surface. Streams flowing into the lake carry sediment, which can fill the depression in a relatively short time, geologically speaking. Soon the lake becomes a swamp, and with time the swamp fills with more sediment and vegetation and becomes a meadow or forest with a stream flowing through it.

If most lakes fill quickly with sediment, why are these bodies of water so abundant today? Most lakes exist in places that were covered by glaciers during the latest ice age. About 18,000 years ago, great continental ice sheets extended well south of the Canadian border, and mountain glaciers scoured alpine valleys as far south as New Mexico and Arizona. Similar ice sheets and alpine glaciers existed in the higher latitudes of the Southern Hemisphere.

The glaciers created lakes in several ways. Flowing ice eroded numerous depressions in the land surface, which then filled with water. The Great Lakes of North America and the Finger Lakes of upper New York State are examples of large lakes occupying glacially scoured depressions.

The glaciers also deposited huge amounts of sediment as they melted and retreated. Some of these great piles of glacial debris formed dams across stream valleys. When the glaciers melted, streams flowed down the valleys but were blocked by the sediment dams. Many modern lakes occupy glacially dammed valleys (●FIGURE 11.26). In addition, large blocks of ice were left behind as the glaciers receded. When these ice blocks melted, they left behind depressions called kettles, which in many cases filled with water to become **kettle lake**.

Most of these glacial lakes formed within the past 10,000 to 20,000 years, and sediment is rapidly filling them. Many of the smallest such lakes have already become swamps. In the next few hundred to few thousand years, many of the remaining lakes will fill with mud. The largest, such as the Great Lakes, may continue to exist for tens of thousands of years. But the life spans of lakes such as these are limited, and it will take another glacial episode to replace them.

Lakes also form by nonglacial means. A volcanic eruption can create a crater that fills with water to form a lake, such as Crater Lake in Oregon. Oxbow lakes form in abandoned river channels. Other lakes, such as Lake Okeechobee in the Florida Everglades, form in flatlands with shallow groundwater. These types of lakes, too, fill with sediment and, as a result, exist for a limited time.

● **FIGURE 11.26** An oligotrophic lake in Canada's Banff National Park. This lake was dammed by sediment deposited by a glacier.

A few lakes, however, form in ways that maintain their existence beyond the span of a "normal" lake. For example, Russia's Lake Baikal is a large, deep lake lying in a depression created by active normal faults. Although rivers pour sediment into the lake, movements of the faults repeatedly deepen the basin. As a result, the lake has existed for more than a million years—so long that indigenous species of seals, fish, and other animals have evolved in its ecosystem.

Nutrient Balance in Lakes

In a deep lake, sunlight is available near the surface, but nutrients are abundant mainly on the bottom. Plankton (small, free-floating organisms) grow poorly on the surface due to low nutrient levels there, and bottom-rooted plants cannot grow due to lack of sunlight there. Thus, deep lakes contain low concentrations of nutrients and insufficient bottom sunlight to sustain aquatic food webs that generate much biomass. The low level of biological productivity often causes these lakes to have a deep-blue color that we associate with a clean, healthy lake. Such a lake is called *oligotrophic*, meaning "poorly nourished." **Oligotrophic lakes** have low productivities, meaning that they sustain relatively few living organisms, although a lake of this type is attractive for recreation and can support limited numbers of large game fish.

As a lake fills with sediment, it becomes shallower and sunlight reaches more and more of the lake bottom. The sunlight allows bottom-rooted plants to grow. As the plants die and rot, their litter adds nutrients to the lake water. Plankton increase in numbers, as do the numbers of fish and other organisms. The lake may become so productive that its surface is covered with a green scum of plankton or a dense mat of rooted plants. The litter from this biomass contributes to sediment filling the lake, and eventually the lake becomes a swamp.

A lake of this kind, with a high nutrient supply and high level of productivity in terms of biomass, is called a

Maksim Oleynik/Shutterstock.com

● **FIGURE 11.27** A eutrophic lake in Florida.

eutrophic lake (●**FIGURE 11.27**); eutrophic means "well nourished." Eutrophication occurs naturally as part of the life cycle of a lake. However, the addition of nutrients in the form of sewage, fertilizer runoff, and other kinds of pollution has greatly accelerated the eutrophication of many lakes.

Temperature Layering and Turnover in Lakes

If you have ever dived into a deep lake on a summer day, you have probably discovered that the top meter or so of lake water can be much warmer than deeper water. This occurs because sunshine warms the upper layer of water, making it less dense than the cooler, deeper water. The warm, less dense water floats on the cooler, denser water. The boundary between the warm and cool layers is called the **thermocline**.

In temperate climates, colder autumn weather cools the surface water to a temperature below that of deeper water, so that the surface water becomes more dense than the deeper water. This density difference and the effects of wind cause

---

lake A large, inland body of standing water that occupies a depression in the land surface.

kettle lake A lake that forms in a depression created by a receding glacier, filled with the water from the melting glacier.

oligotrophic lake A deep lake characterized by nearly pure water but with low concentrations of nutrients, thus sustaining relatively few living organisms.

eutrophic lake A relatively shallow lake characterized by abundant nutrients, thus sustaining multiple living organisms.

thermocline The boundary between the upper warm layers and deeper cool layers of water in a lake.

turnover A process, occurring in fall and spring in temperate climates, in which a lake's surface water changes temperature in response to seasonal weather changes and convection mixes the water to equalize temperature throughout the lake.

wetlands Regions that are water soaked or flooded for all or part of the year; includes *swamps, bogs, marshes, sloughs, mudflats,* and *floodplains*.

---

surface water to sink and mix with the deep water, equalizing the water temperature throughout the lake. This process is called fall **turnover**. In the winter, ice floats on the surface and temperature layering develops again. In spring, as ice melts on the lake, the cold surface water again is denser than deep lake waters, and spring turnover occurs. As summer comes, the lake again develops thermal layering. This seasonal process is illustrated in ●**FIGURE 11.28**.

Turnover in temperate lakes illustrates an important Earth systems interaction among the atmosphere, the hydrosphere, and the biosphere. During summer and winter when the lake water is layered, bottom-dwelling organisms may use up most or all of the oxygen in deep water. At the same time, surface organisms may deplete surface waters of dissolved nutrients. However, surface water is rich in oxygen because it is in contact with the atmosphere, and deep water may be rich in nutrients because it is in contact with bottom sediment. Turnover enriches deep water with oxygen and, at the same time, supplies nutrients to the surface water. The latter effect often becomes evident in the form of an *algal bloom*—a sudden and obvious increase in the amount of floating green algae on a lake's surface—in spring and fall.

WETLANDS

Wetlands are known across North America as swamps, bogs, marshes, sloughs, mudflats, and floodplains. They are regions that are water soaked or flooded for all or part of the year. Some wetlands are wet only during exceptionally wet years and may be dry for several years at a time.

Wetland ecosystems vary so greatly that the concept of a wetland defies a simple definition. All wetlands share certain properties, however: the ground is wet for at least part of the time; the soils reflect anaerobic (lacking oxygen) conditions; and the vegetation consists of plants such as cattails, bulrushes, mangroves, and other species adapted to periodic flooding or water saturation. North American wetlands include all stream floodplains, frozen Arctic tundra, warm Louisiana swamps, coastal Florida mangrove swamps, boggy mountain meadows of the Rockies, and the immense swamps of interior Alaska and Canada (●**FIGURE 11.29**).

Wetlands are among the most biologically productive environments on Earth, are important for degrading pollutants, and serve to mitigate the effects of flooding. Two-thirds of the Atlantic fish and shellfish consumed by humans rely on coastal wetlands for at least part of their life cycles. One-third of the endangered species of both plants and animals in the United States also depend on wetlands for survival. More than 400 of the 800 species of protected migratory birds and one-third of all resident bird species feed, breed, and nest in wetlands. Aquatic organisms consume many

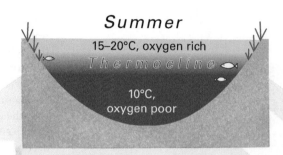

Summer

15–20°C, oxygen rich

Thermocline

10°C,
oxygen poor

Spring turnover   Algal
bloom

3.5–4.5°C, entire lake
oxygenated

Fall turnover   Algal
bloom

3–10°C, entire lake oxygenated

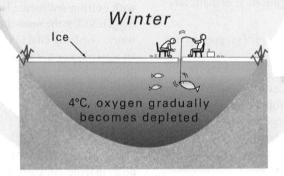

Winter

Ice

4°C, oxygen gradually
becomes depleted

● **FIGURE 11.28** Lakes in temperate climates develop temperature layering in both summer and winter. As a result, bottom waters become depleted in oxygen. In fall and spring, water temperature becomes constant throughout the lake and turnover brings new supplies of oxygen to the deep waters.

(A)

(B)

● **FIGURE 11.29** North American wetlands extend from (A) the Everglades to (B) the immense Alaskan swamps.

pollutants and degrade them to harmless by-products. Because these organisms abound in wetlands, the ecosystems serve as natural sewage treatment systems. Wetlands also mitigate flooding by absorbing excess water that might otherwise overrun towns and farms.

When European settlers first arrived in North America, the continent had 87 million hectares of wetlands (exclusive of those in Alaska). Until relatively recently, most Americans viewed wetlands as mosquito-infested, malarial swamps occupying land that can be farmed or otherwise developed if drained or filled but otherwise is worthless. In the mid-1800s, the federal government passed legislation known as the Swamp Land Acts, which established an official policy for filling and draining wetlands in order to convert them to agricultural uses wherever possible. Over a quarter million square kilometers were officially identified by the acts as swampland available for "reclamation."

Wetlands have been lost or degraded by both humans and natural causes. Many wetlands have been drained, or their sources of water cut off, through the construction of dams or dikes. Others have been filled, logged, or mined, while the introduction of nonnative invasive species or toxic levels of pollution or nutrients continues to cause widespread wetland degradation. Natural wetland loss can occur by subsidence, sea level rise, drought, and erosion by large storms.

According to the U.S. Environmental Protection Agency, over the past several centuries about 900,000 square kilometers of original U.S. wetlands have been degraded or destroyed—an area larger than Texas and Oklahoma combined. California and several upper-midwestern states have lost more than 80 percent of their historic wetlands. Beginning in the 1960s, the importance of wetlands in water purification, flood control, and wildlife habitat began to be recognized, and since the late 1980s, federal policy has been to incur "no net loss" of wetlands. However, it has proven more difficult to reverse practice than policy, and the net loss of wetlands continues. For example, the U.S. Department of the Interior estimated a net loss of 25,000 hectares (250 square kilometers) of wetlands between 2004 and 2009. Today, wetlands make up between 6 and 9 percent of the lower 48 states and as much as 60 percent, or about 80 million hectares, of Alaska (●FIGURE 11.29B). Preserving these wetlands along with those outside the United States will present a serious future challenge as Earth's human population continues to rise and pressure increases to develop, farm, log, or ranch these wetlands.

GROUNDWATER

In most places, if you drill a hole in the ground between a few meters and 100 or so meters deep, the bottom of the hole fills with water, usually within a few minutes to a few

days. The water appears even if no rain falls and no streams flow nearby. The water that seeps into the hole is groundwater, which saturates Earth's crust from a few meters to a few kilometers below the surface.

Groundwater is exploited by digging wells and pumping the water to the surface. It provides drinking water for more than half of the population of North America and is a major source of water for irrigation and industry. However, deep wells and high-speed pumps now extract groundwater more rapidly than natural processes can replace it in many parts of the central and western United States. As a result, groundwater resources that have accumulated over thousands or tens of thousands of years are being permanently lost. In addition, industrial, agricultural, and domestic contaminants seep into groundwater in many parts of the world. Such pollution is often difficult to detect and expensive to clean up.

Porosity and Permeability

Groundwater fills small cracks and voids in soil and bedrock. The volume percentage of these open spaces, called *pores*, is called **porosity**. In unlithified sand or gravel, the network of open pore spaces between the individual sediment clasts results in porosity values that usually are close to 40 percent. In contrast, unlithified mud can have porosity values of up to 90 percent. The generally higher porosity of unlithified mud results from several physical and chemical characteristics of very fine-grained sediment. First, the platy shape of many clay crystals and crystal aggregates creates plentiful open space between sediment particles. Second, water is attracted to silt- and clay-sized particles through capillary action (see Chapter 10). Third, the polar water molecule is attracted to electrically charged surfaces of clay-sized particles. Finally, water is incorporated directly into the crystal structure of some clay minerals.

Most rocks have lower porosities than loose sediment. For example, sandstone and conglomerate commonly have 5 to 30 percent porosity, and in some cases more. Much of this porosity is left behind from the original pore network around the sand and gravel clasts, minus that lost by compaction of the sediment during burial and that lost by the introduction of mineral cements. Pore networks in quartz-dominated sandstone and gravel can stay open during deep burial because the quartz grains provide a rigid framework that resists collapse. In some cases, quartz sandstone and conglomerate with porosity values of 15 percent or higher serve as conventional reservoirs for petroleum or natural gas at depths in excess of two miles.

In contrast, at such depths mud-dominated lithologies—mudstone and claystone in particular—are characterized by porosity values of only a few percent, remarkably lower than the very high porosity of the original unlithified mud. The big loss of porosity during the lithification of mud to mudrock results mainly from compaction during burial, which collapses pore spaces as the platy clay minerals and mineral aggregates become sub-aligned like a collapsed house of cards on a table top. The growth of mineral cements

---

porosity The proportional volume of pores or open space within a material; indicates the maximum possible volume of fluid that could be held within the material.

between the clay and silt-sized grains fills much of the remaining pore space during the lithification of mud, while the transformation of organic matter to kerogen or kerogen to petroleum can further clog available pore space in mudrocks during deep burial. Igneous and metamorphic rocks have very low porosities (a few percent or less) unless they are fractured.

Porosity indicates the amount of water that rock or soil can hold. In contrast, **permeability** refers to how interconnected the pore spaces are and relates to the ability of rock or soil to transmit water or any other fluid including petroleum and natural gas. Water can flow rapidly through material with high permeability, but flows slowly through a low permeability material.

Most materials with high porosity also have high permeability. Sand and sandstone have numerous, relatively large, well-connected pores that allow the water to flow through the material. However, if the pores are very small, as in clay or mudstone, electrical attractions between water and soil or mineral particles retard the passage of water. Although unlithified clay typically has a high number of pores that are so small and tortuous that the electrical attractions with clay particles slow the passage of water, it usually has a very low permeability and transmits water slowly.

The Water Table and Aquifers

When rain falls, much of it soaks into the ground. Water does not descend into the crust indefinitely, however. Below a depth of a few kilometers, the pressure from overlying rock closes the pores, making bedrock both nonporous and impermeable. Water accumulates above this permeability barrier, filling pores in the rock and soil. This completely wet layer of soil and bedrock above the deep, impermeable rock is called the **zone of saturation**. The **water table** is the top of the zone of saturation (●**FIGURE 11.30**). The unsaturated zone, or **zone of aeration**, lies above the water table. In this layer, the rock or soil may be moist but is not saturated, and air occupies some or all of the pore space.

If you dig into the unsaturated zone, the hole does not fill with water. However, if you dig below the water table into the zone of saturation, you have dug a **well**. The water level in a well is at the level of the water table (Figure 11.30). During a wet season, rain seeps into the ground to **recharge** the groundwater, and the water table rises. During a dry season, the water table falls. Thus, the water level in most wells fluctuates with the seasons.

An **aquifer** is any body of rock or soil that can yield economically significant quantities of water. An aquifer must be

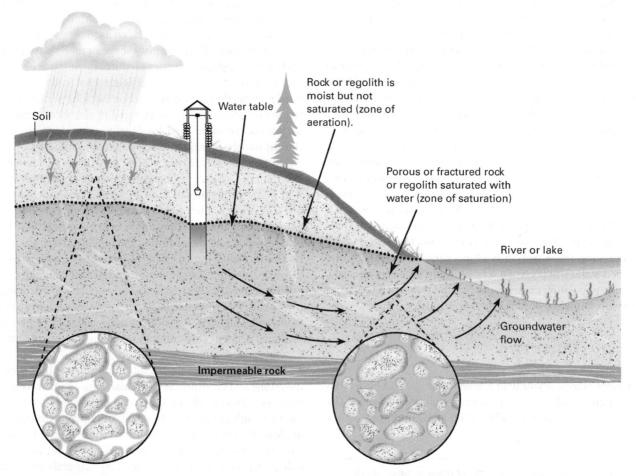

● **FIGURE 11.30** The water table is the top of the zone of saturation near Earth's surface. It intersects the land surface at lakes, springs, and most streams and is the level of standing water in nonartesian wells.

both porous and permeable so that water flows into a well to replenish (recharge) water that is pumped out. Sand and gravel, sandstone, limestone, and highly fractured bedrock of any kind make excellent aquifers. In contrast, mudstone, clay, and unfractured igneous and metamorphic rocks have low porosity and permeability and are called **aquitards**.

Groundwater Movement

Nearly all groundwater seeps slowly through bedrock and soil. Groundwater often flows at about 4 centimeters per day (about 15 meters per year), although flow rates may be much faster or slower depending on permeability. Most aquifers are like sponges through which water seeps, rather than underground pools or streams. However, groundwater can flow very rapidly through large fractures in bedrock, and in a few regions underground rivers flow through caverns.

In general, the water table mimics its surrounding topography. Refer again to Figure 11.30. Groundwater flows from zones where the water table is highest toward areas where it is lowest, and so some groundwater flows roughly parallel the sloping surface of the water table toward the valley. But groundwater also flows from zones of high pressure toward zones of low pressure. Because water pressure is greatest beneath the highest part of the water table, the pressure difference forces much of the groundwater to flow downward beneath the hill,

then laterally toward the valley, and finally upward beneath the lowest part of the valley, where a stream flows. This is how groundwater feeds stream flow, and why streams flow even when no rain has fallen for weeks or months.

SPRINGS AND ARTESIAN WELLS A **spring** occurs where the water table intersects the land surface and water flows or seeps onto the surface. In some places, a layer of impermeable rock or clay lies above the main water table. Porous, saturated rock or sediment above the impermeable layer can form a **perched aquifer** (●FIGURE 11.31A). Hillside springs often flow from perched aquifers. Springs also occur where fractured bedrock or cavern systems intersect the land surface.

FIGURE 11.32 shows a tilted layer of permeable sandstone sandwiched between two layers of impermeable mudstone. An inclined aquifer such as the sandstone layer, bounded top and bottom by impermeable rock, is a **confined aquifer**. Water in the lower part of the aquifer is under pressure from the weight of water above. Therefore, if a well is drilled through the mudstone and into the sandstone, water rises in the well without being pumped. A well of this kind is called an **artesian well**. If pressure is sufficient, the water spurts out onto the land surface.

permeability The ability of a solid material such as a rock to transmit water or another fluid through its pore network; depends on the size, shape, and interconnectedness of the pores within the material.

zone of saturation A subsurface zone below the water table in which all porosity within soil and bedrock is filled with water.

water table The top surface of the zone of saturation; the water table separates this zone from the zone of aeration above.

zone of aeration A subsurface zone located between the ground surface and the water table and in which the pores are mostly filled with air; also called the *unsaturated zone*.

well A hole dug or drilled into Earth, generally for the production of water, petroleum, natural gas, brine, or sulfur, or for exploration.

recharge To replenish an aquifer by the addition of water.

aquifer A body of rock that can yield economically significant quantities of groundwater; should be both porous and permeable.

aquitards A body of sediment or rock that has low porosity and permeability and that inhibits the flow of groundwater.

spring A seep or flow of groundwater onto the surface; commonly occurs where the water table intersects the land surface.

perched aquifer A local aquifer formed where a layer of impermeable rock or sediment exists above the regional water table and creates a locally saturated zone.

confined aquifer An inclined aquifer sandwiched between layers of impermeable rock; typically, the water in the lower part of the aquifer is under pressure from the weight of water above.

artesian well A well drilled into a confined aquifer, in which the water rises without pumping and in some cases flows to the surface.

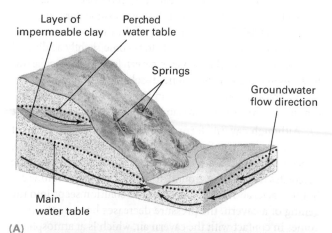

(A)

(B)

● **FIGURE 11.31** (A) Springs can form where a perched water table intersects a hillside. (B) Water flows from a spring on a hillside in British Columbia.

Courtesy of Graham R. Thompson/Jonathan Turk

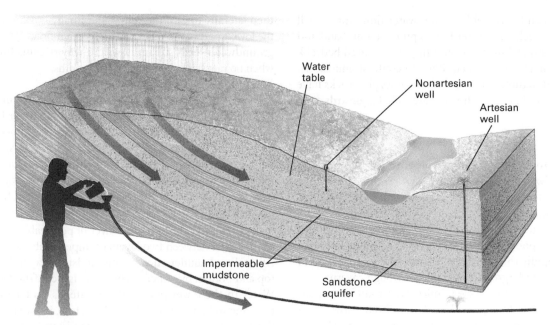

● **FIGURE 11.32** A confined aquifer capable of supporting an artesian well occurs where a tilted layer of permeable rock, such as sandstone, lies sandwiched between layers of impermeable rock, such as mudstone. Water rises in an artesian well without being pumped. A hose with a hole (inset) illustrates the pressure driving the upward flow of water in artesian wells.

CAVERNS Like surface streams, groundwater is capable of eroding rock and leaving behind sedimentary deposits (●**FIGURE 11.33**). Recall from Chapter 10 that rainwater reacts with atmospheric carbon dioxide to become slightly acidic and capable of dissolving limestone. A **cavern** forms when acidic water seeps into cracks in limestone, dissolving the rock and enlarging the cracks. Mammoth Cave in Kentucky and Carlsbad Caverns in New Mexico are two famous caverns formed in this way.

Although caverns form when limestone dissolves, most caverns also contain features formed by deposition of calcite. When a solution of water, dissolved calcite, and carbon dioxide percolates through the ground, it is under pressure from water in the cracks above it. If a drop of this solution seeps into the ceiling of a cavern, the pressure decreases because the drop comes in contact with the cavern air, which is at atmospheric pressure. The high humidity of the cave prevents the water from evaporating rapidly, but the lowered pressure allows some of the carbon dioxide to escape as a gas. When the carbon dioxide escapes, the drop becomes less acidic. This decrease in acidity causes some of the dissolved calcite to precipitate as the water drips from the ceiling. Over time, a beautiful and intricate **stalactite** (from the Greek for "drip") grows to hang icicle-like from the ceiling of the cave (●**FIGURE 11.34**).

Only a portion of the dissolved calcite precipitates as the drop seeps from the ceiling. When the drop falls to the floor, it spatters and releases more carbon dioxide. The acidity of the drop decreases further, and another minute amount of calcite precipitates. Thus, a cone-shaped **stalagmite** (from the Greek for "drop") builds from the floor upward to complement the stalactite. Because stalagmites are formed by splashing water, they tend to be broader than stalactites. As the two features continue to grow, they may eventually meet and fuse together to form a **column**.

SINKHOLES If the roof of a cavern collapses, a **sinkhole** forms on Earth's surface (●**FIGURES 11.33** and **11.35**). A sinkhole can also form as limestone dissolves from the surface downward. In tropical regions, deep sinkholes formed from collapse of limestone caves can expose groundwater, resulting in a deep lake called a **cenote**. Each year, sinkholes cause significant economic damage as buildings, roads, and other structures undergo collapse in densely populated regions underlain by limestone. A well-documented sinkhole formed in May 1981 in Winter Park, Florida. During the initial collapse, a three-bedroom house, half a swimming pool, and six Porsches in a dealer's lot all fell into the underground cavern. Within a few days, the sinkhole was 200 meters wide and 50 meters deep, and it had devoured additional buildings and roads.

Although sinkholes form naturally, human activities can accelerate the process. The Winter Park sinkhole formed when the water table dropped, removing support for the ceiling of the cavern. The water table fell as a result of a severe drought augmented by excessive removal of groundwater by humans.

---

cavern An underground cavity or series of chambers created when groundwater dissolves large volumes of rock, usually limestone; also called a *cave*.

cenote A deep, usually clear lake formed when groundwater is exposed at the surface through collapse of caves in limestone bedrock.

stalactite An icicle-like dripstone of precipitated calcite that hangs from the ceiling of a cavern.

stalagmite A cone-shaped deposit of calcite formed over time as drops of water fall to the same spot on the floor of a cavern, release carbon dioxide gas upon impact, and precipitate the mineral as a result of the increase in pH.

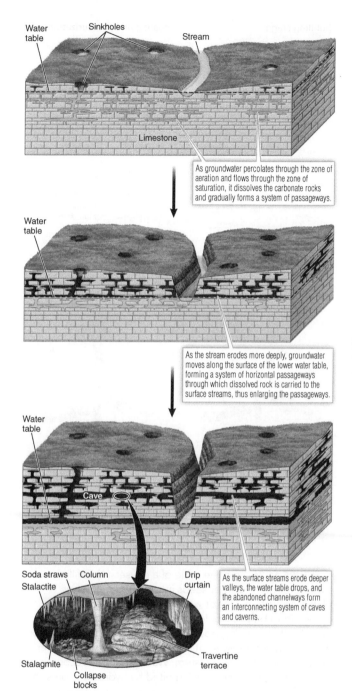

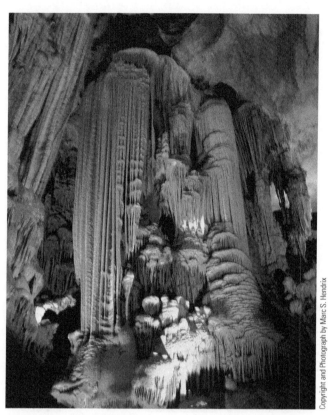

As groundwater percolates through the zone of aeration and flows through the zone of saturation, it dissolves the carbonate rocks and gradually forms a system of passageways.

As the stream erodes more deeply, groundwater moves along the surface of the lower water table, forming a system of horizontal passageways through which dissolved rock is carried to the surface streams, thus enlarging the passageways.

As the surface streams erode deeper valleys, the water table drops, and the abandoned channelways form an interconnecting system of caves and caverns.

● **FIGURE 11.33** Sinkholes and caverns are characteristic of karst topography. Streams commonly disappear into sinkholes and flow through the caverns, to emerge elsewhere.

column A cave deposit formed when a stalactite and a stalagmite meet and grow together.

sinkhole A depression on Earth's surface caused by the collapse of a cavern roof or by the dissolution of surface rocks, usually limestone.

karst topography A landscape that forms over limestone or other soluble rock and is characterized by abundant sinkholes, disappearing streams, and caverns.

hot springs A spring formed where hot groundwater flows to the surface.

● **FIGURE 11.34** Spectacular columns have formed in the Grotte des Demoiselles in southern France as downward-growing stalactites connected with upward-growing stalagmites.

KARST TOPOGRAPHY An irregular landscape called **karst topography** forms in regions underlain by limestone and other readily soluble rocks such as halite (rock salt). Caverns and sinkholes are common features; surface streams often pour into the sinkholes and disappear into the caverns (Figure 11.33). In the area around Mammoth Cave in Kentucky, streams are given names such as Sinking Creek because of their disappearing acts.

HOT SPRINGS, GEYSERS, AND GEOTHERMAL ENERGY

At numerous locations throughout the world, hot water naturally flows to the surface to produce **hot springs**. Most hot springs form from the heating of groundwater in one of two ways:

1. Earth's temperature increases by about 30°C per kilometer of depth in the upper portion of the crust. Therefore, if groundwater descends through cracks to depths of 2 to 3 kilometers, it is heated by 60°C to 90°C. The hot water or steam then can rise because it is less dense than the colder water above it. However, it is unusual for fissures to descend so deep into Earth, and this type of hot spring is uncommon.

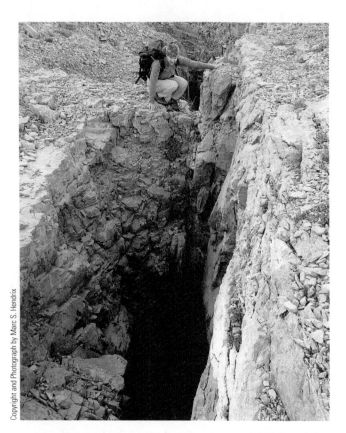

● **FIGURE 11.35** This sinkhole formed on Cambrian limestone in the Scapegoat Wilderness of Montana.

2. In regions of recent volcanism, magma or hot igneous rock near the surface can heat groundwater at relatively shallow depths. Hot springs heated in this way are common throughout western North and South America because these regions have been magmatically active in the recent past and remain so today. For example, magma only a few kilometers below the surface heats the hot springs and geysers of Yellowstone National Park.

Most hot springs form from water rising gently to the surface from cracks in bedrock. However, **geysers** violently erupt hot water and steam. Geysers form from a network of cracks and small caverns in hot underground rock. Before a geyser erupts, hot groundwater seeps into the network, where it continues to be heated by the surrounding rock (●**FIGURE 11.36A**). Usually, cooler water fills the network of cracks and caverns from above, while hotter water fills them from below. As the water fills the caverns, the temperature of the deepest water can rise above the boiling point. Yet this **superheated water** does not boil, because pressure from the overlying water keeps it from doing so. Gradually, steam bubbles form in water within shallower parts of the crack system where the pressure is lower, and these bubbles start to rise just as in a heated teakettle. If part of the channel is constricted, the bubbles can accumulate there, forming a temporary barrier that allows the steam pressure below to increase in a manner similar to a pressure cooker (●**FIGURE 11.36B**). The rising pressure eventually forces some of the bubbles upward past the constriction, pushing water out of the

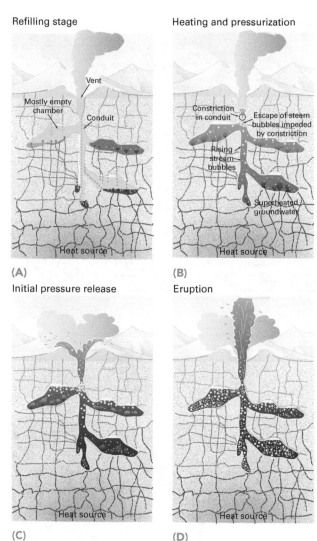

● **FIGURE 11.36** Eruptive cycle of a geyser. (A) Refilling stage: Hot groundwater seeps into the empty geyser plumbing system. Hotter water seeps into the plumbing system from below. Cooler water seeping in from above includes that draining downward from the previous eruption. (B) Heating and pressurization stage: Hot rock heats water in the geyser plumbing system. Water at depth becomes superheated, but boiling is inhibited by hydrostatic pressure. Steam bubbles forming at shallower levels rise upward through the conduit and increase in size because of reduced pressure at shallower depths. Some rising steam bubbles become trapped at a constriction in the conduit, further increasing pressure in the water below. As pressure rises, continued heating causes nearly all of the water in the plumbing system to become superheated. (C) Initial pressure release: With the conduit full, accumulation of upward-rising steam at constriction forces small bursts of water out of the vent. This escape of water decreases the hydrostatic pressure in shallow parts of the plumbing system, allowing superheated water there to flash to steam and forcing more water past the constriction and out of the vent. (D) Eruption: As pressure drops with these initial releases of water, superheated water throughout the entire plumbing system flashes to steam, causing the main eruption. Most of the water in the plumbing system is forced upward past the constriction as an upward-spouting jet of hot water.

geyser vent and lowering the pressure on the superheated water below the constriction (●FIGURE 11.36C). Because it is already above the boiling point, the superheated water suddenly flashes to stream, forcing hot water and steam skyward past the constriction and out of the geyser's vent during the main eruption (●FIGURES 11.36D and 11.37). Eventually, an insufficient volume of superheated water remains in the crack system to sustain the eruption, and it stops.

The most famous geyser in North America is Old Faithful in Yellowstone National Park, which erupts on the average of once every 91 minutes. Old Faithful is not as regular as people like to believe; the intervals between eruptions can vary from about 45 to 125 minutes. These variations in the eruptive interval are caused by the year-to-year changes in the volume of water available to fill the crack system as well as seasonal changes in the temperature of water infiltrating the crack system from the surface. In addition, changes in the eruptive behavior of geysers such as Old Faithful can be caused by earthquakes. As seismic waves pass through rocks forming a geyser's plumbing system, its shape can change as rocks or sediment are dislodged.

Hot groundwater can be used to drive turbines and generate electricity, or it can be used directly to heat homes and other buildings. Energy extracted from Earth's heat is called **geothermal energy**. In 2013, 64 geothermal plants, mostly in California and Nevada, were operating in the United States. These produced a generating capacity of about 2,700 megawatts, enough to supply about 3 million people with electricity.

Unfortunately, compared with the potential of geothermal as a source of energy in the United States, the current generating capacity is miniscule. By the end of 2013, the most

● **FIGURE 11.37** An eruption of Spa Geyser in Upper Geyser Basin, Yellowstone National Park, Wyoming.

recent data available from the International Energy Agency, only about 0.4 percent of U.S. electricity was generated by geothermal sources. In an effort to make the development of additional geothermal areas more economic, some developers are constructing enhanced geothermal systems in which permeability in rock that is already hot is increased through drilling and hydraulic fracturing. The first commercial-scale enhanced geothermal system in the United States is the Desert East pilot project which began operating in 2013.

geyser A type of hot spring that periodically erupts with violent jets of hot water and steam; eruptions occur when groundwater recharging a geyser's subsurface plumbing system becomes superheated and forces a small volume of water out of the geyser's vent. The release of this water lowers pressure at deeper levels, causing water there also to flash to steam and initiating the main eruption.

superheated water Water that has a temperature above the boiling point but that does not boil because it is under pressure.

geothermal energy Energy extracted from Earth's heat.

Key Concepts Review

- Only about 0.64 percent of Earth's water is fresh. The rest is salty seawater and glacial ice. The constant circulation of water among the sea, land, the biosphere, and the atmosphere is called the hydrologic cycle, or the water cycle. Most of the water that evaporates from the seas and from land returns to the surface as rain or snow. The precipitation that falls on the continents returns to the sea via runoff and moving groundwater, or it returns to the atmosphere via evaporation and transpiration.
- A stream is any body of water flowing in a channel. The velocity of a stream is determined by its gradient (steepness), discharge (amount of water), and channel

characteristics (shape and roughness of the stream channel). Streams shape Earth's surface by eroding soil and bedrock. Streams transport sediment as dissolved load, suspended load, and bed load. Downcutting, lateral erosion, and mass wasting combine to form a stream valley. Base level is the lowest elevation to which a stream can erode its bed; it is usually sea level.
- When streams slow down they deposit their bed load first. If water slows sufficiently, the suspended load may also fall to the bottom, but the dissolved load remains in the water until the chemical environment changes. A stream feeding a delta or fan splits into many channels called distributaries.

- A flood occurs when a stream overtops its banks and flows over its flood plain. Artificial levees and channels can contain small floods but they may exacerbate large ones.
- Lakes are short-lived landforms because streams fill them with sediment. Recent glaciers created many modern lakes; as a result, we live in an unusual time of abundant lakes. The life history of a lake commonly involves a progression from an oligotrophic lake to a eutrophic lake to a swamp and, finally, to a flat meadow or forest.
- Wetlands are among the most biologically productive environments on Earth. In addition, they are natural water purification systems, and they mitigate flood effects by absorbing floodwaters. Despite these facts, government and private efforts have eliminated more than half of the wetlands that existed in the lower 48 states when European settlers first arrived.

- Much of the rain that falls on land seeps into soil and bedrock to become groundwater. Ground water saturates the upper few kilometers of soil and bedrock to a level called the water table, which separates the zone of saturation from the zone of aeration. Most groundwater moves slowly, about 4 centimeters per day. Springs occur where the water table intersects the land surface and water flows or seeps onto the surface. An inclined layer of permeable rock sandwiched between layers of impermeable rock can produce an artesian aquifer. Caverns form where groundwater dissolves limestone. A sinkhole forms when the roof of a limestone cavern collapses. Karst topography, with numerous caves, sinkholes, and subterranean streams, is characteristic of limestone regions.
- Hot springs and geysers develop when hot groundwater rises to the surface. Hot groundwater has been tapped to produce geothermal energy.

Important Terms

alluvial fan (p. 263)

alluvium (p. 263)

aquifer (p. 274)

aquitards (p. 275)

artesian well (p. 275)

artificial levee (p. 266)

avulsion (p. 265)

banks (p. 255)

base level (p. 256)

bed load (p. 256)

bed (p. 255)

braided streams (p. 260)

capacity (p. 256)

cavern (p. 276)

channel characteristics (p. 255)

column (p. 276)

competence (p. 256)

confined aquifer (p. 275)

delta front (p. 264)

delta top (p. 264)

delta (p. 264)

discharge (p. 254)

dissolved load (p. 256)

distributary channels (p. 264)

downcutting (p. 256)

drainage basin (p. 261)

drainage divide (p. 261)

eutrophic lake (p. 271)

flood (p. 265)

floodplain (p. 265)

geothermal energy (p. 279)

geysers (p. 278)

graded stream (p. 258)

gradient (p. 254)

hot springs (p. 277)

hydrologic cycle (p. 253)

intermittent streams (p. 254)

karst topography (p. 277)

kettle lake (p. 270)

lake (p. 270)

lateral erosion (p. 258)

meanders (p. 258)

oligotrophic lakes (p. 270)

oxbow lake (p. 260)

perched aquifer (p. 275)

perennial streams (p. 254)

permeability (p. 274)

point bar (p. 259)

porosity (p. 273)

prodelta (p. 264)

recharge (p. 274)

runoff (p. 253)

sinkhole (p. 276)

spring (p. 275)

stalactite (p. 276)

stalagmite (p. 276)

stream (p. 254)

superheated water (p. 278)

suspended load (p. 256)

thermocline (p. 271)

transpiration (p. 254)

tributaries (p. 254)

turnover (p. 271)

water table (p. 274)

well (p. 274)

wetlands (p. 271)

zone of aeration (p. 274)

zone of saturation (p. 274)

Review Questions

1. Describe the movement of water through the hydrologic cycle.

2. Describe the factors that determine the velocity of stream flow.

3. What is karst topography? How can it be recognized? How does it form?

Gary Saxe/Shutterstock.com

12

WATER RESOURCES

The Englebright Dam on California's Yuba River was built in 1941 by the U.S. Army Corps of Engineers. The dam was built primarily to trap sediment from historic and anticipated gold mining operations in the headwaters of the Yuba River. Although the dam generates hydroelectric power and impounds a reservoir that provides various recreational opportunities, it also poses an impassable barrier to ocean-run Chinook salmon and Steelhead trout that spawn in the Yuba River and its headwaters.

INTRODUCTION

More than two-thirds of Earth's surface is covered with water. But most of this water is salty, and most of the freshwater is frozen into the Antarctic and Greenland ice caps. Humans divert half of all the flowing water from rain and snow that falls on the continents, but even that is not enough for our burgeoning population and per capita consumption. Today, as we reach—and in some cases exceed—the limits of our water resources, the well indeed is running dry.

Currently about one-fifth of the world's population lives in a region of **water scarcity** in which the aggregate demand by all sectors exceeds that available. Under the existing climate change scenario, the United Nations has projected that by the year 2030, water scarcity is expected to affect half the world's population and displace between 24 million and 700 million people. Not only will water scarcity increase future levels of human poverty but it also may lead to armed conflict as the competition for the dwindling resource increases.

An important aspect of water scarcity is water quality. In addition to being physically scarce in some regions, much of the surface water and groundwater in industrial areas is heavily polluted, destroying natural habitats and making the resource less useful for humans.

WATER SUPPLY AND DEMAND

Rain and snow continuously replenish freshwater on land, so that the amount of available freshwater is about the same as it was two centuries ago. But the demand for water has risen dramatically, until it has approached or exceeded supply in

many parts of the world (●**FIGURE 12.1**). In the early 1800s, there were 1 billion people on Earth; now there are over 7 billion. Thus, as the human population has grown, the amount of available freshwater per person has diminished. In addition, our technological society uses water at rates that would have been inconceivable to Ben Franklin.

Water use falls into two categories. Any process that uses water and then returns it to Earth locally is called **withdrawal**. Most of the water used by industry and homes is returned to streams or groundwater reservoirs near the place from which it was taken. For example, river water pumped through an electric power-generating station to cool the exhaust is returned almost immediately to the river. Water used to flush a toilet in a city is pumped to a sewage treatment plant, purified, and discharged into a nearby stream.

In contrast, a process that uses water and then returns it to Earth far from its source is called **consumption**. Most irrigation water evaporates, disperses with the wind, and returns to Earth as precipitation hundreds or thousands of kilometers from its source (●**FIGURE 12.2**). Although industry accounts for about half of all water withdrawn in the United States, agriculture accounts for most of the water that is consumed and not returned to its place of origin. Globally, about two-thirds of the total water that is withdrawn is consumed.

Water use (withdrawal plus consumption) is subdivided into domestic, industrial, and agricultural categories. According to the U.S. Geological Survey, by far the largest users of water in the United States are power plants, which in 2015 accounted for about 41 percent of total water withdrawals, and irrigation for agriculture, which accounts for about 38 percent of total consumption (●**FIGURE 12.3**). The specific categories for which the water is used varies by state, with the arid western states consuming more water for irrigation and the more populous eastern states using more water for thermoelectric power (●**FIGURE 12.4**).

●**FIGURE 12.5** shows the total freshwater withdrawals by source (surface water vs. groundwater) from 1950 through 2015, plotted alongside U.S. population. Although the U.S. consumes an extravagant volume of water by global standards, total water withdrawals fell between 2010 and 2015. The drop in total withdrawals was primarily caused by decreases in withdrawals for thermoelectric power, which accounted for 89 percent of the decrease in total withdrawals, and public-supply withdrawals, which accounted for another 9 percent of the decline. Partially offsetting these reductions in water consumption were increases in withdrawals for irrigation and mining.

Domestic Water Use

The average American household uses about 1100 liters (300 gallons) per day—many times the amount needed to maintain a healthy life—although average consumption per household varies significantly by region. Households located in arid regions commonly use more than twice as much water as households located in humid regions, mainly because of differences in landscape irrigation—particularly lawns.

Evolution of Global Water Use
Withdrawal and Consumption by Sector

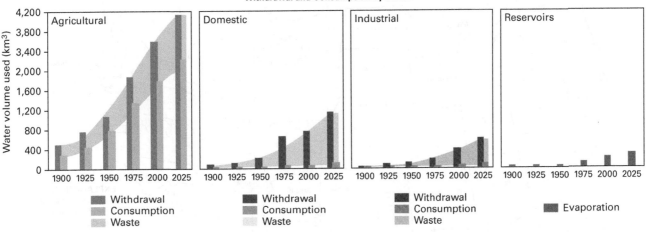

Note: Domestic water consumption in developed countries (500–800 liters per person per day) is about six times greater than in developing countries (60–150 liters per person per day).

● **FIGURE 12.1** Global water use, 1900 to 2025. Between 2000 and 2054, the world's population is expected to increase by 3 billion people, so water demand will continue to rise.

● **FIGURE 12.2** Most irrigation water evaporates or is carried away by wind. Spray irrigation of a spinach field.

Category	Billions gal/day	Percentage
Public supply	39	12
Self-supplied domestic	3.26	1
Irrigation	118	37
Livestock	2	1
Aquaculture	7.55	2
Self-supplied industrial	14.8	5
Mining	4	1
Thermoelectric	133	41
	321.61	100

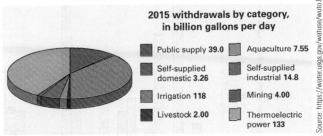

Source: https://water.usgs.gov/watuse/wuto.html

● **FIGURE 12.3** Water withdrawal by category in 2015 for the United States.

Source: Compiled from Worldwatch Institute data

These differences in average household water use among the different states are compounded by the fact that many of the states experiencing recent large population increases are also states in which the arid climate leads to higher average household consumption (●**FIGURE 12.6**). As the arid western states have become more populated, domestic water conservation practices have become more commonplace, leading to a decrease in overall average household consumption between 2005 and 2015. ●**FIGURE 12.7** shows typical volumes

water scarcity A situation in which the aggregate demand for water by all users, including the environment, exceeds the available supply.

withdrawal Any process that uses water and then returns it to Earth locally.

consumption Any process that uses water and then returns it to Earth far from its source.

of water used in a variety of domestic activities in the United States and tips for reducing this volume. Although greater water awareness has reduced the average U.S. household consumption since 2005, it is important to keep in mind that domestic use accounts for only about 12 percent of the water used in the United States.

In contrast to the comparatively extravagant per capita water usage in the United States, the World Health Organization defines "reasonable water access" in less-developed agricultural societies as 20 liters (about 5.28 gallons) of sanitary water per person per day, assuming that the source is less than 1 kilometer from the home. Yet 1.1 billion people, about one-sixth of the global population, lack even this small amount of clean water. Water access can fall short for three reasons:

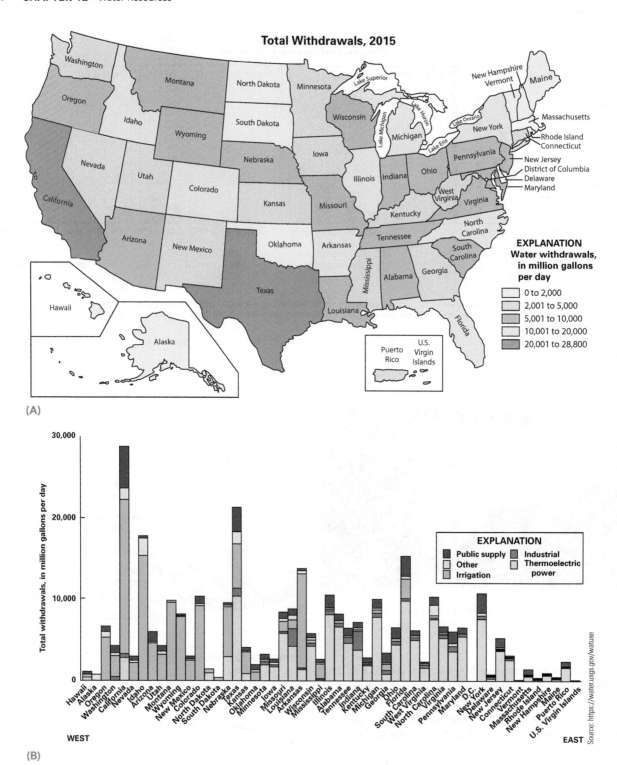

● **FIGURE 12.4** (A) Total water withdrawals for each state in 2015 in millions of gallons per day. (B) Total withdrawals by category for each state in 2015. In this chart, states are arranged from west on the left to east on the right. Note the much higher irrigation usage in the arid western states versus the higher usage for thermoelectric power in the more populous east.

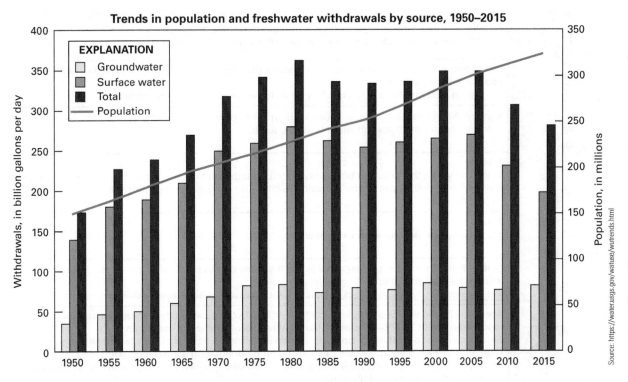

● FIGURE 12.5 Changes in total freshwater withdrawals by source (surface water vs. groundwater) from 1950 through 2015, plotted alongside U.S. population.

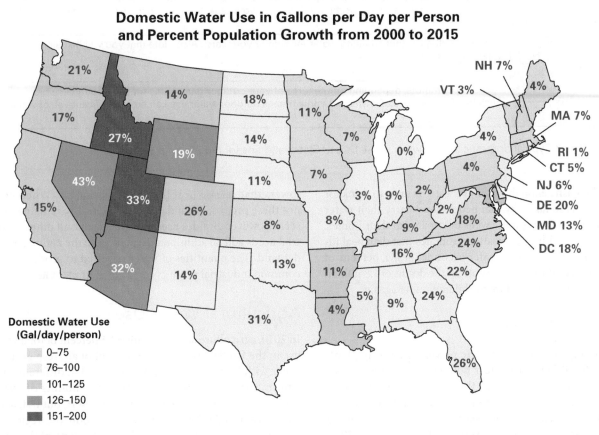

● FIGURE 12.6 Domestic water use per day per person in each of the lower 48 United States, along with population growth between 2000 and 2015 for each state.

Bath	A "full tub" varies, of course, but 36 gallons is good average amount. **Tip:** Taking a shower instead of a bath should save a good bit of water.
Shower	Old showers used to use up to 5 gallons of water per minute. Water-saving shower heads produce about 2 gallons per minute. **Tip:** Taking a shorter shower using a low-flow showerhead saves lots of water.
Teeth brushing	1 gallon. Newer bath faucets use about 1 gallon per minute, whereas older models use over 2 gallons. **Tip:** Simply turn the faucet off when brushing teeth.
Hands/ face washing	1 gallon **Tip:** Simply turn the faucet off before drying your hands and face. If you don't mind a brisk wash, don't run the faucet until it gets hot before using it. Installing a faucet-head aerator will also reduce the water flow rate.
Face/leg shaving	1 gallon **Tip:** Simply turn the faucet off when shaving.
Dishwasher	6–16 gallons. Newer, EnergyStar models use 6 gallons or less per wash cycle, whereas older dishwashers might use up to 16 gallons per cycle. **Tip:** EnergyStar dishwashers not only save a lot of water but also save electricity.
Dish washing by hand:	About 8–27 gallons. This all depends on how efficient you are at hand-washing dishes. Newer kitchen faucets use about 1.5–2 gallons per minutes, whereas older faucets use more. **Tip:** Efficient hand-washing techniques include installing an aerator in your faucet head and scraping food off, soaking dishes in a basin of soapy water before getting started, and not letting the water run while you wash every dish. And it's best to have two basins to work in—one with hot, soapy water and the other with warm water for a rinse.
Clothes washer	25 gallons/load for newer washers. Older models might use about 40 gallons per load. **Tip:** EnergyStar clothes washers not only save a lot of water but also save electricity.
Toilet flush	3 gallons. Most all new toilets use 1.6 gallons per flush, but many older toilets used about 4 gallons. **Tip:** Check for toilet leaks! Adjust the water level in your tank. But, best to install a new low-flow toilet.
Glasses of water you drank	8 oz. per small glass (not counting water for Fido or your cats). Also, note that you will be using water for cooking.
Outdoor watering	2 gallons per minute, depending on the force of your outdoor faucet. This may not sound like too much but the large size of lawns and yards means outdoor water use can be a significant use of water.

Source: https://water.usgs.gov/edu/qa-home-percapita.html

● **FIGURE 12.7** Typical domestic uses of water in the United States and tips for reducing daily usage.

either there isn't enough water, the water is unfit to drink, or the source is so far away that people must carry their water long distances from public wells or streams to their homes. The World Health Organization estimates that diarrheal illnesses caused by unsafe water account for 4.1 percent of global disease and are responsible for the deaths of 1.8 million people each year, most of them children under age five.

Industrial Water Use

The cooling systems in electric power-generating plants (run by fossil fuels or nuclear power) account for roughly 45 percent used each day in the United States. Although much of this water is returned to the stream from which it was taken, it is considerably warmer, which, as we shall see, affects aquatic ecosystems. As a result, in March of 2011, the U.S. Environmental Protection Agency issued proposed standards for closed-cycle cooling systems utilizing cooling

towers that release heat through evaporation. Whether or not these proposed standards become reality remains to be seen, however. Besides use by power plants, all other industrial needs add an additional 10 percent to the national water demand. The quantities of water required to produce some common industrial products are shown in ●**FIGURE 12.8**.

Agricultural Water Use

In 2010, agriculture accounted for 32.5 percent of total water use in the United States and 76 percent of all fresh groundwater withdrawals. ●**FIGURE 12.9** shows that different agricultural products use vastly different amounts of water. California's Central Valley was once a desert, but irrigation has transformed it into an immensely productive region for growing fruits and vegetables. The productivity of the Great Plains in the western United States would decline by one-third to one-half if irrigation were to cease.

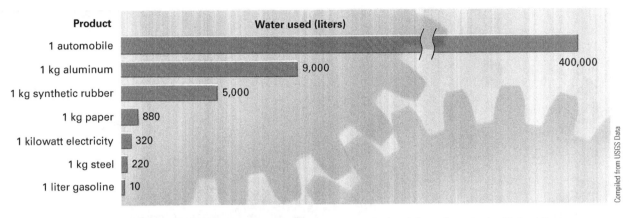

● **FIGURE 12.8** Large quantities of water are used to produce common industrial products in the United States.

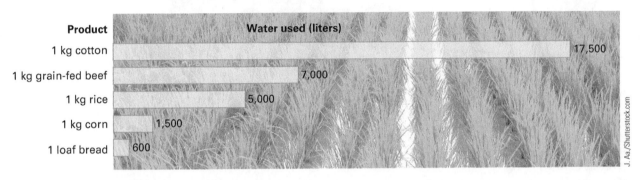

● **FIGURE 12.9** Different agricultural products require vastly different amounts of irrigation water in the United States.

Globally, many nations rely on irrigation for more than half of their food production. Estimates vary, but roughly 70 percent of all global water consumption is used to irrigate crops. In dry regions most if not all farmland must be irrigated, and as a result, both surface and groundwater resources are stressed, leading to a higher risk of insufficient water availability (●**FIGURE 12.10**).

DAMS AND DIVERSION

Much of Earth's rain falls in the wrong places or at the wrong times to be of much use to humans. For thousands of years, people realized that the problem could be ameliorated by building dams to store water that falls at the wrong time and by diverting water that falls in the wrong places. For example, the Tigris and Euphrates Rivers flow through deserts of the Middle East from distant mountains, but ancient Babylonians farmed the desert by irrigating the parched lowlands with diverted river water. In modern times, engineers build huge dams to store water. In 1950, there were 5,700 large dams in the world; as of 2018, there are about 57,000.* Almost half of the new dams were built in China. This unprecedented building boom has increased water availability in many regions and produced large amounts of hydroelectric energy, but as we will see in the discussion that follows, these benefits are offset against significant human and environmental costs.

In many regions, surface water is scarce, but huge reservoirs of groundwater are accessible beneath the surface. This groundwater can be pumped to the surface for human use.

The United States receives about 3 times more water from precipitation than it uses. However, some of the driest regions are those that use the greatest amounts of water (Figure 12.4). For example, some of the most productive agricultural regions are located in the deserts of California, Texas, and Arizona, where crops are entirely supported by irrigation. Desert cities such as Phoenix, Los Angeles, Albuquerque, and Tucson support large populations that use hundreds of times more water than is locally available and renewable.

Surface Water Diversion

A **diversion system** is a pipe or canal constructed to transport water. Many of these systems use gravity to move the water, while others use pumps to lift water uphill. Diversion systems are often augmented by dams that store water. Dams

---

diversion system A pipe, canal, or other infrastructure that transports water from its natural place and path in the hydrologic cycle to a new place and path to serve human needs.

---

* A large dam is defined as rising 15 meters or more from base to top and storing more than 3 million cubic meters of water.

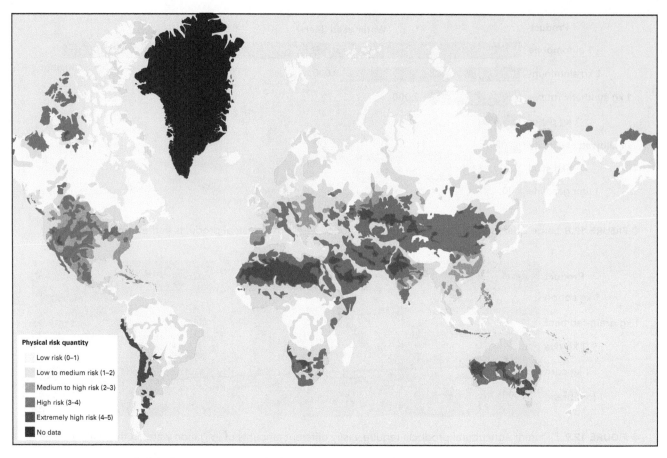

● **FIGURE 12.10** Map of global water stress, in which regions are assigned a physical risk of being impacted by short- or long-term water availability.

Physical risk quantity

Low risk (0–1)

Low to medium risk (1–2)

Medium to high risk (2–3)

High risk (3–4)

Extremely high risk (4–5)

No data

are especially useful in regions of seasonal rainfall. For example, in the arid and semiarid western United States, much of the annual precipitation occurs in winter and spring. Dams store this water for the summer irrigation season, when crops need it. In addition, the potential energy of the water in a reservoir can generate electricity, and because reservoirs can refill with water, hydroelectric power is renewable. In 2016, hydroelectric power from dams accounts for about 16 percent of total global electricity consumption. Thus, dams are beneficial, but they also can create numerous undesirable effects, described next.

LOSS OF WATER The reservoir formed by a dam provides more surface area for evaporation and more bottom area for seepage into bedrock than did the stream that preceded it. (For example, about 270,000 cubic meters of water per year evaporate from Lake Powell, above the Glen Canyon Dam on the Colorado River.) Therefore, less total water flows downstream. In addition, many canals built to carry water from a reservoir to farmland are simply ditches excavated in soil or bedrock, and they leak profusely.

SALINIZATION As described in Chapter 11, all streams contain a dissolved load—ions derived from continental weathering and dissolved in a stream. When farmers use

river water to irrigate cropland, the water evaporates, leaving the ions behind to form various salts in the soil. If desert or semidesert soils are irrigated for long periods of time, salt accumulates and lowers the soil fertility. In the United States, **salinization** lowers crop yields on 25 to 30 percent of irrigated farmland, more than 50,000 square kilometers, an area roughly the size of Vermont and New Hampshire combined (●**FIGURE 12.11**). The problem is particularly acute in the San Joaquin and Imperial Valleys of Southern California. Globally, salinization affects between 10 and 20 percent of irrigated land, some 250,000 square kilometers or about the size of Wyoming, and it is increasing at a rate of roughly 6 percent per year.

SILTING A stream deposits its sediment in a reservoir, where the current slows. Rates of sediment accumulation in reservoirs vary with the sediment load of the dammed river. Lake Mead, impounded behind Hoover Dam on the Colorado River in Arizona and Nevada, lost 6 percent of its capacity in its first 35 years as a result of sediment accumulation. At such a rate, a reservoir lasts for hundreds of years, providing a considerable societal resource in terms of power generation, water storage, and recreation. However, in a few other instances, engineers have made expensive miscalculations of sedimentation rates that have greatly shortened the lifespan of reservoirs. The Tarbela Dam in Pakistan was completed

● **FIGURE 12.11** White salts cover a vast expanse of Nevada desert. Salinization has reduced the productivity of millions of hectares of agricultural land globally.

in 1978 after 9 years of construction, and by 2005, it was already more than 25 percent filled with silt. The reservoir behind the Sanmenxia Dam on the Yellow River in China filled 4 years after the dam was finished, making both the dam and reservoir useless. In 1907, Milltown Dam in western Montana was built to supply local hydroelectric power, but the very next year the reservoir above the dam nearly filled with mine tailings washed in from a large flood. The tailings not only greatly reduced the capacity of the dam to produce hydroelectric power, but they also formed a concentration of heavy metals that polluted groundwater in the area.

EROSION Sediment trapping by dams also can significantly impact the hydrology and ecology of areas downstream that had received sediment during predam floods. Recently, the beaches along the Colorado River in the Grand Canyon were disappearing rapidly because management practices at the Glen Canyon Dam upstream prevented flooding in the Grand Canyon. In a similar way, beaches are vanishing from many other dammed streams.

A classic example of the way dams affect the movement of sediment has occurred in the Colorado River within the Grand Canyon, located directly below the Glen Canyon Dam. Once the dam was built in 1963, all of the sandy sediment being transported downstream in the Colorado River was deposited in Glen Canyon Reservoir. The only sand delivered to the Colorado within Grand Canyon came from tributaries that entered the river below the dam. Between 1963 and 1991, this reduction in sand delivery caused sand bars and beaches within the Grand Canyon to erode, not

salinization A process whereby salts accumulate in soil that is irrigated heavily, lowering soil fertility.

only reducing the availability of prime camping spots within the park but greatly limiting the availability of shallow, slow-flowing aquatic environments critical to the survival of fish. In addition, the lack of floods promoted the widespread establishment of *Tamarix*, a non-native plant that colonized the river banks and further disrupted the river ecosystem (●**FIGURE 12.12**).

In the spring of 1996, and again in the fall of 2004 and spring of 2008, scientists released enough water from Glen Canyon Dam to create a flood in the Grand Canyon. These high flow experiments were designed to mimic the effects of pre-dam flooding on the Colorado River and investigate whether such floods could help rebuild sandbars, benefit aquatic organisms, and limit the spread of *Tamarix*. The first flood in 1996 eroded sediment from existing sand bars and moved it to higher elevations, but it did not move much sand from the river bed to the sandbars as was hoped.

In response, the 2004 and 2008 floods were strategically timed to coincide with tributary floods that brought "new" sand into the Colorado. These floods succeeded in moving much of the newly delivered sand from tributaries onto the banks and sandbars in the Colorado, causing them to enlarge. Research conducted on long-term study sites in the Grand Canyon showed that 75 percent of the sand bars grew over the course of the three artificial floods (●**FIGURE 12.13**). In addition, the spring releases in 1996 and 2008 showed that these floods helped reduce the spread of *Tamarix* not only

● **FIGURE 12.12** Damming of the Colorado River has greatly reduced the size and frequency of large floods, leading to extensive colonization of river banks and sand bars by *Tamarix*, a non-native invasive plant that has negatively impacted the river ecosystem. *Tamarix* is now widespread across many southwestern rivers.

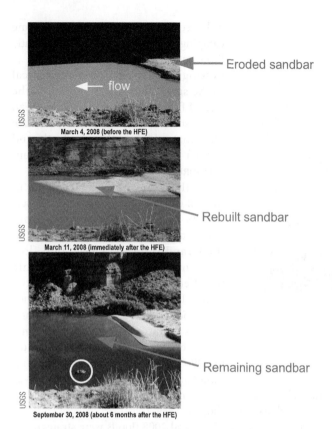

FIGURE 12.13 Photos of a sand bar before an induced flood, immediately after the flood, and about six months after the flood. The lowest photo has an 18-foot boat for scale (circled). The floods have succeeded in partially rebuilding many of the bars in the Grand Canyon that had eroded because the sand that normally would form the bars was being trapped behind the Glen Canyon Dam.

by partially removing existing plants but also by reducing the establishment of new *Tamarix* seedlings because the spring releases came before the plants went to seed. Lastly, the spring floods were found to increase the density of small aquatic invertebrates, especially insects, that are an important food source for fish. Along with the greater availability of shallow-water, slow-flowing environments, the increases in aquatic invertebrates following the floods contributed to increasing the population of rainbow trout, an important fishery below the dam, by 800 percent.

Because a reservoir accumulates silt and sand, it also interrupts the supply of sediment to the seacoast, causing some coastal beaches to erode faster than new sand can be delivered by natural processes. This problem has led to expensive sand replenishment projects on some popular beaches in the eastern United States.

RISK OF DISASTER A dam can break, creating a disaster in the downstream floodplain. One of the greatest floods since the last ice age roared down Idaho's Teton River canyon when the Teton Dam failed on June 5, 1976. The Teton Dam failed because of deficiencies in the design of the dam foundation, construction short-cuts, lack of a low-fill spillway, and rapid

initial filling of the reservoir. In particular, the Teton Dam was built on a foundation of highly permeable bedrock and this bedrock was insufficiently filled with grout prior to construction of the dam. In addition, one of the dam abutments was not sealed against the bedrock, and the dam core was not completely grouted.

The reservoir behind the Teton Dam was first filled in the spring of 1976, and it filled more quickly than expected due to rapid melting of snow in the drainage basin. On June 5, 1976, clear water was observed leaking through the dam foundation and subsequently through ungrouted joints in the bedrock adjacent to the dam abutments. Within a matter of hours, the leaking water became muddy, indicating that internal erosion of the dam itself was beginning to occur, and a sinkhole developed on the downstream side of the dam and began to flow additional muddy water. Although workers attempted to plug the sinkhole with large boulders, it quickly enlarged and propagated to the top of the dam causing a breach. As reservoir water poured through the breach, it quickly enlarged and lead to draining of the entire contents of the reservoir. Although a low-elevation spillway had been designed for controlled lowering of the reservoir, construction delays prevented it from being operational when the reservoir was first filled that spring. Without the spillway, there was no way to fill the reservoir slowly or prevent it from filling to capacity.

Within hours, the entire contents of the Teton Reservoir emptied, inundating about 300 square miles with floodwater and traveling over 150 miles downstream. Floodwaters swept away 154 of the 155 buildings in the town of Wilford, 6 miles below the dam; the town no longer exists. The flood inundated the lower half of Rexburg, a larger town farther downstream, damaging or destroying 4,000 homes and 350 businesses (●**FIGURE 12.14**). Damages totaled about $400 million (about $1.7 billion inflation-adjusted to 2018). Only 11 people died because civil authorities warned and evacuated most of the people. But if the flood had occurred in the middle of the night when warnings and evacuations would have been more difficult, thousands might have been killed.

RECREATIONAL AND AESTHETIC IMPACTS Dams are often built across narrow canyons to minimize engineering and construction costs. But when the canyon upstream of the dam is flooded, unique scenery and ecosystems are destroyed. Glen Canyon Dam on the Colorado River created Lake Powell by flooding one of the most spectacular desert canyons in the American West. Today, Glen Canyon is submerged for more than 300 kilometers above the dam and can be seen only in old photos. Although these environments are now effectively gone, the creation of Lake Powell provides new opportunities for fishing and other water sports that did not exist prior to construction of Glen Canyon Dam.

Disputes often arise among various groups that use water in dammed reservoirs. To prevent flooding, a reservoir should be nearly empty by spring so that it can store water and fill during spring runoff. Thus, a reservoir should be drawn down slowly during summer and fall. Agricultural users also prefer to have water stored during spring

(A)

(B)

(C)

● **FIGURE 12.14** (A) The Teton Dam was brand new in the spring of 1976. (B) Rapidly rising water broke the dam on June 5, 1976, sending lake water pouring through the breach. (C) Floodwaters from the failed Teton Dam inundated the city of Rexburg, Idaho.

runoff and supplied for irrigation during summer months. However, recreational users object to a lowering of reservoir levels during summer, when they visit reservoirs most frequently. Managers of hydroelectric dams prefer to run water through their turbines during times of peak demand for electricity, which rarely correspond to flood-management, irrigation, or recreational schedules.

ECOLOGICAL DISRUPTIONS A river is an integral part of the ecosystem that it flows through, and when the river is altered, the ecosystem changes. Dams interrupt water flow during portions of the year, prevent flooding, change the temperature of the downstream water, and often create unnatural daily fluctuations in stream flow. All of these changes alter relative abundances of aquatic and riparian species and may create specific ecological problems in individual rivers.

For example, before Egypt's Aswan Dam was built, the Nile River flooded every spring, depositing nutrient-rich sediment over the floodplain and delta. When the dam and reservoir cut off the silt supply, however, erosion of the delta front by the sea caused the Nile delta to shrink. In addition, the loss of an annual supply of nutrients eliminated a source of free fertilizer for floodplain farms. Although the energy produced by the dam is used to manufacture commercial fertilizers, many of the poorer farmers on the floodplain cannot afford to purchase what the river once provided for free.

HUMAN COSTS In the past 50 years, as many as 80 million people globally—approximately equal to 26 percent of the U.S. population—have been forcibly removed from their homes and farms to make way for the reservoirs behind dams. When these people moved into the relocation areas, they impacted the lives of the people already living there. Additionally, in many regions waterborne diseases become more prevalent surrounding stagnant water impounded behind a dam than they were near free-flowing streams that existed prior to the dam.

Construction of the Three Gorges Dam in China (●**FIGURE 12.15**), the largest hydroelectric dam in the world, has forced the relocation of about 1.4 million people. Roughly 300,000 more than originally anticipated were relocated because the rise and fall of the immense reservoir has triggered landslides along its banks, while the slower flow has caused harmful pollutants to accumulate.

DAM REMOVAL Although dams continue to provide an indispensable source of hydroelectric power along with water for agriculture and recreation, the societal risks and environmental costs of dams has significantly reduced the rate of new dam construction in the United States over the past 50 years (●**FIGURE 12.16**). In fact, hundreds of dams have been removed in the United States in recent years in part because of the expense required to maintain the aging structures and in part because of the rapidly growing interest in river restoration. In 2017, for example, a total of 86 dams were removed in the United States, reconnecting over 550 miles of river. Between 1912 and 2017, a total of 1,492 dams were removed.

Although few new dams are being constructed presently in the United States and the removal of dams is accelerating, this trend is largely restricted to the United States. Elsewhere around the globe, a major new initiative in hydroelectric

(A)

North

10 km

(B)

● **FIGURE 12.15** (A) Spanning the Yangtze River in China, the Three Gorges Dam is the largest in the world. It is over 2 kilometers across, impounds a reservoir over 600 kilometers long, and took 18 years to fully complete. The dam was built mainly to generate hydroelectric power and mitigate the effects of devastating floods on the lower reaches of the river. Despite these positive outcomes, the dam and reservoir are a source of significant controversy, as roughly 1.4 million people were forced to relocate and a wide variety of environmental problems have resulted. (B) This satellite image was acquired by the Advanced Spaceborne Thermal Emission and Reflection Radiometer (ASTER satellite) on May 15, 2012. The red colors on the image correspond to green colors (vegetation) on the ground.

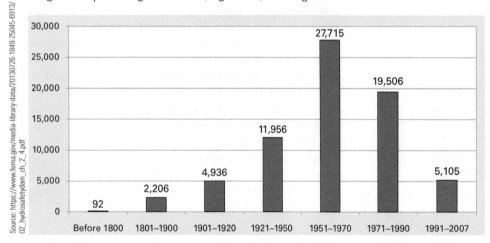

● **FIGURE 12.16** Years during which construction of new dams were completed in the United States.

of large dams for hydropower historically has been viewed as a sign of "human progress," these benefits are offset and may be ultimately outweighed by the negative impacts of dams on river ecosystems, the displacement of humans, and the alteration of regional water balances across watersheds.

Groundwater Diversion

Groundwater provides drinking water for more than half of the population of North America and is a major source of water for irrigation and industry. It is a valuable resource because:

1. It is abundant; 60 times more freshwater exists underground than in streams and lakes combined.
2. Groundwater moves very slowly and remains available during dry periods.
3. In some regions, groundwater flows from wet environments to arid ones, providing a source of water in dry areas.

power generation is underway in an effort to bring reliable electricity to more people (●**FIGURE 12.17**). As of 2014, at least 3700 new large dams with at least one Megawatt of electricity capacity were either planned or were under construction, primarily in countries with emerging economies. The new dams are projected to increase the global capacity of hydroelectric power by 73 percent or more, but even this dramatic increase in generation capacity will fail to keep up with projected increase in electricity demand. In addition, the construction of new dams outside the United States is projected to reduce the number of large, free-flowing rivers by 21 percent and significantly reduce freshwater biodiversity (●**FIGURE 12.18**). Thus, although the construction

cone of depression A cone-shaped depression in the water table, created when water is pumped out of a well more rapidly than it can flow through the aquifer to the well. If water continues to be pumped from the well at such a rate, the water table will drop.

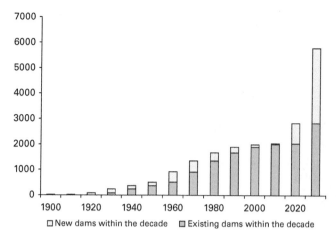

● **FIGURE 12.17** Worldwide the pace of construction of new hydroelectric dams is accelerating and is expected to continue to do so.

Groundwater Depletion

If groundwater is pumped to the surface faster than it can flow through the aquifer to the well, the water table forms a **cone of depression** surrounding the well (●**FIGURE 12.19**). When the pump is turned off, groundwater flows back toward the well, typically in a matter of days or weeks if the aquifer has good permeability, and the cone of depression disappears. Conversely, if water is continuously removed more rapidly than it can flow to the well through the aquifer, the water table drops.

Before the development of advanced drilling and pumping technologies, the human impact on groundwater was minimal. Today, however, deep wells and high-speed pumps can extract groundwater more rapidly than it is recharged. Where such excessive pumping is practiced, the water table falls as the groundwater reservoir becomes depleted.

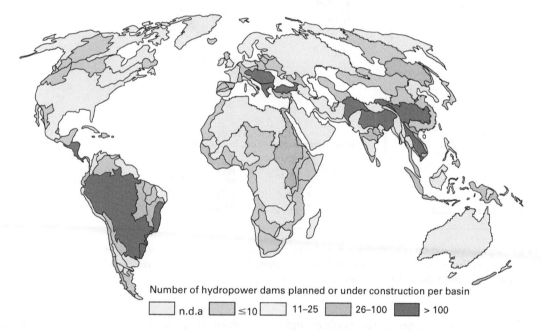

Number of hydropower dams planned or under construction per basin

☐ n.d.a   ☐ ≤10   ☐ 11–25   ☐ 26–100   ■ > 100

● **FIGURE 12.18** Number of future hydroelectric dams projected to be built per major river basin.

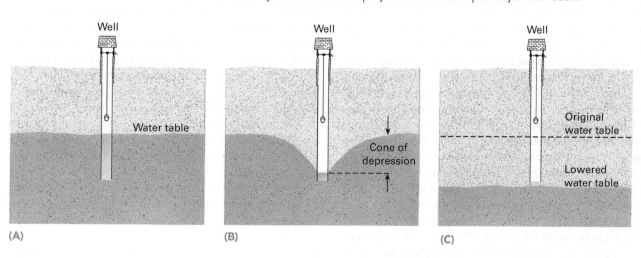

(A)   (B)   (C)

● **FIGURE 12.19** (A) A well is dug or drilled into an aquifer. (B) A cone of depression forms because water is withdrawn from the well at a faster rate than it can flow through the aquifer to the well and replenish the water removed. (C) If water continues to be extracted at the same rate, the water table falls.

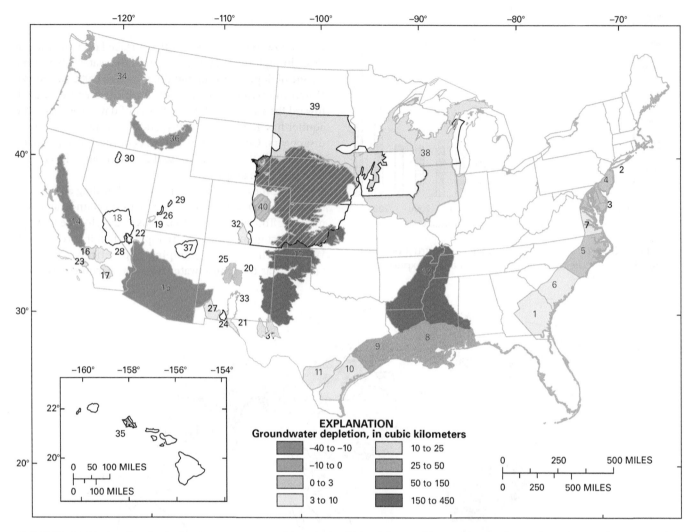

● **FIGURE 12.20** Aquifer depletion is a common problem in the United States. The numbers refer to the volume of water used in excess of the daily recharge rate, in cubic kilometers. The cross-hatched pattern in western Nebraska and Kansas and eastern Colorado indicate that water from more than one aquifer is being depleted.

In some cases, the aquifer is no longer able to supply enough water to support the farms or cities that have overexploited it. This situation is common in the central, western, and southwestern United States and is becoming more common along the eastern coastal plain (●FIGURE 12.20).

The Ogallala aquifer extends almost 900 kilometers from the Rocky Mountains eastward across the prairie, and from Texas to South Dakota (●FIGURE 12.21). It consists of water-saturated porous sandstone and conglomerate deposited as sediment shed eastward off the southern Rocky Mountains, mostly between 6 and 2 million years ago. The aquifer averages about 65 meters in thickness. Its upper surface, the water table, ranges in depth from zero where the aquifer recharges surface streams to about 150 meters. The Ogallala is the world's largest known aquifer and has been heavily exploited since the 1930s. About 200,000 wells were drilled into the aquifer between 1930 and 2000, and extensive irrigation systems were installed throughout the region. Today, the Ogallala aquifer supplies roughly 30 percent of all groundwater pumped for irrigation in the United States.

Prior to development, the total volume of the Ogallala aquifer was about 3,800 cubic kilometers of water, more than exists in Lake Huron and enough to cover the entire lower-48 U.S. landmass with about a half meter of water. Most of the aquifer's water accumulated when the last Pleistocene ice sheet melted, about 15,000 years ago. However, today the High Plains receive little rain, and the aquifer is mostly recharged by rain and snowmelt from the southern Rocky Mountains, hundreds of kilometers to the west.

The problem is that the rate of consumption of groundwater from the Ogallala aquifer is much greater than the rate of recharge from all combined sources. Not only is the total volume of precipitation across the recharge area for the Ogallala aquifer far smaller than the volume of water formerly available from melting of Pleistocene glaciers, but water moves very slowly through the

---

subsidence The irreversible sinking or settling of Earth's surface.

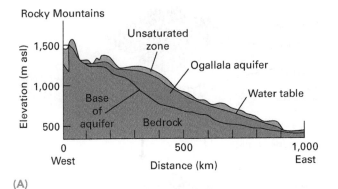

(A)

aquifer, at an average rate of only about 15 meters per year. At this rate, groundwater takes 60,000 years to travel from the southern Rocky Mountains to the eastern edge of the aquifer. Presently, approximately 9.6 billion liters are returned to the aquifer every year through rainfall and groundwater flow. However, farmers remove nearly 10 times this volume annually. Under such conditions, the deep groundwater is, for all practical purposes, non-renewable. Indeed, in some of the drier regions of north-west Texas and west central Kansas, wells drilled into the Ogallala aquifer already are depleted (Figure 12.21B). If the present pattern of water use continues, one-quarter of the original water in the Ogallala aquifer will be consumed by 2020.

Depletion of the Ogallala aquifer has serious economic consequences. About 20 percent of the total U.S. agricultural output, including 40 percent of the feed-lot beef, is produced by water from the aquifer, with a total value of $32 billion. If the aquifer is depleted and another source of irrigation water is not found, productivity in the central High Plains is expected to decline by 80 percent. Farmers will go bankrupt, and food prices will rise throughout the nation.

SUBSIDENCE Excessive removal of groundwater can cause **subsidence**, the sinking or settling of Earth's surface. When water is withdrawn from an aquifer, pore spaces that had been filled with water collapse. As a result, the volume of the aquifer decreases and the overlying ground subsides. In the aquifer underlying the San Joaquin Valley of California, the estimated reduction in water storage capacity by groundwater withdrawal is equal to about 40 percent of the total capacity of all surface reservoirs in the state.

Subsidence rates can reach 5 to 10 centimeters per year, depending on the rate of pumping and the nature of the aquifer. Some areas in the San Joaquin Valley have sunk nearly 10 meters (●**FIGURE 12.22**). The land surface has subsided by as much as 3 meters in the Houston–Galveston area of Texas. The problem is particularly severe in cities. For example, Mexico City is built on an old lake bottom. Over the years, as the weight of buildings and roadways has increased and much of the groundwater has been removed, parts of the city have settled as much as 8.5 meters. Many millions of dollars have been spent to

EXPLANATION
Water-level change, in feet, 1980 to 1997

Declines
More than 40
20 to 40
10 to 20
5 to 10

No significant change
−5 to 5

Rises
5 to 10
10 to 20
20 to 40
More than 40

Area of little or no saturated thickness

Regional subdivision boundary

0   50   100 MILES

0   50   100 KILOMETERS

(B)

U.S. Geological Survey Library

●**FIGURE 12.21** The Ogallala aquifer supplies water to much of the High Plains. (A) A cross-sectional view of the aquifer shows that much of its water originates in the Rocky Mountains and flows slowly toward the east as groundwater beneath the High Plains. (B) This map shows the extent and water level changes in the Ogallala aquifer from 1950 to 2015.

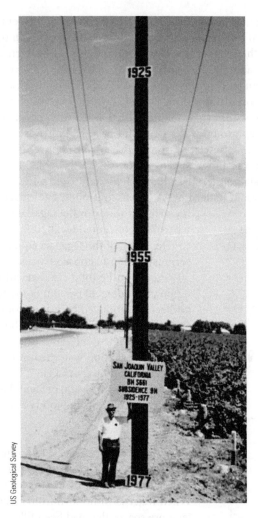

● **FIGURE 12.22** The land surface at this point in San Joaquin, California, was at the height of the 1925 marker in that year. Subsidence resulting from groundwater extraction for irrigation had lowered the ground surface by about 10 meters in 52 years.

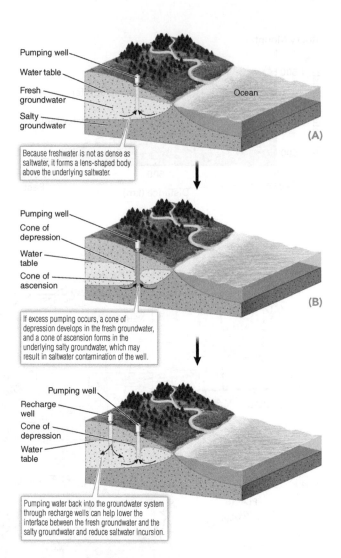

● **FIGURE 12.23** Saltwater intrusion can pollute coastal aquifers. (A) Freshwater lies above salt water, and the water in the well is fit to drink. (B) If too much freshwater is removed, the water table falls. The level of saltwater rises and contaminates the well.

maintain the city on its unstable base. Similar problems are affecting Phoenix, Arizona, and other U.S. cities.

Unfortunately, subsidence is irreversible. Once groundwater is withdrawn and an aquifer compacts, porosity is permanently lost so that groundwater reserves cannot be completely recharged, even if water becomes abundant again.

SALTWATER INTRUSION Two types of groundwater occur in coastal areas: fresh groundwater and salty groundwater that seeps in from the sea. Freshwater floats on top of salty water because it is less dense. If too much freshwater is removed from an aquifer, salty groundwater that is unfit for drinking, irrigation, or industrial use rises to the level of wells (●**FIGURE 12.23**). **Saltwater intrusion** has affected much of south Florida's coastal groundwater reservoirs. Once an area is contaminated by saltwater intrusion, it dramatically reduces freshwater storage in that aquifer.

THE GREAT AMERICAN DESERT

In the years following the Civil War, an American geologist, John Wesley Powell, explored the area between the western mountains (the Sierra Nevada-Cascade crest) and the 100° meridian (the line of longitude running through the Dakotas; Nebraska; Kansas; and Abilene, Texas). He recognized that most of this region is arid or semiarid, and he called it the Great American Desert. Despite Powell's warning about

saltwater intrusion A condition along the coasts of oceans and inland salt water bodies in which excessive pumping of fresh groundwater causes salty groundwater to invade an aquifer.

inadequate water, Americans flocked to settle there. Several parts of this region now use huge amounts of water for cities, industry, and agriculture, although they receive little precipitation. Some desert cities, including Phoenix, El Paso, Reno, and Las Vegas, rely on diversion of surface water and/or overpumping of groundwater for nearly all of their water.

Billions of dollars have been spent on projects to divert rivers and pump groundwater to serve this area of the country. In California alone, the two largest irrigation projects in the world have been built, along with 1,400 dams. More than 80 percent of the water diverted to Southern California is used to irrigate desert and near-desert cropland. The water supplied to these farms commonly irrigates crops such as cotton, rice, and alfalfa. Cotton fields require 17,500 liters of water per kilogram of cotton product; rice and alfalfa also use great amounts of water. The U.S. government pays large subsidies to farmers in naturally wet southeastern and south-central states not to grow those same crops. The irrigation water used to support cattle ranching and other livestock in California is enough to supply the needs of the entire human population of the state. Controversy over water diversion projects is not limited to the American West. New York City obtains much of its water from reservoirs in upstate New York, and the New York City Board of Water Supply controls those reservoirs and watersheds hundreds of miles from the city. Opposing interests of local land-use groups and city water users have provoked heated conflicts in the state.

The Colorado River

The Colorado River runs through the southwestern portion of the Great American Desert. Starting from the snowy mountains of Colorado, Wyoming, Utah, and New Mexico, it flows across the arid Colorado Plateau and southward into Mexico, where it empties into the Gulf of California.

Because the river flows through a desert, farmers, ranchers, cities, and industrial users along the entire length of the river compete for rights to use the water. In the 1920s, the Colorado River discharged 18 billion cubic meters of water per year into the Gulf of California. In 1922, the U.S. government apportioned 9 billion cubic meters for the river's Upper Basin (Colorado, Utah, Wyoming, and New Mexico) and the remaining 9 billion cubic meters for Lower Basin users in Arizona, California, and Nevada. Thus, all the water was allocated and none of it was set aside for maintenance of the river ecosystem or for Mexican users south of the border. No water flowed into the sea. In 1944, an international treaty awarded Mexico 1.5 billion cubic meters of water via the Colorado River, less than one-tenth of the annual volume that had been flowing into the Gulf of California 20 years earlier. Half of this amount was to be taken from the Upper Colorado River Basin, and half from the Lower Basin. In 2001, the U.S. government passed legislation specifying how water surpluses within the Colorado River Basin should be shared among the states involved but did not set aside any surplus water for Mexico. These collective actions created hard feeling among Mexicans living along the lower Colorado River toward U.S. management of river water north of the border. Finally, in 2012, the U.S. government passed legislation that included Mexico among the list recipients of surplus Colorado River water stored in U.S. reservoirs. In return, Mexico agreed to cut down on its use of Colorado River water during periods of drought.

In addition to the challenge of equitably providing enough water from the Colorado River to sustain the environment and meet human demand on both sides of the international border is the challenge of maintaining the quality of the water along the length of the river. The Colorado and many of its tributaries flow across sedimentary rocks, many of which contain soluble salt deposits. As a result, the Colorado is a naturally salty river. In addition, the U.S. government built 10 large dams on the Colorado, and evaporation from the reservoirs has further concentrated the salty water.

In 1961, the water flowing into Mexico contained 27,000 ppm (parts per million) dissolved salts, compared with an average salinity of about 100 ppm for large rivers. For comparison, 27,000 ppm is 77 percent as salty as seawater.

Mexican farmers used this water for irrigation, and their crops died. In 1973, the U.S. government built a desalinization plant to reduce the salinity of Mexico's share of the water.

There are also significant questions about the possibility of a long-term drought in the Southwest. Tree ring studies have shown that more extreme and extended droughts have occurred in the past several centuries than have been experienced in modern times. With current demands on the Colorado River increasing, even a minor drought can have severe consequences. The end of 2004 marked five consecutive years of lower-than-average rainfall and subsequent decline in reservoir levels. In January 2005, Lake Powell was at 35 percent of its capacity, the lowest level since 1969, when the reservoir was being filled. Some of this water since has been replenished, but as of August 2018, the reservoir was only at 48.3 percent of capacity. Inflows over the past water year were only about 50 percent of normal, whereas outflows required by treaty were about 96 percent of normal for the past water year. Clearly, maintaining a reliable supply of water for the growing population of the Colorado River basin is one of the most challenging public policy issues of the Southwest.

WATER AND INTERNATIONAL POLITICS

The Jordan River flows along the border between Israel and its long-time adversaries, Jordan and Syria. In the north, it forms the boundary between Israel and the disputed Golan Heights, and downstream it forms the boundary between Israel and the disputed West Bank. Population in the

The Los Angeles Water Project

In the mid-1800s, Los Angeles was a tiny, neglected settlement on the California coast. The Los Angeles River frequently flooded in winter and diminished to a trickle in summer. The average annual precipitation of the Los Angeles area is 38 centimeters, most of which falls during a few winter weeks.

Among the early settlers were Mormons, who had become experts at irrigating dry farmland from their experience in the Utah desert. By the late 1800s, irrigated farms in the Los Angeles basin were producing a wealth of fruits and vegetables. Suddenly, Los Angeles was an attractive, growing town. In 1848, the town had a population of 1,600. The population passed 100,000 by 1900 and 200,000 by 1904. Then the city ran out of water. Wells dried up, and the river was inadequate to meet demand. Los Angeles was surrounded on three sides by deserts and on the fourth by the Pacific Ocean. There was no nearby source of freshwater.

However, 250 miles to the northeast, the Owens River flowed out of the Sierra Nevada. The city of Los Angeles bought water rights, bit by bit, from farmers and ranchers in the Owens Valley and at the headwaters of the river. The city then spent millions of dollars building the Los Angeles Aqueduct across some of the most difficult, earthquake-prone terrain in North America. When finished, the aqueduct was 357 kilometers long, including 85 kilometers of tunnel. Siphons and pumps carried the water over hills and mountains too treacherous for tunneling (●**FIGURE 12.24**). On November 5, 1913, the first water poured from the aqueduct into the San Fernando Valley, located just north of Los Angeles.

● **FIGURE 12.24** A portion of the California aqueduct that carries water from the southern Sierra Nevada to Los Angeles.

The following 10 to 15 years were unusually rainy in the Los Angeles area and in the Owens Valley. The rain recharged the groundwater under the Los Angeles basin, and the Los Angeles River flowed freely again. As a result, irrigated farmland increased from 1,200 hectares in 1913, when the aqueduct opened, to more than 30,000 hectares in 1918.

In the 1920s, normal dryness and drought returned. Los Angeles and the farms of the San Fernando Valley demanded more water from the Owens River, and the Owens Valley dried up.

By the mid-1930s, Los Angeles owned 95 percent of the farmland and 85 percent of the residential and commercial property in the Owens Valley. Although the city leased some of the farmland back to the farmers and ranchers, the water supply became so unpredictable that agriculture slowly dried up with the land. As Los Angeles continued to grow, the water from the Owens River was not sufficient, so the Los Angeles Department of Water and Power drilled wells into the Owens Valley and began pumping its groundwater aquifer dry.

Today, the Owens Valley is parched. Its remaining citizens pump gas, sell beer, and make up motel beds for tourists driving through on their way to somewhere else. Ironically, the water that once irrigated farms and ranches in the naturally arid Owens Valley was diverted through the Los Angeles Aqueduct, at great cost to taxpayers, to irrigate farms in the naturally arid San Fernando Valley.

water-parched Jordan River basin was about 44 million in 2011 and is expected to increase to about 73 million by 2050.

Compounding the region's political tensions is the fact that nearly all of the renewable surface water runoff is being used and groundwater systems are being pumped at levels that are not sustainable, causing local saltwater intrusion in some of the coastal areas. Additional tensions related to water resources are caused by planned or actual construction of dams on tributaries of the Jordan River that alter the volume of water flowing down the river. In addition, a large volume of water is diverted into irrigation projects in Israel, further decreasing the amount available to Jordan and Syria. Pollution of the very limited surface water and groundwater supplies has led to additional conflict.

Current international water law offers little help in resolving conflicts over water. In general, downstream nations argue that a river or aquifer is a communal resource to be shared equitably among all adjacent nations. Predictably, many upstream nations maintain that they have absolute control over the fate of water within their borders and have no responsibility to downstream neighbors.

Any international code of water use must be based on three fundamental principles:

1. Water users in one country must not cause major harm to water users in other countries downstream.
2. Water users in one country must inform neighbors of actions that may affect them before the actions are taken. (For example, if a dam is built, engineers must inform downstream users that they plan to stop or reduce river flow to fill a reservoir.)
3. People must distribute water equitably from a shared river basin.

Unfortunately, these principles, especially the last one, are open to such subjective and self-interested interpretations that their intent is easily and often subverted. Despite these political challenges, since the 1950s more than 200 agreements have been developed to address issues of water management between individual nations and these agreements cover issues of flood control, hydropower projects, or allocations for consumptive or nonconsumptive uses. Although these are promising developments, many of the agreements specify allocations in fixed amounts, thereby ignoring annual variability in precipitation and changing needs.

WATER POLLUTION

While consumption reduces the quantity of water in a stream, lake, or groundwater reserve, **pollution** is the reduction of the quality of the water by the introduction of impurities (●FIGURE 12.25). A polluted stream may run full to the banks, yet it may be toxic to aquatic wildlife and unfit for human use.

In the early days of the Industrial Revolution, factories and sewage lines dumped untreated wastes into rivers. As the population grew and industry expanded, many of our waterways became fetid, foul smelling, and unhealthy. The first sewage treatment plant in the United States was built in Washington, D.C., in 1889, more than 100 years after the Revolutionary War. Soon other cities followed suit, but few laws regulated industrial waste discharge.

In 1952 and again in 1969, the Cuyahoga River in Cleveland, Ohio, was so heavily polluted with oil and industrial chemicals that it caught fire, spreading flames and smoke across the water. While the highly publicized photographs of the flaming river flowing through downtown Cleveland had a powerful visual impact (●FIGURE 12.26), other water pollution disasters were insidious, such as that of Love Canal in New York State, discussed below.

These and similar incidents made it clear that water pollution was endangering health and reducing the quality

● **FIGURE 12.25** Pollution is the reduction of the quality of a resource by the introduction of impurities.

● **FIGURE 12.26** In 1952, the Cuyahoga River in Ohio was so polluted by oil and industrial chemicals that it caught fire.

---

pollution The reduction of the quality of a resource by the introduction of impurities.

Love Canal

Love Canal in Niagara Falls, New York, was excavated to provide water to an industrial park that was never built. In the 1940s, Hooker Chemical Company purchased part of the old canal as a site to dispose of toxic manufacturing wastes. The engineers at Hooker considered the relatively impermeable ground surrounding the canal to be a reasonably safe disposal area. During the following years, the company disposed of approximately 19,000 tons of chemical wastes by loading them into 55-gallon steel drums and placing the drums in the canal. In 1953, the company covered one of the sites with dirt and sold the land to the Board of Education of Niagara Falls for $1, after warning the city of the buried toxic wastes. The city then built a school and playground on the site.

In the process of installing underground water and sewer systems to serve the school and growing neighborhood, the relatively impermeable soil surrounding the old canal and waste burial site was breached. Gravel used to backfill the water and sewer trenches then provided a permeable path connecting the toxic waste dump to surrounding parts of the city.

During the following decades, the buried drums rusted through, and the chemical wastes seeped into the groundwater. In the spring of 1977, heavy rains raised the water table to the surface, and the area around Love Canal became a muddy swamp. But it was no ordinary swamp—the leaking drums had contaminated the groundwater with toxic and carcinogenic compounds. The poisonous fluids soaked the playground, seeped into basements of nearby homes, and saturated gardens and lawns. Children who attended the school and adults who lived nearby developed epilepsy, liver malfunctions, skin sores, rectal bleeding, and severe headaches. In the years that followed, an abnormal number of pregnant women suffered miscarriages, and large numbers of babies were born with birth defects. Thus, a human tragedy at Love Canal was caused by the deliberate, but legal, burial of industrial poisons, coupled with flawed hydrogeologic analysis and a complete disregard of public health by both Hooker Chemical Company and local officials. In the decades following this public health tragedy, environmental laws have become much stricter and public awareness has increased dramatically. It is almost unimaginable that anyone would build a school over a toxic waste site today. However, less-flagrant releases of toxic and radioactive materials continue to this day, and the Love Canal disaster is a reminder to maintain vigilance.

of life for people throughout the United States. As a result, Congress passed the **Clean Water Act** in 1972. The legislation states that:

1. It is the national goal that the discharge of pollutants into the navigable waters be eliminated by 1985.
2. It is the national goal that, wherever attainable, an interim goal of water quality which provides for the protection and propagation of fish, shellfish, and wildlife and provides for recreation in and on the water be achieved by July 1, 1983.
3. It is the national policy that the discharge of toxic pollutants in toxic amounts be prohibited.

Types of Pollutants

Water pollutants have different levels of toxicity to humans and the environment. In addition, some pollutants are easier and less expensive to remove or stabilize than others, depending on the physical and chemical nature of the pollutant itself. Of particular importance is whether or not a pollutant can be removed through the metabolic actions of bacteria that either already exists or else are introduced into the polluted water as part of a cleanup operation.

BIODEGRADABLE POLLUTANTS A biodegradable material is one that decays naturally, being consumed or destroyed in a reasonable amount of time by organisms that live in soil and water. Several types of contaminants are **biodegradable pollutants**:

1. Sewage is wastewater from toilets and other household drains. It includes biodegradable organic material such as human and food wastes, soaps, and detergents. Sewage also includes some nonbiodegradable chemicals because people often flush paints, solvents, pesticides, and other chemicals down the drain.
2. Disease organisms, such as typhoid and cholera, are carried into waterways in the sewage of infected people.
3. Phosphates and nitrate fertilizers flow into surface water and groundwater mainly from agricultural runoff. Phosphate detergents as well as phosphates and nitrates from feedlots also fall into this category.

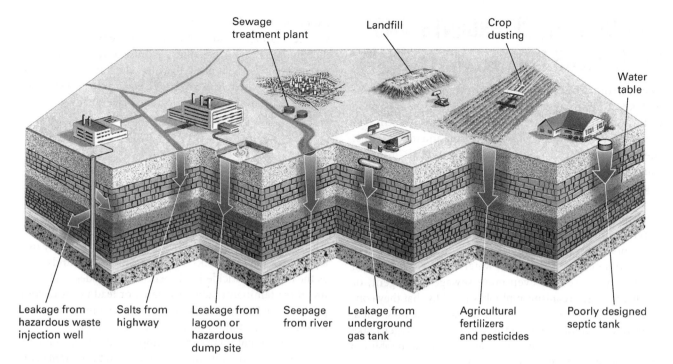

Sewage treatment plant   Landfill   Crop dusting   Water table

Leakage from hazardous waste injection well   Salts from highway   Leakage from lagoon or hazardous dump site   Seepage from river   Leakage from underground gas tank   Agricultural fertilizers and pesticides   Poorly designed septic tank

● **FIGURE 12.27** Many different sources contaminate groundwater.

NONBIODEGRADABLE POLLUTANTS Many materials are not decomposed naturally by environmental chemicals or consumed by decay organisms, and therefore they are nonbiodegradable. Not all of these are poisons, but some are. Nonbiodegradable poisons are called **persistent bioaccumulative toxic chemicals (PBTs)**. The Environmental Protection Agency (EPA) recognizes tens of thousands of PBTs. Yet even nontoxic materials can pollute and adversely affect ecosystems. **Nonbiodegradable pollutants** can be roughly subdivided into three broad categories:

1. Many organic industrial compounds are particularly troublesome because they are toxic in high doses, are suspected carcinogens in low doses, and can survive for decades in aquatic systems. Examples include some pesticides; dioxin; and polychlorinated biphenyl compounds, commonly referred to as PCBs.

2. Toxic inorganic compounds include mine wastes, road salt, and dissolved ions such as those of cadmium, arsenic, and lead.

3. When sediment enters surface waters, it neither fertilizes nor poisons the aquatic system. However, the sediment muddies streams and buries aquatic habitats, thus degrading the quality of an ecosystem.

RADIOACTIVE MATERIALS Radioactive materials include wastes from nuclear power plants, nuclear weapons, the mining of radioactive ores, and medical and scientific applications.

HEAT Heat can also pollute water. In a fossil fuel or nuclear electric generator, an energy source (coal, oil, gas, or nuclear fuel) boils water to form steam. The steam runs a generator to produce electricity. The exhaust steam must then be cooled to maintain efficient operation.

The cheapest cooling agent is often river or ocean water. A 1,000-megawatt power plant heats 10 million liters of water by 35°C every hour. The warm water can kill fish directly, affect their reproductive cycles, and change the aquatic ecosystems to eliminate some native species and favor the introduction of new species. For example, if water is heated, indigenous cold-water species such as trout will die out and be replaced by carp or other warm-water fish.

Any of these categories of pollutants can be released in either of two ways. **Point source pollution** arises from a specific site such as a septic tank, a gasoline spill, or a factory. In contrast, **nonpoint source pollution** is generated over a broad area. Fertilizer and pesticide runoff from lawns and farms falls into this latter category (●**FIGURE 12.27**).

Clean Water Act A federal law passed in 1972 mandating the cleaning of the nation's rivers, lakes, and wetlands and forbidding the discharge of pollutants into waterways.

biodegradable pollutants Pollutants that decay naturally in a reasonable amount of time by being consumed or destroyed by organisms that live in soil or water.

persistent bioaccumulative toxic chemicals (PBTs) Nonbiodegradable toxins that are released into and accumulate in the environment.

nonbiodegradable pollutants Pollutants that do not decay naturally in a reasonable amount of time, including some industrial compounds, toxic inorganic compounds, and nontoxic sediment that muddies streams and habitats.

point source pollution Pollution that arises from a specific site such as a septic tank or a factory.

nonpoint source pollution Pollution that is generated over a broad area, such as fertilizers and pesticides spread over agricultural fields.

HOW SEWAGE, DETERGENTS, AND FERTILIZERS POLLUTE WATERWAYS

Recall from Chapter 11 that an oligotrophic stream or lake contains few nutrients and clear, nearly pure water. Sewage, detergents, and agricultural fertilizers are plant nutrients. They do not poison aquatic systems; they nourish them. If humans dump one or more of these nutrients into an oligotrophic waterway, aquatic organisms multiply so quickly that they consume most of the dissolved oxygen. Many fish, such as trout and salmon, die. Other organisms, such as algae, carp, and water worms, can proliferate. Thus, an increased supply of nutrients can transform a clear, sparkling stream or lake into a slime-covered, eutrophic waterway (●FIGURE 12.28).

If humans release even more sewage, detergent, or fertilizer, decay organisms multiply so rapidly that they consume all the oxygen. As a result, most aquatic life dies. Then, **anaerobic** bacteria (bacteria that live without oxygen) take over, releasing noxious hydrogen sulfide gas that bubbles to the surface and smells like rotten eggs.

When raw sewage consisting of human waste is discharged into a stream, decay organisms immediately begin to break it down. At the same time, the current carries the sewage and decay organisms downstream. Over time, the organisms consume the waste and they eventually die off when their food supply is consumed. Natural turbulence replenishes oxygen in the water and fully aerobic organisms become reestablished. Thus, the river cleanses itself. However, if the original source of sewage is too large, or if every town along the river dumps sewage into the same waterway, then these natural cleaning mechanisms become overwhelmed and the river remains polluted all the way to the sea. Moreover, many sewage treatment plants ultimately release the treated water directly into nearby rivers. Just because water is treated in a sewage plant does not mean that all pollutants are completely removed, and the release of treated water into surface water systems can still significantly degrade water quality.

TOXIC POLLUTANTS, RISK ASSESSMENT, AND COST–BENEFIT ANALYSIS

Although human feces can be broken down naturally over a period of days or weeks, many organic compounds, such as the banned pesticide DDT, persist for decades, and heavy metal contaminants such as arsenic or lead never decompose. These heavy metals may change chemical form and become benign, but the atoms never decompose.

Environmental laws in developed nations regulate the discharge of pollutants into waterways. Yet pollutants continue to enter streams and rivers. Even in situations where waterways are contaminated with very small amounts of toxic nonbiodegradable compounds, a great many different pollutants may be found in a single river, and even small concentrations may cause cancer or birth defects.

Scientists do not know the doses at which many compounds become harmful to human health. The uncertainty results in part from the delayed effects of some chemicals. For example, a contaminant might increase the risk of a certain type of cancer, but the cancer may not develop until 10 to 20 years after exposure. Because scientists cannot perform direct toxicity experiments on humans, and because they cannot wait decades to observe results, they frequently attempt to assess the carcinogenic properties of a contaminant by feeding it to laboratory rats. If rats are fed very high doses, sometimes hundreds of thousands of times more concentrated than environmental pollutants, they may then contract cancer in weeks or months—rather than in years or decades. Suppose that a rat gets cancer after drinking the equivalent of 100,000 glasses of polluted well water each day for a few weeks. Can we say that the same contaminant will cause cancer in humans who drink 10 glasses a day for 20 years? No one knows. It may or may not be legitimate to extrapolate from high doses to low doses and from one species to another.

Scientists also use epidemiological studies to assess the risk of a pollutant. For example, if the drinking water in a city is contaminated with a pesticide and a high proportion of people in the city develop an otherwise rare disease, then the scientists may infer that the pesticide caused the disease and that its presence in drinking water constitutes a high level of risk to human health.

Because neither laboratory nor epidemiological studies can prove that low doses of a pollutant are harmful to humans, scientists are faced with a question: Should businesses and governments spend money to clean up the pollutant? Some argue that such expenditure is unnecessary until we can prove

● **FIGURE 12.28** This stream flowing through La Paz, Bolivia, is an open sewer. Decay organisms have consumed all of the oxygen in the water, and anaerobic bacteria have taken over.

that the contaminant is harmful. Others invoke the precautionary principle that it's better to be safe than sorry.

Pollution control is expensive. However, pollution is also expensive. If a contaminant causes people to sicken, the cost to society can be measured in medical bills and loss of income resulting from missed work. Many contaminants damage structures, crops, and livestock. People in polluted areas also bear the economic repercussions of diminished tourism and land values when people no longer want to visit or live in a contaminated area.

GROUNDWATER POLLUTION

Water in a sponge saturates tiny pores and passages. To clean a contaminant from a sponge, it would be necessary to clean every pore of the sponge that came into contact with contaminated water. Because this is nearly impossible to achieve completely, no one would wash dishes with a sponge that had been used to clean a toilet.

Many different types of sources contaminate groundwater. Because these contaminants permeate an aquifer in a manner analogous to the way in which contaminated water saturates a sponge, the removal of a contaminant from an aquifer is both difficult and expensive. Once a pollutant enters an aquifer, the natural flow of groundwater disperses it, forming a **plume of contamination** (●FIGURE 12.29). Because groundwater flows slowly, usually at a few centimeters per day, the plume spreads slowly.

TABLE 12.1 Average Residence Times for Water in Various Reservoirs	
Atmosphere	8 days
Rivers	16 days
Soil moisture	75 days
Seasonal snow cover	145 days
Glaciers	40 years
Large lakes	100 years
Shallow groundwater	200 years
Deep groundwater	10,000 years

Most contaminants persist in a groundwater aquifer for much longer than in a stream or lake. The rapid flow of water through streams and lakes replenishes their water quickly, but groundwater flushes much more slowly (**TABLE 12.1**). In addition, oxygen, which reacts to decompose many contaminants, is less abundant in groundwater than in surface water thereby allowing the contaminants to persist for years. For

---

anaerobic A condition in which oxygen is not present; anaerobic bacteria can survive in the absence of oxygen.

plume of contamination The slow-growing three-dimensional zone within an aquifer and/or body of surface water affected by a dispersing pollutant.

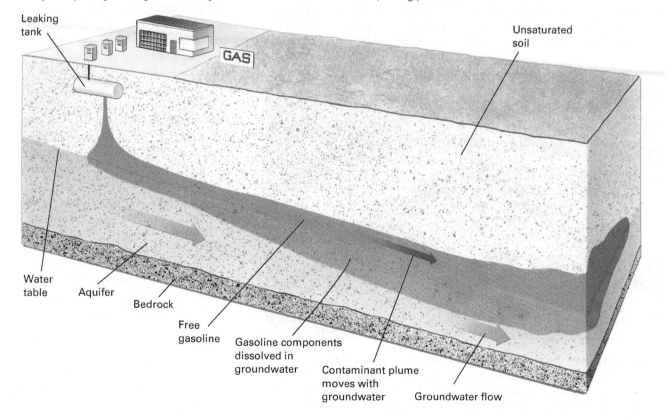

● **FIGURE 12.29** Pollutants disperse as a plume of contamination. Gasoline and many other contaminants are less dense than water. As a result, they float and spread out on top of the water table. Soluble components may dissolve and migrate with groundwater.

example, in 2006, a 25,000 gallon spill of regular unleaded gasoline occurred from underground tanks at an Exxon gas station in Jacksonville, Maryland. The clean-up operation lasted for over five years and involved the drilling of 284 test wells to depth up to 615 feet. Over 77 million gallons of polluted groundwater were treated.

Because groundwater pollution can persist for long periods of time, the following situations exist despite decades of government concern and action:

- Up to 25 percent of the usable groundwater in the United States is contaminated. About 45 percent of

DIGGING DEEPER

Yucca Mountain Controversy

In December 1987, the U.S. Congress chose a site near Yucca Mountain, Nevada, about 175 kilometers from Las Vegas, as the national burial ground for all spent reactor fuel—unless sound environmental objections were found. The Yucca Mountain site is located in the Basin and Range Province, a region noted for faulting and volcanism related to ongoing tectonic extension of that part of the western United States. Bedrock at the Yucca Mountain site is welded tuff, a hard volcanic rock. The tuff erupted from several volcanoes that were active from 16 to 6 million years ago. The last eruption near Yucca Mountain occurred 15,000 to 25,000 years ago. Geologists have mapped 32

faults adjacent to the Yucca Mountain site that have moved during the past 2 million years. The site itself is located within a structural block bounded by parallel faults. On July 29, 1992, a 5.6-magnitude earthquake occurred about 20 kilometers from the Yucca Mountain site and damaged some of the facilities there. Critics of the Yucca Mountain site have argued that recent earthquakes and volcanoes prove that the area is geologically active.

The Yucca Mountain environment is desert dry, and the water table lies 550 meters beneath the surface. In order to isolate the waste from both surface water and groundwater, the repository would consist of a series of

tunnels and caverns dug into the tuff 300 meters beneath the surface and 250 meters above the water table. However, it is impossible to predict future climatological changes that could alter the present hydrologic setting of the site. If the climate becomes appreciably wetter, the water table could rise or surface water could percolate downward. In addition, if an earthquake fractures rocks beneath the site, then contaminated groundwater might disperse more rapidly than predicted. Critics point out that construction of the repository will involve blasting and drilling, and these activities could fracture underlying rock, opening conduits for flowing water.

After decades of political wrangling, Congress designated the Yucca Mountain site to be the official long-term storage site for spent nuclear waste in the summer of 2002, and President George W. Bush approved. This decision was vetoed by the Nevada governor, as allowed by the Nuclear Waste Repository Act of 1982, but Congress subsequently overturned the veto.

Although the Yucca Mountain facility was approved by Congress, the legislative body is not authorized to grant the license for long-term disposal of nuclear waste to the Department of Energy (DOE) which would own and operate the facility. This licensing authority rested solely with the Nuclear Regulatory Commission (NRC).

In 2009, the Obama Administration announced plans to stop the pursuit of the operational license for the

AP Images/JOE CAVARETTA

● **FIGURE 12.31** The entrance to the Yucca Mountain Nuclear Waste Repository site. Despite spending billions of dollars to study the geology of the site and construct a large infrastructure for long-term storage of nuclear waste, no waste is presently stored there and the future of the site as a repository remains mired in political wrangling.

- municipal groundwater supplies in the United States are contaminated with organic chemicals.
- The gasoline additive MTBE is a frequent and widespread contaminant of underground drinking water. According to the EPA, 5 to 10 percent of drinking water in areas using reformulated gasoline shows MTBE contamination.
- Wells in 38 states contain pesticide levels high enough to threaten human health. Every major aquifer in New Jersey is contaminated (●FIGURE 12.30).
- In Florida, where 92 percent of the population drinks groundwater, more than 1,000 wells have been closed because of contamination and over 90 percent of the

Yucca Mountain facility, and in 2011, the administration terminated funding for the project. This action and other constraints caused both the DOE and NRC to suspend their efforts to secure a license operate the Yucca Mountain facility, leaving the United States with no long-term storage facility for nuclear waste generated by power plants and the manufacturing of nuclear armaments for defense.

After the effort to license Yucca Mountain was stopped, a group of plaintiffs that includes the states of Washington and South Carolina—both of which contain large, heavily polluted nuclear development facilities—filed a petition in the U.S. Court of Appeals in Washington, D.C., to restart the process. The petitioners argued that roughly 67,500 metric tons of high-level radioactive waste were being stored in 80 closed and operating nuclear plants across the United States and that the Yucca Mountain site would be safer than these unlicensed, temporary on-site storage sites (●FIGURE 12.32). In 2013, the court sided with the petitioners, and on November 18, 2013, the NRC ordered the licensing proceeding be restarted by completing the Yucca Mountain Safety Evaluation Report. The purpose of the safety evaluation report was to insure that the license application submitted by the DOE complies with existing environmental regulations. In 2015, the report was completed. Although the NRC found the Yucca Mountain site to be "acceptable" in terms of environmental regulations, they recommended that

the license application be denied because the DOE does not have jurisdiction over the land at the site. Rather, the land encompassing the site is under the jurisdiction of five separate federal agencies, and it would require an act of Congress to transfer this jurisdiction solely to the DOE. Moreover, a supplement filed by the NRC in 2015 showed that the DOE does not hold the necessary water rights at the Yucca Mountain site. The water rights are owned by the State of Nevada, which has refused to transfer them to the federal government.

Following publication of the Yucca Mountain Safety Evaluation Report, the site was effectively abandoned. As of 2019, no federal money has been

appropriated to support continued NRC licensing efforts or investigation of the Yucca Mountain site. Even if the land jurisdiction and water rights issues are resolved along with any third-party lawsuits that are filed, the DOE estimates that the cost of actually building the repository and transportation infrastructure needed to get the waste to the site would be about $97 billion. As it stands, roughly $8 billion has been spent since 1987 studying and testing the Yucca Mountain site as a repository, likely making it the most closely studied piece of real estate using public money in the history of geology. Today, the only thing that exists there is a boarded-up single 5-km-long exploratory tunnel.

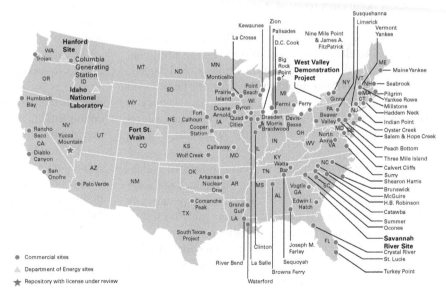

● FIGURE 12.32 Presently, spent nuclear fuel rods and other nuclear waste is stored at 80 operating and closed nuclear plants across the country. This map shows the location of the commercial nuclear power plants, Department of Energy nuclear sites, and the Yucca Mountain nuclear waste repository.

Courtesy of Graham R. Thompson/Jonathan Turk

● **FIGURE 12.30** An oil refinery in New Jersey. New Jersey suffers from some of the worst groundwater pollution in North America as a result of heavy industry.

remaining wells have detectable levels of industrial or agricultural chemicals.

- In 2003, the EPA reported 436,494 confirmed releases of dangerous volatile organic compounds leaking from underground fuel storage tanks in the United States.

Treating a Contaminated Aquifer

The treatment, or **remediation**, of a contaminated aquifer commonly occurs in a series of steps.

ELIMINATING THE SOURCE The first step in treating an aquifer is to eliminate the pollution source so additional contaminants do not escape. If an underground tank is leaking, the remaining liquid in the tank can be pumped out and the tank dug from the ground. If a factory is discharging toxic chemicals into the groundwater, courts may issue an injunction ordering the factory to stop the discharge.

Elimination of the source prevents additional pollutants from entering the groundwater, but it does not solve the problem posed by the pollutants that have already escaped. For example, if a buried gasoline tank has leaked slowly for years, many thousands of gallons of gas may have contaminated the underlying aquifer. Once the tank has been dug up and the source eliminated, people must deal with the gasoline in the aquifer.

MONITORING When aquifer contamination is discovered, a hydrogeologist monitors the contaminants to determine how far, in what direction, and how rapidly the plume is moving and whether the contaminant is becoming dilut-

ed or degraded by bacteria that live in the aquifer. The hydrogeologist may take samples from existing wells and may drill additional wells to monitor the movement of the plume through the aquifer.

MODELING After measuring the rate at which the contaminant plume is spreading, the hydrogeologist develops a computer model to predict future dispersion of the contaminant through the aquifer. To predict dilution effects, the model considers the local geologic structure, the permeability of the aquifer, directions of groundwater flow, and the mixing rates of groundwater.

REMEDIATION Several processes are currently used to clean up a contaminated aquifer. For example, contaminated groundwater can be contained by building an underground barrier to isolate it from other parts of the aquifer. Or, if the trapped contaminant does not decompose by natural processes, hydrogeologists can drill wells into the contaminant plume and pump the fluid to the surface, where it is treated to destroy the pollutant.

Bioremediation uses microorganisms to decompose a contaminant. Specialized microorganisms can be developed by genetic engineers to use a particular contaminant as a source of food or energy without damaging the ecosystem. Once a specialized microorganism is developed, it is relatively inexpensive to breed it in large quantities. The microorganisms are then pumped into the contaminant plume, where they attack the pollutant. When the contaminant is gone, the microorganisms die, leaving a clean aquifer. Bioremediation can be among the cheapest of all cleanup procedures.

Chemical remediation is similar to bioremediation. If a chemical compound reacts with a pollutant to produce harmless products, the compound can be injected into an aquifer to destroy contaminants. Common compounds used in chemical remediation include oxygen and dilute acids and bases. Oxygen may react with a pollutant directly or provide an environment favorable for microorganisms, which then degrade the pollutant. Acids or bases can neutralize certain contaminants or remove pollutants from groundwater by precipitating them to solid forms.

Reclamation teams also can dig up the entire contaminated portion of an aquifer. The contaminated soil is incinerated or treated with chemical processes to destroy the pollutant. The treated soil can then be returned to fill the hole. This process is prohibitively expensive, however, and is used only in extreme cases.

NUCLEAR WASTE DISPOSAL

In a nuclear reactor, radioactive uranium nuclei split into smaller nuclei, many of which are also radioactive. Most of these radioactive waste products are useless and must be disposed of without exposing people to the radioactivity. In the

United States, military processing plants, 104 commercial nuclear reactors, and numerous laboratories and hospitals generate more than 2,000 metric tons of high-level radioactive wastes every year.

Chemical reactions cannot destroy radioactive waste because radioactivity is a nuclear process and atomic nuclei are unaffected by chemical reactions. Therefore, the only feasible method for disposing of radioactive wastes is to store them in a place safe from geologic hazards and human intervention and to allow them to decay naturally. The U.S. Department of Energy defines a permanent repository as one that will isolate radioactive wastes for 10,000 years. However, this number is totally arbitrary, based on the fact that 10,000 years is so far into the future that most people won't worry about pollution occurring at that time. Actually, radioactive wastes will remain harmful for 1 million years or more because of the long half-lives of some radioactive isotopes. As a result, radioactive wastes could become harmful to human or nonhuman ecosystems over the course of geologic time.

For a repository to keep radioactive waste safely isolated for long periods of time, it must meet at least three geologic criteria:

1. It must be safe from disruption by earthquakes and volcanic eruptions.

---

remediation The treatment of a contaminated area, such as an aquifer, to remove or decompose a pollutant.

bioremediation The use of microorganisms to decompose an environmental contaminant.

chemical remediation Treatment of a contaminated area by injecting it with a chemical compound that reacts with the pollutant to produce harmless products.

2. It must be safe from landslides, soil creep, and other forms of mass wasting.
3. It must be free from floods and seeping groundwater that might corrode containers and carry wastes into aquifers.

THE CLEAN WATER ACT: A MODERN PERSPECTIVE

When it was adopted in 1972, the Clean Water Act set an ambitious agenda for cleaning the nation's rivers, lakes, and wetlands. Have we achieved our goals? The good news is that municipalities no longer dump raw sewage and factories no longer discharge untreated waste directly into our waterways. The bad news is that much of our water remains polluted and most documented violations remain unpunished. For example, a 2009 *New York Times* study reported that over a half-million documented violations of the Clean Water Act had occurred over the prior 5 years, yet less than 3 percent of these violations resulted in any fines or other punishment.

In 2004, the EPA reported impaired water quality for 44 percent of the nation's assessed streams, 64 percent of assessed lakes, and 30 percent of assessed estuaries. (An estuary is a former river valley that has been inundated by post-glacial sea-level rise; Chesapeake Bay is a good example.) A 2007 EPA study found that 49 percent of U.S. lakes contain fish that are contaminated with mercury and considered dangerous for human consumption. By 2008, the EPA listed over 8700 water bodies in 43 states, including the District of Columbia and Puerto Rico, that exceeded the mercury concentrations limits established by the Clean Water Act. Clearly, meeting these standards is easier said than done.

Key Concepts Review

- Growing human populations have created demands for water that stretch, and in some cases exceed, the amount of water that is available both globally and in the United States. Most of the water used by homes and industry is withdrawn and then returned to streams or groundwater reservoirs near the site of withdrawal. But most of the water used by agriculture is consumed because it evaporates. U.S. water consumption falls into three categories: (1) domestic use accounts for 10 percent; (2) industrial use accounts for 49 percent; and (3) agricultural use accounts for 41 percent.
- The United States receives about 3 times more water from precipitation than it uses, but some of the driest regions use the greatest amounts of water. Water diversion projects collect and transport surface water and groundwater from places where water is available to places where it is needed. Dams are especially useful in regions that receive seasonal rainfall, and some dams also generate hydroelectric

energy. But they can create undesirable effects, including water loss, salinization, silting, erosion, disaster when a dam fails, recreational and aesthetic losses, and ecological disruptions. Groundwater projects pump water to Earth's surface for human use, but because groundwater flows so slowly, the extraction often creates a cone of depression that may lead to a drop in the water table. The chapter's discussion of the Ogallala aquifer illustrates groundwater depletion. Groundwater withdrawal may also cause subsidence and saltwater intrusion.

- The Great American Desert is a mostly arid region of the western United States. Americans have built great cities and extensive farms and ranches in the region, all supplied by irrigation systems. Diminishing water reserves and increasing costs of water-diversion projects suggest that their future is uncertain. The chapter outlines the Owens River and Colorado River diversion projects.

- Water conflict leads to serious international conflict between hostile nations.
- Water pollution is the reduction in the quality of water by the introduction of impurities. In the United States, the Clean Water Act was passed in 1972 in response to serious water pollution disasters such as the Cuyahoga River fire and the Love Canal disaster. Water pollutants include biodegradable materials such as sewage, disease organisms, and fertilizers; nonbiodegradable materials such as industrial organic compounds, toxic inorganic compounds, and sediment; radioactive materials; and heat. Pollution can originate from both point sources and nonpoint sources.
- Biodegradable pollutants nourish an ecosystem and lead to oxygen depletion and growth of anaerobic bacteria, with resultant degradation of the ecosystem.
- It is difficult to determine the health effects of low doses of water pollutants. Cost–benefit analysis compares the cost of pollution control with the cost of externalities.
- Groundwater pollutants normally spread slowly into an aquifer as a plume of contamination. Because contaminants permeate all the tiny pore spaces in an aquifer, any cleanup, or remediation, is expensive and difficult. Techniques include elimination of the source, monitoring, modeling, bioremediation, chemical remediation, and removal of contaminated rock and soil.
- Radioactive wastes must be isolated from water resources because they are impossible to destroy and persist for long times. In the United States, Yucca Mountain is the designated site for spent fuels from nuclear power plants.
- Since the Clean Water Act was passed in 1972, waterways in the United States have been cleaned considerably—but many of our streams still remain polluted.

Important Terms

anaerobic (p. 302)

biodegradable pollutants (p. 300)

bioremediation (p. 306)

chemical remediation (p. 306)

Clean Water Act (p. 300)

cone of depression (p. 293)

consumption (p. 282)

diversion system (p. 287)

nonbiodegradable pollutants (p. 301)

nonpoint source pollution (p. 301)

persistent bioaccumulative toxic chemicals (PBTs) (p. 301)

plume of contamination (p. 303)

point source pollution (p. 301)

pollution (p. 299)

remediation (p. 306)

salinization (p. 288)

saltwater intrusion (p. 296)

subsidence (p. 295)

water scarcity (p. 282)

withdrawal (p. 282)

Review Questions

1. The United States receives three times more water in the form of precipitation than it uses. Why do water shortages exist in many parts of the country?

2. Describe the three main categories of water use. What proportion of total U.S. water use falls into each category?

3. Explain the differences among water use, water withdrawal, and water consumption.

4. Why is agriculture responsible for the greatest proportion of water consumption?

5. What are the two main sources of water exploited by water diversion projects?

6. Describe the beneficial effects of dams and their associated water delivery systems.

7. Describe the negative effects of dams and their associated water delivery systems.

8. Why does salinization commonly result from desert irrigation?

9. Describe the factors that make groundwater a valuable resource.

10. Describe problems caused by excessive pumping of groundwater.

11. Why does groundwater depletion cause longer-term problems than might result from the draining of a surface reservoir?

12. If land subsides when an underlying aquifer is depleted, will it rise to its original level when pumping stops and the aquifer is recharged? Explain your answer.

13. Why does saltwater intrusion affect only coastal areas?

14. Describe the geographic region that John Wesley Powell called the Great American Desert.

15. List four categories of water pollutants. What is/are the source(s) of each, and what are the harmful effects?

16. How does a nontoxic substance, such as cannery waste, become a water pollutant?

17. Discuss the difficulties in assessing the health risks of low doses of toxic water pollutants.

18. Explain why pollutants persist in groundwater longer than they do in surface water.

19. Outline possible procedures for cleaning polluted groundwater.

20. Discuss the controversy over the Yucca Mountain nuclear waste repository.

13

Exit glacier in Kenai Fjords National Park, Alaska. Deep crevasses have formed in this portion of the glacier as it flows downslope from the Harding Ice Field.

GLACIERS AND GLACIATIONS

LEARNING OBJECTIVES

LO1 Explain how early humans may have been affected by their interactions with glaciers.

LO2 Summarize the glaciation history of the Earth.

LO3 Compare and contrast Alpine and Continental glaciers.

LO4 Discuss the mechanisms that cause glaciers to move.

LO5 Explain the mass balance of a glacier.

LO6 Locate erosional glacial features on a diagram of a previously glaciated area.

LO7 Identify depositional glacial features on a diagram of a previously glaciated area.

LO8 Describe the importance and formation of drumlins.

LO9 Explain the formation of glacial features consisting of stratified drift.

LO10 Discuss climate changes associated with the Earth's orbit, spin axis, and precession.

LO11 Describe the association between the global volume of glacial ice and changes in global temperature.

INTRODUCTION

We often think of glaciers as features of high mountains and the frozen polar regions, yet anyone living in the northern third of the United States is familiar with landscapes created by a vast ice sheet that covered this region as recently as 17,000 years ago. In New England, which was virtually covered by ice at that time, glaciers sculpted Maine's rocky coastline and modified the shapes of the Northern Appalachian Mountains. The ice also deposited sediment that today forms Massachusetts' famous hook-shaped peninsula and the offshore islands around Martha's Vineyard. Further west, glacial ice carved out the depressions now occupied by the Great Lakes and deposited vast blankets of sediment across much of the upper Midwest. Even further west, glacial erosion deepened valleys and sharpened mountain peaks and ridges in the Rocky Mountains and created the subdued lowlands of Puget Sound and the San Juan Islands in western Washington State.

Numerous times during Earth's history, glaciers grew to cover large parts of Earth and then melted away. Before the most recent major glacial advance, beginning about 100,000 years ago, the world was free of ice except for on high mountains and in the polar ice caps of Antarctica and Greenland. Then, in a relatively short time—perhaps only a few thousand years—Earth's climate cooled. As winter snow failed to melt completely in summer, the polar ice caps spread into lower latitudes. At the same time, glaciers formed near mountain summits, even near the equator. They flowed down mountain valleys into nearby lowlands. When the glaciers reached their maximum size between about 24,000 and 17,000 years ago, they covered one-third of Earth's continents.

Humans, who have been on this planet for about 100,000 years, lived through the most recent glaciation. In southwest France and northern Spain, humans developed sophisticated spearheads and carved body ornaments between 40,000 and 30,000 years ago. People first began experimenting with agriculture between about 12,000 and 10,000 years ago.

FORMATION OF GLACIERS

In most temperate regions, winter snow melts completely in spring and summer. However, in certain cold, wet environments, some of the winter snow remains unmelted during the summer and accumulates year after year. During summer, the snow crystals become rounded and denser as the snowpack is compressed and alternately warmed during daytime and cooled at night. If snow survives through one summer, it converts to rounded ice grains called firn. Mountaineers like firn because the sharp points of their ice axes and crampons sink into it easily and hold firmly. If firn is buried deeper in the snowpack, it converts to glacial ice, which consists of closely packed ice crystals (●**FIGURE 13.1**).

A **glacier** is a massive, long-lasting, moving accumulation of compacted snow and ice. Glaciers form only on land, and they derive their ice from regions in which the amount of snow that falls in winter exceeds the amount that melts in summer. Because ice under pressure behaves like plastic and deforms easily, mountain glaciers flow downhill. In many cases, glaciers also slide downslope along their base. Glaciers on level land flow outward under their own weight. Unlike water, glacial ice is capable of moving upslope if it is being pushed from behind by ice flowing downslope from higher elevation portions of the same glacier.

Glaciers form in two environments: Alpine glaciers form at all latitudes on high, snowy mountains. Continental ice sheets form at all elevations in the cold polar regions.

Alpine Glaciers

Mountains are generally colder and wetter than adjacent lowlands. Near the summits, winter snowfall is deep and summers are short and cool. These conditions create alpine glaciers (●**FIGURE 13.2**). With the exception of Australia, alpine glaciers exist on every continent—in the Arctic and Antarctica, in temperate regions, and in the tropics. Glaciers cover the summits of Mount Kenya in Africa and Mount Cayambe in South America, even though both peaks are near the equator.

Some alpine glaciers flow great distances from the peaks into lowland valleys. For example, the Kahiltna Glacier, which flows down the southwest side of Denali (Mount McKinley) in Alaska, is about 65 kilometers long, 12 kilometers across at its widest point, and about 700 meters thick. Most alpine glaciers are smaller than the Kahiltna. For example, Switzerland's Gorner Glacier, shown in Figure 13.2, is about 14 kilometers long, about 1.5 kilometers across at its widest point, and up to 450 meters thick.

The growth of an alpine glacier depends on both temperature and precipitation. The average annual

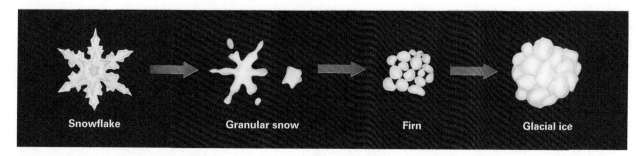

● **FIGURE 13.1** Newly fallen snow changes through several stages to form glacial ice.

● **FIGURE 13.2** The Gorner Glacier in Switzerland is the second-largest alpine glacier in the Alps.

temperature in the state of Washington is warmer than in Montana, yet alpine glaciers in Washington are larger and flow to lower elevations than those in Montana. Winter storms buffet Washington from the moisture-laden Pacific. Consequently, Washington's mountains receive such heavy winter snowfall that even though summer melting is rapid, snow generally accumulates every year. In much drier Montana, snowfall is light enough that most of it melts in the summer, and thus Montana's mountains have only a few small glaciers.

Continental Glaciers

In polar regions, winters are so long and cold and summers so short and cool that glaciers cover most of the land regardless of its elevation. An **ice sheet, or continental glacier,**

---

glacier A massive, long-lasting accumulation of compacted snow and ice that forms on land and moves downslope or spreads outward under its own weight.

ice sheet, or continental glacier A glacier that covers an area of 50,000 square kilometers or more and spreads outward in all directions under its own weight.

basal slip Movement of a glacier in which the entire mass slides over bedrock.

plastic flow Movement of a glacier in which the ice flows as a viscous fluid.

---

is a large accumulation of ice that spreads outward in all directions under its own weight and can cover areas of 50,000 square kilometers or more.

Today, Earth has only two ice sheets, one in Greenland and the other in Antarctica. These two ice sheets contain 99 percent of the world's ice and about three-fourths of Earth's freshwater. The Greenland sheet is more than 3.3 kilometers thick in places and covers 1.7 million square kilometers. Yet it is small compared with the Antarctic ice sheet, which blankets nearly 14 million square kilometers, roughly the size of the contiguous United States and Mexico combined. The Antarctic ice sheet covers entire mountain ranges, and the mountains that rise above its surface are islands of rock in a sea of ice.

Whereas the South Pole lies in the interior of the Antarctic continent, the North Pole is situated in the Arctic Ocean. At the North Pole, only a few meters of ice freeze on the relatively warm sea surface, and the ice fractures and drifts with the currents. As a result, no ice sheet exists there.

GLACIAL MOVEMENT

The rate of glacial movement varies with slope steepness, precipitation, and air temperature. In the coastal ranges of southeast Alaska, where annual precipitation is high and average temperature is relatively warm (for glaciers), some glaciers move 15 centimeters to 1 meter per day and in some instances, individual glaciers have been observed to surge at a speed of 10 to 100 meters per day. In contrast, in the interior of Alaska, where conditions are colder and drier, glaciers move only a few centimeters per day. At these rates, it takes hundreds to thousands of years for ice to flow the length of an alpine glacier.

Glaciers move by two mechanisms: **basal slip** and **plastic flow**. In basal slip, the entire glacier slides over bedrock in the same way that a bar of soap slides down a tilted board. Just as wet soap slides more easily than dry soap, water between bedrock and the base of a glacier accelerates basal slip.

Several factors cause water to accumulate near the base of a glacier. Earth's heat melts ice near bedrock. Friction from glacial movement also generates heat. Water occupies less volume than an equal amount of ice. As a result, pressure from the weight of overlying ice favors melting.

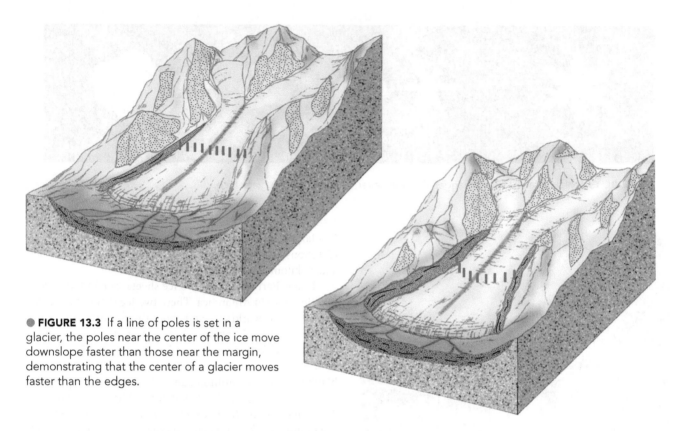

● **FIGURE 13.3** If a line of poles is set in a glacier, the poles near the center of the ice move downslope faster than those near the margin, demonstrating that the center of a glacier moves faster than the edges.

Finally, during the summer, water melted from the surface of a glacier may seep downward to its base.

A glacier also moves by plastic flow, in which the ice flows as a viscous fluid. Plastic flow is demonstrated by two experiments. In one, scientists set a line of poles in the ice (●**FIGURE 13.3**). After a few years, the ice moves downslope so that the poles form a U-shaped array. This experiment shows that the center of the glacier moves faster than the edges. Frictional resistance with the valley walls slows movement along the edges of the glacier. Because glacial ice flows plastically, this resistance slows the edges of the glacier down relative to the center.

In another experiment, scientists drive a straight, flexible pipe downward into a glacier to study the flow of ice at depth (●**FIGURE 13.4**). After several years, the entire pipe has moved downslope and become bent.

This experiment demonstrates that at the surface of a glacier, the ice is brittle, like an ice cube or the ice found on the surface of a lake. In contrast, at depths greater than about 40 meters, pressure is sufficient to allow ice to deform plastically. The curvature in the pipe shows not only that ice moves plastically but also that middle levels of a glacier move faster than lower levels. The base of a glacier is slowed by friction against bedrock, so it moves even more slowly than the plastic portion above it.

The relative rates of basal slip and plastic flow depend mainly on the steepness of the bedrock underlying the glacier, the thickness of the ice, and the temperature of the glacier. Glaciers located on steep slopes are prone to more basal slip than glaciers located on flat ground. The thickness of a glacier determines the pressure at its base. Because pressure depresses the freezing/melting point of ice, a thick glacier

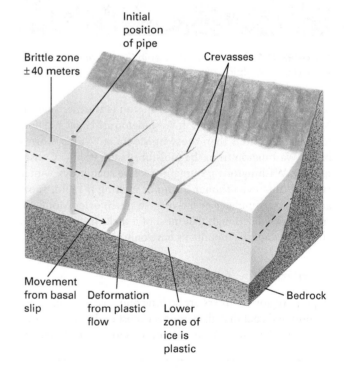

● **FIGURE 13.4** In this experiment, a pipe is driven through a glacier until it reaches bedrock. The entire pipe moves downslope with the ice but also becomes curved. The lower part of the pipe curves because friction with bedrock slows movement of the bottom of the glacier. Middle layers of ice flow more rapidly because they are less affected by the friction. The top 40 meters of ice also move rapidly, but the pipe remains straight there because this section of ice does not flow plastically.

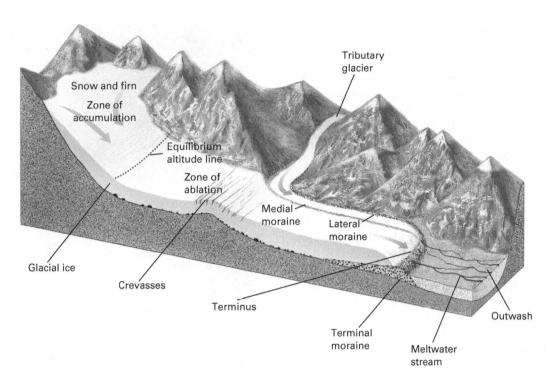

● **FIGURE 13.5** A schematic view of an alpine glacier. Crevasses form in the upper, brittle zone of a glacier where the ice flows over uneven bedrock.

will be more prone to develop a layer of water at its base, which will in turn promote basal sliding. Similarly, warmer temperatures at the base of a glacier will produce a layer of water that acts like a lubricant and promotes basal sliding. In contrast, if the temperature at the base of the glacier is cold enough, the glacier can freeze to the underlying substrate, thereby eliminating basal slip and causing the glacier to move entirely by internal deformation.

When a glacier flows over uneven bedrock, the deeper plastic ice bends and flows over bumps, while the brittle upper layer stretches and cracks, forming a **crevasse** (●FIGURES 13.5 and 13.6). Crevasses form only in the brittle upper 40 meters or so of a glacier, not in the lower plastic zone. Crevasses open and close slowly as a glacier moves. An **icefall** is a section of a glacier consisting of crevasses and towering ice pinnacles. The pinnacles form where ice blocks break away from the crevasse walls and rotate as the glacier moves. With crampons, ropes, and ice axes, a skilled mountaineer might climb into a crevasse. The walls are a pastel blue, and sunlight filters through the narrow opening above. The ice shifts and cracks, making creaking sounds as the glacier advances. Unfortunately, many mountaineers have been crushed by falling ice while traveling into crevasses and through icefalls.

---

crevasse A fracture or crack in the brittle upper 40 meters of a glacier, formed when the glacier flows over uneven bedrock.

icefall A section of a glacier consisting of numerous crevasses and towering ice pinnacles.

zone of accumulation The higher-altitude part of an alpine glacier where snow accumulates from year to year and forms glacial ice.

● **FIGURE 13.6** Crevasses in the Bugaboo Mountain Range of British Columbia. The widest crevasses are about 10 meters across.

The Mass Balance of a Glacier

Consider an alpine glacier flowing from the mountains into a valley. At the upper end of the glacier, snowfall is heavy, temperatures are below freezing for much of the year, and avalanches transport large quantities of snow from the surrounding slopes onto the ice. More snow falls in winter than melts in summer, and snow accumulates from year to year. This higher-elevation part of the glacier is called the **zone of accumulation**.

Lower in the valley, the temperature is higher throughout the year and less snow falls. This lower part of a glacier, where more snow melts in summer than accumulates in winter, is called the **zone of ablation**. When the snow melts, a surface of old, hard glacial ice is left behind. The **equilibrium altitude line, or ELA**, is the boundary between the zone of accumulation and the zone of ablation. The ELA shifts up and down the glacier from year to year, depending on weather. Ice exists in the zone of ablation because the glacier flows downward from the accumulation area. Further down the valley, the rate of glacial flow cannot keep pace with melting, so the glacier ends at its **terminus**.

Glaciers grow and shrink. For example, the zone of accumulation in an alpine glacier will grow thicker and the glacier's ELA will descend to a lower elevation if annual snowfall increases and more firn survives the summer without completely melting. At first the glacial terminus may remain stable, but eventually it will advance farther down the valley. The lag time between a change in climate and the advance of an alpine glacier may range from a few years to several decades, depending on the size of the glacier, its rate of movement, and the magnitude of the climate change. If annual snowfall decreases or the climate warms and less firn survives the summer, the alpine glacier will grow thinner, its ELA will recede upslope, and the glacier terminus will retreat.

When a glacier retreats, its ice continues to flow downhill, but the terminus melts back faster than the rate at which ice reaches the terminus. Typically, the ELA ascends to a higher elevation as more of the glacier undergoes melting. In Glacier Bay, Alaska, glaciers have retreated over 100 kilometers in the past 200 years. Near the terminus, newly exposed rock is bare and lifeless. A few kilometers from the glacier, where the ground has been exposed for a few decades, lichens grow on otherwise bare rock, while mosses and dwarf fireweed can survive on thin soils formed in sheltered cracks where seabird droppings have mixed with windblown silt. Near the head of Glacier Bay, which deglaciated 200 years ago, tidal currents and ocean storms have washed enough sediment over the exposed bedrock to create a thicker soil capable of supporting a spruce-hemlock rain forest.

In equatorial and temperate regions, glaciers commonly terminate at an elevation of 3,000 meters or higher. However, in a cold, wet climate, a glacier may extend into the sea, where giant chunks of ice break off, forming **icebergs** (●FIGURE 13.7).

The largest icebergs in the world are those that break away from the Antarctic **ice shelf**. Between the years 2000 and 2002, two plates of ice the size of Connecticut and at least one the size of Rhode Island broke free from the West Antarctic Ice Sheet and floated into the Antarctic Ocean. In 2010, a 78-kilometer-long iceberg with a surface area greater than that of Massachusetts broke off of the terminus of the Mertz Glacier in Antarctica. In 2012, scientists analyzing 40 years' worth of satellite imagery from the ice shelf in West Antarctica reported that large portions of the ice shelf are slowly beginning to break away from the continent. If this breakaway continues, it will lead to the formation of more very large icebergs.

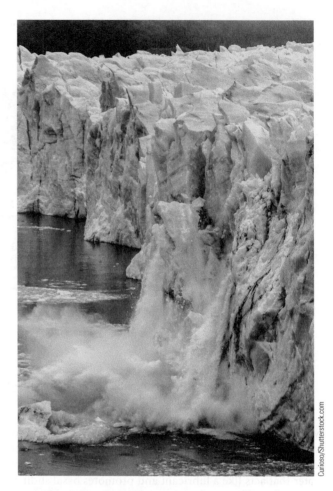

● **FIGURE 13.7** Icebergs form when the terminus of a glacier reaches the sea and pieces of the ice break off. A large mass of ice is shed into the Pacific Ocean from the terminus of the Perito Moreno Glacier in Patagonia.

GLACIAL EROSION

Bedrock at the base and sides of a glacier may have been fractured by tectonic forces, frost wedging, or pressure-release fracturing. Moving glacial ice commonly will dislodge and pluck individual boulders from such fractured bedrock. The plucked boulders are then transported downslope as clasts either within or at the base of the glacier (●FIGURE 13.8). Ice is viscous enough to pick up and carry particles of all sizes, from house-sized boulders to clay-sized grains. Thus, glaciers erode and transport huge quantities of rock and sediment.

Ice itself is not abrasive to bedrock, because it is too soft. However, rocks embedded in the ice scrape across bedrock, gouging deep, parallel grooves and scratches called **glacial striations** (●FIGURE 13.8B). When glaciers melt and striated bedrock is exposed, the markings show the direction of ice movement. Geologists measure the orientation of glacial striations and use them to map the flow directions of ancient glaciers.

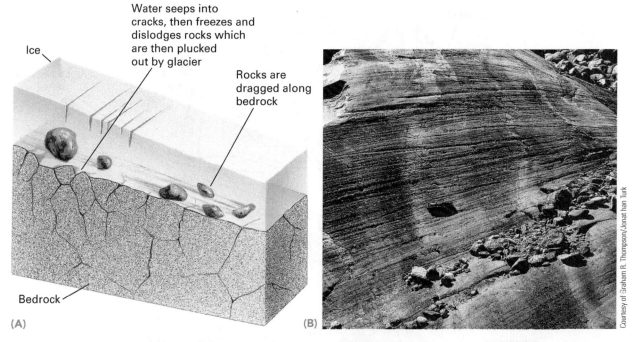

Ice

Water seeps into cracks, then freezes and dislodges rocks which are then plucked out by glacier

Rocks are dragged along bedrock

Bedrock

(A)

(B)

Courtesy of Graham R. Thompson/Jonathan Turk

● **FIGURE 13.8** (A) A glacier plucks rocks from bedrock and then drags them along as clasts, abrading both the clast and the bedrock. (B) Stones embedded in the base of a glacier gouged these striations in bedrock in British Columbia.

Erosional Landforms Created by Alpine Glaciers

Let's take an imaginary journey through a mountain range that was glaciated in the past but is now mostly ice free (●**FIGURE 13.9**). We start with a helicopter ride to the summit of a high, rocky peak. Our first view from the helicopter is of sharp, jagged mountains rising steeply above smooth, rounded valleys carved by glaciers.

We've already seen that mountain streams in unglaciated regions commonly erode downward into their beds,

---

zone of ablation The lower-altitude part of an alpine glacier where more snow melts in summer than accumulates in winter, and where the melting snow leaves behind a surface of old, hard glacial ice.

equilibrium altitude line, or ELA The boundary between the zone of accumulation and the zone of ablation.

terminus The end, or foot, of a glacier.

icebergs A large chunk of ice that breaks from a glacier into a body of water.

ice shelf A thick mass of ice that floats on the ocean surface but is connected to and fed from a glacier on land.

glacial striations Parallel grooves and scratches in bedrock that form as rocks are dragged along at the base of a glacier.

U-shaped valley A glacially eroded valley with a broad, characteristic U-shaped cross section.

cirque A steep-walled, spoon-shaped depression eroded into a mountain peak by a glacier.

tarn A small lake at the base of a cirque.

paternoster lakes A series of lakes in a glacial valley, strung out like beads and connected by short streams and waterfalls.

---

cutting steep-sided, V-shaped valleys (Chapter 11). A glacier, however, is not confined to a narrow streambed, but instead extends across the valley bottom and commonly fills the entire valley itself. As a result, the glacier scours the sides of the valley as well as the bottom, carving a broad, rounded **U-shaped valley** (●**FIGURES 13.9** and **13.10**).

We land on one of the peaks and step out of the helicopter. Beneath us, a steep cliff drops off into a scoop-shaped depression in the mountainside called a **cirque**. A small glacier at the head of the cirque reminds us of the larger mass of ice that existed in a colder, wetter time (●**FIGURE 13.11A**).

To understand how a glacier creates a cirque, imagine snow accumulating and a glacier forming on the side of a mountain (●**FIGURE 13.11B**). As the ice flows down the mountainside, it erodes a small depression that grows slowly as the glacier moves (●**FIGURE 13.11C**). With time and additional erosion, the depression deepens, causing the cirque walls to become steeper and higher. The glacier carries the eroded rock from the cirque to lower parts of the valley (●**FIGURE 13.11D**). When the glacier finally melts, it leaves behind a steep-walled, rounded cirque.

Streams and lakes are common in glaciated mountain valleys. As a cirque forms, the glacier may erode a depression into the bedrock beneath it. When the glacier melts, this depression can fill with water, forming a small lake, or **tarn**, nestled at the base of the cirque. If we hike down the valley below the high cirques, we may encounter a series of **paternoster lakes**. The term paternoster refers to a string of rosary beads connected by a string. Paternoster lakes are formed as a glacier erodes a sequence of small basins that later fill with water when the glacier recedes (●**FIGURE 13.12**). The lakes are connected by a stream that commonly features rapids and waterfalls.

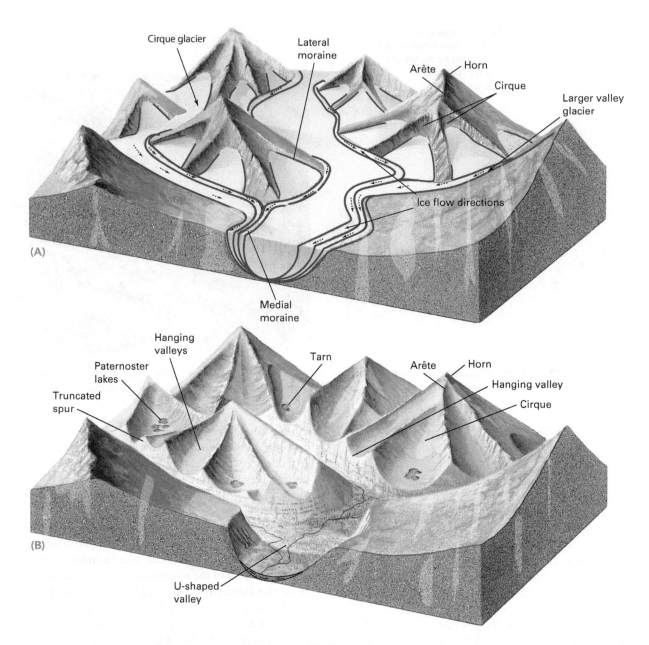

Cirque glacier
Lateral moraine
Arête
Horn
Cirque
Larger valley glacier
Ice flow directions
(A)
Medial moraine

Hanging valleys
Paternoster lakes
Truncated spur
Tarn
Arête
Horn
Hanging valley
Cirque
(B)
U-shaped valley

● **FIGURE 13.9** Two views of the same glacial landscape. (A) The landscape as it appeared when it was mostly covered by glaciers. (B) The same landscape as it appears now, after the glaciers have melted.

Harvepino/Shutterstock.com

● **FIGURE 13.10** A U-shaped valley formed by glacial erosion in the Lofoten archipelago of northern Norway.

If glaciers erode three or more cirques into different sides of a peak, they will create a steep, pyramid-shaped rock summit called a **horn**. The Matterhorn in the Swiss Alps is a famous horn (●**FIGURES 13.9** and **13.13**). Similarly, two alpine glaciers flowing down opposite sides of a mountain ridge will erode both sides, forming a sharp, narrow rib of rock called an

horn A sharp, pyramid-shaped rock summit where three or more cirques intersect near the summit.

arête A sharp narrow ridge of rock between adjacent valleys or between two cirques, created when two alpine glaciers moved along opposite sides of the mountain ridge and eroded both sides.

hanging valley A small glacial valley lying high above the floor of the main valley.

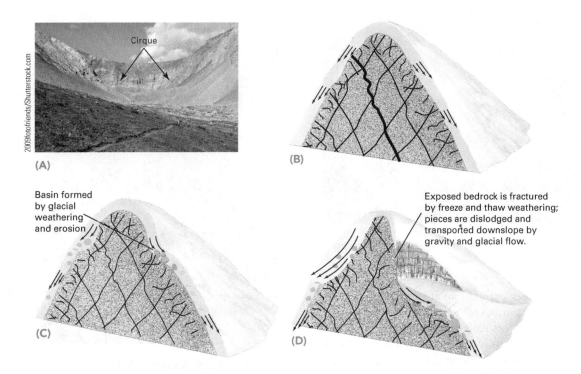

(A)

(B)

Basin formed by glacial weathering and erosion

(C)

Exposed bedrock is fractured by freeze and thaw weathering; pieces are dislodged and transported downslope by gravity and glacial flow.

(D)

● **FIGURE 13.11** (A) A glacier eroded this concave cirque into a mountainside Alberta, Canada, forming Ptarmigan Cirque. (B) To form a cirque, snow accumulates and a glacier begins to flow from the summit of a peak, shown here in cross section. (C) Weathering and glacial erosion form a small depression in the mountainside. (D) Continued glacial movement enlarges the depression. When the glacier melts, it leaves a cirque carved in the side of the peak, as in the photograph.

arête. When the glaciers melt, the arête will form a drainage divide that separates the two adjacent valleys or cirques.

Looking downward from our peak, we may see a waterfall pouring from a small, high valley into a larger, deeper one. A small glacial valley lying high above the floor of the main valley is called a **hanging valley** (●FIGURES 13.9 and **13.14**). The famous waterfalls of Yosemite Valley in California cascade from hanging valleys. A hanging valley forms where a small tributary glacier joined a much larger one. The tributary glacier eroded a shallow valley, while the massive main glacier gouged a deeper one. When the

● **FIGURE 13.12** Glaciers eroded bedrock to form this string of paternoster lakes in the Tatra Mountains of Poland.

● **FIGURE 13.13** The Matterhorn in Switzerland formed as three alpine glaciers eroded cirques into the peak from three different sides. The sharp ridge between two cirques is called an arête.

● **FIGURE 13.14** Yosemite Falls cascades from a hanging valley in Yosemite National Park.

● **FIGURE 13.15** A steep-sided fjord bounded by 1,000-meter-high cliffs on Baffin Island, Canada.

glaciers melted, they exposed an abrupt drop where the small valley joins the main valley.

Deep, narrow inlets called **fjords** extend far inland on many high-latitude seacoasts. Most fjords are glacially carved valleys that were later flooded by encroaching seas as the glaciers melted (●**FIGURE 13.15**).

Erosional Landforms Created by a Continental Glacier

A continental glacier erodes the landscape just as an alpine glacier does. However, a continental glacier is considerably larger and thicker and is not confined to a valley. As a result, it covers a large area that can include entire mountain ranges. The most recent continental glacier in North America stretched from beyond the present New England coastline westward to the high plains of the Dakotas and eastern Montana (●**FIGURE 13.16**).

GLACIAL DEPOSITS

In the 1800s, geologists recognized that the large deposits of sand and gravel found in the Alps and other places had been transported from distant sources. A popular hypothesis at the time explained that this material had drifted in on icebergs during catastrophic floods. The deposits were called drift after this inferred mode of transport.

Today we know that continental glaciers covered vast parts of the land only 10,000 to 20,000 years ago and that these glaciers carried and deposited drift. Although the term "drift" is a misnomer, it remains in use. Now geologists define **drift** as an all-embracing term for sediments of glacial origin, no matter how the sediments were deposited.

Drift is divided into two categories. **Till** was deposited directly by glacial ice. **Stratified drift** refers to sediments first carried by a glacier but ultimately deposited by glacial meltwater.

Landforms Composed of Till

Ice is so much more viscous than water that it carries a wide range of particle sizes. When a glacier melts, it deposits particles of all sizes—from fine clay to huge boulders—in an unsorted, unstratified mass (●**FIGURE 13.17**). Cobbles and boulders carried by a glacier often show scratches or grooves caused by the slow grinding of one clast against another during transport within the ice. However, gravel carried by a glacier is not rounded in the same way as gravel carried by a stream, because the clasts in a glacier do not undergo numerous collisions during transport. Therefore, if you find rounded gravel in a till, it probably became rounded by a stream before the glacier picked it up.

In some places, large boulders or smaller-sized rock fragments consisting of a rock type different from the bedrock in the immediate vicinity can be found in glaciated country (●**FIGURE 13.18**). Rocks of this type are called **erratics**. Most were transported to their present locations

---

fjords A deep, narrow, glacially carved valley on a high-latitude seacoast that was later flooded by encroaching seas as the glaciers melted.

drift Any rock or sediment transported and deposited by a glacier or by glacial meltwater.

till Glacial drift that was deposited directly by glacial ice.

stratified drift Glacial drift that was first carried by a glacier and then transported and deposited in layers by a stream.

erratics Boulders, usually different from bedrock in the immediate vicinity, that were transported to their present location by a glacier.

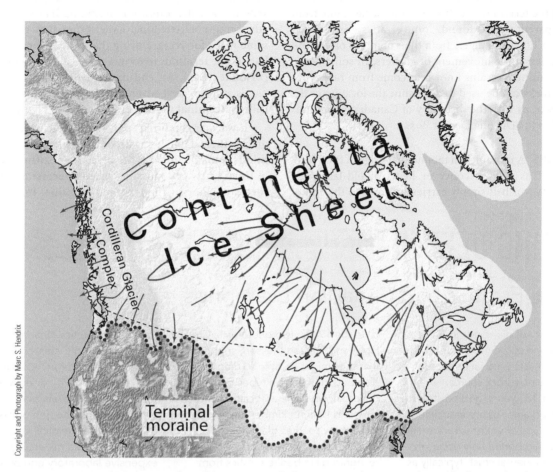

Copyright and Photograph by Marc S. Hendrix

● **FIGURE 13.16** Maximum extent of the continental glaciers in North America during the latest glacial advance, approximately 18,000 years ago. The arrows show directions of ice flow. Terminal moraines associated with the maximum advance stretch from offshore New Jersey to the Puget Sound of Washington State.

Copyright and Photograph by MARC S. HENDRIX

● **FIGURE 13.17** Glacial till consists of sediment deposited directly by ice and typically is unsorted. The boulders were rounded by stream action before being transported by glacial ice and deposited along with sand, silt, and clay in this till found in northwestern Wyoming.

Copyright and Photograph by MARC S. HENDRIX

● **FIGURE 13.18** These large boulders of granite in northern Yellowstone National Park are glacial erratics. The boulders were plucked from the hills seen on the skyline by glacial ice that carried them to this site before depositing them.

by a glacier. The origins of erratics can be determined by exploring the terrain in the direction the glacier came from until the parent rock is found. Some erratics were carried 500 or even 1,000 kilometers from their points of origin and provide clues to the movement of glaciers. In some cases, individual mineral grains known to come from rare types of parent rocks can be used to determine the location of the parent rocks, as with the discovery of Canadian diamonds discussed in the Digging Deeper Box.

MORAINES A moraine is a mound or a ridge of till. Think of a glacier as a giant conveyor belt. An old-fashioned airport conveyor belt simply carries suitcases to the end of

the belt and dumps them in a pile. Similarly, a glacier carries sediment and deposits it at its terminus. If a glacier is neither advancing nor retreating, its terminus may remain in the same place for years. During that time, sediment accumulates at the terminus to form a ridge called an **end moraine** (●**FIGURE 13.20**). An end moraine that forms when a glacier is at its greatest advance, before beginning to retreat, is called a **terminal moraine** (refer again to Figure 13.5).

If warmer conditions prevail, the glacier recedes. If the glacier then stabilizes again during its retreat and the terminus remains in the same place for a sufficient amount of time, a new end moraine, called a **recessional moraine**, forms upslope of the terminal moraine. In some cases,

DIGGING DEEPER

Glacial Erratics and Canadian Diamonds

The discovery of diamonds in Arctic Canada is one of the most storied examples of using glacial erratics to pinpoint their bedrock source.

For over 100 years, geologists have known that diamonds are associated with a rare type of volcanic rock called kimberlite. Kimberlite only occurs in very old continental crust, typically as carrot-shaped bodies, called "pipes," that are only a few hundred meters across at the surface. About 2 percent of kimberlites contain diamonds of mineable quality and quantity.

Prior to 1991, most of the world's diamonds were produced from a few tightly controlled mines associated with ancient continental crust and kimberlites in Africa, Siberia, and Australia. Although the oldest known continental crust on Earth occurs in Canada's Northwest Territories, nearly everything there is covered by a blanket of glacial till ranging up to several tens of meters thick. Until 1989, this vast till sheet and Arctic Canada's harsh climate prevented geologists from locating any kimberlite pipes there, although glacial erratics known to be associated with diamond-bearing kimberlites had been identified in some samples of Canadian till prior to that time.

In the 1980s, two geologists, Chuck Fipke and Stewart Blusson, supposed that they might be able to map the distribution of certain types of mineral grains known to be associated with diamond-bearing kimberlites to locate new sources of diamonds in Arctic Canada. However, previously recognized kimberlite erratics from the Canadian till consisted only of sand-sized mineral grains from a few isolated samples of till collected across a vast region. In addition, the till containing the erratics had been deposited by multiple generations of continental glaciers that formed and melted over the past few million years. Each glacial advance redistributed sediment left behind by the previous glaciation, further complicating the picture. Thus, in addition to the challenge of isolating and recognizing rare, sand-sized diamond-bearing indicator minerals across a till sheet many thousands of square kilometers in area, the geologists had to understand how each separate glacial advance redistributed the erratics before they could use them for locating the position of diamond-bearing kimberlite pipes.

Over the next 8 years, geologists Fipke and Blusson sampled till covering an arc over 1,200 kilometers

long across Arctic Canada and painstakingly isolated the diamond-indicator minerals from each till sample. To correctly identify the indicator minerals associated with diamonds, the geologists had to analyze the chemical composition of the sand-sized minerals using expensive laboratory equipment. They also measured the size and roundness of the grains to estimate how far they had been transported by the ice.

By 1989, Fipke and Blusson were working independently toward the same goal. Fipke was confident that he was near enough to a till-covered diamond-bearing kimberlite source that he began to register a claim with the Canadian government, an act that requires physically driving wooden stakes into the ground to mark the claim boundaries. On April 13, 1989, on the last day of sampling and staking claim, the group returned to collect a sample from the shoreline of an unusually circular, deep lake that reminded Fipke of exposed kimberlite pipes he had seen before in Africa. There, one of Fipke's associates picked up a pea-size piece of chrome diopside, a bright green diamond-indicator mineral that normally breaks down very close to its source. A piece of diopside this big suggested that

multiple recessional moraines are deposited upslope of the terminal moraine.

---

end moraine A ridge of till that forms at the end, or terminus, of a glacier that is neither advancing nor retreating and whose terminus has remained in the same place for years.

terminal moraine An end moraine that forms when a glacier is at its greatest advance before beginning to retreat.

recessional moraine A moraine that forms at the new terminus of a glacier as the glacier stabilizes temporarily during retreat.

ground moraine A moraine formed when a glacier recedes steadily and deposits till in a relatively thin layer over a broad area.

moraine A mound or ridge of till deposited directly by glacial ice.

When a glacier recedes steadily, till is deposited in a relatively thin layer over a broad area, forming a **ground moraine**. Ground moraines fill old stream channels and other low spots. Often this leveling process disrupts drainage patterns. Many of the swamps in the northern Great Lakes region and northern New England lie on ground moraines formed when the most recent continental glaciers receded.

End moraines and ground moraines are characteristic of both alpine and continental glaciers. An end moraine deposited by a large alpine glacier may extend for several kilometers and be so high that even a person in good physical condition would have to climb for an hour to reach the top. Moraines may be dangerous to hike over if

the Fipke team was standing on or very close to the kimberlite pipe. Realizing this, they raced home to find the money to expand their claim and further test their new find.

Within months, they had secured half-billion dollars from BHP Billiton, a global mining company. With this money, they flew in a drill rig that was able to cut a core from the bedrock under the lake. At a depth of 455 feet, the drill core encountered diamond-bearing kimberlite that verified the location of the pipe itself. The lake on top of it had formed because the relatively soft kimberlite

volcanic rock was easily eroded by the glacial ice, so a depression formed that collected water when the ice melted.

In November of 1991, BHP released a small press statement announcing the discovery. A nearly instant modern-day gold rush followed, with large and small companies appearing from all over the world to stake claims nearby. Since then, extensive exploration across the region has resulted in the identification of over 100 kimberlite pipes, roughly one-third of which are diamond bearing. As of 2017, there were five active diamond mines in the area, each recovering diamonds

from a separate kimberlite pipe. In 2015 alone, a total of 11.6 million carats of rough diamonds valued at $1.6 billion were produced from Canadian mines in the Northwest Territories. More recently, in December of 2018, a single 522 carat diamond - the largest North American diamond ever found - was recovered from the Diavik Diamond Mine about 135 miles south of the Arctic Circle. Remarkably, in less than 25 years, Canada has become the third largest world producer of diamonds in terms of dollar value, after Botswana and Russia, and it produces about 15 percent of gem-quality diamonds worldwide.

(A)

(B)

● **FIGURE 13.19** (A) A diamond in kimberlite. (B) Uncut, rough diamonds.

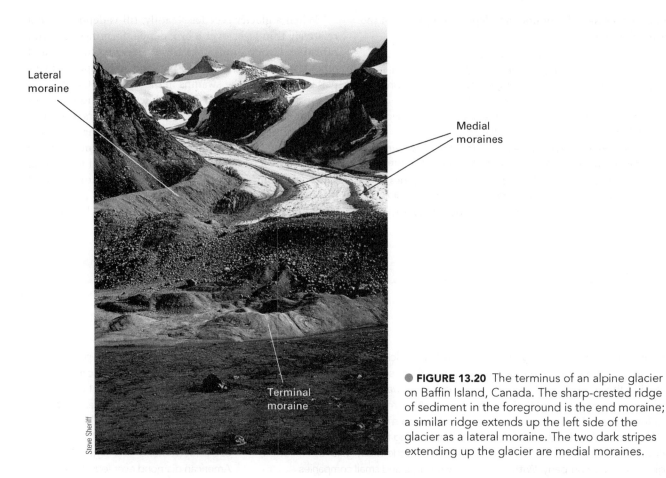

Lateral moraine

Medial moraines

Terminal moraine

Steve Sheriff

● **FIGURE 13.20** The terminus of an alpine glacier on Baffin Island, Canada. The sharp-crested ridge of sediment in the foreground is the end moraine; a similar ridge extends up the left side of the glacier as a lateral moraine. The two dark stripes extending up the glacier are medial moraines.

their sides are steep and the till is loose. Large boulders are mixed randomly with rocks, cobbles, sand, and clay. A careless hiker can dislodge boulders and send them tumbling to the base.

The most recent Pleistocene continental glaciers reached their maximum extent about 18,000 years ago. Their terminal moraines record the southernmost extent of those glaciers. In North America, the terminal moraines lie in a broad, undulating front extending across the northern

United States (Figure 13.16). Enough time has passed since the glaciers retreated that soil has formed on the till and vegetation now covers the terminal moraines (●**FIGURE 13.21**).

When an alpine glacier moves downslope, it erodes the valley walls as well as the valley floor. Additional debris falls or slides down from the valley walls and accumulates directly on mountain glaciers and their margins. Thus, large sediment loads are carried along the lateral edges of glaciers. When the glacier retreats, this sediment is left behind as a ridge of till called a **lateral moraine** (Figure 13.20).

If two alpine glaciers converge, their lateral moraines merge into the middle of the resulting larger glacier. This till forms a visible dark stripe on the surface of the ice, called a **medial moraine** (●**FIGURE 13.5 and** ●**FIGURE 13.20**).

DRUMLINS Elongate parallel hills, called **drumlins**, are common across the northern United States and are well

Science Geaphics/Ward's Natural Science Inc.

● **FIGURE 13.21** This wooded terminal moraine in New York State marks the southernmost extent of glaciers in that region.

lateral moraine A ridgelike moraine that forms from sediment on or adjacent to the sides of a mountain glacier.

medial moraine A moraine formed in or on the middle of a glacier by the merging of lateral moraines as two glaciers flow together.

drumlins Elongate hills, usually occurring in clusters, formed when a glacier flows over and reshapes a mound of till or stratified drift.

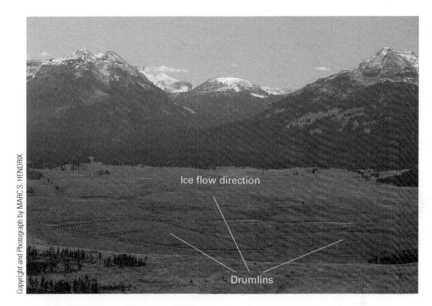

● **FIGURE 13.22** A field of drumlins in northern Yellowstone National Park. The drumlins are the elongate, sage-covered hills in front of the timbered mountains and indicate ice flow from left to right. One of the drumlins is highlighted by a dotted black line.

● **FIGURE 13.23** Streams flowing from the terminus of a glacier filled this valley on Baffin Island with outwash.

outwash streams A stream that emerges from below the snout of a glacier and carries glacial sediment further downslope.

outwash Sediment deposited by streams flowing from the terminus of a melting glacier.

outwash plain A broad, gently-sloping surface formed when outwash spreads onto a wide valley or plain beyond a glacier.

exposed across parts of upstate New York, Michigan, Wisconsin, and Minnesota. Drumlin is from the Old Irish for "back" or "ridge," and each one looks like a whale swimming through the ground, with its humped back in the air. An individual drumlin is typically about 1 to 2 kilometers long and about 15 to 50 meters high.

Drumlins usually occur in clusters, called drumlin fields (●**FIGURE 13.22**). Drumlins typically consist of multiple layers of till that are roughly parallel to the drumlins' upper surface but that also show erosion on the upstream side. A recent study of drumlins in Iceland shows that they are formed near the terminus of a glacier by multiple surges of the ice. According to this study, crevasses in the ice concentrate piles of sediment directly below, along the bed of the glacier. As the glacier surges, the ice sculpts the pile of till into a teardrop shape in which the steeper upstream side is partly eroded and the till is carried to the downstream side, where it is deposited to form the more gently sloping tail of the drumlin. With each successive surge of the glacier, another layer of till is plastered to the outer edge of the drumlin and shaped by the glacial movement.

Landforms Consisting of Stratified Drift

Because a glacier erodes great amounts of sediment, subglacial streams and **outwash streams** are typically laden with silt, sand, and gravel. Most of this sediment is deposited beyond the glacier terminus as **outwash** (●**FIGURES 13.5, 13.23,** and **13.24**). Outwash streams carry such a heavy sediment load that they often become braided, flowing in multiple channels. Over time, outwash deposited in a valley will form a broad, gently-sloping valley floor called a braidplain. Outwash deposited in front of continental glaciers forms a larger **outwash plain**.

During summer, melting of snow and ice causes streams to form on the surface of a glacier. Many of these streams flow off the front or sides of the glacier, while others plunge into tunnels (called moulins) in the ice and flow downward, ultimately to become part of a subglacial stream. Subglacial streams not only can erode the sediment bed below the glacier but they also can carve upside-down channels into the base of the ice. These channels in the ice can become filled with well-sorted stratified drift that forms a long,

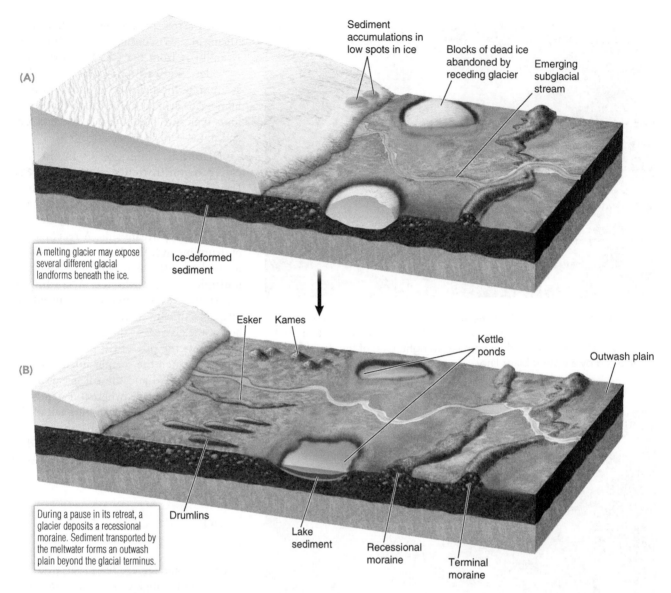

Sediment accumulations in low spots in ice

Blocks of dead ice abandoned by receding glacier

Emerging subglacial stream

(A)

A melting glacier may expose several different glacial landforms beneath the ice.

Ice-deformed sediment

Esker   Kames

Kettle ponds

Outwash plain

(B)

During a pause in its retreat, a glacier deposits a recessional moraine. Sediment transported by the meltwater forms an outwash plain beyond the glacial terminus.

Drumlins

Lake sediment

Recessional moraine

Terminal moraine

● **FIGURE 13.24** (A) A melting glacier may expose several different glacial landforms that were created beneath the ice, including drumlins and eskers. (B) During a pause in its retreat, a glacier deposits a recessional moraine. Sediment transported by the meltwater forms an outwash plain beyond the glacial terminus.

sinuous deposit called an **esker** when the glacier melts (●**FIGURE 13.24A**). Eventually, subglacial streams will reach the end of the glacier and emerge from beneath its terminus as an outwash stream.

Commonly, large pieces of glacial ice will separate from the main glacier during glacial retreat, forming **stagnant ice** that is no longer connected to the glacier's conveyor-belt ice delivery system. Sediment from outwash streams flowing around the stagnant ice will accumulate in a pile called a **kame**. As large blocks of stagnant ice melt, they leave behind depressions, called **kettles**, that often fill with water. Kames and kettles commonly occur together, forming **kame and kettle topography** that is typical of stagnant ice associated with a retreating glacier (●**FIGURE 13.24B**).

Large volumes of glacial meltwater can temporarily collect in lakes that form downslope from a glacier. Proglacial lakes form immediately in front of a retreating glacier and usually are dammed by the glacier's terminal moraine. In addition, large **pluvial lakes** can form in topographic basins that do not have an outflow. During the Pleistocene, several such pluvial lakes formed in the topographic basins of western Utah, Nevada, and southern Idaho. The Great Salt Lake in Utah is a remnant of a once-larger pluvial lake called glacial Lake Bonneaville.

Because kames, eskers, and sediment accumulating in glacial lakes are not deposited directly by ice, they typically show sorting and sedimentary bedding, which distinguishes them from unsorted and unstratified till. In addition, transport by streams usually rounds clasts deposited in these environments, in contrast with the more angular clasts typical of glacial till.

THE PLEISTOCENE GLACIATION

Geologists have found terminal moraines extending across all high-latitude continents. By studying those moraines, as well as lakes, eskers, outwash plains, and other glacial landforms, geologists have concluded that massive glaciers once covered large portions of the continents, altering Earth systems. A time when alpine glaciers descend into lowland valleys and continental glaciers advance across the land surface at high latitudes is called a **glaciation**. During a glaciation, glaciers several kilometers thick and thousands of kilometers across can spread across the land surface. Beneath this burden of ice, the continents sink deeper into the asthenosphere. The ice erodes rock and soil and deposits it elsewhere, completely altering the landscape.

Geologic evidence shows that Earth has been warm and relatively ice free for at least 90 percent of the past 1 billion years. However, at least six major glaciations occurred during that time. Each one lasted from 2 to 10 million years and was separated in time from relatively ice-free **interglacial periods**.

---

esker A long, snakelike ridge formed as the channel deposit of a stream that flowed within or beneath a melting glacier.

stagnant ice Glacial ice that has broken away from the front of a terrestrial glacier and is no longer connected to the glacier's ice-delivery system.

kame A small mound or ridge of stratified drift deposited by a stream that flows on top of, within, or beneath a glacier.

kettles A small depression formed by a block of stagnant ice; many fill with water to become a kettle lake.

kame and kettle topography Rolling hill topography associated with glacial kames and kettles.

pluvial lakes A lake formed in a topographic basin as the result of a moist climate during a glacial interval.

glaciation A time when alpine glaciers descend into lowland valleys and continental glaciers grow over high-latitude continents; usually used in reference to Pleistocene Glaciation.

interglacial periods A relatively warm, ice-free time separating glaciations.

Pleistocene Glaciation The most recent series of glaciations occurring during the Pleistocene Epoch and beginning around 2 million years ago. Most climate models indicate that Earth is still in the Pleistocene Glaciation.

eccentricity A term referring to the elliptical shape of Earth's orbit around the Sun; the more elliptical the orbit, the more eccentric it is said to be.

---

1. Mark Chandler, "Trees Retreat and Ice Advances," Glacial Cycles, *Nature* 381 (June 6, 1996): 477–478.

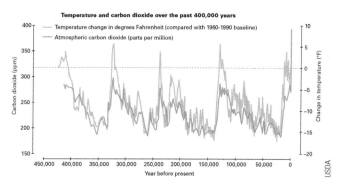

● **FIGURE 13.25** Glacial cycles during the Pleistocene Epoch. Cold temperatures coincided with Pleistocene glacial advances, and warm intervals coincided with glacial melting.

Source: https://www.fs.usda.gov/ccrc/sites/default/files/Figure2_primer_updated2014r2.png

The most recent glaciations have taken place during the Pleistocene Epoch and collectively are called the **Pleistocene Glaciation**. The Pleistocene Glaciation began about 2 million years ago in the Northern Hemisphere, although evidence of an earlier beginning has been found in the Southern Hemisphere. However, Earth was not glaciated continuously during the Pleistocene Epoch; instead, climate fluctuated and continental glaciers grew and then melted away several times (●**FIGURE 13.25**). During the most recent interglacial period, the average temperature was about the same as it is today, or perhaps a little warmer. Then, high-latitude temperature dropped at least 15°C, causing the ice to advance.[1] Many climate models indicate that the conditions leading to the Pleistocene Glaciation still exist and that continental ice sheets may advance again.

Causes of the Pleistocene Glacial Cycles

For reasons that are poorly understood, prior to 2 million years ago Earth's climate had been cooling for tens of millions of years—and then something happened to push the planet over a climate threshold and plunge it into a glaciation.

An increase in volcanic ash 2 million years ago coincided closely with a dramatic increase in sediment grains that showed glacial markings. Although volcanic dust may have triggered the onset of the Pleistocene Glaciation, no such events seem to be associated with the repeated growth and melting of glaciers that characterize the glacial interval. Instead, scientists have found that slight, periodic variations in Earth's orbit and orientation relative to the Sun coincided with glacial expansion and shrinking during the Pleistocene Glaciation.

In the 19th century, astronomers recognized three periodic variations in Earth's orbit and spin axis (●**FIGURE 13.26**):

1. Earth's orbit around the Sun is elliptical rather than circular. The shape of the ellipse is called **eccentricity**. The more elliptical the orbit, the more eccentric it is said to be. Earth's orbit eccentricity varies in a regular cycle lasting about 100,000 years.

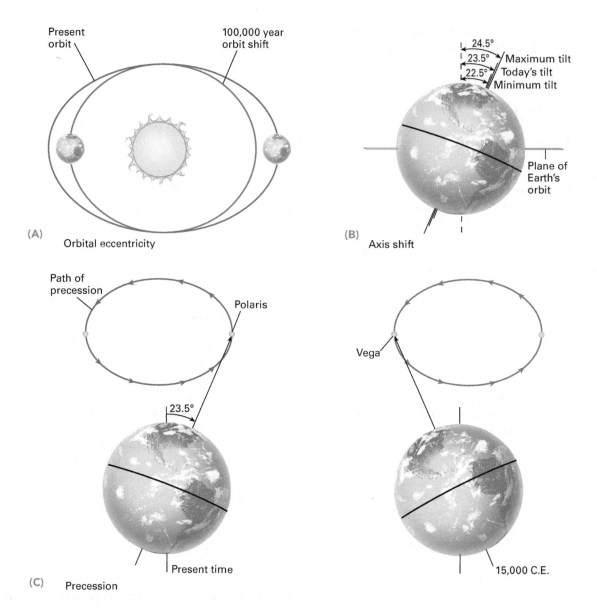

● **FIGURE 13.26** Earth's orbital variations may explain the temperature oscillations and glacial advances and retreats during the Pleistocene Epoch. Earth's orbit and spin axis vary in three ways: (A) the elliptical shape of the orbit changes over a cycle of about 100,000 years; (B) the tilt of Earth's axis of rotation oscillates by about 2 degrees over a cycle of about 41,000 years; and (C) Earth's axis completes a full cycle of precession about every 26,000 years.

2. The **tilt**, or angle of Earth's axis, is currently about 23.5 degrees with respect to a line perpendicular to the plane of its orbit around the Sun. The tilt oscillates by about 2 degrees on about a 41,000-year cycle.
3. Earth's axis, which now points directly toward the North Star (Polaris), circles like that of a wobbling top. This circling, called **precession**, completes a full cycle every 26,000 years.

These changes affect both the total solar radiation received by Earth and the distribution of solar energy with respect to latitude and season. Seasonal changes in sunlight reaching higher latitudes can reduce summer temperatures. If summers are cool and short, winter snow and ice persist, leading to the growth of glaciers.

Early in the 20th century, a Yugoslavian astronomer, Milutin Milanković, calculated that the orbital variations generate alternating cool and warm climates in the mid-latitudes and higher latitudes. Moreover, the timing of the calculated cooling coincided with that of Pleistocene glacial advances. Therefore, he concluded that orbital variations caused Pleistocene glacial cycles.

Modern calculations indicate that orbital cycles by themselves are not sufficient to cause glaciers to advance and retreat. Instead, orbital cycles disturb other Earth systems, which in turn cause additional cooling. Thus, a relatively small initial disturbance is amplified to cause a major climate change. In one recent study, researchers calculated that orbital variations probably caused high-latitude climate to cool enough to kill vast regions of Pleistocene northern forest. Forests control climate by absorbing solar energy and warming the atmosphere. When the forests died, more solar energy reflected back out to space. This loss of solar energy caused Earth to cool even more, and the glaciers advanced.[2] Because ice reflects more solar radiation back into space, a feedback process developed in which the growing glaciers accelerated global cooling. In this way, variations in Earth's orbit could have altered the biosphere enough to trigger a feedback that led to global cooling and a major glacial advance.

Effects of Pleistocene Continental Glaciers

At its maximum extent about 18,000 years ago, the most recent North American ice sheet covered 10 million square kilometers—most of Alaska, Canada, and parts of the northern United States (Figure 13.16). At the same time, alpine glaciers flowed from the mountains into the lowland valleys.

The erosional features and deposits created by these glaciers dominate much of the landscape of the northern United States. Today, terminal moraines form a broad band of rolling hills from Montana across the Midwest and eastward to the Atlantic Ocean. Kettle lakes or lakes dammed by moraines are abundant in northern Minnesota, Wisconsin, and Michigan, while drumlins dot the landscape across much of the northern Midwest. Ground moraines, outwash, and windblown glacial silt, or **loess** (pronounced "luss") cover much of the northern Great Plains. These deposits have weathered to form the fertile soil of North America's breadbasket.

Sea Level Changes with Glaciation

When glaciers grow, they accumulate water that would otherwise be in the oceans, and sea level falls. When glaciers melt, sea level rises again. When the Pleistocene glaciers reached their maximum extent roughly 18,000 years ago, global sea level fell to about 130 meters below its present elevation. As submerged continental shelves became exposed, the global land area increased by 8 percent (although about one-third of the land was ice covered).

---

tilt The angle of Earth's axis with respect to a line perpendicular to the plane of its orbit around the Sun. Earth's axis is tilted by about 23.5 degrees.

precession The circling or wobbling of Earth's axis as the planet travels in its orbit, like that of a wobbling top.

loess Deposits of windblown glacial silt.

tillite Till that was deposited by glaciers so long ago that it became lithified into solid rock.

---

2. R. G. Gallimore and J. E. Kutzbach, "Role of Orbitally Induced Changes in Tundra Area in the Onset of Glaciation," *Nature* 381 (June 6, 1996): 503–505.

When the ice sheets melted, most of the water returned to the oceans, raising sea level again. At the same time, portions of continents rebounded isostatically as the weight of the ice was removed. Thus, while the rising seas submerged some coastlines, others rebounded more than sea level rose, causing the sea to retreat. For example, former beaches in the Canadian Arctic now lie tens of meters to a few hundred meters above the sea.

SNOWBALL EARTH: THE GREATEST GLACIATION IN EARTH'S HISTORY

Research suggests that at least twice, and perhaps as many as five times in late-Precambrian time between 800 and 550 million years ago, massive ice sheets completely covered all continents and the world's oceans froze over—even at the equator—entombing the entire globe in a 1-kilometer-thick shell of ice. This glaciation, called Snowball Earth by the researchers who discovered it, contrasts sharply with the Pleistocene Glaciation, when ice covered just a third of the continents and only the polar seas froze over.[3]

The main evidence for the Precambrian Snowball Earth is based on a unique rock called **tillite**. Recall that till consists of an unsorted mixture of boulders, silt, and clay that was deposited by a glacier. Pleistocene tills are made of loose sediment that can be easily dug up with a shovel. Tillite, however, is hard, solid rock that in every other respect resembles the Pleistocene tills. Simply put, tillite is till that was deposited by glaciers so long ago that it has become cemented into hard rock. Researchers have found at least two thick layers of tillite between 750 and 580 million years old on almost every continent. Recall that continents have moved around Earth through geologic time. Other types of evidence show that some of the continents lay at the equator when the tillites formed. Other continents were nearer to the poles at the same time, showing that the glaciations were global in scope.

THE EARTH'S DISAPPEARING GLACIERS

Glaciers are now shrinking in more places and at more rapid rates than at any other time since scientists began keeping records (●TABLE 13.1). In 2003, during a single, hot summer in Europe, 10 percent of the glacial ice in the Alps melted. Scientists point out that this accelerated loss of glacial ice

3. Paul F. Hoffman, Alan J. Kaufman, Galen P. Halverson, and Daniel P. Schrag, "A Neoproterozoic Snowball Earth," *Science* 281 (August 28, 1998): 1342–1346.

TABLE 13.1 Selected Examples of Ice Melt around the World

Name	Location	Measured Loss
Arctic Sea Ice	Arctic Ocean	The extent of sea ice in September 2012 shrunk to the lowest observed since 1979 and was 49% below the average extent between 1979 and 2000. Largest loss of ice extent between March maximum and September minimum since 1979 also occurred in 2012.
Greenland Ice Sheet	Greenland	In summer of 2012, the largest areal percentage of the ice sheet in which melting was detected and the longest duration of seasonal ice melting since 1979 both occurred. The lowest albedo in 12 years also was observed as a consequence of the melting.
Columbia Glacier	Alaska	Has retreated nearly 20 km between 1980 and 2011; has lost about half its ice volume and has thinned to half its thickness since the 1980s.
Glacier National Park	Rocky Mountains, U.S.	Since 1850, the number of glaciers has dropped from 150 to fewer than 25 in 2010 that were at least 10 hectares (25 acres) in area. Some computer models suggest these remaining glaciers could disappear completely in several decades.
Antarctic Ice Sheet	Southern Ocean	Between 2003 and 2010, the Antarctic ice sheet lost an estimated 1155 km^3 of ice.
Pine Island Glacier	West Antarctica	In 2012, a crevasse 30 km long formed on the ice shelf, signaling the near-future breaking off of a 900 km^2 iceberg.
Larsen B Ice Shelf	Antarctic Peninsula	In 2002, a 3,250 km^2 and 220 m thick section of the ice shelf that had been stable for >10,000 years broke apart and disintegrated.
Carstensz Glaciers, West Meren Glacier	Papua (formerly Irian Jaya) Indonesia	Surface area of Carstensz Glaciers collectively shrank from 11 km^2 in 1942 to 2.4 km^2 by 2000 to 1.8 km^2 in 2005, a cumulative 84% decrease over 63 years. West Meren Glacier alone retreated 2,600 m since first surveyed in 1936 and disappeared completely between 1997 and 1999.
Dokriani Bamak Glacier	Himalayas, India	Between 1962 and 1995, the volume of glacial ice shrunk by roughly 20%, the glacier retreated 550 m.
Tien Shan	Central Asia	Analysis of glaciers within 15 separate parts of the Tien Shan indicate that all 15 regions have experienced loss of ice surface area ranging from <10% to 40% over the past ~50 years.
Caucasus Mountains	Russia	Monitoring of 113 glaciers between 1985 and 2000 revealed that 94% retreated during this period, with the largest glaciers retreating twice as fast as the smaller glaciers.
Alps	Western Europe	Nearly 600 glaciers within the French Alps have lost an average surface area of 26% since the late 1960s with 20% occurring since 1985–1986. Between 2000 and 2005, 230 of 230 glaciers observed in Austria and Switzerland and 50 of 60 Italian glaciers retreated. Researchers have concluded that the rate of glacial retreat today is faster than that of the past few decades.
Mount Kilimanjaro	Tanzania	Has lost 75% of glacial ice cover and 80% of ice volume over the past century.
Mount Kenya	Kenya	Has lost 75% of glacial ice cover over past century; 7 of the mountain's 18 glaciers have completely disappeared over this time frame.
Speka Glacier	Uganda	Retreated by more than 150 m between 1977 and 1990, compared with only 35 m to 45 m between 1958 and 1977.
Upsala Glacier	Argentina	Has retreated throughout the 20th century. Retreat rate has recently accelerated from 400 meters/year between 1990 and 1993 to 960 meters/year between 2008 and 2011.
Quelccaya Glacier	Andes, Peru	Unfossilized, well-preserved rooted plants now exposed along the glacial margins have been dated at roughly 5,000 years before present, indicating that it has been at least 50 centuries since the glacier was smaller than today.

Sources: Arctic Sea Ice: http://www.arctic.noaa.gov/reportcard/sea_ice.html

Greenland Ice Sheet: http://www.arctic.noaa.gov/reportcard/greenland_ice_sheet.html

Columbia Glacier, Alaska: http://earthobservatory.nasa.gov/Features/WorldOfChange/columbia_glacier.php

Glacier National Park: Hall, M. P. and D. B. Fagre. 2003. Modeled climate-induced glacier change in Glacier National Park, 1850–2100. Bioscience 53(2): 131–140

Antarctic Ice Sheet: Jacob, T., Wahr, J., Pfeffer, W.T., and Swenson, S., Recent contributions of glaciers and ice caps to sea level rise: Nature, v. 482, p. 514–518.

Pine Island Glacier: http://www.climatecodered.org/2010/01/pine-island-glacier-loss-must-force.html

Larson B Ice Sheet: http://nsidc.org/news/press/larsen_B/2002.html

Carstensz Glaciers, Indonesia: Kincaid, Joni L., 2007, An assessment of regional climate trends and changes to the Mt. Jaya glaciers of Irian Jaya: unpublished M.S. Thesis, Texas A&M University, 87 p. Prentice, M. L., & Hope, G. S. (2007). Climate of Papua . The Ecology of Papua'. (Eds A. J. Marshall and B. M. Beehler.) pp, 177–195.

Dokriani Bamak Glacier: Dobhal, D. P., Gergan, J. T., and Thayyen, R. J., 2004, Recession and morphogeometrical changes of Dokriani glacier (1962–1995) Garhwal Himalaya, India: CURRENT SCIENCE, VOL. 86, NO. 5, p. 692–696. http://www.iisc.ernet.in/currsci/mar102004/692.pdf

Tien Shan, central Asia: http://www.paleoglaciology.org/regions/TienShan/GlacierRetreat/

Caucasus Mountains: Stokes, C. R., Gurney, S. D., Shahgedanova, M., and Popovnin, V., Late-20th-century changes in glacier extent in the Caucasus Mountains, Russia/Georgia: Journal of Glaciology, vol. 52, no. 17, p. 99–109.

Alps: http://www.bbc.co.uk/news/science-environment-16025568 and http://www.paleoglaciology.org/regions/TienShan/GlacierRetreat/

Mount Kilimanjaro: Thompson, L. G., Mosley-Thompson, E., Davis, M. E., Henderson, K. A., Brecher, H. H., Zagorodnov, V. S., Mashiotta, T. A., Lin, P.-N., Mikhalenko, V. N., Hardy, D. R., and Beer, J. 2002. Kilimanjaro ice core records: evidence of Holocene climate change in tropical Africa. Science 298: 589–593.

Mount Kenya: Hastenrath, S. 1993. Toward the satellite monitoring of glacier changes on Mount Kenya. Annals of Glaciology 17: 245–249.

Speka Glacier: http://www.worldwatch.org/melting-earths-ice-cover-reaches-new-high.

Upsala Glacier: Warren C. R., Greene D. R. and Glasser N. F. (1995) Glaciar Upsala,Patagonia: rapid calving retreat in fresh water: Annals of Glaciology, v. 21, p.311–316. and Sakakibara, D., Sugiyama, S., Sawagaki, T., Marinsek, S., and Skvarca, P., 2013, Rapid retreat, acceleration and thinning of Glacier Upsala, Southern Patagonia Icefield, initiated in 2008: Annals of Glaciology, v. 54, no. 63, p. 131–138.

Quelccaya Glacier: Buffen, A. M., Thompson, L. G., Mosley-Thompson, E., and Huh, K. I., 2009, Recently exposed vegetation reveals Holocene changes in the extent of the Quelccaya Ice Cap, Peru: Quaternary Research, v. 72, p. 157–163.

coincides with an increase in atmospheric carbon dioxide levels of about 50 percent in the past century, and they suggest that the melting may be one of the first symptoms of human-caused global warming.

Alpine glaciers reflect these changes in a highly visible way. The larger glaciers in Glacier National Park, Montana, have shrunk to a third of their size since 1850, and they continue to melt away today. Many of the smaller glaciers have disappeared completely (●FIGURE 13.27). Since 1850, the number of glaciers in the park has decreased from 150 to fewer than 25 in 2010. One computer model predicts that all of the glaciers will be gone from the park by 2030 if global temperatures continue to rise as predicted, though other models suggest that this prediction is overstated.

As demonstrated in Table 13.1, significant loss of glacial ice volumes through melting is a global phenomenon. In some instances, ice disintegration has been particularly dramatic. For example, half of Alaska's Columbia Glacier melted between 1980 and 2011, and the glacier currently releases about 5 cubic kilometers of meltwater into Prince William Sound every year. Elsewhere, scientists predict that glaciers in the European Alps will be all but gone by 2050. Global climate changes do not occur uniformly around the globe. The north polar regions are warming faster than the average rate for all of Earth, and this warming is having a profound effect on Arctic sea ice. During the winter, Arctic sea ice covers an area about the size of the United States, though it is rapidly shrinking in size and thickness. Ice thickness measurements during 2011 and 2012 indicate that the summer ice pack those years was the thinnest, least extensive on record, and many scientists predict that Arctic summer sea ice will completely disappear in the next few decades. Such a loss would greatly increase the amount of solar energy that is absorbed by the Arctic Ocean, potentially accelerating global warming.

Ice shelves are thick masses of ice that are floating in the ocean but are connected to glaciers on land. They gain ice by flow from the land glaciers and lose ice by melting and when large chunks break off and float into the ocean as icebergs. Most of Earth's ice shelves surround Antarctica. Ice shelves respond to rising temperature more sensitively than glaciers do. Since 1974, seven Antarctic ice shelves have shrunk by a total of 13,500 square kilometers. These ice shelves are described further in Chapter 21.

The melting of the Arctic and Antarctic ice has profound effects for the planet. This melting contributes to the rise in sea level, and the freshwater flowing into the ocean may ultimately alter ocean currents and global climate.

(A)   (B)

US Geological Survey, Glacier Field Station

● FIGURE 13.27 (A) Boulder Glacier in Glacier National Park, Montana, in July 1932. (B) The glacier in July 1988, 56 years later, photographed from the same point. The glacier had disappeared completely by 1988.

Key Concepts Review

- If snow survives through one summer, it becomes a relatively hard, dense material called firn. A glacier is a massive, long-lasting accumulation of compacted snow and ice that forms on land and creeps downslope or outward under the influence of gravity and its own weight. Alpine glaciers form in mountainous regions; continental glaciers cover vast regions.

- Glaciers move by two mechanisms: basal slip and plastic flow. The upper 40 meters of a glacier is too brittle to flow, and large cracks called crevasses develop in this layer. In the zone of accumulation, the annual rate of snow accumulation is greater than the rate of melting, whereas in the zone of ablation, melting exceeds accumulation. The equilibrium altitude line (ELA) separates the two zones. The end of a glacier is called its terminus.

- Glaciers erode bedrock, forming U-shaped valleys, cirques, and other landforms.

- Drift is rock or sediment transported and deposited by a glacier. The unsorted drift deposited directly by a glacier is called till. Most glacial terrain is characterized by large mounds of till known as moraines. Terminal moraines, ground moraines, recessional moraines, lateral moraines, medial moraines, and drumlins are all depositional features formed by ice contact. Stratified drift consists of sediment first carried by a glacier and then transported, sorted, and deposited by streams. Outwash plains, kames, and eskers are composed of stratified drift. A kettle is a depression created when a large block of ice left behind by a retreating glacier melts.

- During the past 1 billion years, at least six major glaciations have occurred. The most recent is the Pleistocene Glaciation. One hypothesis contends that Pleistocene advances and retreats were caused by climate change induced by variations in Earth's orbit and the orientation of its rotational axis. Sea level falls when continental ice sheets form and rises again when the ice melts.

- The greatest glaciation in Earth's history occurred during late Precambrian time. Glaciers covered all continents and the seas froze over completely, entombing the globe in a shell of ice.

- Glaciers and sea ice are shrinking in more places and at more rapid rates than at any other time since scientists began keeping records. The melting of the Arctic and Antarctic ice has profound effects for the planet.

Important Terms

arête (p. 317)

basal slip (p. 311)

cirque (p. 315)

crevasse (p. 313)

drift (p. 318)

drumlins (p. 322)

eccentricity (p. 325)

end moraine (p. 320)

equilibrium altitude line, or ELA (p. 314)

erratics (p. 318)

esker (p. 324)

fjords (p. 318)

glacial striations (p. 314)

glaciation (p. 325)

glacier (p. 310)

ground moraine (p. 321)

hanging valley (p. 317)

horn (p. 316)

ice sheet, or continental glacier (p. 311)

ice shelf (p. 314)

icebergs (p. 314)

icefall (p. 313)

interglacial periods (p. 325)

kame and kettle topography (p. 324)

kame (p. 324)

kettles (p. 324)

lateral moraine (p. 322)

loess (p. 327)

medial moraine (p. 322)

outwash plain (p. 323)

outwash streams (p. 323)

outwash (p. 323)

paternoster lakes (p. 315)

plastic flow (p. 311)

Pleistocene Glaciation (p. 325)

pluvial lakes (p. 324)

precession (p. 326)

recessional moraine (p. 320)

stagnant ice (p. 324)

stratified drift (p. 318)

tarn (p. 315)

terminal moraine (p. 320)

terminus (p. 314)

till (p. 318)

tillite (p. 327)

tilt (p. 326)

U-shaped valley (p. 315)

zone of ablation (p. 314)

zone of accumulation (p. 313)

Review Questions

1. List the major steps in the metamorphism of newly fallen snow to glacial ice.

2. Differentiate between alpine glaciers and continental glaciers. Where are alpine glaciers found today? Where are continental glaciers found today?

3. Distinguish between basal slip and plastic flow.

4. Why are crevasses only about 40 meters deep, even though many glaciers are much thicker?

5. Describe the surface of a glacier in the summer and in the winter in the zone of accumulation and in the zone of ablation.

6. Describe how glacial erosion can create a cirque, a paternoster lake, and striated bedrock.

7. Describe the formation of arêtes, horns, and hanging valleys.

8. Distinguish among ground, recessional, terminal, lateral, and medial moraines.

9. Why are kames and eskers features of receding glaciers? How do they form?

10. What topographic features were left behind by the continental ice sheets? Where can they be found in North America today?

11. Discuss the evidence for the glaciation known as Snowball Earth.

14

DESERTS AND WIND

Wind erosion sculpted these bedrock spires in the Sahara Desert of Algeria, while wind deposition formed the sand dunes in the background and rippled sand in the lower left foreground. White-colored deposits of salt remain from a now-evaporated temporary lake to the right of the bedrock outcrop.

INTRODUCTION

Earth's continents are classified into climate zones based primarily on precipitation and temperature. In turn, climate determines the communities of plants and animals that live in a region. A **desert** is any region that receives less than 25 centimeters (10 inches) of rain per year and consequently supports little or no vegetation.[1] Most deserts are surrounded by semiarid zones that receive 25 to 50 centimeters of annual rainfall—more moisture than a true desert but less than adjacent regions.

Deserts cover 25 percent of Earth's land surface outside of the polar regions and make up a significant part of every continent. If you were to visit the great deserts of Earth, you might be surprised by their geologic and topographic variety. You would see coastal deserts along the beaches of Chile; shifting dunes in the Sahara; deep, red sandstone canyons in southern Utah; stark granite mountains in Arizona; and bitter-cold polar deserts with a few lichens clinging tenaciously to the otherwise barren rock. The world's deserts are similar only in that they all receive scant rainfall.

WHY DO DESERTS EXIST?

Rain and snow are unevenly distributed over Earth's surface. For example, Mount Wai'ale'ale, Hawaii, one of the wettest places on Earth, receives an average of 1,168 centimeters (38 feet) of rain annually. In contrast, 10 years or more may pass between rains or snowfalls in the Atacama Desert of Peru and Chile. Several factors—including latitude, mountains, and atmospheric circulation—control rainfall patterns and therefore the global distribution of deserts and semiarid lands.

Latitude

The Sun shines most directly overhead near the equator, warming air near Earth's surface. The air absorbs moisture from the equatorial oceans and rises because it is warmer, and therefore less dense, than surrounding air. Rising air cools as pressure decreases. But cool air cannot hold as much water as warm air, so the water vapor condenses and falls as rain (●**FIGURE 14.1**). For this reason, vast tropical rain forests grow near the equator.

This rising equatorial air, which is now drier because of the loss of moisture, flows

---

desert Any region that receives less than 25 centimeters (10 inches) of rain per year and consequently supports little or no vegetation.

---

1. The definition of desert is linked to soil moisture and depends on temperature and amount of sunlight in addition to rainfall. Therefore, the 25-centimeter criterion is approximate.

LEARNING OBJECTIVES

LO1 Summarize the characteristics of deserts.

LO2 Explain the existence of deserts.

LO3 Illustrate the interaction of the geosphere and atmosphere that creates the arid zone referred to as a rain shadow desert.

LO4 Discuss the role of water in the formation of desert features.

LO5 Distinguish between the formation of a bajada and a pediment.

LO6 Describe the importance of accumulating salt mineral precipitates in the desert.

LO7 Explain the formations of desert erosional features by lateral erosion.

LO8 Discuss the formation and the migration of dunes.

LO9 Compare the formation of different types of dunes to the amount of sand and vegetation.

LO10 Discuss the interactions between humans, weather, and loess.

LO11 Summarize the changes in the perceived causes of desertification in Africa and the latest research on the subject.

---

both northward and southward at high altitudes. There the air cools, becomes denser, and sinks back toward Earth's surface at about 30 degrees north and south latitudes. As the air falls, it is compressed and becomes warmer, which

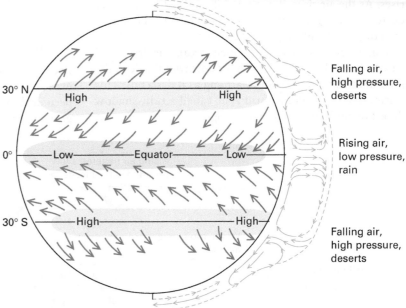

● **FIGURE 14.1** Falling air creates deserts at 30 degrees north and south latitudes. The red arrows indicate surface winds. The blue arrows on the right show airflow on the surface and at higher elevations.

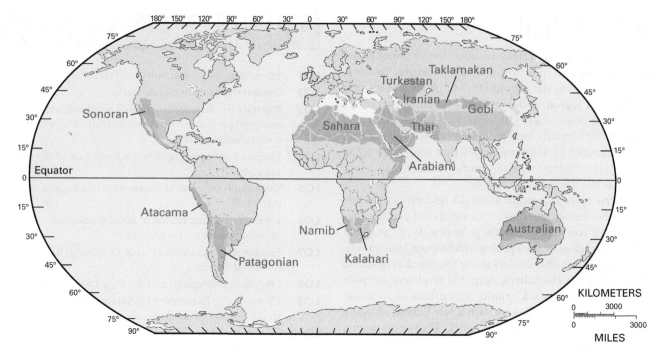

● **FIGURE 14.2** The major deserts of the world are concentrated at approximately 30 degrees north and south latitudes.

enables it to hold more water vapor. As a result, water evaporates from the land surface into the air. Because the sinking air absorbs water, the ground surface is dry and rainfall is infrequent. Thus, many of the world's largest deserts lie at about 30 degrees north and south latitudes (●**FIGURE 14.2**).

Mountains: Rain-Shadow Deserts

When moisture-laden air flows over a mountain range, it rises. As the air rises, it cools and its ability to hold water decreases. As a result, the water vapor condenses into rain or snow, which falls as precipitation on the windward side and on the crest of the range (●**FIGURE 14.3**). The dryer, cool air continues down the leeward (or downwind) side of the mountain, compressing and warming as it descends. This warm, dry air creates an arid zone called a **rain-shadow desert** on the leeward side of the range.

In this way, tectonic forces, which produce mountains, affect very long-term rainfall patterns. In turn, as we learned in Chapter 10, flowing water will weather rock, forming soil and defining the local environmental conditions on which various ecosystems exist. Thus we encounter yet another example of interacting Earth systems, because the building of a mountain range (a tectonic process) alters rainfall (an atmospheric process) and ultimately defines the types of ecosystems that characterize a region (the biosphere).

Coastal and Interior Deserts

Because most evaporation occurs over oceans, one might expect that coastal areas would be moist and climates would become drier with increasing distance from the sea. This is generally true, but a few notable exceptions exist.

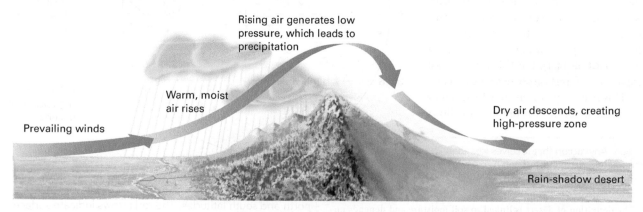

● **FIGURE 14.3** A rain-shadow desert forms where warm, moist air from the ocean rises as it flows over mountains. As it rises, it cools and water vapor condenses to form rain. The dry, descending air on the lee side absorbs moisture, forming a desert.

The Atacama Desert along the west coast of South America is so dry that portions of Peru and Chile often receive no rainfall for a decade or more. Cool ocean currents flow along the west coast of South America. When the cool marine air encounters warm land, the air is heated and expands. As it warms, the air absorbs moisture from the ground, creating a coastal desert.

The Gobi Desert is a broad, arid region in central Asia. The center of the Gobi lies at about 40°N latitude, and its eastern edge is a little more than 400 kilometers from the Yellow Sea. As a comparison, Pittsburgh, Pennsylvania, lies at about the same latitude and is 400 kilometers from the Atlantic Ocean. If latitude and distance from the ocean were the only factors, these regions would have similar climates. However, the Gobi is a barren desert and western Pennsylvania receives enough rainfall to support forests and rich farmland. The Gobi is bounded by the Tibetan Plateau to the south and the Altai and Tien Shan mountain ranges to the west, which shadow it from the prevailing winds. In contrast, winds carry abundant moisture from the Gulf of Mexico, the Great Lakes, and the Atlantic Ocean to western Pennsylvania.

Thus, in some regions deserts extend to the seashore, whereas in other regions the interior of a continent is humid. The climate at any particular place on Earth results from a combination of many factors. Latitude and proximity to the ocean are important, but complex interactions involving the direction of prevailing winds, the direction and temperature of ocean currents, and the positions of mountain ranges also control climate.

WATER AND DESERTS

Although rain and snow rarely fall in deserts, water plays an important role in these dry environments. Water can reach a desert from three sources. Streams flow from adjacent mountains or other wetter regions, bringing surface water to some desert areas. Groundwater may also flow from a wetter source to an aquifer beneath a desert. Finally, rain and snow fall occasionally on deserts. Thus, even in the driest places on Earth, the hydrosphere and geosphere interact.

Vegetation is sparse in most deserts because of the limited water supply; much bare soil is exposed, unprotected from erosion. As a result, rain can easily erode desert soils, and flowing water is an important factor in the evolution of desert landscapes.

Desert Streams

Large rivers flow through some deserts. For example, the Nile River flows through North African deserts and the Colorado River crosses the arid southwestern United

● **FIGURE 14.4** The Colorado River flows from the Rocky Mountains through the arid southwestern United States.

States (●**FIGURE 14.4**). Desert rivers receive most of their water from wetter, mountainous regions bordering the arid lands.

In a desert, the water table is commonly deep below a streambed, so that water seeps downward from the stream into the ground. As a result, many of the smaller desert streams flow only for a short time after a rainstorm, or during the spring when winter snows are melting, before their water seeps below the streambed. A streambed that is dry for most of the year is called a **wash** (●**FIGURE 14.5**).

Desert Lakes

In humid climates, the water table is relatively shallow, and groundwater can flow from saturated sediment or rock in the subsurface directly into a lake. In this situation, the lake surface projects into the subsurface as the water table, which separates the zone of saturation below from the zone of aeration above. In contrast, many desert lakes lie well above the water table.

During the wet season, rain and streams may fill a desert lake. Whereas some desert lakes are drained by outflowing streams, most lose water only by evaporation and seepage. In large desert lakes such as the Great Salt Lake in Utah and the Dead Sea in Israel and Jordan, evaporation has caused the remaining lake water to become very saline and has even resulted in the precipitation of salt within the lake basin (●**FIGURES 14.6** and **14.7**). In smaller desert lakes, inflowing streams may cease to flow during the dry season, causing the lake to dry up completely due to evaporation and seepage. This kind of intermittent desert lake is called a **playa lake**, and the dry lake bed is called a **playa** (●**FIGURE 14.8**).

Recall from Chapter 11 that streams and groundwater contain dissolved salts. When this slightly salty water fills a desert lake and then evaporates, the ions are left behind and salt minerals precipitate on the playa. Over many years, economically valuable mineral deposits, such as those of Death Valley, may accumulate (●**FIGURE 14.9**).

---

rain-shadow desert A desert formed on the downwind side of a mountain range.

wash A streambed that is dry for most of the year.

playa lake A temporary desert lake that dries up during the dry season.

playa The dry desert lake bed of a playa lake.

● **FIGURE 14.5** This small wash in the Grand Canyon is normally dry but prone to violent flash floods, as seen here.

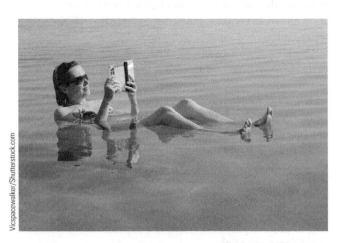

● **FIGURE 14.6** Evaporation has caused the Dead Sea in Israel and Jordan to become many times more salty than the ocean. The dissolved salts substantially increase the density of the lake water, so people swimming in the lake are able to float much higher on its surface than they can in freshwater. This woman is able to comfortably read a book while floating unassisted in the Dead Sea.

● **FIGURE 14.7** Evaporation of water from the Dead Sea has caused these salt terraces to precipitate along the sea's shoreline.

● **FIGURE 14.8** Mud cracks form in a dry playa in the Namib Desert of Namibia. A very large sand dune can be seen on the horizon.

● **FIGURE 14.9** Borax and other valuable minerals are abundant in the evaporite deposits of Death Valley. Mule teams hauled the ore from the valley in the 1800s.

Flash Floods

Bedrock or tightly compacted soil covers the surface of many deserts, and little vegetation is present to help absorb moisture. As a result, rainwater runs over the surface and collects in gullies and washes. During a rainstorm, a dry streambed may fill with water so rapidly that a **flash flood**—a brief, intense local flood—occurs (see Figure 14.5). Occasionally, novices to desert camping pitch their tents in a wash, where they find soft, flat sand to sleep on and shelter from the wind. However, if a thunderstorm occurs upstream during the night, a flash flood may fill the wash with a wall of water mixed with rocks and boulders, creating disaster for the campers. By midmorning of the next day, the wash may contain only a tiny trickle, and within 24 hours it may be completely dry again.

When rainfall is unusually heavy and prolonged, runoff flowing across a sloping desert soil will rapidly pick up sediment and evolve into a slurry-like debris flow. As its

---

flash flood A rapid, intense, local flood of short duration, usually following a rainstorm.

bajada A broad, gently sloping depositional surface formed by the merging of alluvial fans from closely spaced canyons and extending outward into a desert valley.

---

consistency thickens with the addition of the sediment, a debris flow can become powerful enough to carry large boulders and substantially damage or destroy buildings in its path. Some of the most expensive homes in Phoenix, Arizona, and other desert cities are built on alluvial fans and steep mountainsides, where they afford good views but are at risk from the effects of debris flows during wet years.

Pediments and Bajadas

Recall from Chapter 11 that when a steep, flooding mountain stream empties into a flat valley, the water slows abruptly and deposits most of its sediment at the mountain front, forming an alluvial fan. Although fans form in all climates, they are particularly conspicuous in deserts (●FIGURE 14.10). A large fan may be several kilometers across and rise a few hundred meters above the surrounding valley floor.

If the mouths of several canyons are spaced only a few kilometers apart, the alluvial fans extending from each canyon may merge. A **bajada** is a broad, gently sloping depositional surface formed by merging alluvial fans and extending into the center of a desert valley. Typically, the fans merge incompletely, forming an irregular surface that follows the mountain front for tens of kilometers. Over millions of years, the alluvial sediment forming the bajada may accumulate to a thickness of several kilometers.

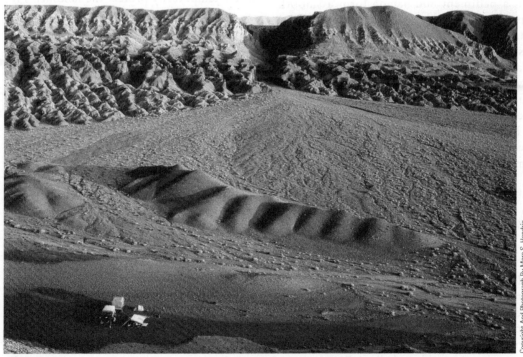

Copyright And Photograph By Marc S. Hendrix

● **FIGURE 14.10** An alluvial fan is a deposit of sediment formed where a relatively steep, confined stream enters a less confined region with a lower slope angle. Streams associated with alluvial fans usually are intermittent, frequently dry, and capable of transporting large volumes of sediment and large clasts quickly. The sediment is distributed across the fan surface as the stream frequently shifts its course. This alluvial fan in Mongolia formed where an intermittent stream (dry in this photograph) passes through a small, confining canyon and emerges into a more open, gently sloping plain.

● **FIGURE 14.11** The bajada in the foreground merges with a gently sloping pediment to form a continuous surface in front of these mountains in Mongolia. This basin is filling with sediment from the surrounding mountains because it has no external drainage.

A **pediment** is a broad, gently sloping erosional surface. Pediments commonly form along the front of desert mountains, because tectonic uplift results in erosion and, hence, development of the pediment. The surface of a pediment is covered with a thin veneer of gravel that is in the process of being transported from the mountains, across the pediment, to the bajada.

Together, a pediment and bajada form a smooth surface from the mountain front to the valley center (●**FIGURE 14.11**). The surface steepens slightly near the mountains, so it is concave up. To distinguish a pediment from a bajada, you would have to dig or drill a hole. If you were on a pediment when making the hole, you would strike the eroded rock or sediment after only a few meters. On a bajada, in contrast, hundreds or even thousands of meters of recent gravel would cover the older rock or sediment.

TWO AMERICAN DESERTS

The Colorado Plateau

The Colorado Plateau covers a broad region encompassing portions of Utah, Colorado, Arizona, and New Mexico (●**FIGURE 14.12**). During the past billion years of Earth's history, this region has been alternately covered by shallow seas, deserts, and broad alluvial plains with rivers and lakes. Sediment accumulated in these environments and slowly lithified to become flat layers of sedimentary rock. Tectonic forces later uplifted the lithified flat layers of rock over a broad region, forming

the Colorado Plateau. Today, much of the Colorado Plateau receives less than 25 centimeters of rainfall per year.

As the plateau rose, the Colorado River cut downward through the bedrock to form the 1.6-kilometer-deep Grand Canyon and its tributary canyons. The modern Colorado River receives most of its water from snowmelt and rain in the high Rocky Mountains east and north of the plateau, and the river then flows southwestward through the heart of the great desert to the Gulf of California, in northern Mexico, where it empties.

In addition to water, the Colorado River and its tributaries transport much sediment toward the Gulf of California. Uplift of the Colorado Plateau causes these streams to erode downward into bedrock. This downcuttting continues until a resistant layer is reached. As discussed in Chapter 11 in section on "Downcutting and Base Level," such a resistant layer will form a temporary base-level. Once the base-level is reached, the stream begins to erode laterally, widening the canyon by undercutting its walls and causing the rock to collapse along vertical joints. Continued lateral erosion and removal of the sediment by the river forms a relatively level plain with flat-topped mesas and buttes that rise above it as erosional remnants that are left behind.

The flat tops of mesas and buttes usually form on a bed of sandstone or other sedimentary rock that is relatively resistant to erosion (●**FIGURE 14.13**). These resistant upper layers temporarily protect the landforms from erosion,

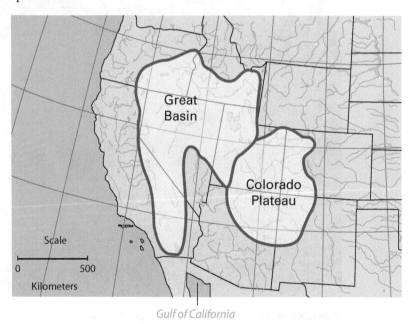

● **FIGURE 14.12** The Colorado Plateau and the Great Basin together make up a large desert and semiarid region of the western United States. The Colorado River flows through the Colorado Plateau to the Gulf of California, but no streams flow out of the Great Basin.

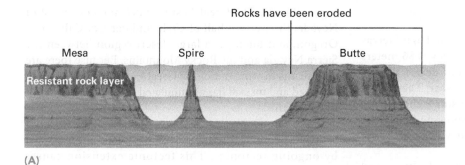

Rocks have been eroded

Mesa   Spire   Butte

Resistant rock layer

(A)

● **FIGURE 14.13** (A) Spires and buttes form when streams reach a temporary base level and erode laterally. The streams transport the eroded sediment away from the region. (B) Spires and buttes in Monument Valley, Arizona.

(B)

somewhat like a roof protects a building, although continued lateral erosion of the sides of the landform ultimately causes it to be removed. In many cases across the Colorado Plateau, extensive lateral erosion above a temporary base-level leaves spectacular pinnacles or spires, isolated remnants of once-continuous rock layers that the streams have all but completely eroded away. The huge volumes of sediment that are produced through this lateral erosion are transported away from the region by streams and ultimately are redeposited near the shoreline as part of a delta or carried further offshore to be deposited on a deep-water submarine fan. We will explore deltas and submarine fans in Chapter 15.

---

pediment A broad, gently sloping erosional surface that forms along the front of desert mountains uphill from a bajada, usually covered by a patchy veneer of gravel only a few meters thick.

plateau A large, elevated area of relatively flat land.

mesa A flat-topped mountain, shaped like a table, that is smaller than a plateau and larger than a butte.

butte A flat-topped mountain, smaller and more towerlike than a mesa, characterized by steep cliff faces.

This style of lateral erosion to a temporary base-level produces a wide variety of spectacular landforms that are common in the desert west where flat-lying layered sedimentary rocks are exposed at the surface. The largest of these features is a **plateau**, a region of fairly flat land that has been uplifted by tectonic forces and usually is made of relatively flat-lying sedimentary or volcanic rock. The Colorado Plateau is a good example, as it consists of relatively flat-lying sedimentary rocks that have been uplifted to elevations of several kilometers. The Colorado Plateau is surrounded by the North American Basin and Range extensional province and this extension is slowly nibbling away at the margins of the plateau. Other plateaus include the Columbia River Plateau—formed by the extensive flat-lying basalt flows of Cenozoic age (see Chapter 8) and the Tibetan Plateau, a large feature in central Asia resulting from the continent–continent collision between India and Asia (see Chapter 9). Smaller than a plateau is a flat-topped mountain shaped like a table and called a **mesa**—Spanish for the word "table." A **butte** is also a flat-topped mountain, smaller and more towerlike than a mesa, and characterized by steep cliff faces. Even smaller is a spire—a single tower with a pointy top. Mesas, buttes, and spires are all erosional landforms that are common in the Colorado Plateau.

Death Valley and the Great Basin

Death Valley lies in the rain shadow of the High Sierras in California. The deepest part of the valley is 86 meters below sea level. It is a classic rain-shadow desert, receiving a scant 5 centimeters of rainfall per year. The mountains to the west receive abundant moisture, and during the winter rainy season and spring snowmelt, streams flow from the mountains into the valley, where a playa lake forms.

On the Colorado Plateau, the Colorado River and its tributaries carry sediment away from the desert to the Gulf of California, leaving deep canyons. In contrast, streams flow into Death Valley from the surrounding mountains, but no streams flow out. Because Death Valley has no external drainage, the valley is filling with sediment eroded from the surrounding mountains. The sediment collects to form vast alluvial fans and bajadas. Stream water flows into a playa lake that dries up under the hot summer sun.

Death Valley is just one small part of a vast desert region in the American West that has no external drainage (see Figure 14.12). The Great Basin (which includes most of Nevada; the western half of Utah; and parts of California, Oregon, and Idaho) is a large desert region between the Sierra Nevada and the Rocky Mountains. Because there are no streams flowing out of the Great Basin, sediment is not removed and instead has accumulated to become thousands of meters thick.

Like the Colorado Plateau, the Great Basin currently is being uplifted, but it is also being pulled apart by ongoing tectonics. This tectonic extension causes normal faults to form that downdrop the valleys and uplift the mountains of the Great Basin. As we discussed in Chapter 9, the downdropped valleys are called *graben*, and the uplifted mountains are called *horsts* (●**FIGURE 14.14A**).

Desert streams have partially eroded the mountains and deposited sediment to form pediments, alluvial fans, and bajadas that today are very common in the Great Basin (●**FIGURE 14.14B**). With no streams carrying deposits out of the Great Basin, the mountains are drowning in their own sediment (●**FIGURE 14.14C**).

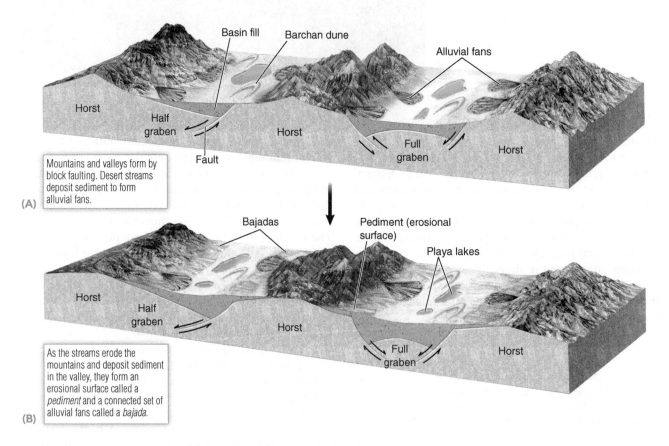

● **FIGURE 14.14** A scenario for the formation of bajadas and pediments. (A) The mountains and valleys form by block faulting. Desert streams deposit sediment to form alluvial fans. (B) As the streams erode the mountains and deposit sediment in the valley, they form both the erosional surface, called a *pediment*, and the depositional surface, called a *bajada*.

WIND

Just as water from the hydrosphere plays an important role in the desert environment, moving air of the atmosphere also shapes and sculpts the desert landscape. When wind blows through a forest or across a prairie, the trees or grasses shield the soil from wind erosion. Water dampens the soil and binds particles together, further protecting it from wind erosion. In contrast, a desert commonly has little or no vegetation and rainfall, so wind erodes bare, unprotected desert soil.

Wind erosion is not limited to deserts. Wind is an important agent of erosion wherever the wind blows over unvegetated soil. Windblown dunes of sand are common along seacoasts, where salty sea spray limits plant growth, and in regions recently left bare by receding glaciers.

Wind Erosion

Wind erosion, called **deflation**, is a selective process. Because air is much less dense than water, wind is capable of moving only small particles, generally those sand-sized and finer. Imagine bare soil containing silt, sand, pebbles, and cobbles. When wind blows, it removes only the silt and sand, leaving the pebbles and cobbles behind to form a continuous armoring of stones called a **desert pavement** (●**FIGURE 14.15A**). A desert pavement prevents the wind from eroding additional sand and silt, even though this finer sediment may be abundant beneath the stoney armor. Approximately 80 percent of the world's desert area is covered by a pavement of pebbles and cobbles (●**FIGURE 14.15B**), whereas only 20 percent is covered by sand.

Transport and Abrasion

Wind erosion is caused by sand-sized grains of sediment that are entrained in wind. When it is strong enough, wind can begin to roll and bounce sand grains across the ground surface. If the wind strengthens, some of the grains will be kicked upward then carried downwind a short distance before falling back to the surface, in a process called **saltation** (from the Latin verb *saltare*, "to dance"). The resulting trajectory of

---

deflation Erosion by wind.

desert pavement A continuous cover of closely packed gravel- or cobble-sized clasts left behind when wind erodes smaller particles such as silt and sand.

saltation The asymmetric jumping movement of sedimentary particles that are ejected off the bed through impact by another particle and are carried downstream by wind or water for some distance before falling back to the bed surface. Most sand grains in desert environments move via saltation by wind.

saltation carpet When a very strong wind blows across dry sand, so many grains saltate at once that a sand-air fluid is formed, usually a few centimeters or decimeters high. This sand-air fluid is dominated by grain-to-grain collisions and is unaffected by turbulence from the wind.

dune A mound or ridge of wind-deposited sand.

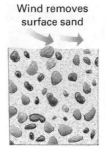
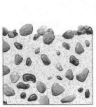

Wind removes surface sand

Formation of desert pavement complete—no further wind erosion

(A)

(B)

● **FIGURE 14.15** (A) Wind erodes silt and sand but leaves larger rocks behind to form desert pavement. (B) Stony desert pavement in Bladensburg National Park, Australia. This region is north of the Stony Sturt Desert in southern Queensland.

each grain is asymmetric, with a steeper ascent angle and a more shallow descent angle. When the wind is strong enough to cause many grains to saltate at one time, a **saltation carpet** is formed (●**FIGURE 14.16**). Such windblown, saltating sand is abrasive and capable of eroding bedrock. In some cases where hard bedrock such as quartzite is exposed, windblown sand can polish the bedrock surface. Because saltation carpets form next to the ground surface, wind erosion usually is most intense close to the ground. In rare cases, oddly shaped temporary spires can result in which a thin rock pedestal supports a larger mass above. Eventually, such "balanced rocks" will collapse as the pedestal continues to be undercut.

In contrast to sand-sized particles, smaller silt- and clay-sized particles can be carried by wind for much longer distances. Large storms in the Sahara, for example, can carry small particles in suspension in the atmosphere for hundreds of miles before they settle back to the surface. Skiers in the Alps commonly encounter a thin layer of silt on the snow surface resulting from wind transport of fine sediment from the Sahara across the Mediterranean Sea.

Dunes

A **dune** is an asymmetric mound or ridge of wind-deposited sand (●**FIGURE 14.17**). As explained earlier, wind removes

● **FIGURE 14.16** Sand being blown from right to left across sand dunes in a strong wind. Circled in red on the up-wind side of one dune is a saltation carpet in which the mixture of sand and air moves along the ground surface as a fluid that is not affected by turbulence from the wind above.

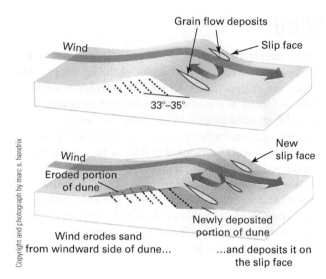

● **FIGURE 14.17** Sand dunes migrate in the downwind direction as sand is eroded off the upwind side of the dune, is transported to the brink of the dune, and then avalanches down the steep slip face on the downwind side in small grain flows.

sand from the surface in many deserts, leaving behind a stoney desert pavement. The wind commonly deposits such sand in a topographic depression or other place where the wind slows down. Dunes commonly grow to heights of 30 to 100 meters, and some giants exceed 500 meters. Most dunes are between about 100 meters and 1 kilometer across, although some dune ridges in the western Sahara are hundreds of kilometers long. The largest dune field on Earth is the Rub al-Khali ("empty quarter") in Arabia, which covers 560,000 square kilometers, larger than the state of California. Most dune fields cover a few square kilometers.

Dunes also form where glaciers have recently receded and along sandy coastlines. Glacial deposits consist of large quantities of bare, unvegetated sediment. A sandy beach is commonly unvegetated because sea salt prevents plant growth. As a result, both of these environments contain the essentials for dune formation: an abundant supply of sand and a windy environment with sparse vegetation.

Most dunes are asymmetrical. Sand eroded by wind from the windward side of the dune saltates up to the dune crest. From there, the sand moves down the steep leeward dune face, called the **slip face**, as a series of small, tongue-like grain flow deposits before coming to rest as an inclined layer of sand (Figure 14.17). Typically, the slip face dips about 35 degrees from horizontal—the angle of repose for dry sand. The inclined layers of sand that result from this process form cross-beds that reflect the original slope of the dune slip face (Figure 14.17).

Migrating dunes can overrun buildings and highways. For example, near the town of Winnemucca, Nevada, dunes advance across U.S. Highway 95 several times a year. Highway crews must remove as much as 4,000 cubic meters of sand (roughly half the volume of an Olympic-sized swimming pool) to reopen the road.

Engineers often attempt to stabilize dunes in inhabited areas. One method is to plant vegetation to reduce deflation and stop dune migration. The main problem with this approach is that desert dunes commonly form in regions that are too dry to support vegetation. Another solution is to build artificial windbreaks to create dunes in places where they do the least harm. For example, a fence traps blowing sand and forms a dune, thereby protecting areas downwind. Fencing is a temporary solution, however, because eventually the dune covers the fence and resumes its migration. In Saudi Arabia, dunes are sometimes stabilized by covering them with tarry wastes from petroleum refining.

TYPES OF SAND DUNES Wind speed and sand supply control the shapes and orientation of dunes. A **barchan dune** is one of four common types of dunes and forms in rocky deserts where there is a limited supply of sand. The center of the dune grows higher than the edges (●**FIGURE 14.18A**) but migrates forward more slowly, causing the dune to become crescent shaped, with its tips pointing downwind (●**FIGURE 14.18B**). Barchan dunes are not generally connected to one another but instead migrate independently.

If sparse desert vegetation is present, the wind may form a small depression called a blowout in a bare area among the desert plants. As sand is carried out of the blowout, it accumulates in a **parabolic dune**, the tips of which are anchored by plants on each side of the blowout (●**FIGURES 14.19A** and **14.19B**). Parabolic dunes are common in moist semidesert regions and along seacoasts, where sparse vegetation grows in the sand.

In environments where sand is plentiful and evenly dispersed, it accumulates in long ridges called **transverse dunes** aligned perpendicular to the prevailing wind (●**FIGURE 14.20**).

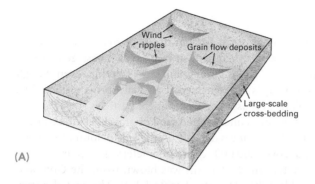

(A)

(B)

Courtesy of Graham R. Thompson/Jonathan Turk

● **FIGURE 14.18** (A) When sand supply is limited, the tips of a barchan dune travel faster than the center and point downwind. Barchan dunes typically are covered by wind ripples on their upwind side and are characterized by a steep slip face in which small grain flows transport sand to the toe of the slip-face. Large-scale cross-beds that typically represent the steeper downwind side of the dunes can be preserved in sediment deposited as part of barchan dunes. (B) Barchan dunes in the Southern California desert. The prevalent wind direction is from right to left.

---

slip face The steep leeward side of a dune, typically at the angle of repose for loose sand, so that the sand flows or slips down the face, where it is deposited.

barchan dune A crescent-shaped dune, highest in the center, with the tips pointing downwind; typically forms in rocky deserts where there is a general shortage of sand.

parabolic dune A crescent-shaped dune with tips pointing into the wind; forms in moist semidesert regions and along seacoasts where sparse vegetation is present to anchor the tips of the dune.

transverse dunes A relatively long, straight dune with a gently sloping windward side and a steep lee face that is perpendicular to the prevailing wind; forms where sand is plentiful and evenly dispersed.

longitudinal dunes A long, symmetrical dune that forms as a result of two different wind directions with comparable magnitude.

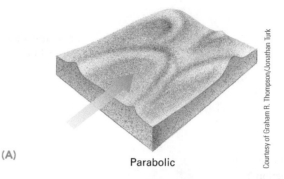

(A)   Parabolic

Courtesy of Graham R. Thompson/Jonathan Turk

(B)

Orxy/Shutterstock.com

● **FIGURE 14.19** (A) A parabolic dune is crescent shaped, with its tips pointing upwind. It forms where wind blows sand from a blowout, and grass or shrubs anchor the dune tips. (B) Small shrubs anchor the tips of these parabolic dunes in the Namib Desert.

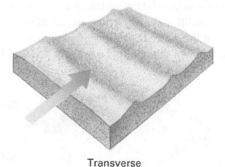

Transverse

● **FIGURE 14.20** Transverse dunes form perpendicular to the prevailing wind direction in regions with abundant sand.

If two different wind directions prevail throughout the year and both are of comparable magnitude, then long, straight **longitudinal dunes** can form. In portions of the western Sahara Desert, longitudinal dunes reach 100 to 200 meters in height and are as much as 100 kilometers long. Depending on the angle between the two prevailing wind directions, longitudinal dunes form either perpendicular or parallel to the average annual wind direction (●**FIGURE 14.21**).

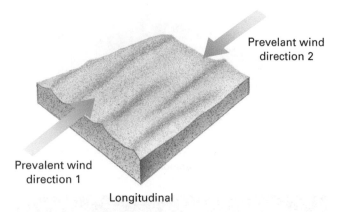

FIGURE 14.21 Longitudinal dunes are long, straight dunes that form when nearly opposite wind directions with comparable magnitude prevail.

FOSSIL DUNES Sedimentary structures are common when sand dunes are buried and lithified over geologic time to become fossil dunes. For example, ●**FIGURE 14.22** shows a rock face in Zion National Park, located on the Colorado Plateau in Utah. The beds in Figure 14.22 appear to be tilted; however, recall from the discussion above that sedimentary rocks on the Colorado Plateau are mostly flat-lying. In this case, the steep layers of sandstone were not produced by tectonic tilting but rather are cross-beds that represent the original layers of the dune slip face. The cross-beds dip in the direction that the wind was blowing when it deposited the sand.

Notice that the cross-beds in Figure 14.22 dip in different directions. The variety of dip directions results in part from the curved shape of each cross-bed surface within the dune. Each cross-bed surface is curved because it is the lithified slip-face of the original dune which itself was curved (Figure 14.18A). In addition, however, the variety of dip directions suggests that the wind direction shifted somewhat through time as the ancient sand dunes were being deposited.

Loess

Wind can carry silt for hundreds or even thousands of kilometers and then deposit it as **loess**. Loess generally is composed of silt, is porous, and lacks layering. Even though loess usually is not cemented, it typically forms vertical cliffs and bluffs because the fine sediment particles stick together more than does uncemented sand or gravel (●**FIGURE 14.23**).

The largest loess deposits in the world, found in central China, cover 800,000 square kilometers and are more than 300 meters thick. The silt was blown from the Gobi and the Taklamakan Deserts of central Asia. The particles stick together so effectively that people have dug caves into the loess cliffs to make their homes. However, in 1920, a great earthquake caused the cave system to collapse, burying and killing an estimated 100,000 people.

Thick loess deposits accumulated in North America during the most recent Pleistocene glacial advance, when continental ice sheets ground bedrock into silt. Streams carried this fine sediment from the melting glaciers and deposited much of it along their banks and floodplains. These environments were cold, windy, and devoid of vegetation, and wind easily picked up and transported the silt, depositing thick layers of loess as far south as Vicksburg, Mississippi. During the Civil War's Battle of Vicksburg, a siege that lasted for over seven weeks, residents of Vicksburg took shelter from federal bombardment in nearby loess caves (●**FIGURE 14.24**).

Loess deposits in the United States range from about 1.5 meters to 30 meters thick (●**FIGURE 14.25**). Soils formed on loess are generally fertile mollisols that support vast natural grassland ecosystems and make good farmland. Much of the rich soil of the central plains of the United States and eastern Washington State formed on loess (●**FIGURE 14.26**).

loess A homogenous, porous deposit of windblown glacial silt, typically unlayered, that forms vertical bluffs and cliffs. Fertile, agriculturally productive soils—usually mollisols—commonly form on loess.

FIGURE 14.22 This cross-bedded sandstone in Zion National Park preserves the sedimentary bedding of ancient sand dunes.

FIGURE 14.23 This home in central China was dug from the thick deposits of loess that occur there.

(A)

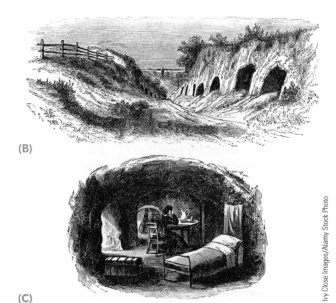

(B)

(C)

● **FIGURE 14.24** During the U.S. Civil War, the city of Vicksburg, Mississippi, was the key to controlling navigation on the Mississippi River. Beginning in May and continuing through July 1863, the city was under siege by Union troops commanded by General U.S. Grant. Union gunboats continually shelled the town, and many of its residents dug caves in the soft loess to seek refuge. (A) A small neighborhood of caves at Vicksburg as they appeared during the siege. B) Sketches of several caves dug and used during the Vicksburg siege. C) A domestic scene inside one of the hand-dug caves during the siege.

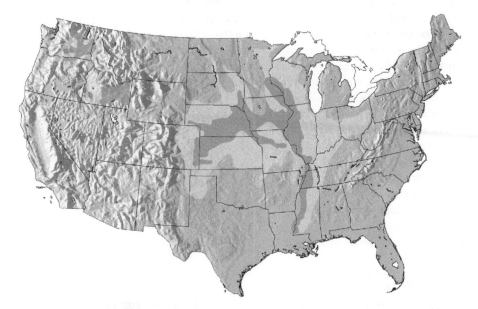

● **FIGURE 14.25** Loess deposits cover large areas of the United States.

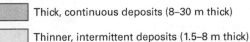 Thick, continuous deposits (8–30 m thick)

Thinner, intermittent deposits (1.5–8 m thick)

● **FIGURE 14.26** Much of central and eastern Washington State is covered by several meters of loess. The rolling hills seen here are characteristic of silt dunes formed by wind deposition and subsequently modified by surface erosion. Agriculturally productive mollisol soils have formed on the loess. These soils and the rich, silty character of loess on which they form support the production of large wheat and legume crops.

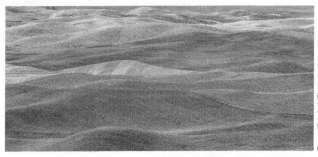

The North American "Dust Bowl"

Although loess can form rich farmland, its uniformly fine size makes it capable of being easily eroded by wind, especially when farming breaks up the topsoil and exposes the loess. Arguably the worst human-caused environmental disasters in U.S. history, the infamous dust bowl of the 1930s resulted from years of dryland farming practices in the North American mid-continent region, followed by a severe drought.

In the early part of the twentieth century, thousands of Americans were drawn to the southern Great Plains to begin a new life in what the U.S. government described as "the last frontier of agriculture." The migration resulted from a combination of generous federal farm policies intended to settle the southern Plains, unusually high rainfall, and high wheat prices brought on by World War I. The result was "the Great Plow-Up" in which an estimated 5.2 million acres of native grassland—roughly the combined size of Connecticut and Delaware —was converted from thick, native grassland into wheat fields (●FIGURE 14.27). The native grasslands and mollisol soils that were plowed under to create the wheat fields took thousands of years to become established and were very effective at stabilizing the fine-grained soil and loess against the wild weather variations and high winds typical of the Great Plains.

Following the stock market crash of 1929, the nation sank into depression and wheat prices plummeted from about $2/bushel to about 40 cents/bushel. Many farmers responded by increasing the size of their wheat fields in the hopes of overcoming the reductions in price by expanding the volume of wheat

they produced. As the increased wheat supply further eroded prices, many farmers simply abandoned their farms, leaving the former grasslands behind as exposed, dry fields.

Compounding the situation was a long-term, severe drought across the region. Not only did the drought prevent the native grasslands from re-establishing but the accompanying winds rapidly swept up the fine particles of soil and loess, producing massive dust storms that buried farms, choked livestock and humans, and removed much of the soil that had taken so long to develop. Thousands of farmers and their families from the Dakotas to Texas went bankrupt and left (●FIGURE 14.28).

Although farming practices in the Great Plains and other loess-covered regions such as central Washington State have changed since the 1930s, these

regions are still at risk of losing additional valuable soil and a repeat of the Dust Bowl is not out of the question. (●FIGURE 14.29). Today, dryland crops are rotated and the fields are terraced so that the furrows do not run downhill. Fields are regularly rested to help the soil recover between crops. Irrigation helps to stabilize the soil, although as described in Chapter 12, the primary Great Plains Aquifer—the Ogallala Aquifer —is rapidly being depleted and irrigation in other dryland farming regions is under similar pressure. Average annual temperatures continue to warm, requiring more irrigation and longer recovery periods between crops. As more pressure to produce falls on the "breadbasket of the nation," these headwinds will only intensify and it will be up to future generations to come up with solutions to avoid a repeat of the Great Dust Bowl of the 1930s.

● **FIGURE 14.27** A farmer drives his tractor along with his young son near Cland, New Mexico in 1938.

(A)

(B)

(C)

(D)

● **FIGURE 14.28** Scenes from the North American "Dust Bowl" of the 1930s. (A) A dust storm rolls into Rolla, Kansas, in 1935. (B) Children preparing to head off to school, Lakin, Kansas, 1935. (C) Farming equipment buried by wind-blown sand and silt, Dallas, South Dakota, May 1936. (D) A Dust Bowl farmer and his son raising a fence so it does not become buried by drifting sand in Oklahoma.

● **FIGURE 14.29** A wheat farmer disks his field in the fall in Washington State. The soil in this region is made of loess and supports a native grassland ecosystem. Despite modern farming practices designed to minimize soil erosion, the dry conditions, frequently strong winds, and agricultural need to disk the soil result in erosion as seen here.

DESERTIFICATION

The Sahara is the largest desert on the planet. South of the Sahara lies the semiarid Sahel (●**FIGURE 14.30**). During the 1960s, unusually heavy rains caused the Sahel to bloom. People expanded their flocks to take advantage of the additional forage. With foreign aid from rich countries, medical attention and sanitation in the area improved, and the human population grew dramatically. Many people predicted a new era of prosperity for the Sahel, but the favorable rains were an anomaly. In the late 1960s and early 1970s, drought destroyed the region. During this period, governments in North Africa began to enforce national borders more strictly, curtailing nomadism. When people settled in specific regions, their flocks grazed the same area throughout the year. Plants did not have time to regenerate, and the hungry animals chewed the grasses down to the roots. Civil and international war brought instability to the region, and famine struck.

Reports issued in the 1970s and 1980s claimed that the Sahara was expanding southward into the Sahel at a rate of 5 kilometers per year. Scientists argued that overgrazing, farming, and firewood gathering had caused the desert expansion. This growth of the desert caused by human mismanagement has been called **desertification**.

More recent research has shown that overgrazing a semiarid region causes land degradation but does not cause a desert to expand. The Sahel–Sahara desert boundary is clearly visible on satellite photographs as a boundary separating a region of sparse vegetation from one with almost no plants. Researchers plotted changes in the size of the desert by studying satellite photographs taken between 1980 and 1990. As shown in ●**FIGURE 14.31**, the desert expanded (blue line) when rainfall declined (red line).[2] Thus, decreasing rainfall—not overgrazing—may have been responsible for expansion of the Sahara.

A more recent study in 2018[3] confirmed this conclusion by showing that over the period 1902–2013, the size of the region receiving less than 100 mm of rain per year—the official definition of a desert—grew by 16% during the winter and 11% during the summer. For reference, the 16% wintertime increase in the size of the Sahara between 1902 and 2013 exceeds the combined size of Alaska and California!

The 2018 study found that during the summer, the rainfall boundary between the southern Sahara and the Sahel expanded southward, whereas in the winter the northern boundary of the Sahara expanded northward. The study also found that over the course of the twentieth century, precipitation during the rainy season (spring to fall) in the headwaters of the Niger River declined by as much as 25%. ●**FIGURE 14.32** shows the resulting average monthly declines in stream discharge (vertical black bars) for a stream monitoring station located in the Niger River headwaters. Similarly, significant drops in rainy season precipitation over the twentieth century were observed in the headwaters of the Blue Nile and Congo Rivers.

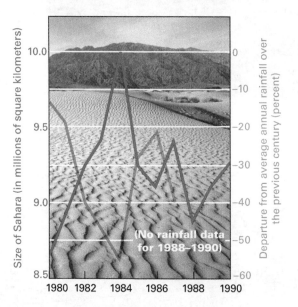

● **FIGURE 14.31** The Sahara Desert expanded and contracted between 1980 and 1990. Note that the Sahara expanded (blue line) when rainfall decreased (red line).

2. William H. Schlesinger, James F. Reynolds, Gary L. Cunningham, Laura F. Huenneke, Wesley M. Jarrell, Ross A. Virginia, and Walter G. Whitford, "Biological Feedbacks in Global Desertification," *Science* 247 (March 1990): 1043–1048.

3. Natalie Thomas and Sumant Niga, "Twentieth-Century Climate Change over Africa: Seasonal Hydroclimate Trends and Sahara Desert Expansion," *Journal of Climate* 31 (1 May 2018): 3349-3370.

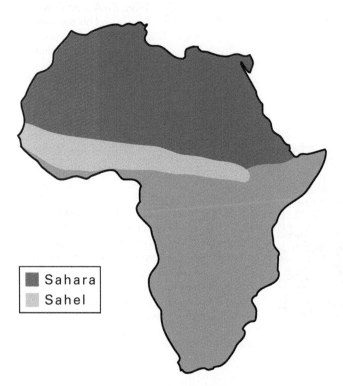

● **FIGURE 14.30** The Sahara Desert and the semiarid Sahel region dominate the ecosystems of northern Africa.

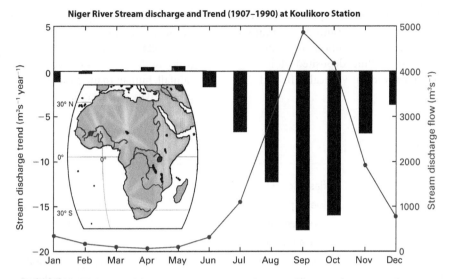

● FIGURE 14.32 Monthly stream discharge values (red line, right axis) and discharge trends (black bars, right axis) for the period 1907–1990 at a monitoring station located in the headwaters of the Niger River (red dot).

Source: Natalie Thomas and Sumant Niga, "Twentieth-Century Climate Change over Africa: Seasonal Hydroclimate Trends and Sahara Desert Expansion," *Journal of Climate* 31 (1 May 2018): 3349-3370.

Despite these century-long trends, which the authors of the 2018 study ascribe to global warming and associated changes in ocean and atmospheric circulation patterns, massive agricultural aid to the region has led to improved farming and herding practices in some regions. Farmers have learned land management techniques specific to desert environments, which has helped keep the land usable despite fluctuations in rainfall. As a result, thousands of acres of land have been rehabilitated, and millet and sorghum yields have increased significantly. This story reminds us once again that, while a desert ecosystem is created by low rainfall, human management can significantly alter the productivity of the land.

desertification A process by which semiarid land is converted to desert by human mismanagement or by climate change.

Key Concepts Review

- Deserts have an annual precipitation of less than 25 centimeters. The world's largest deserts occur near 30 degrees north and south latitudes, where warm, dry, descending air absorbs moisture from the land. Deserts also occur on the leeward side (in rain shadows) of mountains, in continental interiors, and in coastal regions adjacent to cold ocean currents.
- Desert streams are often dry for much of the year, but flash floods may occur during a rainstorm. Playa lakes are desert lakes that dry up periodically, leaving abandoned lake beds called playas. Alluvial fans in desert environments may be several kilometers across and rise a few hundred meters above the surrounding valley floor. A bajada is a broad depositional surface formed by merging alluvial fans. A pediment is a broad, gently sloping erosional surface that forms along the front of desert mountains and that merges imperceptibly with a bajada.

- The Colorado River drains the Colorado Plateau desert. Thus, streams carry sediment away from the region, forming canyons and eroding the Colorado Plateau to form mesas and buttes. Death Valley and the Great Basin have no external drainage and, as a result, the valleys are filling with sediment eroded from the surrounding mountains.
- Deflation is erosion by wind. Silt and sand are removed selectively, leaving larger stones on the surface and creating desert pavement. Sand grains are relatively large and heavy and are carried only short distances by saltation, seldom rising more than a meter above the ground. Silt can be transported great distances at higher elevations. Wind erosion forms blowouts. Windblown particles are abrasive, but because the heaviest grains travel close to the surface, abrasion occurs mainly near ground level. A mound or ridge of wind-deposited sand is called a dune. Most dunes are asymmetrical, with

gently sloping, windward sides and steeper slip faces on the lee sides. Dunes migrate. The various types of dunes include barchan dunes, transverse dunes, longitudinal dunes, and parabolic dunes. Wind-deposited silt is called loess, which typically forms vertical cliffs and bluffs.

- The Sahara is the largest desert on the planet. South of the Sahara lies the semiarid Sahel. Low rainfall has been responsible for the expansion of the Sahara into the Sahel. Still, while a desert ecosystem is created by low rainfall, human management can significantly alter the productivity of the land.

Important Terms

bajada (p. 337)
barchan dune (p. 342)
butte (p. 339)
deflation (p. 341)
desert (p. 333)
desertification (p. 348)
desert pavement (p. 341)
dune (p. 341)

flash flood (p. 337)
loess (p. 344)
longitudinal dunes (p. 343)
mesa (p. 339)
parabolic dune (p. 342)
pediment (p. 338)
plateau (p. 339)
playa (p. 335)

playa lake (p. 335)
rain-shadow desert (p. 334)
saltation (p. 341)
saltation carpet (p. 341)
slip face (p. 342)
transverse dunes (p. 342)
wash (p. 335)

Review Questions

1. Why are many deserts concentrated along zones at 30 degrees latitude in both the Northern Hemisphere and the Southern Hemisphere?
2. List three factors that control global distribution of deserts.
3. Why do flash floods and mudflows occur in deserts?
4. Why are alluvial fans more conspicuous in deserts than in humid environments?
5. Compare and contrast floods in deserts with floods in more humid environments.
6. Compare and contrast pediments and bajadas.
7. Why is wind erosion more prominent in desert environments than it is in humid regions?
8. Describe the formation of desert pavement.
9. Describe the evolution and shape of a dune.
10. Describe the differences among barchan dunes, transverse dunes, parabolic dunes, and longitudinal dunes. Under what conditions does each type of dune form?
11. Compare and contrast desert plateaus, mesas, and buttes. Describe the formation of each.
12. Compare the effects of stream erosion and deposition in the Colorado Plateau and Death Valley.

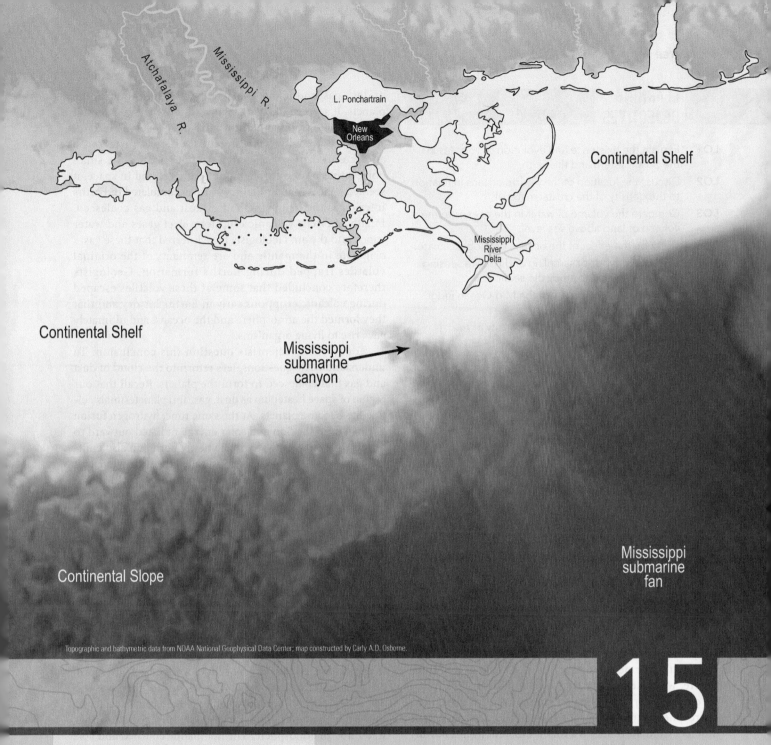

Atchafalaya R.

Mississippi R.

L. Ponchartrain

New Orleans

Continental Shelf

Mississippi River Delta

Continental Shelf

Mississippi submarine canyon

Continental Slope

Mississippi submarine fan

Topographic and bathymetric data from NOAA National Geophysical Data Center; map constructed by Carly A.D. Osborne.

15

OCEAN BASINS

An image of the seafloor extending from the Mississippi River Delta shoreline southward to the abyssal plain in the central Gulf of Mexico. The broad, relatively shallow water of the continental shelf passes further offshore into the deeper water of the continental slope. A major submarine canyon cuts across the slope and funnels sediment into even deeper water. The hummocky seafloor topography of the continental slope is the result of numerous kilometer-scale blobs of rock salt (halite), called diapirs, that are slowly flowing upward toward the seafloor and deforming it in the process. Offshore of this zone of salt movement is the deeper water of the abyssal plain.

THE ORIGIN OF OCEANS

The primordial Earth, heated by the impacts of colliding planetesimals and the decay of radioactive isotopes, was molten, or near molten. The sky, without an atmosphere, was black. There were no oceans—no life. Today, we consider Earth in terms of four spheres: the geosphere, hydrosphere, atmosphere, and biosphere. Each sphere is as different from the others as a rock is different from a flowing stream, a breath of air, or a butterfly.

For the moment, let's abandon our view of Earth's four spheres and think of only two kinds of Earth compounds: compounds that are volatile, or can evaporate rapidly and easily escape into the atmosphere, and compounds that are not. Water is a good example of a volatile compound; carbon dioxide is another. Most scientists agree that the surface of primordial Earth consisted mainly of rock and partially molten rock and contained few volatile compounds. How, then, did our thick atmosphere, vast oceans, and a global biosphere of living organisms form?

For many years, geologists hypothesized that abundant volatiles, including water and carbon dioxide, were trapped within early Earth's interior. This reasoning was based on three observations and inferences. First, cosmogenic models showed that volatiles were evenly dispersed in the cloud of dust, gas, and planetesimals that coalesced to form the planets. It seemed likely that some of those volatiles would have become trapped within Earth as it formed. Second, scientists detected volatiles in modern comets, meteoroids, and asteroids. If volatiles were trapped within the small objects that passed through our neighborhood in space, it seemed logical to infer that they also accumulated in Earth's interior as the original cloud of dust and gas coalesced. Finally, modern volcanic eruptions eject gases and water vapor into the air. Geologists have inferred that these gases originate in the mantle and are remnants of the original volatiles trapped during Earth's formation. Geologists therefore concluded that some of these volatiles escaped during volcanic eruptions early in Earth's history and that they formed the atmosphere and the oceans, and ultimately gave rise to living organisms.

Today, many scientists question this conclusion. To understand their questions, let's return to the cloud of dust and gas that coalesced to form the planets. Recall that our region of space heated up as dust, gas, and planetesimals collided to become planets. At the same time, hydrogen fusion began within the Sun and solar energy radiated outward to heat the inner Solar System. The newly born Sun also emitted a stream of ions and electrons, called the solar wind, that swept across the inner planets, blowing their volatile compounds into outer regions of the Solar System. As a result, most of Earth's volatile compounds essentially boiled off and were swept by the solar wind into the cold outer regions of the Solar System (●**FIGURE 15.1**).

According to a currently popular hypothesis, shortly after our planet formed and lost its volatiles, a Mars-sized object smashed into Earth. The cataclysmic impact blasted through the crust and deep into the mantle, ejecting huge quantities of pulverized rock into orbit. The fragments eventually coalesced to form the Moon. The impact also ejected most of Earth's remaining volatiles with enough velocity that they escaped Earth's gravity and disappeared into space. According to this hypothesis, Earth's surface then was left barren and rocky, with few volatiles either on the surface or in the deep mantle. Thus, Earth's early surface had neither water nor an atmosphere. The hot mantle churned and volcanic eruptions repaved the planet with lava, but these events brought relatively few volatiles to Earth's surface. According to one estimate, outgassing of the deep mantle accounted for no more than 10 percent of Earth's hydrosphere, atmosphere, and biosphere.[1]

If this scenario is correct, why do modern volcanoes release volatiles? According to one hypothesis, most of the gases given off by modern volcanoes are recycled from the surface. Water, carbon (in the form of carbonate rocks such as limestone), and other light compounds are carried into shallow parts of the mantle by subducting plates. There, the volatiles are incorporated into magma that forms within the

---

1. Paul J. Thomas, Christopher F. Chyba, and Christopher P. McKay, *Comets and the Origin and Evolution of Life* (New York: Springer-Verlag, 1997).

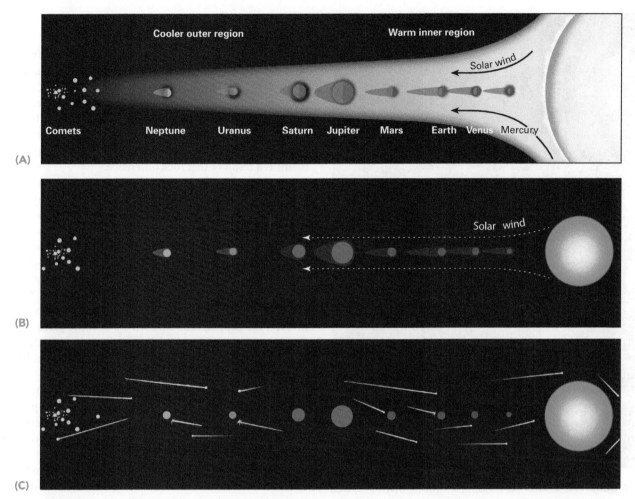

● **FIGURE 15.1** (A) As the Solar System formed, volatiles boiled away from the inner planets. The solar wind then blew them into the cooler region beyond Mars. (B) The outer planets captured some of the volatiles, and some condensed to form comets in the frigid zone beyond Neptune. As a result, the inner four planets were left with little or no atmosphere or ocean and the outer planets grew to become gaseous giants. (C) Comets and meteorites crashed into Earth and other planets, returning some of the volatiles to their surfaces. Over time, the volatiles accumulated to form the atmosphere and oceans and therefore the foundation of life on Earth. (Sizes and distances in this drawing are not to scale.)

volcanic arc associated with the subduction zone. The volatiles are returned to the surface when volcanoes forming the arc erupt. Thus, modern volcanic eruptions, like their primordial ancestors, do not deliver significant quantities of volatiles from the deep mantle to Earth's surface.

Now let's return to the volatiles that streamed away from the hot, inner Solar System. As they flew away from the Sun, volatiles entered a cooler region beyond Mars. Most of the volatiles were captured by the outer planets—Jupiter, Saturn, Uranus, and Neptune—but some continued their journey toward the outer fringe of the Solar System (●FIGURE 15.1B). There, beyond the orbits of the known planets, volatiles from the inner Solar System combined with residual dust and gas to form comets (●FIGURE 15.2). A comet's nucleus has been

compared to a dirty snowball because it is composed mainly of ice and rock. However, other compounds not common in snowballs exist in comets as well. These include frozen carbon dioxide, ammonia, and simple organic molecules. Volatiles are also abundant in certain types of meteoroids and asteroids in the region between Mars and Jupiter.

Astronomers calculate that the early Solar System was crowded with comets, meteoroids, and asteroids—space debris left over from planetary formation. Much of this material contained volatile compounds. When a large piece of space debris crashes into a planet, it is called a **bolide**. Early in the formation of the Solar System, a large number of bolides crashed into Earth, nearby planets, and moons. In this way, the bolides delivered and deposited a large mass of volatile compounds to the inner Solar System (●FIGURE 15.1C). While falling space debris added less than 1 percent to Earth's total mass, it imported roughly 90 percent of its modern reservoir of volatiles.

bolide A large piece of space debris, such as an asteroid, that crashes into a planet.

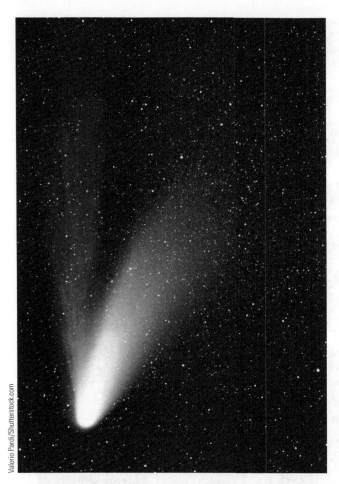

Valerio Pardi/Shutterstock.com

● **FIGURE 15.2** Comet Hale–Bopp, shown here in 1997, was probably the most widely observed comet of the twentieth century. The early Solar System was crowded with comets, meteoroids, and asteroids; these bodies are composed of rock and condensed volatile materials such as ice and solid carbon dioxide.

Upon entry and impact, the frozen volatiles in the bolides vaporized, releasing water vapor, carbon dioxide, ammonia, simple organic molecules, and other volatile compounds. As the planet cooled and atmospheric pressure increased, the water vapor condensed to liquid, forming the first oceans. The light molecules transported to Earth in bolides also provided gases that formed the atmosphere and the raw materials for life. (The formation and evolution of the atmosphere is discussed in Chapter 17.)

Thus, at least some, and probably most, of the compounds necessary to produce the hydrosphere, the atmosphere, and the biosphere traveled to Earth from outer regions of the Solar System. The water that fills Earth's oceans came from interplanetary space.

Later in Earth's history, impacts from outer space blasted rock and dust into the sky, causing mass extinctions and killing large portions of life on Earth. (See Chapter 4 for more on mass extinctions.) Ironically, extraterrestrial impacts may have provided the raw materials for the oceans, the atmosphere, and for life—but they later caused mass extinctions.

THE EARTH'S OCEANS

If you were to ask most people to describe the difference between a continent and an ocean, they would almost certainly reply, "Why, obviously, a continent is land and an ocean is water!" This observation is true, of course, but to a geologist what is more important is that the rock beneath the ocean water is different from the rock beneath the land surface. The accumulation of seawater in the world's ocean basins is a result of that rock difference.

Modern oceanic crust is dense basalt and varies from 4 to 7 kilometers thick. Continental crust is made of lower-density granite and averages 20 to 40 kilometers in thickness. In addition, the entire continental lithosphere is both thicker and less dense than oceanic lithosphere. As a result of these differences, the thick, lower-density continental lithosphere floats isostatically at high elevations, whereas oceanic lithosphere sinks to low elevations. Most of Earth's water flows downhill, to collect in the vast depressions formed by oceanic lithosphere. Even if no water existed on Earth's surface, oceanic crust would form deep basins and continental crust would form regions of higher elevation.

Oceans cover about 71 percent of Earth's surface. The seafloor is about 5 kilometers deep in the central parts of the ocean basins, although it is only 2 to 3 kilometers deep above the Mid-Oceanic Ridge and plunges to 11 kilometers in the Mariana Trench (●**FIGURE 15.3**).

The ocean basins contain 1.4 billion cubic kilometers of water—18 times more than the volume of all land above sea level. So much water exists at Earth's surface that if Earth were a perfectly smooth sphere, it would be covered by a global ocean 2,000 meters deep.

The size and shape of Earth's ocean basins have changed over geologic time. At present, the Atlantic Ocean is growing wider at a rate of a few centimeters each year as the seafloor spreads apart at the Mid-Atlantic Ridge and as the Americas move away from Europe and Africa. At the same time, the Pacific is shrinking at a similar rate, as oceanic crust sinks into subduction zones around its edges. In short, the Atlantic Ocean basin is now expanding at the expense of the Pacific.

Global climate is profoundly influenced by the immense volume of Earth's oceans and the fact that the oceans consist of liquid water. On average, the ocean is about 3,800 meters deep, yet the thermal capacity of Earth's entire atmosphere is equivalent only to the uppermost 2.5 meters of ocean water (less than 0.05%). Ocean currents transport heat from the equator toward the poles, cooling equatorial climates and warming polar environments. The seas also absorb and store solar heat more efficiently than do rocks and soil. As a result, oceans are commonly warmer in winter and cooler in summer than adjacent land is. Most of the water that falls as rain or snow is water that evaporated from the seas. In these and other ways, the oceans have played a major role in controlling Earth's climate and the distribution of different climate zones through geologic time.

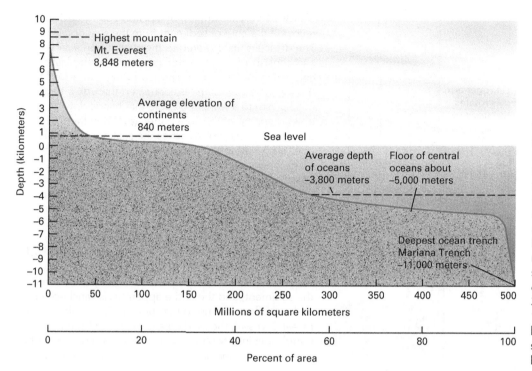

● **FIGURE 15.3** A schematic cross section of the continents and ocean basins. The vertical axis shows elevations relative to sea level. The horizontal axis shows the basic breakdown of Earth's main topographic surfaces. Thus, for example, roughly 30 percent, or about 150 million square kilometers, of Earth's surface lies above sea level.

STUDYING THE SEAFLOOR

Seventy-five years ago, scientists had better maps of the Moon than of the seafloor. The Moon is clearly visible in the night sky, and we can view its surface with a telescope. The seafloor, however, is deep, dark, and inhospitable to humans. Modern oceanographers use a variety of techniques to study the seafloor, including several types of sampling and remote sensing.

Sampling

Several devices collect sediment and rock directly from the ocean floor. A **rock dredge** is an open-mouthed steel net dragged along the seafloor behind a research ship. The dredge breaks rocks from submarine outcrops and hauls them to the surface along with whatever other sediment and biota might be scooped up. Oceanographers sample seafloor mud by lowering a piston coring device to the bottom (●**FIGURE 15.4**). This device consists of a steel core barrel with a sharp bit on the lower end, removable plastic tubing that fits inside the hollow core barrel, a large weight attached to the top of the barrel, and a trigger-and-piston mechanism. The trigger is released when the tip of the core barrel is only a meter or two above the seafloor, allowing the weighted core barrel to free-fall to the bottom, where it plunges bit-first into the mud. As the weight drives the core barrel downward, the piston inside the tubing is simultaneously drawn upward by a cable, in order to make way for and suck in the mud core. Once the barrel has stopped settling into the mud (usually less than a minute), the entire device is yanked out of the bottom sediments, winched to the surface, and brought on-board. There, scientists extracted, split, and describe the core.

Because piston coring involves driving the core barrel downward into the mud in one swift action, the length of the core itself is always limited by the strength of the barrel and its ability to penetrate downward into the sediment rather than fold over as the weight drives it downward. Most piston cores are not over 10 meters long. In contrast, **seafloor drilling** methods developed for oil exploration can recover continuous cores from seafloor sediment as well as lithified rock that are hundreds of meters long. Large drill rigs are mounted on offshore platforms and on research vessels. The drill cuts cylindrical cores from sediment and rock, which are then brought to the surface for study.

Remote Sensing

Remote sensing methods do not require direct physical contact with the ocean floor, and for some studies this approach is both effective and economical. The **echo sounder** is an instrument commonly used to map seafloor topography. It emits a sound signal from a research ship and then records

---

rock dredge An open-mouthed steel net dragged along the seafloor behind a research ship for the purpose of sampling rocks and sediment from the ocean floor.

seafloor drilling A process in which drill rigs mounted on offshore platforms or on research vessels cut cylindrical cores from both the sediment and rock of the seafloor. Following their extraction, the cores are brought to the surface for study.

echo sounder An instrument that emits timed sound waves that reflect off the seafloor, return, and are recorded; the data are used to measure water depth and define the topography of the seafloor.

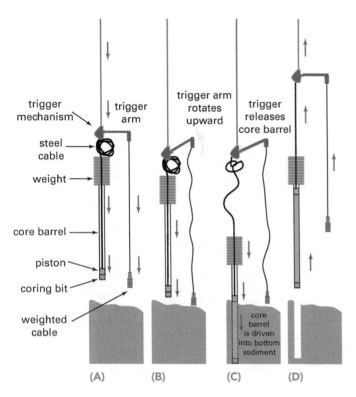

trigger
mechanism

trigger
arm

trigger arm
rotates
upward

trigger
releases
core barrel

steel
cable

weight

core barrel

piston

coring bit

weighted
cable

core
barrel
is driven
into bottom
sediment

(A)   (B)   (C)   (D)

● **FIGURE 15.4** Illustration of a piston-coring device recovering a core of sediment from the seafloor. (A) The device is first lowered to the seafloor on a cable and consists of a steel piston core barrel with plastic core lining inside and a heavy lead weight attached to the top. The core barrel is connected through a trigger mechanism to a weighted cable that hangs down beyond the tip of the core barrel. (B) When the end of the weighted cable reaches the seafloor, the cable goes slack, allowing the spring-loaded trigger arm to rotate upward and release the core barrel. The core barrel and weight free fall a few meters to the seafloor. (C) The inertia of the falling, weighted core barrel drives it into the sediment of the seafloor. As the core barrel penetrates downward into the sediment, the piston within the core lining is pulled upward by a cable, making room for the sediment core and helping to suck it into the core barrel. (D) After a minute or so, friction stops the core barrel from penetrating any further into the sediment, and the entire apparatus is winched out of the seafloor and up to the surface. Once onboard the research vessel, the mud core is extracted from the core barrel, split into two halves, and described by scientists.

Source: Adapted from http://c2fn.dt.insu.cnrs.fr/spip/IMG/jpg/carottier_kullenberg_et_schema.jpg

DIGGING DEEPER

Invaluable sediment archives recently recovered from the deep ocean floor

Japanese scientists set world record for longest sediment-core recovery in response to 2011 Tōhoku earthquake/ tsunami disaster

Although the use of drilling platforms and research vessels to extract continuous cores of sediment from the deep ocean floor is expensive, the core sediments themselves represent the physical record of geologic history for that location and, as such, can be invaluable. Far from simply recovering a long tube of sediment or cylinder of rock for study, modern state-of-the-art deep-sea drilling involves simultaneously conducting numerous measurements of the physical conditions of the sediment and rock being cut by the drill bit. The simultaneous drilling and measuring or *logging* of these physical conditions, called logging while drilling (LWD) technology, provides real-time

information about the sediment and rock being drilled, including its temperature and pressure, natural radioactivity, and ability to conduct electricity. LWD technology also provides real-time information about the drilling process, such as the drilling rate, orientation of the drill string, and circulation of fluids needed to drill the hole.

An excellent example of the tremendous potential scientific and societal value of seafloor drilling, logging, and coring is the spring 2012 LWD campaign of the Japan Trench conducted by scientists aboard the Japanese research vessel *Chikyu*. The LWD work that took place in April 2012 was in direct response to the devastating Tōhoku earthquake and tsunami that struck only 13 months earlier on March 9, 2011. (See Chapter 7 for more on this earthquake and tsunami and

the fault rupture that produced both.) The overarching research question behind the LWD program carried out aboard the *Chikyu* was this: "What mechanisms control the occurrence of destructive earthquakes, landslides, and tsunami?" More specifically, the research team sought to understand what controls the state of stress across faults that rupture to produce very large earthquakes and to assess whether all of this stress had been released during the 2011 Tōhoku Earthquake. In addition, the researchers sought to characterize the physical conditions associated with the main plate boundary fault that ruptured during the 2011 seismic event, using LWD technology in combination with recovery of continuous sediment/rock core and deployment of a permanent suite of temperature and pressure data sensors.

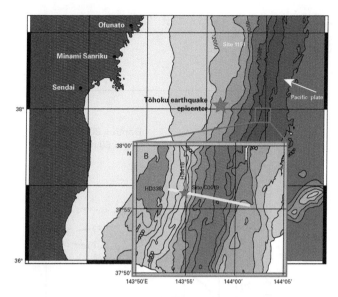

● **FIGURE 15.5** Location map of the 2012 *Chikyu* deep-sea drilling site—the Tōhoku earthquake epicenter of March 9, 2011—and the seismic reflection profile shown in Figure 15.7.

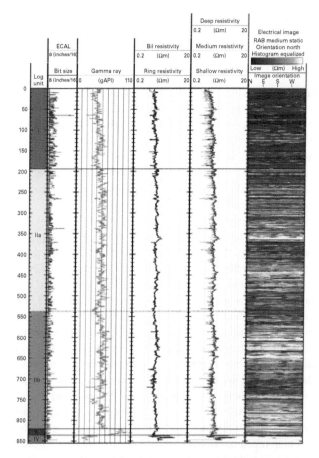

● **FIGURE 15.6** Example of data collected from sediment and rock cut during the drilling process. This "logging while drilling" (LWD) technology provides onboard scientists with real-time information about the geology being reached by the tip of the drill string as drilling is taking place.

After departing from the Japanese port of Shimizu on April 1, 2012, the *Chikyu* arrived at the drilling site (located about 250 kilometers east of the Sendai peninsula) two days later (●**FIGURE 15.5**). While en route, the crew used the time to organize and prepare for deployment of the many sections of drill pipe that would be strung together to recover the core. After arriving on site and waiting out two days of bad weather, the crew deployed the ship's powerful stabilizers, which were needed to maintain the ship at a constant location and position throughout the drilling process. The stabilizers are controlled by computers that constantly monitor the ship's absolute position using a high-precision GPS system.

On April 25, 2012, Japanese scientists aboard the *Chikyu* used a deep-sea drilling rig to recover a core 850 meters long from the outer portion of the subduction complex containing the fault that ruptured to produce the 2011 earthquake and tsunami. In doing so, the scientists set a world record by using a drill pipe with a total underwater length of 7.7 kilometers to recover the core in water 5.9 kilometers deep.

Both the hole drilled and logged by the Japanese team and the core recovered from the hole provide direct access to the fault that ruptured to produce the devastating Tōhoku earthquake. An east–west-oriented seismic reflection profile shot before the drilling campaign clearly shows the main fault rupture surface within sediments of the subduction complex, in addition to imaging the basaltic top of the downgoing Pacific Plate and several normal faults that offset it (●**FIGURE 15.6**). The normal faults are produced by extensional stress in the top of the downgoing slab as it is being bent downward into the subduction zone and are unrelated to the Tōhoku earthquake.

Examples of the data collected during and after the drilling process are shown on the following page.

Figure 15.6 is an example of the graphical data resulting from the logging process. Not only were various physical attributes of the sediment measured, such as its natural radioactivity, but an image of the sediment itself was captured by a camera system used in the logging process.

The fault that ruptured on March 9, 2011, was reached by the *Chikyu*'s drill bit at a total depth of ~820 meters below the seafloor. The fault puts a sequence of deformed, clayey to silty mudstones containing abundant terrigenous sediment and volcanic ash over a relatively undeformed sequence of pelagic mudstones inferred to have

(Continued)

Invaluable sediment archives recently recovered from the deep ocean floor (Continued)

been deposited in deep water on the Pacific Plate (●FIGURE 15.7). The fault itself is actually a 5-meter-thick fault zone consisting of highly sheared clay with numerous centimeter-scale curved fault surfaces that give the clay a scaly appearance.

After extracting the sediment core from the drillhole, the scientists installed an array of extremely sensitive temperature and pressure sensors in order to measure the residual frictional heat left over from the fault rupture, to better quantify the amount of energy released by the earthquake, and to assess whether all energy was released by the fault rupture—which was estimated to involve 50 meters of offset at the

drill site. An example of the temperature data is shown in ●FIGURE 15.8. The temperature peak coincides with the fault surface and represents heat still dissipating from the rupture.

Though it will take scientists many years to slowly sift through the data generated by the Chikyu's LWD campaign and analyze the cores extracted from deep within the subduction complex, the information generated by this historic expedition is sure to reveal numerous additional insights into the geology behind the Tōhoku earthquake while furthering the scientific understanding of how extremely large earthquakes occur.

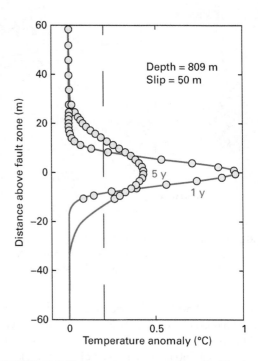

● **FIGURE 15.8** Temperature sensors installed in the borehole show a clear temperature spike across the ruptured fault zone. The higher temperatures result from heat still dissipating from frictional resistance during the 2011 fault rupture. These data help geoscientists understand how energy is released during fault ruptures that produce very large earthquakes.

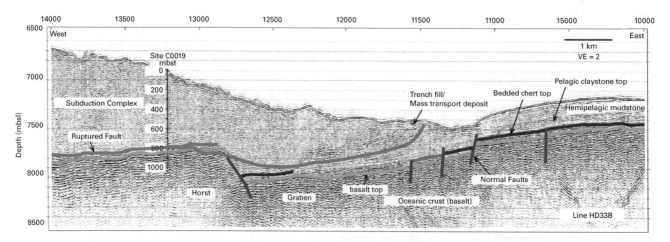

● **FIGURE 15.7** Seismic reflection profile showing the main geologic features associated with the Chikyu's drill site. The top of the basaltic oceanic crust of the downgoing Pacific Plate is clearly imaged on the profile, as is the ruptured fault surface that produced the Tōhoku earthquake. The fault was intersected by the Chikyu's drill bit approximately 820 meters below the seafloor.

the signal after it bounces off the seafloor and travels back up to the ship. The water depth is calculated from the time required for the sound to make the round trip. A topographic map of the seafloor is constructed as the ship steers a carefully navigated course with the echo sounder operating continuously. Modern echo sounders, called sonar, transmit 1,000 signals at a time to create more-complete and accurate maps.

A **seismic reflection profiler** works in the same way but uses a higher-energy signal that penetrates and reflects from layers of sediment and rock below the seafloor (●**FIGURE 15.9**). This geophysical technique provides an image of the layering and structure of seafloor sediments, the rock layers and crust below, and major structures such as folds and faults that might be present.

●**FIGURE 15.10** is an example of a modern seismic reflection profile that was shot by the USGS in shallow water off the coast of central California. The red and blue layers in the image represent layers of sediment and sedimentary rock below the seafloor. Notice the presence of two major strike-slip fault zones, associated with the transform plate boundary between the North American and Pacific Plates.

---

seismic reflection profiler A device that emits a high-energy timed seismic signal that penetrates into and reflects from layers of sediment and rock beneath the seafloor; the data are used to construct an image of rock and sediment layers along with other geologic structures below the seafloor.

A different type of remote sensing that involves the reflection of compression waves off the seafloor is side scan sonar imaging. This technique involves towing a heavy, torpedo-shaped vessel called a "tow-fish" behind a powerful ship. Inside the tow-fish is an array of transducers that emit, then receive, an acoustic pulse of energy of a specific frequency. ●**FIGURE 15.11A** is a cartoon showing a tow-fish emitting and receiving sonar information as it is being towed behind the research ship. Notice that the zone through which the acoustic pulse is sent and received is very narrow in the towing direction but very wide in the direction perpendicular to the towing track. This configuration is designed to maximize the return of sonar energy along the seafloor. As the ship travels and the tow-fish sends and receives sonar pulses with this configuration, an image called a sonogram is stitched together showing the topography along the seafloor (●**FIGURE 15.11B**). Rougher areas return more sonar energy and so are bright on the sonogram, whereas smooth areas or shadows result in little sonar energy return and so are darker on the sonogram.

Very similar to side scan, sonar imaging is multibeam imaging. Instead of involving a tow-fish, multibeam imaging systems are mounted directly next to the hull of the research vessel (●**FIGURE 15.12A**). Multiple frequencies of acoustic sonar beams are emitted and received, and the resulting signal is processed on-board and shows the topographic information directly below the ship as it moves (●**FIGURE 15.12B**). A major advantage of the multibeam system is that bubbles that form in the wake of the ship do not affect the sonar sending and receiving system because it is not being towed behind the ship.

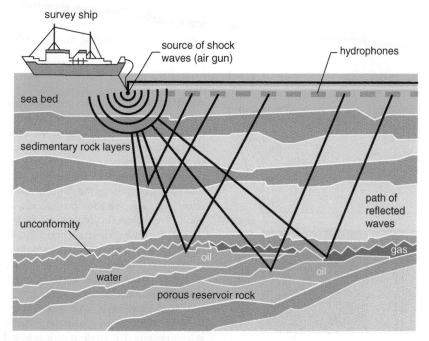

● **FIGURE 15.9** Seismic reflection profiling involves emitting source energy in the form of sound waves and recording that portion of the sound energy that reaches an array of detectors. The length of time required for the sound energy to travel downward to a particular layer of rock below the seafloor, bounce off of it, and travel farther to the detector depends on the depth of the rock layer and the velocity with which the sound waves travel to and from the layer. By using highly engineered source emitters and detectors, and carefully processing the returned sound recording through a computer, a seismic reflection profile is produced that is a visual representation of the geology below the seafloor as shown by example in Figures 5.7 and 5.10.

Source: Adapted from http://www.corelab.com/ps/cms/images/low-frequency-measurements.jpg

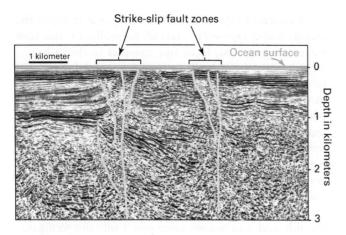

FIGURE 15.10 An example seismic reflection profile; note the horizontal and vertical scales. This profile was shot by the USGS offshore central California on the continental shelf. The red and blue lines in the profile reflect subsurface layers of sediment or rock from which significant returning sound energy was detected. Notice that the profile crosses two fault zones, both of which are associated with the transform boundary between the North American and Pacific Plates.

Source: http://walrus.wr.usgs.gov/mapping/csmp/images/seismic_multichannel.jpg

A **magnetometer** is an instrument that measures a magnetic field. Magnetometers towed behind research ships measure the magnetism of seafloor sediment and rock. Data collected by shipborne magnetometers resulted in the now-famous discovery of symmetric magnetic stripes on the seafloor. That discovery rapidly led to the development of the seafloor spreading hypothesis and of the theory of plate tectonics shortly thereafter, as described in Chapter 6.

Satellite-based **microwave radar** instruments measure the echo of microwave pulses to detect subtle swells and depressions on the sea surface. These features reflect seafloor topography. For example, the mass of a seafloor mountain 4,000 meters high creates sufficient gravitational attraction to produce a gentle 6-meter-high swell on the sea surface directly above it. The microwave data are used to make seafloor maps.

FEATURES OF THE SEAFLOOR
The Mid-Oceanic Ridge System

The discovery of the Mid-Oceanic Ridge system took place over the course of several decades and was possible only by scientific analysis of large volumes of oceanographic data, much of which was collected for military purposes. Following World War I (1914–1918), oceanographers began using early versions of echo-sounding devices to measure ocean depths. Those surveys showed that the seafloor was much more rugged than previously thought, and they further identified the continuity and size of the Mid-Atlantic Ridge.

During World War II, naval commanders needed topographic maps of the seafloor to support submarine warfare. Those detailed maps, made with early versions of

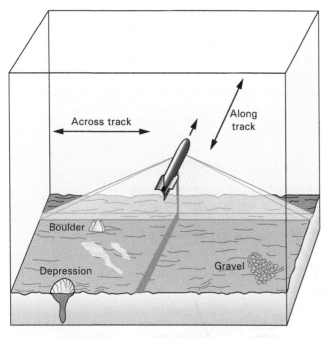

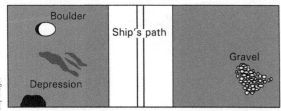

(A)

Ralph White/Corbis Documentary/Getty Images

(B)

FIGURE 15.11 Side scan sonar imaging. (A) A torpedo-shaped "tow-fish" is towed behind a research vessel. As both the vessel and tow-ship move, a pulse of acoustic sonar energy (sometimes called a "chirp") is emitted, then received, by an array of transducers inside the tow-fish. Some of the sonar energy reflects off the seafloor and is returned to the tow-fish that measures the direction and magnitude of the returning pulse. The zone covered in a single "chirp-listen" cycle—highlighted in yellow in the figure to the left—is very short in the towing direction and very wide in the direction perpendicular to towing. (B) A sonogram is the image made by stitching together multiple individual cycles of emitting and receiving the sonar pulse. The sonogram shows the topography along the seafloor. Rougher surfaces reflect more sonar energy back to the tow-fish and are shown in bright shades. Smoother surfaces and shadows are shown in darker shades. This image is a composite of numerous sonograms shot by a remotely controlled submersible that used side scan sonar. Shown in the image is the bow section of the RMS *Titanic*, which sunk in the North Atlantic in the early hours of April 15, 1912, after hitting an ice berg. Also shown in the image is a portion of the debris field surrounding the wreck.

Source: https://woodshole.er.usgs.gov/operations/sfmapping/sonar.htm

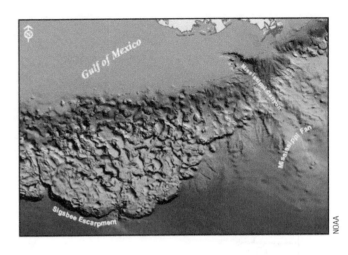

(A)

(B)

● **FIGURE 15.12** Multibeam sonar imaging. (A) Pulses of sonar energy of multiple different frequencies are emitted, then received, by an array of transducers located directly against the hull of the research vessel. Some of the sonar energy reflects off the seafloor and is returned to the ship and detected. As with side scan sonar, the zone covered in a single "chirp-listen" cycle— highlighted in yellow above—is very short in the towing direction and very wide in the direction perpendicular to towing. (B) By assembling the results from numerous multibeam surveys, scientists can produce a terrain model of the seafloor as shown in this example from the Gulf of Mexico. The tip of the Mississippi River Delta is depicted in the upper right, and the Mississippi Canyon and Mississippi Fan are shown on the right side of the image. The rest of the image shows the transition from the continental shelf shown in warm brown colors down the continental slope to the deep abyss on the floor of the Gulf of Mexico, shown in dark blue colors. The Sigsbee Escarpment marks the toe of the continental slope. The numerous pock-marks on the continental slope show mounds and circular basins caused by the mobilization of salt within the slope sediments.

the echo sounder, were kept secret by the military. When they became available to the public after peace was restored, scientists were surprised to learn that the ocean floor has at least as much topographic diversity and relief as the continents do. Broad plains, high peaks, and deep valleys form a varied and fascinating submarine landscape. In the 1950s, oceanographic surveys conducted by several nations led to the discovery that the Mid-Atlantic Ridge is just part of a great submarine mountain range, now called the Mid-Oceanic Ridge system.

Recall from Chapter 6 that the Mid-Oceanic Ridge system is a continuous submarine mountain chain formed by the rifting apart of two oceanic plates. The Mid-Ocean Ridge system encircles the globe and has a total length exceeding 80,000 km (●**FIGURE 15.13**). In some places it is more than 1,500 kilometers wide. The ridge system rises an average of 2 to 3 kilometers above the surrounding deep seafloor because heat from the rifting causes rocks in and around the plate boundary to be hot. The hot rocks of the ridge system take up

more volume so float higher in the underlying asthenosphere. Although the Mid-Ocean Ridge lies almost exclusively beneath the sea surface, it is Earth's longest continuous mountain chain.

A **rift valley** is an elongated depression that develops at a divergent plate boundary (●**FIGURE 15.14**). In the Mid-Oceanic Ridge system, a rift valley 1 to 2 kilometers deep and several kilometers wide splits many segments of the ridge crest. Oceanographers in small research submarines have dived into the rift valley, where they have documented gaping vertical cracks up to 3 meters wide on the valley floor. In some cases, extremely hot hydrothermal water highly concentrated with dissolved metals and metal-bearing compounds streams out of the cracks.

To understand the geologic significance of such cracks, recall that the Mid-Oceanic Ridge system is a spreading center, where two lithospheric plates are moving apart from each other. The cracks form as tensile stress at the ridge axis causes brittle oceanic crust to separate. Basaltic magma then rises through the resulting crack and flows onto the floor of the rift valley. This basalt becomes new oceanic crust as two lithospheric plates spread outward from the ridge axis.

The new crust (and the underlying lithosphere) at the ridge axis is warmer, and therefore of relatively lower density, than older crust and lithosphere located farther from the ridge axis. Its buoyancy causes it to float high above the surrounding seafloor, elevating the Mid-Oceanic Ridge system 2 to 3 kilometers above the deep seafloor. The new lithosphere cools as it spreads away from the ridge. As a result of cooling, the lithosphere shrinks and cracks as it becomes denser and sinks to lower elevations, forming the deeper seafloor on both sides of the ridge (●**FIGURE 15.15**).

magnetometer An instrument that measures the strength and, in some cases, the direction of a magnetic field.

microwave radar A satellite-based instrument that measures the travel time of microwave pulses that reflect off of the sea surface. The processed data allow for the detection of subtle swells and depressions on the sea surface, which are controlled by seafloor topography and can be used to map the seafloor.

rift valley An elongate depression that develops at a divergent plate boundary. Examples include continental rift valleys and the rift valley along the center of the Mid-Oceanic Ridge system.

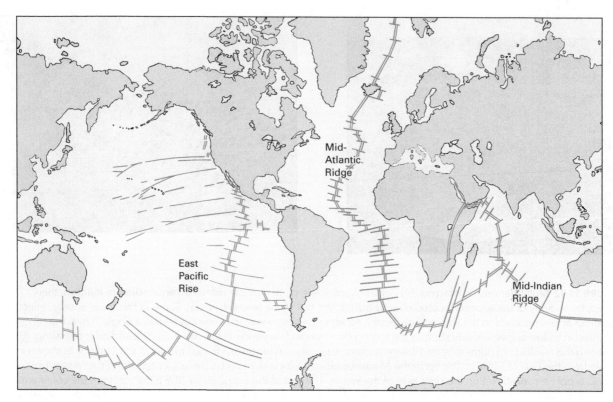

● **FIGURE 15.13** This map shows that divergent plate boundaries, or spreading centers, coincide exactly with the Mid-Oceanic Ridge system in the world's oceans. The spreading centers are shown in double red lines; the single red lines are transform faults.

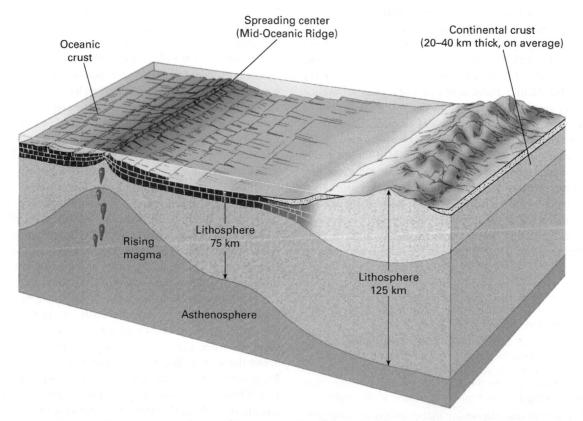

● **FIGURE 15.14** A cross-sectional view of the central rift valley in the Mid-Oceanic Ridge. As the plates separate, blocks of rock drop down along the fractures to form the rift valley, bounded by normal faults. Movements across these faults cause earthquakes.

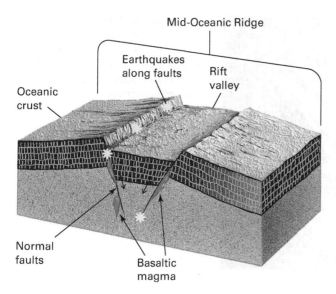

Mid-Oceanic Ridge

Earthquakes along faults

Rift valley

Oceanic crust

Normal faults

Basaltic magma

● **FIGURE 15.15** The seafloor sinks as it grows older. At the Mid-Oceanic Ridge, new lithosphere is buoyant because it is hot and of low density. As it moves away from the ridge, the lithosphere cools, thickens, and becomes denser, causing it to sink. On average, the seafloor lies at a depth of about 4 kilometers, relative to 2 to 3 kilometers at the Mid-Ocean Ridge.

Normal faults and shallow earthquakes are common along the Mid-Oceanic Ridge system because oceanic crust fractures as the two plates separate. Blocks of crust drop downward along some of the seafloor cracks, forming faults that bound the rift valley.

Thus, the cracks documented along the mid-ocean spreading ridge by oceanographers can be explained as the result of three primary causes: (1) separation of rock by tectonic tension associated with the spreading ridge; (2) decrease of rock volume by the cooling and contraction of newly formed basalt crust on the ocean floor; and (3) normal faulting in which rock is physically displaced across a crack.

On the volcanically active Mid-Oceanic Ridge system, hot rocks heat seawater as it circulates through fractures in oceanic crust. The hot water dissolves metals and sulfur from the rocks. Eventually, the hot metal-and-sulfur-laden water rises back to the seafloor surface, spouting from fractures as a jet of cloudy, black water called a black smoker (described in Chapter 5). The black color is caused by precipitation of fine-grained metal sulfide minerals as the solutions cool on contact with seawater (●**FIGURE 15.16**).

These scalding, sulfurous waters are as hot as 400°C, and seemingly would produce a sterile environment in which nothing could survive. Yet, remarkably, the deep seafloor around a black smoker teems with life. At the vents, bacteria produce energy from hydrogen sulfide in a process called **chemosynthesis**. Thus, the bacteria release energy from

OAR/National Undersea Research Program (NURP); NOAA

● **FIGURE 15.16** A black smoker spouts from the East Pacific Rise. Seawater is heated as it circulates through the hot rocks of the rift zone, and it dissolves metals and sulfur from the rocks. The ions precipitate as "smoke," consisting of tiny mineral grains, when the hot suspension spews into cold ocean water.

chemicals and are not dependent on photosynthesis. The chemosynthetic bacteria are the foundation of a deep-sea food chain: either larger vent organisms eat them or the larger organisms live symbiotically with the bacteria. For example, instead of a digestive tract, the red-tipped tube worm (●**FIGURE 15.17**) has a special organ that hosts the chemosynthetic bacteria. In return for providing a home for the bacteria, the worm receives nutrition from the bacteria's wastes. Other vent organisms in this unique food chain include giant clams and mussels, eyeless shrimp, crabs, and fish.

In addition to cracks or faults located along the axis of the mid-oceanic ridges and rift valleys are hundreds of fractures, called **transform faults**, that cut across the rift valleys and ridges (●**FIGURE 15.18**). These fractures extend through the entire thickness of the lithosphere and develop because the Mid-Oceanic Ridge system consists of many short segments. Each segment is slightly offset from adjacent segments by a transform fault. Transform faults are original features of the Mid-Oceanic Ridge; they form as an accommodation to Earth's spherical shape when lithospheric spreading begins.

Some transform faults displace the ridge by less than a kilometer, but others offset the ridge by hundreds of kilometers. In some cases, a transform fault can grow so large that it forms a transform plate boundary. The San Andreas Fault in California is a transform plate boundary that offsets both oceanic and continental crust.

---

chemosynthesis A process in which bacteria produce energy from hydrogen sulfide or other inorganic compounds and thus are not dependent on photosynthesis.

transform faults A strike-slip fault between two offset segments of a mid-ocean ridge or along a strike-slip plate boundary.

● **FIGURE 15.17** These red tubeworms are part of a thriving plant and animal community living near a submarine hydrothermal vent in the Guaymas Basin in the Gulf of California.

NOAA Okeanos Explorer Program, Galapagos Rift Expedition 2011

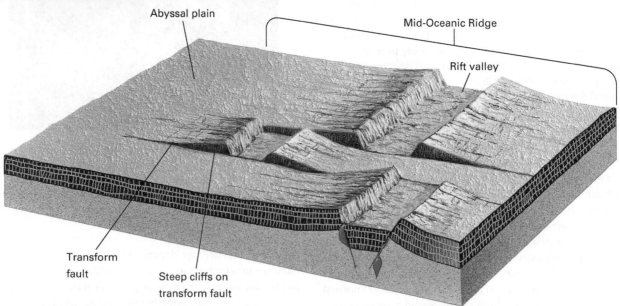

Abyssal plain

Mid-Oceanic Ridge

Rift valley

Transform fault

Steep cliffs on transform fault

● **FIGURE 15.18** Transform faults offset segments of the Mid-Oceanic Ridge. Adjacent segments of the ridge may be separated by steep cliffs 3 kilometers high. Note the flat abyssal plain far from the ridge.

Global Sea-Level Changes and the Mid-Oceanic Ridge System

A thin layer of marine sedimentary rocks blankets large areas of Earth's continents. These rocks tell us that those places must have been below sea level when the sediment accumulated.

Tectonic activity can cause a continent to sink, allowing the sea to flood a large area. However, at particular times in the past (most notably during the Cambrian, Carboniferous, and Cretaceous Periods), marine sediments accumulated on low-lying portions of all continents simultaneously, indicating simultaneous global flooding of low-elevation regions on all continents. Although our plate tectonics model explains the sinking of individual continents, or parts of continents, it does not explain why all continents should sink at the same time. Therefore, we need to explain how sea

level could rise globally by hundreds of meters to flood all continents simultaneously.

Continental glaciers have advanced and melted numerous times in Earth's history. During the growth of continental glaciers, seawater evaporates and is frozen into the ice that rests on land. As a result, sea level drops. When glaciers melt, the water runs back into the oceans and sea level rises. The alternating growth and melting of glaciers during the Pleistocene Epoch caused sea level to fluctuate by as much as 150 meters. However, the ages of most marine sedimentary rocks on continents do not coincide with times of glacial melting. Therefore, we must look for a different cause to explain continental flooding.

Recall that the new, hot lithosphere at a spreading center is buoyant, causing the Mid-Oceanic Ridge system to rise above the surrounding seafloor. This submarine mountain chain displaces a huge volume of seawater. If the Mid-Oceanic Ridge system were smaller, it would displace less seawater and sea level would fall. If it were larger, sea level would rise.

The Mid-Oceanic Ridge rises highest at the spreading center, where new lithosphere rock is hottest and has the lowest density. The elevation of the ridge decreases on both sides of the spreading center because the lithosphere cools and shrinks as it moves outward.

Now consider a spreading center where spreading is very slow (perhaps 1 to 2 centimeters per year). At such a slow rate, the newly formed lithosphere would cool before it migrated far from the spreading center. As a result, the ridge would be narrow and of low volume, as shown in ●FIGURE 15.19A. In contrast, rapid seafloor spreading of 10 to 20 centimeters per year would create a high-volume ridge because the newly formed, hot lithosphere would be carried a considerable distance away from the spreading center before it cooled and shrank (●FIGURE 15.19B). This high-volume ridge would displace considerably more seawater than would a low-volume ridge, and the high-volume ridge would cause global sea level to rise. To summarize, rapid spreading produces a larger volume of hot rock and pushes aside more seawater, causing global sea level to go up. If spreading then slows down, a smaller volume of hot rock is produced, displacing less seawater and causing global sea level to fall.

Seafloor age data indicate that the rate of seafloor spreading has varied from about 2 to 16 centimeters per year over the past 200 million years ago, since the Jurassic Period. During Late Cretaceous times, between 110 and 85 million years ago, seafloor spreading was unusually rapid, and that rapid spreading caused the formation of an unusually high-volume Mid-Oceanic Ridge. As the total volume of the Mid-Oceanic Ridge grew with the increase in average global spreading rate, more and more seawater was displaced and low-lying portions of continents were inundated. Geologists have found marine sedimentary rocks of Late Cretaceous age on nearly all continents, indicating that Late Cretaceous time was, in fact, one of abnormally high global sea level. Thus, a process initiated by heat transfer deep within the geosphere profoundly affected the sea level of the hydrosphere and life throughout the biosphere. Unfortunately, because no oceanic crust is older than about 200 million years, the specific relationships between seafloor spreading rates and global sea level cannot be tested for earlier times when extensive marine sedimentary rocks accumulated on continents.

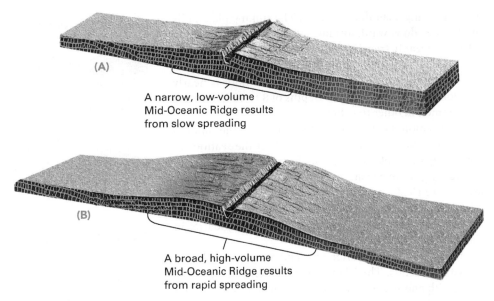

● **FIGURE 15.19** (A) Slow seafloor spreading creates a narrow, low-volume Mid-Oceanic Ridge that displaces less seawater and lowers sea level. (B) Rapid seafloor spreading creates a wide, high-volume ridge that displaces more seawater and raises sea level.

Oceanic Trenches and Island Arcs

In many parts of the Pacific Ocean, and in some other ocean basins, two oceanic plates converge. One dives beneath the

oceanic trench A long, narrow, steep-sided depression of the seafloor formed where a subducting oceanic plate sinks into the mantle, causing the seafloor to bend downward like a flexed diving board.

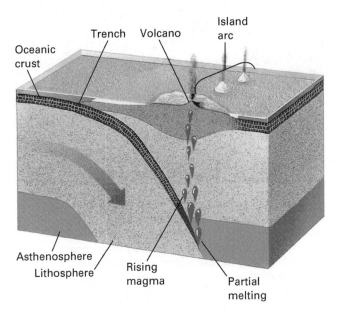

● **FIGURE 15.20** An oceanic trench forms at a convergent boundary between two oceanic plates. One of the plates sinks and heats, generating magma that rises to form a chain of volcanic islands called an island arc.

other, forming a subduction zone. The sinking plate pulls the seafloor downward, forming a long, narrow depression called an **oceanic trench** (●FIGURE 15.20). The deepest place on Earth is in the Mariana Trench, north of New Guinea in the southwestern Pacific, where the ocean floor is nearly 11 kilometers below sea level. Depths of 8 to 10 kilometers are common in other trenches.

Huge amounts of magma are generated in the subduction zone. The magma rises and erupts at the seafloor to form submarine volcanoes next to the trench. The volcanoes eventually grow to become a chain of islands called an island arc (Figure 15.20), as we learned in Chapter 9. The western Aleutian Islands are an example of an island arc. Many others occur at the numerous convergent plate boundaries in the western Pacific (●FIGURE 15.21).

If subduction stops after an island arc forms, volcanic activity also ends. The island arc may then ride passively along with another tectonic plate until it arrives at another subduction zone. However, the density of island arc rocks is relatively low, making them too buoyant to sink into the mantle (●FIGURE 15.22A and 15.22B). Instead, the island arc is mashed against the side of the overriding plate.

When an island arc and continent begin to slowly mash together, or "collide," the subducting plate commonly fractures on the seaward side of the island arc to form a new subduction zone. In this way, the island arc breaks away from the ocean plate and becomes part of the continent, enlarging it (●FIGURE 15.22C). Much of the crust now underlying western California, Oregon, Washington, and western British Columbia was added to North America in this way between 180 million and about 50 million years

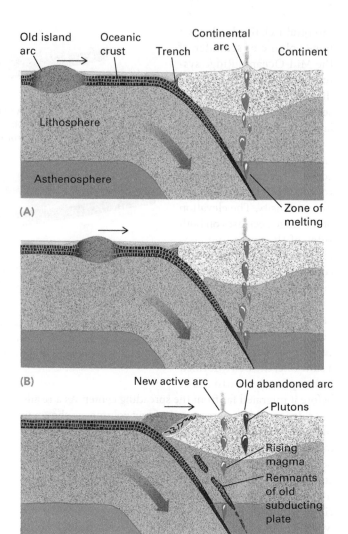

(A)

(B)

(C)

● **FIGURE 15.22** (A) An island arc forms on a plate that is itself being subducted beneath a continent. (B) The island arc reaches the subduction zone but the arc's low density prevents it from subducting into the mantle. (C) The island arc is jammed onto the continental margin and becomes part of the continent. Both the subduction zone and trench jump to the seaward side of the island arc, thereby enlarging the continent.

● **FIGURE 15.21** Photograph of an active stratovolcano on Pagan Island, part of the Commonwealth of the Northern Marianas in the western Pacific Ocean. The photo was taken on March 6, 2012, by an astronaut on the International Space Station, located about 480 kilometers southwest of the island at an altitude of about 400 kilometers. Pagan Island is part of the island arc formed by melting of old, cold subducting oceanic lithosphere of the Pacific Plate westward beneath oceanic crust of the Philippine Plate. The Marianas Trench is located immediately to the east of the arc and contains the deepest point of Earth's modern ocean at 10.9 kilometers.

ago, when several island arcs formerly in the eastern Pacific Ocean slowly collided with the western margin of the North American Plate. A simplified geologic map of these late additions to the North American continent, in general called **accreted terranes**, is shown in ●FIGURE 15.23.

Seamounts, Oceanic Islands, and Atolls

A **seamount** is a submarine mountain that rises 1 kilometer or more above the surrounding seafloor. An **oceanic island** is a seamount that rises above sea level. Both are common in all ocean basins but are particularly abundant in the southwestern Pacific Ocean. Seamounts and oceanic islands sometimes occur as isolated peaks, but they are more commonly found in chains. Dredge samples show that seamounts, oceanic islands, and the ocean floor itself are all made of basalt. Many seamounts and oceanic islands are volcanoes that formed at a hot spot above a mantle plume, and most form within a tectonic plate rather than at a plate boundary. An isolated seamount or short chain of small seamounts probably formed over a plume that lasted for only a short time. In contrast, a long chain of large islands, such as the Hawaiian Island-Emperor Seamount chain, formed over a long-lasting plume. In this case, the lithospheric plate migrated over the plume as the magma continued to rise from a source beneath the lithosphere. Each volcano formed directly over the plume and then became extinct as the moving plate carried it away from the plume. As a result, the seamounts and oceanic islands become progressively younger toward the Island of Hawaii, located at the end of the chain and currently located over the active mantle plume (●FIGURE 15.24).

After a volcanic island forms, it begins to sink, or subside. Three factors contribute to the sinking:

1. If the mantle plume stops rising, it stops producing magma. Then the lithosphere beneath the island cools and becomes denser, and the island sinks. Alternatively, a moving plate may carry the island away from the hot spot. This also results in the cooling, contraction, and sinking of the island.
2. The weight of the newly formed volcano causes isostatic sinking.
3. Erosion lowers the top of the volcano.

These three factors gradually transform a volcanic island to a seamount over geologic time (●FIGURE 15.25). Calculations suggest that if the Pacific Ocean Plate continues to move at its present rate, the island of Hawaii may sink beneath the sea within 10 to 15 million years. In this case, sea waves likely will erode a horizontal upper surface on the sinking island, forming a flat-topped seamount called a **guyot** (pronounced "gee-o," after Swiss-born American geologist Arnold Henri Guyot), as illustrated in ●FIGURE 15.26.

The South Pacific and portions of the Indian Ocean are dotted with numerous other islands called atolls. An **atoll**

---

accreted terranes A mappable, fault-bounded landmass that originates as an island arc or a microcontinent that is later added onto a continent.

seamount A submarine mountain, usually of volcanic origin, that rises 1 kilometer or more above the surrounding seafloor.

oceanic island A submarine mountain (seamount) that rises above sea level.

guyot A flat-topped seamount, formed when the top of a sinking island, usually of volcanic origin, is eroded by wave energy.

atoll A circular coral reef that forms a ring of islands around a central lagoon and that is bounded on the outside by the deep water of the open sea. Usually forms on top of a subsided seamount.

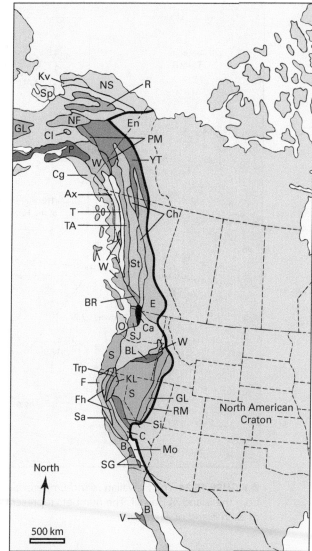

Symbol	Name	Symbol	Name
Ax	Alexander	O	Olympic
B	Baja	P	Peninsular
BL	Blue Mountains	PM	Pingston and McKinley
BR	Bridge River	R	Ruby
C	Calaveras	RM	Roberts Mountain
Ca	Cascade	S	Siletzia
Cg	Chugach	Sa	Salinian
Ch	Cache Creek	SG	San Gabriel
Cl	Chulitna	Si	Northern Sierra
E	Eastern terranes	SJ	San Juan
En	Endicott	Sp	Seward Peninsula
F	Franciscan and Great Valley	St	Stikine
Fh	Foothills belt	T	Taku
GL	Golconda	TA	Tracy Arm
I	Innoko	Trp	Western Klamath Mountains
KL	Klamath Mountains	V	Vizcaino
Kv	Kagvik	W	Wrangellia
Mo	Mohave	YT	Yukon-Tanana
NF	Nixon Fork		
NS	North Slope		

● **FIGURE 15.23** The accreted terranes of western North America are microcontinents and island arcs that were added to the continent from the Pacific Ocean.

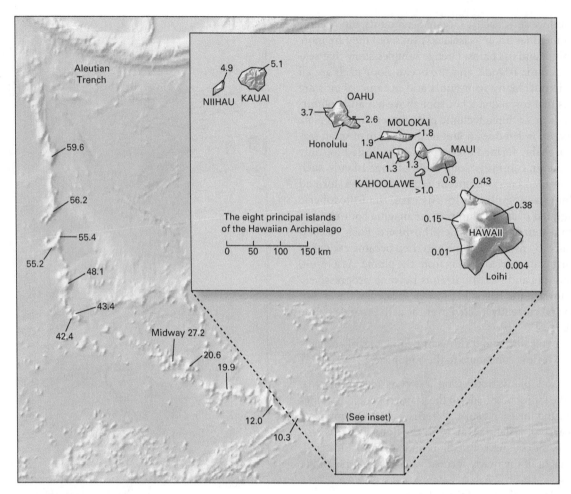

● **FIGURE 15.24** The Hawaiian Island-Emperor Seamount chain becomes older in a direction going away from the island of Hawaii. The numbers represent ages, in millions of years, of the oldest volcanic rocks of each island or seamount.

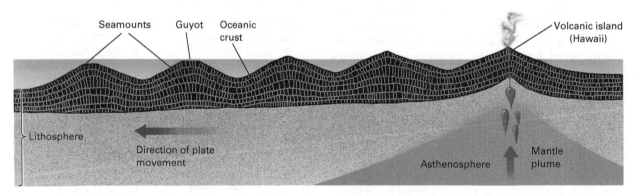

● **FIGURE 15.25** The Hawaiian Islands and Emperor Seamounts sink as they move away from the mantle plume currently located under the volcanically active Island of Hawaii.

is a circular coral reef that forms a ring of islands around a central lagoon (●FIGURE 15.27). Atolls vary from 1 to 130 kilometers in diameter and are surrounded by the deep water of the open sea.

If corals live only in shallow water, how did atolls form in the deep sea? Charles Darwin studied this question during his famous voyage on the Beagle from 1831 to 1836. He reasoned that a coral reef must have formed in shallow water on the flanks of a volcanic island. Eventually the island sank, but the reef continued to grow upward, so that the living portion always remained in shallow water (●FIGURE 15.28). This proposal was not accepted at first because scientists could not explain how a volcanic island could sink. However, when scientists drilled into a Pacific atoll shortly after

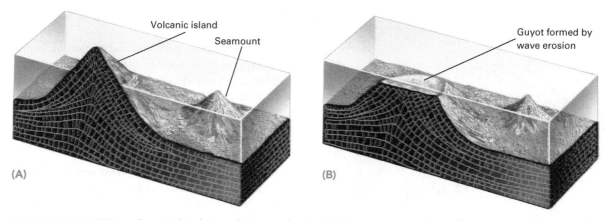

● **FIGURE 15.26** (A) A volcanic island rises above sea level. (B) Wave energy erodes a flat top on a sinking island to form a guyot.

Tahiti Tourist Board

● **FIGURE 15.27** The Tetiaroa Atoll in French Polynesia formed by the process described in Figure 15.26. Over time, storm waves wash coral sand on top of the reef and vegetation grows on the sand, forming the individual islands in the atoll.

World War II and found volcanic rock hundreds of meters beneath the reef, Darwin's hypothesis was revived. It is considered accurate today, in light of our ability to explain why volcanic islands sink.

SEDIMENT AND ROCKS OF THE SEAFLOOR

Early oceanographers had believed that the oceans are 4 billion years old, so mud on the ancient seafloor should have been very thick, having had so much time to accumulate. In 1947, however, scientists on the U.S. research ship *Atlantis* discovered that the mud layer on the bottom of the Atlantic Ocean is much thinner than they expected. Why is there so little mud on the seafloor? The answer to this question would be a crucial piece of evidence in the development of the theory of plate tectonics.

Earth is 4.6 billion years old, and rocks as old as 4.04 billion years have been found on the North American continent. Because of its buoyancy, most continental crust remains near Earth's surface. In contrast, no parts of the seafloor are older than about 200 million years, because oceanic crust forms continuously at the Mid-Oceanic Ridge and then recycles into the mantle at subduction zones. As

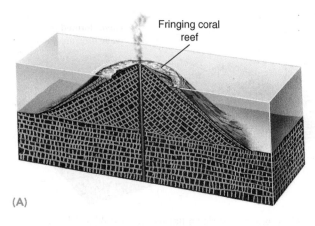

(A)

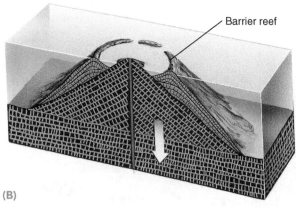

(B)

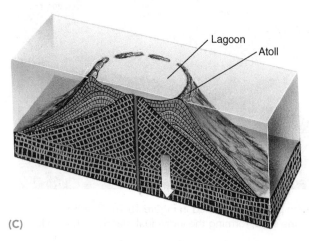

(C)

● **FIGURE 15.28** (A) A fringing reef grows along the shore of a young volcanic island. (B) As the island sinks, the reef continues to grow upward to form a barrier reef that encircles the island. (C) Finally, the island sinks below sea level and the reef forms a circular atoll. This model of atoll formation was proposed by Charles Darwin in 1842, following his voyage on the *Beagle*.

the theory of plate tectonics was emerging in the early 1960s, it became apparent that the presence of only a thin layer of mud covering the oceanic crust could be easily explained if the oldest seafloor is less than 200 million years old, not 4 billion years old as oceanographers had once advocated.

This evidence supported the fledgling plate tectonics theory in its early days.

Seismic reflection profiling and seafloor drilling show that oceanic crust is made of three layers. The uppermost layer consists of sediment, and the lower two are basalt (●**FIGURE 15.29**).

Ocean-Floor Sediment

The uppermost layer of oceanic crust consists of two types of sediment. **Terrigenous sediment** is sand, silt, and clay eroded from the continents and delivered to deep water mainly through submarine canyons that extend outward from the continental shelf. Thus, most terrigenous sediment is found close to the continents. **Pelagic sediment**, however, collects even on the deep seafloor far from continents. It is a gray and red-brown mixture of silt- and clay-sized particles that were transported by wind as dust derived from continental weathering and that lived in the ocean surface waters as sand-sized and smaller phytoplankton that photosynthesized, died, and sank (●**FIGURE 15.30**).

Pelagic sediment accumulates at a rate of about 2 to 10 millimeters per 1,000 years. Near the Mid-Oceanic Ridge system there is very little to virtually no sediment because the seafloor is so young. The sediment thickness increases with distance from the ridge because the seafloor becomes older as it spreads away from the ridge and, consequently, the sediment has had more time to accumulate. Close to shore, pelagic sediment becomes diluted by more and more terrigenous sediment, forming accumulations that can exceed 10 kilometers in thickness, though more typically are around four kilometers in thickness. The observation that the thickness of seafloor mud increased with distance from the ridge also supported the plate tectonics theory in its early days.

Parts of the ocean floor beyond the Mid-Oceanic Ridge system are flat, level, essentially featureless surfaces called the **abyssal plains**. They are the flattest surfaces on Earth. Seismic profiling shows that the basaltic crust is rough and jagged throughout the ocean. On the abyssal plains, however, pelagic sediment buries this rugged surface, forming smooth surfaces. If you were to remove all of the sediment, you would see rugged topography similar to that of the Mid-Oceanic Ridge.

BASALTIC OCEANIC CRUST Below the superficial layer of sediment on the ocean floor is a layer of basalt about 1 to 2 kilometers thick. It consists mostly of **pillow basalt**, which forms as hot magma oozes onto the seafloor. Contact with cold seawater causes newly erupting molten lava to rapidly form a solidified outer rind. Still-molten and pressurized magma immediately below the rind causes it to deform plastically into bulbous, pillow-shaped spheroids, called pillows (●**FIGURE 15.31**).

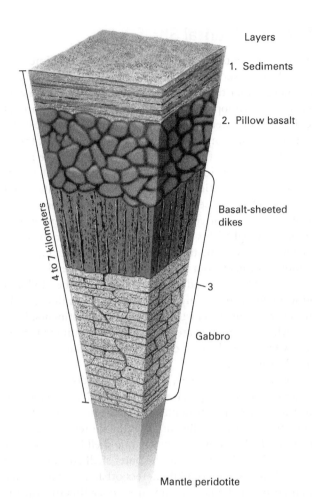

Layers

1. Sediments

2. Pillow basalt

Basalt-sheeted dikes

3

Gabbro

4 to 7 kilometers

Mantle peridotite

● **FIGURE 15.29** The three layers of oceanic crust. The uppermost layer consists of sediment. The middle layer consists of pillow basalt. The deepest layer is made of vertical dikes of basalt that merge downward into gabbro. Below this lowermost layer of oceanic crust is the upper mantle.

Beneath the pillow basalt is the deepest and thickest layer of oceanic crust, consisting of between 3 and 5 kilometers of basalt that did not erupt onto the seafloor. This basalt directly overlies the mantle. The upper portion consists of vertical basalt dikes that formed as basaltic magma oozing toward the surface froze in the cracks of the rift valley. The lower portion consists of gabbro, the coarse-grained

---

terrigenous sediment Sediment composed of sand, silt, and clay eroded from the continents.

pelagic sediment Muddy ocean sediment made up of the skeletons of tiny marine organisms.

abyssal plains The flat, level, largely featureless parts of the ocean floor between the Mid-Oceanic Ridge and the continental rise.

pillow basalt Molten basaltic lava that solidified under water, forming spheroidal lumps of basalt that resemble a stack of pillows.

● **FIGURE 15.30** This scanning electron microscope image shows foraminifera, tiny organisms that float near the ocean surface. When these organisms die, their remains sink to the seafloor, to become part of the pelagic mud layer. Each of the fossils is the size of a fine sand grain.

equivalent of basalt. The gabbro forms as pools of magma cool slowly, insulated by the basalt dikes above them.

The pillow basalt, vertical basalt dikes, and gabbro all form at the Mid-Oceanic Ridge. These rocks make up the foundation of all oceanic crust because all oceanic crust forms at a ridge axis and then spreads outward. In some places, chemical reactions with seawater have altered the basalt to a soft, green rock called serpentinite that contains up to 13 percent water. Serpentinite is a fairly common rock type in parts of California and the Pacific Northwest due to the presence of abundant basalt and altered basalt that represents oceanic crust associated with the accreted terranes found there. Superb outcrops of serpentinite are found on U.S. Highway 1 south of Big Sur, California (●**FIGURE 15.32**).

● **FIGURE 15.31** Underwater photo of pillow basalt off the island of Hawaii.

Copyright: and Photograph by Marc S. Hendrix

● **FIGURE 15.32** Close-up of strongly sheared serpentinite exposed along the shoreline south of Big Sur, California. This serpentinite formed through the alteration of basalt that once was part of oceanic crust. The basalt was deformed as it was slowly mashed into the western part of the North American Plate, and that deformation continued after the basalt was altered to serpentinite. The car keys are for scale.

CONTINENTAL MARGINS

A continental margin is a place where continental lithosphere meets oceanic lithosphere. Two types of continental margins exist. A **passive continental margin** occurs where continental and oceanic lithosphere are firmly joined together. Because it is not a plate boundary, little tectonic activity occurs at a passive margin. Continental margins on both sides of the Atlantic Ocean are passive margins. In contrast, an **active continental margin** occurs at a convergent plate boundary, where oceanic lithosphere sinks beneath the continent in a subduction zone. The west coast of South America is an active continental margin, also called an Andean margin, as you learned in Chapter 9.

Passive Continental Margins

Recall from Chapter 6 that about 250 million years ago all of Earth's continents were joined into the supercontinent called Pangea. Shortly thereafter, Pangea began to rift apart into the continents as we know them today. The Atlantic Ocean opened as the east coast of North America separated from Europe and Africa. As Pangea broke up, the continental crust fractured and thinned near the fractures (●**FIGURE 15.33A**). Basaltic magma rose at the new spreading center, forming oceanic crust between North America and Africa (●**FIGURE 15.33B**). All tectonic activity then focused on the continually spreading Mid-Atlantic Ridge, and no further tectonic activity occurred at the continental margins; hence the term "passive continental margin" (●**FIGURE 15.33C**).

The Continental Shelf

On all continents, streams and rivers deposit sediment in coastal deltas such as the Mississippi Delta. Much of this sediment is redistributed along the coastline by ocean currents and downslope as sediment-gravity flows—mixtures of sediment and water that flow downslope as a fluid, usually as a bottom-hugging current. Much of the sediment transported downslope from the shoreline and offshore forms a shallow, gently sloping underwater surface that marks the edge of the continent and is called a **continental shelf** (●**FIGURE 15.34**). As sediment accumulates on a continental shelf, the edge of the continent sinks isostatically because of the added weight. This effect keeps the shelf generally below sea level. At the same time, the delivery and accumulation of more sediment from the weathering of the nearby continent causes the outer edge of the continental shelf to build outward, ultimately enlarging the size of the continent itself. In this way, mass is redistributed from the continental interior, where mountains are eroded, to the continental margins, which actively grow outward due to the accumulation of sediment there.

Over millions of years, several kilometers of sediment accumulated on the passive east coast of North America, forming a broad continental shelf that projects outward along the entire coast. Beginning at the shoreline, the water depth of the North American Atlantic shelf increases gradually to about 200 meters at the outer shelf edge, which in most places is over 100 kilometers offshore. The average inclination of the continental shelf over this distance is about 0.1 degree.

A continental shelf on a passive margin can be a very large feature. The shelf off the coast of southeastern Canada is about 500 kilometers wide (●**FIGURE 15.35**), and parts of the shelves of Siberia and northwestern Europe are even wider.

In some places, a supply of sediment may be lacking, either because no rivers bring sand, silt, or clay to the shelf or because ocean currents do not deliver sediment to the particular region. In warm regions where terrigenous sediment is either lacking or is delivered in very small quantities, carbonate reef-building organisms such as corals, mollusks, bryozoa, calcareous algae, and numerous others can thrive. When environmental conditions are optimal, the carbonate-generating ecosystem can produce very large volumes of carbonate sediment in very short periods of time. As a result, in tropical and subtropical latitudes where the rate of clastic sediment is low enough and water temperature and salinity is optimal, thick beds of limestone can accumulate. Limestone accumulations of this type may be hundreds of meters thick and hundreds of kilometers across and are called **carbonate platforms**. The Florida Keys and the Bahamas are modern-day examples of carbonate platforms on continental shelves (●**FIGURE 15.36**).

Some of the world's richest petroleum reserves occur on the continental shelves of the North Sea between England and Scandinavia, in the Gulf of Mexico, and in the Beaufort

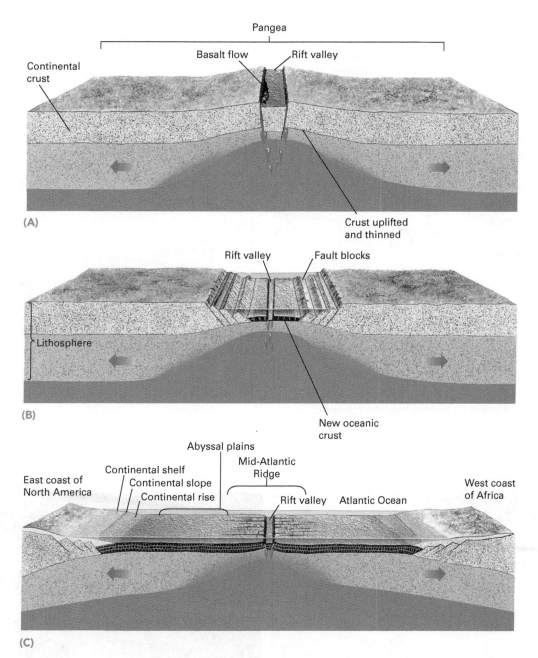

● **FIGURE 15.33** (A) Continental crust fractured as Pangea began to rift. (B) Faulting and erosion thinned the crust as it separated. Rising basaltic magma formed new oceanic crust in the rift zone. (C) Sediment eroded from the continents formed broad continental shelves on the passive margins of North America and Africa.

passive continental margin A margin that occurs where continental and oceanic crust are firmly joined together and where little tectonic activity occurs. Not a plate boundary.

active continental margin A continental margin that occurs at a convergent or transform plate boundary.

continental shelf A shallow, very gently sloping portion of the seafloor that extends from the shoreline to ~200 meters water depth at the top of the continental slope.

carbonate platforms Extensive accumulations of limestone, such as the Florida Keys and the Bahamas, formed in warm regions on an isolated continental shelf where terrigenous clastic sediment does not muddy the water and reef-building organisms thrive.

Sea on the northern coast of Alaska and western Canada. In recent years, oil companies have explored and developed these offshore reserves. Deep drilling has revealed that granitic continental crust lies beneath the sedimentary rocks, confirming that the continental shelves are truly parts of the continents despite the fact that they are covered by seawater.

THE CONTINENTAL SLOPE AND RISE At the outer edge of a shelf, the seafloor gradually steepens to an average slope of about 3 degrees (though it can vary from 1 degree to 10 degrees) as it increases in depth from 200 meters to

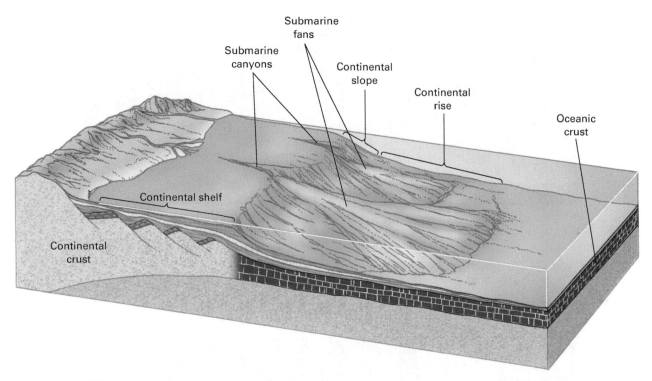

● **FIGURE 15.34** A passive continental margin consists of a broad continental shelf, a slope, and a rise formed by accumulation of sediment eroded from the continent.

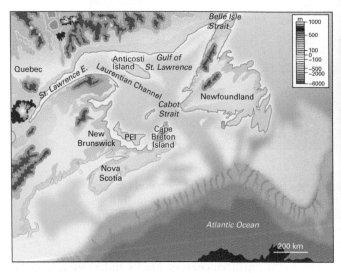

● **FIGURE 15.35** Color-coded map showing both onshore topography and seafloor bathymetry for part of New England, United States, as well as New Brunswick, Nova Scotia, and Newfoundland, Canada. See elevation/depth color key in the upper right. Notice that the St. Lawrence River flows into the Laurentian Channel which itself cuts across the continental shelf all the way to where it ends against the top of the continental slope. Sediment carried down the river and across the Laurentian Channel subsequently moves as turbidity currents and other forms of sediment gravity flows through one or more submarine canyons down across the slope to the continental rise.

● **FIGURE 15.36** Clear skies and calm waters allowed the MODIS satellite to capture this stunning image of southern Florida, the Bahamas, and Cuba. In the center, Andros Island is surrounded by the bright blue halo of the Great Bahama Bank, a carbonate platform that was inundated by a rising sea level between 10,000 and 2,500 years ago as the last ice-age glaciers were melting. In most places, the water above the platform doesn't exceed 6 meters.

about 3 kilometers. This steep region of the seafloor averages about 50 kilometers wide and is called the **continental slope**. It is a surface formed by sediment accumulation, much like the shelf. Its steeper angle is due primarily to thinning of continental crust where it nears the junction with oceanic crust. Offshore seismic reflection profiling shows that the sedimentary layering is commonly disrupted where sediment has slumped and slid down the steep incline.

A continental slope becomes less steep as it gradually merges with the deep ocean floor. This region, called the **continental rise**, consists of an apron of terrigenous sediment that was transported across the continental shelf and deposited on the deep ocean floor at the foot of the slope. The continental rise averages a few hundred kilometers wide. Typically, it joins the deep seafloor at a depth of about 5 kilometers.

In essence, then, the shelf-slope-rise complex is a smoothly sloping, submarine surface on the edge of a continent formed by accumulation of sediment eroded from the continent.

SUBMARINE CANYONS AND SUBMARINE FANS In many places, seafloor maps show deep valleys, called **submarine canyons**, eroded into the continental shelf and slope (Figures 15.34 and 15.35). They look like submarine stream valleys. A canyon typically starts on the outer edge of a continental shelf, usually beyond the outer reaches of a major river delta on the inner shelf, and continues across the slope to the rise. At its lower end, a submarine canyon commonly leads into a **submarine fan** (Figures 15.34 and 15.35), a large, fan-shaped accumulation of terrigenous sediment delivered down the submarine canyon to the continental rise by sediment gravity flows.

Most submarine canyons occur downslope of a region where large rivers enter the sea. When they were first discovered, geologists thought the canyons had been eroded by rivers during the Pleistocene Epoch, when accumulation of glacial ice on land lowered sea level by as much as

150 meters. However, this explanation cannot account for the deeper portions of submarine canyons that cut erosionally into the lower continental slopes at depths of a kilometer or more. These deeper parts of the submarine canyons must have formed through erosion, and a submarine mechanism must be found to explain them.

Geologists subsequently discovered that **sediment gravity flows**—underwater mixtures of sediment and water that flow downslope—can erode the continental shelf and slope to create or deepen the submarine canyons. Particularly efficient at eroding rock and sediment underwater are **turbidity currents**, a form of sediment gravity flow in which a turbulent mixture of sediment and seawater that is more dense than the seawater alone and so flows downslope along the bottom (•FIGURE 15.37). You can create and observe a turbidity current at home by slowly pouring a cup of muddy water into a sloping basin of clear water.

Turbidity currents can be triggered by an earthquake, a large storm, or simply by the oversteepening of the slope as sediment accumulates. (Recall the angle of repose for loose sediment, described in Chapter 10.) When the sediment starts to move, it mixes with water and flows across the shelf and down the slope as a turbulent, chaotic fluid. A turbidity current can travel at speeds greater than 100 kilometers per hour and for distances in excess of 1,200 kilometers.

Sediment-laden water traveling at such speed has tremendous erosive power. Once a turbidity current cuts a small channel into the shelf and slope, subsequent currents follow the same channel, just as a stream uses the same channel year after year. Over time, the currents erode a deep submarine canyon into the shelf and slope. Turbidity currents slow down when they reach the deep seafloor. The sediment accumulates there to form a submarine fan. Most submarine canyons

---

continental slope The relatively steep (averaging 3 degrees but varying between 1 degree and 10 degrees) submarine slope between the continental shelf and the continental rise.

continental rise An apron of sediment at the foot of the continental slope that merges with the deep seafloor.

submarine canyons A deep, V-shaped, steep-walled trough eroded into a continental slope and, in some cases, outer shelf. Funnels sediment from the continental shelf across the slope to the continental rise.

submarine fan A large, fan-shaped accumulation of sediment deposited on the deep seafloor, usually within and beyond the mouth of submarine canyons.

sediment gravity flows Underwater mixtures of sediment and water that flow downslope.

turbidity currents A highly turbulent mixture of sediment and water that flows rapidly downslope in a subaqueous setting. Capable of causing substantial subaqueous erosion.

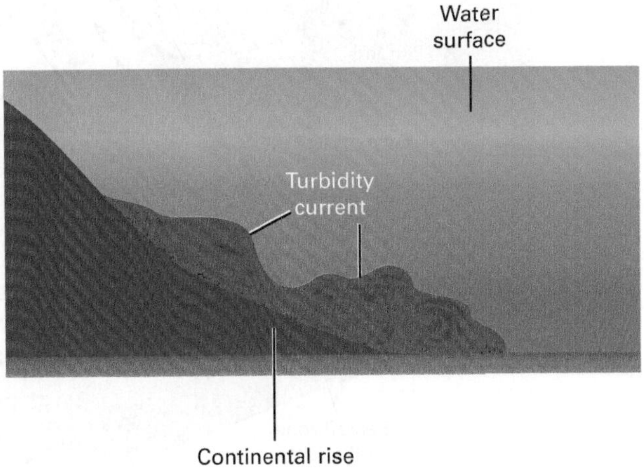

● **FIGURE 15.37** A turbidity current flowing down a laboratory flume filled with water. The metal hardware and wiring measure the velocity of the turbidity current and the density of the sediment-water mixture at different depths above the bed. Turbidity currents such as this deliver a tremendous volume of sediment to the continental rise and are responsible for much erosion in submarine canyons that cut across the continental slope.

and fans form near the mouths of large rivers because the rivers supply the great amount of sediment needed to create turbidity currents and other types of sediment gravity flows. Most of the largest submarine fans form on passive continental margins. Submarine fans that form on active margins typically are much smaller because the trench swallows much of the sediment. Furthermore, most of the world's largest rivers drain toward passive margins. The largest known fan is the Bengal Fan, which covers about 4 million square kilometers beyond the mouth of the Ganges River in the Indian Ocean east of India. More than half of the sediment eroded from the rapidly rising Himalayas ends up in this fan. Interestingly, the Bengal Fan has no associated submarine canyon, perhaps because the sediment supply is so great that the rapid accumulation of sediment prevents erosion of a canyon.

Active Continental Margins

Active continental margins form along Andean-style subduction zones, where an oceanic plate converges with a continent, and along less common transform plate boundaries where oceanic crust is sliding past continental crust, as with the San Andreas Fault in California. In convergent margins, most of the sediment transported from a continent is swallowed up in the trench. As a result, an active margin commonly has a narrow continental shelf or none at all. The landward wall of the trench (the side toward the continent) is the continental slope of an active margin. It typically inclines at 4 degrees or 5 degrees in its upper part and steepens to 15 degrees or more near the bottom of the trench. The continental rise is absent or relatively small because sediment gravity flows generally transport sediment into the trench instead of across it to the ocean floor located over subducting plate (●FIGURE 15.38).

Note that a trench can form wherever subduction occurs—where oceanic crust sinks beneath the edge of a continent, or where it sinks beneath another oceanic plate. However, the trench will be associated with a continental margin only when the overriding plate is made of continental lithosphere.

At convergent boundaries, the oceanic plate sinks into the mantle at descent angles ranging from about 15 degrees to

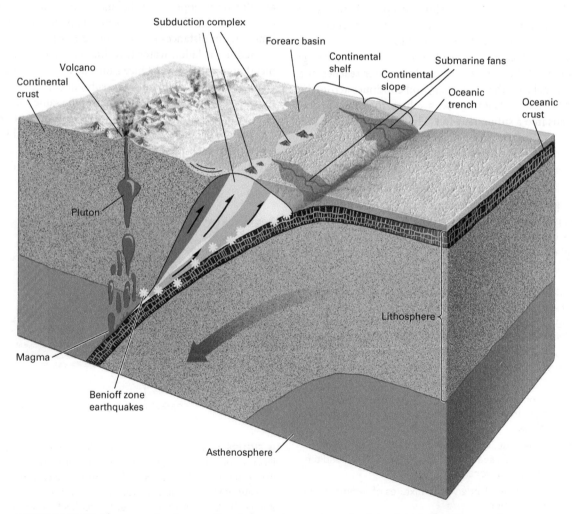

●**FIGURE 15.38** Along most active continental margins, an oceanic plate sinks beneath a continent, forming an oceanic trench. The continental shelf is narrow, the slope is steep, and the continental rise is small to nonexistent.

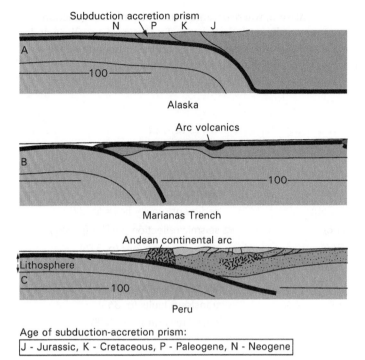

Subduction accretion prism

N   P   K   J

A

——100——

Alaska

Arc volcanics

B

——————100——

Marianas Trench

Andean continental arc

Lithosphere

C

——100——

Peru

Age of subduction-accretion prism:

J - Jurassic, K - Cretaceous, P - Paleogene, N - Neogene

● **FIGURE 15.39** A subducting slab will not usually descend into the mantle at the same angle as its dip angle. However, when it does occur (A above), as off the southern coast of Alaska, a subduction complex forms but the region behind the arc is neither compressed nor extended. (B) In situations where an old, cold ocean slab is sinking steeply into the mantle at an angle greater than its dip, the arc is subject to extensional stress, causing the overriding plate to break apart, as in the volcanic arc adjacent to the Marianas trench. When a young, hot and relatively buoyant oceanic plate subducts, it does so at an angle that is shallower than its dip, causing compression in the overriding plate and contractile deformation there as off the coast of Peru.

beyond vertical (●**FIGURE 15.39**). Regardless of this remarkably wide variation in the descent angle of subducting slabs around Earth's ocean basins, an ocean trench is formed although in some regions the trench is filled with sediment and is not as obvious as elsewhere. For example, offshore of the Pacific Northwest, the Juan de Fuca plate is subducting below the North America plate, but the subduction is filled with sediment resulting from draining of the Cascades and northern Rockies. The Columbia River has been the largest deliverer of sediment to the trench, as huge volumes of water and sediment poured out of the Columbia River during and following each glaciation over the past 3 million years or so. In contrast, the Marianas trench off the eastern coast of the Philippines includes Earth's deepest ocean depth at 10,994 km—6.8 miles.

Key Concepts Review

- Ocean water, atmospheric gases, and the molecular building blocks of life were carried to Earth by bolides early in Earth's history.
- Oceans cover about 71 percent of Earth's surface. Continental lithosphere is thicker and less dense than oceanic lithosphere. Consequently, continents float isostatically to high elevations, whereas oceanic lithosphere sinks to low elevations. The ocean basins form topographic depressions on Earth's surface, which fill with water to form oceans.
- Because of the great depth and remoteness of the ocean floor and oceanic crust, knowledge of them comes mainly from sampling and remote sensing.
- The Mid-Oceanic Ridge system is a submarine mountain chain that extends through all of Earth's major ocean basins. A rift valley runs down the center of many parts of the ridge, and the ridge and rift valley are both offset by numerous transform faults. The Mid-Oceanic Ridge forms at a spreading center where new oceanic crust is added to the seafloor. The Mid-Oceanic Ridge system supports thriving ecosystems based on chemosynthesis. Oceanic trenches are the deepest parts

of ocean basins. Island arcs (chains of volcanoes formed at subduction zones where two oceanic plates collide) are common features of some ocean basins, particularly the southwestern Pacific. Seamounts and oceanic islands form in oceanic crust as a result of volcanic activity over mantle plumes. An atoll is a circular coral reef growing on a sinking volcanic island.

- Oceanic crust varies from about 4 to 7 kilometers thick and consists of three layers. The top layer (layer 1) is sediment, which varies from 0 to 3 or more kilometers thick. Beneath this, in layer 2, lies about 1 to 2 kilometers of pillow basalt. Layer 3, the deepest layer of oceanic crust, is from 3 to 5 kilometers thick and consists of basalt dikes on top of gabbro. The base of this layer is the boundary between oceanic crust and mantle. The age of seafloor rocks increases regularly away from the Mid-Oceanic Ridge. No oceanic crust is older than about 200 million years because it recycles into the mantle at subduction zones.

- A passive continental margin includes a continental shelf, a continental slope, and a continental rise formed by accumulation of terrigenous sediment.

Submarine canyons eroded by turbidity currents cut into continental margins and commonly lead downslope to submarine fans, where the turbidity currents deposit sediments on the continental rise. An active continental margin, where oceanic crust sinks into a subduction zone beneath the margin of a continent, usually includes a narrow continental shelf and a continental slope that steepens abruptly into an oceanic trench.

Important Terms

abyssal plains (p. 370)	echo sounder (p. 355)	rock dredge (p. 355)
accreted terranes (p. 366)	guyot (p. 367)	seafloor drilling (p. 355)
active continental margin (p. 372)	magnetometer (p. 360)	seamount (p. 366)
atoll (p. 367)	microwave radar (p. 360)	sediment gravity flows (p. 375)
bolide (p. 353)	oceanic island (p. 366)	seismic reflection profiler (p. 359)
carbonate platforms (p. 372)	oceanic trench (p. 366)	submarine canyons (p. 375)
chemosynthesis (p. 363)	passive continental margin (p. 372)	submarine fan (p. 375)
continental rise (p. 375)	pelagic sediment (p. 370)	terrigenous sediment (p. 370)
continental shelf (p. 372)	pillow basalt (p. 370)	transform faults (p. 363)
continental slope (p. 375)	rift valley (p. 361)	turbidity currents (p. 375)

Review Questions

1. Describe the main differences between oceans and continents.

2. Sketch a cross section of the central rift valley in the Mid-Oceanic Ridge.

3. Describe the dimensions of the Mid-Oceanic Ridge system.

4. Explain why the Mid-Oceanic Ridge is topographically elevated above the surrounding ocean floor. Why does its elevation gradually decrease away from the ridge axis?

5. Explain the origin of the rift valley in the center of the Mid-Oceanic Ridge.

6. Why are the abyssal plains characterized by such low relief?

7. Sketch a cross section of oceanic crust from the seafloor. Label, describe, and indicate the approximate thickness of each layer.

8. Describe the two main types of seafloor sediment. What is the origin of each type?

9. Compare the ages of oceanic crust with the ages of continental rocks. Why are they so different?

10. Sketch a cross section of both an active continental margin and a passive continental margin. Label the features of each. Give approximate depths below sea level of each of the features.

11. Explain how a continental shelf–slope–rise complex forms on a continental margin.

12. Why does an active continental margin typically have a steeper continental slope than a passive margin?

13. Why do oceanic islands sink after they form?

16

The Dingle Peninsula, located on the southwest coast of Ireland, is a classic rocky coastline formed because the land surface has been submerged by rising sea level since the Pleistocene Glaciation. Although this area is undergoing isostatic uplift because of the disappearance of the glacial ice, the rate of uplift has been slower than the rate of sea level rise.

OCEANS AND COASTLINES

LEARNING OBJECTIVES

LO1   Give examples of dissolved ions and gases in sea water.

LO2   Discuss how acidification of seawater can affect marine ecosystems.

LO3   Explain the relationship between the location of the Earth, Moon, and **Sun with respect to both the size** and occurrence of tides.

LO4   Illustrate a sea wave including major features.

LO5   Describe the relationship between surface and deep water currents and climate.

LO6   Discuss the Great Pacific Garbage Patch and its composition in relation to ocean gyres.

LO7   Describe the Coriolis effect.

LO8   Explain fluctuations in the salinity of water by location.

LO9   Summarize the effect of sea water on the weathering and erosion of coastal areas.

LO10   Discuss the transport of sand along a coastline.

LO11   Compare the sequence of sediments deposited in an emerging coastline with those of a submerging coastline.

LO12   Contrast features of sandy and rocky coastlines.

LO13   Describe life forms that are found in the oceans.

LO14   Explain the relationship between sea level and global climate.

INTRODUCTION

About 60 percent of the world's human population lives within 100 kilometers of the coast. One million people live on low coral islands, and many millions more live on low-lying coastal land vulnerable to coastal flooding. At risk are not just individuals and villages but unique human cultures. Faced with changing coastal environments, these people are threatened with forced abandonment of their nations.

Popular media coverage has linked damage along the American Gulf Coast to sea level rise induced by global warming. However, many geologists and oceanographers argue that other factors such as crustal subsidence, described in Chapter 15, may be responsible for the submergence of this region. In a similar manner, several Pacific island nations risk elimination due to the combined effects of global sea level rise and cooling-related subsidence of the volcanic rock on which they are built.

Regardless of the cause of coastal sea level rise, for many people, it is disturbing that solid land should sink beneath the waves or that a modern city such as New York could be rendered nearly defenseless against a major storm, as was the case in 2012 with Hurricane Sandy. Yet geologists know that continents and islands have risen from the sea and sunk back beneath the waves throughout Earth's history and that, even viewed through the much shorter period of written human history, changes in sea level have directly affected

● **FIGURE 16.1** This structure, located on the Pechora Peninsula in the eastern Gulf of Corinth, Greece, was built to receive ships when sea level was several meters higher. This portion of coastline within the Gulf of Corinth is uplifting roughly half a millimeter per year because it is in the footwall of a major normal fault. The uplifted wave-cut notch, highlighted by the arrow, contains the remains of clamshells that were dated at about 6,390 years ago.

the economic activities of humans living and working near the coast (●**FIGURE 16.1**). In this chapter, we will continue studying Earth's oceans, focusing specifically on the processes that form, shape, or destroy islands and coastlines.

Coastlines are among the most geologically active environments on Earth. Sea level rises and falls, flooding shallow parts of continents and stranding beaches high above the sea surface. Regions of continental and oceanic crust also rise and sink, with results similar to those caused by fluctuating sea level. Rivers deposit great quantities of sand and mud on coastal deltas. Waves and currents erode beaches and transport sand along hundreds or even thousands of kilometers of shoreline. Converging tectonic plates buckle coastal regions, creating mountain ranges, earthquakes, and volcanic eruptions.

GEOGRAPHY OF THE OCEANS

All of Earth's oceans are connected, and water flows from one to another, so in one sense, Earth has just one global ocean. However, several distinct ocean basins exist within the global ocean (●**FIGURE 16.2**). The largest and deepest is the Pacific. It covers one-third of Earth's surface, more than all land combined, and contains more than half of the world's water. The Atlantic Ocean has about half the surface area of the Pacific. The Indian Ocean is slightly smaller than the Atlantic. The Arctic Ocean surrounds the North Pole and extends southward to the shores of North America, Europe, and Asia. Therefore, it is bounded by land, with only a few straits and channels connecting it to the Atlantic and Pacific Oceans. The surface of the Arctic Ocean freezes in winter, and parts of it melt for a few months during summer and early fall. The Antarctic Ocean, feared by sailors for its cold and ferocious

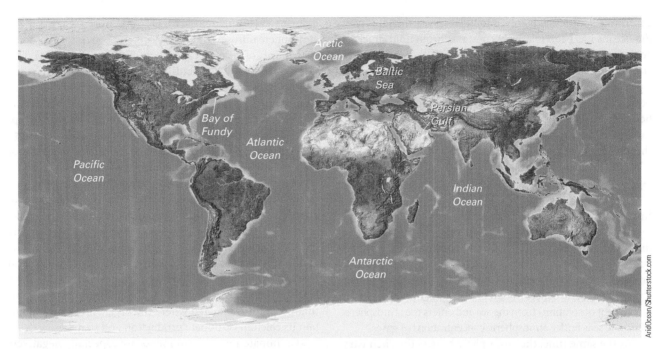

● **FIGURE 16.2** The oceans of the world. The "seven seas" are the North Atlantic, South Atlantic, North Pacific, South Pacific, Indian, Arctic, and Antarctic. However, these designations are related more closely to commerce than to the geology and oceanography of the ocean basins. Geologists and oceanographers recognize four major ocean basins: the Atlantic, Pacific, Indian, and Arctic. Also labeled on this map are a few smaller seas and bays that are mentioned in the text.

winds, has no sharp northern boundary. The northernmost limit of the Antarctic Ocean is the zone where warm currents from the north converge with cold Antarctic water.

SEAWATER

Salts and Trace Elements

The **salinity** of seawater is the total quantity of dissolved salts, expressed as a percentage. In this case, the term salts refers to all dissolved ions, not only sodium and chloride.

Dissolved ions make up about 3.5 percent of the weight of ocean water. The six ions listed in ●**FIGURE 16.3** make up 99 percent of the ocean's dissolved compounds. However, almost every other element found on land is also found dissolved in seawater, albeit in trace amounts. For example, seawater contains about 0.000000004 (4×10^{-9}) percent gold. Although the concentration is small, the oceans are large and therefore contain a lot of gold. About 4.4 kilograms of gold is dissolved in each cubic kilometer of seawater. Because the oceans contain about 1.3 billion cubic kilometers of water, about 5.7 billion kilograms of gold exists in the oceans. Unfortunately, it would be hopelessly expensive to extract even a small portion of this amount.

The world's rivers carry more than 2.5 billion tons of dissolved salts to the oceans every year. Underwater volcanoes

salinity The total quantity of dissolved salts in seawater, expressed as a percentage.

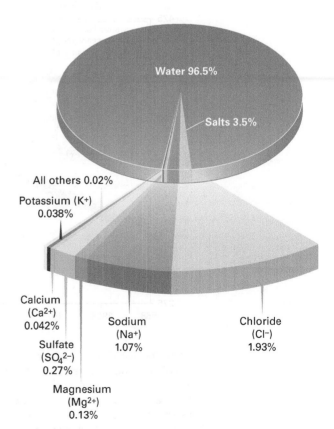

● **FIGURE 16.3** Six common ions form most of the salts in seawater.

contribute additional dissolved ions. However, the salinity of the oceans has been relatively constant throughout much of geologic time because salt has been removed from seawater at roughly the same rate at which it has been added. When a portion of a marine basin becomes cut off from the open ocean, the water evaporates, precipitating thick sedimentary beds of salt. Additionally, large quantities of salt become incorporated into mudstone and other sedimentary rocks through the precipitation of salt minerals within these rocks.

Dissolved Gases

In addition to trace elements and salts, seawater also contains dissolved gases, especially carbon dioxide and oxygen. These dissolved gases exchange freely and continuously with the atmosphere. If the atmospheric concentration of carbon dioxide or oxygen rises, much of the gas quickly dissolves into seawater; if the atmospheric concentration of a gas falls, that gas exsolves (comes out of solution) from the sea and enters the atmosphere. Thus, the seas buffer atmospheric concentrations of gases.

At the same time, the atmosphere buffers the chemistry of seawater. When carbon dioxide dissolves in seawater, the seawater becomes more acidic. In recent years, increasing concentrations of carbon dioxide gas from the burning of fossil fuels has caused not only the atmosphere and oceans to become warmer but also the seawater to become more acidic. That is, when carbon dioxide (CO_2) dissolves in seawater,

it forms carbonic acid (H_2CO_3). As more CO_2 from the atmosphere diffuses into the oceans and converts to carbonic acid, the pH of the ocean as a whole drops (●**FIGURE 16.4**). This process is called **ocean acidification**. Over the past 28 years, oceanographers have measured a drop in the pH of the oceans from about 8.1 to about 8.0. Although this small change may seem insignificant, the pH scale is not linear, and small drops in pH correspond to large increases in acidity (●**FIGURE 16.5**). A pH drop from 8.1 to 8.0 represents an increase in acidity, or the concentration of hydrogen ions, of about 58%.

Recent and ongoing ocean acidification has very seriously affected marine ecosystems. Corals, many kinds of algae, shellfish, foraminifera, and other organisms that make hard body parts out of calcium carbonate are particularly hard hit because calcium carbonate dissolves under acidic conditions. As the oceans acidify, these organisms spend more of their metabolic energy trying to precipitate their hard parts and so are stressed. As the acidity increases, eventually a point is reached where the organism cannot maintain its calcium carbonate production and so it dies.

Carbonate reef environments are especially negatively affected by recent increases in ocean acidification, and the recent loss of coral reef ecosystems is really alarming. From 2016 to 2018, half of the coral reefs in Australia's Great Barrier Reef died and no longer exist as a reef ecosystem. Left behind are vast underwater thickets of bleached, white calcium carbonate skeletons from dead corals (●**FIGURE 16.6**)

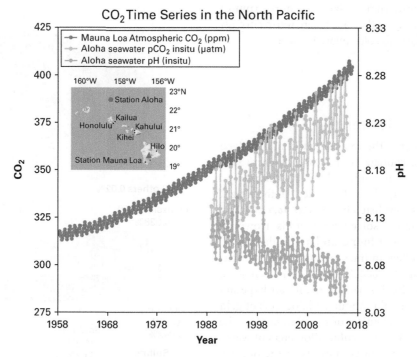

● **FIGURE 16.4** Concentrations of CO_2 gas in the atmosphere (red line) and dissolved CO_2 in seawater (green line) at the Aloha research station, about a hundred miles north of the Hawaiian Islands (see inset map). Notice that the red and green curves track each other. The blue curve shows the measured pH at the Aloha station from 1990 to 2018. The drop in pH from 8.1 to 8.0 represents an increase in acidity (the concentration of hydrogen ions in the water) of about 50%.

Source: NOAA PMEL Carbon Program: www.pmel.noaa.gov/co2/ .Mauna Loa data from NOAA ESRL. ALOHA data adapted from Dore et al. 2009

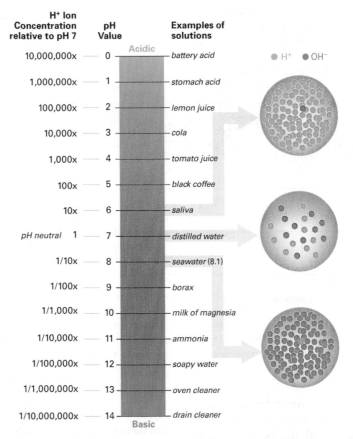

H⁺ Ion Concentration relative to pH 7	pH Value	Examples of solutions
10,000,000x	0	battery acid
1,000,000x	1	stomach acid
100,000x	2	lemon juice
10,000x	3	cola
1,000x	4	tomato juice
100x	5	black coffee
10x	6	saliva
pH neutral 1	7	distilled water
1/10x	8	seawater (8.1)
1/100x	9	borax
1/1,000x	10	milk of magnesia
1/10,000x	11	ammonia
1/100,000x	12	soapy water
1/1,000,000x	13	oven cleaner
1/10,000,000x	14	drain cleaner

Acidic

Basic

● H⁺   ● OH⁻

● **FIGURE 16.5** The pH scale is not linear, and small changes in pH correspond to large changes in acidity or alkalinity.

Source: http://www.whoi.edu/cms/images/oceanus/phScale_99988_286053_300859.gif

(A)

(B)

● **FIGURE 16.6** (A) Under normal conditions, the Great Barrier Reef in Australia is one of the most biologically diverse ecosystems on Earth. (B) Bleached, dead coral thickets in Douglas Reef offshore northern Queensland. Between 2016 and 2017, an estimated 50% of the entire Great Barrier Reef died, transforming from the vibrant ecosystem shown in (A) to the pile of calcium carbonate coral skeletons shown in (B).

ocean acidification A decrease in pH of the world's oceans resulting from higher atmospheric carbon dioxide concentrations that, in turn, cause more carbon dioxide to dissolve in ocean water. Once dissolved, the carbon dioxide forms carbonic acid, decreasing the pH.

Temperature

Recall from Chapter 11 that temperate lakes are comfortably warm for swimming during the summer because warm water floats on the surface and does not readily mix with the deep, cooler water. Much of Earth's oceans develop a similar temperature layering, but because sea waves and currents stir the surface water, the warm layer can extend from the surface to a depth as great as 450 meters (●**FIGURE 16.7**).

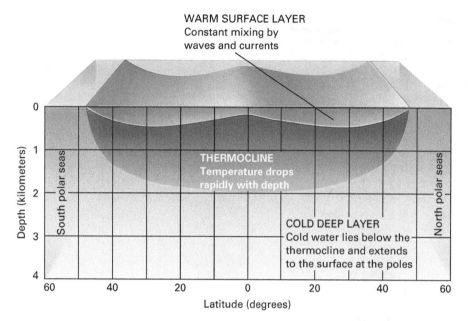

WARM SURFACE LAYER
Constant mixing by
waves and currents

THERMOCLINE
Temperature drops
rapidly with depth

COLD DEEP LAYER
Cold water lies below the
thermocline and extends
to the surface at the poles

South polar seas

North polar seas

Depth (kilometers)

Latitude (degrees)

● **FIGURE 16.7** There are three temperature layers in the ocean. The warm, upper layer exists from the surface downward to depths as great as 450 meters; temperature cools rapidly with depth in the thermocline, and seawater of the deep ocean is cold.

Below this layer is the thermocline, where the temperature drops rapidly with depth. The thermocline extends to a depth of about 2 kilometers. Beneath the thermocline, the temperature of ocean water varies from about 1°C to 2.5°C. The cold, dense water in the ocean depths mixes very little with the surface. Thus, there are three distinct temperature zones in the ocean, as shown in ●**FIGURE 16.7**. This layered structure does not exist in the polar seas because cold surface water sinks, causing vertical mixing.

TIDES

Even the most casual observer will notice that on any beach the level of the ocean rises and falls on a cyclical basis. If the water level is low at noon, it will reach its maximum height at about 6:13 p.m. and be low again at about 12:26 a.m. These vertical displacements are called **tides**. Most coastlines experience two high tides and two low tides during an interval of about 24 hours and 53 minutes.

Tides are caused by the gravitational pull of the Moon and Sun on the sea surface. Although the Moon is much smaller than the Sun, it is so much closer to Earth that its influence predominates. At any given time, one region of Earth (point A in ●**FIGURE 16.8**) lies directly closest to (i.e., "under") the Moon. Because gravitational force is greater for objects that are closer together, the part of the ocean nearest to the Moon is attracted with the strongest force. There, the ocean surface bulges upward toward the Moon, forming a tidal bulge. As a particular geographic point on Earth's coastlines rotates underneath a **tidal bulge**, that coastline experiences a high tide.

But so far our explanation is incomplete. As Earth spins on its axis, a given point on Earth passes directly under the Moon approximately once every 24 hours and 53 minutes, but the period between successive high tides is only 12 hours and 26 minutes. Why are there ordinarily two high tides in a day? The tide is high not only when a point on Earth is directly under the Moon but also when it is 180 degrees away. That is, the tidal bulge created by the gravitational effect of the Moon has two ends: one end faces the Moon and one end faces directly opposite the Moon on the other side of Earth.

To understand how a tidal bulge forms on the side of Earth facing away from the Moon, we must consider the Earth–Moon orbital system. Most people visualize the Moon orbiting around Earth, but it is more accurate to say that Earth and the Moon orbit around a common center of gravity. The two celestial partners are locked together like dancers spinning around in each other's arms. Just as the back of a dancer's dress flies outward as she twirls, the ocean surface on the opposite side of Earth from the Moon bulges outward. This bulge forms the high tide 180 degrees away from the Moon (point B in ●**FIGURE 16.8**). Thus, the tides rise and fall twice daily.

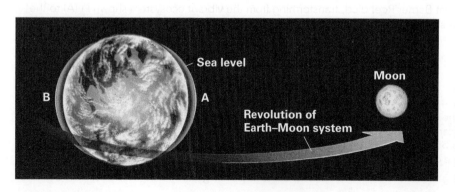

Sea level

B

A

Moon

Revolution of
Earth–Moon system

● **FIGURE 16.8** The Moon's gravity causes a high tide at point A, directly under the Moon. The rotation of the Earth–Moon pair around its common center of gravity causes a high tide at point B, opposite point A—in the same way the body of a child can be suspended above the ground surface by a strong adult as the two rotate around their common center of mass.

High and low tides do not occur at the same time each day but are delayed by approximately 53 minutes every 24 hours. Earth makes one complete rotation on its axis in 24 hours, but at the same time, the Earth–Moon pair is rotating around its common center of gravity. After a point on Earth makes one complete rotation in 24 hours, that point must continue to rotate for an additional 53 minutes before it catches up with the new position of the Moon resulting from rotation of the Earth–Moon pair. This is why the Moon rises approximately 53 minutes later each day. In the same manner, the tides are approximately 53 minutes later each day (●FIGURE 16.9).

Although the Sun's gravitational pull on the oceans is smaller than the Moon's, it does affect tides. When the Sun and Moon are directly in line with Earth, their gravitational fields combine to create a large tidal bulge. During these times, the variation between high and low tides is large, producing **spring tides** (●FIGURE 16.10A). When the Moon is 90 degrees out of alignment with the Sun and Earth, each partially offsets the effect of the other and the differences between the levels of high and low tide are smaller. These relatively small tides are called **neap tides** (●FIGURE 16.10B).

Tidal range is strongly affected by the shape of the coastline. For example, the Bay of Fundy is shaped like a giant funnel (Figure 16.2). The shape of the bay concentrates the rising and falling tides, and as a result the tidal range is as much as 15 meters during a spring tide. In contrast, tides vary by less than 2 meters along a straight stretch of coast such as that in Santa Barbara, California. In the central oceans, the tides average about 1 meter. Mariners consult tide tables that give the time and height of the tides in any area on any day.

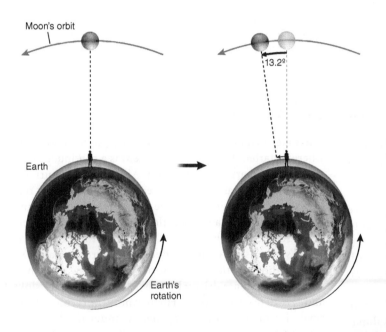

● **FIGURE 16.9** The Moon moves 13.2 degrees every day as a result of the difference in the time it takes Earth to complete one full rotation and the time it takes the Earth–Moon pair to complete one full rotation around its common center of gravity. (A) The Moon is directly above an observer on Earth. (B) One day later, Earth has completed one rotation, but the Moon has traveled 13.2 degrees farther than it was at that same time the previous day. Thus, Earth must now continue to rotate for another 53 minutes before the observer is again directly under the Moon.

tides The cyclic rise and fall of ocean water caused by the gravitational force of the Moon and, to a lesser extent, of the Sun.

tidal bulge A bulge that forms on the surface of Earth's oceans facing directly toward and away from the moon. The bulge facing the moon forms due to gravitational attraction of the ocean water by the moon, whereas the bulge facing away from the moon forms to compensate for the displacement of mass in the opposite bulge.

spring tides The very high and very low tides that occur when the Sun, Moon, and Earth are aligned and their gravitational fields combine to create a strong tidal bulge.

neap tides The relatively small tides that occur when the Moon is 90 degrees out of alignment with the Sun and Earth.

tidal range The vertical distance between low and high tide at a particular point on Earth's oceans.

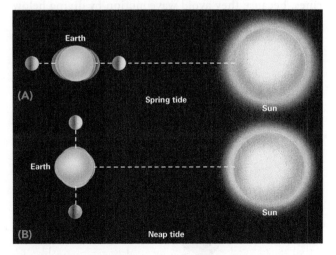

● **FIGURE 16.10** (A) Spring tides occur when Earth, Sun, and Moon are lined up, with their gravitational fields combining to create a strong tidal bulge. (B) Neap tides occur when the Moon lies at right angles to a line drawn between the Sun and the Earth.

SEA WAVES

Most ocean waves develop when wind blows across water. Waves vary from tiny ripples on an otherwise smooth surface to destructive giants that can topple beach houses and sink ships. In deep water, the size of a wave depends on (1) the wind speed, (2) the length of time that the wind has blown, and (3) the distance that the wind has traveled (sailors call this last factor fetch). A 25-kilometer-per-hour wind blowing for 2 to 3 hours across a 15-kilometer-wide bay generates waves about 0.5 meters high. But if a storm blows at 90 kilometers per hour for several days over a fetch of 3,550 kilometers, it can generate 30-meter-high waves, as tall as a ship's mast.

The highest part of a wave is called the **crest**; the lowest is the **trough** (●FIGURE 16.11). The **wavelength** is the distance between successive crests. The **wave height** is the vertical distance from the crest to the trough.

Recall from our discussion of earthquakes in Chapter 7 that when a wave travels through rock, the energy of the wave is transmitted rapidly over large distances, but the rock itself moves only slightly. In a similar manner, a single water molecule in a water wave does not travel with the wave. While the wave moves horizontally, the water molecules move in small circles, called **wave orbitals**, as shown in ●FIGURE 16.12. That is why a ball on the ocean bobs up and down and sways back and forth as the waves pass, but it does not travel along with the waves.

The circles of water movement produced as a wave passes by become smaller with depth. At a depth equal to about half the wavelength, the disturbance becomes negligible. This depth is called **wave base**, and it varies with the size of the waves. If you dive deep enough, you can escape wave motion from even the biggest storm waves that have a wavelength of around 500 meters and a wave base of about 250 meters. No one gets seasick in submarines, because they avoid wave motion by diving below local wave base.

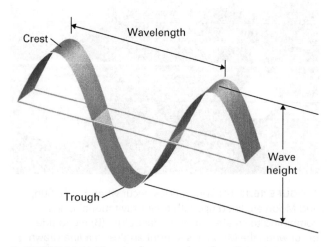

● **FIGURE 16.11** Terminology used to describe a wave.

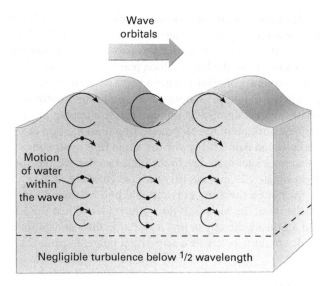

● **FIGURE 16.12** While a wave moves horizontally across the sea surface, the water itself moves only in small circles, called wave orbitals.

OCEAN CURRENTS

The water in an ocean wave oscillates in circles, ending up where it started. In contrast, a **current** is a continuous flow of water in a particular direction. Although river currents are familiar and easily observed, early mariners did not recognize ocean currents. **Surface currents**, caused by wind blowing over the sea surface, flow in the upper 400 meters of the seas and involve about 10 percent of the water in the world's oceans. In contrast, **deep-sea currents** transport seawater both vertically and horizontally below a depth of 400 meters and are driven by gravity as denser water sinks and less-dense water rises. Furthermore, in certain places several processes (discussed below under the headings "Deep-Sea Currents" and "Upwelling") cause surface water to sink to great depths, and in other places deep water rises to the sea surface. Thus, the entire global ocean circulates continuously. ●FIGURE 16.13 shows major surface currents and deep-sea currents in the Atlantic Ocean.

Surface Currents

In the mid-1700s, Benjamin Franklin lived in London, where he was deputy postmaster general for the American colonies. He noticed that mail ships took 2 weeks longer to sail from England to North America than did merchant ships. Franklin learned that the captains of the merchant ships had discovered a current flowing northward along the east coast of North America and then across the Atlantic to England. When sailing from Europe to North America, the merchant ships saved time by avoiding the current. The captains of the mail ships were unaware of this current and lost time sailing against it on their westward journeys. In 1769, Franklin and his cousin, Timothy Folger, a merchant captain, charted the current and named it the **Gulf Stream** (●FIGURE 16.14).

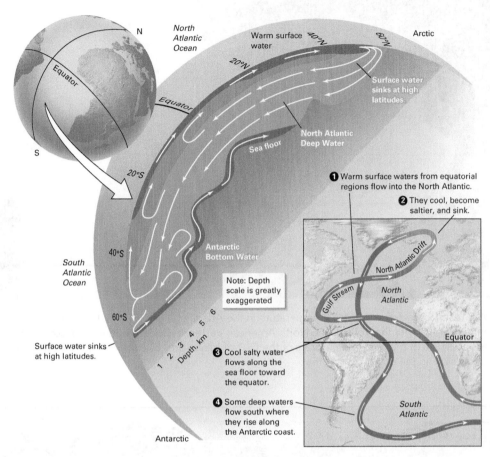

● **FIGURE 16.13** Surface and deep-sea currents of the Atlantic Ocean. Both types of currents profoundly affect climate and other components of Earth systems. This artist's rendition has simplified the currents to emphasize the major flow patterns.

crest The highest part of a wave.

trough The lowest part of a wave.

wavelength The distance between successive wave crests (or troughs).

wave height The vertical distance from the crest to the trough of a wave.

wave orbitals The circular motion of water that occurs as a surface wave passes by. Wave orbitals decrease in diameter with depth, ultimately disappearing altogether at wave base.

wave base The depth below which surface wave energy does not reach. Roughly as deep as half a single wavelength and limited by wind-driven surface waves to about 200 meters on the outer continental shelf.

current A continuous flow of water in a particular direction.

surface currents Horizontal flow of water in the upper 400 meters of the oceans, caused by wind blowing over the sea surface.

deep-sea currents Vertical and horizontal flow of water below a depth of 400 meters in the oceans, caused mainly by gravity-driven differences in water density.

Gulf Stream One of many oceanic surface currents. It begins near the Gulf of Mexico and travels northward up the Atlantic coast, grows wider and slower as it moves to the northeast, and bathes the coastlines of western Europe and Scandinavia with exceptionally warm water.

As ship traffic increased and navigators searched for the quickest routes around the globe, they discovered other ocean currents (●**FIGURE 16.15**). Ocean currents have been described as rivers in the sea. The analogy is only partially correct; ocean currents have no well-defined banks, and they carry much more water than even the largest river. The Gulf Stream is 80 kilometers wide and 650 meters deep near the east coast of Florida and moves at approximately 5 kilometers per hour, a moderate walking speed. As it moves northward and eastward, the current widens and slows; east of New York City, it is more than 500 kilometers wide and travels at less than 8 kilometers per day.

Another important difference between rivers and surface currents is that rivers flow in response to gravity, whereas ocean surface currents are driven primarily by wind. When wind blows across water in a constant direction for a long time, it drags surface water along with it, forming a current. In many regions, the wind blows in the same direction throughout the year, forming currents that vary little from season to season. However, in other places, changing winds cause currents also to change direction. For example, in the Indian Ocean prevailing winds shift on a seasonal basis. When the winds shift, the ocean currents follow suit.

● **FIGURE 16.14** A satellite image of sea surface temperature in the Atlantic Ocean. The Gulf Stream is clearly shown by the streak of dark orange (warm) water that follows the North American eastern seaboard and then passes eastward across the North Atlantic, passing south of the cold (blue) waters of the Gulf of Maine and Labrador Sea. Ireland and the southwest coast of England are visible in the upper right corner.

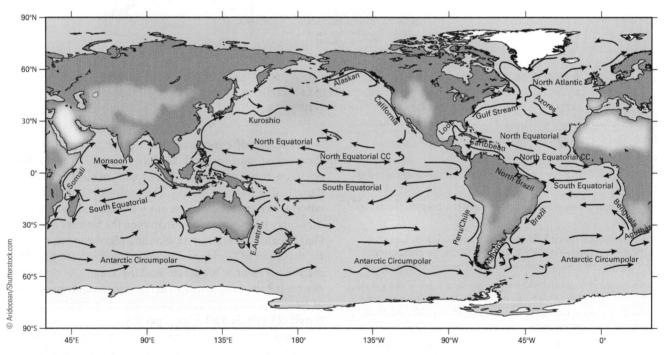

● **FIGURE 16.15** Major oceanic surface currents of the world.

Ocean currents profoundly affect Earth's climate. The Gulf Stream transports 1 million cubic meters of warm water northward past any point every second, warming both North America and Europe. For example, Churchill, Manitoba, is a village in the interior of Canada. It is frigid and icebound for much of the year. Polar bears regularly migrate through town. Yet it is at about the same latitude as Glasgow, Scotland, which is warmed by the Gulf Stream and therefore experiences relatively mild winters. In general, the climate in Western Europe is warmer than that at similar latitudes in other regions not heated by tropical ocean currents.

Why Surface Currents Flow in the Oceans

Surface currents are driven primarily by friction between wind blowing over the sea surface and surface water. The winds simply drag the sea surface along in the same direction that the winds are blowing. In ●FIGURE 16.16, the green arrows show the major prevailing winds over the North and South Atlantic Oceans. The orange arrows show the elliptical surface currents, called **gyres**, in the same regions, simplified from Figure 16.13. The North Atlantic gyre circulates in a clockwise direction, and the southern gyre circulates counterclockwise.

The circular motion characteristic of gyres tends to trap and accumulate floating debris, especially human garbage. First discovered in the early 1990s, the Great Pacific Garbage Patch is a region of floating garbage that has accumulated in the center of the eastern Pacific Ocean between California and Hawaii (●FIGURE 16.17A). Because the gyres move in a circular fashion, floating human garbage in the oceans tends to accumulate in the center. Although the size of the accumulation is difficult to measure and it is possible to sail directly through the garbage patch without noticing an obvious accumulation, the size of the Great Pacific Garbage Patch is estimated to be about twice the size of Texas. Similar but smaller floating garbage patches exist in the western Pacific and north Atlantic.

Although the name conjures up images of a floating garbage heap, these gyre-produced concentrations of garbage consists mostly of small pieces of floating or suspended plastic (●FIGURE 16.17B). Most commonly used plastics do not decompose in the ocean but rather break down into smaller and smaller pieces. This **microplastic debris** is mixed and moved about by wave and wind energy and disperses over huge surface areas throughout the upper part of the ocean while also mixing with ocean water below the surface. At present, scientists understand very little about the effects these dispersing pieces of plastic have on the ocean ecosystem.

Notice that the currents do not flow in exactly the same directions as the prevailing winds. Instead, the east–west surface currents in the Northern Hemisphere are deflected to the right of the winds. Where their flow is blocked by a continent, the currents veer clockwise to continue their circuit. In the Southern Hemisphere, the currents are deflected to the left of the winds and turn counterclockwise when they

gyres A circular or elliptical current in either water or air.

microplastic debris Small pieces of plastic, typically a few millimeters or centimeters long, that result from accumulation and concentration of plastic garbage in parts of the ocean. The debris results from wave and wind energy that breaks plastic litter down into smaller and smaller pieces. The effects of microplastics in the ocean ecosystem are presently unknown.

Coriolis effect A deflection of air or water currents caused by the rotation of Earth.

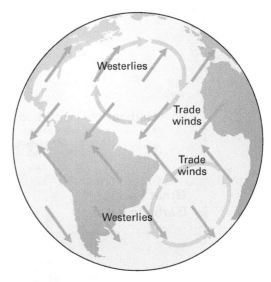

●**FIGURE 16.16** The green arrows show major prevailing winds over the North and South Atlantic Oceans—the trade winds (easterlies) and the westerlies. The orange arrows show the surface oceanic currents in the same regions.

encounter land. These differences between prevailing wind directions and the flow directions of the surface currents suggest that forces other than wind also affect the great gyres.

One force that deflects the gyres away from the prevailing winds is the **Coriolis effect**, named for Gaspard-Gustave de Coriolis, the 19th-century French scientist who described it. To understand this effect, consider the rotating Earth: The circumference of Earth is greatest at the equator and decreases to zero at the poles. But all parts of the planet make one complete rotation every day. Therefore, a point on the equator must travel farther and faster than any point closer to the poles. At the equator, all objects move eastward with a velocity of about 1,600 kilometers per hour; at the poles there is no eastward movement at all and the velocity is 0 kilometers per hour.

Now imagine a rocket fired from the equator toward the North Pole. Before it was launched, it was traveling eastward at 1,600 kilometers per hour with the rotating Earth. As it takes off, it is moving eastward at 1,600 kilometers per hour and northward at its launch speed. As it moves north from the equator, it is still traveling eastward at 1,600 kilometers per hour, but points on Earth beneath it move eastward at a slower and slower speed as the rocket approaches the pole. As a result, the rocket curves toward the east, or the right. In a similar manner, a mass of water or air deflects in an easterly direction as it moves north from the equator, as shown in ●FIGURE 16.18A.

Conversely, consider an ocean current flowing southward from the Arctic Ocean toward the equator. Since it started near the North Pole, this water moves more slowly to the east than Earth's surface near the equator, and therefore the current veers toward the west, or to the right, as shown in ●FIGURE 16.18B. Thus, north–south currents always veer to the right in the Northern Hemisphere. In the Southern Hemisphere, currents turn toward the left for the same reason.

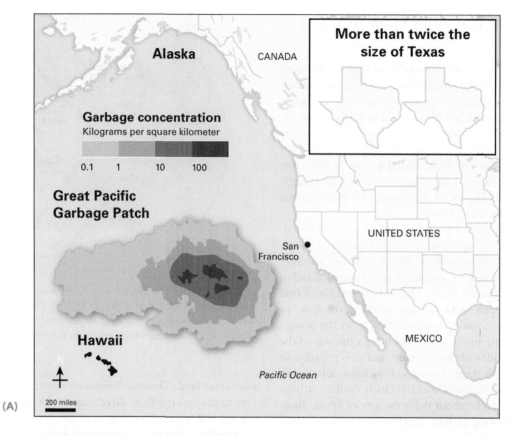

● **FIGURE 16.17** (A) The Great Pacific Garbage Patch is a zone in the eastern Pacific Ocean where the North Pacific subtropical gyre concentrates floating and submerged pieces of plastic garbage. Similar floating "garbage patches" exist in the western Pacific and north Atlantic. (B) Many commonly used plastics do not decompose in the ocean but instead break down into smaller and smaller pieces that make up most of the litter in the Great Pacific Garbage Patch.

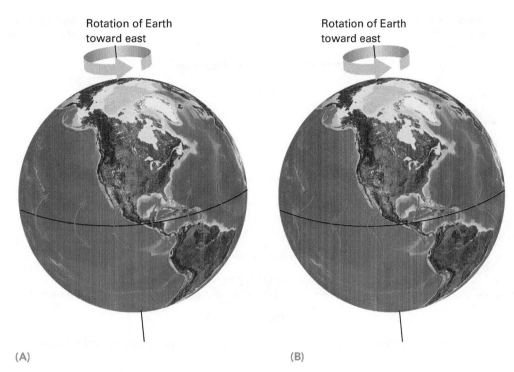

Rotation of Earth
toward east

Rotation of Earth
toward east

(A)   (B)

● **FIGURE 16.18** The Coriolis effect deflects water and wind currents. (A) Water or air moving poleward from the equator is traveling in an easterly direction faster than the surface of the Earth directly below and veers to the east (turns right in the Northern Hemisphere and left in the Southern Hemisphere). (B) Water or air moving toward the equator is traveling in an easterly direction slower than the surface of the Earth directly below and veers to the west (turns right in the Northern Hemisphere and left in the Southern Hemisphere).

At the sea surface, both the prevailing winds and the Coriolis effect affect current directions. However, below the surface, the water doesn't "feel" the wind; it "feels" only the movement of water directly above. The Coriolis effect is as strong at depth as it is at the sea surface. As a result, each successively deeper layer is less affected by the wind and more affected by the Coriolis effect. Consequently, deeper layers of water in the surface currents are deflected even more to the right in the Northern Hemisphere, and to the left in the Southern Hemisphere, than is the shallowest water. The net effect of this process on the flow directions of the gyres extends down to a depth of about 100 meters and is called **Ekman transport**, after Vagn Walfrid Ekman, the Swedish scientist who developed the mathematics that describes the process.

Ekman transport The natural process by which surface water moved by wind drags the layer of ocean water below it, which in turn drags the layer below it, and so forth, to a depth that depends on the wind strength. Only the surface layer responds directly to the wind. The layers below the surface respond both to the directional movement of the water layer directly above and to the Coriolis effect. Deeper layers are deflected more by the Coriolis effect than shallower levels.

thermohaline circulation The force behind deep-sea currents, created by differences in water temperature and salinity and therefore differences in water density.

Deep-Sea Currents

Wind does not affect the ocean depths, and oceanographers once thought, therefore, that water of the deep ocean was almost motionless and the seafloor topography changed little over time as a result. In her 1951 book, *The Sea around Us*, Rachael Carson wrote that the ocean depths are "a place where change comes slowly, if at all." However, in 1962, ripples and small dunes were photographed on the floor of the North Atlantic. Because flowing water forms these features, the photographs demonstrated that water must be moving in the ocean depths. More recently, oceanographers have measured deep-sea currents directly with flow meters and photographed moving sand and mud with underwater video cameras.

Wind drives surface currents, but deep-sea currents are driven by differences in water density. Dense water sinks and flows horizontally along the seafloor, forming a deep-sea current. Two factors cause water to become dense and sink: decreasing temperature and increasing salinity. The global deep-sea circulation shown in ●**FIGURES 16.18** and **16.19** is caused by these two factors and is called **thermohaline circulation** (thermo for temperature and haline for salinity).

Recall that water is most dense when it is cold, close to freezing. Therefore, as tropical surface water moves poleward and cools, it becomes denser and sinks. In addition, water density increases as salinity increases. So water sinks

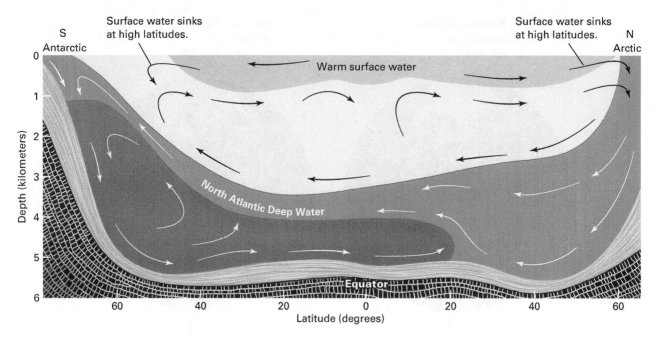

● **FIGURE 16.19** A north (right) to south (left) profile of the Atlantic Ocean showing surface and subsurface currents.

when it becomes saltier. Seawater can become saltier if surface water evaporates. Polar seas also become saltier when the surface freezes, because salt does not become incorporated into the ice. Thus, Arctic and Antarctic water is dense because the water is both cold and salty.

In contrast, addition of freshwater makes seawater less salty and less dense. This effect is pronounced in enclosed bays with abundant, inflowing fresh river water and also in the polar regions when icebergs float into the ocean and melt. As we will see in Chapter 21, recent rapid melting of Greenland glaciers has introduced enough freshwater into the North Atlantic to reduce the density of surface water and alter its buoyancy.

Figure 16.15 shows that the Gulf Stream originates in the subtropics. When this water reaches the northern part of the Atlantic Ocean near the tip of Greenland, it is cool and salty enough to sink. When the sinking water reaches the seafloor, it is deflected southward to form the North Atlantic Deep Water (●**FIGURE 16.19**), which flows along the seafloor all the way to Antarctica. An individual water molecule that sinks near Greenland may travel for 500 to 2,000 years before resurfacing half a world away in the south polar sea.

Upwelling

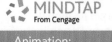

Animation: Coastal Upwelling

If water sinks in some places, it must rise in others to maintain mass balance. This upward flow of water is called **upwelling**. Upwelling carries cold water from the ocean depths to the surface. Upwelling also brings nutrients from the deep ocean to the surface, creating rich fisheries along the coasts of California and Peru. Several processes can cause upwelling, both in the open oceans and along coastlines.

In Figure 16.15, note that the California Current flows southward along the coast of California. In the Southern Hemisphere, the Humboldt Current moves northward along the west coast of South America. Both currents are deflected westward—away from shore—by the Coriolis effect modified by Ekman transport. As these surface currents veer away from shore, water from the ocean depths rises toward the surface along the California coast and the west coast of South America. In August, water in the mid-Atlantic coast of the United States is warmed by the Gulf Stream and may be a comfortable 21°C. However, at the same latitude on the central California coast, the cool California Current combines with the upwelling deep water to produce water that is only 15°C, and surfers and swimmers must wear wetsuits to stay in the water for long.

The frictional drag of a prevailing offshore wind can pull surface water away from a coast. Deep water then upwells along the edge of the continental shelf to replace the surface water flowing away from shore. Winds blowing parallel to shore can also create upwelling. The wind causes water to flow parallel to shore, but the Coriolis effect and Ekman transport deflect the current, driving it away from the coast. Deep water then rises to replace the surface water.

In most years, an offshore wind drives surface water away from the coast of Peru and adjacent portions of western South America, creating a strong, nutrient-rich upwelling current that produces rich fisheries along that coast. However, about every 3 to 7 years—in El Niño years—the offshore wind weakens and the upwelling does not occur or is much reduced in magnitude. As a result, unusually warm, nutrient-poor water accumulates along the west coast of South America, displacing the cold Humboldt Current. El Niño's effects last for about a year before conditions return to normal. Many meteorologists now think that El Niño affects weather patterns for nearly three-quarters of Earth. The causes and effects of El Niño are described in Chapter 19.

Figure 16.15 shows that the south equatorial currents of both the Atlantic and Pacific flow near the equator. Although the Coriolis effect is weak near the equator, in the open oceans, water in equatorial currents is deflected poleward. As equatorial surface water moves toward each pole, it diverges, causing deeper, colder water to rise and replace the surface water, in what is called **equatorial upwelling**.

THE SEACOAST

As we mentioned in the introduction to this chapter, coastlines are among the most geologically active zones on Earth, where atmosphere, geosphere, hydrosphere, and biosphere all affect the local environments. Subduction occurs along many continental coasts. Waves and currents weather, erode, transport, and deposit sediment continuously on all coastlines. In addition, the shallow waters of continental shelves are among the most productive biological ecosystems on Earth. Many organisms that live in these regions build hard shells or skeletons of calcium carbonate. If they do not disintegrate following the organism's death, the remains of these shells and skeletons can become incorporated into mudstone, sandstone, or limestone as clasts. Thus, living organisms become part of the rock cycle.

Weathering and Erosion on the Seacoast

Waves batter most coastlines. If you walk down to the shore, you can watch turbulent water carry sand grains or even small cobbles along the beach. Even on a calm day, waves steepen as they approach shoreline, crash in the surf zone, and wash up against the beach.

When a wave enters water shallower than its wave base, the bottom of the wave drags against the seafloor. This drag compresses the circular motion of the

---

upwelling A rising ocean current that transports cold water and nutrients from the depths to the surface.

equatorial upwelling Oceanic upwelling in which surface currents flowing westward on both sides of the equator are deflected poleward and are replaced by upward flow of deeper, nutrient-rich waters.

breaks To collapse or crash, as when a wave approaches a beach and the front of the wave rises over its base, growing steeper until it collapses forward.

surf The chaotic, turbulent waves breaking along the shore.

refraction The bending of a wave that occurs when it approaches the shore at an angle; the end of the wave in shallow water slows down, while the end in deeper water continues at a faster speed.

---

lowermost wave orbitals into ellipses and slows down the lower part of the wave. At first, the incoming wave simply builds upward from a swell to a wave as the continually shallowing bottom causes the incoming water to bunch up, forcing the surface of the wave upward. Eventually, however, the bottom of the wave becomes so deformed by drag and the upper part of the growing wave gets so far ahead of the lower part that the wave collapses forward, or **breaks** (●FIGURE 16.20). Chaotic, turbulent waves breaking along a shore are called **surf**.

Most coastal erosion occurs during intense storms because storm waves are much larger and more energetic than normal waves (●FIGURE 16.21). A 6-meter-high wave strikes shore with 40 times the force of a 1.5-meter-high wave. A giant 10-meter-high storm wave strikes a 10-meter-wide seawall with 4 times the thrust energy of the three main orbiter engines of a space shuttle.

Seawater weathers and erodes coastlines by hydraulic action, abrasion, solution, and salt cracking—processes that are familiar from our earlier discussions of weathering and streams.

A wave striking a rocky cliff drives water into cracks or crevices in the rock, compressing air in the cracks. The air and water combine to create hydraulic forces strong enough to dislodge rock fragments or even huge boulders. Storm waves create forces as great as 25 to 30 tons per square meter. Engineers built a breakwater in Wick Bay, Scotland, of car-sized rocks weighing 80 to 100 tons each. The rocks were bound together with steel rods set in concrete, and the seawall was topped by a steel-and-concrete cap weighing more than 800 tons. A large storm broke the cap and scattered the rocks about the beach. The breakwater was rebuilt, reinforced, and strengthened, but a second storm destroyed this wall as well. On the Oregon coast, the impact of a storm wave tossed a 60-kilogram rock over a 25-meter-high lighthouse. After sailing over the lighthouse, it crashed through the roof of the keeper's cottage, startling the inhabitants.

While images of flying boulders are spectacular, most wave erosion occurs gradually, by abrasion. Water is too soft to abrade rock, but waves carry large quantities of silt, sand, and gravel. Breaking waves roll this sediment back and forth over bedrock, acting like liquid sandpaper, eroding the rock. At the same time, smaller cobbles are abraded as they roll back and forth in the surf zone.

Seawater slowly dissolves rock and carries ions in solution. Saltwater also soaks into bedrock; when the water evaporates, the growing salt crystals pry the rock apart.

Sediment Transport along Coastlines

Most waves approach the shore at an angle rather than head-on. When this happens, one end of the wave encounters shallow water and slows down, while the rest of the wave is still in deeper water and continues to advance at a relatively faster speed. As a result, the wave bends (●FIGURE 16.22A and 16.22B). This effect is called **refraction**. Consider the analogy of a sled gliding down a snowy hill onto a cleared, paved pathway. If the sled hits the paved surface at an angle, one

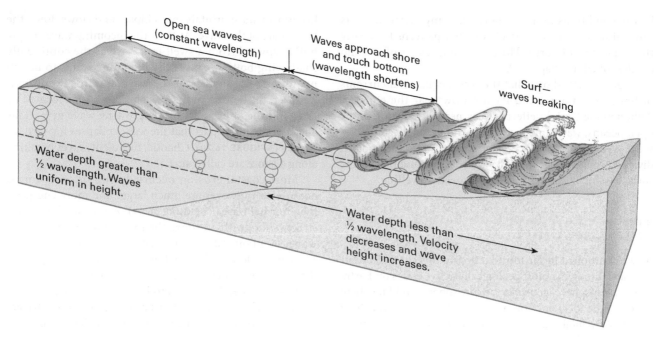

● **FIGURE 16.20** When a wave approaches the shore, wave orbitals at the bottom of the wave flatten out and becomes elliptical. The bottom of the wave drags against the seafloor. As a result, the wavelength shortens and the wave steepens until it finally breaks, creating surf.

runner will reach it before the other. The runner that hits the pavement first slows down, while the other, which is still on the snow, continues to travel rapidly (●**FIGURE 16.22C**). As a result, the sled turns abruptly.

As waves approach an irregular coast, they reach the headlands first, breaking against the point and eroding it. The waves then refract around the headland and travel parallel to its sides. Surfers seek such refracted waves because these waves move nearly parallel to the coast and therefore travel for a long distance before breaking. For example, the classic big-wave mecca at Waimea Bay of Oahu, Hawaii,

forms along a point of land that juts into the sea. Waves build when they strike an offshore coral reef, then refract along the point and charge into the bay.

Refracted waves transport sand and other sediment toward the interior of a bay. As the headlands erode and the interiors of bays fill with sand, an irregular coastline eventually smooths out (●**FIGURE 16.23**).

When waves strike the coast at an angle, they form a **longshore current** that flows parallel to the shore (●**FIGURE 16.24**). Longshore currents involve water from the surf zone and a little further offshore and may travel for tens or even hundreds of kilometers parallel to the coastline. They transport sand for great distances. Longshore sediment transport also occurs by **beach drift**. If a wave strikes the beach obliquely, it pushes sand up and along the beach in the direction that the wave is traveling. When water recedes, the sand flows straight down the beach, as shown in ●**FIGURE 16.24**. Thus, at the end of one complete wave cycle, the sand has moved a short distance parallel to the shoreline. The next wave transports the sand a little farther, and so on. Over time, sediment can move many kilometers down a coastline in this manner.

Longshore currents and beach drift work together to transport and deposit huge amounts of sand along a coast. Much of the sand found at Cape Hatteras, North Carolina, originated hundreds of kilometers away, from the mouth of the Hudson River in New York

● **FIGURE 16.21** Expensive seashore houses such as these along the California coast are often threatened by coastal erosion that occurs during intense storms.

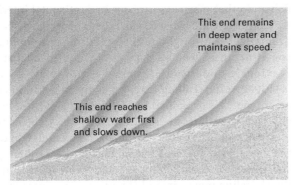

(A)

Sled analogy

Snowy hill

Road

(C)

(B)

● **FIGURE 16.22** (A) When a water wave strikes the shore at an angle, the end in shallow water slows down, causing the wave to bend, or refract. (B) Wave refraction on a lake surface. (C) A sled turns when it strikes a paved surface at an angle because one of the sled's runners hits the roadway and slows down before the other does.

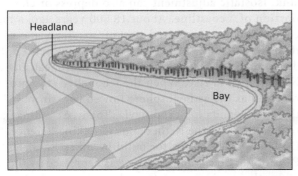

(A)

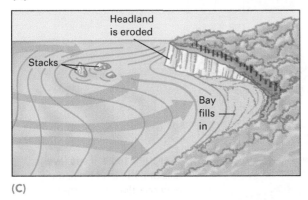

(C)

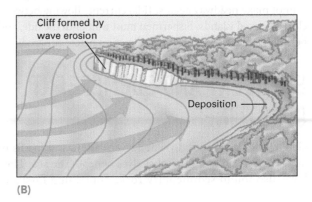

(B)

● **FIGURE 16.23** (A) When a wave strikes a headland, the shallow water causes that portion of the wave to slow down. (B) Part of the wave breaks against the headland, weathering the rock. A portion of the wave refracts, transporting sediment and depositing it inside the bay. (C) Eventually, this selective weathering, erosion, and deposition will smooth out an irregular coastline.

longshore current A current generated when waves strike a shore at an angle, producing flow parallel and close to the coast. Some longshore currents are capable of transporting sand for hundreds of kilometers along the coastline.

beach drift The gradual movement of sediment (usually sand) along a beach, parallel to the shoreline, when waves strike the beach obliquely but return to the sea directly.

and from glacial deposits on Long Island and southern New England. Midway along this coast, at Sandy Hook, New Jersey, an average of 2,000 tons of sandy sediment a day will move past any point on the beach. As a result of these longshore currents, some beaches have been called rivers of sand.

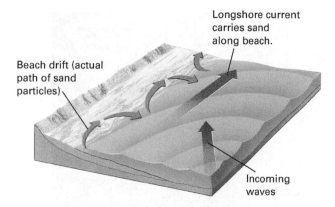

● **FIGURE 16.24** Longshore currents and beach drift transport sediment along a coast.

Tidal Currents

When tides rise and fall along an open coastline, water moves in and out from the shore as a broad sheet. If the flow is channeled by a bay with a narrow entrance or by islands, the moving water funnels into a **tidal current**, which is a flow of ocean water caused by tides. Tidal currents can be intense where large differences exist between high and low tides and narrow constrictions occur in the shoreline. On parts of the west coast of British Columbia, a diesel-powered fishing boat cannot make headway against tidal currents flowing between closely spaced islands. Fishermen must wait until the tide, and hence the tidal current, reverses direction before proceeding.

EMERGENT AND SUBMERGENT COASTLINES

Geologists have found drowned river valleys and fossils of land animals on continental shelves beneath the sea. They have also found sedimentary rocks containing fossils of fish and other marine organisms in continental interiors. As a result, we infer that sea level has changed, sometimes dramatically, throughout geologic time. An **emergent coastline** forms when a portion of a continent that was previously underwater becomes exposed as dry land (●**FIGURE 16.25**). Falling sea level or rising land can cause emergence. As explained in section on "Beaches," many emergent coastlines are sandy. In such cases, the delivery of sandy sediment to the coastline is high enough to cause the shoreline to migrate seaward. In contrast, a **submergent coastline** develops when the sea floods low-lying land and the shoreline moves inland (Figure 16.25). Submergence occurs when sea level rises or coastal land sinks. A submergent coast is commonly irregular, with many bays, sea cliffs, and headlands. The coast of Maine, with its numerous fjords, inlets, and rocky bluffs, is a classic submergent coastline (●**FIGURE 16.26**). Small, sandy beaches form in some protected coves, but most of the shoreline is rocky and steep because it consists of eroding bedrock and glacial sediment.

Factors That Cause Coastal Emergence and Submergence

Tectonic processes can cause a coastline to rise or sink. Isostatic adjustment can also depress or elevate a portion of a coastline. About 18,000 years ago, a huge continental glacier covered most of Scandinavia, causing it to sink isostatically. As the lithosphere settled, the displaced asthenosphere flowed southward, causing the Netherlands to rise. When the ice melted, the process reversed as the asthenosphere flowed back from below the Netherlands to Scandinavia. Today, Scandinavia is rebounding and the Netherlands is sinking. (Hence, the Dutch are well known for building dikes.) During the Pleistocene Glaciation, Canada was depressed by the ice, and asthenosphere rock flowed southward. Today, the

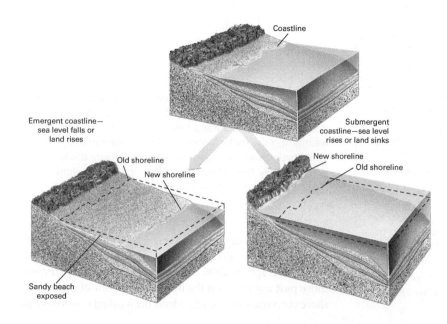

● **FIGURE 16.25** If sea level falls or if the land rises, the new coastline is emergent. Offshore sand is exposed to form a sandy beach. If coastal land sinks or sea level rises, the new coastline is submergent. Areas that were once land are flooded. Irregular shorelines develop, and beaches are commonly rocky.

● **FIGURE 16.26** The Maine coast is a rocky, irregular, submergent coastline.

asthenosphere is flowing back north, much of Canada is rebounding, and much of the continental United States is sinking.

Sea level can also change globally. A global sea level change, called **eustatic sea level change**, occurs by three mechanisms: the growth or melting of glaciers, changes in water temperature, and changes in the volume of the Mid-Oceanic Ridge.

During an ice age, vast amounts of water are withdrawn from the sea to form continental glaciers, causing sea level to drop and coastlines around the world to emerge. Similarly, when glaciers melt, sea level rises globally, causing global submergence of coastlines.

---

tidal current A current, channeled by a bay with a narrow entrance or by closely spaced islands, caused by the rise and fall of the tides.

emergent coastline A coastline that was previously underwater but has become exposed to air, because either the land has risen or sea level has fallen.

submergent coastline A coastline that was previously above sea level but has been drowned, because either the land has sunk or sea level has risen.

eustatic sea level change A global sea level change caused by three different processes: the growth or melting of glaciers, changes in water temperature, and changes in the volume of the Mid-Oceanic Ridge.

beach Any strip of shoreline washed by waves and tides.

shoreface The seaward-sloping seafloor surface extending from the mean low tide line to the mean fair-weather wave base. Includes that part of a beach in which incoming waves reach the bottom, steepen, and break, releasing much mechanical energy.

foreshore or intertidal zone The part of a beach that lies between the high-tide and low-tide lines and is exposed to the air at low tide but covered by water at high tide.

backshore The uppermost zone of a beach, consisting typically of a dry sandy surface that slopes gently landward but that is washed over by waves during large storms.

Seawater expands when it is heated and contracts when it is cooled. Although this change is not noticeable in a glass of water, the volume of the oceans is so great that a small temperature change can alter sea level measurably. As a result, global warming causes sea level rise, and global cooling leads to falling sea level.

Temperature changes and glaciation are linked. When global temperature rises, seawater expands and glaciers melt. Thus, even minor global warming can lead to a large sea level rise. The opposite effect is also true: when temperature falls, seawater contracts, glaciers grow, and sea level falls.

As explained in Chapter 15, changes in the volume of the Mid-Oceanic Ridge can also affect sea level. The ridge displaces seawater. When lithospheric plates spread slowly from the Mid-Oceanic Ridge, they create a narrow ridge that displaces relatively little seawater, resulting in low sea level. In contrast, rapidly spreading plates produce a high-volume ridge that displaces more water, causing a global sea level rise. At times in Earth's history, spreading has been relatively rapid, and as a result, global sea level has been high.

BEACHES

When most people think about going to the **beach**, they think of gently sloping expanses of sand. However, a beach is any strip of shoreline that is washed by waves and tides. Although many beaches are sandy, others are swampy or rocky (●**FIGURE 16.27**).

A beach is divided into three zones, although only two of these are exposed above the sea surface. The **shoreface** extends from mean low tide to mean fair-weather wave base. The shoreface typically slopes seaward and is the zone of wave shoaling, in which waves slow down, steepen, and break, expending mechanical energy against the bottom. The **foreshore**, also called the **intertidal zone**, lies between mean high and low tides and is alternately exposed to the air at low tide and covered by water at high tide. The **backshore** exists above the foreshore and is usually dry but is washed over by waves during storms. It commonly slopes gently landward as a result of sediment delivered and deposited by storm waves losing energy as they wash inland. Many terrestrial plants cannot survive in saltwater, so specialized, salt-resistant plants live in the backshore. The backshore can be wide or narrow, depending on its topography, the local tidal difference, and the frequency and intensity of storms. In a region where the land rises steeply, the backshore may be a narrow strip. In contrast, if the coast consists of low-lying plains and if coastal storms occur regularly, the backshore may extend several kilometers inland.

If weathering and erosion occur along all coastlines, why are some beaches sandy and others rocky? The answer lies partly in the fact that most sand is not formed by weathering and erosion at the beach itself. Instead, several processes

● **FIGURE 16.27** Only very small beaches form on the emergent central California coastline between Monterrey and Big Sur.

transport sand to a seacoast. Rivers carry large quantities of sand, silt, and clay to the sea and deposit it on deltas that may cover thousands of square kilometers. In some coastal regions, glaciers deposited large quantities of sandy till along coastlines during the Pleistocene Ice Age. In tropical and subtropical latitudes, eroding reefs supply carbonate sand to nearby beaches. A sandy coastline is one with abundant sediment from one or more of these sources.

Longshore currents transport and deposit the sand along the coast. Much of the sand carried by these currents accumulates on underwater offshore bars. Thus, a great deal of sand may be stored offshore from a beach. If such a coastline emerges, this vast supply of sand becomes exposed as dry land and is available, if eroded, for building more beaches and more offshore bars. Thus, sandy beaches are abundant on emergent coastlines.

In contrast, rocky coastlines occur where sediment from any of these sources is scarce. With no abundant sources of sand, small sandy beaches may form in protected bays but most of the coast will be rocky. On a submergent coastline, rising sea level puts the stored offshore sand even farther out to sea and below the depth of waves. As a result, submergent coastlines commonly have beaches that are rocky, not sandy.

Sandy Coastlines

A **spit** is a small, fingerlike ridge of sand or gravel that extends outward from a beach (●**FIGURE 16.28**). As sediment migrates along a coast, the spit may continue to grow. A well-developed spit may rise several meters above high-tide level and may be tens of kilometers long. A spit may block the entrance to a bay, forming a **baymouth bar**. A spit may also extend outward into the sea, creating a trap for other moving sediment.

A **barrier island** is a long, low-lying sandy island that extends parallel to the shoreline. It looks like a beach or spit and is separated from the mainland by a sheltered body of water called a **lagoon**. Barrier islands extend along the east coast of the United States from New York to Florida. They are so nearly continuous that a sailor in a small boat can navigate the entire coast inside the barrier island system and remain protected from the open ocean most of the time. Barrier islands also line the Texas Gulf Coast.

Barrier islands form in several ways. The two essential ingredients are a large supply of sand and the waves or currents to transport it. If a coast is shallow for several kilometers outward from shore, breaking storm waves may carry sand toward shore and deposit it just offshore as a barrier island. Alternatively, if a longshore current veers out to sea, it slows down and deposits sand where it reaches deeper water. Waves may then pile up the sand, forming a barrier island.

Other mechanisms that create barrier islands involve sea level change. Underwater sand bars may be exposed as a coastline emerges. Alternatively, sand dunes or beaches may form barrier islands if a coastline sinks.

DEVELOPMENT ON SANDY COASTLINES The Atlantic coast of the United States is fringed with the longest chain of barrier islands in the world. Many seaside resorts are built

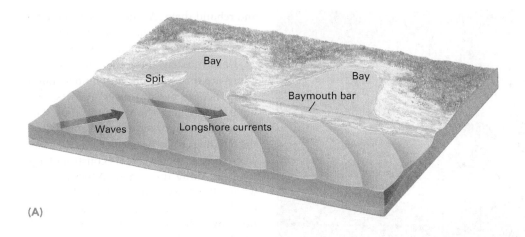

(A)

(B)

(C)

● **FIGURE 16.28** (A) Spits and baymouth bars are common features of sandy emergent coastlines with substantial longshore drift. (B) Aerial photograph of a spit that formed along a low-lying coast in northern Siberia. (C) Longshore drift has produced this baymouth bar across the mouth of a small stream along the Northern California coastline.

on these islands, and developers often ignore the fact that they are transient and changing landforms (●**FIGURE 16.29**). If the rate of erosion exceeds that of deposition for a few years in a row, a barrier island can shrink or disappear completely, leading to destruction of beach homes and resorts. In addition, barrier islands are especially vulnerable to hurricanes, which can wash over low-lying islands and move enormous amounts of sediment in a very brief time. In September 1996, Hurricane Fran flattened much of Topsail Island, a low-lying barrier island in North Carolina. Geologists were not surprised, because the homes were not only built on sand but they were built on sand that was virtually guaranteed to move.

---

spit A small ridge of sand or gravel extending from a beach into a body of water.

baymouth bar A spit that extends partially or completely across the entrance to a bay.

barrier island A long, narrow, low-lying island that extends parallel to the shoreline.

lagoon A sheltered body of water separated from the sea by a reef or barrier island.

As a second example, Long Island extends eastward from New York City and is separated from Connecticut by Long Island Sound. Longshore currents flow westward, eroding sand from glacial deposits at the eastern end of the island and depositing it to form beaches and barrier islands on the south side of the island (●**FIGURE 16.30**). At any point along the beach, the currents erode and deposit sand at approximately the same rates (●**FIGURE 16.31A**). Geologists calculate that the supply of sand at the eastern end of the island is large enough to last a few hundred years. When the glacial deposits at the eastern end of the island become exhausted, however, the flow of sand will cease. Then the entire coastline will erode and the barrier islands and beaches will disappear.

Now let's narrow our time perspective and look at a Long Island beach over a season or during a single storm. Over this time frame, the rates of erosion and deposition are not equal. Thus, beaches shrink and expand with the seasons or the passage of violent gales. In the winter, violent waves and currents erode beaches, whereas sand accumulates on the beaches during the calmer summer months. In an effort to prevent these seasonal fluctuations and to protect their

● **FIGURE 16.29** Many resorts, condos, and homes are built on transient and changing barrier islands, such as these on Hutchinson Island, Florida.

personal beaches, Long Island property owners have built stone barriers called **groins** from shore out into the water (●**FIGURE 16.32**). The groin intercepts the steady flow of sand moving from the east and keeps that particular part of the beach from eroding. But the groin impedes the overall flow of sand. West of the groin, the beach erodes as usual, but the sand is not replenished because the upstream groin traps it. As a result, beaches downcurrent from the groin erode away (●**FIGURE 16.31B**). The landowners living downcurrent from groins may then decide to build other groins to protect their beaches (●**FIGURE 16.31C**). The situation has a domino effect, with the net result that millions of dollars are spent in ultimately futile attempts to stabilize a system that was naturally stable in its own dynamic manner (●**FIGURE 16.33**).

Storms pose another dilemma. Hurricanes commonly strike Long Island in the late summer and fall, generating storm waves that flatten dunes, erode beaches, and transport large volumes of sand into nearby salt marshes, disrupting the ecosystem there. When the storms are over, gentler waves and longshore currents carry sediment back to the beaches

and rebuild them. As the sand accumulates again, salt marshes rejuvenate and the dune grasses grow back within a few months.

Unfortunately, these short-term fluctuations are incompatible with human ambitions. People build houses, resorts, and hotels on or near the shifting sands. The owner of a home or resort hotel cannot allow the buildings to be flooded or washed away. Therefore, property owners construct large seawalls along the beach. When a storm wave rolls across an undeveloped low-lying beach, it dissipates its energy gradually as it flows over the dunes and transports sand. The beach is like a judo master who defeats an opponent by yielding to the attack, not countering it head-on. A seawall interrupts this gradual absorption of wave energy. The waves crash violently against the barrier and erode sediment at its base until the wall collapses. It may seem surprising that a reinforced concrete seawall is more likely to be permanently destroyed than a beach of grasses and sand dunes, yet this is often the case (●**FIGURE 16.34**).

Rocky Coastlines

A rocky coastline is one without any of the abundant sediment sources described previously. In many areas on land, bedrock is exposed or covered by only a thin layer of soil. If this type of sediment-poor terrain is submerged, and if there are no other sources of sand, the coastline is rocky.

A **wave-cut cliff** forms when waves erode the headland into a steep profile. As the cliff erodes, it leaves a flat or gently sloping **wave-cut platform** (●**FIGURES 16.35**). If waves cut a cave into a narrow headland, the cave may eventually erode all the way through the headland, forming a scenic **sea arch**. When an arch collapses or when the inshore part of a headland erodes faster than the tip, it leaves behind a pillar of rock called a **sea stack** (●**FIGURE 16.36**). As waves continue to batter the rock, eventually the sea stack crumbles.

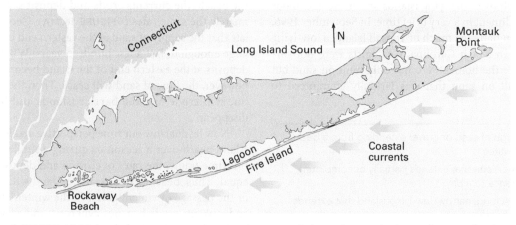

● **FIGURE 16.30** Longshore currents carry sand westward along the south shore of Long Island creating a series of barrier islands.

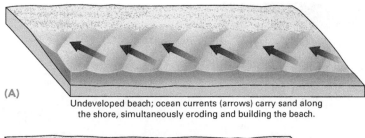

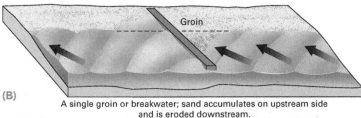

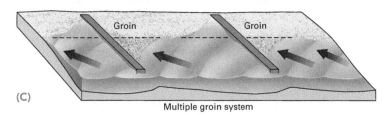

● **FIGURE 16.31** (A) Longshore currents simultaneously erode and deposit sand along an undeveloped beach. (B) A single groin or breakwater traps sand on the upstream side, resulting in erosion on the downstream side. (C) A multiple-groin system propagates the uneven distribution of sand along the entire beach.

(A)

(B)

● **FIGURE 16.32** (A) Several groins (red circles) are shown on this aerial shot of western Long Island. (B) Wooden groins used to trap sand on the beach along the coast of Wales, UK.

groins A narrow barrier or wall built on a beach, perpendicular to the shoreline, to trap sand transported by currents and waves.

wave-cut cliff A cliff created when waves erode the headland of a rocky coastline into a steep profile.

wave-cut platform The horizontal or gently sloping platform left when a wave-cut cliff is eroded back.

sea arch An arch created when a short cave is eroded all the way through a narrow headland.

sea stack A pillar of rock left when a sea arch collapses or when the inshore portion of a headland erodes faster than the tip.

If the sea floods a long, narrow, steep-sided coastal valley, a sinuous bay called a fjord is formed (Chapter 13). Fjords are common at high latitudes, where rising sea level has flooded coastal U-shaped valleys scoured by Pleistocene glaciers. Fjords may be hundreds of meters deep, and often the cliffs at the shoreline drop straight into the sea.

(A)

(B)

● **FIGURE 16.33** (A) This aerial photograph of a Long Island beach shows sand accumulating on the upstream side of a groin and erosion on the downstream side. (B) A close-up of one house on that Long Island beach shows waves lapping against the foundation.

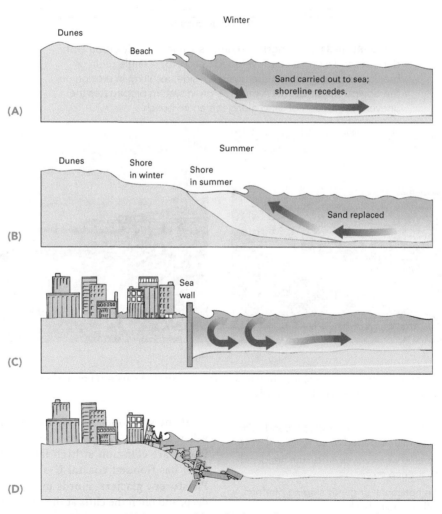

● **FIGURE 16.34** (A) In a natural beach, the violent winter waves often move sand out to sea. (B) The gentler summer waves push sand toward shore and rebuild the beach. (C) Wave energy concentrates against a seawall and (D) may eventually destroy it.

● **FIGURE 16.35** Waves hurl sand and gravel against solid rock, eroding cliffs and creating a wave-cut platform along the Oregon coast.

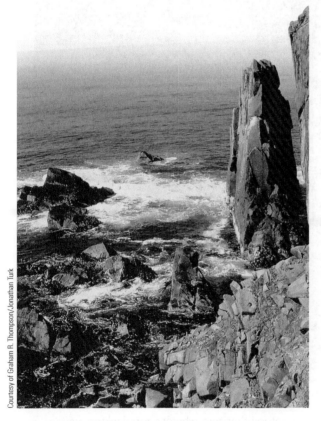

● **FIGURE 16.36** Massive waves of the Antarctic Ocean eroded cliffs, forming these sea stacks near Cape Horn.

plankton Small marine organisms that live mostly within a few meters of the sea surface, where sunlight is available, and that conduct most of the photosynthesis and nutrient consumption in the ocean and form the base of the marine food web.

phytoplankton Plankton that conduct photosynthesis like land-based plants and that are the base of the food chain for marine animals.

zooplankton Tiny marine animals that live mostly within a few meters of the sea surface and feed on phytoplankton.

LIFE IN THE SEA

On land, most photosynthesis is conducted by multicellular plants including mosses, ferns, grasses, shrubs, and trees. Large animals such as cows, deer, elephants, and bison consume the plants. In contrast, most of the photosynthesis and consumption in the ocean is carried out by small organisms called **plankton**. Many plankton are single-celled and microscopic; others are more complex and are up to a few centimeters long. One major difference between terrestrial and aquatic ecosystems is that on land soil nutrients are abundant on the surface, where light is also abundant. However, in the oceans, light is available only in the photic zone which ranges from the surface downward up to 200 meters in the open ocean but typically is shallower closer to shore where the water is more turbid. In contrast to light, nutrients in the oceans tend to settle to the dark depths. Plankton live mostly within a few meters of the sea surface, where light is available. However, the growth and productivity of these organisms is limited by the fact that most of the nutrients—such as nitrogen, iron, and phosphorus—are incorporated into the tissue of planktonic organisms that die and sink to the bottom, carrying the nutrients with them. If the surface waters are not replenished by a fresh supply of nutrients, their productivity will be limited.

Phytoplankton conduct photosynthesis like land-based plants do. Therefore, they are the base of the food chain for aquatic animals. Although phytoplankton are not readily visible, they are so abundant that they supply about 50 percent of the oxygen in our atmosphere. **Zooplankton** are tiny animals that feed on the phytoplankton (●**FIGURE 16.37**). The larger and more familiar marine plants (such as seaweed) and animals (such as fish, sharks, and whales) play a relatively small role in overall oceanic photosynthesis and consumption. However, many of these organisms are important to human as a major source of protein.

World Fisheries

The shallow water of a continental shelf supports large populations of marine organisms. In addition, many deep-sea fish spawn in shallow water within 1 or 2 kilometers of the shoreline. Shallow zones in bays, lagoons, and estuaries are especially hospitable to life because they have (1) easy access to the deep sea, (2) lower salinity than the open ocean, (3) a high concentration of nutrients originating from land and sea, (4) shelter, and (5) abundant plant life rooted to the seafloor in addition to the phytoplankton floating on the surface. As a result, about 99 percent of the marine fish caught every year are harvested from the shallow waters of the world's continental shelves.

Historically, fishing pressure by humans has impacted fish populations, causing a shift in the type of fish targeted, which typically has moved down the food chain. For example, in 1970, fishermen harvested 3 million tons of cod, a predatory fish. This large catch turned out to be biologically

(A)

(B)

(C)

(D)

● **FIGURE 16.37** Plankton are microscopic, usually single-celled photosynthesizers, whereas zooplankton are tiny animals that live near the sea surface and feed on phytoplankton. (A) *A variety of diatoms, single-celled photosynthesizing algae that secrete a siliceous skeleton.* (B) A photosynthesizing nannoplankton called a coccolithophorid, in this case *Emiliania huxleyi*. (C) Slipper lobster phyllosoma larva, a zooplankton. (D) Mantis shrimp larvae, a zooplankton.

unsustainable and had declined to 1 million tons by 1993. In 1995, biologists suspended cod fishing in many areas to allow the populations to recover. In 1978, stocks of herring and mackerel, both of which eat mostly plankton and therefore are lower on the food chain, began to decline; so boats switched nets to trawl for squid. In the 1980s, however, the squid population began to decline. Thus, fishing pressure has worked through the food chain, disrupting the entire ecosystem.

In one study, oceanographers documented research showing that industrial fishing fleets have caused a 90 percent reduction of large predatory fish such as tuna, marlin, swordfish, cod, halibut, and flounder. At the advent of intensive commercial fishing 100 years ago, most fleets caught 6 to 12 fish for every 100 baited hooks. Today, despite sophisticated electronic and aerial fish-finding techniques, the catch rate has plummeted to 1 fish per 100 hooks. The Worldwatch Institute has pointed out that as industrial fishing depletes "large, long-lived predatory species … that occupy the highest levels of the food chain, [fishing operations] move down to the next level—to species that tend to be smaller, shorter-lived, and less valuable. As a result, fishers worldwide now fill their nets with plankton-eating species such as squid, jacks, mackerel, sardines, and invertebrates including oysters, mussels, and shrimp." Commercial fishermen now work harder, spend more time, and consume more fuel to capture smaller quantities of "less valuable species—they are essentially fishing down the marine food web." The authors of the study continue:

> But the cycle of fishing down the marine food web can't go on forever. (At lower trophic levels, the species are so small and diluted that it is no longer economically feasible to fish.) At the current rate of descent, it will take only 30 to 40 years to fish down to the level of plankton.[1]

The UN's Food and Agricultural Organization (FAO) predicted that world fish harvests would remain fairly constant at around 90 million tons per year until the year 2010, but thereafter, ecosystem destruction and overfishing would lead to sharp declines in fish populations and harvests. One 2018 study analyzed the FOA data and compared it to a reconstructed data set from its own analysis of 10 years locally collected fishing data from around the world. The 2018 study indicates that, because the FOA relies on official government numbers for fish caught, it is missing the large size of the fish catches for subsistence, artisanal (hand-made using traditional techniques), or recreational purposes. Many countries are neither interested in nor have the capacity to tally catches from these fisheries, and some nations have altered the reported catch numbers for political reasons. According to the 2018 study, the global catch related to subsistence, artisanal, and recreational fisheries represented 25 percent of the total catch over the past decade. Although the FOA has defended its data gathering methodology for global fish stocks, many wild stocks continue to decline without globally agreed catch limits designed to replenish fish stocks. Collective international adherence to a set of global catch limits, as well as a broader data-gathering system for determining what fish species are harvested, along with the location and date of the catch, is needed.

Reefs

A **reef** is a wave-resistant ridge or mound built by corals, oysters, algae, or other marine organisms. Because corals need sunlight and warm, clear water to thrive, coral reefs develop in shallow tropical seas where little suspended clay or silt muddies the water (●FIGURE 16.38). As the corals die, their offspring grow on their remains. Oyster reefs, on the other hand, form in temperate estuaries and can grow in more turbid water.

(A)

(B)

(C)

● **FIGURE 16.38** (A) Coral reefs form abundantly in clear, shallow, tropical water. An aerial view of the Great Barrier Reef, Australia. Notice on the horizon the change to deep blue water, marking the steep frontal edge of the reef. The much shallower water in the foreground consists of a network of reefs (brown-red) growing upward from the bottom, with areas of unconsolidated carbonate sand in between (lighter-colored areas). (B) Reefs grow in the warm, clear, shallow water and are extremely diverse ecologically. A sea fan and a variety of corals and fish can be seen in this photograph from the Great Barrier Reef. (C) A barrier reef exposed at low tide along the southern coast of Fiji in the South Pacific Ocean. The surf breaks on the outer edge of the reef but dissipates before reaching the top of the reef.

---

reef A wave-resistant ridge or mound built by corals or other marine organisms.

1. From Anne Platt McGinn, "Freefall in Global Fish Stocks," *World Watch Magazine* 11 (May–June 1998), 10.

Globally, coral reefs cover about 600,000 square kilometers—about the area of France—but they spread out in long, thin lines. They are extraordinarily productive ecosystems because the corals provide shelter for many invertebrates, fish, and other marine organisms. Within the past 50 years, 10 percent of the world's coral reefs have been destroyed and an additional 30 percent are in critical condition. Several factors have contributed to this destruction:

- Increased levels of carbon dioxide in Earth's atmosphere have caused higher concentrations of the gas to be dissolved in the world's oceans, causing ocean acidification and making it more difficult for calcium carbonate–secreting organisms, such as corals, to live.
- Corals thrive best in a narrow temperature range. In recent years, oceanographers have compiled considerable evidence that sea surface temperatures have become warmer in recent decades, and the warm water is leading to massive death of corals.
- Silt from cities, urban roadways, farms, and improper logging smother the delicate reef organisms.
- Fertilizer runoff from farms and sewage runoff from cities have added nutrients to coastal waters, feeding coral predators and lowering the levels of dissolved oxygen levels below the level required for oxygen-respiring organisms, including corals, to survive. For example, a 2017 study by National Geographic reported that nutrient-rich runoff from the Mississippi River caused the development of New Jersey-sized "dead zone" along the Louisiana and much of the Texas Coasts. The "dead zone" is characterized

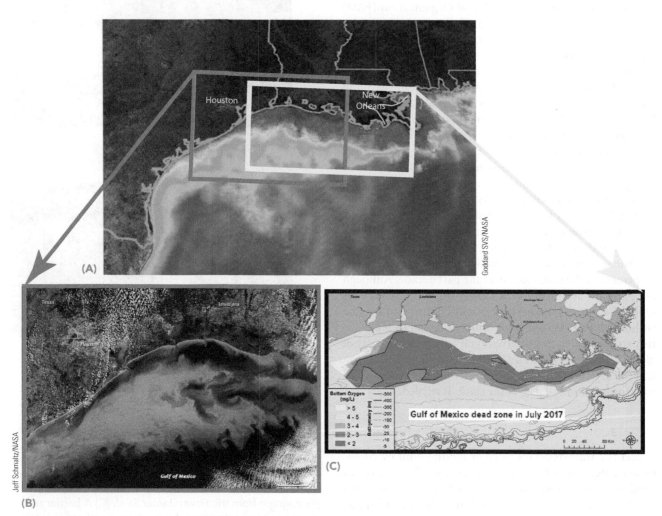

● **FIGURE 16.39** A) This satellite image of the Gulf of Mexico was taken during the summer of 2017 by the Moderate Resolution Imaging Spectroradiometer (MODIS) instrument on NASA's Aqua satellite. The hot colors (red, orange) represent blooms of phytoplankton feeding off nutrients being delivered primarily from the Mississippi River drainage. (B) Satellite image of the red-outlined inset in (A). The lighter colors indicate nutrient-rich sediment being carried offshore into deeper water. The nutrients produce blooms of phytoplankton which use up oxygen when they decay, leading to hypoxic conditions. (C) Map of actual measurements of dissolved oxygen content in the portion of the Gulf of Mexico highlighted by the yellow inset box in (A). The red color indicates dissolved oxygen levels that are low enough to cause hypoxia, corresponding to the dead zone.

by levels of dissolved oxygen too low to sustain life and is formed because high concentrations of nutrients—especially nitrogen and phosphorus—are delivered by the Mississippi River (●FIGURE 16.39). The Mississippi River drainage is by far the largest in North America and includes nearly the entirety of the "American breadbasket" from Montana to Pennsylvania. Runoff in the Mississippi drainage picks up fertilizers, soil particles, animal wastes, and sewage. These sources combine to produce huge algal blooms in the northern Gulf of Mexico, altering the food chain and depleting dissolved oxygen in the water. The size and dimensions of the dead zone fluctuate along with seasonal discharge from the Mississippi River, large rainfall events, and hurricanes.

GLOBAL WARMING AND RISING SEA LEVEL

Sea level has risen and fallen repeatedly in the geologic past, and coastlines have emerged and submerged throughout Earth's history. During the past 40,000 years, sea level has fluctuated by about 150 meters, primarily in response to the growth and melting of glaciers (●FIGURE 16.40). The rapid sea level rise that started about 18,000 years ago began to level off about 7,000 years ago. By coincidence, humans began to build cities about 7,000 years ago. Thus, civilization developed during a short time when sea level was relatively constant.

Shore-based gauging stations and satellite radar studies agree that global sea level is presently rising at about 3 millimeters (about the thickness of a nickel) per year. (Records from the past century indicate an average rise in sea level of 1 to 2 millimeters per year.) Thus, if present rates continue, sea level will rise about 20 centimeters in 100 years. Such a rise would be significant along very low-lying areas such as the Netherlands, Bangladesh, and many Pacific islands.

Consequences of rising sea level vary with location and economics. The wealthy, developed nations could build massive barriers to protect cities and harbors from a small sea level rise. In regions where global sea level rise is compounded by local tectonic sinking, dikes are already in place or planned. Portions of Holland lie below sea level, and the land is protected by a massive system of dikes. In London, where the high-tide level has risen by 1 meter in the past century, multimillion-dollar storm gates have been built on the Thames River. Venice, Italy, which is built over canals at sea level, has flooded frequently in recent years, and here too expensive engineering projects are under way. Rising sea level has resulted in erosion along much of the coastline along the Gulf Coast and eastern seaboard, reducing the size of nearshore wetlands which are needed to help decompose wastes and protect low-lying coastal areas from the effects of hurricanes.

However impressive the engineering of modern urban coastlines becomes, it is unlikely that coastal cities will ever be fully protected against a dramatic sea level rise of several meters. Coastal cities worldwide would be inundated. Moreover, many poor countries cannot afford coastal protection even for a modest sea level rise. For example, a 1-meter rise in sea level would flood roughly 17 percent of the land area of Bangladesh, displacing tens of millions of inhabitants.

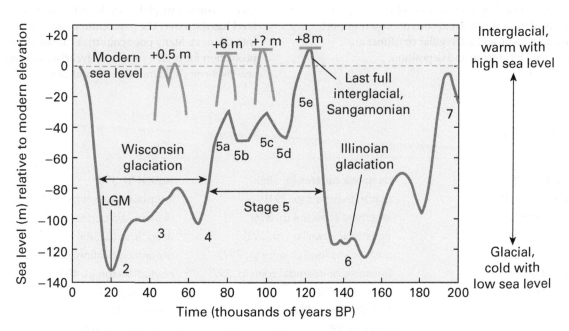

● FIGURE 16.40   Sea level has fluctuated by roughly 150 meters during the past 200,000 years.

Key Concepts Review

- All of Earth's oceans are connected, so Earth has a single global ocean with several distinct ocean basins existing within it.
- Seawater contains about 3.5 percent dissolved salts. The upper layer of the ocean, about 450 meters thick, is relatively warm. In the thermocline, below the warm surface layer, temperature drops rapidly with depth. Deep ocean water is consistently around 1°C to 2.5°C.
- Tides are caused by the gravitational pull of the Moon and Sun. Two high tides and two low tides occur approximately every day.
- Wave size depends on (1) wind speed, (2) the length of time the wind has blown, and (3) the distance that the wind has traveled. The highest part of a wave is the crest; the lowest is the trough. The distance between successive crests is called the wavelength. Wave height is the vertical distance from the crest to the trough. The water in a wave moves in a circular path.
- A current is a continuous flow of water in a particular direction. Surface currents are driven by wind and deflected by the Coriolis effect, Ekman transport, and sea-surface topography. Deep-sea currents are driven by differences in seawater density. Cold, salty water is dense and therefore sinks and flows along the seafloor. When ocean water sinks in some places, it must rise in others to maintain mass balance. This upward flow of water is called upwelling.
- When a wave nears the shore, the bottom of the wave slows and the wave breaks, creating surf. Ocean waves weather and erode coastlines by hydraulic action, abrasion, solution, and salt cracking. The bending of a wave as it strikes shore is called refraction. Refracted waves often form longshore currents that transport sediment along a shore. Tidal currents also transport sediment in some areas. Irregular coastlines are straightened by erosion and deposition.
- If land rises or sea level falls, the coastline migrates seaward and old beaches are abandoned above sea level, forming an emergent coastline. In contrast, a submergent coastline forms when land sinks or sea level rises.
- A beach is a strip of shoreline washed by waves and tides. Most coastal sediment is transported to the sea by rivers. Glacial drift, reefs, and local erosion also add sand in certain areas. Coastal emergence may expose large amounts of sand. Spits, baymouth bars, and barrier islands are common on sandy coastlines. Human intervention such as the building of groins may upset the natural movement of coastal sediment and alter patterns of erosion and deposition on beaches. A rocky coast is dominated by wave-cut cliffs, wave-cut platforms, sea arches, and sea stacks. A fjord is a steep-sided, narrow, glacially carved, submerged valley on a high-latitude seacoast.
- Most of the photosynthesis and nutrient consumption in the ocean is carried out by plankton, tiny organisms that live mostly within a few meters of the sea surface where light is available. Phytoplankton conduct photosynthesis, just as land-based plants do, and are the base of the food chain for aquatic animals. Zooplankton are tiny animals that feed on the phytoplankton. The shallow water of a continental shelf supports large populations of marine organisms. Commercial overfishing and habitat destruction threaten world fish harvests. A reef is a wave-resistant ridge or mound built by corals, oysters, algae, or other organisms.
- Ocean water expands when it warms, and when glaciers melt large volumes of water flow into the sea—both mechanisms cause global sea-level to rise. Global sea level has risen over the past century and may continue into the next. Many poor countries cannot afford coastal protection for even a small sea-level rise; a dramatic sea-level rise could inundate coastal cities.

Important Terms

backshore (p. 397)

barrier island (p. 398)

baymouth bar (p. 398)

beach (p. 397)

beach drift (p. 394)

breaks (p. 393)

Coriolis effect (p. 389)

crest (p. 386)

current (p. 386)

deep-sea currents (p. 386)

Ekman transport (p. 391)

emergent coastline (p. 396)

equatorial upwelling (p. 393)

eustatic sea level change (p. 397)

foreshore or intertidal zone (p. 397)

groins (p. 400)

Gulf Stream (p. 386)

gyres (p. 389)

lagoon (p. 398)

longshore current (p. 394)

microplastic debris (p. 389)

neap tides (p. 385)

ocean acidification (p. 382)

phytoplankton (p. 403)

plankton (p. 403)

reef (p. 405)

refraction (p. 393)

salinity (p. 381)

sea arch (p. 400)

sea stack (p. 400)

shoreface (p. 397)

spit (p. 398)

spring tides (p. 385)

submergent coastline (p. 396)

surf (p. 393)

surface currents (p. 386)

thermohaline circulation (p. 391)

tidal bulge (p. 384)

tidal current (p. 396)

tidal range (p. 385)

tides (p. 384)

trough (p. 386)

upwelling (p. 392)

wave base (p. 386)

wave height (p. 386)

wave orbitals (p. 386)

wave-cut cliff (p. 400)

wave-cut platform (p. 400)

wavelength (p. 386)

zooplankton (p. 403)

Review Questions

1. Name the major ocean basins and give their locations.
2. Describe the temperature profile of the open oceans.
3. Explain why two high tides occur every day, even though the Moon lies directly above any portion of Earth only once a day.
4. List the three factors that determine the size of a wave.
5. Explain the Coriolis effect. What is a gyre?
6. What is refraction? How does it affect coastal erosion?
7. Explain how coastal processes straighten an irregular coastline.

17

THE ATMOSPHERE

Tropical Storm Toraji on September 2, 2013. The eye of the storm is in the East China Sea. To the west of the storm's eye, the light blue, shallow continental shelf of eastern mainland China is visible, while suspended sediment muddies the water closer to the shore. Tropical Storm Toraji spawned tornadoes in Japan, the outline of which is shown in the top center of the image. To the east is the deep blue water of the Pacific Ocean.

INTRODUCTION

Nearly every multicellular organism needs oxygen to survive. If the oxygen abundance in the atmosphere were to drop below 44 percent of its current value, life on Earth as we know it would perish. If oxygen is essential to most life, would we be better off if we had an even greater supply? The answer is yes, to a limit. Even at sea level, athletes can enhance their performance by breathing a small amount of bottled oxygen. But, paradoxically, too much oxygen is poisonous. If you breathe air that has 55 percent or more oxygen than is found at sea level, your body metabolism becomes so rapid that essential molecules and enzymes decompose. In addition, fires burn more rapidly with increased oxygen concentration. If the oxygen level in the atmosphere were to rise significantly, extremely large wildfires would engulf the planet, altering ecosystems as we know them.

Earth is the near-perfect size and distance from a stable and medium-temperature star to permit optimal atmospheric conditions for life. But Earth's atmospheric composition and temperature are not determined solely by planetary size and distance from the Sun. Rather, our planet's environment is finely regulated by interactions among Earth systems. Over the past 4 billion years, the Sun's output has slowly increased (although there were numerous fluctuations during this period). However, Earth's atmospheric temperature has remained remarkably constant because of interactions among Earth's systems. In this chapter, we will explore these interactions and see how they led to the structure and dynamics of Earth's atmosphere today.

EARTH'S EARLY ATMOSPHERES

Scientists have revealed a nearly continuous record of rocks of different ages, from 4.04 billion years ago to the present. Therefore, when they study the history of Earth's continental crust, they can analyze the chemical composition of these ancient rocks for clues.

Unfortunately, there are no samples of very old atmospheres. So how can we determine atmospheric composition millions to billions of years ago? Many gaps in understanding exist, but the history described next comes from information derived from computer modeling as well as the direct study of rocks. Modeling involves calculations about how atmospheric gases would have behaved under the presumed environment of early Earth. To test these models, scientists study the geochemistry of ancient rocks.

Rocks react with water and air, and the nature of these reactions depends on the geochemistry of the entire system. By studying the rocks that existed in a specific time period, scientists deduce the other components of the system that would have produced those reactions. For example, as we will discuss, iron reacts with oxygen to produce

LEARNING OBJECTIVES

LO1 Summarize the composition and evolution of the atmosphere 4.6 billion years ago to 2.6 billion years ago.

LO2 Discuss the earliest life forms on Earth.

LO3 Explain the early interactions among the geosphere, atmosphere, and biosphere that resulted in the deposition of banded iron formation.

LO4 Describe the importance of ozone in the stratosphere.

LO5 Explain the relationship between elevation and temperature in the troposphere.

LO6 Discuss the layers of the atmosphere in terms of temperature and elevation above sea level.

LO7 Describe the incident that occurred in Donora, Pennsylvania and its relationship to the Clean Air Act of 1963.

LO8 Explain the role of ozone in the troposphere.

LO9 Describe how dioxin enters the food chain.

LO10 Discuss the role that CFC's play in depletion of the ozone layer.

iron oxides. Thus, if we find iron oxides in certain types of sedimentary rocks that formed 2.6 billion years ago, we deduce that oxygen must have been present in the air and water at that time.

The First Atmospheres: 4.6 to 4.0 Billion Years Ago

Our Solar System formed from a cold, diffuse cloud of interstellar gas and dust. About 99.8 percent of this cloud was composed of the two lightest elements: hydrogen and helium. Consequently, when Earth formed, its primordial atmosphere was composed almost entirely of these two light elements. But because Earth is relatively close to the Sun and its gravitational force is relatively weak, its primordial hydrogen and helium atmosphere rapidly boiled off into space and escaped. **TABLE 17.1** shows this and the subsequent atmospheres of Earth described in this chapter.

In Chapter 15, we learned that most of the volatile compounds that form Earth's hydrosphere, atmosphere, and biosphere originated from outer parts of the Solar System. Recall that, in its infancy, the Solar System was crowded with bits of rock, comets, ice chunks, and other debris left over from the initial coalescence of the planets. These bolides crashed into the planet in a near-continuous rain that lasted almost 800 million years. Carbonate compounds and carbon-rich rocks reacted under the heat and pressure of impact to form carbon dioxide. Ice quickly melted into water. Ammonia, common in the icy tail of comets, reacted to form nitrogen (**FIGURE 17.1**).

TABLE 17.1 The Earth's Atmosphere through Time

Events That Formed the Atmosphere	Age of Atmosphere	Atmospheric Composition and/or Changes in Composition
Primordial atmosphere: From initial accretion of planets	4.6 billion years ago	Hydrogen (H_2) and helium (He)
Secondary atmosphere: Bolide impact from outer space	4.5 billion years ago	Carbon dioxide (CO_2), water (H_2O), and nitrogen (N_2)
Atmosphere formed by outgassing and modified by reactions of gases with geosphere	4.5 to 2.7 billion years ago	Predominantly hydrogen (H_2) and carbon dioxide (CO_2); some water (H_2O) and nitrogen (N_2)
Evolution of cyanobacteria, which begin producing oxygen	2.7 billion years ago	Oxygen (O_2) produced by cyanobacteria is removed as quickly as it is produced
First Great Oxidation Event	2.4 billion years ago	Oxygen (O_2) accumulates in the atmosphere; hydrogen (H_2) becomes a trace gas
Second Great Oxidation Event	600 million years ago	Atmospheric oxygen concentrations increase as biological and geological processes slow decay
Modern atmosphere: High oxygen concentration maintained by biological photosynthesis	Today	Primarily nitrogen (N_2) and oxygen (O_2), with smaller concentrations of other gases

When Life Began: 4.0 to 2.6 Billion Years Ago

The carbon dioxide, water, and nitrogen atmosphere changed as volatiles escaped from Earth's mantle to the surface in volcanic eruptions, in a process called **outgassing**. In 1953, Stanley Miller and Harold Urey hypothesized—on the basis of little direct modeling—that by 4 billion years ago, Earth's atmosphere consisted primarily of methane (CH_4), ammonia (NH_3), hydrogen (H_2), and water (H_2O). They mixed these gases in a glass flask and fired sparks across the flask to simulate Earth's early atmosphere, beset by lightning storms. Amino acids, the building blocks of proteins, formed in the flask. The Miller–Urey model immediately became popular because scientists speculated that the first living organisms formed by accretion of these abiotic (nonliving) amino acids.

In the early 1970s, the idea of a methane-ammonia-hydrogen atmosphere was largely discredited, and most scientists postulated that the Hadean atmosphere was composed mainly of carbon dioxide (CO_2), with smaller amounts of nitrogen (N_2), water (H_2O), and other gases. However, in 2002, geologists analyzed the isotopic geochemistry of single grains of zircon that are the oldest known preserved mineral grains on Earth, dating between 4.5 and 4.0 billion years before present. Zircon is a silicate mineral that only forms in rocks with a silica-rich, felsic composition; it does not exist in the silica-poor mafic and ultramafic rocks, so the geologists concluded that granitic crust had formed within the first 500 million years of Earth's history. In addition, the isotopic composition of oxygen making up the silicate anionic groups within the zircon mineral crystal was high enough for the geologists to conclude that the surface of the early Earth was cold, not

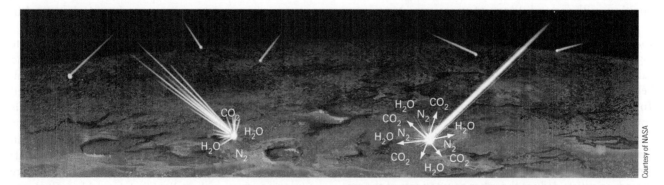

● **FIGURE 17.1**   Comets, meteoroids, and asteroids imported Earth's volatiles from outer parts of the Solar System.

hot, and that water would have been mostly liquid. Subsequently, in 2005, scientists used models of Earth's Hadean geosphere to calculate how gases trapped in the interior would react with rocks and minerals in the planet's interior and surface. These results also suggested that early Earth had a cool surface that included liquid water.

From this work, one unobvious, yet pivotal, question emerged: When did Earth's core form? Earth's core is the deepest layer of the geosphere. The atmosphere is a thin veneer surrounding the crust. How could one affect the other? Recall that the core is composed primarily of iron and nickel. Before Earth's interior segregated into a layered core and mantle, the mantle contained more iron than it does today. Then, when iron settled into the core, the mantle became relatively iron-poor.

Modern hypotheses state that Earth was hot at the time of its formation and that most of the planet's iron was sequestered in its core shortly after the planet evolved, leaving an iron-poor mantle behind. Because mantle rocks rise to the surface through volcanic eruptions, the composition of the mantle directly influences the composition of erupted material, including the types of volatiles that are emitted.

By modeling the reactions of volatiles in an iron-poor mantle, researchers concluded in 2005 that the delivery of these volatiles to the atmosphere resulted in a composition that included not only large amounts of hydrogen, as Miller and Urey postulated, but also high concentrations of carbon dioxide, as later models had proposed.[1] Even more recently, a different set of scientists argued that the Earth's early surface temperatures were significantly warmer than today, with estimates of surface temperatures between 45°C and 85°C.

Models of Earth and its evolution are constantly changing. Scientists propose hypotheses and theories, consider new ideas—and often other scientists disagree. This is the nature and the joy of science. In a review article, Christopher Chyba of SETI Institute and Stanford University wrote that current disagreement about the composition of Earth's earliest atmospheres "makes it a great time for young scientists to enter the field, but it also reminds us that some humility regarding our favorite models is in order."[2] Further constraining the composition and evolution of Earth's earliest atmosphere continues to be a major scientific challenge that

will involve much additional research, discussion, and—quite likely—disagreement among different scientists.

Life, Iron, and the Evolution of the Modern Atmosphere

As explained previously, the first living organisms may have formed by the accretion of complex abiotic (nonliving) organic molecules formed in early Earth's surficial environments. However, complex organic molecules, such as those postulated to have led to the first organisms, become oxidized and destroyed in an oxygen-rich environment. (This oxidation process is analogous to slow burning.) These chemical reactions suggest that if large amounts of oxygen were present in Earth's early atmosphere, the abiotic precursors to living organisms could not have formed. A geochemist can determine whether a rock formed in an oxygen-rich or an oxygen-poor environment by analyzing the mineral assemblage and geochemistry of the rock and applying what is currently known about the thermal and chemical conditions needed to form that suite of minerals. Recent studies of Earth's oldest rocks indicate that the atmospheric oxygen concentration in early Precambrian time was extremely low. These results suggest that the molecules necessary for the emergence of living organisms—namely methane (CH_4), ammonia (NH_3), hydrogen (H_2), and water (H_2O)—would have been preserved in the primordial atmosphere.

Although life could not have emerged in an oxygen-rich environment, complex multicellular life requires an oxygen-rich atmosphere to survive. How, then, did oxygen become abundant in our atmosphere?

The world's earliest organisms probably were **chemoautotrophs** that obtained their energy from reactions with minerals such as iron and sulfur in an extremely inefficient process. (Recall from Chapter 15 that chemoautotrophic bacteria are common today around vents of extremely hot hydrothermal water emanating from parts of the Mid-Oceanic Ridge.) Later, organisms subsisted, in part, by eating each other. But these food chains were limited because there were only a few organisms on Earth. A crucial step in evolution occurred when primitive bacteria evolved the ability to harness the energy in sunlight and produce organic tissue. This process, known as **photosynthesis**, is the foundation for virtually all modern life. During photosynthesis, organisms convert carbon dioxide and water to organic sugars. They release oxygen as a by-product. In 1972, an English chemist named James Lovelock hypothesized that the oxygen produced by primitive organisms gradually accumulated, creating the modern atmosphere. By Late Precambrian time, atmospheric oxygen concentration had reached the critical level needed to sustain efficient metabolism. As a result, multicellular organisms evolved and the biosphere as we know it was born. Lovelock was so overwhelmed by the intimate connection between living and nonliving

---

outgassing The release of volatiles from Earth's mantle and crust during volcanic eruptions at the surface

photosynthesis The process by which chlorophyll-bearing plant cells convert carbon dioxide and water to organic sugars, using sunlight as an energy source; oxygen is released in the process.

chemoautotroph an organism, typically a bacterium, that derives its metabolic energy by oxidizing inorganic compounds.

---

1. Feng Tian, Owen B. Toon, Alexander A. Pavlov, and H. De Sterck, "A Hydrogen-Rich Early Earth Atmosphere," *Science* 308 (May 13, 2005), 1014–1017.

2. Christopher F. Chyba, "Rethinking Earth's Early Atmosphere," *Science* 308 (May 13, 2005), 962–963.

components of Earth's systems that he likened our planet to a living creature, which he called **Gaia** (Greek for "Earth").

The Lovelock hypothesis, now over 45 years old, remains generally accepted, but scientists are still investigating many of the details. For example, blue-green algae called **cyanobacteria** began producing oxygen 2.7 billion years ago, but appreciable quantities of oxygen didn't begin to appear in the atmosphere until 2.4 billion years ago, in the **First Great Oxidation Event** (●**FIGURE 17.2**), when the concentration of atmospheric oxygen rose abruptly.

The production of oxygen by cyanobacteria beginning 2.7 billion years ago started slowly, and for about 700 million years thereafter, atmospheric oxygen concentration remained far below today's level. At times during this 700-million-year period, Earth's atmosphere may have been anoxic, meaning there was no oxygen.

These abrupt variations in the concentration of oxygen in Earth's earliest atmosphere—from periods of anoxia on the one hand to the First Great Oxidation Event 2.4 billion years ago on the other—strongly suggest that the presence of oxygen was driven by a mechanism involving a chemical threshold. To understand the mechanism behind this threshold reaction, we must study atmosphere—geosphere and atmosphere—biosphere systems interactions.

Systems Interactions That Affected Oxygen Concentration in Earth's Early Atmosphere

GEOSPHERE–ATMOSPHERE INTERACTIONS Even though large quantities of iron coalesced into Earth's core, appreciable quantities remained in the mantle and crust. Today's crust, for example, is about 5 percent iron by weight. The iron that remained near the surface during early Earth played an important role in concentrating oxygen in Earth's early atmosphere.

Oxygen gas dissolves in water. As a result, when early cyanobacteria released oxygen, some of the gas dissolved in seawater. In an oxygen-poor environment, iron also dissolves in water. However, when oxygen is abundant, the dissolved oxygen reacts with dissolved iron and causes the iron to precipitate rapidly as iron-oxide minerals.

Because iron is soluble in water lacking dissolved oxygen, large amounts of iron dissolved in the early Precambrian Ocean at times when its concentration of dissolved oxygen was below the threshold level. However, when the concentration of dissolved oxygen from cyanobacteria rose beyond the threshold, iron that had been dissolved in the seawater precipitated rapidly as iron-oxide minerals. These minerals settled to the seafloor, forming a layer there

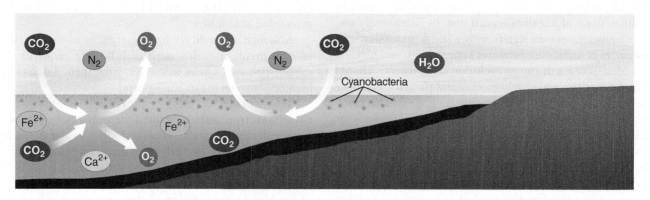

● **FIGURE 17.2**   The oxygen content of the atmosphere slowly and steadily increased after photosynthesizing cyanobacteria began to release oxygen 2.7 billion years ago.

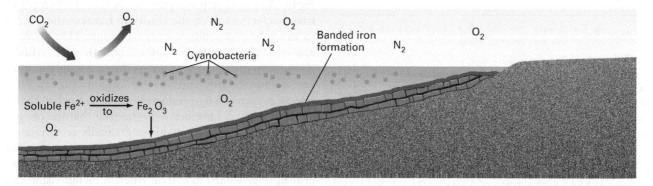

● **FIGURE 17.3**   When the oxygen concentration in the atmosphere and seawater reached a threshold, the oxygen combined with dissolved iron to form iron oxide minerals. The iron oxide minerals precipitated from the water and settled to the seafloor, forming the iron-rich layer of a banded iron formation.

● **FIGURE 17.4** In this banded iron formation from Michigan, the red bands are iron oxide minerals and the dark layers are chert (silica).

(●**FIGURE 17.3**). This process also removed oxygen from the water, explaining why the oxygen concentration in the atmosphere didn't rise at this time in Earth's early history, even though cyanobacteria were releasing oxygen as a gas.

Roughly 90 percent of the iron ore that is mined globally comes from **banded iron formation**, chemical sedimentary rock that consists of alternating layers ("bands") of iron oxide and chert a few centimeters thick (●**FIGURE 17.4**). Most of the banded iron formation on Earth is between 2.6 and 1.9 billion years old and is inferred to have originated when the oxygen level in the seas hovered near the threshold at which soluble iron combined with oxygen to form iron-oxide minerals. At times when dissolved oxygen concentrations were high enough, the oxygen would combine with dissolved iron to precipitate iron-oxide minerals, which formed a layer on the seafloor. The formation of these minerals extracted oxygen as well as iron from the seawater, lowering its oxygen concentration below the threshold. Then

---

Gaia The term (Greek for "Earth") used by James Lovelock to refer to our planet, which he likened to a living creature due to the interconnectivity of all of Earth's systems.

cyanobacteria Blue-green algae that were among the earliest photosynthetic life-forms on Earth.

First Great Oxidation Event The sudden increase in Earth's atmospheric oxygen concentration from trace amounts to appreciable quantities that occurred approximately 2.4 billion years ago, probably because of a combination of biological and geochemical processes.

banded iron formation A marine chemical sedimentary rock formed mostly between 2.7 and 1.9 billion years ago and consisting of centimeter-scale interbeds of iron oxide and chert. Formed as atmospheric oxygen alternately went through periods of accumulation as a waste product from photosynthesizing cyanobacteria and periods of withdrawal during widespread precipitation of iron oxide minerals.

---

dissolved iron accumulated again in the seas from chemical weathering of exposed rock, while the oxygen was slowly replenished by photosynthesizing cyanobacteria. During that time, clay and other minerals washed from the continents and accumulated on the seafloor as they do today, forming the thin layers of silicate minerals that lie between the iron-rich layers in banded iron formation. When the oxygen concentration rose above the threshold again, another layer of iron oxide minerals formed.

Banded iron formations contain thousands of alternating layers of iron minerals and silicates and can cover tens of square kilometers. The great volume of the iron formations, coupled with the fact that they continued to form from 2.6 to 1.9 billion years ago, suggests that the reactions that formed them must have kept the levels of dissolved oxygen close to the threshold for about 700 million years. Thus the iron-rich rocks that support our industrial society were formed by interactions among early photosynthesizing organisms, sunlight, air, and the oceans.

BIOSPHERE–ATMOSPHERE INTERACTIONS In the primordial atmosphere, free oxygen also reacted with hydrogen to form water. This process helped keep the oxygen concentration in the atmosphere low. However, after life evolved, bacteria in the oceans removed atmospheric hydrogen in a process that produced methane. When the hydrogen concentration decreased sufficiently, the concentration of free oxygen in the atmosphere could rise again.

Whatever the exact combination of biological and geochemical processes, evidence in the rocks indicates that the oxygen concentration in the atmosphere remained low and then jumped suddenly from trace to appreciable quantities approximately 2.4 billion years ago.

Evolution of the Modern Atmosphere

Several deposits of banded iron formed after the First Great Oxidation Event, and this process continued to remove oxygen that was released during photosynthesis. The last, major banded iron layer was deposited about 1.9 billion years ago, but the oxygen concentration in the atmosphere did not increase dramatically. At least two more critical steps were required before efficient multicellular organisms could evolve.

The Sun emits energy largely in the form of high-energy ultraviolet light. These rays are energetic enough to break apart complex molecules and kill evolving multicellular organisms. But high-altitude oxygen absorbs ultraviolet radiation in a process that forms ozone (O_3). Thus, the oxygen concentration couldn't increase in the lower atmosphere until appreciable concentrations first accumulated in the upper atmosphere. To summarize: oxygen, largely produced by the earliest photosynthetic organisms, was not only necessary for life as we know it today but as ozone it also protected multicellular life by filtering out harmful solar rays.

Multicellular plants and animals emerged in Late Precambrian time, between 1 billion and 543 million years ago. About 600 million years ago, the oxygen level in the atmosphere increased rapidly a second time, in a process called the **Second Great Oxidation Event**. What changed abruptly 1.3 billion years after the last banded iron layers were deposited to allow oxygen to accumulate? Scientists propose that prior to 600 million years ago, biological decay was almost as rapid as photosynthesis. Therefore, the oxygen that was released into the atmosphere was immediately consumed during respiration, according to the following reactions:

During photosynthesis:
Carbon dioxide + Water → Sugars + Oxygen

During respiration and decay:
Sugars + Oxygen → Carbon dioxide + Water

Then, beginning abruptly 600 million years ago, several processes occurred that allowed preservation of organic matter in sediments before it could decay. These processes effectively removed the organic carbon (the sugars) from the system, thereby causing the opposing reactions above to lean in the direction favoring photosynthesis. Stated differently, environmental conditions that allowed the generation but subsequent removal and storage of organic carbon in sediment also were conducive to the expansion of photosynthetic organisms and the generation of much oxygen as a by-product. All the proposed processes for sequestering organic matter involve complex reactions among Earth's four spheres:

- Geochemical processes produced an abundance of clays 600 million years ago. These clays buried and preserved organic matter on the seafloor.
- Zooplankton evolved in the seas. These organisms produced dense, organic-laden feces that sank rapidly to the seafloor, where they accumulated in the clays (mentioned above) and were subsequently buried, thereby removing organic carbon from the system.
- Simple lichens evolved on land. The lichens accelerated weathering, and the weathered ions washed into the sea and provided nutrients for phytoplankton. In turn, the phytoplankton fed the zooplankton, which sequestered nutrients as described previously.

Thus, numerous complex chemical, physical, and biological processes combined to set the stage for the Second Great Oxidation Event.

Earth's atmosphere not only sustains us but it insulates Earth's surface as winds distribute the Sun's heat around the globe so that the surface is neither too hot nor too cold for life to exist. Clouds form from water vapor in the atmosphere and rain falls from clouds. In addition, the atmosphere filters out much of the Sun's ultraviolet radiation, which can destroy living tissue and cause cancer. The atmosphere carries sound; without air we would live in silence. Without an atmosphere, airplanes and birds could not fly; wind would not transport pollen and seeds; the sky would be black rather than blue; and no reds, purples, and pinks would color the sunset. When we understand this complex web of interacting processes and realize that we are the only planet in the Solar System to be so fortunate, we can only marvel at the fragility of Earth's atmosphere.

THE MODERN ATMOSPHERE

The modern atmosphere is mostly gas, but also contains droplets of liquid water and suspended particles of dust. The gaseous composition of dry air is roughly 78 percent nitrogen, 21 percent oxygen, and 1 percent other gases (●**FIGURE 17.5**). Nitrogen, the most abundant gas, does not react readily with other substances. Oxygen, though, reacts chemically as fires burn, iron rusts, and plants and animals respire. Carbon dioxide, which by some models formed as much as 80 percent of Earth's early atmosphere, is a trace gas in the modern atmosphere, with a concentration of only 0.035 percent.

The types and quantities of gases, water vapor, droplets, and dust vary with both location and altitude. In a hot, steamy jungle, air may contain 5 percent water vapor by weight, whereas in a desert or cold polar region only a small fraction of a percent may be present.

If you sit in a house on a sunny day, you may see a sunbeam passing through a window. The visible beam is light reflected from tiny specks of suspended dust. Clay, salt, pollen, bacteria, viruses, bits of cloth, hair, and skin are all components of dust. People travel to the seaside to enjoy the "salt air." Visitors to the Great Smoky Mountains in Tennessee view the bluish, hazy air formed by sunlight reflecting from pollen and other dust particles.

Within the past century, humans have altered the chemical composition of the atmosphere in many different

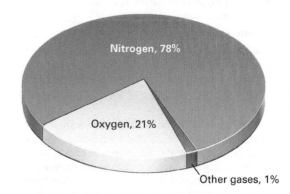

Composition of the Modern Atmosphere

● **FIGURE 17.5** Composition of the modern atmosphere.

ways. We have increased the carbon dioxide concentration by burning fossil fuels and igniting wildfires. Factories release chemicals into the air—some benign, others poisonous. Smoke and soot change the clarity of the atmosphere. These changes are discussed in the section on "Air Pollution."

ATMOSPHERIC PRESSURE

The molecules in a gas zoom about in a random manner. For example, at 20°C an average oxygen molecule is traveling at 425 meters per second (950 miles per hour). In the absence of gravity, or where there are temperature differences or other perturbations, a gas will fill a space homogeneously.

Thus, if you float a cylinder of gas in space, the gas would disperse until there is an equal density of molecules and equal pressure throughout the cylinder. In contrast, gases that surround Earth are perturbed by many influences, which ultimately create the complex and turbulent atmosphere that helps shape the world we live in.

Within our atmosphere, gas molecules zoom around, as in the imaginary cylinder, but in addition, gravity pulls them downward. As a result of this downward force, more molecules concentrate near the surface of Earth than at higher elevations. Therefore, the atmosphere is denser at sea level than it is at higher elevations—and the pressure is higher. Density and pressure then decrease exponentially with elevation (●FIGURE 17.6). At an elevation of about 5,000 meters, the atmosphere contains about half as much oxygen as it does at sea level. If you ascended in a balloon to 16,000 meters (16 kilometers) above sea level, you would be above 90 percent of the atmosphere and would need an oxygen mask to survive. At an elevation of 100 kilometers, pressure is only 0.00003 that of sea level, approaching the vacuum of outer space. There is no absolute upper boundary to the atmosphere.

Atmospheric pressure, often called barometric pressure, is measured with a **barometer**. A simple but accurate barometer is constructed from a glass tube that is sealed at one end. The tube is evacuated and the unsealed end placed in a dish of a liquid such as mercury. The mercury rises in the tube because atmospheric pressure depresses the level of mercury in the dish, but there is no air in the tube to counter that pressure so the surface of the mercury rises

---

Second Great Oxidation Event The process, occurring about 600 million years ago, when the oxygen concentration in Earth's atmosphere increased abruptly a second time in response to interactions among chemical, physical, and biological processes.

atmospheric pressure, often called barometric pressure The pressure of the atmosphere at any given location and time.

barometer A device used to measure barometric pressure.

bar Unit of measurement for atmospheric pressure. One bar is roughly equal to atmospheric pressure at sea level.

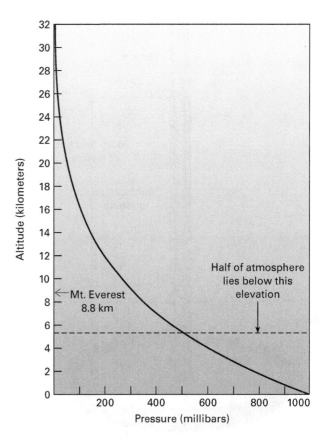

● **FIGURE 17.6** Atmospheric pressure decreases with altitude. One-half of the atmosphere lies below an altitude of 5,600 meters.

(●FIGURE 17.7). At sea level mercury rises approximately 76 centimeters, or 760 millimeters (about 30 inches), into an evacuated tube.

Meteorologists express pressure in inches or millimeters of mercury, referring to the height of the column of mercury in a barometer. They also express pressure in bars and millibars. A **bar** is approximately equal to sea level atmospheric pressure. A millibar is 0.001 of a bar.

A mercury barometer is a cumbersome device nearly a meter tall, and mercury vapor is poisonous. A safer and more portable instrument for measuring pressure, called an aneroid barometer, consists of a partially evacuated metal chamber connected to a pointer. When atmospheric pressure increases, it compresses the chamber and the pointer moves in one direction. When pressure decreases, the chamber expands, directing the pointer the other way (●FIGURE 17.8).

Changing weather can also affect barometric pressure. On a stormy day at sea level, pressure may drop to 980 millibars (28.94 inches), although barometric pressures below 900 millibars (26.58 inches) have been reported during some hurricanes. In contrast, during a period of clear, dry weather, a typical high-pressure reading may be 1,025 millibars (30.27 inches). These changes are discussed more in Chapter 19.

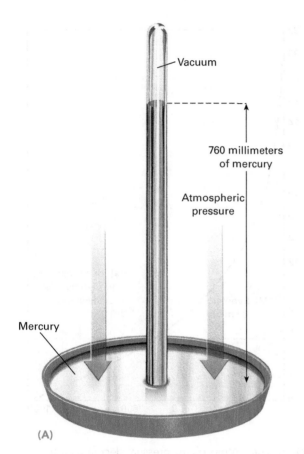

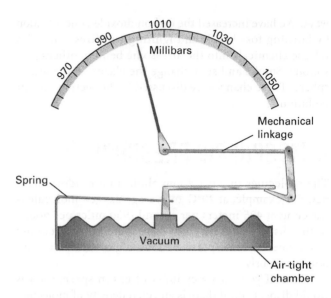

● **FIGURE 17.8**   In an aneroid barometer, increasing air pressure compresses the air-tight chamber and causes the connected pointer to move in one direction. When the pressure decreases, the chamber expands, deflecting the pointer the other way.

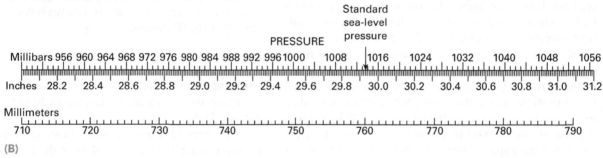

● **FIGURE 17.7**   (A) Atmospheric pressure forces mercury upward in an evacuated glass tube. The height of the mercury in the tube is the measure of air pressure. (B) Three common scales for reporting atmospheric pressure and the conversion among them.

ATMOSPHERIC TEMPERATURE

The temperature of the atmosphere changes with altitude (●**FIGURE 17.9**). The layer of air closest to Earth—the layer we live in—is the **troposphere**. Virtually all of the water vapor and clouds exist in this layer, and almost all weather occurs here. Earth's surface absorbs solar energy, and thus the surface of the planet is warm. But, as explained earlier, continents and oceans also radiate heat, and some of this energy is absorbed by the troposphere. Lower parts of the troposphere absorb most of the heat radiating from Earth's surface; in contrast, at higher elevations in the troposphere, the atmosphere is thinner and absorbs less energy. Consequently, temperature decreases with increased elevation in the troposphere; mountaintops are generally colder than valley floors, and commercial jet airliners must heat their cabins once reaching cruising altitudes, which are generally between 10 and 11 kilometers elevation.

The top part of the troposphere is the **tropopause**, which lies at an altitude of about 17 kilometers at the equator, although it is lower at the poles. The tropopause is the boundary between the troposphere and the **stratosphere** above. It is characterized by an abrupt cessation in the steady decline in temperature with altitude, because cold air from the upper troposphere is too dense to rise higher. As a result, little mixing of air molecules occurs across the tropopause. The tropopause forms the floor of the stratosphere, in which temperature remains constant to 35 kilometers and then increases with altitude until, at about 50 kilometers, it is as warm as air at the Earth's surface.

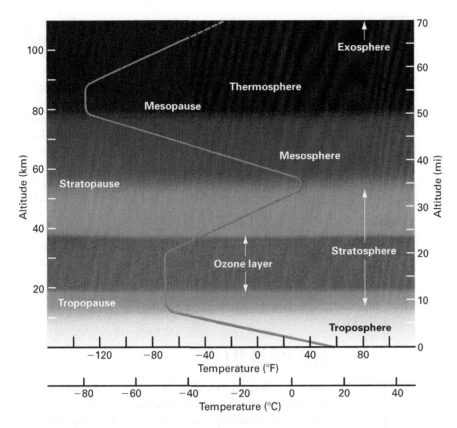

● **FIGURE 17.9** Atmospheric temperature varies with altitude. The atmospheric layers are zones in which different factors control the temperature.

Although the stratosphere is above elevations at which commercial air liners can fly, many military jets can reach it if they really try; the stratosphere is not too high to become polluted by radioactive elements from atmospheric tests of atomic and hydrogen bombs. Large levels of anthropogenic radionuclides (specific isotopes of radioactive elements known only to come from human-caused nuclear reactions) were released into the atmosphere during atmospheric tests of nuclear weapons conducted mostly in the 1950s and 1960s by the United States and former Soviet Union and in the 1970s by France and China. Most of the radioactive debris from atmospheric test of hydrogen bombs reached the stratosphere and accumulated there, forming a reservoir of radioactive radionuclides.

For many years, global fallout of these radionuclides from the stratosphere downward through the troposphere to Earth's surface was observed during the late spring when heating of the ground surface and lower troposphere formed rising hot air that broke through the tropopause and disrupted the lower stratosphere. This disruption caused cold, radionuclide-rich air from the lower stratosphere to rapidly descend, causing the nuclear fallout. Each spring beginning prior to the 1963 moratorium on atmospheric testing of nuclear weapon until the early 1990s, the annual rate of radionuclide fallout peaked, although the annual peak concentration fell through this period.

Since the 1990s, the biggest fluctuations in the concentration of radionuclides in the troposphere are due mostly to their resuspension from contaminated soil (●FIGURE 17.10). Thus, although the annual springtime "radionuclide fall-out event" no longer is significant in terms of providing the source of radionuclides in the troposphere, the deposition of radioactive elements during those events and the physical reworking of those deposits today provide the main source of atmospheric radionuclides in the troposphere.

The reversal in the temperature profile between the troposphere and the stratosphere occurs because the two atmospheric layers are heated by different mechanisms. As already explained, the troposphere is heated primarily from below, by Earth. The stratosphere, however, is heated primarily from above, by direct incoming solar radiation.

Oxygen molecules (O_2) in the stratosphere absorb energetic ultraviolet rays from the Sun. The radiant energy breaks the oxygen molecules apart, releasing free oxygen atoms. These free oxygen atoms then recombine to form ozone (O_3). Ozone absorbs ultraviolet energy more efficiently than oxygen does, and the absorption of UV radiation by the ozone warms the upper stratosphere.

Ultraviolet radiation is energetic enough to affect organisms. Small quantities give us a suntan, but large doses cause skin cancer and cataracts of the eye, inhibit the growth of many plants, and otherwise harm living tissue. The ozone in the upper atmosphere protects life on Earth by absorbing much of this high-energy radiation before it reaches Earth's surface.

---

troposphere The layer of air that lies closest to Earth's surface and extends upward to about 17 kilometers.

tropopause The top of the troposphere; the boundary between the troposphere and the stratosphere.

stratosphere The layer of air above the tropopause, extending upward to about 55 kilometers.

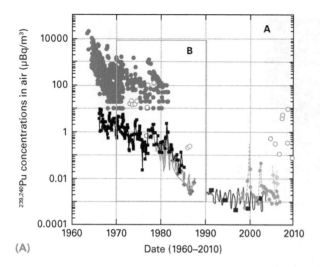

(A)

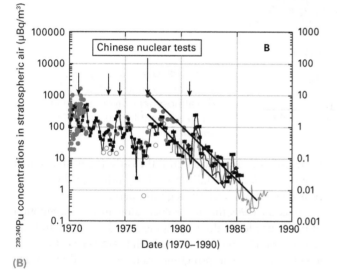

(B)

● **FIGURE 17.10**   (A) Measured levels of Plutonium-239 and Plutonium-240 concentrations in the troposphere between 1960 and 2010.[3] Following the moratorium on atmospheric testing of nuclear weapons in 1963, the concentration of the two radionuclides dropped in the troposphere and the stratosphere. (B) Atmospheric nuclear weapons tests by the Chinese during the 1970s and early 1980s led to high sustained and high concentrations of Plutonium 239 and Plutonium-240 in the stratosphere. Following the elimination of these tests, stratospheric concentrations of the Plutonium radionuclides dropped.

Ozone concentration declines in the upper portion of the stratosphere, and therefore at about 55 kilometers above Earth, temperature once more begins to fall rapidly with elevation. This boundary between rising and falling temperature is the **stratopause**, the ceiling of the stratosphere. The second zone of declining temperature in Earth's modern atmosphere is the **mesosphere**. Little radiation is absorbed in the mesosphere, and the thin air is extremely cold. The ceiling of the mesosphere

3. Katsumi Hirose and Pavel P. Povinec, "Sources of plutonium in the atmosphere and stratosphere-troposphere mixing", www.nature.com/scientificreports, 28 October 2015.

is the **mesopause**. Starting at about 80 kilometers above Earth, the temperature again remains constant and then rises rapidly in the **thermosphere**. Here the atmosphere absorbs high-energy X-rays and ultraviolet radiation from the Sun. High-energy reactions strip electrons from atoms and molecules, producing ions. The temperature in the upper portion of the thermosphere is just below freezing—not extremely cold by surface standards.

The uppermost layer of the atmosphere is the **exosphere**, a layer of very diffuse gas that overlies the thermosphere and thins upward into the vacuum of space. In the exosphere, gas molecules are gravitationally influenced by to the Earth, but the density of gas is too low for the individual molecules to collide with each other, as gas normally does at lower atmospheric levels. Instead of colliding with other molecules, gas molecules in the exosphere travel along 'ballistic trajectories' similar to the arc of a thrown ball. As they are pulled by Earth's gravity, slower gas molecules arc back into the thermosphere while some of the faster-moving molecules continue into outer space and are lost.

AIR POLLUTION

Ever since the first cave dwellers huddled around a smoky fire, people have introduced impurities into the air. The total quantity of these impurities is minuscule compared with the great mass of our atmosphere and with the monumental changes that occurred during the evolution of the planet. Yet air pollution remains a significant health, ecological, and climatological problem for modern industrial society.

In 1948, Donora was an industrial town with a population of about 14,000 located 50 kilometers south of Pittsburgh, Pennsylvania. One large factory in town manufactured structural steel and wire, and another produced zinc and sulfuric acid. During the last week of October 1948, dense fog settled over the town. But it was no ordinary fog; the moisture contained pollutants from the two factories. After four days, visibility became so poor that people could not see well enough to drive, even at noon with their headlights on. Gradually at first, and then in increasing numbers, residents sought medical attention for nausea, shortness of breath, and constrictions in the throat and chest. Within a week, 20 people had died and about half of the town was seriously ill.

stratopause The ceiling of the stratosphere; the boundary between the stratosphere and the mesosphere.

mesosphere The layer of air that lies above the stratopause, extending upward from about 55 kilometers to about 80 kilometers above Earth's surface.

mesopause The ceiling of the mesosphere; the boundary between the mesosphere and the thermosphere.

thermosphere An extremely high and diffuse region of the atmosphere lying above the mesosphere, from about 80 kilometers upward.

exosphere The outermost layer of the in which gas molecules are so diffuse they do not collide. Merges outward into interstellar space.

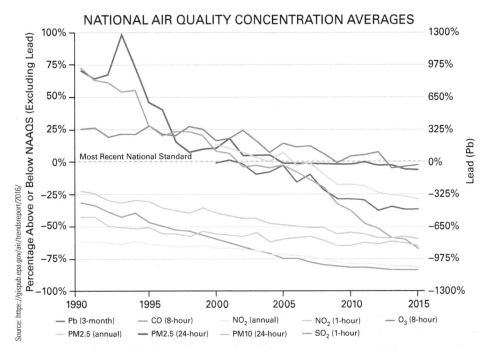

NATIONAL AIR QUALITY CONCENTRATION AVERAGES

Source: https://gispub.epa.gov/air/trendsreport/2016/

— Pb (3-month)   — CO (8-hour)   — NO₂ (annual)   — NO₂ (1-hour)   — O₃ (8-hour)
— PM2.5 (annual) — PM2.5 (24-hour) — PM10 (24-hour) — SO₂ (1-hour)

● FIGURE 17.11 Measurements of air pollutant concentrations in the United States between 1990 and 2015. This plot, from the U.S. EPA, shows average national concentrations of each air pollutant has fallen through the period, reflecting increased awareness of air quality and implementation of air quality measures.

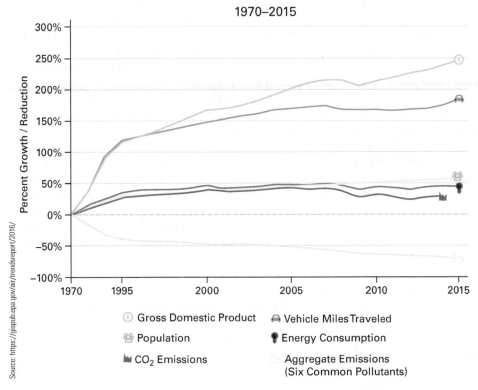

1970–2015

Source: https://gispub.epa.gov/air/trendsreport/2016/

Gross Domestic Product   Vehicle Miles Traveled
Population   Energy Consumption
CO₂ Emissions   Aggregate Emissions
                         (Six Common Pollutants)

● FIGURE 17.12 Between 1970 and 2015, a continuous decrease in the aggregate emissions of six common pollutants occurred while gross domestic product increased along with population and total vehicle miles traveled. The six air pollutants included in this plot are: particulate matter of 2.5 and 10 microns diameter, sulfur dioxide, nitrous oxides, volatile organic compounds, carbon monoxide, and lead.

Other incidents similar to what happened in Donora occurred worldwide. In response to the growing problem, the United States enacted the Clean Air Act in 1963. As a result of the Clean Air Act and its amendments, total emissions of air pollutants have decreased and air quality across the country has improved (●**FIGURE 17.11**). It is even more encouraging to note that this decrease in emissions has occurred at a time when population, energy consumption, vehicle miles traveled, and gross domestic product (GDP) have increased dramatically (●**FIGURE 17.12**). Donora-type incidents in the United States have not been repeated. Smog has decreased, and rain has become less acidic. Yet some people believe that we have not gone far enough and that air pollution regulations should be strengthened further.

Sources and types of air pollution are listed in ●**FIGURE 17.13** and discussed in the following section.

Gases Released When Fossil Fuels Are Burned

Coal is largely carbon, which, when burned completely, produces carbon dioxide. Petroleum is a mixture of hydrocarbons, compounds composed of carbon and hydrogen. When hydrocarbons burn completely, they produce carbon dioxide and water as the only combustion products. Neither is poisonous, but both are greenhouse gases. If fuels were composed purely of compounds of carbon and hydrogen, and if they always burned completely, air pollution from the burning of fossil fuels would pose little direct threat to our health (although combustion of fossil fuels would still contribute to global warming). However, fossil fuels contain impurities, and combustion is usually incomplete. As a result, other products form—most of which are harmful.

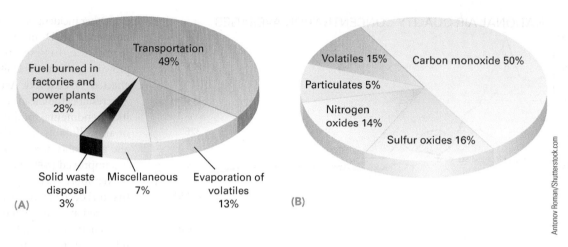

● **FIGURE 17.13**   (A) Sources of air pollution in the United States. (B) Types of air pollutants in the United States. (Although carbon dioxide is a greenhouse gas, it is not listed as a pollutant because it is not toxic.)

Products of incomplete combustion include hydrocarbons such as benzene and methane. Benzene is a carcinogen (a compound that causes cancer), and methane is another greenhouse gas. Incomplete combustion of fossil fuels releases many other pollutants, including carbon monoxide (CO), which is colorless and odorless yet very toxic.

Additional problems arise because coal and petroleum contain impurities that generate other kinds of pollution when they are burned. Small amounts of sulfur are present in coal and, to a lesser extent, in petroleum. When these fuels burn, the sulfur forms oxides, mainly sulfur dioxide (SO_2) and sulfur trioxide (SO_3). High sulfur dioxide concentrations have been associated with major air pollution disasters of the type that occurred in Donora. Today, the primary global source of sulfur dioxide pollution is electricity generation by coal-fired power plants.

Nitrogen, like sulfur, is common in living tissue and therefore is found in all fossil fuels. This nitrogen, together with a small amount of atmospheric nitrogen, reacts when coal or petroleum is burned. The products are mostly nitrogen oxide (NO) and nitrogen dioxide (NO_2). Nitrogen dioxide is a reddish-brown gas with a strong odor. It contributes to the "browning" and odor of some polluted urban atmospheres. Automobile exhaust is the primary source of nitrogen oxide pollution.

Acid Rain

As explained earlier, sulfur and nitrogen oxides are released when coal and petroleum burn. These oxides are also released when metal ores are refined. In moist air, sulfur dioxide reacts to produce sulfuric acid and nitrogen oxides react to form nitric and nitrous acid. These strong atmospheric acids dissolve in water droplets and fall as **acid precipitation, also called acid rain** (●**FIGURE 17.14**).

Acidity is expressed on the **pH scale**. A solution with a pH of 7 is neutral, neither acidic nor basic. On a pH scale, numbers lower than 7 represent acidic solutions and numbers higher than 7 represent basic ones. For example, soapy water is basic and has a pH of about 10, whereas vinegar is an acid with a pH of 2.4.

Rain reacts with carbon dioxide in the atmosphere to produce a weak acid. As a result, natural rainfall has a pH of about 5.7. However, in the "bad old days" before the Clean Air Act was properly enforced, rain was much more acidic. A fog in Southern California in 1986 reached a pH of 1.7, which approaches the acidity of toilet bowl cleaners.

Consequences of Acid Rain

Sulfur and nitrogen oxides impair lung function, aggravating diseases such as asthma and emphysema. They also affect the heart and liver and have been shown to increase vulnerability to viral infections such as influenza.

Acid rain corrodes metal and rock. Limestone and marble are especially susceptible because they dissolve rapidly in mild acid. In the United States, the cost of damage and deterioration to buildings and building materials caused by acid precipitation is estimated at several billion dollars per year.

FIGURE 17.15A is a map of average pH for rainfall across the conterminous United States in 1992. The northeastern United States was particularly susceptible to acid rain because it was downwind from the interior "Rust Belt," the large region of the northern interior United States in which large-scale industry had developed since the nineteenth century (●**FIGURE 17.15B**). More recently, a 2016 study involved the analysis of 27 sites in the northeastern United States and eastern Canada over a period ranging between 8 and 24 years. The results show that long-term increases in pH occurred in the O and B soil horizons at most sites, trends that the study's authors attributed to reversal of the effects of acid rain on North American soils.

Smog and Ozone in the Troposphere

Imagine that your great-grandfather had entered the exciting new business of making moving pictures. Old-time photographic film was "slow" and required lots of sunlight, so

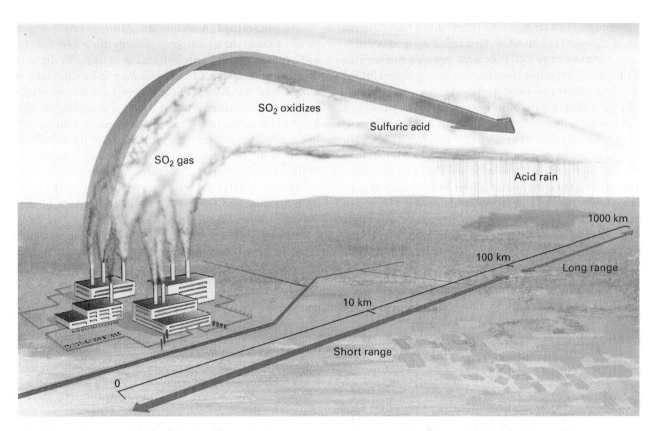

● **FIGURE 17.14**   Acid rain develops from the addition of sulfur compounds to the atmosphere by industrial smokestacks.

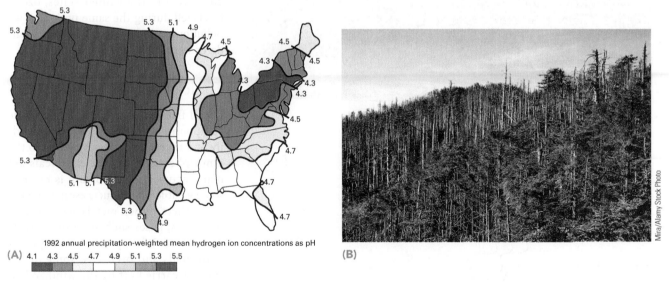

● **FIGURE 17.15**   (A) Annual average measurements of the pH of precipitation during 1992. The highest levels of acidity in precipitation occurred in the northeastern United States, which is generally downwind of the industrial "Rust Belt." (B) Effects of acid rain damage in North Carolina.

acid precipitation, also called acid rain Rain, snow, fog, or mist that has become acidic after reacting with air pollutants.

pH scale A logarithmic scale that measures the acidity of a solution. A pH of 7 is neutral; numbers lower than 7 represent acidic solutions, and numbers higher than 7 represent basic ones.

many filmmakers left the polluted, overcast, industrial Northeast. Southern California, with its warm, sunny climate and little need for coal, was preferable. Thus, a district of Los Angeles called Hollywood became the center of the movie industry. Its population boomed, and after World War II, automobiles became about as numerous as people. Then the quality of the air deteriorated in a strange way. People noted four different kinds of changes: (1) a brownish haze called **smog** settled over the city (●**FIGURE 17.16**); (2) people felt irritation in their eyes and throats; (3) vegetable crops became damaged; and (4) the sidewalls of rubber tires developed cracks.

In the 1950s, air pollution experts worked mostly in the industrialized cities of the East Coast and the Midwest. When they were called to diagnose the problem in Southern California, they looked for the sources of air pollution they knew well, especially sulfur dioxide. But the smog was nothing like the pollution they were familiar with. These researchers eventually learned that incompletely burned gasoline in automobile exhaust reacts with nitrogen oxides and atmospheric oxygen in the presence of sunlight to form ozone (O_3). The ozone then reacts further with automobile exhaust to form smog (●**FIGURE 17.17**).

● **FIGURE 17.16**   Smog in the Los Angeles basin.

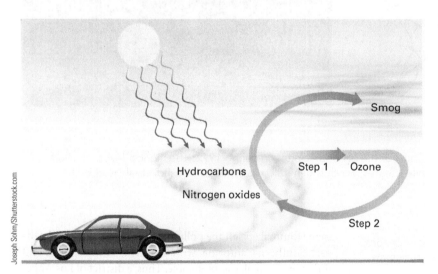

● **FIGURE 17.17**   Smog forms in a sequential process. Step 1: Automobile exhaust reacts with air in the presence of sunlight to form ozone. Step 2: Ozone reacts with automobile exhaust to form smog.

As you learned in section on "Atmospheric Temperature," ozone in the stratosphere absorbs ultraviolet radiation and protects life on Earth. Yet, excessive ozone in the air we breathe is harmful. Is ozone a pollutant to be eliminated, or a beneficial component of the atmosphere that we want to preserve? The answer is that it is both, depending on where it is found: ozone in the troposphere reacts with automobile exhaust to produce smog, and therefore it is a pollutant. Ozone in the stratosphere is beneficial, and the destruction of the ozone layer there creates serious problems.

Ozone irritates the respiratory system, causing loss of lung function and aggravating asthma in susceptible individuals. Ozone also increases susceptibility to heart disease and is a suspected carcinogen. High ozone concentrations slow the growth of plants, which is a particularly serious problem in the rich, agricultural areas of California.

Ozone pollution ranks as the 33rd leading cause of human death. As of 2018, over 95 percent of Earth's human population lives with unsafe levels of air pollution, using air quality guidelines from the World Health Organization. Using population-adjusted measurements of ozone, with greater weight given to ozone concentrated in densely populated areas, the State of Global Air/2018 reported that global levels of ozone pollution increased between 1990 and 2016. Within the United States, ozone levels fell during the same period, as air pollution controls and fuel emission standards were implemented.

Toxic Volatiles

Recall from Chapter 15 that a volatile compound is one that evaporates readily and therefore easily escapes into the atmosphere. Whenever chemicals are manufactured or petroleum is refined, some volatile by-products escape into the atmosphere. When metals are extracted from ores, gases such as sulfur dioxide are released. When pesticides are sprayed onto fields and orchards, some of the spray is carried off by the wind. When you paint your house, the volatile parts of the paint evaporate into the air. As a result of all these processes, tens of thousands of volatile compounds are present in polluted air: some are harmless, others are poisonous, and many have not been studied.

Consider the case of dioxin. Very little dioxin is intentionally manufactured. It is not an ingredient in any herbicide, pesticide, or other industrial formulation.

You cannot buy dioxin at your local hardware store or pharmacy. Dioxin forms as an unwanted by-product in the production of certain chemicals and when specific chemicals are burned. For example, in the United States today, garbage incineration is the most common source of dioxin. When a compound containing chlorine, such as the plastic polyvinyl chloride (PVC), is burned, some of the chlorine reacts with organic compounds to form dioxin. The dioxin then goes up the smokestack of the incinerator, diffuses into the air, and eventually falls to Earth. Cattle eat grass lightly dusted with dioxin and store the dioxin in their fat. Humans ingest the compound mostly in meat and dairy products. The EPA estimates that the average U.S. citizen ingests about 0.0000000001 gram (100 picograms) of dioxin in food every day. Although this is a minuscule amount, the EPA has argued that dioxin is the most toxic chemical known and that even these low background levels may cause adverse effects such as cancer, disruption of regulatory hormones, reproductive and immune system disorders, and birth defects. Others disagree. The Chemical Manufacturers Association has written: "There is no direct evidence to show that any of the effects of dioxins occur in humans in everyday levels."

A 2016 study published in the *Journal of Epidemiology* presented results from an 18 year-long study in which dioxin levels were measured from the milk of first-time Japanese mothers. The study showed that dioxin levels fell between 1998 and 2014 from 20.8 to 7.2 picograms of dioxin toxic equivalence in milk fat, despite a trend toward Japanese women having children later in life during this study period. Because dioxin accumulates in humans through time, milk from older mothers has higher dioxin levels. Thus, the lower overall dioxin levels over the course of the study suggests that concentrations in Japanese first-time mothers fell over the study period.

No one knows whether very small doses of potent poisons ingested over long periods of time are harmful. Environmentalists argue that it is "better to be safe than sorry" and that therefore we should reduce ambient concentrations of volatiles such as dioxin. This argument has helped spur the movement toward increased recycling and decreasing incineration of waste plastic as a means of disposal. Others counter that because the harmful effects of compounds like dioxin are unproven, we should not burden our economy with the costs of control.

Particulates and Aerosols

A **particle, or particulate**, is any small piece of solid matter, such as dust or soot. An **aerosol** is any small particle that is larger than a molecule and suspended in air. In the context of air pollution, all three terms are used interchangeably. Many natural processes release aerosols. Windblown silt, pollen, volcanic ash, salt spray from the oceans, and smoke and soot from wildfires are all aerosols. Industrial emissions add to these natural sources.

Smoke and soot are carcinogenic aerosols formed whenever fuels are burned. Coal always contains clay and other noncombustible minerals that have accumulated along with the organic matter in the depositional environment in which the coal formed, commonly a swamp. When the coal burns, some of these minerals escape from the chimney as **fly ash**, which settles as gritty dust. When metals are mined, the drilling, blasting, and digging raise dust, and this too adds to the total load of aerosols.

In 1988, EPA epidemiologists noted that whenever atmospheric aerosol levels rose above a critical level in Steubenville, Ohio, the number of fatalities from all causes—car accidents to heart attacks—rose. After several studies substantiated the Steubenville report, the EPA proposed additional reductions of ambient aerosol levels in the United States. Opponents argued that it is unfair to target all aerosols, because the term covers a wide range of substances from a benign grain of salt to a deadly mist of toxic volatiles.

DEPLETION OF THE OZONE LAYER

As described in section on "Atmospheric Temperature," solar energy breaks apart oxygen molecules (O_2) in the stratosphere, releasing free oxygen atoms (O). The free oxygen atoms combine with oxygen molecules to form ozone (O_3). Ozone absorbs high-energy ultraviolet light. This absorption protects life on Earth because ultraviolet light causes skin cancer, inhibits plant growth, and otherwise harms living tissue.

In the 1970s, scientists learned that organic compounds containing chlorine and fluorine, called **chlorofluorocarbons (CFCs)**, and compounds containing bromine and chlorine, called **halons**, rise into the upper atmosphere, react with, and destroy ozone (●**FIGURE 17.18**). CFCs have no natural source, but were entirely synthesized for such diverse uses that included cooling agents in almost all refrigerators and air conditioners, propellants in aerosol cans, and as an expanding agent in styrofoam coffee cups and some polystyrene building insulation manufactured before the mid-1980s.

---

smog Visible, brownish air pollution formed through chemical reactions that involve incompletely combusted automobile gasoline, nitrogen dioxide, oxygen, ozone, and sunlight.

particle or particulate In pollution terminology, any small piece of solid matter larger than a molecule, such as dust or soot.

aerosol In pollution terminology, a particle or particulate that is suspended in air.

fly ash Noncombustible minerals that escape into the atmosphere when coal burns, eventually settling as gritty dust.

chlorofluorocarbons (CFCs) Organic compounds containing chlorine and fluorine, which rise into the upper atmosphere and destroy the ozone layer there.

halons Compounds containing bromine and chlorine, which rise into the upper atmosphere and destroy the ozone layer there.

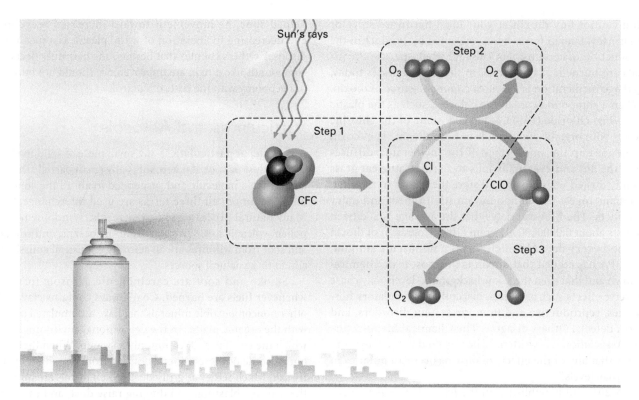

● **FIGURE 17.18** CFCs destroy the ozone layer in a three-step reaction. Step 1: CFCs rise into the stratosphere. Ultraviolet radiation breaks the CFC molecules apart, releasing chlorine atoms. Step 2: Chlorine atoms react with ozone, O_3, to destroy the ozone molecule and release oxygen, O_2. The extra oxygen atom combines with chlorine to produce ClO. Step 3: The ClO sheds its oxygen, producing another free chlorine atom. Thus, chlorine is not used up in the reaction, and one chlorine atom reacts over and over again, to destroy many ozone molecules.

In 1985, scientists observed an unusually low ozone concentration in the stratosphere over Antarctica, a phenomenon called the **ozone hole**. Since it was discovered, the low concentration of ozone over Antarctica declined further until, by 1993, it was 65 percent below normal over 23 million square kilometers, an area almost the size of North America. Research groups also reported significant increases in ultraviolet radiation from the Sun at ground level in the region. In addition, scientists recorded ozone depletion in the Northern Hemisphere. In March 1995, ozone concentration above the United States was 15 to 20 percent lower than during March 1979.

Data on global ozone depletion persuaded the industrial nations of the world to limit the use of CFCs and other ozone-destroying compounds. In a series of international agreements signed between 1978 and 1992, many nations of the world agreed to reduce or curtail production of compounds that destroy atmospheric ozone. Most industrialized countries stopped production of CFCs on January 1, 1996.

The international bans have had positive results. The concentration of ozone-destroying chemicals peaked in the troposphere (lower atmosphere) in 1994 and has been declining ever since, although CFCs are projected to remain in the atmosphere for over 100 years. The CFCs and halons already in the atmosphere break down slowly, but at least the measured concentrations of these ozone-destroying chemicals are declining. In a review article published in May 2006, the authors concluded that "ozone abundances have at least not decreased. . .for most of the world," although natural cycles make it difficult to judge the relative effects of human and natural influences. In summary, they write, "it is therefore unlikely that ozone will stabilize at levels observed before 1980, when a decline in ozone concentrations was first observed"[4] (●**FIGURE 17.19**). As of 2016, eleven different CFC compounds known to be completely synthetic still occur in Earth's atmosphere (**TABLE 17.2**).

---

ozone hole An unusually low ozone concentration in the stratosphere that is centered roughly over Antarctica.

---

4. Elizabeth C. Weatherhead and Signe Bech Andersen, "The Search for Signs of Recovery of the Ozone Layer," *Nature* 441 (May 4, 2006), 39; online at http://www.nature.com/nature/journal/v441/n7089/abs/nature04746.html.

TABLE 17.2 CFC compounds in the atmosphere

Gas	Pre-1750 Tropospheric Concentration	Recent Tropospheric Concentration	Atmospheric Lifetime (Years)
Concentrations in parts per million (ppm)			
Carbon dioxide (CO_2)	~280	399.5	~ 100-300
Concentrations in parts per billion (ppb)			
Methane (CH_4)	722	1834	12.4
Nitrous oxide (N_2O)	270	328	121
Tropospheric ozone (O_3)	237	337	hours-days
Concentrations in parts per trillion (ppt)			
CFC-11 (CCl_3F)	zero	232	45
CFC-12 (CCl_2F_2)	zero	516	100
CFC-113(CCl_2CClF_2)	zero	72	85
HCFC-22($CHClF_2$)	zero	233	11.9
HCFC-141b(CH_3CCl_2F)	zero	24	9.2
HCFC-142b(CH_3CClF_2)	zero	22	17.2
Halon 1211 ($CBrClF_2$)	zero	3.6	16
Halon 1301 ($CBrClF_3$)	zero	3.3	65
HFC-134a(CH_2FCF_3)	zero	84	13.4
Carbon tetrachloride (CCl_4)	zero	82	26
Sulfur hexafluoride (SF_6)	zero	8.6	3200

Source: http://cdiac.ess-dive.lbl.gov/pns/current_ghg.html

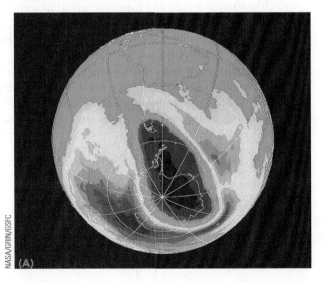 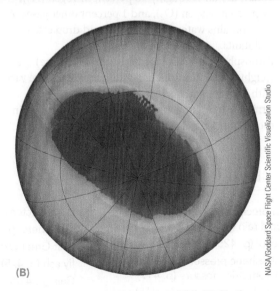

● **FIGURE 17.19** (A) A satellite image of stratospheric ozone concentrations over Antarctica in 1994. (B) Similar data for 2004. In both images, dark purple shows the lowest ozone concentration.

Key Concepts Review

• Earth's primordial atmosphere was composed almost entirely of hydrogen and helium. After these light elements boiled off into space, bolides carried gases to create a secondary atmosphere, which was composed predominantly of carbon dioxide, with lesser amounts of water and nitrogen, and trace gases. Four billion years ago, outgassing, combined with reactions of gases with rocks of the geosphere, created an atmosphere rich in hydrogen and carbon dioxide.

• Photosynthesis by primitive cyanobacteria began producing oxygen about 2.7 billion years ago. However, for 300 million years, the oxygen concentration did not increase dramatically in the atmosphere due to two processes: (1) as the oxygen concentration increased, it initially reacted with dissolved iron in seawater to form banded iron formation; oxygen concentration rose to its present level only after nearly all the dissolved iron had been removed by precipitation; and (2) when life first evolved, the hydrogen concentration of the atmosphere remained high and oxygen reacted with the hydrogen to produce water; bacteria converted the hydrogen to methane, and the oxygen concentration in the atmosphere rose only after the hydrogen concentration declined. The First Great Oxidation Event occurred about 2.4 billion years ago. After that time, further production of oxygen was roughly balanced by respiration. About 600 million years ago, numerous processes reduced decay and respiration, leading to the Second Great Oxidation Event.

• Today, dry air is roughly 78 percent nitrogen (N_2), 21 percent oxygen (O_2), and 1 percent other gases. Air also contains water vapor, dust, liquid droplets, and pollutants.

• Atmospheric pressure (or barometric pressure) is the weight of the atmosphere per unit area. Pressure varies with weather and decreases with altitude.

• Atmospheric temperature decreases with altitude in the troposphere. Then the temperature rises in the stratosphere because ozone absorbs solar radiation; at an altitude of about 50 kilometers, the stratosphere is as warm as at Earth's surface. The temperature decreases again in the mesosphere where it is extremely cold, and then in the uppermost layer, the thermosphere, temperature increases to just below freezing as high-energy radiation is absorbed.

• The increasing ill effects of air pollution prior to and just after World War II convinced lawmakers to pass legislation such as the 1963 Clean Air Act. Incomplete combustion of coal and petroleum produces carcinogenic hydrocarbons such as benzene as well as carbon monoxide and methane, which is a greenhouse gas. Impurities in these fuels burn to produce oxides of nitrogen and sulfur. Nitrogen and sulfur oxides react in the atmosphere to produce acid rain, which damages health, weathers materials, and reduces growth of crops and forests.

• Incompletely burned fuels in automobile exhaust react with nitrogen oxides in the presence of sunlight and atmospheric oxygen to form ozone; ozone then further reacts with automobile exhaust to form smog. Dioxin is an example of a compound that is produced inadvertently during chemical manufacture and when certain materials are burned. Some scientists argue that even tiny amounts of dioxin and other toxic volatiles may be harmful to human health, but others disagree. Scientific studies show that industrial aerosols are harmful to health, but aerosols are so varied that it is difficult to know which ones are most harmful.

• Chlorofluorocarbons and halons, compounds containing chlorine and bromine, rise into the stratosphere and deplete the ozone that filters out harmful UV radiation and protects Earth. Ozone-destroying chemicals have been regulated by international treaty, and their concentration in the troposphere is slowly diminishing. Overall, Earth's stratospheric ozone level has been increasing, but serious ozone holes remain at the poles.

Important Terms

acid precipitation, also called acid rain (p. 422)

aerosol (p. 425)

atmospheric pressure, often called barometric pressure (p. 417)

banded iron formation (p. 415)

bar (p. 417)

barometer (p. 417)

chemoautotroph (p. 413)

chlorofluorocarbons (CFCs) (p. 425)

cyanobacteria (p. 414)

exosphere (p. 420)

First Great Oxidation Event (p. 414)

fly ash (p. 425)

Gaia (p. 414)

halons (p. 425)

mesopause (p. 420)

mesosphere (p. 420)

outgassing (p. 412)

ozone hole (p. 426)

particle, or particulate (p. 425)

pH scale (p. 422)

photosynthesis (p. 413)

Second Great Oxidation Event (p. 416)

smog (p. 424)

stratopause (p. 420)

stratosphere (p. 418)

thermosphere (p. 420)

tropopause (p. 418)

troposphere (p. 418)

Review Questions

1. Discuss the formation and composition of Earth's earliest atmosphere.

2. How and when did Earth's secondary atmosphere form? Compare the composition of this atmosphere with that of the modern one.

3. How did outgassing affect the composition of the Hadean atmosphere?

4. What process first began producing oxygen about 2.7 billion years ago? Why did the oxygen concentration remain low for 300 million years?

5. Briefly describe the First Great Oxidation Event.

6. How did banded iron formation form? How did the deposition of banded iron formation affect the atmosphere?

7. Briefly describe the Second Great Oxidation Event.

8. List the two most abundant gases in the atmosphere. List three other, less abundant, gases. List three nongaseous components of natural air.

9. Draw a graph of the change in pressure with altitude. Explain why the pressure changes as you have shown.

10. What is a barometer and how does it work?

11. Draw a figure showing temperature changes with altitude. Label all the significant layers in Earth's atmosphere.

12. Discuss the air pollution disaster in Donora. What pollutants were involved? Where did they come from? How did they become concentrated?

13. Briefly list the major sources of air pollution.

14. What air pollutants are generated when coal and gasoline burn?

15. What is acid rain, how does it form, and how does it affect people, crops, and materials?

16. What is smog, how does it form, and how does it differ from automobile exhaust?

17. What is a volatile? Why are volatiles hard to control?

18. Discuss the production, dispersal, and health effects of dioxin.

19. What is an aerosol? Briefly discuss the health effects of aerosols.

20. Explain how CFCs deplete the ozone layer.

21. Why is ozone in the troposphere harmful to humans, while ozone in the stratosphere is beneficial?

22. Discuss the effects of the international ban on ozone-destroying chemicals.

NASA/GSFC/Jeff Schmaltz/MODIS Land Rapid Response Team

18

ENERGY BALANCE IN THE ATMOSPHERE

Strong winds blow across Bristol Bay, Alaska, southward through the Aleutian Islands, and into the northern Pacific Ocean, stretching stratocumulus clouds into long parallel bands in the process. Sea ice with numerous fractures floats in the northern part of Bristol Bay and also has accumulated on the windward (northern) side of the Aleutian Islands. As it passes over and around the tall volcanic peaks of the Aleutian island arc, the wind became turbulent, forming symmetric eddies and intricate patterns in the clouds on the leeward side of the islands in the northern Pacific Ocean. Notice that the cloud bands are thin to nonexistent over the sea ice in northern Bristol Bay and thicken over the open water of Bristol Bay and the northern Pacific, where row upon row of clouds are visible. This true-color image was captured by the Moderate Resolution Imaging Spectroradiometer (MODIS) aboard NASA's Terra satellite on January 11, 2012.

INTRODUCTION

Weather is the state of the atmosphere at a given place and time. Temperature, wind, cloudiness, humidity, and precipitation are all components of weather. The weather changes frequently, from day to day or even from hour to hour.

Climate is the characteristic weather of a region, particularly the temperature and precipitation, averaged over several decades. Miami and Los Angeles have warm climates; summers are hot and even the winters are warm. In contrast, New York and Chicago experience much greater temperature extremes. Even though winters are cool and often snowy, summers can be almost as hot as those in Miami. Seattle experiences moderate temperatures with foggy, cloudy winters. In this and the following two chapters, we discuss the atmosphere, weather, and climate. In Chapter 21, we will consider global climate change.

Many of the processes that drive energy balance in the atmosphere vary in gradual and predictable manners. But the weather and climate don't always vary gradually and predictably because they are affected by many complex, overlapping system interactions, and by threshold and feedback mechanisms. As a simple example, incoming solar radiation is most intense at the equator and decreases predictably toward the poles. But some high-latitude locations are warmer than regions closer to the equator, and on any given day, it can be warmer in Montreal than in Houston. In this and subsequent chapters, we will explain the fundamental processes that drive weather and climate, and how they interact.

INCOMING SOLAR RADIATION

Solar energy streams from the Sun in all directions, and Earth receives only one two-billionth of the total solar output. However, even this tiny fraction warms Earth's surface and makes it habitable.

The space between Earth and the Sun is nearly empty. How does sunlight travel through a vacuum? In the late 1600s and early 1700s, light was poorly understood. Isaac Newton postulated that light consists of streams of particles that he called "packets" of light. Two other physicists, Robert Hooke and Christian Huygens, argued that light travels in waves. Today we know that Hooke and Huygens were correct—light behaves as a wave; but Newton was also right—light acts as if it is composed of particles. But how can light

---

weather The state of the atmosphere at a given place and time as characterized by temperature, wind, cloudiness, humidity, and precipitation.

climate The characteristic weather of a region, averaged over several decades; refers to yearly cycles of temperature, wind, rainfall, and so on, and not to daily variations.

photons A particle of light; the smallest particle or packet of electromagnetic energy.

electromagnetic radiation Radiation consisting of an oscillating electric and magnetic field, including radio waves, infrared, visible light, ultraviolet, x-rays, and gamma rays.

LEARNING OBJECTIVES

LO1 Contrast weather and climate.

LO2 Compare absorption and emission of energy from light.

LO3 Give examples of albedo from dark- and light-colored objects.

LO4 Explain the scattering of light in relationship to its wavelength.

LO5 Discuss the use of the term "greenhouse gas" in relation to the Earth's atmosphere.

LO6 Contrast the terms heat and temperature.

LO7 Compare heat transfer by conduction and convection.

LO8 Illustrate the changes of state for water including calories absorbed and released.

LO9 Explain the effect of latitude on temperature.

LO10 Explain the occurrence of seasons on the Earth.

LO11 Explain the relationship of geography to temperature.

---

be both a wave and a particle at the same time? In a sense, this is an unfair question because light is fundamentally different from familiar objects. Light is unique; it behaves as a wave *and* a particle simultaneously.

Particles of light are called **photons**. In a vacuum, photons travel only at one speed, the speed of light, never faster and never slower. The speed of light is 3×10^8 meters per second. At that rate a photon covers the 150 million kilometers between the Sun and Earth in about 8 minutes. Photons are unlike ordinary matter in that they appear when they are emitted and disappear when they are absorbed.

Light also behaves as an electrical and magnetic wave, called **electromagnetic radiation**. Terminology describing a light wave is the same as for sound waves or ocean waves (**●FIGURE 18.1**). Its wavelength is the distance between

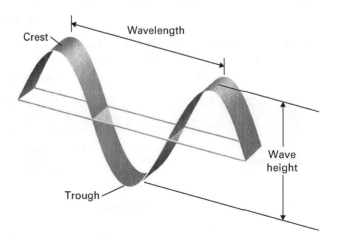

● FIGURE 18.1 The terms used to describe a light wave are identical to those used for water, sound, and other types of waves.

successive wave crests. The **frequency** of a wave is the number of complete wave cycles, from crest to crest, that pass by any point in a second. (Think of how *frequently* the waves pass by.) Electromagnetic radiation occurs in a wide range of wavelengths and frequencies. The **electromagnetic spectrum** is the continuum of radiation of different wavelengths and frequencies (●FIGURE 18.2). At one end of the spectrum, radiation given off by ordinary household current has a long wavelength (5,000 kilometers) and low frequency (60 cycles per second). At the other end, cosmic rays from outer space have a short wavelength (about one-trillionth of a centimeter, or 10^{-14} meter) and very high frequency (10^{22} cycles per second). Visible light is only a tiny portion (about one-millionth of 1 percent) of the electromagnetic spectrum.

Absorption and Emission

If you go outside on a cold, snowy, February day, you wear a heavy jacket, hat, and gloves. Yet the sunlight warms your face. If the sky is clear, even though the temperature may be well below freezing, you must wear sunglasses to prevent snow blindness, and if you have unprotected fair skin, you will suffer sunburn.

Each photon is a tiny packet of concentrated energy. The energy of a single photon is related only to the frequency of the light, and *not* to the ambient temperature. When a photon strikes your face, it may be absorbed. During **absorption of radiation**, the energy of the photon may initiate chemical

and physical reactions in the molecules of your skin. One possible reaction is roughly analogous to cooking, and this reaction causes sunburn or tan. A photon may also cook a molecule in the retina of your eye, and if this reaction occurs frequently enough, you will become snow blind.

Some absorbed photons do not cause chemical reactions; instead they cause molecules to vibrate or rotate more rapidly. This rapid motion makes your skin feel warm.

Alternatively, a photon's energy may be transferred to an electron in a molecule. In this case, the electron jumps to what scientists call an **excited state**—it has a higher level of energy than its *ground state* (lowest level). But electrons do not remain in their excited states forever. Frequently they fall back to a lower energy state, initiating **emission of radiation** in the process. In other words, the photon's energy was transferred to the electron and eventually the electron releases that energy back in the form of a photon. All objects emit radiant energy at some wavelength (except at absolute zero—the theoretically lowest possible temperature at which all molecular motions stop.)

Let's shift our focus from your skin to an iron bar. An iron bar at room temperature emits *infrared* radiation. This radiation has low energy and long wavelengths. It is invisible because the energy is too low to activate the sensors in our eyes. Infrared radiation is sometimes called *radiant heat*, or *heat rays*. Thus, even at room temperature, an iron bar radiates heat. If you place the bar in a hot flame, it begins to glow with a dull, red color when it becomes

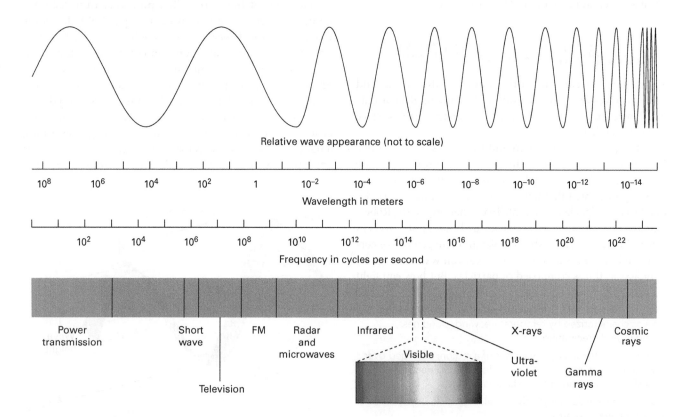

● **FIGURE 18.2** The electromagnetic spectrum. The wave shown is not to scale. In reality the wavelength varies by a factor of 10^{22}, and this huge difference cannot be shown.

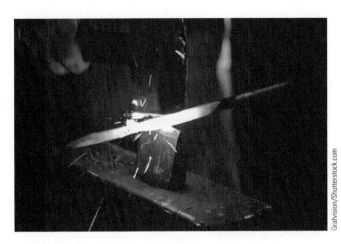

● **FIGURE 18.3** Iron glows red, then orange, then heats to white as a blacksmith works in the shop.

hot enough. The heat has excited electrons in the iron bar, and the excited electrons then emit visible, red, electromagnetic radiation (●**FIGURE 18.3**). If you heat the bar further, it gradually changes color until it becomes white. This demonstration shows another property of emitted electromagnetic radiation: the wavelength (color) of the radiation is determined by the temperature of the source. With increasing temperature, the energy level of the radiation increases, the wavelength decreases, and the color changes progressively.

The Sun's surface temperature is about 6,000°C. Because of its high temperature, the Sun emits relatively high-energy (short-wavelength) radiation, primarily in the ultraviolet and visible portions of the spectrum. When this radiation strikes Earth, it is absorbed by rock and soil. After the radiant energy is absorbed, the rock and soil reemit it. But Earth's surface is much cooler than the Sun's. Therefore, Earth emits low-energy, infrared heat radiation, which has a relatively long wavelength and low frequency. Thus, Earth absorbs high-energy, visible light and emits low-energy, invisible, infrared heat radiation.

Reflection

Radiation reflects from many surfaces. We are familiar with the images reflected by a mirror or the surface of a still lake. Some surfaces are better reflectors than others. The reflectivity of a surface is referred to as its **albedo** (from the Latin for "whiteness") and is often expressed as a percentage. A mirror reflects nearly 100 percent of the light that strikes it and has an albedo close to 100 percent. Even some dull-looking objects are efficient reflectors. Light is bouncing back to your eye from the white paper of this page, although very little is reflected from the black letters.

Snowfields and glaciers have high albedos and reflect 80 to 90 percent of sunlight. Clouds have the second-highest albedo and reflect 50 to 55 percent of sunlight. On the other hand, city buildings and dark pavement have albedos of only 10 to 15 percent (●**FIGURE 18.4**). Forests, with many independent surfaces of dark leaves, have an even lower albedo of about 5 percent. The oceans, which cover about two-thirds of Earth's surface, also have a low albedo. As a result, they absorb considerable solar energy and strongly affect Earth's radiation balance. Thus, the temperature balance of the atmosphere is profoundly affected by the albedos of the hydrosphere, geosphere, and biosphere. If Earth's albedo were to rise by growth of glaciers or cloud cover, the surface of our planet would cool. Alternatively, a decrease in albedo (caused by the melting of glaciers) would cause warming. Thus, the growth and shrinking of Earth's snow and ice is a classic example of a feedback mechanism. If Earth were to cool by a small amount, snowfields and glaciers would grow.

---

frequency The number of complete wave cycles, from crest to crest, that pass by any point in a second.

electromagnetic spectrum The entire range of electromagnetic radiation from very-long-wavelength (low frequency) radiation to very-short-wavelength (high frequency) radiation.

absorption of radiation The process that occurs when energy is absorbed: the energy of a photon is converted to electrical, chemical, vibrational, or heat energy, and the photon disappears.

excited state A state of physical energy higher than the lowest energy level (or *ground state*) of an electron in an atom or molecule.

emission of radiation The process that occurs when energy, in the form of a photon, is emitted as an electron falls out of an excited state, with the equivalent loss of energy from the emitting substance.

albedo The reflectivity of a surface; surfaces that reflect more light have a higher albedo.

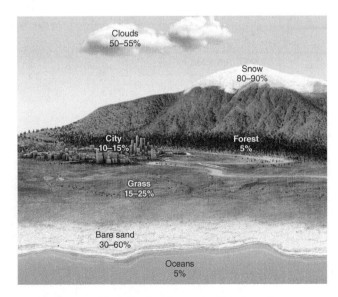

● **FIGURE 18.4** The albedo of common Earth surfaces varies greatly.

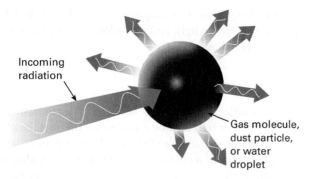

● FIGURE 18.5 Atmospheric gases, water droplets, and dust scatter incoming solar radiation. When radiation scatters, its direction changes but the wavelength remains constant.

But snow and ice reflect solar radiation back into space and lead to additional cooling—which causes further expansion of the snow and ice—and so on.

Scattering

On a clear day, the Sun shines directly through windows on the south side of a building, but if you look through a north-facing window, you cannot see the Sun. Even so, light enters through the window, and the sky outside is blue. If sunlight were only transmitted directly, a room with north-facing windows would be dark and the sky outside the window would be black. Atmospheric gases, water droplets, and dust particles scatter sunlight in all directions, as shown in ●**FIGURE 18.5**. It is this scattered light that illuminates a room with north-facing windows and turns the sky blue.

The amount of scattering is inversely proportional to the wavelength of light. Short-wavelength blue light, therefore, scatters more than longer wavelength red light. The Sun emits light of all wavelengths, which combine to make up white light. Consequently, in space the Sun appears white. The sky appears blue from Earth's surface because the blue component of sunlight scatters more than other frequencies and colors the atmosphere. The Sun appears yellow from Earth because yellow is the color of white light with most of the blue light removed.

THE RADIATION BALANCE

With this background, let us examine the fate of sunlight as it reaches Earth (●**FIGURE 18.6**). Of all the sunlight that reaches Earth, 50 percent is absorbed, scattered, or reflected by clouds and atmosphere, 3 percent is reflected by Earth's surface, and 47 percent is absorbed by Earth's surface. The absorbed radiation warms rocks, soil, and water.

If Earth absorbs radiant energy from the Sun, why doesn't Earth's surface get hotter and hotter until the oceans boil and the rocks melt? The answer is that rocks, soil, and water reemit virtually all the energy they absorb. As explained previously,

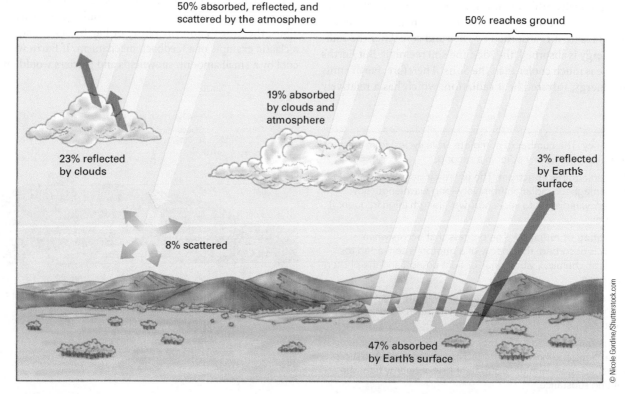

● FIGURE 18.6 Half of the incoming solar radiation reaches Earth's surface. The atmosphere scatters, reflects, and absorbs the other half. All of the radiation absorbed by Earth's surface is reradiated as long-wavelength heat radiation.

most solar energy that reaches Earth is short-wavelength, visible, and ultraviolet radiation. Earth's surface absorbs this radiation and then reemits the energy mostly as long-wavelength, invisible, infrared (heat) radiation. Some of this infrared heat escapes directly into space, but some is absorbed by the atmosphere. The atmosphere traps this heat radiating from Earth and acts as an insulating blanket.

If Earth had no atmosphere, radiant heat loss would be so rapid that Earth's surface would cool drastically at night. Earth remains warm at night because the atmosphere absorbs and retains much of the radiation emitted by the ground. If the atmosphere were to absorb even more of the long-wavelength radiant heat from Earth, the atmosphere and Earth's surface would become warmer. This warming process is called the **greenhouse effect** (●FIGURE 18.7).[1]

Some gases in the atmosphere absorb infrared radiation and others do not. Oxygen and nitrogen, which together make up almost 99 percent of dry air at ground level, do not absorb infrared radiation; water, carbon dioxide, methane, and a few other gases do. Thus, they are called *greenhouse gases*. Water is the most abundant and most powerful greenhouse gas in Earth's atmosphere and for 50 to 75 percent of the greenhouse effect today. The abundance of water vapor at the tropopause and in the lower stratosphere is particularly important because it determines how much energy escapes the atmosphere into outer space. Water passes from the troposphere across the tropopause to the lower stratosphere mostly at the low-latitude tropics, where hot, wet air wells upward through the troposphere and beyond into the stratosphere. In addition to water, the upwelling also carries hydrogen and methane across the tropopause and into the lower stratosphere where they react to form more water. Once in the stratosphere, the water moves to higher latitudes and eventually downwells back into the troposphere. This circulation of water back and forth across the tropopause is called Brewer Dobson Circulation (●FIGURE 18.8).

Stratospheric water is important because under the bombardment of radiation from the sun, water in the stratosphere breaks down, releasing OH− molecules that destroy ozone. In addition, the OH− molecules react with gases containing Cl− to produce compounds that also destroy ozone. Thus, in addition to being a major contributor to global warming as a greenhouse gas, water in the stratosphere

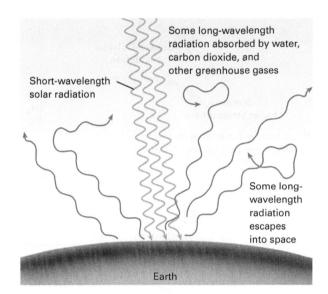

● **FIGURE 18.7** The greenhouse effect can be viewed as a three-step process. Step 1: Rocks, soil, and water absorb short-wavelength solar radiation, and become warmer (orange lines). Step 2: The Earth reradiates the energy as long-wavelength infrared heat rays (red lines). Step 3: Molecules in the atmosphere absorb some of the heat, and the atmosphere becomes warmer.

contributes to ozone loss which allows more harmful radiation to reach the surface.

One study published in *Nature* suggested that decreased concentrations of stratospheric water between 2000 and 2010 partially offset the global temperature increases caused by steadily increased concentrations of carbon dioxide during that time. The study also showed that relatively high levels of stratospheric water during the 1980s and 1990s likely accelerated global warming during those periods beyond that produced from increased carbon dioxide emissions alone.

Carbon dioxide is an important greenhouse gas because its abundance in the atmosphere can vary as a result of several natural and industrial processes. Methane has become important because large quantities are released by industry and agriculture. The greenhouse effect and global climate change are discussed in more detail in Chapter 21.

Dust, cloud cover, aerosols, and other particulate air pollutants also affect Earth's atmospheric temperature by altering the amount of sunlight that is absorbed or reflected. Large volcanic eruptions can inject enough ash and dust into the troposphere to partially block sunlight from reaching Earth, causing several years of global cooling as the dust settles. In addition, large volcanic eruptions in which ash clouds reach the tropopause (the boundary between the troposphere and stratosphere) can inject sulfur dioxide gas into the overlying stratosphere (●FIGURE 18.9). Within about a month, the sulfur dioxide gas converts to tiny droplets of sulfuric acid about one micrometer in diameter. The droplets, called stratospheric aerosols, also reflect incoming solar radiation

---

greenhouse effect An increase in the temperature of the planet's surface caused when infrared-absorbing gases in the atmosphere trap energy from the Sun.

---

1. The comparison between the atmosphere and a greenhouse is only partially correct. Both the glass in a greenhouse and Earth's atmosphere are transparent to incoming, short-wavelength radiation and partially opaque to emitted, long-wavelength, heat radiation. However, the glass in a greenhouse is also a physical barrier that prevents heat loss through air movement, whereas the atmosphere is not. The absorption of radiation is sufficient to warm the atmosphere and cause significant changes in Earth's climate. We use the term *greenhouse effect* because it has become common, both in atmospheric science and in everyday use.

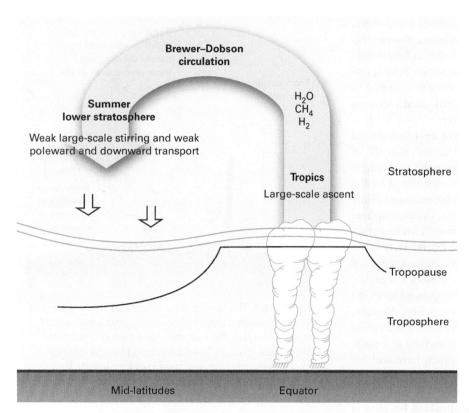

● **FIGURE 18.8** Water exchange between the troposphere and stratosphere (Brewer–Dobson circulation). At the tropics, wet hot air rises quickly and can inject water vapor, methane, and hydrogen gas through the tropopause and into the stratosphere. There, the methane and hydrogen react to form additional water. Once in the stratosphere, the water is transported away from the equator and eventually passes back down through the tropopause and reenters the troposphere. Stratospheric water is important because it causes a strong greenhouse effect and its presence controls how much energy escapes beyond the stratosphere into outer space.

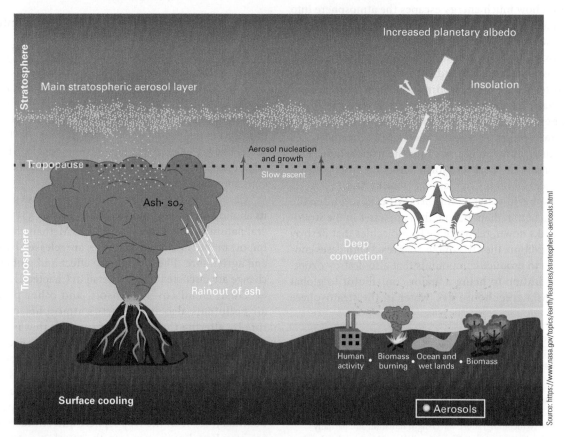

● **FIGURE 18.9** Large volcanic eruptions inject tons of ash into the troposphere, and sulfur dioxide gas from the eruption continues upward beyond the tropopause and into the overlying stratosphere. Once there, the sulfur dioxide converts to droplets of sulfuric acid to form a layer of aerosols that partially block solar insolation, causing cooling of the surface. Sulfur dioxide is also introduced into the troposphere through the burning of fossil fuels and through natural environmental processes, but these aerosol sources are minor compared to that from large volcanic eruptions.

CALIPSO AEROSOL 17–21 km

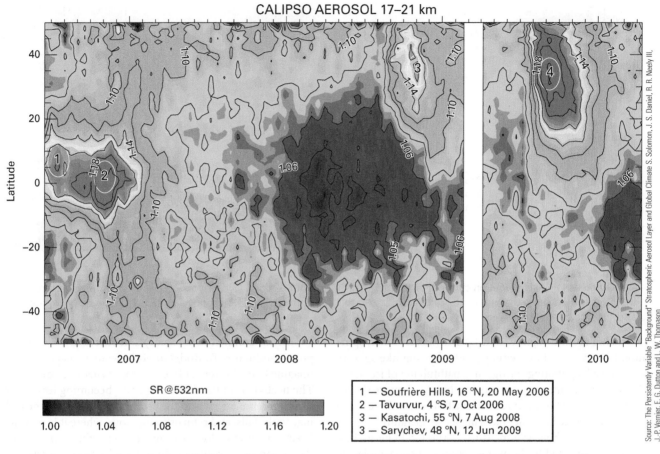

Source: The Persistently Variable "Background" Stratospheric Aerosol Layer and Global Climate S. Solomon, J. S. Daniel, R. R. Neely III, J.-P. Vernier, E. G. Dutton and L. W. Thomason

● **FIGURE 18.10** This plot shows sulfur aerosol levels in the stratosphere between 2006 and 2010, as measured between 17 and 21 kilometer elevation by the CALIPSO (Cloud-Aerosol Lidar and Infrared Pathfinder Satellite Observation) satellite. The y-axis is Earth latitude between 50 degrees north and south. The x-axis is time, from July 2006 on the left to March 2010 on the right. The colors refer to the degree to which light in the stratosphere is scattered by sulfur aerosols, with hot colors corresponding to high aerosol levels and scattering and cool colors corresponding to low aerosol levels and scattering. The numbered "bulls eyes" with hot colors indicating high sulfur stratospheric aerosol levels that are attributed to the four large volcanic eruptions numbered inside the black box with their latitude and date.

back into outer space, causing cooling. One 2011 study estimated that, between 2000 and 2010, stratospheric sulfur aerosols from volcanic eruptions offset the effects of CO_2-caused global warming by roughly one-third (●**FIGURE 18.10**).

The effect of human-caused pollution on global climate change is harder to interpret. Dust and aerosols close to the surface of Earth not only reflect incoming solar radiation but also absorb infrared radiation emitted by the ground. The net result depends on a complex balance of factors, including particle size, particle composition, and natural cloudiness. Whatever the outcome, it is certain that whenever people change the composition of air, they run the risk of altering weather and climate.

ENERGY STORAGE AND TRANSFER: THE DRIVING MECHANISMS FOR WEATHER AND CLIMATE

Heat and Temperature

All matter consists of atoms and molecules that are in constant motion. They fly through space, they rotate, and they vibrate. The **temperature**, or measure of heat in a substance, is proportional to the average speed of the atoms or molecules in a sample.[2] In a teacup full of boiling water, the water molecules

---

temperature A measure of heat in a substance, proportional to the average speed of atoms and molecules in a sample.

---

2. More precisely, temperature is proportional to the kinetic energy of the atoms and molecules. In turn, the kinetic energy equals (mass) × (velocity)^2 for a monatomic gas.

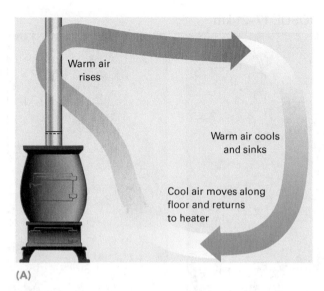

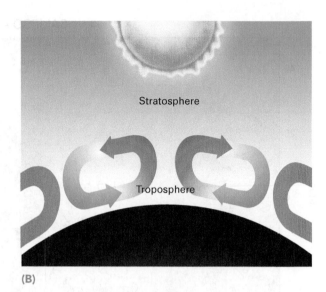

● **FIGURE 18.11** (A) Convection currents distribute heat throughout a room. (B) Convection also distributes heat through the atmosphere when the Sun heats Earth's surface. In this case, the ceiling is the boundary between the troposphere and the stratosphere.

are racing around rapidly, smashing into each other, spinning like so many boomerangs, and vibrating like spheres connected by pulsating springs. In a bathtub full of ice water, the water molecules are also moving, but much more slowly. Molecules in the hot water are moving faster than those in the cold water, so the hot water has a higher temperature.

In contrast, **heat** is a measure of the total energy in a sample. It is related to the average energy of every molecule multiplied by the total number of molecules. It may seem counterintuitive, but there is more heat in a bathtub full of ice water than in a teacup full of boiling water. The average molecule in the ice water is moving slower and therefore has less energy than the average molecule in the boiling water. But there are so many more molecules in the bathtub than in the teacup that the total heat energy is greater.

Heat Transport by Conduction and Convection

If you place a metal frying pan on the stove, the handle gets hot, even though it is not in contact with the burner, because the metal conducts heat from the bottom of the pan to the handle. **Conduction** is the transport of heat by direct collisions among atoms or molecules. When the frying pan is heated from below, the metal atoms on the bottom of the pan move more rapidly. They then collide with their neighbors and transfer energy to them. Like a falling row of dominoes, energy is passed from one atom to another throughout the pan until the handle becomes hot.

Metals conduct heat rapidly and efficiently, but air is a poor conductor. To understand how air transports heat, imagine that a heater is placed in one corner of a cold room. The heated air in the corner expands, becoming less dense. This light, hot air rises to the ceiling. It flows along the ceiling, cools, falls, and returns to the stove, where it is reheated (●**FIGURE 18.11A**). Recall from Chapter 6 that *convection* is the upward and downward flow of fluid material in response to heating and cooling.

In addition to heat-driven movement of fluids through convection, movements of hot or cold fluids can be driven by processes other than heating and/or cooling and resulting changes in density. For example, the Gulf Stream moves heat horizontally from the Caribbean region northward to the coast of Western Europe which is warmed by the current. Such lateral movement of heat is called *advection*. Although advection is commonly confused with convection because both involve moving fluids that carry heat, it is the heat itself that drives convection (think of a convecting pot of tomato soup on the stovetop), whereas advection involves a fluid that is moving for some other reason and that carries heat from one place to another in the process. For example in a forced-air furnace, air heated by the furnace is advected into the living space by a fan and series of ducts.

Both advection and convection also occur in the atmosphere (●**FIGURE 18.11B**). If air in one region is heated above the temperature of surrounding air, this warm air becomes less dense and rises—or convects—just like a hot-air balloon. Eventually, the warm air moves laterally, causing advection of the heat as the air mass moves from one location to another. Similarly, cool, dense air from another part of the atmosphere might sink or convect downward. As it does so, it will displace other air laterally, and this air will advect whatever heat it has with it as it moves. In common speech, this horizontal airflow is called *wind*.

---

heat A measure of the total energy in a sample.

conduction The transport of heat by direct collision among atoms or molecules.

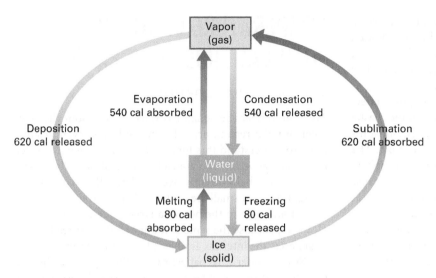

● **FIGURE 18.12** Water releases or absorbs latent heat as it changes among its liquid, solid, and vapor states. (Calories are given per gram at 0°C and 100°C. The values vary with temperature. Red arrows show processes that absorb heat; blue arrows show those that release heat.)

Changes of State

Given the proper temperature and pressure, most substances can exist in three states: solid, liquid, and gas. However, at Earth's surface, many substances commonly exist in only one state. In our experience, rock is almost always solid and molecular oxygen is almost always a gas. Water commonly exists in all three states—as solid ice, as liquid, and as gaseous water vapor.

Latent heat (stored heat) is the energy released or absorbed when a substance changes from one state to another. As shown in ●**FIGURE 18.12**, about 80 calories are required to melt a gram of ice at a constant temperature of 0°C. Another 540 to 600 calories are needed to evaporate a gram of water at constant temperature and pressure.[3] The energy transfers also work in reverse. When 1 gram of water vapor condenses to liquid, 540 to 600 calories are released. When 1 gram of water freezes, 80 calories are released. *Sublimation* is the transformation directly from solid ice to water vapor, without passing through an intermediate the liquid phase. For water, this process requires about 620 calories per gram.

If you walk onto the beach after swimming, you feel cool, even on a hot day. Your skin temperature drops because water on your body is evaporating, and evaporation absorbs heat, cooling your skin. Similarly, evaporation from any body of water cools the water and the air around it. Conversely, condensation releases heat. The energy released when water condenses into rain during a single hurricane can be as great as the energy released by several atomic bombs.

The energy absorbed and released during freezing, melting, evaporation, and condensation of water is important in the atmospheric energy balance. For example, in the northern latitudes, March is usually colder than September, even though equal amounts of sunlight are received in both months. However, in March much of the solar energy is absorbed by melting snow. Snow also has a high albedo and reflects sunlight efficiently. Evaporation cools seacoasts, and the energy of a hurricane comes, in part, from the condensation of massive amounts of water vapor.

Heat Storage

If you place a pan of water and a rock outside on a hot summer day, the rock becomes hotter than the water. Both have received identical quantities of solar radiation.[4]

Why is the rock hotter?

1. **Specific heat** is the amount of energy needed to raise the temperature of 1 gram of material by 1°C. Specific heat is different for every substance. Water has an unusually high specific heat: the amount of energy needed to raise one cubic centimeter (one milliliter) of water by 1°C is one calorie. In contrast, the specific heat of basalt is only about one-fifth that of water, so only about 20 percent of the heat required to heat water by 1°C is needed to heat the same volume of basalt by 1°C. Thus, if water and rock absorb equal amounts of energy, the rock becomes hotter than the water.

2. Rock absorbs heat only at its surface, and the heat travels slowly through the rock. As a result, heat concentrates at the surface. Heat disperses more effectively through water for two reasons. First, solar radiation penetrates several meters below the surface of the water, warming it to this depth. Second, water is a fluid and transports heat by convection.

3. Evaporation is a cooling process. Water loses heat and cools by evaporation, but rock does not.

Think of the consequences of the temperature difference between rock and water. On a hot summer day you may burn your feet walking across dry sand or rock, but the surface of a lake or ocean is never burning hot. Suppose that both the ocean and the adjacent coastline are at the same temperature

---

latent heat Stored heat; the energy released or absorbed when a substance changes from one state to another, by melting, freezing, vaporization, condensation, or sublimation.

specific heat The amount of energy required to raise the temperature of 1 gram of a substance by 1°C.

---

3. The heat of vaporization varies with temperature and pressure; at 100°C and 1 atmosphere pressure, the value is 539.5 calories per gram.

4. Water actually absorbs a bit more solar energy than rock because water has a lower albedo than rock, but this difference is overwhelmed by the other factors.

in spring. As summer approaches, both land and sea receive equal amounts of solar energy. But the land becomes hotter, just as rock becomes hotter than water. Along the seacoast the cool sea moderates the temperature of the land. The interior of a continent is not cooled in this manner and is generally hotter than the coast. In winter, the opposite effect occurs, and inland areas are generally colder than the coastal regions. Thus, coastal areas are commonly cooler in summer and warmer in winter than continental interiors. The coldest temperatures recorded in the Northern Hemisphere occurred in central Siberia and not at the North Pole, because Siberia is landlocked, whereas the North Pole lies in the middle of the Arctic Ocean.[5] In summer, however, Siberia is considerably warmer than the North Pole. In fact, the average temperature in some places in Siberia ranges from −50°C in winter to +20°C in summer, the greatest range on Earth.

---

5. Even though the Arctic Ocean is covered with ice, water lies only a few meters below the surface and it still influences climate.

TEMPERATURE CHANGES WITH LATITUDE AND SEASON

Temperature Changes with Latitude

The region near the equator is warm throughout the year, whereas polar regions are cold and ice-bound even in summer. To understand this temperature difference, consider first what happens if you hold a flashlight above a flat board. If the light is held directly overhead and the beam shines straight down, a small area is brightly lit. If the flashlight is held at an angle to the board, a larger area is illuminated. However, because the same amount of light is spread over a larger area, the intensity is reduced (●FIGURE 18.14).

Now consider what happens when the Sun shines directly over the equator. The equator, analogous to the flat board under a direct light, receives the most concentrated radiation. The Sun strikes the rest of the globe at an angle and thus radiation is

DIGGING DEEPER

Latitude and Longitude

If someone handed you a perfectly smooth ball with a dot on it and asked you to describe the location of the dot, you would be at a loss to do so because all positions on the surface of a sphere are equal. How, then, can locations on a spherical Earth be described? Even if we ignore irregularities, continents, and oceans, Earth has points of reference because it rotates on its axis and has a magnetic field that nearly (but not exactly) coincides with the axis of rotation.

The North Pole and South Pole lie on the rotational axis, and lines of latitude form imaginary horizontal rings around the axis. Mathematicians measure distance on a sphere in degrees. Using this system, the equator is defined as 0 degree latitude, the North Pole is 90 degrees north latitude, and the South Pole is 90 degrees south latitude (●FIGURE 18.13).

No natural east–west reference exists, so a line running through Greenwich, England, was arbitrarily chosen as the 0 degree line. The planet was then divided by lines of longitude, also measured in degrees. On a globe with the rotational axis vertically oriented, lines of longitude also run vertically. To a navigator they measure east–west angular distance from Greenwich, England.

The system is easy to use. Minneapolis–St. Paul lies at 45 degrees north latitude and 93 degrees west longitude. The latitude tells us that the city lies halfway between the equator (0°) and the North Pole (90°). Because a circle has 360 degrees, the longitude tells us that Minneapolis–St. Paul is about one-quarter of the way around the world in a westward direction from Greenwich, England.

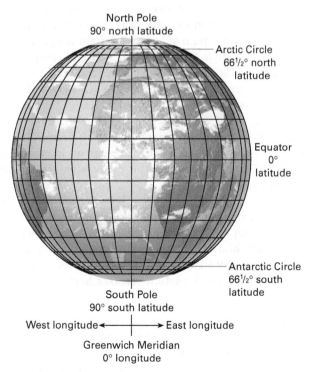

● **FIGURE 18.13** Latitude and longitude allow navigators to identify a location on a spherical Earth.

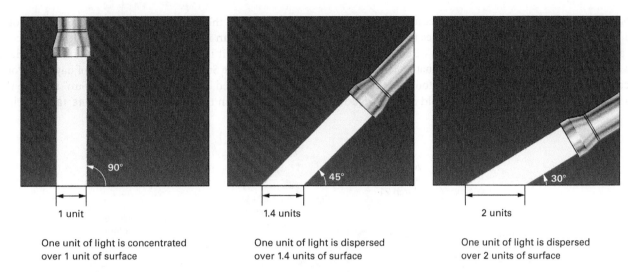

One unit of light is concentrated
over 1 unit of surface

One unit of light is dispersed
over 1.4 units of surface

One unit of light is dispersed
over 2 units of surface

● **FIGURE 18.14** If a light shines from directly overhead, the radiation is concentrated on a small circular area. However, if the light shines at an angle, or if the surface is tilted, the radiant energy is dispersed over a larger, elliptical area.

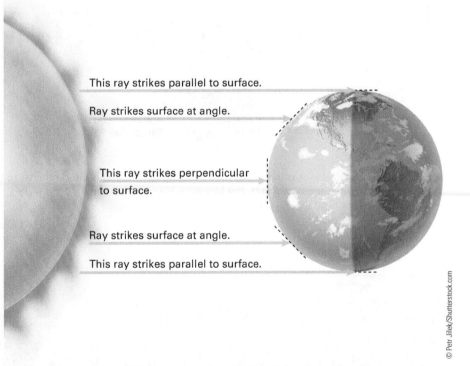

This ray strikes parallel to surface.

Ray strikes surface at angle.

This ray strikes perpendicular to surface.

Ray strikes surface at angle.

This ray strikes parallel to surface.

● **FIGURE 18.15** When the Sun shines directly over the equator, the equator receives the most intense solar radiation, and the poles receive little.

less concentrated at higher latitudes (●**FIGURE 18.15**). Because the equator receives the most concentrated solar energy, it is generally warm throughout the year. Average atmospheric temperature becomes progressively cooler poleward (north and south of the equator). But, as mentioned in the introduction to this chapter, the average temperature does not change steadily with latitude, because many other factors—such as winds, ocean currents, albedo, and proximity to the oceans—also affect atmospheric temperature in any given region.

The Seasons

Earth circles the Sun in a planar orbit, while simultaneously spinning on its axis. This axis is tilted at 23.5 degrees from a line drawn perpendicular to the orbital plane.

Earth revolves around the Sun once a year. As shown in ●**FIGURE 18.16**, the North Pole tilts toward the Sun in summer and away from it in winter. June 21 is the summer

solstice in the Northern Hemisphere because at this time, the North Pole leans the full 23.5 degrees toward the Sun. As a result, sunlight strikes Earth from directly overhead at a latitude 23.5 degrees north of the equator. This latitude is called the **tropic of Cancer**. If you stood on the tropic of Cancer at noon on June 21, you would cast no shadow.

June is warm in the Northern Hemisphere for two reasons: (1) when the Sun is high in the sky, sunlight is more concentrated than it is in winter; (2) when the North Pole is tilted toward the Sun, it receives 24 hours of daylight. Polar regions are called "lands of the midnight Sun" because the Sun never sets in the summertime (●FIGURE 18.17). Below

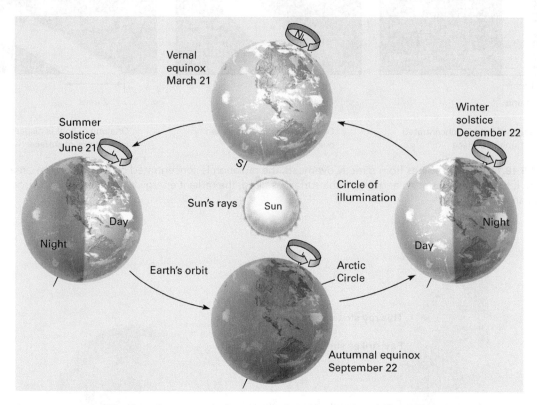

● **FIGURE 18.16**   Weather changes with the seasons because Earth's axis is tilted relative to the plane of its orbit around the Sun. As a result, the Northern Hemisphere receives more direct sunlight during summer but less during winter.

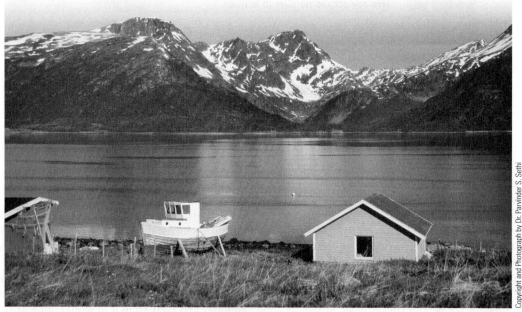

● **FIGURE 18.17**   The Sun shines down brightly on this Norwegian lake at 2 O'clock in the morning.

TABLE 18.1 Hours of Sunlight per Day

Latitude	Geographic Reference	Summer Solstice	Winter Solstice	Equinoxes
0° north	Equator	12 hr	12 hr	12 hr
30° north	New Orleans	13 hr 56 min	10 hr 04 min	12 hr
40° north	Denver	14 hr 52 min	9 hr 08 min	12 hr
50° north	Vancouver	16 hr 18 min	7 hr 42 min	12 hr
90° north	North Pole	24 hr	0 hr 00 min	12 hr

the Arctic Circle, the Sun sets in the summer, but the days are always longer than they are in winter (**TABLE 18.1**).

When it is summer in the Northern Hemisphere, the South Pole tilts away from the Sun and thus the Southern Hemisphere receives low-intensity sunlight and has short days. June 21 is the first day of winter in the Southern Hemisphere. Six months later, on December 21 or 22, the seasons are reversed. The North Pole tilts away from the Sun, giving rise to the winter solstice in the Northern Hemisphere, while it is summer in the Southern Hemisphere. On this day, sunlight strikes Earth directly overhead at the **tropic of Capricorn**, latitude 23.5 degrees south. At the North Pole, the Sun never rises and it is continuously dark, while the South Pole is bathed in continuous daylight.

On March 21 and September 22, Earth's axis lies at right angles to a line drawn between Earth and the Sun. As a result, the poles are not tilted toward or away from the Sun and the Sun shines directly overhead at the equator at noon. If you stand at the equator at noon on either of these two dates, you would cast no shadow. But north or south of the equator, a person casts a shadow even at noon. In the Northern Hemisphere, March 21 is the first day of spring and September 22 is the first day of autumn, whereas the seasons are reversed in the Southern Hemisphere. On the first days of spring and autumn, every portion of the globe receives 12 hours of direct sunlight and 12 hours of darkness. For this reason, March 21 and September 22 are called the **equinoxes**, meaning "equal nights."

All areas of the globe receive the same total number of hours of sunlight every year. The North Pole and South Pole receive direct sunlight in dramatic opposition, 6 months of continuous light and 6 months of continuous darkness, whereas at the equator, each day and night are close to 12 hours long throughout the year. Although the poles receive the same number of sunlight hours as do the equatorial regions, the sunlight reaches the poles at a much lower angle and therefore delivers much less total energy per unit of surface area.

TEMPERATURE CHANGES WITH GEOGRAPHY

Even though all locations at a given latitude receive equal amounts of solar radiation, some places have cooler climates than others at the same latitude. ●**FIGURE 18.18** shows temperatures around Earth in January and in July. Lines called **isotherms** connect areas of the same average temperature. Note that the isotherms loop and dip across lines of latitude. For example, the January 0°C line runs through Seattle, Washington, dips southward across the center of the United States, and then swings northward to northern Norway. Such variations occur with latitude because winds and ocean currents transport heat from one region of Earth to another. Other factors, discussed in section on "Energy Storage and Transfer: The Driving Mechanisms for Weather and Climate," include the different heat storage properties of oceans and continents, evaporative cooling of the oceans, and the latent heat of snow and ice.

solstice Either of two times per year when the Sun is furthest from the equator and shines directly overhead at either 23.5 degrees north latitude (on or about June 21) or 23.5 degrees south latitude (on or about December 22). The solstice on or about June 21 marks the longest day of the year in the Northern Hemisphere and the shortest day in the Southern Hemisphere. The solstice on or about December 22 marks the longest day in the Southern Hemisphere and the shortest day in the Northern Hemisphere.

tropic of Cancer The latitude 23.5 degrees north of the equator. On or about June 21, the summer solstice in the Northern Hemisphere, sunlight strikes Earth from directly overhead at noon at this latitude.

tropic of Capricorn The latitude 23.5 degrees south of the equator. On or about December 22, the summer solstice in the Southern Hemisphere, sunlight strikes Earth from directly overhead at noon at this latitude.

equinoxes Either of two times during the year—on or about March 21 and September 22—when the Sun shines directly overhead at the equator and every portion of Earth receives 12 hours of daylight and 12 hours of darkness.

isotherms Lines on a weather map connecting areas with the same average temperature.

Altitude

Recall from Chapter 17 that at higher elevations in the troposphere, the atmosphere is thinner and absorbs less energy. In addition, lower parts of the troposphere have absorbed much of the heat radiating from Earth's surface. Consequently, temperature decreases with elevation.

January

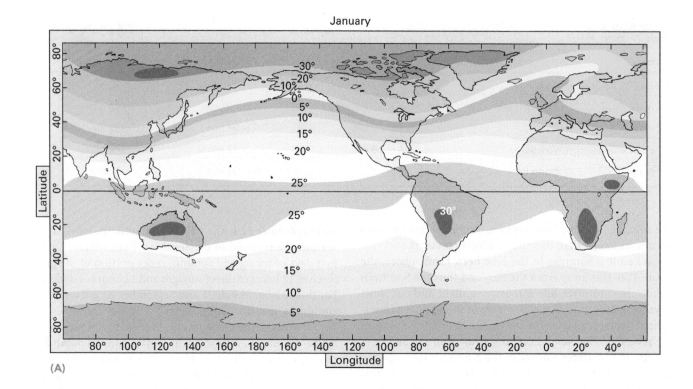

(A)

July

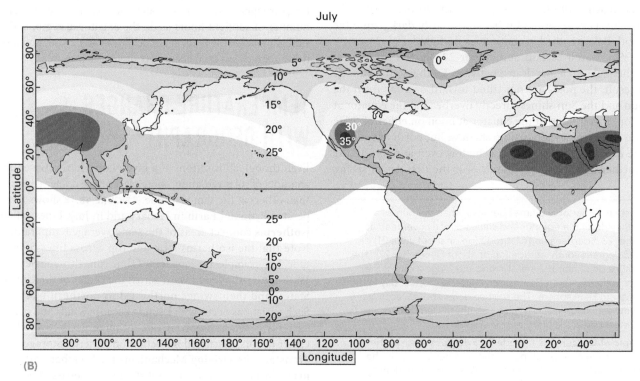

(B)

● **FIGURE 18.18** (A) Global temperature distributions in January. Isotherm lines connect places with the same average temperatures. (B) Global temperature distributions in July. Isotherm lines connect places with the same average temperatures.

Mount Everest lies at 28 degrees north, about the same latitude as Tampa, Florida. Yet, at 8,000 meters, climbers on Everest wear heavy down clothing to keep from freezing to death, while during the same season sunbathers in Tampa lounge on the beach in bikinis.

Ocean Effects

Recall from section on "Energy Storage and Transfer: The Driving Mechanisms for Weather and Climate" that land heats more quickly in summer and cools more quickly

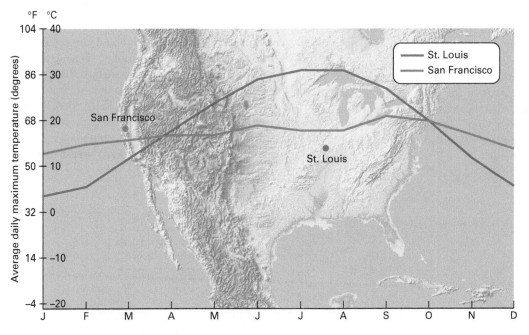

●**FIGURE 18.19** The average temperature of continental St. Louis (red line) is colder in winter and warmer in summer than that of coastal San Francisco (blue line).

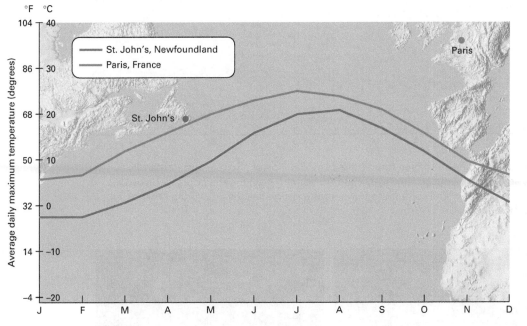

●**FIGURE 18.20** Paris is warmed by the Gulf Stream and the North Atlantic Drift. On the other hand, St. John's, Newfoundland, is alternately warmed by the Gulf Stream and cooled by the Labrador Current. The cooling effect of the Labrador Current depresses the temperature of St. John's year-round.

in winter than ocean surfaces do. As a result, continental interiors show greater seasonal extremes of temperature than coastal regions. For example, San Francisco and St. Louis both lie at approximately 38 degrees latitude, but St. Louis is in the middle of the continent and San Francisco lies on the Pacific coast. On average, St. Louis is 9°C cooler in winter and 9°C warmer in summer than San Francisco (●**FIGURE 18.19**).

Ocean currents also play a major role in determining average temperature. Paris, France, lies at 48 degrees north latitude, north of the U.S.–Canadian border on the other side of the Atlantic Ocean. Although winters are dark because the Sun is low in the sky, the warm Gulf Stream carries enough heat northward, so the average minimum temperature in Paris in January is around 6°C, comfortably above freezing. St. John's, Newfoundland, is at about the same latitude but under the influence of the Labrador Current that flows from the North Pole. As a result, the average minimum January temperature in St. John's is around –3°C, considerably colder than it is in Paris (●**FIGURE 18.20**).

Wind Direction

Winds carry heat from region to region just as ocean currents do. Vladivostok is a Russian city on the west coast of the Pacific Ocean at about 43 degrees north latitude. It has a near-Arctic climate with frigid winters and heavy snow. Forests are stunted by the cold, and trees are generally small.

In comparison, Portland, Oregon, lies at 45 degrees north latitude on the east coast of the Pacific Ocean. Its temperate climate supports majestic cedar forests, and rain is more common than snow in winter (●**FIGURE 18.21**). The main difference in this case is that Vladivostok is cooled by frigid Arctic winds from Siberia.

Cloud Cover and Albedo

If you are sunbathing on the beach and a cloud passes in front of the Sun, you feel a sudden cooling. During the day, clouds reflect sunlight and cool the surface. In contrast, clouds have the opposite effect and warm the surface during the night. Recall that after the Sun sets, Earth cools because radiant heat from soil and rock escapes into the air. Clouds act as an insulating blanket by absorbing outgoing radiation and reradiating some of it back downward (●**FIGURE 18.22**).

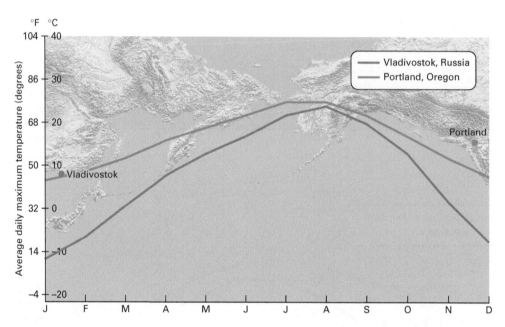

● **FIGURE 18.21** During summer, temperatures in Vladivostok, Russia, and Portland, Oregon, are nearly identical. However, frigid Arctic wind from Siberia cools Vladivostok during the winter (red line) so that the temperature is significantly colder than that of Portland (blue line).

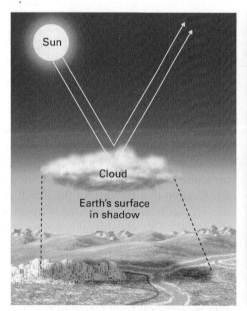

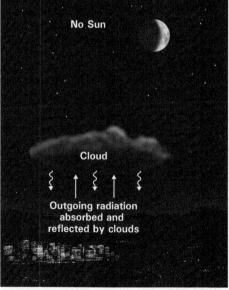

● **FIGURE 18.22** Clouds cool Earth's surface during the day but warm it during the night.

Thus, cloudy nights are generally warmer than clear nights. Because clouds cool Earth during the day and warm it at night, cloudy regions generally have a narrower daily temperature range than regions with predominantly blue sky and starry nights.

Snow also plays a major role in affecting regional temperature. Imagine that it is raining and the temperature is 0.5°C, or just above freezing. After the rain stops, the Sun comes out. The bare ground absorbs solar radiation and the surface heats up. Now imagine that the temperature had dropped to −0.5°C during the storm. In this case, the precipitation would fall as snow. After the storm passed and the Sun came out, the surface would be covered with a sparkling white cover with 80 to 90 percent albedo. Most of the incoming solar radiation would reflect back into the atmosphere and the surface would remain much cooler. If the surface temperature did rise above freezing, the snow would have to melt before the ground temperature would rise. But snow has a very high latent heat; it takes a lot of energy to melt the snow without raising the temperature at all. In this example, a seemingly insignificant 1 degree initial temperature difference during a storm would cause a much larger temperature difference when the Sun came out after the storm.

This type of threshold mechanism is very important in weather and climate and will be discussed in Chapters 19 and 20.

Key Concepts Review

- Light is a form of electromagnetic radiation and exhibits properties of both waves and particles. Wavelength is the distance between wave crests, and frequency is the number of cycles that pass in a second. The electromagnetic spectrum is the continuum of radiation of different wavelengths and frequencies.

- When radiation is absorbed by matter, the electrons may be promoted into an excited state in which they have an increased level of energy. When they fall back to their lower energy state, they reemit the energy, but often at a different frequency from the incoming photons. The Sun's high-energy (short-wavelength) radiation is absorbed by rock and soil, which reemit low-energy (long-wavelength) radiation. Radiation reflects from many surfaces; the reflectivity of a surface is referred to as its albedo. Scattering of sunlight makes the sky appear blue.

- About 8 percent of solar radiation is scattered back into space; 19 percent is absorbed in the atmosphere; 26 percent is reflected; and 47 percent is absorbed by soil, rocks, and water. However, the absorbed radiation is eventually reemitted into space. The greenhouse effect is the warming of the atmosphere due to the absorption of this long-wavelength radiation by water vapor, carbon dioxide, methane, and other green-house gases.

- Temperature is a measure of heat proportional to the average speed of atoms or molecules in a sample, while heat measures the total thermal energy in a sample. Conduction is the transport of heat by direct collisions among atoms or molecules. Winds and ocean currents transfer heat from one region of the globe to another by convection and advection. Latent heat (stored heat) is the energy released or absorbed when a substance changes from one state to another. Large quantities of latent heat are absorbed or emitted when water freezes, melts, vaporizes, or condenses.

- Oceans affect weather and climate because water has low albedo and a high specific heat. In addition, currents transport heat both vertically and horizontally. Finally, evaporation cools the sea surface. As a result of all of these factors, coastal areas are generally cooler in summer and warmer in winter than continental interiors at the same latitude.

- The general decrease in temperature from the equator to the poles results from the decreasing intensity of solar radiation from the equator to the poles. Changes of seasons are caused by the tilt of Earth's axis relative to the Earth–Sun plane.

- Temperature changes with geography as a result of altitude, differences in temperature between ocean and land, ocean currents, wind direction, cloud cover, and albedo.

Important Terms

absorption of radiation (p. 432)	equinoxes (p. 443)	photons (p. 431)
albedo (p. 433)	excited state (p. 432)	solstice (p. 442)
climate (p. 431)	frequency (p. 432)	specific heat (p. 439)
conduction (p. 438)	greenhouse effect (p. 435)	temperature (p. 437)
electromagnetic radiation (p. 431)	heat (p. 438)	tropic of Cancer (p. 442)
electromagnetic spectrum (p. 432)	isotherms (p. 443)	tropic of Capricorn (p. 443)
emission of radiation (p. 432)	latent heat (p. 439)	weather (p. 431)

Review Questions

1. Briefly outline the nature of light. What is a photon? What is the electromagnetic spectrum?

2. What happens to light when it is absorbed? When it is emitted? Give an example of each of these phenomena.

3. What is albedo? How does the albedo of a surface change when snow melts? When forests are converted to agricultural fields? When fields are paved into parking lots?

4. What is the fate of the solar radiation that reaches Earth?

5. Explain the greenhouse effect.

6. Explain the difference between heat and temperature.

7. Explain the difference between conduction and convection.

8. What is latent heat? How does it affect Earth's surface temperature?

9. What is specific heat? How does it affect Earth's surface temperature?

10. Explain how the tilt of Earth's axis affects climate in the temperate and polar regions.

11. Discuss temperature and lengths of days at the poles, the midlatitudes, and the equator at the following times of year: June 21, December 21, and the equinoxes.

12. Explain how altitude, wind, the ocean, cloud cover, and albedo affect regional temperature.

Serkan Senturk/Shutterstock.com

19

Heavy rain falls in the distance from a thick cumulonimbus cloud cover.

MOISTURE, CLOUDS, AND WEATHER

LEARNING OBJECTIVES

LO1 Distinguish between absolute and relative humidity.

LO2 Explain dew point using saturation level.

LO3 Describe radiation and contact cooling.

LO4 Compare dry and wet adiabatic rates.

LO5 Describe the three ways that air can be lifted.

LO6 Explain the characteristics of stable and unstable air.

LO7 Describe the formation of the three main types of clouds.

LO8 Compare the different types of precipitation.

LO9 Distinguish between the types of fog.

LO10 Describe the formation and intensity of winds near the Earth's surface.

LO11 Discuss air masses and fronts associated with weather systems.

LO12 Describe the formation of thunderstorms.

LO13 Compare hurricane and tornados, including formation, intensity, and size.

LO14 Explain the effects of El Niño on weather systems around the Pacific Ocean.

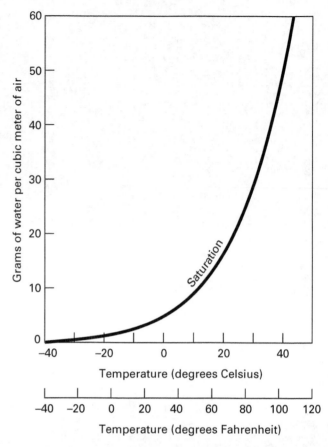

● **FIGURE 19.1** Warm air can hold more water vapor than cold air can.

MOISTURE IN AIR

Precipitation occurs only when there is moisture in the air. Therefore to understand precipitation, we must first understand how moisture collects in the atmosphere and how it behaves.

Humidity

When water boils on a stove, a steamy mist rises above the pan and then disappears into the air. The water molecules have not been lost; they have simply become invisible. In the pan, water is liquid, and in the mist above, the water exists as tiny droplets. These droplets then evaporate, and the invisible water vapor mixes with air. Water also evaporates into air from the seas, streams, lakes, and soil. Winds then distribute this moisture throughout the atmosphere. Thus, all air contains some water vapor—even over the driest deserts.

Humidity is the amount of water vapor in air. **Absolute humidity** is the mass of water vapor in a given volume of air, expressed in grams per cubic meter (g/m^3).

Air can hold only a certain amount of water vapor, and warm air can hold more water vapor than cold air can. For example, air at 25°C can hold 23 g/m^3 of water vapor, but at 12°C, it can hold only half that quantity, 11.5 g/m^3 (●**FIGURE 19.1**). **Relative humidity** is the amount of water

vapor in air relative to the maximum it can hold at a given temperature. It is expressed as a percentage:

$$\text{Relative humidity (\%)} = \frac{\text{Actual quantity of water per unit of air}}{\text{maximum quantity at the same temperature}} \times 100$$

If air contains half as much water vapor as it can hold, its relative humidity is 50 percent. Suppose that air at 25°C contains 11.5 g/m^3 of water vapor. Since air at that temperature can hold 23 g/m^3, it is carrying half of its maximum, and the relative humidity is 11.5 g/23 g × 100 = 50 percent.

Now let us take some of this air and cool it without adding or removing any water vapor. Because cold air holds less water vapor than warm air holds, the relative humidity increases even though the *amount* of water vapor remains constant. If the air cools to 12°C, and it still contains 11.5 g/m^3, the relative humidity reaches 100 percent because air at that temperature can hold only 11.5 g/m^3.

When relative humidity reaches 100 percent, the air is *saturated*. The temperature at which **saturation** occurs, 12°C in this example, is the **dew point**. If saturated air cools below the dew point, some of the water vapor may condense into liquid droplets (although, as discussed next, under special conditions in the atmosphere, the relative humidity can rise above 100 percent).

Supersaturation and Supercooling

When the relative humidity reaches 100 percent (at the dew point), water vapor condenses quickly onto solid surfaces such as rocks, soil, and airborne particles. Airborne particles such as dust, smoke, and pollen are abundant in the lower atmosphere. Consequently, water vapor may condense easily at the dew point in the lower atmosphere, and there the relative humidity rarely exceeds 100 percent. However, in the clear, particulate-free air high in the troposphere, condensation occurs so slowly that for all practical purposes it does not happen. As a result, the air commonly cools below its dew point but water remains as vapor. In that case, the relative humidity rises above 100 percent, and the air reaches a point of **supersaturation**.

Similarly, liquid water does not always freeze at its freezing point. Small droplets can remain liquid in a cloud even when the temperature is −40°C. Such water has undergone **supercooling**.

COOLING AND CONDENSATION

Moisture condenses to form water droplets or ice crystals when moist air cools below its dew point. Clouds and fog are visible concentrations of this airborne water and ice. Three atmospheric processes cool air to its dew point and cause condensation: (1) Air cools when it loses heat by radiation. (2) Air cools by contact with a cool surface such as water, ice, rock, soil, or vegetation. (3) Air cools when it rises.

---

humidity The amount of water vapor in the air.

absolute humidity The mass of water vapor in a given volume of air, expressed in grams per cubic meter (g/m^3).

relative humidity The ratio of the amount of water vapor in a given volume of air divided by the maximum amount of water vapor that can be held by that air at a given temperature, expressed as a percentage.

saturation The maximum amount of water vapor that air can hold.

dew point The temperature at which the relative humidity of air reaches 100 percent and the air becomes saturated.

supersaturation A condition in which the relative humidity of the air exceeds 100 percent.

supercooling A condition in which water droplets in air do not freeze even when the air cools below the freezing point.

dew Moisture that is condensed onto objects from the atmosphere, usually during the night, when the ground and leaf surfaces become cooler than the surrounding air.

frost Ice crystals formed directly from vapor when the dew point is below freezing.

Radiation Cooling

As described in Chapter 18, the atmosphere, rocks, soil, and water absorb the Sun's heat during the day and then radiate some of this heat back out toward space at night. As a result of heat lost by radiation, air, land, and water become cooler at night, and condensation may occur.

Contact Cooling: Dew and Frost

You can observe condensation on a cool surface with a simple demonstration. Heat water on a stove until it boils and then hold a cool drinking glass in the clear air just above the steam. Water droplets will condense on the surface of the glass because the glass cools the hot, moist air to its dew point. The same effect occurs in a house on a cold day. Water droplets or ice crystals appear on windows as warm, moist, indoor air cools on the glass (●FIGURE 19.2).

In some regions, the air on a typical summer evening is warm and humid. After the Sun sets, plants, houses, windows, and most other objects lose heat by radiation and therefore become cool. During the night, water vapor condenses on the cool objects. This condensation is called **dew**. If the dew point is below freezing, **frost** forms. Frost is not frozen dew, but ice crystals formed directly from vapor.

Cooling of Rising Air

Radiation and contact cooling close to Earth's surface form dew, frost, and some types of fog. However, clouds and precipitation normally form at higher elevations where the air is not cooled by direct contact with the ground. Almost all cloud formation and precipitation occur when air cools as it rises (●FIGURE 19.3).

Work and heat are both forms of energy. Work can be converted to heat or heat can be converted to work, but energy is never lost. If you pump up a bicycle tire, you are performing work to compress the air. This energy is not lost; much of it converts to heat. Therefore, both the pump and

● **FIGURE 19.2** Ice crystals condense on a window on a frosty morning.

Veter Sergey/Shutterstock.com

● **FIGURE 19.3** Most clouds form as rising air cools. The cooling causes invisible water vapor to condense into visible water droplets, or ice crystals, which we see as a cloud.

the newly filled tire feel warm. Conversely, if you puncture a tire, the air rushes out. It must perform work to expand, so the air rushing from a punctured tire cools. Variations in temperature caused by compression or expansion of gas are called **adiabatic temperature changes.** *Adiabatic* means without gain or loss of heat. During adiabatic warming, air warms up because work is done on it, not because heat is added. During adiabatic cooling, air cools because it performs work, not because heat is removed.

Air pressure decreases with elevation. When dense surface air rises, it expands because the atmosphere around it is now of lower density, just as air expands when it rushes out of a punctured tire. Rising air performs work to expand,

and therefore it cools adiabatically. Dry air cools by 10°C for every 1,000 meters it rises (5.5°F/1,000 ft). This cooling rate is called the **dry adiabatic lapse rate**. Thus, if dry air were to rise from sea level to 9,000 meters (about the height of Mount Everest), it would cool by 90°C (162°F).

Almost all air contains some water vapor. As moist air rises and cools adiabatically, its temperature may eventually decrease to the dew point. At the dew point, moisture may condense as droplets, and a cloud forms. But recall that condensing vapor releases latent heat. As the air rises through the cloud, its temperature now is affected by two opposing processes. It cools adiabatically, but at the same time it is heated by the latent heat released by condensation. However, the warming caused by latent heat is generally less than the amount of adiabatic cooling. The net result is that the rising air continues to cool, but more slowly than at the dry adiabatic lapse rate. The **wet adiabatic lapse rate** is the cooling rate after condensation has begun. It varies from 5°C for every 1,000 meters it rises (2.7°F/1,000 ft) for air with a high moisture content to 9°C for every 1,000 meters it rises (5°F/1,000 ft) for relatively dry air (●**FIGURE 19.4**). Thus, once clouds start to form, rising air no longer cools as rapidly as it did lower in the atmosphere. Rising air cools at the dry adiabatic lapse rate until it cools to its dew point and condensation begins. Then, as it continues to rise, it cools at the lesser, wet adiabatic lapse rate as condensation continues.

In contrast, sinking air becomes warmer because of adiabatic compression. Warm air can hold more water vapor than cool air can. Consequently, water does not condense from sinking, warming air, and the latent heat of condensation does not affect the rate of temperature rise. As a result, sinking air always becomes warmer at the dry adiabatic rate.

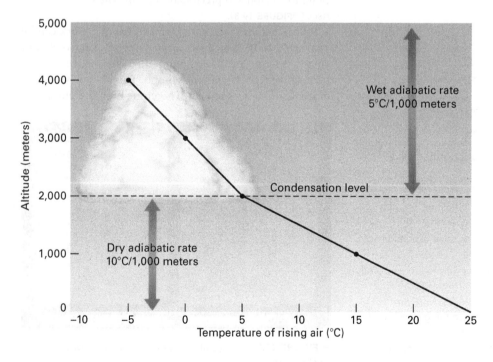

● **FIGURE 19.4** A rising air mass initially cools rapidly at the dry adiabatic lapse rate. Then, after condensation begins and clouds start to form, it cools more slowly at the wet adiabatic lapse rate.

RISING AIR AND PRECIPITATION

To summarize: When moist air rises, it cools and forms clouds. Three mechanisms cause air to rise (●FIGURE 19.5): orographic lifting, frontal wedging, and convection–convergence.

Orographic Lifting

When air flows over mountains, it is forced to rise by a mechanism called **orographic lifting**. This rising air frequently causes rain or snow over the mountains, as will be explained in section on "How the Earth's Surface Features Affect Weather."

Frontal Wedging

A moving mass of cool, dense air may encounter a mass of warm, less-dense air. When this occurs, the cool, denser air slides under the warm air mass, forcing the warm air upward to create a weather front. This process is called **frontal wedging**. We will discuss weather fronts in more detail in sections on "Fronts and Frontal Weather," "Thunderstorms," and "Tornadoes and Tropical Cyclones."

Convection–Convergence

Recall that *convection* is the upward, downward, and horizontal flow of fluids in response to heating and cooling. If one portion of the atmosphere becomes warmer than the surrounding air, the warm air expands, becomes less dense, and rises. Thus, a hot-air balloon rises because it contains air that is warmer and less dense than the air around it. If the Sun heats one parcel of air near Earth's surface to a warmer temperature than that of surrounding air, the warm air will rise, just as the hot-air balloon rises.

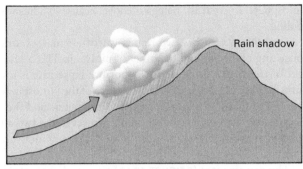

(A)   Orographic lifting

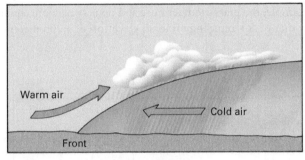

(B)   Frontal wedging

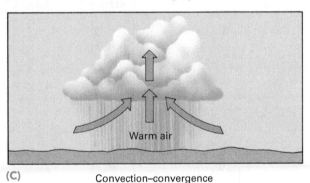

(C)   Convection–convergence

● **FIGURE 19.5** Three mechanisms cause air to rise and cool: (A) orographic lifting, (B) frontal wedging, and (C) convection–convergence.

adiabatic temperature changes Temperature changes caused by compression or expansion of gas without gain or loss of heat.

dry adiabatic lapse rate The rate at which dry air cools adiabatically as it rises—10°C for every 1,000 meters above sea level. Rising air cools at the dry adiabatic lapse rate until it cools to its dew point and condensation begins.

wet adiabatic lapse rate The rate at which rising moist air cools adiabatically after it has reached its dew point and condensation has begun—varies depending on moisture content from 5°C to 9°C for every 1,000 meters it rises.

orographic lifting Lifting of air that occurs when air flows over a mountain.

frontal wedging A process by which a moving mass of cool, dense air encounters a mass of warm, less-dense air; the cool, denser air slides under the warm air mass, forcing the warm air upward to create a weather front.

normal lapse rate The vertical temperature structure of the atmosphere; in other words, the rate at which air that is neither rising nor falling cools with elevation.

Convective Processes and Clouds

On some days, clouds hang low over the land and obscure nearby hills. At other times, clouds float high in the sky, well above the mountain peaks. What factors determine the height and shape of a cloud?

Recall that air is generally warmest at Earth's surface and cools with elevation throughout the troposphere. The rate at which air that is neither rising nor falling cools with elevation is called the **normal lapse rate**. The average normal lapse rate is 6°C for every 1,000 meters of elevation (3.3°F/1,000 ft) and thus is less than the dry adiabatic lapse rate. However, the normal lapse rate is variable. Typically, it is greatest near Earth's surface and decreases with altitude. The normal lapse rate also varies with latitude, the time of day, and the seasons. It is important to note that the normal lapse rate is simply the vertical temperature structure of the atmosphere. In contrast, rising air cools because of adiabatic cooling.

●**FIGURE 19.6** shows two rising, warm-air masses, one consisting of dry air and the other of moist air. The central part of the figure shows that the normal lapse rate is the same for both air masses: the temperature of the atmosphere decreases rapidly in the first few thousand meters and then more slowly with increasing elevation. However, the two air masses behave differently because of their different moisture contents.

The dry air mass (●**FIGURE 19.6A**) rises and cools at the dry adiabatic lapse rate of 10°C/1,000 m. As a result, in this example, the mass's temperature and density become equal to that of surrounding air at an elevation of 3,000 meters.

Because the density of the rising air is the same as that of the surrounding air, the rising air is no longer buoyant, and it stops rising. No clouds form because the air has not cooled to its dew point.

In ●**FIGURE 19.6B**, the rising moist air initially cools at the dry adiabatic rate of 10°C/1,000 m, but only until the air cools to its dew point, at an elevation of 1,000 meters. At that point, moisture begins to condense, and clouds form. But the condensing moisture releases latent heat of condensation. This additional heat causes the rising air to cool more slowly, at the wet adiabatic rate of 5°C/1,000 m. As a result, the rising air remains warmer and more buoyant than surrounding air, and it continues to rise for thousands of meters, creating a towering, billowing cloud with the potential for heavy precipitation.

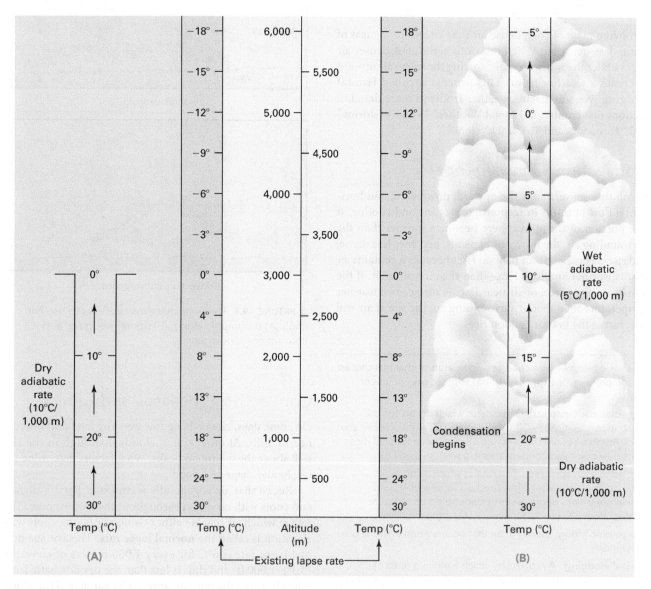

● **FIGURE 19.6** (A) As dry air rises, it expands and cools at the dry adiabatic lapse rate. Thus, it soon cools to the temperature of the surrounding air, and it stops rising. (B) As moist air rises, initially it cools at the dry adiabatic lapse rate. It soon cools to its dew point, and clouds form. Then, it cools more slowly at the wet adiabatic lapse rate. As a result, it remains warmer than surrounding air and continues to rise for thousands of meters. It stops rising when all moisture has condensed, and the air again cools at its dry adiabatic rate.

In simple terms, warm, moist air is **unstable air** because it rises rapidly, forming towering clouds and heavy rainfall. Also, as shown in ●**FIGURE 19.5C**, air rushes along the ground to replace the rising air, thus generating surface winds. Most of us have experienced a violent thunderstorm on a hot, summer day. Puffy clouds seem to appear out of nowhere in a blue sky. These clouds grow vertically and darken as the afternoon progresses. Suddenly, gusts of wind race across the land, and shortly thereafter, heavy rain falls. These events, to be described in more detail in section on "Fronts and Frontal Weather," are all caused by unstable, rising, moist air.

In contrast, warm, dry air doesn't rise rapidly, doesn't ascend to high elevations, and doesn't lead to cloud formation and precipitation. Thus warm, dry air is said to be **stable air**. Yet, keep in mind that convection is only one of the three processes that leads to rising air. Orographic lifting and frontal wedging also lead to rising air, cloud formation, and rain, as explained earlier.

TYPES OF CLOUDS

Even a casual observer of the daily weather will notice that clouds are quite different from day to day. Different meteorological conditions create the various cloud types and, in turn, a look at the clouds can provide useful information about the daily weather.

Cirrus (Latin for "wisp of hair") clouds are wispy clouds that look like hair blowing in the wind or feathers floating across the sky. Cirrus clouds form at high altitudes, 6,000 to 15,000 meters (20,000 to 50,000 feet). The air is so cold at these elevations that cirrus clouds are composed of ice crystals rather than water droplets. High winds aloft blow them out into long, gently curved streamers (●**FIGURE 19.7**).

Stratus (Latin for "layer") clouds are horizontally layered, sheetlike clouds. They form when condensation occurs at the same elevation at which air stops rising and the clouds

spread out into a broad sheet. Stratus clouds form the dark, dull-gray, overcast skies that may persist for days and bring steady rain (●**FIGURE 19.8**).

Cumulus (Latin for "heap" or "pile") clouds are fluffy, white clouds that typically display flat bottoms and billowy tops (●**FIGURE 19.9**). On a hot, summer day, the top of a cumulus cloud may rise 10 kilometers or more above its base in cauliflowerlike masses. The base of the cloud forms at the altitude at which the rising air cools to its dew point and condensation starts. However, in this situation the rising air remains warmer than the surrounding air and therefore continues to rise. As it rises, more vapor condenses, forming the billowing columns.

Other types of clouds are named by combining these three basic terms (●**FIGURE 19.10**). **Stratocumulus** clouds are low, sheetlike clouds with some vertical structure. The term *nimbo* refers to a cloud that precipitates; thus a **cumulonimbus** cloud is a towering rain cloud. If you see one, you should seek shelter, because cumulonimbus clouds commonly produce intense rain, thunder, lightning, and sometimes hail. A **nimbostratus** cloud is a stratus cloud from which rain or snow falls. Other prefixes are also added to cloud names. For example, *Alti* is derived from the Latin root *altus*, meaning "high." An **altostratus** cloud is simply a high stratus cloud.

---

unstable air A parcel of warm, moist air that rises rapidly, ascends to high elevations, and leads to formation of towering clouds and heavy rainfall.

stable air A parcel of warm, dry air that does not rise rapidly, does not ascend to high elevations, and does not lead to cloud formation and precipitation.

cirrus Wispy, high-altitude clouds composed of ice crystals.

stratus Horizontally layered clouds that spread out into a broad sheet, usually creating dark, overcast skies.

cumulus Fluffy white clouds with flat bottoms and billowy tops.

stratocumulus Low, sheetlike clouds with some vertical structure.

cumulonimbus Towering storm clouds that form in columns and produce intense rain, thunder, lightning, and sometimes hail.

nimbostratus Stratus clouds from which rain or snow falls.

altostratus High-altitude stratus clouds.

● **FIGURE 19.7** Cirrus clouds are high, wispy clouds composed of ice crystals.

● **FIGURE 19.8** Stratus clouds spread out across the sky in a low, flat layer.

● **FIGURE 19.9** Cumulus clouds are fluffy white clouds with flat bottoms and billowy tops.

Why does rain fall from some clouds, whereas other clouds float across a blue sky on a sunny day and produce no rain? The droplets in a cloud are small, about 0.01 millimeter in diameter (about one-seventh the diameter of a human hair). In still air, such a droplet would require 48 hours to fall from a cloud 1,000 meters above Earth. But these tiny droplets never reach Earth because they evaporate faster than they fall.

If the air temperature in a cloud is above freezing, the tiny droplets may collide and coalesce. You can observe similar behavior in droplets sliding down a window pane on a rainy day. If two droplets collide, they merge to become one large drop. If the droplets in a cloud grow large enough, they fall as drizzle (0.1 to 0.5 millimeter in diameter) or light rain (0.5 to 2 millimeters in diameter). About 1 million cloud droplets must combine to form an average-size raindrop.

In many clouds, however, water vapor initially forms ice crystals rather than condensing as tiny droplets of supercooled water. Part of the reason for this is that the temperature in clouds is commonly below freezing, but another factor also favors ice formation. At temperatures near or below freezing, air that is slightly undersaturated with respect to water is slightly supersaturated with respect to ice. For example, if the relative humidity of air is 95 percent with respect to water, it is about 105 percent with respect to ice. Thus, as air cools toward its dew point, all the vapor forms ice crystals rather than supercooled water droplets. The tiny ice crystals then grow larger as more water vapor condenses on them, until they are large enough to fall. The

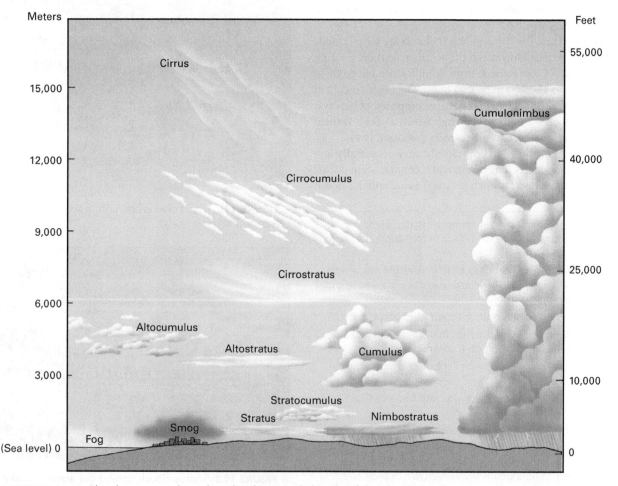

● **FIGURE 19.10** Cloud names are based on the shape and altitude of the clouds.

ice then melts to form raindrops as it falls through warmer layers of air.

If you have ever been caught in a thunderstorm, you may remember raindrops large enough to be painful as they struck your face or hands. Recall that a cumulus cloud forms from rising air and that its top may be several kilometers above its base. The temperature in the upper part of the cloud is commonly below freezing. As a result, ice crystals form and begin to fall.

Condensation continues as the crystal falls through the towering cloud, and the crystal grows. If the lower atmosphere is warm enough, the ice melts before it reaches the surface. Raindrops formed in this manner may be 3 to 5 millimeters in diameter, large enough to hurt when they hit.

SNOW, SLEET, AND GLAZE As explained previously, when the temperature in a cloud is below freezing, the cloud is composed of ice crystals rather than water droplets. If the temperature near the ground is also below freezing, the crystals remain frozen and fall as snow. In contrast, if raindrops form in a warm cloud and fall through a layer of cold air at lower elevation, the drops freeze and fall as small spheres of ice called **sleet**. Sometimes the freezing zone near the ground is so thin that raindrops do not have time to freeze before they reach Earth. However, when they land on subfreezing surfaces, they form a coating of ice called **glaze** (●FIGURE 19.11).

● **FIGURE 19.11** Glaze forms when rain falls on a surface that is colder than the freezing temperature of water.

---

sleet Small spheres of ice that develop when raindrops form in a warm cloud and freeze as they fall through a layer of cold air at lower elevation.

glaze An ice coating that forms when rain falls on subfreezing surfaces.

hail Large ice globules varying from 5 millimeters to a record 14 centimeters in diameter that fall from cumulonimbus clouds.

advection fog Fog that forms when warm, moist air from the sea blows onto cooler land, where the air cools and water vapor condenses at ground level.

radiation fog Fog that occurs when Earth's surface and the air near the surface cool by radiation during the night, and water vapor in the air condenses because it cools below its dew point.

Glaze can be heavy enough to break tree limbs and electrical transmission lines. It also coats highways with a dangerous icy veneer. In the winter of 1997–1998, a sleet and glaze storm in eastern Canada and the northeastern United States caused billions of dollars in damage. The ice damaged so many electric lines and power poles that many people were without electricity for a few weeks.

HAIL Occasionally, precipitation takes the form of very large ice globules called **hail**. Hailstones vary from 5 millimeters in diameter to a record 14 centimeters in diameter; that record breaker weighed 765 grams (more than 1.5 pounds) and fell in Kansas. A 500-gram (1-pound) hailstone crashing to Earth at 160 kilometers (100 miles) per hour can shatter windows, dent car roofs, and kill people and livestock. Even small hailstones can damage crops. Hail falls only from cumulonimbus clouds. Because cumulonimbus clouds form in columns with distinct boundaries, hailstorms occur in local, well-defined areas. Thus, one farmer may lose an entire crop while a neighbor is unaffected.

A hailstone consists of ice in concentric shells, like the layers of an onion. Two mechanisms have been proposed for their formation. In one, turbulent winds blow falling ice crystals back upward in the cloud. New layers of ice accumulate as additional vapor condenses on the recirculating ice grain. An individual particle may rise and fall several times until it grows so large and heavy that it drops out of the cloud. In the second mechanism, hailstones form in a single pass through the cloud. During their descent, supercooled water freezes onto the ice crystals. The layering develops because different temperatures and amounts of supercooled water exist in different portions of the cloud, and each layer forms in a different part of the cloud.

FOG

Fog is a cloud that forms at or very close to ground level, although most fog forms by processes different from those that create higher-level clouds. **Advection fog** occurs when warm, moist air from the sea blows onto cooler land. The air cools to its dew point, and water vapor condenses at ground level. San Francisco and Seattle, as well as Vancouver, British Columbia, all experience foggy winters as warm, moist air from the Pacific Ocean is cooled first by the cold California current and then by land. The foggiest location in the United States is Cape Disappointment, Washington, on the Pacific Ocean at the mouth of the Columbia River, where visibility is obscured by fog 29 percent of the time.

Radiation fog occurs when Earth's surface and air near the surface cool by radiation during the night (●FIGURE 19.12). Water vapor condenses as fog when the air cools below its dew point. Often the cool, dense, foggy air settles into valleys. If you are driving late at night in hilly terrain, beware, because a sudden dip in the roadway may lead you into a thick fog where visibility is low. A ground fog of this type typically

● **FIGURE 19.12** Radiation fog is seen as a morning mist in this field in Idaho.

low-pressure region, at a rate of about 1 kilometer per day. In contrast, if air in the upper atmosphere cools, it becomes denser than the air beneath it and sinks (●**FIGURE 19.13B**).

Air must flow inward over Earth's surface toward a low-pressure region to replace a rising air mass. But a sinking air mass displaces surface air, pushing it outward from a high-pressure region. Thus, vertical airflow in both high- and low-pressure regions is accompanied by horizontal airflow, called **wind**. Winds near Earth's surface always flow away from a region of high pressure and toward a low-pressure region. Ultimately, all wind is caused by the pressure differences resulting from unequal heating of Earth's atmosphere.

"burns off" in the morning. The rising Sun warms the land or water surface which, in turn, warms the low-lying air. As the air becomes warmer, its capacity to hold water vapor increases, and the fog droplets evaporate. Radiation fog is particularly common in areas where the air is polluted because water vapor condenses readily on the tiny particles suspended in the air.

Recall that vaporization of water absorbs heat and, therefore, cools both the surface and the surrounding air. In addition, vaporization adds moisture to the air. The cooling and the addition of moisture combine to form conditions conducive to fog. **Evaporation fog** occurs when air is cooled by evaporation from a body of water, commonly a lake or river. Evaporation fogs are common in late fall and early winter, when the air has become cool but the water is still warm. The water evaporates, but the vapor cools and condenses to fog almost immediately upon contact with the cold air.

Upslope fog occurs when air cools as it rises along a land surface. Upslope fogs occur both on gradually sloping plains and on steep mountains. For example, the Great Plains rise from sea level at the Mississippi Delta to 1,500 meters (5,000 feet) at the Rocky Mountain front. When humid air moves northwest from the Gulf of Mexico toward the Rockies, it rises and cools adiabatically to form upslope fog. The rapid rise at the mountain front also forms fog.

PRESSURE AND WIND

Warm air is less dense than cold air. Thus, warm air exerts a relatively low atmospheric pressure and cold air exerts a relatively high atmospheric pressure. Warm air rises because it is less dense than the surrounding cool air (●**FIGURE 19.13A**). Air rises slowly above a typical

Pressure Gradient

Wind blows in response to differences in pressure. Imagine that you are sitting in a room and the air is still. Now you open a can of vacuum-packed coffee and hear the hissing as air rushes into the can. Because the pressure in the room is higher than that inside the coffee can, wind blows from the room into the can. But if you blow up a balloon, the air inside the balloon is at higher pressure than the air in the room. When the balloon is punctured, wind blows from the high-pressure zone of the balloon into the lower-pressure zone of the room.

Wind speed is determined by the magnitude of the pressure difference over distance, called the **pressure gradient**. Thus, wind blows rapidly if a large pressure difference exists over a short distance. A steep pressure gradient is analogous to a steep hill. Just as a ball rolls quickly down a steep hill, wind flows rapidly across a steep pressure gradient. To create a pressure-gradient map, air pressure is measured at hundreds of different weather stations. Points of equal pressure are connected by map lines called **isobars**. A steep pressure gradient is shown by closely spaced isobars, whereas a weak pressure gradient is indicated by widely spaced isobars (●**FIGURE 19.14**). Pressure gradients change daily, or sometimes hourly, as high- and low-pressure zones move. Therefore, maps are updated frequently.

Coriolis Effect

Recall from Chapter 16 that the Coriolis effect, caused by Earth's spin, deflects ocean currents. The Coriolis effect similarly deflects winds. In the Northern Hemisphere, wind is deflected toward the right, and in the Southern Hemisphere, to the left (●**FIGURE 19.15**). The Coriolis effect alters wind direction but not its speed.

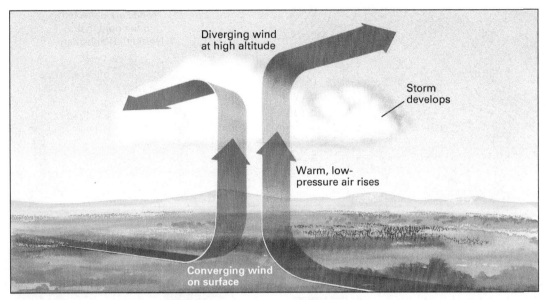

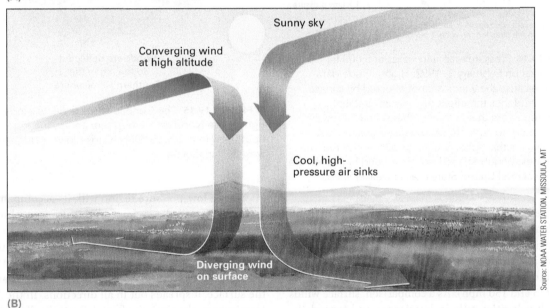

● **FIGURE 19.13** (A) Rising, low-pressure air creates clouds and precipitation. Air flows inward toward the low-pressure zone, creating surface winds. (B) Sinking, high-pressure air creates clear skies. Air flows outward from the high-pressure zone and also creates surface winds.

evaporation fog Fog that forms when air is cooled by evaporation from a body of water, commonly a lake or river and typically in late fall or early winter when the air is cool but the water is still warm. The water evaporates, but the vapor cools and condenses to fog.

upslope fog Fog that forms when air cools as it rises along a land surface.

wind Horizontal airflow caused by pressure differences resulting from unequal heating of Earth's atmosphere. Winds near Earth's surface always flow from a region of high pressure toward a low-pressure region.

pressure gradient A measure of the change in air pressure over distance, used to determine wind speed.

isobars Lines on a weather map connecting points of equal air pressure.

Friction

Rising and falling air generates wind both along Earth's surface and at higher elevations. Surface winds are affected by friction with Earth's surface, whereas high-altitude winds are not. As a result, wind speed normally increases with elevation. This effect was first noted during World War II. On November 24, 1944, U.S. bombers were approaching Tokyo for the first mass bombing of the Japanese capital. Flying between 8,000 and 10,000 meters (27,000 to 33,000 feet), the pilots suddenly found themselves roaring past landmarks 140 kilometers (90 miles) per hour faster than the theoretical top speed of their airplanes! Amid the confusion,

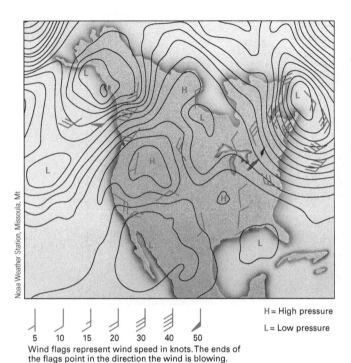

Noaa Weather Station, Missoula, Mt

H = High pressure

L = Low pressure

| 5 | 10 | 15 | 20 | 30 | 40 | 50 |

Wind flags represent wind speed in knots. The ends of the flags point in the direction the wind is blowing.

● **FIGURE 19.14** Pressure map and winds at 5,000 feet in North America on February 3, 1992. High-altitude data are shown because the winds are not affected by surface topography and thus the effect of pressure gradient is well illustrated. Note that in the Northeast and Northwest, steep pressure gradients, shown by closely spaced isobars, cause high winds that spiral counterclockwise into the low-pressure zones. Widely spaced isobars around high-pressure zones in the central United States cause weaker winds.

most of the bombs missed their targets, and the mission was a military failure. However, this experience introduced meteorologists to **jet streams**, narrow bands of high-altitude wind. The jet stream in the Northern Hemisphere flows from west to east at speeds between 120 and 240 kilometers per hour (75 and 150 mph). As a comparison, surface winds attain such velocities only in hurricanes and tornadoes. Airplane pilots traveling from Los Angeles to New York fly with the jet stream to gain speed and save fuel, whereas pilots moving from east to west try to avoid it.

Jet-stream influence on weather and climate will be discussed again later in this chapter and in more detail in Chapter 20.

Cyclones and Anticyclones

● **FIGURE 19.16A** shows the movement of air in the Northern Hemisphere as it converges toward a low-pressure area. If Earth did not spin, wind would flow directly across the isobars, as shown by the black arrows. However, Earth does spin, and the Coriolis effect deflects wind to the right, as shown by the small blue arrows. This rightward deflection creates a counterclockwise vortex near the center of the low-pressure region, as shown by the large, magenta arrows. In the Southern Hemisphere, the direction is clockwise.

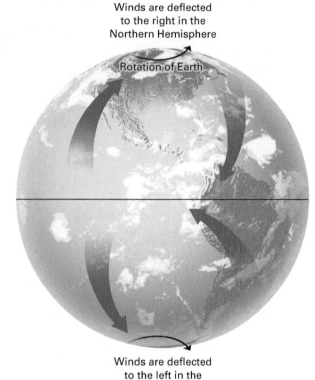

Winds are deflected to the right in the Northern Hemisphere

Rotation of Earth

Winds are deflected to the left in the Southern Hemisphere

● **FIGURE 19.15** The Coriolis effect deflects winds to the right in the Northern Hemisphere and to the left in the Southern Hemisphere. Only winds blowing due east or west are unaffected.

Such a low-pressure region with its accompanying surface wind is called a *cyclone*. In this usage, **cyclone** means a system of inwardly directed rotating winds, not the violent storms that are sometimes called cyclones, hurricanes, or typhoons. The opposite mechanism forms an **anticyclone** around a high-pressure region. When descending air reaches the surface, it spreads out in all directions. In the Northern Hemisphere, the Coriolis effect deflects the diverging winds of an anticyclone to the right, forming a pinwheel pattern, with the wind spiraling clockwise (●**FIGURE 19.16B**). In the Southern Hemisphere, the Coriolis effect deflects winds leftward and creates a counterclockwise spiral.

Pressure Changes and Weather

As explained earlier, wind blows in response to any difference in pressure. However, low pressure generally brings clouds and precipitation with the wind, and sunny days predominate during high pressure. To understand this distinction, recall that warm air is less dense than cold air. If warm and cold air are in contact, the less dense, and therefore buoyant, warm air rises.

Rising air forms a region of low pressure. But rising air also cools adiabatically. If the cooling is sufficient, clouds form and rain or snow may fall. Thus, low barometric pressure is an indication of wet weather. Alternatively, when

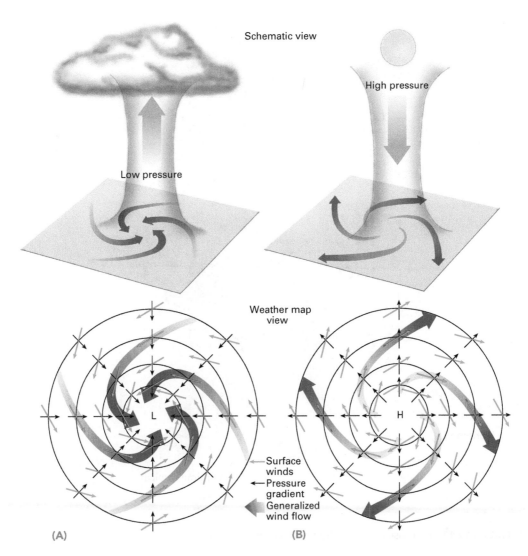

Schematic view

High pressure

Low pressure

Weather map view

← Surface winds
← Pressure gradient
Generalized wind flow

(A)   (B)

● **FIGURE 19.16** (A) In the Northern Hemisphere, a cyclone consists of winds spiraling counterclockwise into a low-pressure region. (B) An anticyclone in the Northern Hemisphere consists of winds spiraling clockwise out from a high-pressure zone.

cool air sinks, it is compressed and the pressure rises. In addition, sinking air is heated adiabatically. Because warm air can hold more water vapor than cold air can, the sinking air absorbs moisture, and thus clouds generally do not form over a high-pressure region. Thus, fair, dry weather generally accompanies high pressure.

---

jet streams Narrow bands of high-altitude, fast-moving wind.

cyclone A low-pressure region with its accompanying system of inwardly directed rotating winds. In common, nonscientific usage the term often refers to a variety of different violent storms including hurricanes and tornados.

anticyclone A high-pressure region with its accompanying system of outwardly directed rotating winds that develop where descending air spreads over Earth's surface.

air mass A large body of air that has approximately the same temperature and humidity at any given altitude throughout.

front In meteorology, the boundary between a warmer air mass and a cooler one.

FRONTS AND FRONTAL WEATHER

An **air mass** is a large body of air with approximately uniform temperature and humidity at any given altitude. Typically, an air mass is 1,500 kilometers or more across and several kilometers thick. Because air acquires both heat and moisture from Earth's surface, an air mass is classified by its place of origin. Temperature can be either polar (cold) or tropical (warm). Maritime air originates over water and has high moisture content, whereas continental air has low moisture content (●**FIGURE 19.17, TABLE 19.1**).

Air masses move and collide. The boundary between a warmer air mass and a cooler one is a **front**. The term was first used during World War I because weather systems were considered analogous to armies that advance and clash along battle lines. When two air masses collide, each may retain its integrity for days before the two mix. During a collision, one of

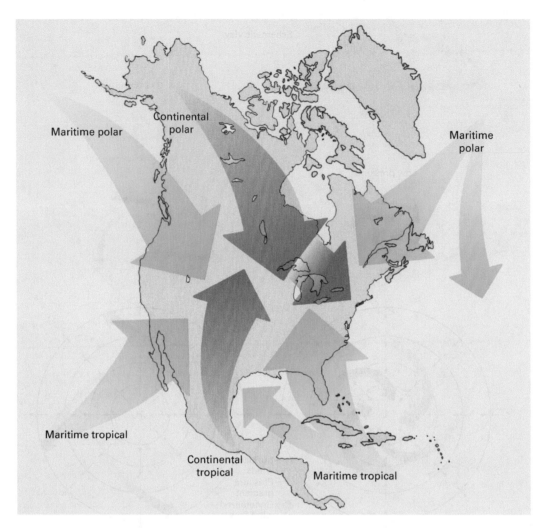

● **FIGURE 19.17** Air masses are classified by their source regions.

the air masses is forced to rise, which often results in cloudiness and precipitation. Frontal weather patterns are determined by the types of air masses that collide and their relative speeds and directions. The symbols commonly used on weather maps to describe fronts are shown in ●**FIGURE 19.18.**

TABLE 19.1 Classification of Air Masses

Classification according to temperature (latitude):

Polar (P) air masses originate in high latitudes and are cold.

Tropical (T) air masses originate in low latitudes and are warm.

Classification according to moisture content:

Continental (c) air masses originate over land and are dry.

Maritime (m) air masses originate over water and are moist.

Symbol	Name	Characteristics
Mp	Maritime polar	Moist and cold
cP	Continental polar	Dry and cold
mT	Maritime tropical	Moist and warm
cT	Continental tropical	Dry and warm

Warm Fronts and Cold Fronts

Fronts are classified by whether a warm air mass moves toward a stationary (or more slowly moving) cold mass, or vice versa. A **warm front** forms when moving warm air collides with a stationary or slower-moving cold air mass. A **cold front** forms when moving cold air collides with stationary or slower-moving warm air.

In a warm front, the moving warm air rises over the denser cold air as the two masses collide (●**FIGURE 19.19**). The rising warm air cools adiabatically and the cooling generates clouds and precipitation. Precipitation is generally light because the air rises slowly along the gently sloping frontal boundary. Figure 19.19 shows that a characteristic sequence of clouds accompanies a warm front. High, wispy cirrus and cirrostratus clouds develop near the leading edge of the rising warm air. These high clouds commonly precede a storm. They form as much as 1,000 kilometers ahead of an advancing band of precipitation that falls from thick, low-lying nimbostratus and stratus clouds near the trailing edge of the front. The cloudy weather may last for several days because of the gentle slope and broad extent of the frontal boundary.

Showers   Rain   Snow

Fronts

Warm   Cold   Occluded   Stationary

● **FIGURE 19.18** Symbols commonly used in weather maps. Warm and cold are relative terms. Air over the central plains of Montana at a temperature of 0°C may be warm relative to polar air above northern Canada but cold relative to a 20°C air mass over the southeastern United States.

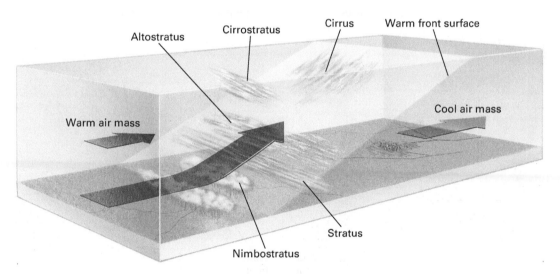

Altostratus   Cirrostratus   Cirrus   Warm front surface

Warm air mass   Cool air mass

Nimbostratus   Stratus

● **FIGURE 19.19** In a warm front, moving warm air rises gradually over cold air. High, wispy cirrus and cirrostratus clouds typically develop near the leading edge of the rising warm air, followed by descending altostratus, stratus, and nimbostratus clouds near the trailing edge of the front.

warm front A front that forms when moving warm air collides with a stationary or slower-moving cold air mass. The moving warm air rises over the denser cold air, cools adiabatically, and the cooling generates clouds and precipitation.

cold front A front that forms when moving cold air collides with stationary or slower-moving warm air. The dense cold air distorts into a blunt wedge and pushes under the warmer air, creating a narrow band of violent weather commonly accompanied by cumulus and cumulonimbus clouds.

occluded front A front that forms when a faster-moving cold air mass traps a warm air mass against a second mass of cold air. Precipitation occurs along both frontal boundaries, resulting in a large zone of inclement weather.

A cold front forms when faster-moving cold air overtakes and displaces warm air. The dense cold air forms a blunt wedge and pushes under the warmer air (●**FIGURE 19.20**). Thus, the leading edge of a cold front is much steeper than that of a warm front. The steep contact between the two air masses causes the warm air to rise rapidly, creating a narrow band of violent weather commonly accompanied by cumulus and cumulonimbus clouds. The storm system may be only 25 to 100 kilometers wide, but within this zone downpours, thunderstorms, and violent winds are common.

Occluded Front

An **occluded front** forms when a faster-moving cold air mass traps a warm air mass against a second mass of cold

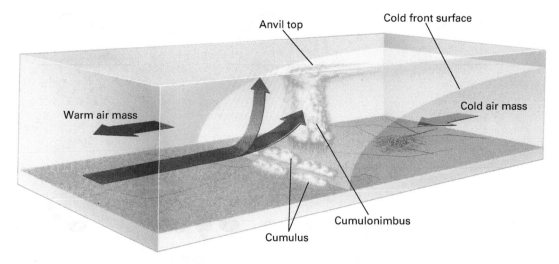

● **FIGURE 19.20** In a cold front, moving cold air slides abruptly beneath warm air, forcing it steeply upward and creating a narrow band of violent weather commonly accompanied by cumulus and cumulonimbus clouds.

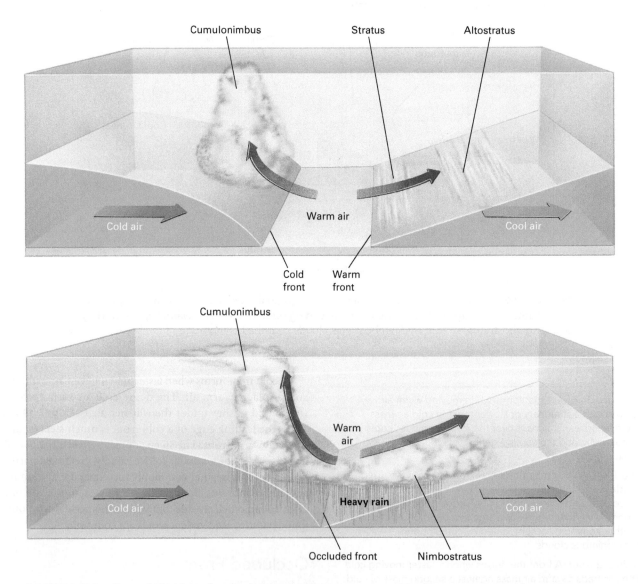

● **FIGURE 19.21** An occluded front forms where warm air is trapped and lifted between two cold air masses. Precipitation occurs along both frontal boundaries in a large zone of short-lived inclement weather.

air. Thus, the warm air mass becomes trapped between two colder air masses (●FIGURE 19.21). The faster-moving cold air mass then slides beneath the warm air, lifting it completely off the ground. Precipitation occurs along both frontal boundaries, combining the narrow band of heavy precipitation of a cold front with the wider band of lighter precipitation of a warm front. The net result is a large zone of inclement weather. A storm of this type is commonly short-lived because the warm air mass is cut off from its supply of moisture evaporating from Earth's surface.

Stationary Front

A **stationary front** occurs along the boundary between two stationary air masses. Under these conditions, the front can remain over an area for several days. Warm air rises, forming conditions similar to those in a warm front. As a result, rain, drizzle, and fog may occur.

The Life Cycle of a Midlatitude Cyclone

As you learned in section on "Pressure and Wind," cyclone is a low-pressure system with rotating winds.[1] Most cyclones in the middle latitudes of the Northern Hemisphere develop along a front between polar and tropical air masses. The storm often starts with winds blowing in opposite directions along a stationary front between the two air masses (●FIGURE 19.22A). In the figure, a warm air mass was moving northward and was deflected to the east by the Coriolis effect. At the same time, a cold air mass traveling southward was deflected to the west.

In ●FIGURE 19.22B, the cold polar air continues to push southward, creating a cold front and lifting the warm air off the ground. Then, some small disturbance—a topographic feature such as a mountain range, airflow from

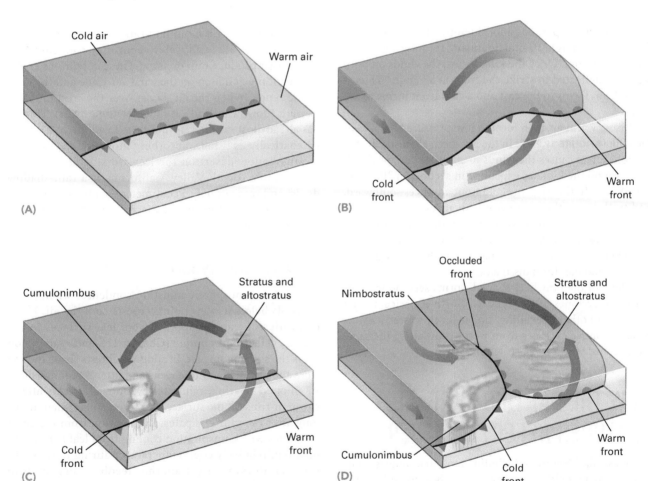

● **FIGURE 19.22** A midlatitude cyclone develops along a front between polar air and a tropical air mass. (A) The front develops. (B) Some small disturbance such as a topographical feature, nearby storm, or local temperature variation creates a kink in the front. (C) A low-pressure region and cyclonic circulation develop near the kink. (D) An occluded front forms.

stationary front A front at the boundary between two stationary air masses of different temperatures.

1. In common usage, the world cyclone is applied loosely to a wide variety of storm systems, and the usage varies from place to place. In this textbook, we apply the scientific usage.

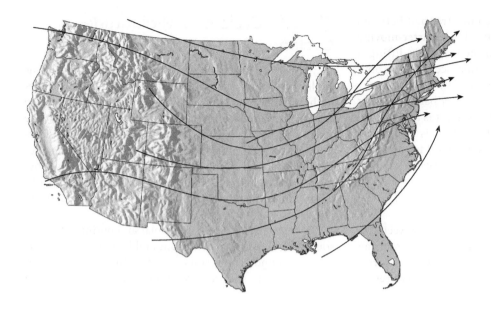

● **FIGURE 19.23** Most North American cyclones follow certain paths, called storm tracks, from west to east.

a local storm, or perhaps a local temperature variation—deforms the straight frontal boundary, forming a wavelike kink in the front. Once the kink forms, the winds on both sides are deflected to strike the front at an angle. Thus, a warm front forms to the east and a cold front forms to the west.

Rising warm air then forms a low-pressure region near the kink (●**FIGURE 19.22C**). In the Northern Hemisphere, the Coriolis effect causes the winds to circulate counterclockwise around the kink, as explained in section on "Pressure and Wind." To the west, the cold front advances southward, and to the east, the warm front advances northward. At the same time, rain or snow falls from the rising warm air (●**FIGURE 19.22D**). Over a period of one to three days, the air rushing into the low-pressure region equalizes pressure differences, and the storm dissipates.

Many of the pinwheel-shaped storms seen on weather maps are cyclones of this type. In North America, the jet stream and other prevailing, upper-level, westerly winds generally move cyclones from west to east along the same paths, called **storm tracks** (●**FIGURE 19.23**).

HOW THE EARTH'S SURFACE FEATURES AFFECT WEATHER

Earth's surface features—including mountain ranges, rainforests, proximity to the sea, and uneven heating and cooling of continents—can create conditions that affect the weather of a region.

Mountain Ranges and Rain-Shadow Deserts

As we described in section on "Rising Air and Precipitation," air rises in a process called *orographic lifting* when it flows over a mountain range. As the air rises, it cools adiabatically, and water vapor may condense into clouds that produce rain or snow. These conditions create abundant precipitation on the windward side and the crest of the range. When the air passes over the crest onto the leeward (downwind) side, it sinks (●**FIGURE 19.24**). This air has already lost much of its moisture. In addition, it warms adiabatically as it falls, absorbing moisture and creating a rain-shadow desert on the leeward side of the range. For example, Death Valley, California, is a rain-shadow desert and receives only 5 centimeters of rain a year, while the nearby west slope of the Sierra Nevada receives 178 centimeters of rain a year.

Forests and Weather

Recall that atmospheric moisture condenses when moist air cools below its dew point. Forests cool the air. Large quantities of water evaporate from leaf surfaces in the process called *transpiration* (Chapter 11), and evaporation cools the surrounding air. In addition, forests shade the soil from the hot sun, and tree roots and litter retain moisture. In an open clearing, rainwater evaporates quickly after a storm or runs off the surface. But forest soils remain moist long after the rain dissipates. Evaporation from soil litter combines with transpiration cooling from leaf surfaces to maintain relatively cool temperatures during times when there is no rain. In yet another feedback mechanism: Forests cool the air. Cool air promotes rainfall. Rainfall supports forests.

In today's tropical rainforests, local rainfall has decreased by as much as 50 percent when the forests were cut and replaced by farmland or pasture. When the rainfall decreases, wildfires ravage the boundary between the logged area and the remaining virgin forests. More forest is destroyed, establishing a negative feedback mechanism of increasing drought, fire, and forest loss.

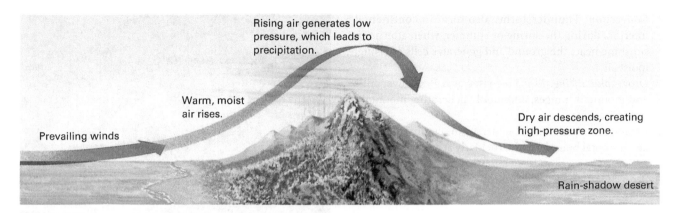

● **FIGURE 19.24** A rain-shadow desert forms where moist air rises over a mountain range and precipitates most of its moisture on the windward side and crest of the range. The dry, descending air on the leeward side absorbs moisture, forming a desert.

Sea and Land Breezes

Anyone who has lived near an ocean or large lake would have encountered winds blowing from water to land and from land to water. Sea and land breezes are caused by uneven heating and cooling of land and water. Recall that land surfaces heat up faster than adjacent bodies of water and cool more quickly. If land and sea are at nearly the same temperature on a summer morning, during the day the land warms and heats the air above it. Hot air then rises over the land, producing a local low-pressure area. Cooler air from the sea flows inland to replace the rising air. Thus, on a hot sunny day, winds generally blow from the sea onto land. The rising air is good for flying kites or hang-gliding but often brings afternoon thunderstorms.

At night the reverse process occurs. The land cools faster than the sea, and descending air creates a local high-pressure area over the land. Then the winds reverse, and breezes blow from the shore out toward the sea.

Monsoons

A **monsoon** is a seasonal wind and weather system caused by uneven heating and cooling of continents and adjacent oceans. Just as sea and land breezes reverse direction with day and night, monsoons reverse direction with the seasons. In the summertime, the continents become warmer than the sea. Warm air rises over land, creating a large low-pressure area and drawing moisture-laden maritime air inland. When the moist air rises as it flows over the land, clouds form and heavy monsoon rains fall. In winter, the process is reversed.

storm tracks Paths repeatedly followed by storms.

monsoon A seasonal wind and weather system caused by uneven heating and cooling of land and adjacent sea, generally blowing from the sea to the land in the summer when the continents are warmer than the ocean, and from land to sea in winter when the ocean is warmer than the land.

The land cools below the sea temperature, and as a result air descends over land, producing dry, continental, high pressure. At the same time, air rises over the ocean and the prevailing winds blow from land to sea. More than half of the inhabitants of Earth depend on monsoons because the predictable, heavy, summer rains bring water to the fields of Africa and Asia. If the monsoons fail to arrive, crops cannot grow and people starve.

THUNDERSTORMS

An estimated 16 million thunderstorms occur every year, and at any given moment, about 27,000 thunderstorms are in progress over different parts of Earth. A single bolt of lightning can involve several hundred million volts of energy and for a few seconds produces as much power as a nuclear power plant. It heats the surrounding air to 25,000°C or more, much hotter than the surface of the Sun. The heated air expands instantaneously to create a shock wave that we hear as thunder.

Despite their violence, thunderstorms are local systems, often too small to be included on national weather maps. A typical thunderstorm forms and then dissipates in a few hours and covers from about 10 to a few hundred square kilometers. It is not unusual to stand on a hilltop in the sunshine and watch rain squalls and lightning a few kilometers away. All thunderstorms develop when warm, moist air rises, forming cumulus clouds that develop into towering cumulonimbus clouds. Different conditions cause these local regions of rising air:

1. *Wind convergence.* Central Florida is the most active thunderstorm region in the United States. As the subtropical Sun heats the Florida peninsula, rising air draws moist air from its both east and west coasts. Where the two air masses converge, the moist air rises rapidly to create a thunderstorm. Thunderstorms also occur in other environments where moist air masses converge.

2. *Convection.* Thunderstorms also form in continental interiors during the spring or summer, when afternoon sunshine heats the ground and generates cells of rising, moist air.

3. *Orographic lifting.* Moist air rises as it flows over hills and mountain ranges, commonly generating mountain thunderstorms.

4. *Frontal thunderstorms.* Thunderstorms commonly occur along frontal boundaries, particularly at cold fronts.

Lightning

Lightning is an intense discharge of electricity that occurs when the buildup of static electricity overwhelms the insulating properties of air (●FIGURE 19.25). If you walk across a carpet on a dry day, the friction between your feet and the rug shears electrons off the atoms on the rug. The electrons migrate into your body and concentrate there. If you then touch a metal doorknob, a spark consisting of many electrons jumps from your finger to the metal knob.

In 1752, Benjamin Franklin showed that lightning is an electrical spark. He suggested that charges separate within cumulonimbus clouds and build until a bolt of lightning jumps from the cloud. In the more than 250 years since Franklin, atmospheric physicists have been unable to agree upon the exact mechanism of lightning. According to one hypothesis, friction between the intense winds and moving ice crystals in a cumulonimbus cloud generates both positive and negative electrical charges in the cloud, and the two types of charges become physically separated (●FIGURE 19.26A). The positive charges tend to accumulate in the upper portion of the cloud, and the negative charges build up in the lower reaches of the cloud. When enough charge accumulates, the electrical potential exceeds the insulating properties of air, and a spark jumps from the cloud to the ground, from the ground to the cloud, or from one cloud to another.

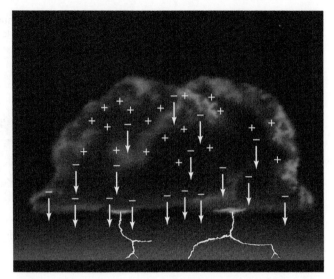

(A)

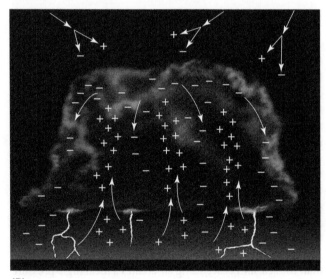

(B)

● **FIGURE 19.26** Two hypotheses for the origin of lightning. (A) Friction between intense winds and ice particles generates charge separation. (B) Charged particles are produced from above by cosmic rays and below by interactions with the ground. The particles are then distributed by convection currents.

Another hypothesis suggests that cosmic rays bombarding the cloud from outer space produce ions at the top of the cloud. Other ions form on the ground as winds blow over Earth's surface. The electrical discharge occurs when the potential difference between the two groups of electrical charges exceeds the insulating properties of air (●FIGURE 19.26B).

Perhaps neither hypothesis is entirely correct and some combination of the mechanisms causes lightning.

Svet_Feo/ShutterStock.com

● **FIGURE 19.25** This time-lapse photo captures multiple cloud-to-ground lightning strikes in the Black Sea.

TORNADOES AND TROPICAL CYCLONES

Tornadoes and tropical cyclones are both intense, low-pressure centers. Strong winds follow the steep pressure gradients and spiral inward toward a central column of rising air.

Tornadoes

A **tornado** is a small, short-lived, funnel-shaped storm that protrudes from the base of a cumulonimbus cloud (●FIGURE 19.27A). The base of the funnel can be from 2 meters to 3 kilometers in diameter. Some tornadoes remain suspended in air while others touch the ground. After a tornado touches ground, it may travel for a few meters to a few hundred kilometers across the surface. The funnel travels at 40 to 65 kilometers per hour, and in some cases as much as 110 kilometers per hour, but the spiraling winds within the funnel are much faster. Few direct measurements have been made of pressure and wind speed inside a tornado. However, we know that a large pressure difference occurs over a very short distance. Meteorologists estimate that winds in tornadoes may reach 500 kilometers per hour or greater. These winds rush into the narrow, low-pressure zone and then spiral upward. After a few seconds to a few hours, the tornado lifts off the ground and dissipates.

Tornadoes are the most violent of all storms. One tornado in 1910 lifted a team of horses and then deposited it, unhurt, several hundred meters away. They were lucky. In the past, an average of 120 Americans were killed every year by these storms, and property damage costs millions of dollars. The death toll has decreased in recent years because effective warning systems allow people to seek shelter, but the property damage has continued to increase (●FIGURE 19.27B). Tornado winds can lift the roof off a house and then flatten the walls. Flying debris kills people and livestock caught in the open. Even so, the total destruction from tornadoes is not as great as that from hurricanes because the path of a tornado is narrow and its duration is short.

Although tornadoes can occur anywhere in the world, 75 percent of the world's twisters concentrate in the Great Plains, east of the Rocky Mountains. Approximately 700 to 1,000 tornadoes occur in the United States each year.

They frequently form in the spring or early summer. At that time, continental polar (dry, cold) air from Canada collides with maritime tropical (warm, moist) air from the Gulf of Mexico. As explained previously, these conditions commonly create thunderstorms. Meteorologists cannot explain why most thunderstorms dissipate harmlessly but a few develop tornadoes. However, one fact is apparent: tornadoes are most likely to occur when large differences in temperature and moisture exist between the two air masses and the boundary between them is sharp.

Tropical Cyclones

A **tropical cyclone or tropical storm** is less intense than a tornado but much larger and longer-lived (TABLE 19.2). Tropical cyclones are circular disturbances that average 600 kilometers in diameter and persist for days or weeks. If the wind exceeds 120 kilometers per hour, a tropical cyclone is called a **hurricane** in North America and the Caribbean, a **typhoon** in the western Pacific, and a *cyclone* in the Indian Ocean (●FIGURE 19.28). Intense low pressure in the center of a hurricane can generate wind as strong as 300 kilometers per hour.

(A)

(B)

tornado A small, intense, short-lived, funnel-shaped storm that protrudes from the base of a cumulonimbus cloud.

tropical cyclone or tropical storm A broad, circular storm with intense low pressure that forms over warm oceans.

hurricane A tropical storm occurring in North America or the Caribbean whose wind exceeds 120 kilometers per hour; called a *typhoon* in the western Pacific and a *cyclone* in the Indian Ocean.

typhoon A tropical storm occurring in the western Pacific Ocean whose wind exceeds 120 kilometers per hour; called a hurricane in North America and the Caribbean and a cyclone in the Indian Ocean.

●FIGURE 19.27 (A) The dark funnel cloud of a tornado descends during an evening storm. (B) This house was lifted off its foundation by a tornado that swept through Lapeer County, Michigan on March 12, 2012.

TABLE 19.2 Comparison of Tornadoes and Tropical Cyclones		
	Range	
Feature	**Tornado**	**Tropical Cyclone**
Diameter	2 to 3 km	400 to 800 km
Path length (distance traveled across terrain)	A few meters to hundreds of kilometers	A few hundred to a few thousand kilometers
Duration	A few seconds to a few hours	A few days to a week
Wind speed	300 to 511* km/hr	60 to 120 km/hr (tropical storm); 120 to 371† km/hr (hurricane)
Speed of motion	40 to 110 km/hr	20 to 30 km/hr
Pressure fall	20 to 200 millibars	20 to 60 millibars

*Doppler radar systems recorded a maximum wind speed of 511 km/hr during the tornado that destroyed Moore, Oklahoma on May 3, 2013.
†A wind gust of 371 km/hr was recorded by a wind anemometer on April 12, 1934, on the summit of Mount Washington, New Hampshire. The wind gust occurred during a hurricane.

The low atmospheric pressure created by a tropical cyclone can raise the sea surface by several meters. Often, as a tropical cyclone strikes shore, strong onshore winds combine with the abnormally high water level created by low pressure to create a **storm surge** that floods coastal areas. In 2005, during Hurricane Katrina, sea level rose about 8.5 meters above normal on the Gulf Coast as a result of a storm surge.

Tropical cyclones form only over warm oceans, never over cold oceans or land. Thus, moist, warm air is crucial to the development of this type of storm. A midlatitude cyclone develops when a small disturbance produces a wavelike kink in a previously linear front. A similar mechanism initiates a

tropical cyclone. In late summer, the Sun warms tropical air. The rising hot air creates a belt of low pressure that encircles the globe over the tropics. In addition, many local low-pressure disturbances move across the tropical oceans at this time of year. If a local disturbance intersects the global tropical low, it creates a bulge in the isobars. Winds are deflected by the bulge and, directed by the Coriolis effect, begin to spiral inward. Warm, moist air rises from the low. Water vapor condenses from the rising air, and the latent heat warms the air further, which causes even more air to rise. As the low pressure becomes more intense, strong surface winds blow inward to replace the rising air. This surface air also rises, and more condensation and precipitation occur. But the

● **FIGURE 19.28** This image shows Hurricane Katrina making landfall near New Orleans on August 28, 2005.

additional condensation releases more heat, which continues to add energy to the storm.

The center of the storm is a region of vertical airflow, called the *eye*. In the outer, and larger, part of the eye, the air that has been rushing inward spirals upward. In the inner eye, air sinks. Thus, the horizontal wind speed in the eye is reduced to near zero (●**FIGURE 19.29**). Survivors who have been in the eye of a hurricane report an eerie calm. Rain stops, and the Sun may even shine weakly through scattered clouds. But this is only a momentary reprieve. A typical eye is only 20 kilometers in diameter, and after it passes, the hurricane rages again in full intensity.

---

storm surge Abnormally high coastal waters and flooding, created by a combination of strong onshore winds and the low atmospheric pressure of a storm that raises the sea surface by several meters.

Thus, a hurricane is powered by a classic feedback mechanism: the low-pressure storm causes condensation; condensation releases heat; heat powers the continued low pressure. The entire storm is pushed by prevailing winds, and its path is deflected by the Coriolis effect. A hurricane usually dissipates only after it reaches land or passes over colder water because the supply of moist, warm air is cut off. Condensing water vapor in a single tropical cyclone releases as much latent heat energy as that produced by all the electric generators in the United States in a six-month period.

The numerical "category" of tropical cyclones is based on a rating scheme called the *Saffir–Simpson scale*, after its developers (**TABLE 19.3**). This scale, commonly mentioned in weather reports, rates the damage potential of a hurricane or other tropical storm, and typical values of atmospheric pressure, wind speed, and height of storm surge associated with storms of increasing intensity.

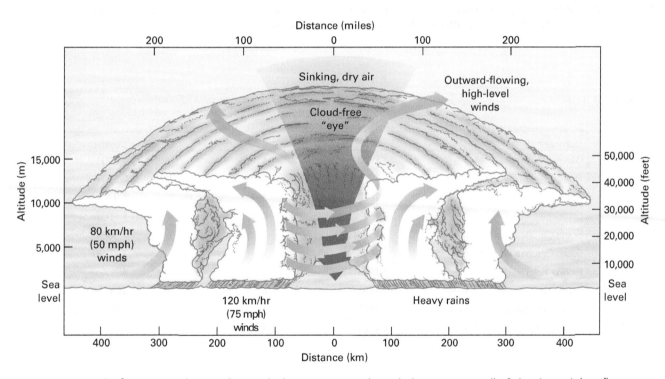

● **FIGURE 19.29** Surface air spirals inward toward a hurricane, rises through the towering wall of clouds, and then flows outward above the storm. Falling air near the storm's center creates the eerie calm in the eye of the hurricane.

TABLE 19.3 The Saffir–Simpson Hurricane Damage Potential Scale

Type	Category	Damage	Pressure (millibars)	Winds (km/h)	Storm surge (m)
Depression				>56	
Tropical storm				63 to 117	
Hurricane	1	Minimal	980	119 to 152	1.2 to 1.5
Hurricane	2	Moderate	965 to 979	154 to 179	1.8 to 2.4
Hurricane	3	Extensive	945 to 964	179 to 209	2.7 to 3.7
Hurricane	4	Extreme	920 to 944	211 to 249	4.0 to 5.5
Hurricane	5	Catastrophic	<920	>249	>5.5

HURRICANE KATRINA

Scope of the Disaster

Hurricane Katrina formed in the South Atlantic on August 23, 2005, made landfall as a category 1 hurricane just north of Miami, then crossed the Florida peninsula and grew to a category 5 storm in the Gulf of Mexico with sustained winds of 280 kilometers per hour. By the time it made a second landfall on the Mississippi Delta, Hurricane Katrina had diminished to a category 3–4 storm, with winds hovering around 200 kilometers per hour. Yet despite the reduced winds, Katrina quickly became the most costly and one of the deadliest natural disasters in U.S. history. Parts of New Orleans were flooded to the rooftops of houses. In addition, a record 8.5-meter storm surge inundated the New Orleans and Mississippi coastlines, flattening and flooding homes over 233,000 square kilometers, an area almost as large as the United Kingdom. Approximately 1.3 million people were driven from their homes. The official death toll was 1,836 with more than 700 people still unaccounted for. The direct structural damage was estimated to cost the United States as much as $110 billion. Of course, indirect costs such as lost wages and disruption of human life are much higher still, running into the hundreds of billions of dollars.

Brief History of Gulf Hurricanes

Hurricanes have ravaged the southeast United States numerous times in the past and will certainly do so again. But while hurricanes are not preventable, hurricane deaths and costs result from a combination of factors, including the severity of the hurricane, building practices, and human responses. For example, from **TABLE 19.4** we see that prior to Hurricane Katrina, the deadliest U.S. hurricanes all occurred before 1957. However, the costliest storms all occurred after 1955. With the exception of Katrina, none of the 10 deadliest storms are on the list of costliest storms, and vice versa. Why is there so little correlation between structural damage (cost) and death toll?

TABLE 19.4 The Deadliest and Costliest Storms in U.S. History

Deadliest Storms				
Rank	Hurricane	Year	Category	Deaths*
1	Galveston, Texas	1900	4	8,000+
2	Florida	1928	4	2,500+
3	Katrina	2005	3–4	1,836+
4	Louisiana	1893	4	1,100+
5	South Carolina and Georgia	1893	3	1,000+
6	South Carolina and Georgia	1891	2	700
7	Florida Keys	1935	5	408
8	Louisiana	1856	4	400
9	Texas and Louisiana (Audrey)	1957	4	390
10	Florida, Alabama, and Mississippi	1926	4	372

Costliest Storms				
Rank	Hurricane	Year	Category	Damage (U.S.)[†]
1	Katrina (SE FL, LA, MS)	2005	3	$160,000,000,000
2	Harvey (TX, LA)	2017	4	$125,000,000,000
3	Maria (PR, USVI)	2017	4	90,000,000,000
4	Sandy (Mid-Atlantic & NE US)	2012	1	70,200,000,000
5	Irma (FL)	2017	4	50,000,000,000
6	Andrew (SE FL/LA)	1992	5	47,790,000,000
7	Ike (TX, LA)	2008	2	34,800,000,000
8	Ivan (AL/NW FL)	2004	3	27,060,000,000
9	Wilma (S FL)	2005	3	24,320,000,000
10	Rita (SW LA, N TX)	2005	3	23,680,000,000

*Does not include offshore deaths.
[†]2017 dollars adjusted for inflation.

In the last half-century, structural damage has been high for three main reasons: (1) more Americans lived near the Atlantic and Gulf coasts in the late 1990s and early 2000s than they had previously; (2) homes and other structures have had an increased inflation-adjusted value; and (3) Americans now own more things than ever before and, consequently, the dollar value of their possessions is at an all-time high.

But while structures are immobile, people can evacuate if given ample warning. The death toll has been low in the past half-century because accurate forecasting warns people of impending hurricanes. The deadliest hurricane in the United States struck Galveston, Texas, in September 1900. More than 8,000 people died because the population was caught unaware. A tropical cyclone has a sharp boundary, and even a few hundred kilometers outside that boundary, fluffy white clouds may be floating in a blue sky. Today, satellites track hurricanes and news reports give people ample time to evacuate.

EL NIÑO

Hurricanes are common in the southeastern United States and on the Gulf Coast, but they rarely strike California. Consequently, Californians were taken by surprise in late September 1997, when Hurricane Nora ravaged Baja California and then, somewhat diminished by landfall, struck San Diego and Los Angeles. The storm brought the first rain to Los Angeles after a record 219 days of drought, then spread eastward to flood parts of Arizona, where it caused the evacuation of 1,000 people.

Other parts of the world also experienced unusual weather during the autumn of 1997. In Indonesia and Malaysia, fall monsoon rains normally douse fires intentionally set in late summer to clear the rainforest. The rains were delayed for two months in 1997, and as a result the fires raged out of control, filling cities with such dense smoke that visibility at times was no more than a few meters. Even an airliner crash was attributed to the smoke. Severe drought in nearby Australia caused ranchers to slaughter entire herds of cattle for lack of water and feed. At the same time, far fewer hurricanes than usual threatened Florida and the U.S. Gulf Coast. Floods soaked northern Chile's Atacama Desert, a region that commonly receives no rain at all for a decade at a time, while record snowfalls blanketed the Andes and heavy rains caused floods in Peru and Ecuador.

All of these weather anomalies have been attributed to **El Niño**, an ocean current that brings unusually warm water to the west coast of South America. But the current does not flow every year; instead, it occurs about every 3 to 7 years,

---

El Niño An episodic weather pattern occurring every 3 to 7 years in which the trade winds slacken in the Pacific Ocean and warm water accumulates off the coast of South America and causes unusual rains and heavy snowfall in the Andes.

and its effects last for about a year before conditions return to normal. Although meteorologists paid little attention to the phenomenon until the El Niño year of 1982–1983, many now think that El Niño affects weather patterns for nearly three-quarters of Earth.

Meteorologists first began recording El Niños in 1982–1983, although they were known to Peruvian fishermen much earlier because they warm coastal waters and diminish fish harvests. The fishermen called the warming events *El Niño* because they commonly occur around Christmas, the birthday of El Niño—the Christ Child.

To understand the El Niño effects, first consider interactions between southern Pacific sea currents and weather in a normal, non–El Niño year (●**FIGURE 19.30**). Normally in fall and winter, strong trade winds blow westward from South America across the Pacific Ocean. The winds drag the warm, tropical surface water away from Peru and Chile and pile it up in the western Pacific near Indonesia and Australia. In the western Pacific, the warm water forms a low mound thousands of kilometers across. The water is up to 10°C warmer and as much as 60 centimeters higher than the surface of the ocean near Peru and Chile.

As the wind-driven surface water flows away from the South American coast, cold, nutrient-rich water rises from the depths to replace the surface water. The nutrients support a thriving fishing industry along the coasts of Peru and Chile.

Abundant moisture evaporates from the surface of a warm ocean. In a normal year, much of this water condenses to bring rain to Australia, Indonesia, and other lands in the southwestern Pacific, which are adjacent to the mound of warm water. On the eastern side of the Pacific, the cold, upwelling ocean currents cool the air above the coasts of Peru and northern Chile. This cool air becomes warmer as it flows over land. The warming lowers the relative humidity and creates the coastal Atacama Desert.

In an El Niño year, for reasons poorly understood by meteorologists, the trade winds slacken (●**FIGURE 19.31**). The mound of warm water near Indonesia and Australia then flows downslope—eastward across the Pacific Ocean toward Peru and Chile. The anomalous accumulation of warm water off South America causes unusual rains in normally dry coastal regions, and heavy snowfall in the Andes. At the same time, the cooler water near Indonesia, Australia, and nearby regions causes drought. The mass of warm water that caused the 2014–2016 El Nino was one of the strongest El Nino weather disturbances in history, causing widespread flooding and drought.

El Niño has global effects that go far beyond regional rainfall patterns. For example, El Niño deflects the jet stream from its normal path as it flows over North America, directing one branch northward over Canada and the other across southern California and Arizona. Consequently, those regions receive more winter precipitation and storms than usual, while fewer storms and warmer winter temperatures affect the Pacific Northwest, the northern plains, the Ohio River valley, the mid-Atlantic states, and New England. Southern Africa

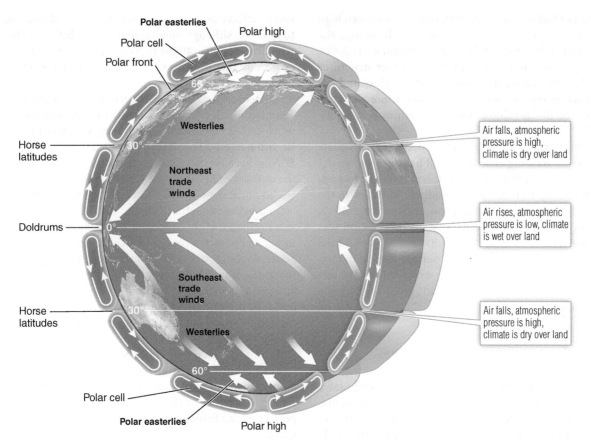

● **FIGURE 19.30** In a normal year, trade winds drag warm surface water westward across the Pacific and pile it up in a low mound near Indonesia and Australia, where the warm water causes rain. The surface flow creates upwelling of cold, deep, nutrient-rich waters along the coast of South America.

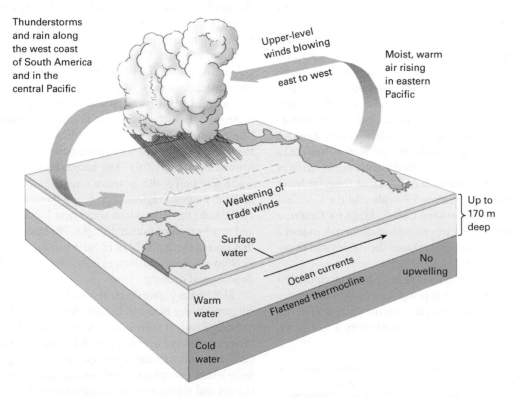

● **FIGURE 19.31** In an El Niño year, the trade winds slacken and the warm water flows eastward toward South America, causing the storms and rain to move over South America, and diminishing the upwelling currents.

experiences drought, while Ecuador, Peru, Chile, southern Brazil, and Argentina receive more rain than usual.

Altered weather patterns created by El Niño wreak havoc. Globally, 2,000 deaths and more than $13 billion in damage are attributed to the 1982–1983 El Niño effects. In the United States alone, more than 160 deaths and $2 billion in damage occurred, mostly from storm and flood damage. In southern Africa, economic losses of $1 billion and uncounted deaths due to disease and starvation have been attributed to the 1982–1983 El Niño.

Key Concepts Review

- Relative humidity is the amount of water vapor in air compared to the amount the air could hold at that temperature. When relative humidity reaches 100 percent, the air is saturated. Condensation occurs when saturated air cools below its dew point.
- Three atmospheric processes cool air to its dew point and cause condensation: (1) radiation, (2) contact with a cool surface, and (3) adiabatic cooling of rising air. Radiation and contact cooling cause the formation of dew, frost, and some types of fog. However, clouds and precipitation normally form as a result of the cooling that occurs when air rises.
- Three mechanisms cause air to rise: orographic lifting, frontal wedging, and convection–convergence. Warm, moist air is said to be unstable because it rises rapidly, forming towering clouds and heavy rainfall. Warm, dry air is said to be stable because it doesn't rise rapidly, doesn't ascend to high elevations, and doesn't lead to cloud formation and precipitation.
- A cloud is a concentration of water droplets or ice crystals in air. The three fundamental types of clouds are cirrus, stratus, and cumulus. Precipitation occurs when the water droplets or ice crystals in a cloud coalesce until they become large enough to fall.
- Fog is a cloud that forms at or very close to ground level, although most fog forms by processes different from those that create higher level clouds. Types of fog include advection fog, radiation fog, evaporation fog, and upslope fog.
- When air is heated, it expands and rises, creating low pressure. Cool air sinks, exerting a downward force and forming high pressure. Uneven heating of Earth's surface causes pressure differences, which, in turn, cause wind. Wind speed is determined by the pressure gradient.
- A front is the boundary between a warmer air mass and a cooler one. When two air masses collide, the warmer air rises along the front, forming clouds and often precipitation.
- Air cools adiabatically when it rises over a mountain range, often causing precipitation. Sea breezes and monsoons arise because ocean temperature changes slowly in response to daily and seasonal changes in solar radiation, whereas land temperature changes quickly. In the case of forests, a feedback mechanism operates: forests cool the air, cool air promotes rainfall, and rainfall supports forests.
- A thunderstorm is a small, short-lived storm from a cumulonimbus cloud. Lightning occurs when charged particles separate within the cloud.
- A tornado is a small, short-lived, funnel-shaped storm that protrudes from the bottom of a cumulonimbus cloud. A tropical cyclone (or tropical storm) is a larger, longer-lived storm that forms over warm oceans and is powered by the energy released when water vapor condenses to form clouds and rain.
- Death and property loss during Hurricane Katrina were caused both by the intensity of the storm and by a variety of human factors, including human-caused erosion of the delta and levee failure.
- El Niño is a weather pattern caused by shifting motion of ocean currents every 3 to 7 years.

Important Terms

absolute humidity (p. 450)

adiabatic temperature changes (p.452)

advection fog (p. 457)

air mass (p. 461)

altostratus (p. 455)

anticyclone (p. 460)

cirrus (p. 455)

cold front (p. 462)

cumulonimbus (p. 455)

cumulus (p. 455)

cyclone (p. 460)

dew (p. 451)

dew point (p. 450)

dry adiabatic lapse rate (p. 452)

El Niño (p. 473)

evaporation fog (p. 458)

front (p. 461)

frontal wedging (p. 453)

frost (p. 451)

glaze (p. 457)

hail (p. 457)

humidity (p. 450)

hurricane (p. 469)

isobars (p. 458)

jet streams (p. 460)

monsoon (p. 467)

nimbostratus (p. 455)

normal lapse rate (p. 453)

occluded front (p. 463)

orographic lifting (p. 453)

pressure gradient (p. 458)

radiation fog (p. 457)

relative humidity (p. 450)

saturation (p. 450)

sleet (p. 457)

stable air (p. 455)

stationary front (p. 465)

storm surge (p. 470)

storm tracks (p. 466)

stratocumulus (p. 455)

stratus (p. 455)

supercooling (p. 451)

supersaturation (p. 451)

tornado (p. 469)

tropical cyclone or tropical storm (p. 469)

typhoon (p. 469)

unstable air (p. 455)

upslope fog (p. 458)

warm front (p. 462)

wet adiabatic lapse rate (p. 452)

wind (p. 458)

20

CLIMATE

Summer in the Smoky Mountains, Tennessee. The name of the mountain range comes from the Cherokee words for 'land of blue smoke'. The 'smoke' is formed by scattering of blue and purple light by volatile organic compounds - various organic molecules released by the abundant vegetation into the humid air. The light scattering and 'smoky' appearance is most pronounced around sunrise and sunset.

LEARNING OBJECTIVES

LO1   Distinguish between weather and climate.

LO2   Describe how the movement of the jet stream affects the weather.

LO3   Diagram the three cell model of global winds.

LO4   Explain the migration of global winds.

LO5   Discuss the Koeppen Climate Classification system criteria.

LO6   Contrast the major climate zones of the Koeppen Climate Classification system based upon their climatographs.

LO7   Discuss the effect of both vertical and horizontal ocean currents on temperature.

LO8   Describe major climate zones of the Koeppen Climate Classification system.

LO9   Explain the urban heat island effect.

INTRODUCTION

If you are planning a picnic for next Sunday, you hope for bright sunshine, but at the same time, you understand that your plans may be spoiled by rain. Next Sunday's temperature and precipitation are determined by daily fluctuations of weather. But weather varies only within a relatively narrow range. You are certain that it is not going to snow on the Fourth of July in Texas, and that if you live in Montana, you are not going to bask outside in shorts and a T-shirt in January.

Over the time span of generations or centuries, climate is stable enough so that we plan our lives around it. Farmers in Kansas plant wheat and never attempt to raise bananas. In winter, hotel owners in Florida prepare for an influx of tourists escaping the northern winter. As a result, climate strongly influences many aspects of our lives: our outdoor recreation, the houses we live in, and the clothes we wear. Humans tend to migrate toward warm, sunny regions. In the United States, the populations of California and Texas have increased rapidly in the past generation, while North Dakota and Montana have seen little growth.

Yet, such stability is only guaranteed over relatively short periods of geologic time. Dinosaur bones and coal deposits indicate that parts of Antarctica were once covered by warm, humid swamps. Desert dunes lie buried beneath the fertile wheat fields of Colorado. When geologists study the climate record in a specific region, they must separate the effects of tectonic movement from global climate change. For example, the coal deposits in Antarctica do not tell us that Earth was once warm enough for coal-producing swamps to grow near the South Pole. Instead, other evidence indicates that the continent of Antarctica was once close to the equator, as explained in our study of tectonics. Yet, even when the effects of tectonic plate movements are factored out, it is clear that global climate has changed dramatically and often abruptly over the long spans of Earth's history. In this

chapter, we will examine today's global climate. Climate is regulated by many mechanisms that have already been discussed, including latitude (Chapter 18), wind (Chapters 18 and 19), oceans and ocean currents (Chapters 16 and 18), altitude (Chapter 17), and albedo (Chapter 18). We discuss global wind systems in more detail below.

GLOBAL WINDS AND CLIMATE

The Sun is the ultimate energy source for winds and evaporation. The Sun shines most directly at or near the equator and warms the air near Earth's surface. The warm air gathers moisture from the equatorial oceans. The warm, moist, rising air forms a vast region of low pressure near the equator, with little horizontal airflow. As the rising air cools adiabatically, the water vapor condenses and falls as rain. Therefore, local squalls and thunderstorms are common, but steady winds are rare. This hot, still region was a serious barrier in the age of sailing ships. Mariners called the equatorial region the **doldrums** (from the Middle English for "dull"), and the old sailing literature is filled with stories recounting the despair and hardship of being becalmed on the vast, windless seas. On land, the frequent rains near the equatorial low-pressure zone nurture lush tropical rainforests.

The air rising at the equator splits to flow north and south at high altitudes. However, these high-altitude winds do not continue to flow due north and south, but rather are deflected by the Coriolis effect. Thus, their poleward movement is interrupted.

In both the Northern Hemisphere and Southern Hemisphere, this air veers until it flows due east at about 30 degrees north and south latitudes (●**FIGURE 20.1**). The air then cools enough to sink to the surface, creating subtropical high-pressure zones at 30 degrees north and south latitudes. The sinking air warms adiabatically, absorbing water and forming clear, blue skies. At the center of the high-pressure area, the air moves vertically and not horizontally, and therefore few steady surface winds blow. This calm, high-pressure belt circling the globe is called the **horse latitudes**. The region was so named because sailing ships were becalmed and horses transported as cargo often died of thirst and hunger. The warm, dry, descending air in this high-pressure zone forms many of the world's great deserts, including the Sahara in northern Africa, the Kalahari in southern Africa, and the Australian interior desert.

Descending air at the horse latitudes splits and flows over Earth's surface in two directions: toward the equator and toward the poles. The surface winds moving toward the equator are deflected by the Coriolis effect, so they blow from the northeast in the Northern Hemisphere and from the southeast in the Southern Hemisphere. Sailors depended on these reliable winds and called them the **trade winds**. The winds moving toward the poles are also deflected by the Coriolis effect, forming the **prevailing westerlies**. They flow from the southwest in the Northern Hemisphere and from the northwest in the Southern Hemisphere (●**FIGURE 20.1A**).

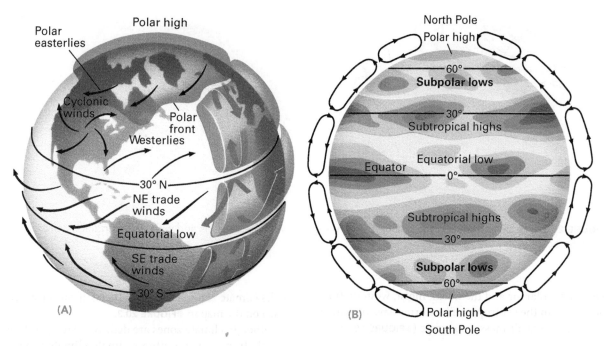

● **FIGURE 20.1** Global wind patterns predicted by the three-cell model. (A) Air rising at the equator moves poleward at high elevations, falls at about 30 degrees north and south latitudes, and returns to the equator, forming the trade winds. The orange arrows show both upper-level and surface wind patterns. The black arrows show only surface winds. (B) High- and low-pressure belts are indicated on the sphere, with surface and upper-level wind patterns shown on the edges.

The poles are cold year-round. The cold polar air sinks, creating yet another band of high pressure. The sinking air flows over the surface toward lower latitudes. In the Northern Hemisphere, these surface winds are deflected by

the Coriolis effect to form the **polar easterlies**. The polar easterlies and prevailing westerlies converge at about 60 degrees latitude. Warm air rises at the convergence, forming a low-pressure boundary zone called the **polar front**.

These global wind patterns follow the **three-cell model**, which depicts three convection cells in each hemisphere. The cells are bordered by alternating bands of high and low pressure (●**FIGURE 20.1B**). In the three-cell model, global winds are generated by heat-driven convection currents, and then their direction is altered by Earth's rotation (Coriolis effect).

Because the rising and falling air of global wind systems is dependent on temperature, these boundaries migrate north and south with the seasons. They are also distorted by surface topography and local air movement. For example, in the Northern Hemisphere, cyclones and anticyclones develop along the polar front as explained in Chapter 19. These storms bring alternating rain and sunshine, conditions that are favorable for agriculture. Thus, the great wheat belts of the United States, Canada, and Russia all lie between 30 and 60 degrees north latitude.

Recall from Chapter 19 that a jet stream is a narrow band of fast-moving, high-altitude air. Jet streams form at boundaries between Earth's climate cells as high-altitude air is deflected by the Coriolis effect. The **subtropical jet stream** flows between the trade winds and the westerlies, and the **polar jet stream** forms along the polar front. When you watch a weather forecast on TV, the meteorologist may show the movement and direction of the polar jet stream as it snakes across North America. Storms commonly occur along this line because the jet stream marks the boundary

---

doldrums A vast low-pressure region of Earth near the equator with hot, humid air and where local squalls and rainstorms are common but steady winds are rare.

horse latitudes A calm, high-pressure region of Earth lying at about 30 degrees north and south latitudes, in which generally dry conditions prevail and steady winds are rare.

trade winds The winds that blow steadily toward the equator from the northeast in the Northern Hemisphere and from the southeast in the Southern Hemisphere, between 5 and 30 degrees north and south latitudes.

prevailing westerlies The winds that blow steadily toward the poles from the southwest in the Northern Hemisphere and from the northwest in the Southern Hemisphere, between 30 and 60 degrees north and south latitudes.

polar easterlies Persistent polar surface winds in the Northern Hemisphere that flow from east to west.

polar front The low-pressure boundary at about 60 degrees latitude formed by warm air rising at the convergence of the polar easterlies and prevailing westerlies.

three-cell model A model of global wind patterns that depicts three convection cells in each hemisphere, bordered by alternating bands of high and low pressure.

subtropical jet stream A jet stream that flows between the trade winds and the westerlies.

polar jet stream A jet stream that flows along the polar front.

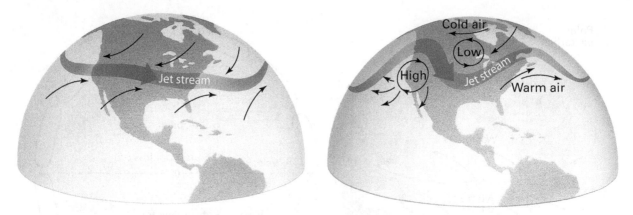

● **FIGURE 20.2** The polar front and the polar jet stream migrate with the seasons and with local conditions. Storms commonly occur along the jet stream.

between cold, polar air and the warm, moist, westerly flow that originates in the subtropics. The storms develop where the two contrasting air masses converge (●**FIGURE 20.2**).

CLIMATE ZONES OF EARTH

Earth's major climate zones are classified primarily by temperature and precipitation. But an area with both wet and dry seasons has a different climate from one with moderate rainfall all year long, even though the two areas may have identical total annual precipitation. Therefore, climatic zones are also classified on the basis of seasonal variations in temperature and precipitation.

The **Koeppen Climate Classification**, used by climatologists throughout the world, is illustrated in **TABLE 20.1**. Subclassifications of these groups are shown in **TABLE 20.2**, and

Earth's climate zones based on this classification system are shown on the map in ●**FIGURE 20.3**.

Although climate zones are defined by temperature and precipitation, you can often estimate climate types from a photograph of an area. Visual classification is possible because specific plant communities grow in specific climates. For instance, cacti grow in the desert, and trees grow where moisture is more abundant. A **biome** is a community of plants living in a large geographic area characterized by a particular climate.

The climate in any location is summarized by a *climograph* that records annual and seasonal temperature and precipitation. ●**FIGURE 20.4** is a model climograph for Nashville, Tennessee.

Next we will explore the principal zones of the Koeppen Climate Classification system.

Climate Type	Name	Description
TABLE 20.1 Koeppen Climate Classification		
A	Humid tropical	In A climates, every month is warm with a mean temperature over 18°C (64°F). The temperature difference between day and night is greater than the difference between December and June averages. There is enough moisture to support abundant plant communities.
B	Arid	B climates have a chronic water deficiency; in most months evaporation exceeds precipitation. Temperatures vary according to latitude: some B climates are hot while others are frigid.
C	Humid mesothermal	C climates occur in midlatitudes, with distinct winter and summer seasons and enough moisture to support abundant plant communities. The winters are mild. Snow may fall but snow cover does not persist, with the average temperature in the coldest month above 3°C.
D	Humid microthermal	D climates are similar to C climates, with distinct summer and winter seasons, but D climates are colder. Winters are more severe, with persistent winter snow cover and an average temperature in the coldest month below –3°C.
E	Polar	In E climates, winters are extremely cold and even the summers are cool, with the average temperature in the warmest month below 10°C (50°F).
F	Highlands	F climates, found at latitudes worldwide, are determined by elevation above Earth's surface. Temperature changes with altitude, ranging from about –18°C to +10°C. Precipitation tends to decrease with altitude; windward sides of mountains usually receive more precipitation than leeward sides.

TABLE 20.2 World Climates: Subtypes of the Koeppen Climate Classification

	Climate Type and Subtype	Description	Vegetation
A	Humid tropical	No winter	
	• Tropical rainforest	Abundant year-round rainfall	Rainforest
	• Tropical monsoon	Wet, with a short dry season	Rainforest
	• Tropical savanna	Summer rains, winter dry season	Grassland savanna
B	Arid	Dry; evaporation greater than precipitation	
	• Tropical steppe	Semiarid	Steppe grassland
	• Midlatitude steppe	Cool and dry; continental interior or rain-shadow location	Steppe grassland
	• Tropical desert	Very warm	Desert
	• Midlatitude desert	Cool and dry; continental interior or rain-shadow location	Desert
C	Humid mesothermal	Midlatitude, mild winter	
	• Humid subtropical	Warm, wet summer with winter cyclonic storms	Forest
	• Mediterranean	Dry summer and wet winter with cyclonic storms; subtropical	Shrubby plants and oak savanna
	• Marine west coast	Cool summer and mild winter; maritime, westerlies, and cyclonic storms	Forest to temperate rainforest
D	Humid microthermal	Midlatitude; severe winter	
	• Humid continental, hot summer	Midlatitude continental; warm summer, cold winter	Prairie and forest
	• Humid continental, mild summer	High-midlatitude continental; cool summer, cold winter	Prairie and forest
	• Subarctic	High-latitude continental	Forest taiga
E	Polar	Little warmth, even in summer	
	• Tundra	High latitude; short summer, harsh and long winter	Mosses, grasses, and flowers; some bushes and even a few small trees in protected areas on southern boundary of tundra
	• Ice cap	High latitude; short summer, harsh and long winter	No vegetation
F	Highlands	Cool to cold; found in mountain or high plateau areas; varies according to elevation	Changes as elevation increases

Koeppen Climate Classification A climate classification system describing Earth's principal climate zones, used by climatologists throughout the world.

biome A community of plants growing in a large geographic area characterized by a particular climate.

tropical rainforests A forest growing in a tropical climate with abundant, year-round rainfall; characterized by tall trees that branch only near the top, creating a dense canopy of leaves that block out most of the light, and with soggy ground and water dripping everywhere.

Humid Tropical Climates: No Winter

TROPICAL CLIMATES WITH ABUNDANT YEAR-ROUND RAINFALL The large, low-pressure zone near the equator causes abundant rainfall, often exceeding 400 centimeters per year, which supports **tropical rainforests** (●FIGURE 20.5). The dominant plants in a tropical rainforest are tall trees with slender trunks. These trees branch only near the top, covering the forest with a dense canopy of leaves. The canopy blocks out most of the light, so as little

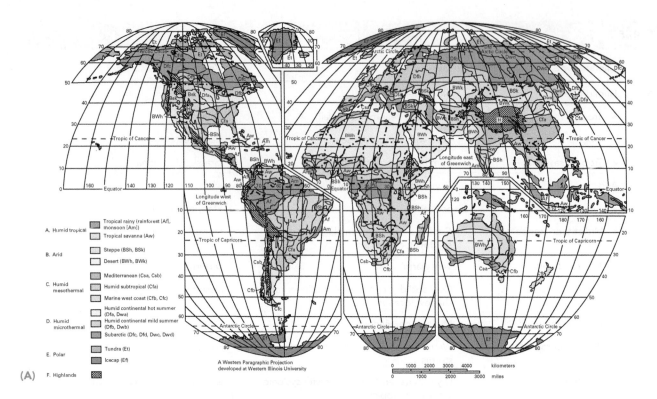

A. Humid tropical

Tropical rainy (rainforest [Af], monsoon [Am])

Tropical savanna (Aw)

B. Arid

Steppe (BSh, BSk)

Desert (BWh, BWk)

C. Humid mesothermal

Mediterranean (Csa, Csb)

Humid subtropical (Cfa)

Marine west coast (Cfb, Cfc)

D. Humid microthermal

Humid continental hot summer (Dfa, Dwa)

Humid continental mild summer (Dfb, Dwb)

Subarctic (Dfc, Dfd, Dwc, Dwd)

E. Polar

Tundra (Et)

Icecap (Ef)

(A)   **F. Highlands**

A Western Paragraphic Projection developed at Western Illinois University

0 1000 2000 3000 4000   kilometers
0 1000 2000 3000   miles

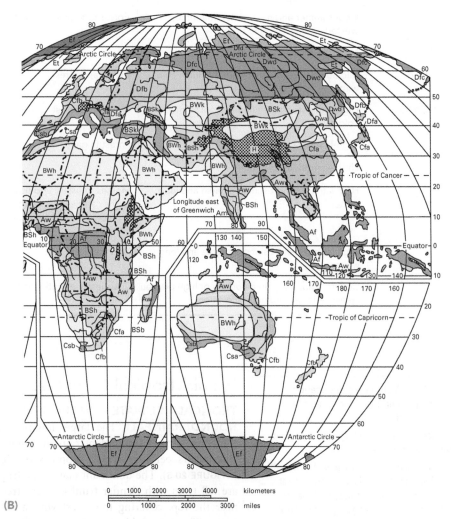

(B)

0 1000 2000 3000 4000   kilometers
0 1000 2000 3000   miles

● **FIGURE 20.3** (A) Global climate zones. Each climate zone supports a unique biome. (B) Global climate zones. Each climate zone supports a unique biome.

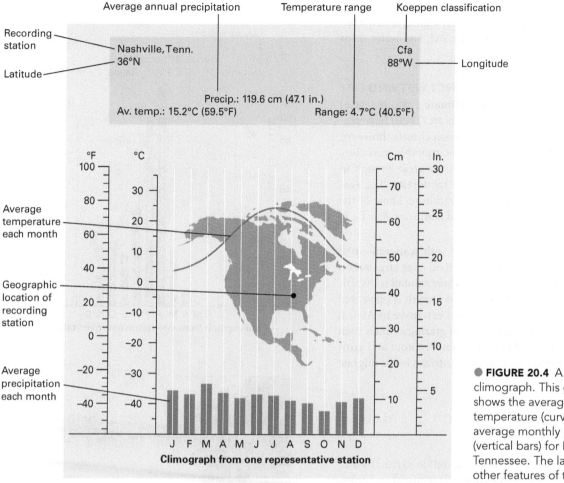

Recording station

Latitude

Average annual precipitation

Temperature range

Koeppen classification

Nashville, Tenn.
36°N

Cfa
88°W

Longitude

Precip.: 119.6 cm (47.1 in.)
Av. temp.: 15.2°C (59.5°F)   Range: 4.7°C (40.5°F)

Average temperature each month

Geographic location of recording station

Average precipitation each month

Climograph from one representative station

● **FIGURE 20.4** A model climograph. This graph shows the average monthly temperature (curved line) and average monthly precipitation (vertical bars) for Nashville, Tennessee. The labels highlight other features of the climograph.

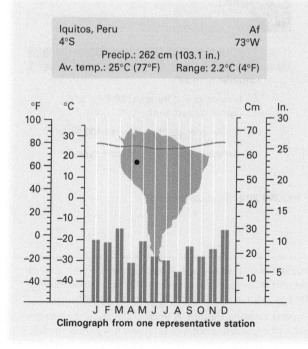

Iquitos, Peru   Af
4°S   73°W
Precip.: 262 cm (103.1 in.)
Av. temp.: 25°C (77°F)   Range: 2.2°C (4°F)

Climograph from one representative station

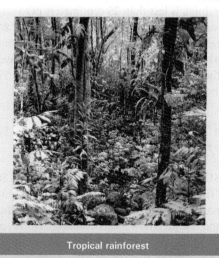

Tropical rainforest

Latitude: 0° to 15°

Av. temperature difference: 23°C to 28°C (winter to summer)

Av. annual precip.: 200 cm/yr. to 500 cm/yr.

General statistics for climate type

● **FIGURE 20.5** Tropical rainforest.

as 0.1 percent of the sunlight reaches the forest floor, which consequently has relatively few plants. The ground in a tropical forest is soggy, the tree trunks are wet, and water drips everywhere.

TROPICAL CLIMATES WITH DISTINCT WET AND DRY SEASONS A **tropical monsoon climate** (●**FIGURE 20.6**) and **tropical savanna climate** (●**FIGURE 20.7**) both have seasonal variations in rainfall. The monsoon climate, however, has greater total precipitation, greater monthly variation, and a shorter dry season. Precipitation is great enough in tropical monsoon biomes to support rainforests. The seasonal precipitation is also ideal for agriculture. Some of the great rice-growing regions in India and Southeast Asia lie in tropical monsoon climates.

A tropical savanna is a grassland with scattered small trees and shrubs. Such grasslands extend over large areas, often in the interiors of continents, where rainfall is insufficient to support forests or where forest growth is prevented by recurrent fires. Savannas are most extensive in Africa, where they support a rich collection of grazing animals such as zebras, wildebeest, and gazelles. Grasses sprout and grow with the seasonal rains, and the great African herds migrate with the foliage.

Dry Climates: Evaporation Greater Than Precipitation

In dry zones where the annual precipitation varies from 25 to 35 centimeters per year, the climate is semiarid and grassland **steppes** predominate. The great steppe grasslands of Central Asia fall into these categories.

If the rainfall is less than 25 centimeters per year, deserts form and support only sparse vegetation (●**FIGURE 20.8**). The world's largest deserts lie along the high-pressure zones of the 30 degrees latitude, although rain-shadow and coastal deserts exist in other latitudes. Thus, some deserts are torridly hot, while Arctic deserts are frigid for much of the year.

Humid Midlatitude Climates with Mild Winters

HUMID SUBTROPICS The southeastern United States has a **humid subtropical climate** (●**FIGURE 20.9**). During the summer, conditions can be as hot and humid as in the tropics, and rain and thundershowers are common. However, during the winter, arctic air pushes southward, forming cyclonic storms. Although the average monthly temperature seldom falls below 7°C, cold fronts occasionally bring frost and snow. Precipitation is relatively constant year-round due to convection-driven thunderstorms in summer and cyclonic storms in winter. This zone supports both trees with needles (conifers) and trees with broad leaves (deciduous) as well as valuable crops such as vegetables, cotton, tobacco, and citrus fruits.

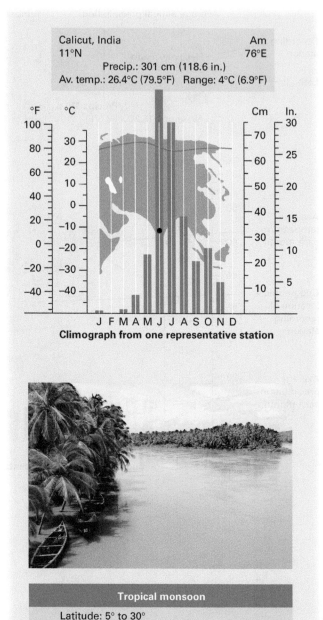

● **FIGURE 20.6** Tropical monsoon.

tropical monsoon climate A tropical climate with distinct wet and dry seasons, characterized by high rainfall during the wet months and a relatively short dry season.

tropical savanna climate A tropical climate with distinct wet and dry seasons but with relatively low annual rainfall, characterized by large areas of grassland supporting herds of grazing animals.

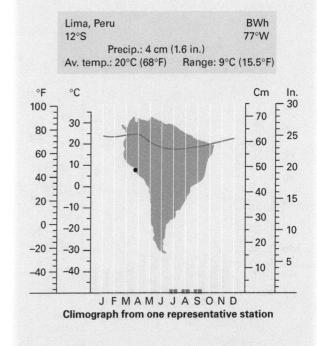

Kano, Nigeria   Aw
12°N   8°E
            Precip.: 86.5 cm (34 in.)
Av. temp.: 26.7°C (80°F)   Range: 9.5°C (17°F)

Climograph from one representative station

Tropical savanna

Latitude: 5° to 25°

Av. temperature difference: 20°C to 30°C
(winter to summer)

Av. annual precip.: 100 cm/yr. to 180 cm/yr.

General statistics for climate type

● **FIGURE 20.7** Tropical savanna.

Lima, Peru   BWh
12°S   77°W
            Precip.: 4 cm (1.6 in.)
Av. temp.: 20°C (68°F)   Range: 9°C (15.5°F)

Climograph from one representative station

Desert

Latitude: Variable

Av. temperature difference: Variable
(winter to summer)

Av. annual precip.: Less than 25 cm/yr.

General statistics for climate type

● **FIGURE 20.8** Desert.

Graeme Shannon/Shutterstock.com

Copyright and Photograph by Dr. Parvinder S. Sethi

steppes A vast, semiarid, grass-covered plain found in dry climates such as those in southeast Europe, central Asia, and parts of central North America.

humid subtropical climate A midlatitude climate characterized by hot, humid summers but cooler winters and rainfall that falls throughout the year.

Mediterranean climate A midlatitude climate characterized by dry summers, rainy winters, and moderate temperatures.

MEDITERRANEAN The **Mediterranean climate** is characterized by dry summers, rainy winters, and moderate temperature (●**FIGURE 20.10**). These conditions occur on the west coasts of all continents between latitudes 30 and 40 degrees. In summer, the subtropical high migrates to higher latitudes, producing near-desert conditions with clear skies as much as 90 percent of the time. In winter, the prevailing westerlies bring warm, moist air from the ocean, leading to fog and rain.

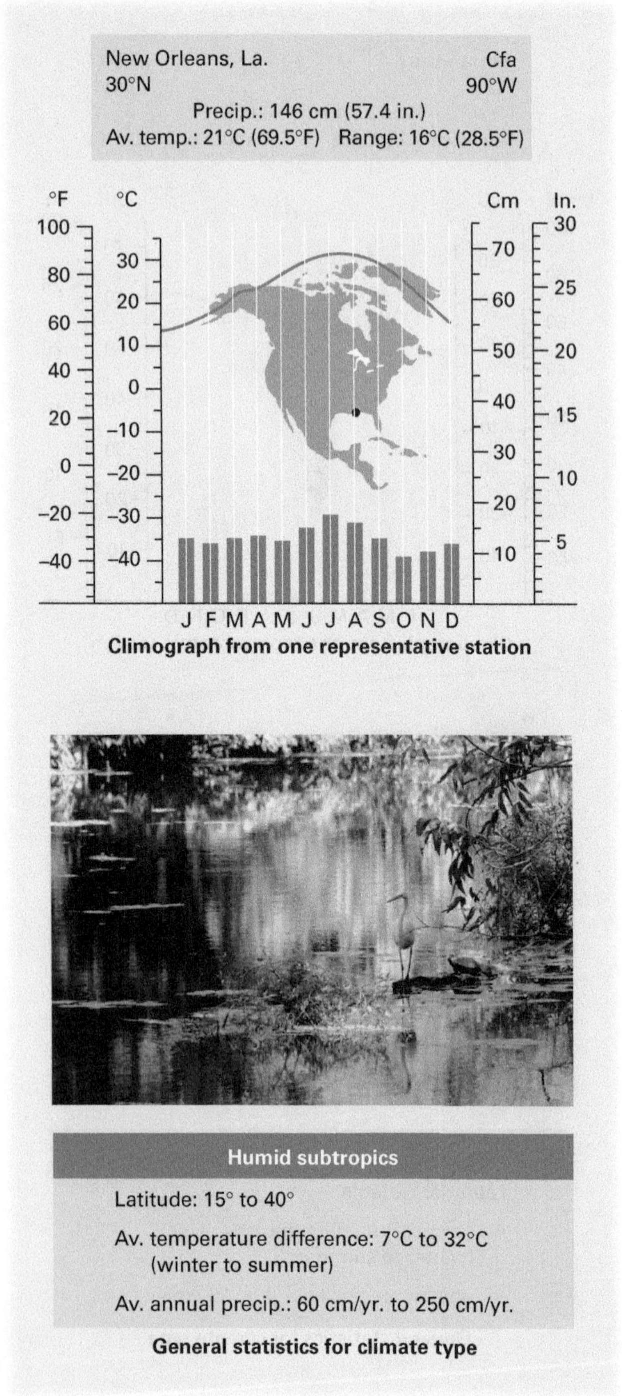

New Orleans, La.
Cfa
30°N
90°W
Precip.: 146 cm (57.4 in.)
Av. temp.: 21°C (69.5°F) Range: 16°C (28.5°F)
°F
°C
Cm
In.
J F M A M J J A S O N D
Climograph from one representative station
Humid subtropics
Latitude: 15° to 40°
Av. temperature difference: 7°C to 32°C (winter to summer)
Av. annual precip.: 60 cm/yr. to 250 cm/yr.
General statistics for climate type

Dustie/Shutterstock.com

● **FIGURE 20.9** Humid subtropics.

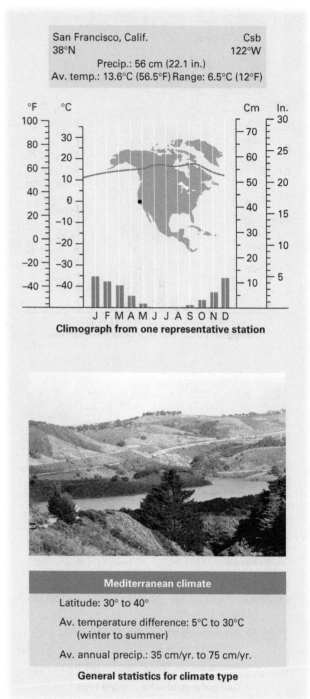

San Francisco, Calif.
Csb
38°N
122°W
Precip.: 56 cm (22.1 in.)
Av. temp.: 13.6°C (56.5°F) Range: 6.5°C (12°F)
°F
°C
Cm
In.
J F M A M J J A S O N D
Climograph from one representative station
Mediterranean climate
Latitude: 30° to 40°
Av. temperature difference: 5°C to 30°C (winter to summer)
Av. annual precip.: 35 cm/yr. to 75 cm/yr.
General statistics for climate type

Copyright and Photograph by Dr. Parvinder S. Sethi

● **FIGURE 20.10** Mediterranean climate.

Thus, 75 percent or more of the annual rainfall occurs in winter. Although redwoods, the largest trees on Earth, grow in specific environments in central California, the summer heat and drought of Mediterranean climates generally retard the growth of large trees. Instead, shrubs and scattered trees dominate. Fires occur frequently during the dry summers and spread rapidly through the dense shrubbery. During the dry wildfire season of 2018, California experienced both the highest number of wildfires in a single year (7,571) and the most destructive, burning an estimated 1.7 million acres, an area slightly bigger than the state of Delaware. The biggest single fire in California's history - the Camp Fire which killed 88 people - also occurred in 2018. If vegetation is destroyed by fire, landslides often occur when the winter rains return. Torrential winter rains frequently bring extensive flooding and landslides to southern California.

http://cdfdata.fire.ca.gov/incidents/incidents_stats?year=2018

MARINE WEST COAST **Marine west coast climate** zones border the Mediterranean zones and extend poleward to 65 degrees (●**FIGURE 20.11**). They are severely influenced by ocean currents that moderate temperature and bring abundant precipitation. Thus, summers are cool and winters warm, and the temperature difference between the seasons is small. For example, average monthly temperatures vary by only 15.5°C in Portland, Oregon. In contrast, Eau Claire, Wisconsin, which is a continental city at the same latitude, experiences a 31.5°C annual temperature range. Seattle and other northwestern coastal cities experience rain and drizzle for days at a time, especially during the winter when the warm, moist, maritime air from the Pacific flows first over cool currents close to shore and then over cool land surfaces. The total rainfall varies from moderate, 50 centimeters per year (20 inches per year), to wet, 250 centimeters per year (100 inches per year). The wettest climates occur where mountains interrupt the maritime air. **Temperate rainforests** grow where rainfall is greater than 100 centimeters per year and is constant throughout the year. Temperate rainforests are common along the northwest coast of North America, from Oregon to Alaska.

Humid Midlatitude Climate with Severe Winters

Continental interiors in the midlatitudes are characterized by hot summers and cold winters, giving rise to **humid continental climates** (●FIGURE 20.12). Thus, in the northern Great Plains the temperature can drop to −40°C in winter and soar to 38°C in summer. Even in a given season, the temperature may vary greatly as the polar front moves northward or southward. For example, in winter the northern continental United States may experience arctic cold one day and rain a few days later. If rainfall is sufficient, this climate supports abundant coniferous forests, whereas grasslands dominate the drier regions. Millions of bison once roamed the continental grasslands of North America, and today wheat and other grains grow from horizon to horizon. The northernmost portion of this climate zone is the **subarctic**, which supports the **taiga** biome, a forest of conifers that can survive extremely cold winters.

---

marine west coast climate A midlatitude climate characterized by relatively mild but wet winters and cool summers, with little temperature difference between seasons.

temperate rainforests A rainforest that grows in marine west coast climates where rainfall is more than 100 centimeters per year and is relatively constant throughout the year, particularly along the northwest coast of North America from Oregon to Alaska.

humid continental climate A midlatitude climate characterized by hot summers, cold winters, and precipitation throughout the year.

subarctic The northernmost edge of the humid continental climate zone, lying just south of the arctic.

taiga A biome of conifers that can survive the extreme winters of the subarctic.

tundra A biome dominated by low-lying grasses, mosses, flowers, and a few small bushes that grow in the Arctic and high alpine regions.

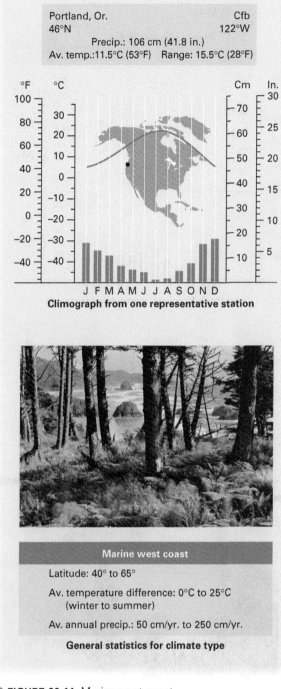

Marine west coast

Latitude: 40° to 65°

Av. temperature difference: 0°C to 25°C (winter to summer)

Av. annual precip.: 50 cm/yr. to 250 cm/yr.

General statistics for climate type

● **FIGURE 20.11** Marine west coast.

Polar Climate

In the Arctic and Antarctic, winters are harsh and long, and the temperature remains above freezing only during a short summer. The climate is classified as a *polar climate*. Trees cannot survive, and low-lying plants such as mosses, grasses, flowers, and a few small bushes cover the land. This biome is called **tundra** (●FIGURE 20.13).

Duluth, Minn.   Dfb
46°N   92°W
Precip.: 76 cm (29.9 in.)
Av. temp.: 4°C (39°F)   Range: 31°C (56°F)

Climograph from one representative station

Midlatitude with severe winters

Latitude: 40° to 65°

Av. temperature difference: −15°C to 25°C (winter to summer)

Av. annual precip.: 50 cm/yr. to 150 cm/yr.

General statistics for climate type

● **FIGURE 20.12** Midlatitude with severe winters.

https://www.un.org/en/development/desa/population/publications/pdf/urbanization/the_worlds_cities_in_2016_data_booklet.pdf

URBAN CLIMATES

If you ride a bicycle from the center of a city toward the countryside, you may notice that the air gradually feels cooler and more refreshing as you leave the city streets and enter the green fields or hills of the outlying area. This feeling is not entirely psychological; the climate of a city is measurably different from that of the surrounding rural regions (**TABLE 20.3**).

As shown in ●**FIGURE 20.14**, the average winter minimum temperature inside the beltway of Washington, D.C., is more than 3°C warmer than in outlying areas. This temperature difference, called the **urban heat island effect**, is caused by numerous factors:

- Stone and concrete buildings and asphalt roadways absorb solar radiation and reradiate it as infrared heat.
- Cities are warmer because little surface water exists and, as a result, little evaporative cooling occurs. In contrast, in the countryside, water collects in the soil and evaporates for days after a storm. Roots draw water from deeper in the soil, and this water evaporates from leaf surfaces.
- Urban environments are warmed by the heat released when fuels are burned. In New York City in winter, the combined heat output of all the vehicles, buildings, factories, and electrical generators is 2.5 times the solar energy reaching the ground.
- Tall buildings block winds that might otherwise disperse the warm air.
- Air pollutants absorb long-wave radiation (heat rays) emitted from the ground and produce a local greenhouse effect.

As warm air rises over a city, a local low-pressure zone develops, and rainfall is generally greater over the city than in the surrounding areas (●**FIGURE 20.15**). Water condenses on dust particles, which are abundant in polluted urban air. Weather systems collide with the city buildings and linger, much as they do on the windward side of mountains. Thus, a front that might pass quickly over rural farmland remains longer over a city and releases more precipitation.

In 1600, less than 1 percent of the global population lived in cities. By 1950, 30 percent of the world's population was urban, and by 2016, that number had grown to 55 percent. By 2030, roughly 60% of Earth's human population is projected to live in urban areas. Therefore, although urban climate change may not affect global climate, it affects the lives of the many people living in cities.

urban heat island effect Warmer temperatures in a city compared to the surrounding countryside, caused by factors in the urban environment itself.

Courtesy of Graham R. Thompson/Jonathan Turk

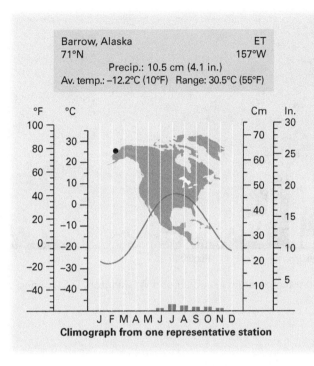

Barrow, Alaska   ET
71°N   157°W
Precip.: 10.5 cm (4.1 in.)
Av. temp.: −12.2°C (10°F)   Range: 30.5°C (55°F)

Climograph from one representative station

Arctic tundra

Latitude: 65° to 80°

Av. temperature difference: −35°C to +10°C
(winter to summer)

Av. annual precip.: generally low, 10 cm/yr.
to 35 cm/yr.

General statistics for climate type

Ekaterina Baranova/Shutterstock.com

● **FIGURE 20.13** Arctic tundra.

TABLE 20.3 Average Changes in Climatic Elements Caused by Urbanization

Weather Element	Comparison with Rural Environment
Cloudiness:	
Cloud cover	5% to 10% more
Fog, winter	100% more
Fog, summer	30% more
Precipitation, total	5% to 10% more
Relative humidity:	
Winter	2% lower
Summer	8% lower
Radiation:	
Total	15% to 20% less
Direct sunshine	5% to 15% less
Temperature:	
Annual mean	0.5°C to 1.0°C higher
Winter minimum (average)	1.0°C to 3.0°C higher
Wind speed:	
Annual mean	20% to 30% lower
Extreme gusts	10% to 20% lower
Calms	5% to 20% higher

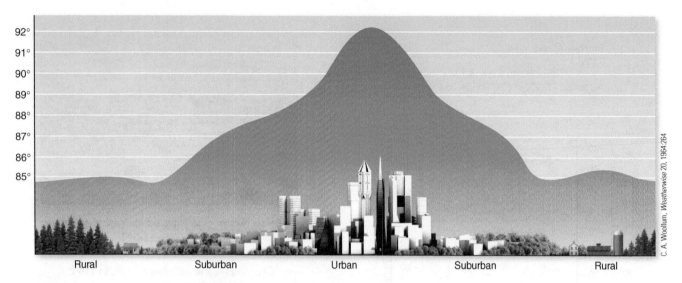

● **FIGURE 20.14** The urban heat island effect. The average minimum temperature in and around Washington, D.C., during the winter; contours represent relative T minima in degrees Celsius.

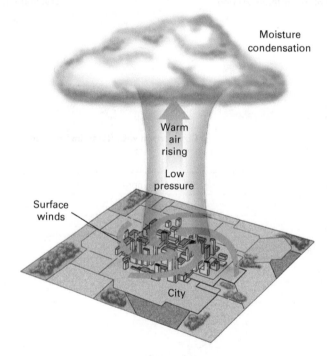

● **FIGURE 20.15** Warm air rising over a city creates a low-pressure zone. As a result, precipitation is greater in the city than over the surrounding countryside.

Key Concepts Review

- The three-cell model of global wind circulation shows three cells of global airflow bordered by alternating bands of high and low pressure. Warm air rises near the equator, forming a low-pressure region called the doldrums. This air flows north and south at high altitude until it

cools and falls at 30 degrees latitude and forms a high-pressure region called the horse latitudes, where most of the world's largest deserts prevail. The falling air splits. A portion flows back toward the equator along the surface, forming the trade winds and completing the tropical cell.

- The remaining air that falls at 30 degrees north and south latitudes flows poleward along the surface. In the Northern Hemisphere, this air is deflected to form the prevailing westerlies. Air rises at a low-pressure region near 60 degrees latitude and returns at high altitude to complete the cell. A second region of high pressure exists over the poles, where the descending air spreads outward to form the polar easterlies. The polar front forms where the polar easterlies and the prevailing westerlies converge. The jet stream blows at high altitude along the polar front and along the boundaries between cells at the horse latitudes.

- Climate zones are classified according to annual temperature, annual precipitation, and variability in either of these factors from month to month or season to season. World climate types include humid tropical, arid, humid mesothermal, humid microthermal, polar, and highland climates.

- Urban areas are generally warmer and wetter than surrounding countryside.

Important Terms

biome (p. 480)

doldrums (p. 478)

horse latitudes (p. 478)

humid continental climate (p. 487)

humid subtropical climate (p. 484)

Koeppen Climate Classification (p. 480)

marine west coast climate (p. 486)

Mediterranean climate (p. 485)

polar easterlies (p. 479)

polar front (p. 479)

polar jet stream (p. 479)

prevailing westerlies (p. 478)

steppes (p. 484)

subarctic (p. 487)

subtropical jet stream (p. 479)

taiga (p. 487)

temperate rainforests (p. 487)

three-cell model (p. 479)

trade winds (p. 478)

tropical monsoon climate (p. 484)

tropical rainforests (p. 481)

tropical savanna climate (p. 484)

tundra (p. 487)

urban heat island effect (p. 488)

Review Questions

1. Discuss the factors that control the average temperature in any region.

2. Discuss how oceans affect coastal climate.

3. Explain the three-cell model for global wind circulation.

4. Describe the trade winds. Why are they so predictable?

5. Why is the doldrums region relatively calm and rainy? Why are the horse latitudes calm and dry?

6. Describe the polar front and the jet stream. How do they affect weather?

7. Explain how horizontal and vertical ocean currents affect temperature.

8. Discuss and describe the major climate zones of the Koeppen Climate Classification and the plant communities associated with each.

9. Explain how urban climate differs from that of the surrounding countryside.

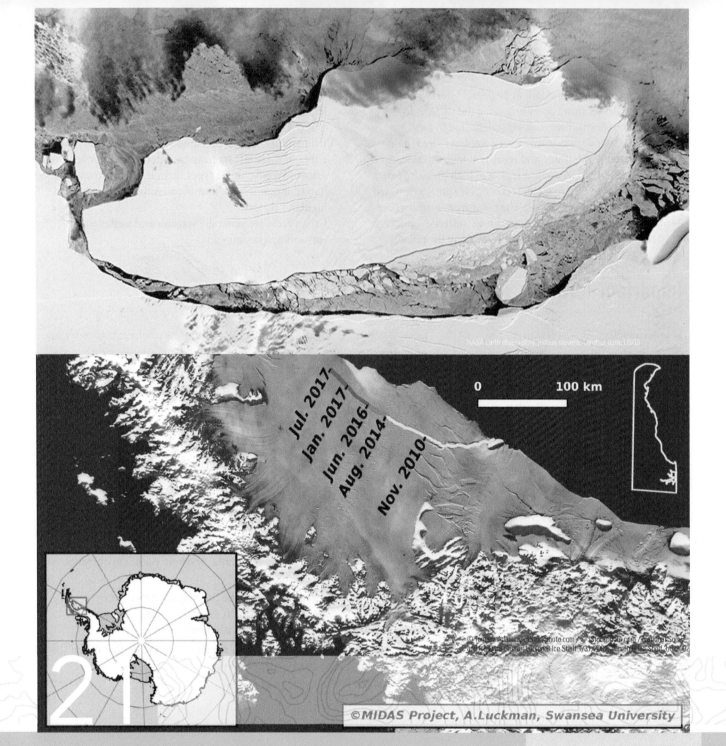

NASA Earth observatory, Joshua Stevens, Landsat data, USGS

Jul. 2017.
Jan. 2017.
Jun. 2016.
Aug. 2014.
Nov. 2010.

0   100 km

© Tomasz Adamczyk/iStockPhoto.com • © iStockphoto.com / National Snow and Ice Data Center, Larson B Ice Shelf 1/31/2002, Larsen B Ice Shelf 3/5/2002

©MIDAS Project, A.Luckman, Swansea University

CLIMATE CHANGE

The Larsen C Ice Shelf rims a small portion of the eastern coastline of the Antarctic Peninsula. Between 2010 and 2017, roughly 5,800 square kilometers of ice, about the size of the state of Delaware, broke off the ice shelf and floated northward.

INTRODUCTION

In March 2000, an 11,000-square-kilometer iceberg as large as the state of Connecticut broke free from the Ross Ice Shelf in Antarctica and drifted north, where it melted in warmer water. Two months later, a massive chunk of ice cracked off the nearby Ronne Ice Shelf. In September, a second Connecticut-sized chunk of the Ross Ice Shelf disintegrated. Then in January of 2002, a Rhode Island–sized section of the Larsen Ice Shelf splintered into millions of small fragments. More recently, a Delaware-sized chunk of the Larson Ice Shelf—12% of the entire Ice Shelf by area—detached after developing a crack that propagated for over 100 km over the course of at least seven years. This huge mass of ice is now drifting northward as seen in the satellite image to the left. Ultimately, it will break apart. Since the height of the last ice age, 18,000 years ago, the Ross Ice Shelf has receded 700 kilometers, shedding 5.3 million cubic kilometers of ice. Although the Ross Ice Shelf has been slowly melting for thousands of years, the disintegration of the Ice Shelf is accelerating as the recent large calving events demonstrate.

In order to have a balanced perspective on what is happening today, we must first study historical climate change. The geological record in almost every locality on Earth provides evidence that past regional climates were different from modern climates. Geologists have discovered sand dunes beneath prairie grasslands near Denver, Colorado, indicating that this semiarid region was recently desert. Moraines on Long Island, New York, tell us that this temperate region was once glaciated. Fossil ferns in nearby Connecticut indicate that, before the glaciers, the northeastern United States was warm and wet.

Many of these regional climatic fluctuations resulted from global climate change. Thus, 18,000 years ago, Earth was cooler than it is today and glaciers descended to lower latitudes and altitudes. During Mississippian time, from 360 to 325 million years ago, Earth was warmer than it is today. Vegetation grew abundantly, and some of it collected in huge swamps to form coal.

During the past 10,000 years, global climate has been mild and stable as compared with the preceding 100,000 years. During this time, humans have developed from widely separated bands of hunter-gatherers to agrarian farmers, and then moved into crowded communities in huge industrial megalopolises. Today, with a global population of more than 6 billion, people have stressed the food-producing capabilities of the planet. If temperature or rainfall patterns were to change, even slightly, crop failures could lead to famines. In addition, cities and farmlands on low-lying coasts could be flooded under rising sea level. As a result, climate change may be one of the most important issues that we face in the 21st century.

CLIMATE CHANGE IN EARTH'S HISTORY

Recall from Chapter 17 that Earth's primordial atmosphere contained high concentrations of carbon dioxide (CO_2) and

LEARNING OBJECTIVES

LO1 Discuss changes in the size of the ice shelves in terms of global temperature change.

LO2 Summarize the changes in climate from 100,000 to 10,000 years ago.

LO3 Describe the effect of a stable climate on the development of civilization.

LO4 Discuss at least four methods used to determine the climate on Earth prior to written temperature records.

LO5 Explain the three basic processes that affect the atmospheric temperature of the Earth.

LO6 Discuss the relationship between the hydrologic cycle and global climate.

LO7 Analyze the relationship between carbon and global climate.

LO8 Describe the position of the continents and the effects that the positions of continents have on climate.

LO9 Examine the connection between tectonics, sea-level, volcanoes, weathering, and global climate.

LO10 Explain the greenhouse effect.

LO11 Describe the consequences of global warming.

water vapor (H_2O). Both greenhouse gases absorb infrared radiation in the atmosphere. Astronomers have calculated that the Sun was 20 to 30 percent fainter early in Earth's history than it is today, yet oceans did not freeze. The high concentrations of atmospheric carbon dioxide and water vapor retained enough of the Sun's diminished radiation to warm Earth's atmosphere and surface to temperatures that kept the oceans liquid. Luckily for us, the concentration of carbon dioxide and water in the atmosphere declined gradually as the Sun warmed.

●**FIGURE 21.1A** are graphs of estimated mean global temperature throughout Earth's history. Note, for example, that the planet plunged into a deep freeze on at least five occasions. There is evidence that the cold periods 600 to 700 million years ago were so extreme that the oceans are thought to have frozen from pole to pole, accompanied by near-total land coverage of glaciers, leading to a "Snowball Earth," described in Chapter 13. In contrast, our planet was relatively warm for 248 million years, from the start of the Mesozoic Era almost to the present. The last 2 million years have witnessed the most recent ice age. According to many climatologists, the current period is an interglacial warming episode and the ice sheets are likely to return in the geologically near future.

Many additional climate changes occurred during Earth's history, but they were too short in duration to be apparent on the graph. For example, ice core records from Greenland, Antarctica, and alpine glaciers show that between 110,000 and 10,000 years ago the mean annual global temperature changed frequently and dramatically. The blue segment of the global temperature estimates in

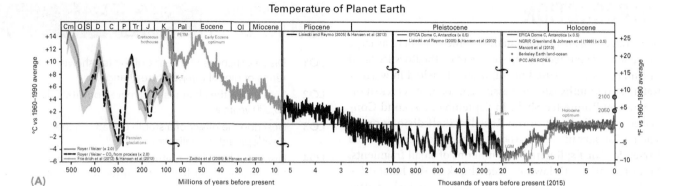

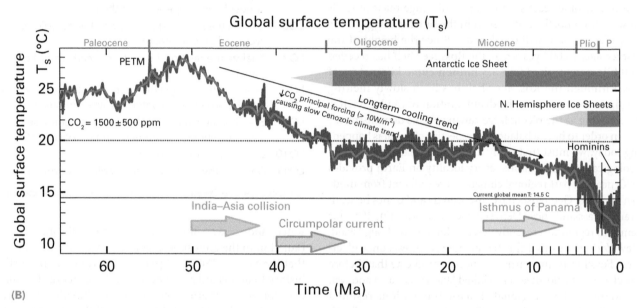

● **FIGURE 21.1** Estimates of global surface temperature over the past 500 million years. (A) The upper curve shows the entire record for 500 million years in four separate panels each with a different colored segment of the overall record. Notice that each panel has a separate time scale on the x-axis and that progressing from left to right the number of years represented gets smaller (i.e., the resolution of the sea-level record increases). (B) The lower curve shows global surface temperature for the past 65 million years (Mesozoic-Cenozoic boundary) using a consistent time scale for the entire curve. Shown also are key tectonic and oceanographic events at the bottom and along with the growth of both polar ice sheets.

Source information and graphs: James Hansen publications. Slide assembly and annotation: Root Routledge, *Deep Time: The Story of Mother Earth.* http://alpineanalytics.com/Climate/DeepTime.html

●**FIGURE 21.1** provides the closest look at portion of the temperature data, corresponding to the most recent one million years. The graph shows that the Earth underwent around a dozen rapid swings in surface temperature of around 10°C and that most of the swings include very steep segments during warming episodes, suggesting that when Earth's surface temperature warmed up, it did so very quickly. Many of the cold intervals shown by the blue curve in Figure 21.1 persisted for 1,000 years or more. As mentioned earlier, the past 10,000 years, during which civilization developed, has witnessed anomalously stable climate.

●**FIGURE 21.2** gives us an even more detailed look at a 128-year span from 1880 to 2008. Note that the temperature

rose slowly from 1880 to 1970 and then increased dramatically. During the past century, people burned large quantities of fuel, thereby injecting carbon dioxide into the atmosphere. The correlation between carbon dioxide emission and global temperature implies—but does not prove—that human activities have caused the current global warming. The temperature rise has been a little less than 0.8°C; temperature changes much larger than this have occurred repeatedly throughout Earth's history. For the past 100 years, meteorologists have used instruments to measure temperature, precipitation, wind speed, and humidity. But how do we interpret prehistoric climates? ●**FIGURE 21.3** reviews several techniques for determining past climate, as represented in Figure 21.1.

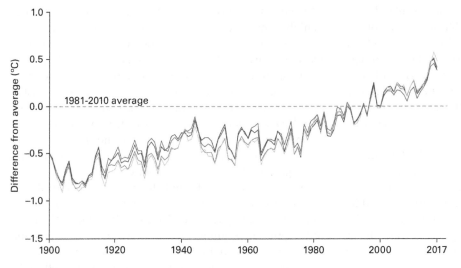

● **FIGURE 21.2** This graph shows the annual global surface temperature from 1900–2017 compared to the average temperature for the period 1981–2010 (dashed line). The different colored curves represent analysis of the historical temperature record by independent research groups.

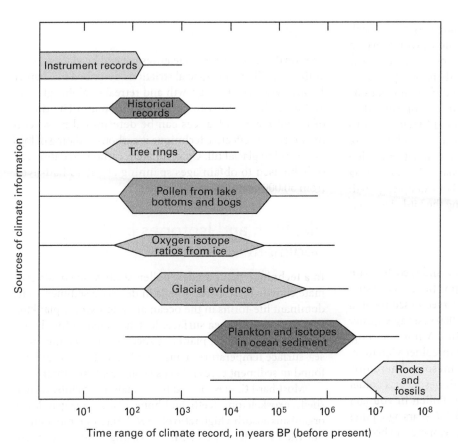

● **FIGURE 21.3** Several methods, each with its own useful time range, allow scientists to determine historical and ancient climates.

Source: Adapted from Cambridge University Press, from T. Webb III, J. Kutzbach, and F. A. Street-Perrott, in *Global Change*, Eds. T. F. Malone and J. D. Roederer (Cambridge, UK: Cambridge University Press, 1985), 212–218.

MEASURING CLIMATE CHANGE

This chapter is devoted to understanding and interpreting the 128-year warming trend shown in Figure 21.2.

Historical Records

Historians search for written records or archeological data that chronicle climate change. In 985 CE, the Viking explorer Eric the Red sailed to southwest Greenland with a few hundred immigrants. They established two colonies, from which they exported butter and cheese to Iceland and Europe, and the population flourished. Some Vikings sailed farther west, colonized the Labrador Coast of North America, and visited Ellesmere Island, near 80 degrees north latitude. Then within 300 to 400 years, the colonies vanished. Sagas tell of heavy sea ice in summer, crop failures, starvation, and death.

At approximately the same time that the Greenland colonies vanished, European glaciers descended into lowland valleys. This period of global cooling, called the *Little Ice Age*, lasted from about 1450 to 1850 and is documented by old landscape paintings and writings that depict a glacial advance between the 15th and 19th centuries. Other historical and archeological evidence chronicles climate changes at different times and places.

Tree Rings

Growth rings in trees also record climatic variations. Each year, a tree's growth is recorded as a new layer of wood called a *tree ring*. Trees grow slowly during a cool, dry year and more quickly in a warm, wet year; therefore, tree rings grow wider during favorable years than during unfavorable ones. Paleoclimatologists date ancient logs preserved in ice, permafrost, or glacial till by carbon-14 techniques to determine when trees died. They then count and measure the rings to reconstruct the history of past climate recorded in the wood. Interpretations of climate change from tree-ring data coincide well with historical data. For example, growth rings are narrow in trees that lived during the Little Ice Age.

Plant Pollen

Plant pollen is widely distributed by wind and is coated with a hard, waxy cover that resists decomposition. As a result, pollen grains are abundant and well preserved in sediment in lake bottoms and bogs. For example, 11,000 years ago, spruce was the most abundant tree species in a Minnesota bog. In modern forests, spruce dominates in colder Canadian climates but is less abundant in Minnesota. Therefore, scientists deduce that the climate in Minnesota was colder 11,000 years ago than it is at present. Pollen in younger layers of sediment shows that about 10,500 years ago, pines displaced the spruce, indicating that the temperature became warmer.

Oxygen Isotope Ratios in Glacial Ice

Oxygen consists mainly of two isotopes, abundant ^{16}O and rare ^{18}O. Both isotopes are incorporated into water, H_2O. Water molecules containing ^{16}O are lighter and evaporate more easily than those containing ^{18}O. At high temperatures, however, evaporating water vapor contains a higher proportion of ^{18}O than it does at lower temperatures.

Therefore, the ratio of $^{18}O/^{16}O$ in vapor from warm water is higher than that in vapor from cool water. Some of the water vapor condenses as snow, which accumulates in glaciers. Thus, the $^{18}O/^{16}O$ ratios in glacial ice reflect water temperature at the time the water evaporated. Because most of the atmospheric water vapor that falls as snow originated from evaporation of ocean water, scientists then use the $^{18}O/^{16}O$ data from glacial ice to estimate mean ocean surface temperatures. Because the sea surface and the atmosphere are in close contact, mean ocean surface temperature reflects mean global atmospheric temperature.

Geologists have drilled deep into Greenland and Antarctic glaciers, where the ice is up to 110,000 years old, and have carefully removed ice cores. The age of the ice at any depth is determined by counting annual ice deposition layers or by carbon-14 dating of windblown pollen within the glacier. The oxygen isotope ratios in each layer reflect the air temperature at the time the snow fell. The temperature data in Figure 21.2 were obtained from Greenland ice cores.

Glacial Evidence

Erosional and depositional features created by glaciers, such as the tills, tillites, and glacial striations described in Chapter 13, are evidence of the growth and retreat of alpine glaciers and ice sheets, which in turn reflect climate. The timing of recent glacial advances can be determined by several methods. One effective technique is carbon-14 dating of logs preserved in glacial till. Unfortunately, carbon-14 dating can only be used to obtain ages spanning about 10 half-lives, from 50,000 years ago to the present.

Plankton and Isotopes in Ocean Sediment

In a technique that parallels pollen studies, scientists estimate climate by studying fossils in deep-sea sediment. The dominant life-forms in the ocean are microscopic plankton that float near the sea surface. Just as pollen ratios change with air temperature, plankton species ratios change with sea-surface temperature. Thus, fossil plankton assemblages found in sediment cores reflect sea-surface temperature.

Most hard tissues formed by animals and plants, such as shells, exoskeletons, teeth, and bone, contain oxygen. Many organisms absorb a high ratio of $^{18}O/^{16}O$ at low temperatures, but the ratio decreases with increasing temperature. For example, consider foraminifera, tiny marine organisms. During a time of Pleistocene cooling and glacial growth, their shells contain an average of 2 percent more ^{18}O than similar shells formed during a warm interglacial interval. Thus, just as scientists estimate paleoclimate by measuring oxygen isotope ratios in glacial ice, they can estimate ancient climate by measuring oxygen isotope ratios in fossil corals, plankton, teeth, and the remains of other organisms. Oxygen is also incorporated into soil minerals, so isotope ratios in soil and seafloor sediment also reflect paleoclimate.

Courtesy of Graham R. Thompson/Jonathan Turk

● **FIGURE 21.4** A fossil fern indicates that a region was wet and warm at the time the fern grew.

The Rock and Fossil Record

Fossils are abundant in many sedimentary rocks of Cambrian age and younger. Geologists can approximate climate in ancient ecosystems by comparing fossils with modern relatives of the ancient organisms (●FIGURE 21.4). For example, modern coral reefs grow only in tropical water. Therefore, we infer that fossil reefs also formed in the tropics. Coal deposits and ferns, like the one in Figure 21.4, formed in moist tropical environments; cacti indicate that the region was once desert.

Looking backward even farther—into the Proterozoic Eon, before life became abundant—it is difficult to measure climate with fossils. Thus, geologists search for clues in rocks. Tillite is a sedimentary rock formed from glacial debris and thus indicates a cold climate. Lithified dunes formed in deserts or along coasts.

Sedimentary rocks form in water, so their existence tells us that the temperature was above freezing and below boiling. Carbonate rocks precipitate from carbon dioxide dissolved in seawater. Geochemists know the chemical conditions under which carbon dioxide dissolves and precipitates, so they can calculate a range of atmospheric and oceanic compositions and temperatures that would have produced limestone and other carbonate rocks. Most thick limestones formed in warm, shallow seas. Some ancient mineral deposits, such as the banded iron deposits described in Chapter 17, also reflect the chemistry of the ancient atmosphere.

ASTRONOMICAL CAUSES OF CLIMATE CHANGE

After scientists learned that climates have changed, they began to search further to understand how climates change. Despite its immense complexity, Earth's atmospheric temperature is determined by three basic processes:

- *The heat of the Sun.* Solar output fluctuates, so in the absence of mitigating factors, when the Sun becomes hotter, Earth warms. When the Sun cools, so does our planet.

- *Albedo.* If the incoming solar radiation is absorbed at Earth's surface, the planet becomes warmer; if heat is reflected back into space, the planet becomes cooler. Therefore, the reflectivity, or albedo, of Earth's surface is a critical factor in determining the planet's temperature.

- *Heat retention by the atmosphere.* Heat that is reflected back into space may be absorbed by gases in the atmosphere, in a process called the *greenhouse effect.* Since different gases retain heat differently, atmospheric composition is the third basic factor that regulates our planet's atmospheric temperature.

These three factors determine the amount of heat available to drive the planet's climate and weather engines. Multitudinous feedback loops and threshold effects—involving atmosphere, hydrosphere, geosphere, and biosphere—then perturb the basic system to produce the resultant climate.

Changes in Solar Radiation

Recall from Chapter 13 that variations in Earth's orbit and rotational pattern may have caused the climate fluctuations responsible for the glacial advances and retreats of the Pleistocene Ice Age. Other astronomical factors may also cause climate change.

A star the size of our Sun produces energy by hydrogen fusion for about 10 billion years. During this time, its energy output increases slowly. When the Earth first formed, the Sun produced only 70 percent of the energy that it produces today. If sunlight were the only factor that influenced temperature, our planet would have been much cooler and the earth would have been covered with ice. However, sedimentary rocks formed in this early earth, indicating that water was present. The Earth's primordial atmosphere contained abundant methane, which is a potent heat-trapping greenhouse gas that warmed the early Earth. As the sun heated up, the concentration of methane declined and the composition of the atmosphere changed. Ultimately, processes leading to warming balanced those leading to cooling, and the Earth's surface remained mostly at temperatures where water in its liquid form was abundant.

Within the past few hundred million years, solar output has changed by only one fifty-millionth of 1 percent per century, and therefore the variation had no measurable influence on climate change over thousands, to even millions, of years.

Bolide Impacts

Evidence strongly suggests that a bolide crashed to Earth about 65 million years ago. The impact blasted enough rock and dust into the sky to block out sunlight and cool the planet. According to one current hypothesis, this cooling led to the extinction of the dinosaurs. Other bolide impacts may have caused rapid and catastrophic climate changes throughout Earth's history.

WATER AND CLIMATE

Water is abundant in all four of Earth's spheres. It occurs in rocks and soil of the geosphere, comprises over 90 percent of the living organisms of the biosphere, constitutes essentially the entire hydrosphere, and exists in the atmosphere as vapor, liquid droplets, and solid crystals of snow and ice. Moreover, water moves freely from one system to another. As a result, "water acts as the Venetian blind of our planet, as its central heating system, and as its refrigerator, all at the same time."[1] We have already discussed many of the components of the complex, and often conflicting, relationship between water and climate. As a brief review:

- Water vapor is the most abundant greenhouse gas. It warms the atmosphere and Earth's surface.
- Clouds reflect sunlight and therefore cool the atmosphere and Earth's surface. But clouds also absorb heat radiating from Earth's surface. Thus, water in the atmosphere causes both warming and cooling.
- Glaciers and snowfields have a high albedo (80 to 90 percent); they reflect sunlight and also cool Earth's climate. Conversely, surface water has a very low albedo, only about 5 percent. As a result, any change from surface water to glaciers, or vice versa, can have a dramatic effect on the planet's albedo and temperature.
- When water evaporates from the ocean surface, solar energy is stored as latent heat of the resultant vapor. The vapor moves vast distances, transporting this heat. The heat is then released when the water vapor condenses to form rain or snow.
- Similarly, heat is released when water freezes to form snow and ice. But once a glacier forms or a portion of Earth's surface becomes snow covered, a lot of heat is required to melt the ice.

- Flowing water weathers rocks and initiates chemical reactions that alter the carbon dioxide concentration in the atmosphere. Carbon dioxide is a greenhouse gas that warms Earth's surface.
- Ocean currents move heat to and from the polar regions. But because more currents flow from the equator toward the poles than the other way around, there is a net transport of heat toward the polar regions. This polar warming effect is counterbalanced by the fact that water evaporates from warm ocean currents. Some of this vapor condenses and falls as snow, thereby increasing the albedo and cooling the polar regions.

Climate models attempt to quantify all of these factors, but clearly the balances are challenging to unravel. For this reason, it is difficult to predict climate changes.

THE NATURAL CARBON CYCLE AND CLIMATE

Carbon circulates among the atmosphere, the hydrosphere, the biosphere, and the geosphere and is stored in each of these reservoirs, in proportions you can see in ●**FIGURE 21.5**.

Carbon in the Atmosphere

Oxygen and nitrogen, the most abundant gases in the atmosphere, are transparent to infrared radiation and are not greenhouse gases. Carbon exists in the atmosphere mostly as carbon dioxide (CO_2), and in smaller amounts as methane (CH_4).

Although only 0.1 percent of the total carbon near Earth's surface is in the atmosphere, this reservoir plays

---

1. Heike Langenberg, commentary introducing the special Insight section "Climate and Water," *Nature* 419 (September 12, 2002): 187.

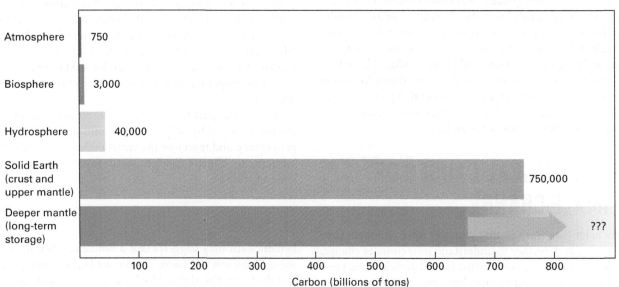

● **FIGURE 21.5** Carbon reservoirs in the atmosphere, biosphere, hydrosphere, and solid Earth. The numbers represent billions of tons of carbon.

Source: Adapted from W. M. Post, T. H. Peng, W. R. Emanuel, A. W. King, V. H. Dale, and D. L. DeAngelis, "The Global Carbon Cycle," *American Scientist* 78 (1990): 310–326.

an important role in controlling atmospheric temperature because carbon dioxide and methane are greenhouse gases; they absorb infrared radiation and heat the lower atmosphere. If either of these compounds is removed from the atmosphere, the atmosphere cools; if they are released into the atmosphere, the air becomes warmer.

Carbon in the Biosphere

Carbon is the fundamental building block for all organic tissue. Plants extract carbon dioxide from the atmosphere and build their body parts predominantly of carbon and hydrogen. This process occurs both on land and in the sea. Most of the aquatic fixation of carbon is conducted by microscopic photoplankton. Therefore, healthy terrestrial and aquatic ecosystems play a vital role in removing carbon from the atmosphere.

Most of the carbon is released back into the atmosphere by natural processes, such as respiration, fire, or decomposition, which are all part of the carbon cycle as demonstrated in ●**FIGURE 21.6**. However, at certain times and places, organic material does not decay completely and is stored as fossil fuels—coal, oil, and gas. Thus, plants transfer carbon from the biosphere to rocks of the upper crust.

Carbon in the Hydrosphere

Carbon dioxide dissolves in seawater. Most of it then reacts to form bicarbonate, HCO_3^- (commonly found in your kitchen as baking soda or bicarbonate of soda, $NaHCO_3$) and carbonate (CO_3^{2-}).

The amount of carbon dioxide dissolved in the oceans depends in part on the temperature of the atmosphere and of the oceans. When seawater warms, it releases dissolved carbon dioxide into the atmosphere, causing greenhouse warming. In turn, greenhouse warming further heats the oceans, causing more carbon dioxide to escape. Warmth evaporates seawater as well, and water vapor also absorbs infrared radiation. Clearly, such a feedback mechanism can escalate. A runaway greenhouse effect may be responsible for the high temperature on Venus.

Carbon in the Crust and Upper Mantle

As shown in Figure 21.5, the atmosphere contains about 750 billion tons of carbon. In contrast, the crust and upper mantle contain 1,000 times as much, or 750 trillion tons of carbon. The upper geosphere, combined with the

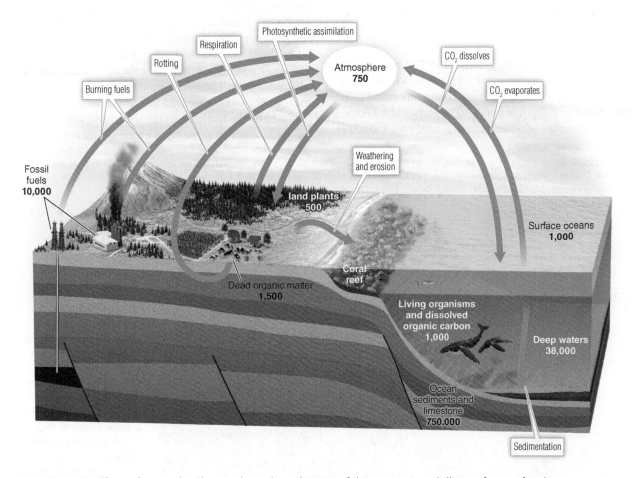

● **FIGURE 21.6** The carbon cycle. The numbers show the size of the reservoirs in billions of tons of carbon.

Data taken from U. Siegenthaler and J. L. Sarmiento, "Atmospheric Carbon Dioxide and the Ocean," *Nature* 365 (September 9, 1993): 119.

hydrosphere and biosphere, contain almost 800 trillion tons of carbon. Thus, if only a minute fraction of the carbon in the geosphere, hydrosphere, and biosphere is released, the atmospheric concentration of carbon dioxide can change dramatically and have a severe impact on climate.

CARBONATE ROCKS Marine organisms absorb calcium and carbonate ions from seawater and convert them into calcium carbonate ($CaCO_3$) in shells and other hard parts. This process removes carbon from seawater and causes more atmospheric carbon dioxide to dissolve into the seawater. Thus, formation of shells removes carbon dioxide from the atmosphere. The shells and skeletons of these organisms gradually collect to form limestone.

When sea level falls or tectonic processes raise portions of the sea floor above sea level, limestone and silicate rocks weather by processes that extract additional carbon dioxide from the atmosphere.[2] The exception to that is with acid rain reactions like $CaCO_3 + H_2SO_4$ (aq) → Ca (aq) + (SO_4) + H_2O + CO_2 (gas), which return carbon dioxide to the air.

CARBON IN FOSSIL FUELS Carbon is stored in fossil fuels, and carbon dioxide is released when these fuels are burned. Recoverable fossil fuels contain about 4 trillion tons of carbon, 5 times the amount in the atmosphere today. For this reason, scientists are concerned that burning fossil fuels will raise atmospheric carbon dioxide levels.

METHANE IN SEAFLOOR SEDIMENT When organic material falls to the sea floor and is buried with mud, bacteria decompose it, releasing methane, commonly called natural gas. Between a depth of about 500 meters and 1 kilometer, the temperature of water-saturated mud on continental shelves is low enough and the pressure is favorable to convert methane gas to a frozen solid called methane hydrate.

Methane hydrate then gradually collects in mud on the continental shelves. After studying both drill samples and seismic data, geochemist Keith Kvenvolden of the United States Geological Survey estimated that methane hydrate deposits hold twice as much carbon as all conventional fossil fuels— 10 times more than is in the atmosphere.

At present, the commercial extraction of methane hydrates to produce natural gas is impractical. It is expensive to drill in deep water, and a thin layer spread throughout the continental shelves would be prohibitively expensive to exploit. However, scientists are studying links between methane hydrates and climate. Tectonic activity at subduction zones or landslides on continental slopes could release methane from hydrate deposits. Changes in bottom temperatures on continental shelves resulting from warming of seawater could also release methane from the frozen hydrates. In turn, increased atmospheric methane could trigger global warming. Ice core studies show that global atmospheric methane concentration has changed rapidly in the past, perhaps by sudden releases of oceanic methane hydrates.

About 55 million years ago, near the end of the Paleocene Epoch, climate suddenly warmed and many aquatic and terrestrial species became extinct. According to one model, a change in sea-surface circulation caused equatorial waters to remain in low latitudes. High equatorial temperatures evaporated enough water to increase the salinity of the sea surface. When the salinity reached a threshold value where the surface water was denser than the cold deep water, the warm, salty water sank. The warm, sinking water melted the methane hydrates and released the methane. Aquatic species were poisoned by the methane in the water, and many terrestrial species succumbed to the rapid greenhouse warming.[3]

2. The net reaction is: $CaCO_3 + CO_2 + H_2O = Ca(HCO_3)^2$ Limestone plus carbon dioxide plus water react to form calcium bicarbonate (soluble).
3. Gerald R. Dickens, Maria M. Castillo, and James C. G. Walker, "A Blast of Gas in the Latest Paleocene: Simulating First-Order Effects of Massive Dissociation of Oceanic Methane Hydrate," *Geology* (March 1997): 259–262.

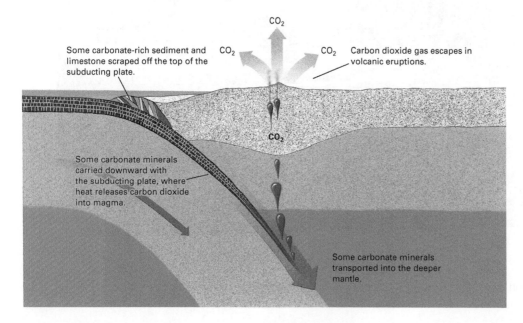

● FIGURE 21.7 A subducting oceanic plate carries limestone and other carbonate-rich sediment into the mantle. Some of the carbonate minerals are heated to produce carbon dioxide, which escapes during volcanic eruptions. Some of the carbonate minerals may be stored in the mantle.

CARBON IN THE DEEPER MANTLE During subduction, oceanic crust sinks into the mantle (●FIGURE 21.7). The descending plate may carry carbonate rocks and sediment. As this material sinks to greater depths, the carbonate minerals become hot and release carbon dioxide, which is carried back to the surface by volcanic eruptions. Some carbonate rock may be carried into deeper regions of the mantle during subduction. Large quantities of carbon were trapped within Earth during its formation. Much of this carbon escaped early in Earth's history, but some remains in the deep mantle and may rise from the mantle to the surface during volcanic eruptions. Carbon exchanges between the deep mantle and the surface are an important topic of current research in Earth Science.

TECTONICS AND CLIMATE CHANGE

Positions of the Continents

A map of Pangea shows that 200 million years ago Africa, South America, India, and Australia were all clustered near the South Pole (●FIGURE 21.8). Because climate is colder at high latitudes than near the equator, continental position alters continental climate.

In addition, continental interiors generally experience colder winters and hotter summers than coastal areas do. When all the continents were joined into a supercontinent, the continental interior was huge, and regional climates must have been different from the climates on many smaller continents with extensive coastlines.

The positions of the continents also influence wind and sea currents, which, in turn, affect climate. For example,

today the Arctic Ocean is nearly landlocked, with three straits connecting it with the Atlantic and Pacific oceans. The Bering Strait between Alaska and Siberia is 80 kilometers across, Kennedy Channel between Ellesmere and Greenland is only 40 kilometers across, and a third, wider seaway runs along the east coast of Greenland. Presently, cold currents run southward through Kennedy Channel and the Bering Strait, and the North Atlantic Drift carries warm water northward along the coast of Norway. If any of these straits were to widen or close, global heat transfer would be affected. Deep-sea currents also transport heat and are affected by continental positions.

Tectonic plates move from 1 to 16 centimeters per year. A plate that moves 5 centimeters per year travels 50 kilometers in a million years. Thus, continental motion can change global climate within a geologically short period of time by opening or closing a crucial strait. (However, even this geologically "short" period of time is extremely long when compared with the rise of human civilization.) Much longer times are required to modify climate by altering the proximity of a continent to the poles or by creating a supercontinent.

Mountains and Climate

Global cooling during the past 40 million years coincided with the formation of the Himalayas and the North American Cordillera,[4] the system of mountain ranges that form the backbone of the continent. Mountains interrupt airflow, altering regional winds. Air cools as it rises and passes over high, snow-covered peaks. However, it is unclear whether this regional cooling could account for the global cooling that accompanied this episode of mountain formation.

Large portions of the Himalayas and the North American Cordillera are composed of marine limestone. Recall from section on "The Natural Carbon Cycle and Climate" that when marine limestone weathers, carbon dioxide is removed from the atmosphere. When seafloor rocks are thrust upward to form mountains, they become exposed to the air. Rapid weathering then may remove enough atmospheric carbon dioxide to cause global cooling.

Volcanoes and Climate

Volcanoes emit ash and sulfur compounds that reflect sunlight and cool the atmosphere. For two years after Mount Pinatubo erupted in 1991, Earth cooled by a few tenths of a degree Celsius. Temperature rose again in 1994 after the ash and sulfur had settled out.

Volcanoes also emit carbon dioxide that warms the atmosphere by absorbing infrared radiation. The net result—warming or cooling—depends on the size

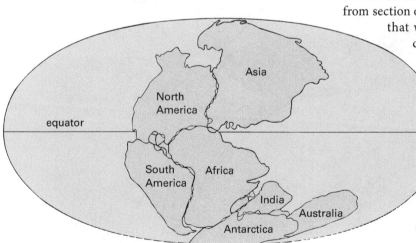

● **FIGURE 21.8** Two hundred million years ago, when the Pangea supercontinent was assembled, Africa, South America, India, and Australia were all positioned close to the South Pole.

4. William F. Ruddiman and John Kutzbach, "Plateau Uplift and Climatic Change," *Scientific American* (March 1991): 66ff.

of the eruption, its violence, and the proportion of solids and gases released. Some scientists believe that a great eruption in Siberia 250 million years ago cooled the atmosphere enough to cause or contribute to the Permian extinction. A huge sequence of eruptions 120 million years ago, called the *mid-Cretaceous superplume*, may have emitted enough carbon dioxide to warm the atmosphere by 7°C to 10°C. Dinosaurs flourished in huge swamps, and some of the abundant vegetation collected to form massive coal deposits.

How Tectonics, Sea Level, Volcanoes, and Weathering Interact to Regulate Climate

Tectonics, sea level, volcanoes, and weathering are all part of a tightly interconnected Earth system that affects both global and regional climates. When tectonic plates spread slowly, the Mid-Oceanic Ridge system is so narrow that it displaces relatively small amounts of seawater. As a result, sea level falls. When sea level falls, large marine limestone deposits on the continental shelves are exposed as dry land. The limestone weathers. Weathering of limestone removes carbon dioxide from the atmosphere, leading to global cooling. At the same time, when seafloor spreading is slow, subduction is also slow. Volcanic activity at both the spreading centers and the subduction zones slows down, so relatively

small amounts of carbon dioxide are emitted. With small additions of carbon dioxide from volcanic eruptions and removal of atmospheric carbon dioxide by weathering, the atmospheric carbon dioxide concentration decreases and the global temperature cools. In addition, dropping sea level decreases the surface area of the oceans and increases the surface area of the higher albedo continents. This results in an increase of average global albedo and, consequently, reinforces the global cooling (●**FIGURE 21.9**). These conditions may have caused the cooling at the end of the Carboniferous Period shown in Figure 21.1.

In contrast, during periods of rapid seafloor spreading, a high-volume Mid-Oceanic Ridge system raises sea level. Marine limestone beds are submerged, weathering slows, and weathering removes less carbon dioxide from the atmosphere. Volcanic activity is high during periods of rapid plate movement, so large amounts of carbon dioxide are released into the atmosphere. Rising sea level decreases continental area and therefore decreases the average global albedo. All of these factors lead to global warming (●**FIGURE 21.10**). But rapid spreading also coincides with rapid subduction and accelerated mountain-building, leading to accelerated weathering on the continents, which consumes carbon dioxide. Once again, climate systems are driven by so many opposing mechanisms that it is often difficult to determine which will prevail.

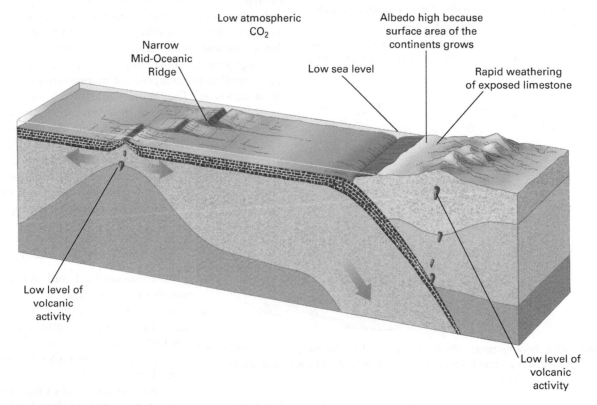

● **FIGURE 21.9** Slow subduction = Narrow Mid-Oceanic Ridge = Cooler Earth.

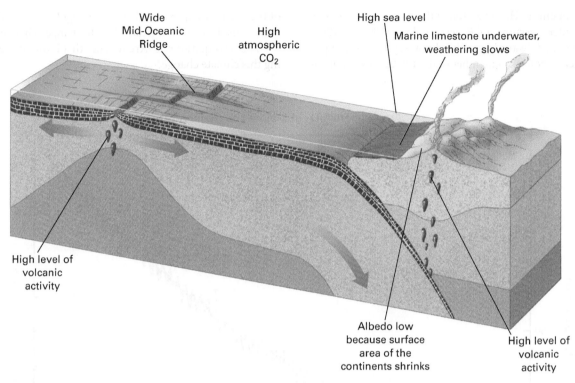

High sea level

Wide Mid-Oceanic Ridge

High atmospheric CO_2

Marine limestone underwater, weathering slows

High level of volcanic activity

Albedo low because surface area of the continents shrinks

High level of volcanic activity

● **FIGURE 21.10** Fast subduction = Wide Mid-Oceanic Ridge = Warmer Earth.

GREENHOUSE EFFECT: THE CARBON CYCLE AND GLOBAL WARMING

We have learned that the amount of carbon in the atmosphere is determined by many natural factors, including rates of plant growth, mixing of surface ocean water and deep ocean water, growth rates of marine organisms, weathering, the movement of tectonic plates, and volcanic activity. Within the past few hundred years, humans have become an important part of the carbon cycle. About 20 percent of this change occurs by cutting forests and other urbanization of land surfaces, thus reducing the carbon uptake from the atmosphere to plants. The remaining 80 percent occurs when gases are emitted by industry and agriculture. Modern industry releases four greenhouse gases—carbon dioxide, methane, chlorofluorocarbons (CFCs), and nitrogen oxides.

People release carbon dioxide whenever they burn fossil fuels or biofuels. This release is inherent in the chemistry of combustion. Carbon in the fuel reacts with oxygen in the air to produce carbon dioxide. Furthermore, once carbon dioxide is released, it is, for all practical purposes, impossible to remove this gas from the atmosphere. If you drive your car to town today, the carbon dioxide released will remain in the atmosphere for centuries. Logging also frees carbon dioxide because stems and leaves are frequently burned and forest litter rots more quickly when it is disturbed by heavy machinery. The recent rise in the concentration of atmospheric carbon dioxide has attracted considerable attention

because it is the most abundant industrial greenhouse gas (●**FIGURE 21.11**).

In addition, several other greenhouse gases are released by modern agriculture and industry. Small amounts of methane are released during some industrial processes. Larger amounts are released from the guts of cows, other animals, and termites, and from rotting that occurs in rice paddies. Today, industry and agriculture combined add about 37×10^{12} grams of methane into the atmosphere every year. N_2O, yet another greenhouse gas, is released from the manufacture and use of nitrogen fertilizer, some industrial chemical syntheses, and from the exhaust of high-flying jet aircraft.

CFCs were used widely in the 1960s and 1970s as the propellant in aerosol cans. By the early 1970s, worldwide production of the compounds had reached nearly one million tons per year and was valued at about half a billion dollars. Meanwhile, evidence linking the use of CFCs to loss of ozone in the stratosphere was published, and by 1978, the United States no longer permitted use of the compounds as aerosol propellants. CFCs were used as a refrigerant beyond 1978, but by 1994 production of new CFC stocks had effectively stopped worldwide. Today, CFCs are only used for certain specialized uses for which an alternative has yet to be found, and the existing stock of CFCs is recycled through a set of "halon banks" designed to eliminate additional loss to the atmosphere. Although CFCs still exist in the stratosphere from their former widespread use, their concentration has decreased as the result of the international agreement to ban CFC production.

Over the past few decades, a few scientists and some politicians and writers have argued that climate change is not really happening. However, today the data are overwhelming; the Intergovernmental Panel on Climate Change (IPCC) concluded that "warming in the climate system is unequivocal." Henry Pollack, a member of the IPCC, wrote that "use of the word 'unequivocal' leaves no wiggle room. 'Maybe, maybe not' is over. Significant climate change is happening."

The next question is: Are we sure that humans are causing this climate change?

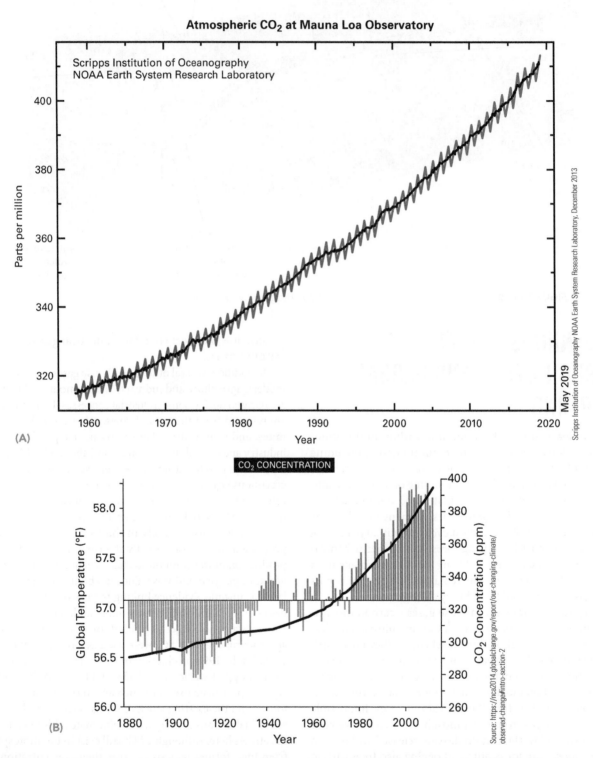

Atmospheric CO$_2$ at Mauna Loa Observatory

Scripps Institution of Oceanography
NOAA Earth System Research Laboratory

(A)

CO$_2$ CONCENTRATION

(B)

● **FIGURE 21.11** (A) Atmospheric carbon dioxide concentration has risen by about 20 percent since 1958. The short-term fluctuations are caused by seasonal changes in carbon dioxide absorption by plants. (B) Global average surface temperature above (red) and below (blue) the long-term average between 1880 and 2010. The left y-axis shows global temperature. The right y-axis shows atmospheric CO$_2$ concentration (parts per million) and corresponds to the black line.

CONSEQUENCES OF GLOBAL WARMING

Agriculture

- Shifts in food-growing areas
- Changes in crop yields
- Increased irrigation demands
- Increased pests, crop diseases, and weeds in warmer areas

Biodiversity

- Extinction of some plant and animal species
- Loss of habitats
- Disruption of aquatic life

Weather Extremes

- Prolonged heat waves and droughts
- Increased flooding
- More intense hurricanes, typhoons, tornadoes, and violent storms

Water Resources

- Changes in water supply
- Decreased water quality

- Increased drought
- Increased flooding

Melting of Arctic Sea Ice and Glaciers

- Changes in albedo
- Rise in sea level
- Changes in temperature and salinity of ocean
- Altered water balance
- Species extinction

Human Population

- Increased deaths
- More environmental refugees
- Increased migration

Forests

- Changes in forest composition and locations
- Disappearance of some forests
- Increased fires from drying
- Loss of wildlife habitats and species

Sea Level and Coastal Areas

- Rising sea levels
- Flooding of low-lying islands and coastal cities
- Flooding of coastal estuaries, wetlands, and coral reefs
- Beach erosion
- Disruption of coastal fisheries
- Contamination of coastal aquifers with salt water

Human Health

- Increased deaths from heat and disease
- Disruption of food and water supplies
- Spread of tropical diseases to temperate areas
- Increased respiratory disease
- Increased water pollution from coastal flooding

Let us first summarize the data:

- Human activities release greenhouse gases.
- The concentration of these gases in the atmosphere has risen since the beginning of the Industrial Revolution.
- Greenhouse gases absorb infrared radiation and trap heat.
- The atmosphere has warmed by almost 0.8°C during the last 50 years.

Eighteen of the last 19 warmest years have occurred since 2001, with 2016 being the hottest year on record.

The obvious connection is that rising atmospheric concentrations of industrial greenhouse gases have caused the recent global temperature rise. Again, some government and industry sources have disagreed, stating that the current warming trend may be an unrelated natural event that just happened to coincide with the atmospheric increase in carbon dioxide. Although this debate has raged for a few decades, today scientists are essentially unanimous in concluding that industrial emissions are causing global warming. Thousands of technical peer-reviewed scientific papers that address recent climate change and that have been published since 2003 have concluded that anthropogenic greenhouse gas emissions are causing the observed warmer temperatures.

Consequences of Greenhouse Warming

Many people ask, "What's the big deal? What difference will it make if the planet is a few degrees warmer than it is today?"

TEMPERATURE EFFECTS ON AGRICULTURE A warmer global climate would mean a longer frost-free period in the high latitudes, which would benefit agriculture. In parts of North America, the growing season is now a week longer than it was a few decades ago. On the negative side, warmth affects plants in many ways that decrease crop yields. In one study, researchers found that rice yields decrease by 10 percent for every 1°C increase in nighttime temperatures. The scientists concluded that warmer nighttime temperatures increased plant respiration and therefore decreased the energy available for storage in the rice kernels. In another study, scientists estimated that heat stress, decreased soil moisture, and increased susceptibility to parasites and disease combined during the 2003 European drought to cause a 30 percent reduction in forest and grassland growth. The effect on crops was variable. Winter wheat yields were barely affected, because the maximum growth period occurred before the heat wave. On the other end of the spectrum, the Italian corn yields decreased by 36 percent, despite extensive irrigation.

PRECIPITATION AND SOIL-MOISTURE EFFECTS ON AGRICULTURE In a warmer world, both precipitation and evaporation would increase. Computer models show that the effects would differ from region to region. Between 1900 and 2000, rainfall increased in Pakistan and in the great grain-growing regions in North America and Russia. However, parts of North Africa, India, and Southeast Asia received less rainfall over the same time. The resulting drought has led to famine during recent decades.

In a hotter world, more soil moisture will evaporate. In the Northern Hemisphere, scientists predict that global warming will lead to increased rainfall, adding moisture to the soil. At the same time, warming will increase evaporation that removes soil moisture. Most computer models forecast a net loss of soil moisture and depletion of groundwater, despite the prediction that rainfall will increase. Drought and soil moisture depletion would increase demands for irrigation. But, irrigation systems are already stressing global water resources, causing both shortages and political instability in many parts of the world. As a result of all these factors, most scientists predict that higher mean global temperature would decrease global food production, perhaps dramatically.

In many regions of the world, winter precipitation accumulates as snow in the high mountains. Snowmelt feeds rivers during the summer and this river water is used to irrigate crops. In a warmer world, more precipitation falls as rain, and the snow that does fall is more likely to melt in early

spring rather than in late summer. Also, glaciers recede, thus diminishing the amount of water in long-term storage. As a result of all these factors, river levels during mid- to late summer are lower in a warmer world than in a cooler world. Thus, unless people build expensive dams, there is less water available for irrigation.

EXTREME WEATHER EVENTS Computer models predict that weather extremes such as intense rainstorms, flooding, heat waves, prolonged droughts, and violent storms—hurricanes, typhoons, and tornadoes—will become more common in a warmer, wetter world. Over the past several decades, this scenario has begun to play out in reality. In the United States, the number of heavy precipitation events (defined as a two-day precipitation total exceeded once every five years) increased between the first and second halves of the 1900s and accelerated into the new millennia (●FIG-URE 21.12). Worldwide, large floods became more frequent during the twentieth century (●FIGURE 21.13). Since 2005 to the time of this writing (April, 2019), a total of 17 hurricanes

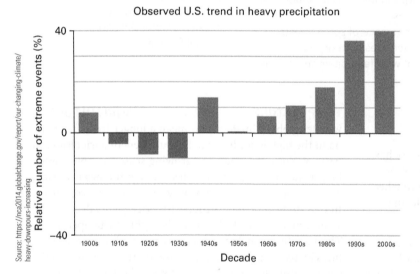

Source: https://nca2014.globalchange.gov/report/our-changing-climate/heavy-downpours-increasing

● **FIGURE 21.12** Observed trends in heavy precipitation events between 1900 and 2010 in the conterminous United States. The y-axis shows the relative number of events as a percentage of the average number of events between 1901 and 1960. A distinct increase in the number of events has been observed since the 1950s.

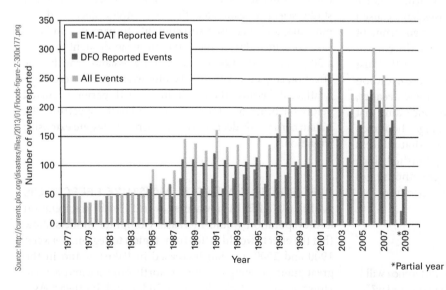

Source: http://currents.plos.org/disasters/files/2013/01/Floods-figure-2-300x177.png

● **FIGURE 21.13** Number of flood events for the period 1977–2009 as reported globally. EM-DAT = International Disaster Database; DFO = Dartmouth Flood Observatory Global Archive of Large Flood Events.

have hit the U.S. These include Hurricanes Katrina (2005) and Sandy (2010) in addition to three devastating storms of 2017: Hurricanes Harvey, Irma, and Maria and two major storms in 2018: Michael and Florence.

Hurricanes are driven by warm sea surface temperatures, and many climate models predict that extreme tropical storms and hurricanes, like Katrina, will become more frequent in a warmer world. Thus, we face the paradox of both more floods and more droughts. Hurricane intensity is likely to increase in a warmer world.

CHANGES IN BIODIVERSITY According to a study published by the World Wildlife Fund in 2000, global warming could alter one-third of the world's wildlife habitats by 2100. In some northern-latitude regions, 70 percent of habitats will be significantly affected. As soil moisture decreases, trees will die and wildfires will become more common, changing forests into savannas. Plants and animals that thrive in cold temperatures will die off. Several recent studies have shown that many plant and animal pathogens thrive better in warmer temperatures than in a colder world. Thus, the pine beetle in northern North America has flourished in recent years, decimating huge swaths of forests in the Rocky Mountains of western United States and Canada. Wildfire frequency and size also has increased significantly in the past several decades with devastating and deadly wildfires occurring in virtually each one of the western United States. The 2018 fire season alone included the Mendocino Complex Fire, the largest in California's history. Populations of frogs have declined dramatically as they have succumbed to virulent disease organisms. In extreme cases, species that are unable to adapt to the warmer temperatures might become extinct. For example, as Arctic sea ice diminishes, polar bears are losing their hunting grounds and migration pathways. As a result, these majestic creatures are threatened. Undoubtedly, other species will flourish. Ecological systems will change, but it is important to remember that terrestrial ecosystems have adjusted to far greater perturbations than we are experiencing today.

MELTING OF ARCTIC SEA ICE AND GLACIERS About 3.5 million years ago, in the mid-Pliocene, the average global temperature was about 2°C to 3.3°C warmer than it is today. During this time, the Northern Hemisphere was ice free and there was significantly less ice in Antarctica than exists there today. Recall that in the past 50 years, Earth's temperature has increased by 0.8°C, and now is only 1.5°C to 2.5°C cooler than that prehistorical warm period. In recent decades, as the Earth has warmed, glacial and sea ice has melted. Between 1979 and 2009, the Arctic Ocean lost approximately 7 percent of its spring ice cover. But more alarming, the ice that remains is significantly thinner than it was 30 years ago, causing concern that the rate of loss could quickly accelerate with a small additional temperature increase. Ice loss in Antarctica in-

creased by 75 percent in the last 10 years due to a speed-up in the flow of its glaciers. Greenland Ice Sheet's annual loss rose from 90 cubic kilometers in 1996 to 150 cubic kilometers in 2007. Stephen Schneider, a climatologist at Stanford, says that there is a "few percent chance" that meltwater that has already percolated into the ice cap will "obliterate the Greenland ice cover irrevocably."[5] He goes on to add that if the Northern Hemisphere warms by 3°C, there is a 90 percent chance that the Greenland ice cap will disappear. Note that this estimate corresponds with what actually happened 3.5 million years ago.

When land-based glaciers melt, sea level rises, affecting civilizations throughout the world. (Melting of sea ice has no effect on sea level.) Changes of sea level and loss of ice cover also affect the Earth's climate balance in many ways, as will be discussed in section on "Feedback and Threshold Mechanisms in Climate Change."

ACIDIFICATION OF THE OCEANS About one-third of all the carbon dioxide that enters the atmosphere dissolves in the ocean. On one hand, this buffer mechanism reduces the concentration of carbon dioxide in the atmosphere, thus maintaining a cooler planet. On the other hand, the dissolved carbon dioxide alters the chemistry of the oceans. When carbon dioxide dissolves in seawater, it reacts to form carbonic acid, H_2CO_3. Since preindustrial times, the oceans have become more acidic, showing a decline of about 0.1 pH units. When the oceans become more acidic, marine organisms have reduced ability to build hard shells or exoskeletons out of calcium carbonate. For example, along the Great Barrier Reef in Australia, the calcification rate decreased by 14 percent between 1990 and 2008. Between 2016 and 2017, half of the Great Barrier Reef died (●**FIGURE 21.14**). Thus, a small change in pH leads directly to a large biological disruption. The resultant changes affect species distributions that could adversely affect ocean food webs.

● **FIGURE 21.14** The remnants of a dead coral reef. The coral that formed this reef died and their calcium carbonate skeletons were bleached and broken down by wave energy.

Rich Carey/Shutterstock.com

SEA-LEVEL CHANGE When water is warmed, it expands slightly. From 1900 to 2000, mean global sea level rose 10 to 25 millimeters, or roughly the thickness of a dime, every year. Oceanographers suggest that even this modest rise may be affecting coastal estuaries, wetlands, and coral reefs. However, sea-level rise is accelerating dramatically because the polar ice sheets of Antarctica and Greenland are melting. In 2005, 150 cubic kilometers of Antarctic ice melted and the water flowed into the oceans. To exacerbate the problem, the flow of ice from Greenland has more than doubled over the past decade (●**FIGURE 21.15**).

Projections of global sea-level rise by the IPCC through the year 2100—82 years into the future—suggest that sea level is very likely to rise somewhere more than half a meter and perhaps more than a meter (3.3 feet; ●**FIGURE 21.16A**). However, actual measurements of sea-level rise from satellite measurements and tidal gages between 1993 and 2013 show that the projections by the IPCC during this same time frame underestimate the magnitude of sea-level rise (●**FIGURES 21.16B** and **16C**). If the IPCC projections for the rest of this century continue to underestimate the actual rate of sea-level rise, over 1 meter of rise could result.

●**FIGURE 21.17** shows results from a sea-level rise simulator available online through the National Aeronautics and Oceanographic Association (NOAA). It shows five feet (1.5 meters) of sea-level rise in the Boston region and includes a hypothetical view of Harvard's Dunster House with this level of rise.

EFFECTS ON PEOPLE Humans evolved during a period of rapid climate change in the savannas and forests of Africa. In fact, many anthropologists argue that we flourished because we adapted to climate change faster and more efficiently than other species in the African ecosystem. A few million years later, we survived the Pleistocene Ice Age. We are a clever and resourceful species. Yet the consequences of global warming could be severe. In a warmer world, tropical diseases such as malaria have been spreading to higher latitudes. But the biggest problem arises because expanding human population is already stressing global systems. Because of high population density, land ownership, and political boundaries, people cannot migrate as freely as they once could. As a result, small shifts in food production or water availability could lead to mass starvation, water scarcity, and political instability. In addition, rising sea level could flood coastal cities and farmlands, causing trillions of dollars worth of damage throughout the world. There is already great stress on our water supply, as you read about in Chapter 12.

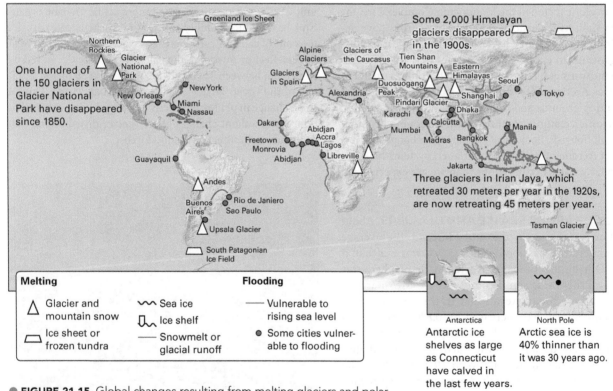

●**FIGURE 21.15** Global changes resulting from melting glaciers and polar ice caps.

From "The Mercury's Rising," from *Newsweek*, 12/4/2000, p.52. Reprinted by permission of the publisher.

5. Stephen H. Schneider, "The Worst-Case Scenario," *Nature* 458 (April 30, 2009): 1104–1105.

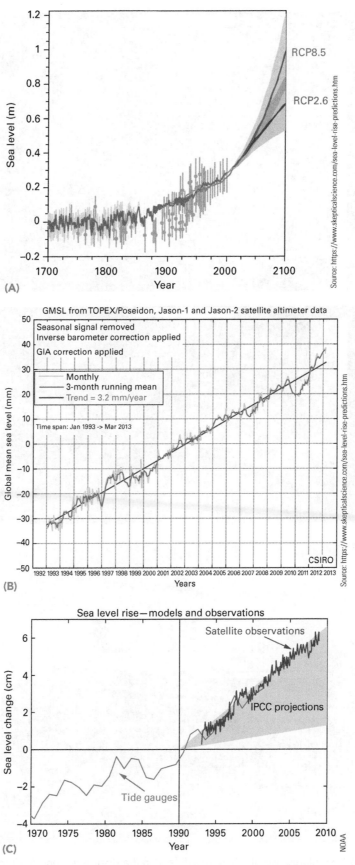

Source: https://www.skepticalscience.com/sea-level-rise-predictions.htm

Source: https://www.skepticalscience.com/sea-level-rise-predictions.htm

NOAA

● **FIGURE 21.16** (A) Observed (pre-2009) and projected (2009–2100) sea-level change in meters from the International Panel on Climate Change (IPCC). The blue projection (RCP2.6) anticipates a lower rate of sea-level rise than the red projection (RCP8.5). (B) Actually observed sea-level rise between 1993 and 2013. (C) Observed global mean sea level between 1970 and 2010 from tide gauges (orange line) and satellite data (light blue curve). Shown in gray is the range of sea-level projections by the IPCC for the same period.

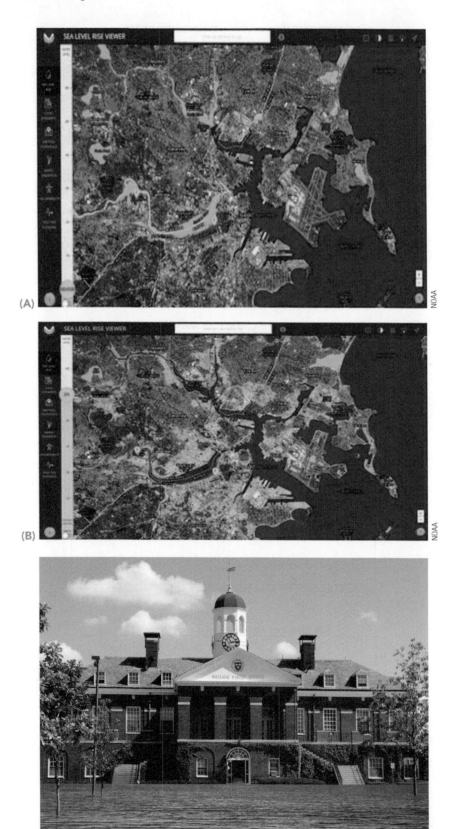

● **FIGURE 21.17** Simulated rise of 1.5 meters (5 feet) in the Boston region. (A) Current (2018) scenario. The bright green highlights inland waterways. (B) Same region after 1.5 meters of simulated sea-level rise. Everything with light blue overlay is below sea level. (C) A hypothetical photo of Harvard University's Dunster House after 1.5 meters of sea-level rise.

FEEDBACK AND THRESHOLD MECHANISMS IN CLIMATE CHANGE

If the concentration of greenhouse gases and the mean global temperature were related linearly, a small change in one would produce a corresponding small change in the other. However, climate change does not follow linear relationships.

For example, the melting of ice is a threshold phenomenon. If the air above a glacier warms from −1.5°C to −0.5°C, the ice does not melt and the warming may cause only minimal environmental change. However, if the air warms another degree to +0.5°C, the ice warms beyond the threshold defined by its melting point. Melting ice can cause sea-level rise and a host of cascading effects.

Recall that a feedback mechanism occurs when a small initial perturbation affects another component of Earth's systems, which amplifies the original effect, which perturbs the system even more, which leads to a greater effect, and so on. Four important climate feedback mechanisms are discussed next.

Albedo Effects

Sparkling snowfields and icy glaciers have high albedos and cool Earth by reflecting sunlight. When Earth warms a little, snow and ice covers decrease, and the albedo decreases. But when the albedo decreases, the atmosphere gets warmer, which melts more snow, which warms the atmosphere even more. A recent study has shown that a second, interlocking feedback mechanism exacerbates the first. In the Arctic, warmer temperatures have promoted the growth of bushes that often protrude above the spring snow and trap solar heat—thereby depressing the albedo. Thus snow melts faster, providing favorable conditions for bushes to grow, which decreases the albedo even further.

The accelerating nature of this double feedback loop has contributed to the fact that between 1960 and 1990, Arctic landscapes warmed by roughly 0.15°C per decade. However, between 1990 and 2004, this warming more than doubled to 0.3–0.4 degrees Centigrade per decade and Arctic surface air temperatures today are warming twice as fast as those across the rest of the globe. Arctic surface air temperatures between 2014 and 2018 exceeded all previous records.

Imbalances in Rates of Plant Respiration and Photosynthesis

Any process that increases the total photosynthesis in an ecosystem will remove carbon dioxide from the air and sequester the carbon in plant tissue. In contrast, any process that increases respiration over photosynthesis will introduce more carbon dioxide into the air and increase global warming. Numerous studies have shown that decay respiration in forest and grassland soils accelerates in a warmer world, thus warming the world more by introducing carbon dioxide into the atmosphere.

In another study, scientists showed that both respiration and photosynthesis decreased during the 2003 European drought, but photosynthesis declined more than respiration, leading to a net increase in carbon emissions into the atmosphere.

Changes in Ocean Currents

Approximately 12,800 years ago, air temperature over the North Atlantic dropped suddenly by more than 10°C, a climate change called the Younger Dryas Event, which you can see on the graph in ●FIGURE 21.18.

The speed of this change implies that some threshold or feedback mechanisms occurred, most probably relating to changes in ocean currents. Recall that in the Atlantic Ocean, warm-water currents carry heat northward, thus heating southern Greenland and northern Europe. One argument proposes that after the ice age maximum, as the planet warmed, massive amounts of ice melted, creating a huge lake, called Lake Agassiz in North America, about where the Great Lakes are now. Initially, Lake Agassiz flowed southward into the Gulf of Mexico. But when the North American ice sheet retreated sufficiently, a huge volume of water suddenly flowed through what is now the St. Lawrence River valley into the North Atlantic. By one estimate, 9,500 cubic kilometers of water cascaded into the ocean. Many scientists argue that the huge influx of freshwater caused the Younger Dryas Event. However, there is significant debate over how this freshwater influx caused the recorded temperature change. According to one argument, the freshwater lowered the surface density of the seawater. Due to this lowered density, seawater in the northeast Atlantic failed to sink, as it had previously. This disruption of vertical movement disrupted the horizontal currents. Others argue that ocean currents are driven primarily by wind systems, and the change in water flow must have altered wind patterns, which in turn altered the currents in the Atlantic. A third argument proposes that the cold freshwater was simply a physical barrier, just as a tributary will alter the flow of the main current in a river system. On the other hand, maybe the emptying of Lake Agassiz was not the cause of the temperature change at all. This argument is supported by the observation that a second meltwater pulse from central North America occurred at the end of the Younger Dryas, when the planet was warming and this influx of meltwater did not cause a cooling trend.

While the mechanism remains uncertain, ice core data tell us that the Younger Dryas cooling, while the most dramatic, was not the only rapid cooling in the Pleistocene Epoch. In fact it was only the most recent of a series of large and abrupt climate swings that occurred

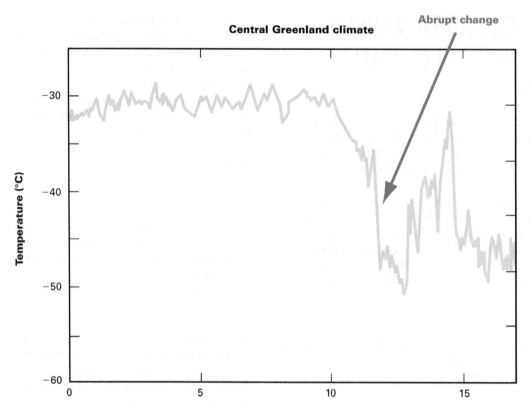

● **FIGURE 21.18** The Younger Dryas Event as an example of abrupt climate change.

repeatedly during the last ice age, and were recorded in many places around the globe. Thus, it is clear that sequences of threshold and feedback mechanisms have changed climate rapidly and dramatically in the recent geologic past.

Permafrost and Deep-Sea Methane Deposits

Large deposits of methane gas are trapped both in permafrost and in sediments beneath oceans. Methane is 20 times as effective in trapping heat in the atmosphere as carbon dioxide. Warming temperatures can release methane from both sources, thus causing a rapid and possibly catastrophic feedback warming. Vast areas of permafrost contain stores of carbon, which accumulated during multiple glacial advances and retreats over millions of years. Some of this carbon already exists as methane, produced by microbial decay in the past. Additional amounts can be released quickly when the ice melts and microbes have lush, carbon-rich material to feast on. In a similar manner, methane is also trapped in seafloor sediments as methane hydrates, in which methane is trapped in a molecular cage of ice. If ocean water warms sufficiently to melt the ice, then the methane

bubbles to the surface. In both of these cases, initial warming from an unrelated cause, such as human release of carbon dioxide, can release the methane, which leads to additional and more rapid warming. Many scientists hypothesize that about 55 million years ago, a 100,000-year hot spell called the Paleocene-Eocene Thermal Maximum was caused by such a release of methane-bearing seafloor deposits. This was the last time that the Earth was entirely without ice.

No one knows what the future will bring, but we do know that climates have changed radically in the past, from the frigid cold of Snowball Earth to the warmth of the Cretaceous. Today, by altering the atmospheric composition, we are undergoing a great and unintentional experiment with the climate system that sustains us. By making artificial changes to the natural order, such as the flow of water, we are altering not only the Earth's topography but also its natural cycles in ways that we may not be prepared to handle. Furthermore, global climate change has already affected the Earth, its ecosystems, and the humans who call this planet home. Are we prepared to account for the consequences linked to the way we use our planet? By understanding how human activity can affect the Earth's natural cycles, perhaps we can work with our environment and not against it to maintain a habitable planet.

Key Concepts Review

- Global climate has changed throughout Earth's history, from extreme cold with extensive glaciation to a 248-million-year warm period from the start of the Mesozoic Era almost to the present. During the last 100,000 years, the mean annual global temperature has changed frequently and dramatically, although the past 10,000 years, during which civilization developed, witnessed an anomalously stable climate.

- Past climates can be studied through historical records, tree rings, pollen assemblages in sediment, oxygen isotope ratios in glacial ice, erosional and depositional features created by glaciers, plankton assemblages and isotope studies in ocean sediment, and fossils in sedimentary rocks.

- Despite its immense complexity, Earth's atmospheric temperature is determined by the heat of the Sun, albedo, and heat retention by the atmosphere, which had great impact on early Earth in particular. When the Earth first formed, the Sun produced about 70 percent of the energy it produces today. Other components of the primordial atmosphere warmed early Earth; as the composition of the atmosphere changed, so did climate. Bolide impacts may also have great impact and caused rapid, catastrophic changes in Earth's climate.

- Water in its various forms can cause warming or cooling. Water vapor warms the atmosphere, but clouds can cause either warming or cooling. Glaciers and snowfields reflect sunlight and cause cooling. Evaporation is a cooling process, while condensation releases heat. The evaporation–condensation cycle transports heat from the equatorial regions to the poles. Weathering alters the carbon dioxide concentration in the atmosphere. Ocean currents transport heat and moisture.

- Carbon circulates among all four of Earth's spheres. Carbon dioxide is a greenhouse gas that warms the atmosphere. Carbon exists in the biosphere as the fundamental building block for organic tissue. Carbon dioxide dissolves in seawater to form bicarbonate and carbonate ions. Carbon exists in the crust and mantle in several forms: (1) Marine organisms absorb calcium and carbonate ions and convert them to solid calcium carbonate (shells and skeletons), and as a result large quantities of carbon exist in limestone and marine sediment. (2) Fossil fuels are largely composed of carbon. (3) Large quantities of carbon exist as methane hydrates on the sea-floor. (4) Carbon also exists in the deeper mantle; some of this is primordial carbon trapped during Earth's formation and some is carried into the mantle on subducting plates.

- Natural climate change also results from changes in the positions of continents, the growth of mountains, and volcanic eruptions. The positions of the continents influence wind and sea currents, which affect climate. Mountains interrupt airflow, changing regional winds. The limestone in several large mountain chains removes carbon dioxide from the air in the weathering process, possibly causing global cooling. Volcanic eruptions emit ash and sulfur compounds that reflect sunlight and cool the atmosphere, but also emit carbon dioxide and ash into the atmosphere, which warms the planet by absorbing infrared radiation. All of these factors interact to determine a specific global temperature.

- People release carbon dioxide into the atmosphere when they burn fuel. Logging, some industrial processes, and some aspects of agriculture contribute greenhouse gases to the atmosphere. Most scientists agree that human introduction of greenhouse gases has altered climate in the past century. Global warming could lead to a longer growing season in the high latitudes, but warmth affects plants in ways to decrease crop yields. One study has shown that rice yields decline in a warmer world, and the ratio of respiration to photosynthesis increases, leading to more carbon dioxide in the atmosphere. Global warming will affect precipitation and snow-melt, cause extreme weather, alter biodiversity, melt arctic sea ice and glaciers, lead to acidification of the oceans, and cause sea level to rise.

- Climate feedback loops include snowmelt and albedo and changes in rates of plant respiration and photosynthesis. The Younger Dryas Event was a sudden cool climate period that may have been caused by changes in the ocean currents due to freshwater influx. The collapse of the West Antarctic Ice Sheet could lead to rising sea level and further global warming. Melting permafrost could release trapped methane gas, which is 20 times as effective at trapping heat as carbon dioxide. Such release could create catastrophic feedback warming.

Review Questions

1. Explain why early Earth was warm, even though the Sun emitted about 30 percent less energy at that time.

2. Briefly outline climate changes both from 100,000 to 10,000 years ago and from 10,000 years ago to the present. List some of the methods used to gather data to determine this paleoclimate history.

3. Explain how volcanic eruptions can cause either a cooling or a warming of the atmosphere.

4. List and explain four feedback loops that affect climate. For each one, outline the contributions of geosphere, atmosphere, biosphere, and hydrosphere.

5. Discuss the consequences of a warmer Earth. How would temperature changes affect precipitation? How would changes in temperature and precipitation affect agriculture, ecosystems, and human society?

Jacques Descloitres, MODIS Rapid Response Team, NASA/GSFC

22

A Hubble Space Telescope view of a star-forming region in the 30 Doradus Nebula. The massive, young stellar grouping, called R136, is only a few million years old.

MOTIONS IN THE HEAVENS

● **FIGURE 22.1** A time exposure of the night sky shows the rotation of the stars around the Pole Star, which is nearly motionless.

THE MOTIONS OF THE HEAVENLY BODIES

Even a casual observer notices that both the Sun and Moon rise in the east and set in the west. In the midlatitudes, summer days are long and the Sun rises high in the sky. Winter days are shorter, and at midlatitudes in the Northern Hemisphere, the Sun never rises very high above the southern horizon, even at noon. In contrast to the yearly cycle of seasons, the Moon completes its cycle once a month. It is full and round one night, then darkens slowly until it is a thin crescent. It disappears completely after two weeks, then reappears as a sliver that grows again, completing the entire cycle in about 29.5 days.

MINDTAP
From Cengage

Animation:
Retrograde Motion:
Mars Retrograde

If you had been a cowboy in the 1800s, the foreman might have told you to guard the herd at night until "the dipper holds water." This phrase refers to two features of the night sky. First, stars remain in fixed positions relative to one another. This fact led the ancients to identify groups of stars, which they called **constellations**—"the dipper" being one of them. Second, in the Northern Hemisphere, the Pole Star, or North Star, is a motionless fixed point in the sky, and all other stars appear to revolve around it (●**FIGURE 22.1**). This motion provided the clock for the cowboy on night watch, because the dipper alternately holds and spills water as it revolves around the Pole Star.

Constellations appear and disappear with the seasons. For example, the Egyptians noted that Sirius, the brightest star in the sky, became visible at dawn just before the Nile began to flood. Farmers would therefore plant crops when this star first appeared, with the assurance that the high water soon would irrigate their fields.

Ancient astronomers noted several objects that appeared to be stars but were different because they changed position with respect to the stars. The ancient Greeks called these objects **planets**, from the word meaning "wanderers." For most of the year, planets appear to drift eastward with respect to the stars, but sometimes they seem to reverse direction and drift westward. This apparent reverse movement is called **retrograde motion**, a concept that we will explore further in sections on "Aristotle and the Earth-Centered Universe" and "The Renaissance and the Heliocentric Solar System."

ARISTOTLE AND THE EARTH-CENTERED UNIVERSE

The Greek philosopher and scientist Aristotle proposed a **geocentric**, or Earth-centered, Universe. In this model, Earth is stationary and positioned at the center of the Universe. A series of concentric **celestial spheres**, made of transparent crystal, surrounds Earth. The Sun, Moon, planets, and stars are imbedded in the spheres. At any one time, a person can see only a portion of each sphere, but as the sphere revolves around Earth, objects appear and disappear.

Aristotle based his conclusions on two observations. First, he reasoned that Earth must be stationary because people have no sensation of motion. People tend to fall off the back of a chariot when horses start to gallop, but they do not fall off Earth, so Aristotle reasoned that Earth must not be moving.

Aristotle's second observation was based on **parallax**, the apparent change in position of an object due to the change in position of the observer. To understand parallax, consider the fence posts in ●FIGURE 22.2. They appear to stand one in front of the other when the photographer is in the first position (A) and seem offset when the photographer steps to the side (B). Thus the relative positions of the stationary posts appear to change with the movement of the observer. The ancients correctly reasoned that if Earth moved around the Sun, the stars should change position relative to one another (●FIGURE 22.3). Because they observed no parallax shift, they concluded that Earth must be stationary. The mistake arose not out of faulty reasoning but because stars are so far away that their parallax shift is too small to be detected with the naked eye.

Aristotle incorporated philosophical concepts, in addition to observation, into scientific theory. He argued that the gods would create only perfection in the heavens and that a sphere is a perfectly symmetrical shape. Therefore, the celestial spheres must be a natural expression of the will of the gods. In addition, the Sun, Moon, planets, and stars must also be unblemished spheres.

Aristotle's theory, although incorrect, did explain the motions of the Sun, Moon, and stars (●FIGURE 22.4). However, it failed to explain the retrograde motion of the planets. In about 150 CE, Claudius Ptolemy modified the celestial sphere model to incorporate retrograde motion. In Ptolemy's model, each planet moves in small circles as it follows its larger orbit around Earth (●FIGURE 22.5). Ptolemy's sophisticated mathematics accurately described planetary motion

and therefore his model was accepted. Still, he retained Aristotle's erroneous idea that the Sun and the planets orbit a stationary Earth.

(A)

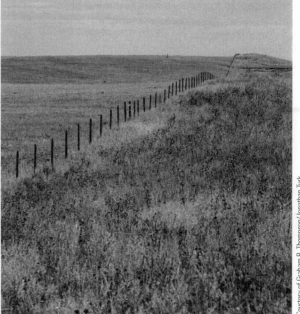

(B)

● **FIGURE 22.2** Parallax is illustrated by two photographs of a fence on the Montana prairie. (A) The photographer is nearly in line with the fence, so the distant posts appear to be in line. (B) When the photographer moved, the posts appear to have shifted position. Now we can see spaces between the more distant posts. Of course, the posts have not moved, only the photographer has. The same effect has been observed in astronomical studies. As Earth revolves about the Sun, the stars appear to shift position relative to one another.

---

constellations A group of stars that seem to form a pattern when viewed from Earth. Many of the patterns have been named by astronomers.

planets A celestial body that revolves in a fixed orbit around a star.

retrograde motion When viewed from Earth, the apparent motions of the planets in which they temporarily move backward (westward) with respect to the stars, before resuming their original eastward motion.

geocentric A model that places Earth at the center of the Universe.

celestial spheres The hypothetical series of concentric transparent spheres surrounding Earth in Aristotle's model of the Universe. Aristotle postulated that the Sun, Moon, planets, and stars are embedded in the spheres.

parallax The apparent change in position of an object due to the change in position of the observer.

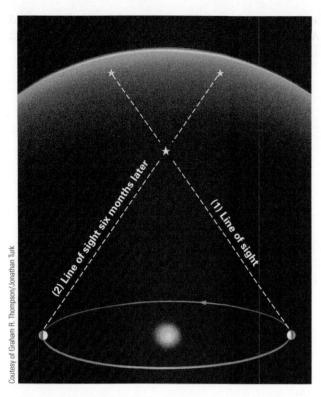

● **FIGURE 22.3** A nearby star appears to change position with respect to the distant stars as Earth orbits around the Sun. This drawing is greatly exaggerated; in reality, the distance to the nearest stars is so much greater than the diameter of Earth's orbit that the parallax angle is only a small fraction of a degree.

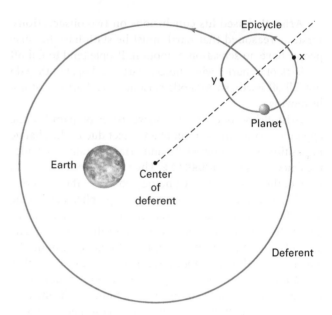

● **FIGURE 22.5** Ptolemy's explanation of retrograde motion. Each planet revolves in a small orbit (the epicycle) around the larger orbit (the deferent). When the planet is in position x, it appears to be moving eastward. When it is in position y, it appears to reverse direction and to be moving westward. Ptolemy did not realize that the planets moved in elliptical orbits. To compensate for this error, he placed Earth away from the center of the deferent.

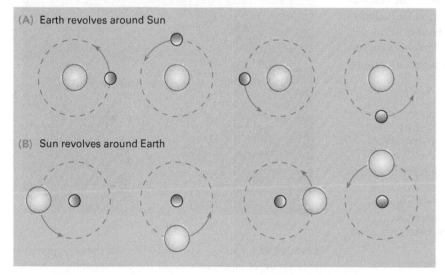

● **FIGURE 22.4** Series A shows Earth revolving around the Sun, and series B shows the Sun revolving around Earth. Now lay some thin paper over series A and trace the outlines of the Sun and Earth but not the arrows or orbits. Lay this tracing over series B and note that they match exactly, after you shift the paper for each sketch to make sure the Sun and Earth superimpose. Conclusion: There is no apparent difference between Earth revolving around the Sun and vice versa, provided you do not refer to anything else, such as the outline of these diagrams or another star.

THE RENAISSANCE AND THE HELIOCENTRIC SOLAR SYSTEM

Aristotle's and Ptolemy's ideas remained essentially unchallenged for 1,400 years. Then, in the 120 years from 1530 to 1650, several Renaissance scholars changed our understanding of motion in the Solar System and, in the process, revolutionized scientific thought.

MINDTAP From Cengage

Animation:
Retrograde Motion:
Geocentric and
Heliocentric Models
of the Solar System

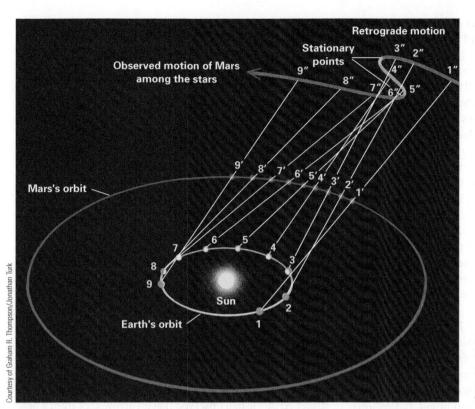

Courtesy of Graham R. Thompson/Jonathan Turk

● **FIGURE 22.6** Retrograde motion is explained within the heliocentric model by changes in the relative positions of Earth and the planet we are observing.

Copernicus

In 1530, a Polish astronomer and cleric, Nicolaus Copernicus, proposed that the Sun, not Earth, is the center of the Solar System and that Earth is a planet, like the other "wanderers" in the sky. Copernicus based his hypothesis on the philosophical premise that the Universe must operate by the simplest possible laws. Copernicus believed Ptolemy's model was too complex and that the motions of the heavenly bodies could be explained more concisely by a **heliocentric** model with the Sun at the center of the Solar System.

●**FIGURE 22.6** shows how the heliocentric model explains retrograde motion. Assume that initially Earth is in position 1 and Mars is in position 1′. An observer on Earth looks past Mars (as shown by the white line) and records its position relative to more distant stars. Mars appears to be in position 1″ in the night sky. After a few weeks, Earth has moved to position 2, Mars has moved to 2′, and Mars's position relative to the stars is indicated by 2″. In the Copernican model, Earth moves faster than Mars because Earth is closer to the Sun. Therefore, Earth catches up to and eventually passes Mars, as shown by positions 3–3′ and 4–4′. During this passage, Mars appears to turn around and move backward, although, of course, this appearance is merely an illusion against the backdrop of the stars. Mars appears to reverse direction again through positions 5 and 6, thus completing one cycle of retrograde motion. In reality, Mars never reverses direction, it just appears to behave in this way as Earth catches up to it and then passes it.

heliocentric A model that places the Sun at the center of the Solar System.

Brahe and Kepler

In the late 1500s, Tycho Brahe, a Danish astronomer, accurately mapped the positions and motions of all known bodies in the Solar System. His maps enabled him to predict where any planet would be seen at any time in the near future, but he never explained their motions. Brahe died somewhat mysteriously in 1601, and his student Johannes Kepler ended up with the vast amount of data that Brahe had collected. Kepler calculated that the planets moved in elliptical orbits, not circular ones, and he derived a set of mathematical formulas to describe their paths. However, Kepler never answered the important question: Why do the planets move in orbits around the Sun rather than flying off into space in straight lines?

Galileo

Galileo was an Italian mathematician, astronomer, and physicist who made so many contributions to science that he is often called the "Father of Modern Science." Perhaps most significant was his realization that the laws of nature must be understood through observation, experimentation, and mathematical analysis. This concept freed scientists from the confines of Aristotelian dogma.

Recall that Aristotle's geocentric theory was based on two observations and one philosophical premise: (1) If Earth were moving, people should fall off, but they do not. (2) Aristotle was unable to observe parallax shift of the stars. (3) The gods would create only an unblemished, symmetrical Universe.

Galileo's experiments and observations showed that Aristotle's model was incorrect and led him to support Copernicus's heliocentric model.

Galileo studied the motion of balls rolling across a smooth marble floor, and legend tells us he also dropped objects from the leaning Tower of Pisa. He organized the results of these experiments into laws of motion that were later expanded and quantified by Isaac Newton. One of these laws states that "an object at rest remains at rest and an object in uniform motion remains in uniform motion until forced to change." This corresponds to Newton's first law of motion, the law of **inertia**. Inertia is the tendency of an object to resist a change in motion.

According to the law of inertia, if Earth were in uniform motion, a person on its surface would be in uniform motion along with it. The person therefore would travel with Earth and would not fall off and be left behind, as Aristotle had assumed. In fact, the person could not even feel the motion.

Galileo built his first telescope in 1609 and turned it to the heavens soon afterward. He learned that the hazy white line across the sky called the Milky Way was not a cloud of light as Aristotle had proposed but a vast collection of individual stars. Next, he trained his telescope at the Moon and saw hills, mountains, giant craters, and broad, flat regions on its surface, which he thought were seas and therefore named *maria* (Latin for "seas"). Looking at the Sun, Galileo recorded dark regions, called sunspots, that appeared and then vanished.

Although these discoveries had no direct bearing on the controversy of a geocentric versus a heliocentric Universe, they were important because they led Galileo to question Aristotle's views. The prevailing scientific opinion at the time was that if the Milky Way were a collection of stars, Aristotle would have known about it. Furthermore, Galileo's observations of the Sun and the Moon did not agree with Aristotle's philosophical assumptions that the heavenly bodies were perfectly homogeneous and unblemished. Galileo reasoned that if Aristotle was wrong about the structures of the Milky Way, the Sun, and the Moon, perhaps he was also wrong about the motions of these celestial bodies.

When Galileo studied Jupiter, he saw four moons orbiting the giant planet. (Today we know that Jupiter has 63 moons, but only four are large enough to have been seen with Galileo's telescope.) According to the geocentric model, every celestial body orbits Earth. However, Jupiter's moons clearly orbited Jupiter, not Earth. The contradiction increased Galileo's doubt about Aristotelian theories.

Finally, Galileo observed that the planet Venus passes through phases as the Moon does. Such cyclical phases could not be readily explained in the geocentric model, even with Ptolemy's modifications (●**FIGURE 22.7**). As a result of his observations, Galileo proposed that the Sun is the center of the Solar System and that the planets orbit around it.

Isaac Newton and the Glue of the Universe

Galileo successfully described the motions in the Solar System, but he never addressed the question: Why do the planets orbit the Sun rather than fly off in straight lines into space? Aristotle had observed that an arrow shot from a bow flies in a straight line, but celestial bodies move in curved paths. He reasoned that arrows have an essence that compels them to move in straight lines, and planets and stars have an essence that compels them to move in circles. In the Renaissance, however, this answer was no longer acceptable.

Isaac Newton was born in 1643, the year Galileo died. During his lifetime, Newton made important contributions to physics and developed calculus. A popular legend tells that Newton was sitting under an apple tree one day when an apple fell on his head and—presto—he discovered gravity. Of course, people knew that unsupported objects fall to the ground long before Newton was born. But he was the first to recognize that gravity is a universal force that governs

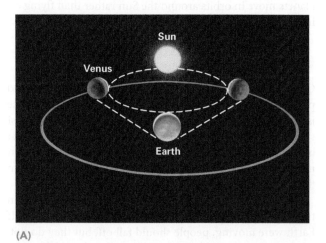

(A)

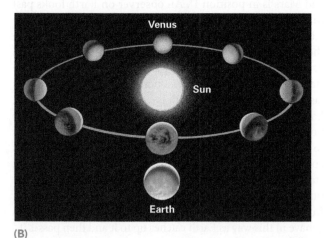

(B)

● **FIGURE 22.7** (A) In Ptolemy's theory, Venus could never move farther from the Sun than is shown by the dotted lines. Therefore it would always appear as a crescent. (B) In the heliocentric theory, Venus passes through phases like the Moon. Galileo observed phases of Venus through a telescope and concluded that the planets must orbit the Sun.

all objects, including a falling apple, a flying arrow, and an orbiting planet.

According to the laws of motion introduced by Galileo and expanded by Newton, a moving body travels in a straight path unless it is acted on by an outside force. Just as a ball rolls in a straight line unless it is forced to change direction, a planet also moves in a straight line unless a force is exerted on it. The gravitational attraction between the Sun and a planet forces the planet to change direction and move in an elliptical orbit. Gravity is the glue of the Universe and affects the motions of all celestial bodies.

THE MOTIONS OF THE EARTH AND THE MOON

By about 1700, astronomers knew that the Sun is the center of our Solar System and that planets revolve around it in elliptical orbits. In addition to revolving around the Sun, the planets simultaneously spin on their axes. Earth spins approximately 365 times for each complete orbit around the Sun. Each complete **rotation** of Earth represents one day. As Earth rotates about its axis, the Sun, Moon, and stars appear to move across the sky from east to west. We explained in Chapter 18 that Earth's axis is tilted and that this tilt combined with Earth's orbit around the Sun produces the seasons.

---

inertia The tendency of an object to resist a change in motion.

rotation Turning or spinning on an axis. Tops and planets rotate on their axes.

revolution Orbiting around a central point. A satellite revolves around Earth, and Earth revolves around the Sun.

As explained in section on "The Motions of the Heavenly Bodies," different stars and constellations are visible during different seasons. Earth's **revolution**, or orbiting, around the Sun causes this seasonal change in our view of the night sky (●**FIGURE 22.8**). Part of the sky is visible on a winter night when Earth is on one side of the Sun, and a different part is visible on a summer night six months later.

More recent measurements have revealed several additional types of planetary motion. As you learned in Chapter 13, as Earth rotates, its axis wobbles like a spinning top in a motion called *precession* (●**FIGURE 22.9**). Presently, Earth's axis points toward Polaris, the North Star. In 12,000 years, the axis will point toward Vega, and Vega will become the North Star. Because precession cycles over a 26,000-year period, Earth's axis will point toward Polaris again by the year 28,000. Recall from Chapter 13 that precession is one factor that contributed to glacial advances and retreats within the Pleistocene Ice Age.

In addition, the Moon's gravity pulls Earth slightly out of its orbit, causing Earth to spiral slightly as it circles the Sun. Our Sun also orbits the center of the Milky Way galaxy. Traveling at a speed of 220 kilometers per second, it completes its orbit in about 200 million years. The entire Milky Way galaxy carries our Sun and planets through intergalactic space.

Motion of the Moon

The gravitational attraction between the Moon and Earth not only holds the two in orbit around each other but it also affects the surfaces and interiors of both. We learned that the Moon's gravitation causes tides on Earth. In turn, Earth's gravitation pulls on the Moon sufficiently to cause it to bulge, despite the fact that it is solid rock. Earth's gravity attracts the bulge, so the side of the Moon with the bulge always faces Earth. As a result, the Moon rotates on

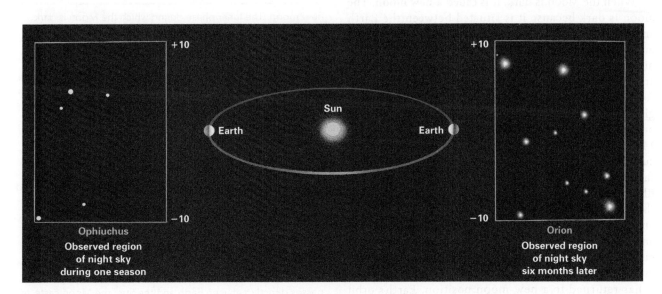

●**FIGURE 22.8** The night sky changes with the seasons because Earth is continuously changing positions as it orbits the Sun.

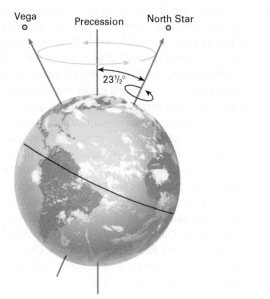

● **FIGURE 22.9** Earth's axis wobbles or precesses like a top, completing one precession cycle every 26,000 years.

its axis at the same rate at which it orbits Earth. Thus, we always see the same side of the Moon, and the other side was invisible to us until it was first photographed by the Soviet Union's Luna 3 space probe in 1959. To understand the phases of the Moon, we must first realize that the Moon does not emit its own light but reflects light from the Sun. The half of the Moon facing the Sun is always bathed in sunlight, while the other half is always dark. The phases of the Moon as we see them depend on how much of the Moon's sunlit area is visible from Earth. In turn, this visible area depends on the relative positions of the Sun, Moon, and Earth.

When the Moon is dark, it is called a **new moon**. The moon is dark because it is situated between the Earth and the Sun, and the Moon's unlit side faces us during our daytime when we can't see it. A few days after a new moon, a thin **crescent moon** appears. The crescent grows and the Moon is described as **waxing**, or becoming full. As the moon waxes, it changes from a crescent moon to a **gibbous moon**, in which the moon is over half-full with only a thin crescent of the moon remaining dark. About 14 to 15 days after a new moon, the Moon appears circular and is a **full moon** (●FIGURE 22.10). At this point, the Earth is situated between the Moon and Sun, so the illuminated side of the Moon is facing directly towards the Earth. A few evenings later, part of the disk is darkened. As the days progress, the visible portion shrinks and the Moon is said to be **waning**. Eventually, only a tiny curved sliver, the waning crescent, is left. After a total cycle of about 29.5 Earth days, the Moon is dark because it has returned to a new-moon position. Earth's orbit around the Sun forms an elliptical plane, with the Sun in

the same plane and near its center. In a similar way, the Moon's orbit around Earth describes another plane with Earth at its center. But the plane of the Moon's orbit is tilted 5.2 degrees with respect to Earth's orbital plane. As a result, the Moon is not usually in the same plane as that of Earth and the Sun (●FIGURE 22.11). Therefore Earth's shadow does not normally fall on the Moon, and the Moon's shadow does not normally fall on Earth.

new moon The lunar phase during which the Moon is dark when viewed from Earth, because it has moved to a position between Earth and the Sun and its sunlit side faces away from us.

crescent moon A lunar phase in which the Moon appears as a thin crescent, when either waxing or waning.

waxing Becoming full; the 14 to 15 days after a new moon and before a full moon when the visible portion of the Moon increases every day.

gibbous moon A bright moon, either waxing or waning, with only a sliver of dark visible.

full moon The lunar phase when the Moon appears round and fully illuminated, because it is on the opposite side of Earth from the Sun and its entire sunlit area is visible.

waning Becoming smaller; the 14 to 15 days after a full moon and before a new moon when the visible portion of the Moon decreases every day.

solar eclipse A phenomenon that occurs when the Moon passes directly between Earth and the Sun; the Moon casts its shadow on Earth, thus blocking the Sun's light.

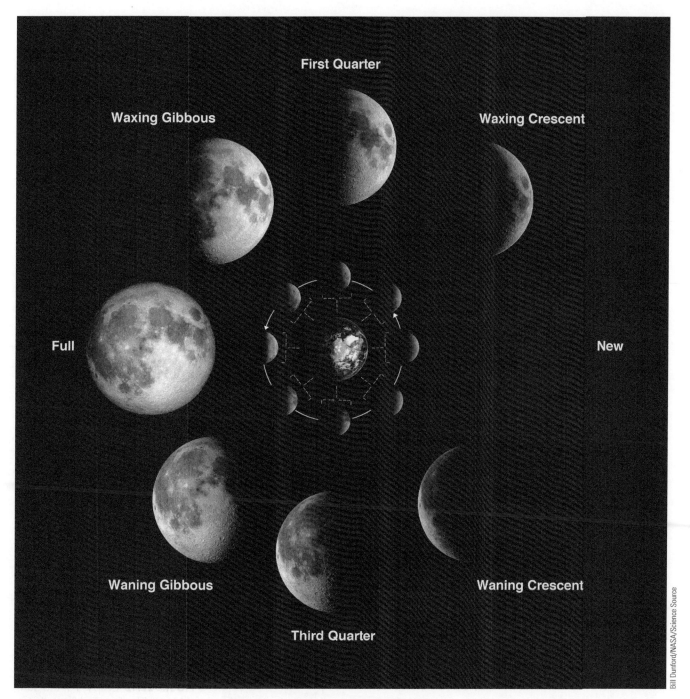

● **FIGURE 22.10** Approximately every 29.5 Earth days, the Moon passes through a complete cycle of phases. The center of the drawing shows the Earth–Moon orbital system viewed over the course of a month. The larger images of the moon show its view from Earth during each phase shown on the central drawing.

Eclipses of the Sun and the Moon

Recall that the plane of the Moon's orbit around Earth is tilted with respect to that of Earth's orbit about the Sun, so normally the Moon lies slightly out of the plane of the Earth–Sun orbit. As a result, during a new moon, the Moon's shadow misses the Earth (●**FIGURE 22.12A**) and at a full moon Earth's shadow misses the Moon (●**FIGURE 22.12B**).

However, on rare occasions, the new moon passes through the Earth–Sun orbital plane. At these times, the Moon passes directly between Earth and the Sun. When this happens, the Moon's shadow falls on Earth, producing a **solar eclipse** (●**FIGURE 22.12C**). As the Moon slides in front of the Sun, an eerie darkness descends, and Earth becomes still and quiet. Birds return to their nests and stop singing. During a *total eclipse*, the Moon blocks out

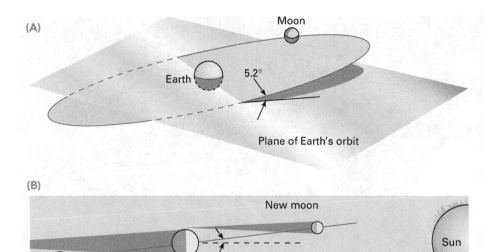

(A)

Moon

Earth

5.2°

Plane of Earth's orbit

(B)

New moon

Sun

Earth 5.2°

Full moon

● **FIGURE 22.11** (A) The Sun and Earth lie in one plane, while the Moon's orbit around Earth lies in another. (B) A sideways look at the Sun, Moon, and Earth shows that most of the time the Moon's shadow misses Earth and Earth's shadow misses the Moon. Scales are exaggerated for emphasis.

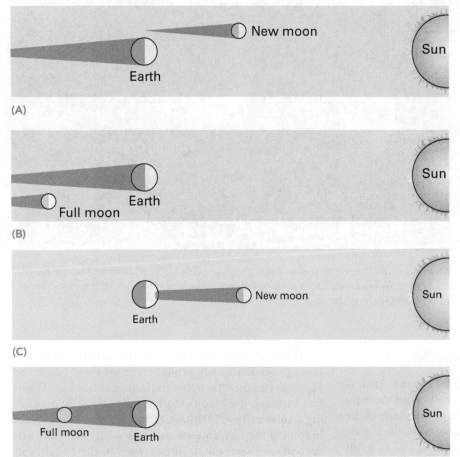

New moon

Sun

Earth

(A)

Sun

Earth

Full moon

(B)

New moon

Sun

Earth

(C)

Sun

Full moon Earth

(D)

the entire surface of the Sun, but the outer solar atmosphere, or **corona**—normally invisible because of the Sun's brilliance—appears as a halo around the black Moon (●**FIGURE 22.13**). Due to the relative distances between Sun, Moon, and Earth and their respective sizes, the Moon's shadow is only a narrow band on Earth (●**FIGURE 22.14**). The band where the Sun is totally eclipsed, called the **umbra**, is never wider than 275 kilometers. In the **penumbra**, a wider band outside of the umbra, only a portion of the Sun is eclipsed. During a *partial eclipse* of the Sun, the sky loses some of its brilliance but does not become dark. Viewed through a dark filter, a semicircular shadow cuts across the Sun.

If the Moon passes through the Earth–Sun orbital plane when it is full, then Earth lies directly between the Sun and the Moon. At these times, Earth's shadow falls on the Moon and the Moon temporarily darkens to produce a **lunar eclipse** (●**FIGURE 22.12D**). Lunar eclipses are more common and last longer than solar eclipses because Earth is larger than the Moon and therefore its shadow is more likely to cover the entire lunar surface. A lunar eclipse can last a few hours.

● **FIGURE 22.12** Eclipses of the Sun and Moon. The Sun and Earth lie in one plane, while the Moon's orbit around Earth lies in another. Scales are exaggerated for emphasis. (A) During a new moon, the Moon's shadow misses Earth. (B) During a full moon, Earth's shadow misses the Moon. (C) An eclipse of the Sun occurs when the Moon is directly between the Sun and Earth, and the Moon's shadow is cast on Earth. (D) An eclipse of the Moon occurs when Earth's shadow is cast on the Moon.

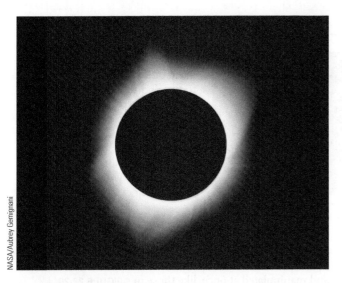

● **FIGURE 22.13** The solar corona appears as a bright halo around the eclipsed Sun. This photograph was taken during the total eclipse of the Sun on July 11, 1991, in La Paz, Baja California.

---

corona The outer atmosphere of the Sun, normally invisible but appearing as a halo around a black Moon during a solar eclipse.

umbra The narrow band of the Moon's shadow where the Sun is completely blocked out during a solar eclipse.

penumbra A wide band outside of the umbra, where only a portion of the Sun is hidden from view during a solar eclipse.

lunar eclipse A phenomenon that occurs when Earth lies directly between the Sun and the Moon, causing Earth's shadow to fall on the Moon and darken it.

resolution The capability to which an instrument can detect or measure detail, such as the capability of an optical telescope to define individual craters on the Moon.

refracting telescope A type of telescope that uses two lenses: one to collect light from a distant object and another to magnify the image.

objective lens The lens, farthest from the eyepiece in a refracting telescope, that collects light from a distant object.

eyepiece The lens, closest to the eye in a refracting telescope, that magnifies the image.

reflecting telescopes A type of telescope that uses a mirror or mirrors to collect and focus an image, which is then reflected to the eyepiece by another mirror.

MODERN ASTRONOMY

Once astronomers understood the relative motions of the Sun, Moon, and planets, they began to ask questions about the nature and composition of these bodies and to probe more deeply into space to study stars and other objects in the Universe. How do we gather data about such distant objects?

If you stand outside at night and look at a speck of light in the sky, the information you receive is limited by several factors. For one, your eye detects only visible light, which is just one-millionth of 1 percent of the electromagnetic spectrum. Thus, more than 99.99 percent of the spectrum is invisible to the naked eye. In addition, the naked eye collects little light, and you may not see faint or distant objects at all. Your eye also has poor **resolution**; it may see one dot when two actually exist (●**FIGURE 22.15**). Finally, the light you see has been distorted by Earth's atmosphere. Modern astronomers attempt to overcome these difficulties with telescopes and other instruments.

Optical Telescopes

A telescope is a device that collects light from a wide area and then focuses it where it can be detected. Detection devices may be simple, such as your eye, or complex such as telescopes that stay trained on one section of the sky for several hours, collecting enough light to form an image from a weak signal.

Galileo and many other early astronomers used a **refracting telescope**, which utilizes two lenses. The first, called the **objective lens**, collects light from a distant object. The second is a small magnifying lens, called the **eyepiece**. Light bends, or refracts, when it passes through the curved surface of the objective lens (●**FIGURE 22.16**). The bent light rays converge on the focus, forming an image of a distant object. The eyepiece then magnifies the image.

The problem with refracting telescopes is that different colors in the spectrum refract by different amounts. Therefore, if you focus the telescope to collect blue light sharply, red light will be fuzzy. As a result, most modern optical telescopes are **reflecting telescopes**. They collect light with a large curved mirror and reflect it to an eyepiece (●**FIGURE 22.17**). Modern telescopes are outfitted with both a camera and electronic detectors at the eyepiece.

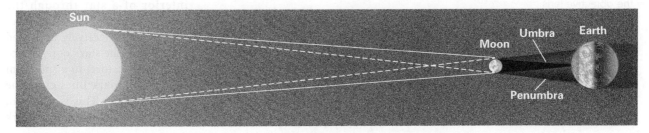

● **FIGURE 22.14** A total solar eclipse is viewed in the narrow band, called the umbra, formed by the projection of the Moon's shadow on Earth. The penumbra is the wider band where a partial eclipse is visible. (Drawing is not to scale.)

Increasing resolution

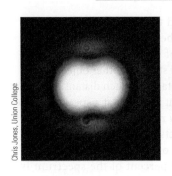

Chris Jones, Union College

● **FIGURE 22.15** Resolution refers to the degree to which details are distinguishable in an image. The photograph on the left has poor resolution and appears to show a single object. With increasing resolution (center and right photos), we clearly see two distinct objects.

As cities and suburban sprawl have grown, electric lights associated with them compete with the lights from distant stars, making them much less visible than in the countryside far from electric lights. This human-generated background light is called **light pollution,** and it reduces the distances and clarity with which telescopes operate.

In the past few decades, astronomers have built more powerful telescopes, both on land and in space. In 1990, the Hubble Space Telescope (HST) was launched into orbit around Earth (●**FIGURE 22.18**). In the vacuum of space, the HST isn't adversely affected by either light pollution or atmospheric interference. The Hubble's high-resolution images have altered our understanding of many celestial bodies (●**FIGURE 22.19**). The HST is outfitted with sensors to collect visible light and radiation in other portions of the spectrum.

Projects like Hubble are enormously expensive, but with developments in technology, astronomers can also build increasingly powerful observatories on land. Because a mirror much larger than 600 centimeters sags under its own weight, recent telescopes use an array of smaller mirrors to increase the mirror area. The mirrors are focused by computer, meaning that a large number of mirrors can be focused and manipulated at once, like those in ●**FIGURE 22.20**.

Telescopes Using Other Wavelengths

Visible light is only a small portion of the electromagnetic spectrum. The wavelengths of electromagnetic radiation emitted by a star are determined by several factors, including the types of nuclear reactions that occur in the star, its chemical composition, and its temperature.

In recent years, astronomers have enhanced our knowledge of stars and other objects in space by studying many different wavelengths, from low-energy radio and infrared signals to high-energy gamma rays and X-rays. The telescopes used in these studies often do not look like conventional optical telescopes but rather are tailored to the characteristics of the wavelength being studied.

Emission and Absorption Spectra

If light passes through a prism, it separates into a **spectrum**, an ordered array of colors[1] (●**FIGURE 22.21**). A rainbow is such an array, with white sunlight separated into its individual colors. Each color is formed by a band of wavelengths.

As light passes from the hot interior of a star through the cooler, outer layers, some wavelengths are selectively absorbed by atoms in the star's outer atmosphere. Therefore, in a spectrum of starlight, dark lines cross the band of colors. This is called an **absorption spectrum**. Each dark

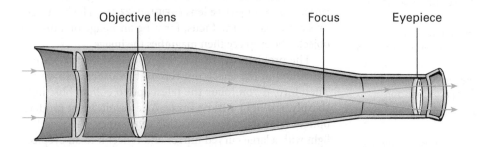

● **FIGURE 22.16** In a refracting telescope, light is collected and focused by a large objective lens. A second lens, called the eyepiece, magnifies the image produced by the objective lens.

Objective lens · Focus · Eyepiece

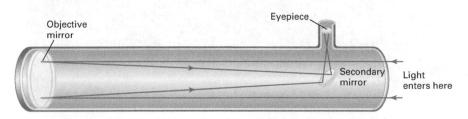

Objective mirror · Eyepiece · Secondary mirror · Light enters here

● **FIGURE 22.17** A Newtonian reflecting telescope. Incoming light (right) is collected and focused by a curved objective mirror. The light is then reflected back to the eyepiece by a secondary mirror.

1. In modern instruments light is dispersed by a diffraction grating, not a prism, but the effect is the same.

● **FIGURE 22.18** The orbiting Hubble Space Telescope has enabled astronomers to make many new discoveries of the Solar System, our galaxy, and intergalactic space.

● **FIGURE 22.19** The Hubble Space Telescope has helped astronomers solve the mystery of this loner starburst galaxy, called NGC 1569, by showing that it is one and a half times farther away than astronomers had thought.

line represents a wavelength that is absorbed by atoms of a particular element. Thus an absorption spectrum enables us to determine the chemical composition of a star. This type of analysis is so effective that astronomers discovered helium in the Sun 27 years before chemists detected helium

---

light pollution The nighttime glow of city lights that competes with the light of distant stars and reduces the vision of a telescope.

spectrum An ordered array of colors; a pattern of wavelengths into which a beam of light or other electromagnetic radiation is separated.

absorption spectrum A spectrum of radiation wavelengths that are absorbed when light passes through a substance, such as a star; absorbed wavelengths appear as dark lines crossing a full-color spectrum.

emission spectrum A spectrum created when absorbed radiation is reemitted from its source.

Doppler effect The observed change in frequency of light or sound that occurs when the source of the wave is moving either toward or away from the observer.

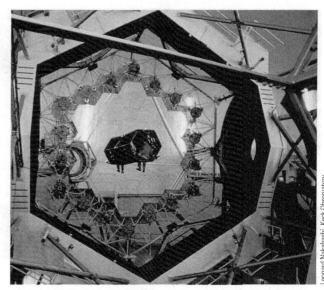

● **FIGURE 22.20** The 10-meter mirror of the Keck Telescope is composed of 36 separate hexagonal sections, each adjusted by computer. This view shows the mirror under construction, with 18 of the hexagonal sections installed. The worker with the orange shirt is installing one hexagonal section; notice how much larger the overall mirror is.

in material on Earth. Because an atom's spectrum changes with temperature and pressure, spectra can also be used to determine surface temperatures and pressures of stars.

Whenever an atom absorbs radiation, it must eventually reemit it. Such an **emission spectrum** is often hard to detect against the bright background of starlight, but these spectra can be seen along the outer edges of some stars and in large clouds of dust and gas in space.

Doppler Measurements

Have you ever stood by a train track and listened to a train speed by, blowing its whistle? As it approaches, the pitch of the whistle sounds higher than usual, and after it passes, the pitch lowers. This change, called the **Doppler effect**, was first explained for both light waves and sound waves by the Austrian physicist Johann Christian Doppler in 1842.

A stationary object remains in the center of the circular waves it generates (●**FIGURE 22.22A**). The waves from a moving object crowd each other in the direction of the object's motion. The object, in effect, is catching up with its own waves (●**FIGURE 22.22B**). If the object is moving toward you, you receive more waves per second (higher frequency) than you would if it were stationary, and if it is moving away from you, you receive fewer waves per second (lower frequency).

In the same way, the frequency of light waves changes with relative motion. The Doppler effect causes light from an object moving away from Earth to reach us at a lower frequency than it had when it was emitted. Lower frequency light is closer to the red end of the spectrum. Thus, a Doppler

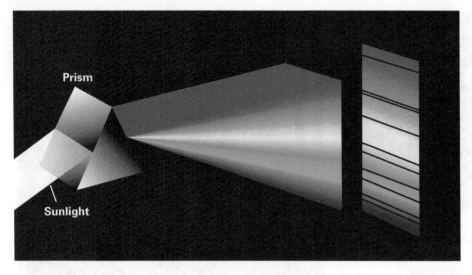

shift to lower frequency is called a **red shift**. Alternatively, light from an object traveling toward us reaches Earth at higher frequency, called a **blue shift**. Using these principles, astronomers measure the relative velocities of stars, galaxies, and other celestial objects millions or billions of light-years away.

● **FIGURE 22.21** A prism disperses a beam of sunlight into a spectrum of component colors. Each color represents a different band of wavelengths.

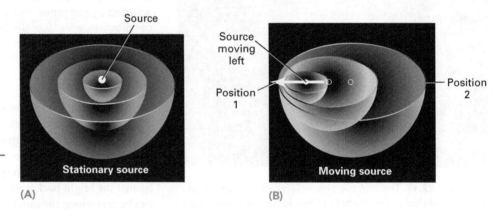

red shift A shift toward longer (red) wavelengths, observed in the spectrum of a distant galaxy or other object that is moving away from Earth; caused by the Doppler effect.

blue shift A shift toward shorter (blue) wavelengths, observed in the spectrum of a distant galaxy or other object that is moving toward Earth; caused by the Doppler effect.

● **FIGURE 22.22** Doppler effect. (A) When a stationary source emits a signal, the frequency of the signal is unaffected by the source, and observers in any direction detect the same frequency. (B) The Doppler effect is the change in the frequency of a signal when the source is moving towards or away from the observer. If the source is moving toward an observer (position 1), the observer detects waves squeezed close together and therefore of higher frequency than waves detected from the same source when it is stationary. If the source is moving away from an observer (position 2), the observer detects waves stretched farther apart and therefore of lower frequency.

Key Concepts Review

- Constellations are groups of stars that seem to form a pattern when viewed from Earth and serve as points of reference for the positions of objects in the night sky. For most of the year, planets appear to drift eastward with respect to the stars, but sometimes they seem to reverse direction and drift westward. This apparent reverse movement is called retrograde motion.

- Aristotle proposed a geocentric, or Earth-centered, Universe in which a stationary, central Earth is surrounded by celestial spheres that contain the Sun, the Moon, the planets, and the stars. Ptolemy modified the

geocentric model to explain the retrograde motion of the planets. In Ptolemy's model, each planet moves in small circles within its larger orbit.

- Copernicus believed that the Universe should operate in the simplest manner possible and showed that a heliocentric Solar System best explains the movement of planets including their retrograde motions. Kepler calculated the elliptical orbits of the planets. Galileo used observation and experimentation to discredit the geocentric model and show that the planets revolve around the Sun. Newton proved that gravity holds the planets in elliptical orbits, keeping them from flying off into space.

- The revolution of the Moon and Earth around their common center of gravity causes the phases of the Moon. A lunar eclipse occurs when Earth lies directly between the Sun and the Moon. A solar eclipse occurs when the Moon lies directly between the Sun and Earth.
- Objects in space are studied with both optical telescopes and telescopes sensitive to invisible wavelengths. Emission and absorption spectra provide information about the chemical compositions and temperatures of stars and other objects. The Doppler effect causes light from an object moving away from Earth to reach us with a lower frequency than it had when it was emitted and light from an object moving toward Earth to have a higher frequency than it had when it was emitted. Instruments carried aloft by spacecraft eliminate interference by Earth's atmosphere and allow closer inspection of objects in the Solar System.

Important Terms

absorption spectrum (p. 526)

blue shift (p. 528)

celestial spheres (p. 516)

constellations (p. 516)

corona (p. 524)

crescent moon (p. 522)

Doppler effect (p. 527)

emission spectrum (p. 527)

eyepiece (p. 525)

full moon (p. 522)

geocentric (p. 516)

gibbous moon (p. 522)

heliocentric (p. 519)

inertia (p. 520)

light pollution (p. 526)

lunar eclipse (p. 524)

new moon (p. 522)

objective lens (p. 525)

parallax (p. 517)

penumbra (p. 524)

planets (p. 516)

red shift (p. 528)

reflecting telescopes (p. 525)

refracting telescope (p. 525)

resolution (p. 525)

retrograde motion (p. 516)

revolution (p. 521)

rotation (p. 521)

solar eclipse (p. 523)

spectrum (p. 526)

umbra (p. 524)

waning (p. 522)

waxing (p. 522)

Review Questions

1. List and explain Aristotle's observations and the reasoning he used to support his geocentric model of the Solar System.

2. Why did Aristotle and other ancient astronomers fail to notice a parallax shift when they observed the stars?

3. What is retrograde motion? How is it explained in Ptolemy's and Copernicus's models?

4. Explain how Galileo's studies of physics contributed to his rejection of Aristotle's geocentric model.

5. Explain how Galileo's observations of the Milky Way, the Sun, and the Moon led him to question Aristotle's geocentric model.

6. What evidence convinced Galileo that Earth revolves around the Sun?

7. What was the astronomical significance of Newton's studies of gravity?

8. How is the Moon positioned with respect to Earth and the Sun when it is full, new, gibbous, and crescent?

9. Explain the difference between rotation and revolution as they relate to planetary motion.

10. Draw a picture of the Sun, Earth, and Moon as they appear during an eclipse of the Sun. Draw a picture of the Sun, Earth, and Moon as they are arranged during an eclipse of the Moon. Explain how these different arrangements produce each type of eclipse.

11. Explain how optical telescopes work. How is a refracting telescope different from a reflecting telescope?

12. Discuss the relative advantages and disadvantages of space telescopes versus ground-based observatories.

13. Describe the differences between absorption spectra and emission spectra.

14. Explain the Doppler effect.

NASA/JPL/Space Science Institute

23

PLANETS AND THEIR MOONS

While cruising around Saturn at a distance of about 6.3 million kilometers in early October 2004, Cassini orbiter captured a series of images that have been composed into the most detailed natural color view of Saturn and its rings ever made.

THE SOLAR SYSTEM: A BRIEF OVERVIEW

The Solar System formed about 4.6 billion years ago from a cold, diffuse cloud of dust and gas rotating slowly in space. The cloud was composed of about 92 percent hydrogen and 7.8 percent helium, about the same elemental composition as the Universe.[1] Most of this hydrogen and helium originated during the early moments of the Big Bang, as smaller atomic particles cooled enough to form these simplest of atoms. The remaining 0.2 percent of cloud forming the early Solar System consisted of all naturally-occurring elements heavier than helium, including lithium, carbon, oxygen, iron, gold, uranium, etc. These heavier elements had formed by fusion of hydrogen and helium along with some of the heavier elements themselves in the centers of stars that had since exploded.

A portion of the cloud gravitated toward its center to form the Sun. Here the pressure became so intense that hydrogen fused, producing energy as some of the hydrogen converted to helium. Hydrogen fusion is still the source of the Sun's energy and will be discussed in Chapter 24. The remaining matter in the original cloud formed a disk-shaped, rotating *nebula*, or cloud of interstellar dust, that eventually coalesced into separate spheres to produce the planets. The evolution of the planets is an example of a feedback mechanism occurring in space. Any object—including a rocket ship or a gas molecule—can escape from a planet's gravity when it reaches a speed known as the **escape velocity**. The escape velocity is proportional to the mass, and hence the gravitational force, of the planet.

Let us now compare the evolution of Mercury, the closest planet to the Sun, with that of Jupiter, one of the more distant planets. In their primordial states, all planets were composed mainly of gases. Gases in a planetary atmosphere are in constant motion, and the higher the temperature, the higher the average speed. Mercury, being closer to the Sun, was originally hotter than Jupiter. Its gases were moving faster, so they were more likely to escape the planet's gravity and fly off into space. As gases escaped, the planet lost mass, so the escape velocity decreased, making it easier for gases to escape. In addition, the **solar wind**, a stream of electrons and positive ions radiating outward from the Sun at high speed, blew even more gases away from the primordial planets. Combining all processes, the inner planets—Mercury, Venus, Earth, and Mars—lost most of their gases, leaving behind spheres composed mostly of nonvolatile metals and silicate rocks. These four are now called the **terrestrial planets**. In contrast, the

protoplanets in the outer reaches of the Solar System were so far from the Sun that they were initially cool. As a result, the outer **Jovian planets**—Jupiter, Saturn, Uranus, and Neptune—retained large amounts of hydrogen, helium, and other light elements. In fact, they actually grew larger as they captured gases that escaped from the terrestrial planets. As the mass increased, the escape velocity also increased, and gas escape became very slow. Through time, both the terrestrial and jovian planets underwent **planetary differentiation** in which heavier elements migrated towards the center of the planets and lighter elements migrated towards the planets outer edges. Here on Earth, this differentiation has produced the solid iron core, liquid outer core consisting of a molten alloy of iron and nickel, a mantle consisting of iron- and magnesium-rich ultramafic minerals, and a crust made of much

LO1 Summarize the formation of the Solar system.
LO2 Distinguish between the terrestrial and Jovian planets.
LO3 Discuss the characteristics of Mercury.
LO4 Explain the effect of atmosphere on the climate of a planet.
LO5 Review the findings of "Sprit" and "Opportunity" on Mars.
LO6 Compare the appearance of the craters on Mercury to those on the Earth and Venus.
LO7 Give examples of tectonic activity on Venus.
LO8 Discuss the origin and formation of the moon.
LO9 Describe the composition of the rings of Saturn.
LO10 Discuss the classification of Pluto as a dwarf planet.
LO11 Compare asteroids, comets, and meteoroids.

1. Composition is given here in percentage of number of atoms. Percentage by mass is significantly different.

About five billion years ago, exploding stars accelerated dust and gas into the great void of space.

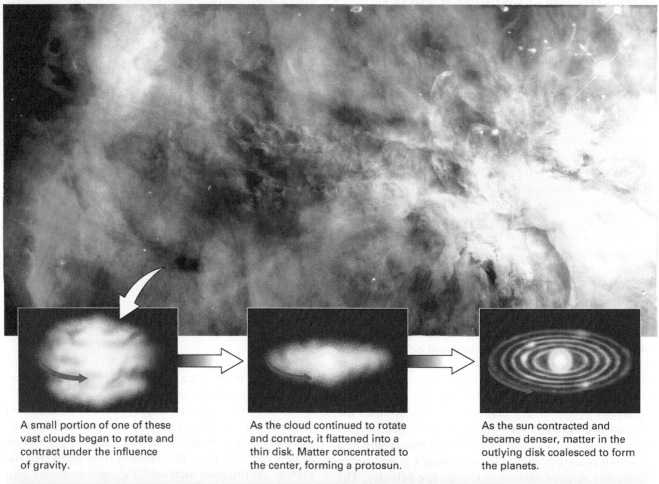

A small portion of one of these vast clouds began to rotate and contract under the influence of gravity.

As the cloud continued to rotate and contract, it flattened into a thin disk. Matter concentrated to the center, forming a protosun.

As the sun contracted and became denser, matter in the outlying disk coalesced to form the planets.

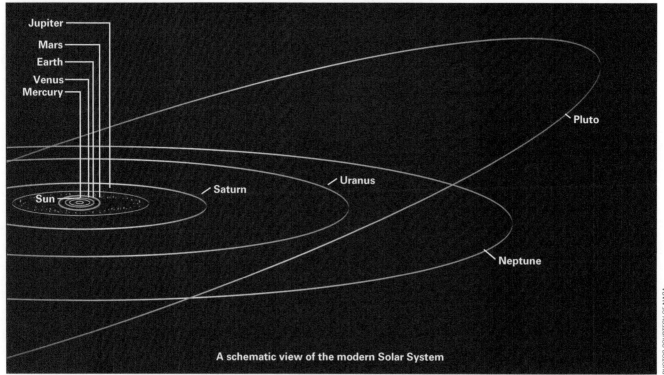

A schematic view of the modern Solar System

The planets drawn to scale. The terrestrial planets are so small on this scale that they are enlarged below. The Jovian planets (right) are predominantly composed of gases and liquids with small rocky and metallic cores, while the terrestrial planets (below) are composed primarily of rock and metal.

Jupiter

Neptune

Saturn

Uranus

Terrestrial planets

Mercury

Moon

Venus (radar image)

Mars

The planets and the Sun drawn to scale. Note that the Earth–Moon orbital system would easily fit into a portion of the Sun's surface. The Earth is the most massive of the terrestrial planets, but is only 1/300 the mass of Jupiter. Saturn's rings would reach from the Earth to the Moon.

Mercury

Sun

Venus

Earth

Moon

Mars

Jupiter

Saturn

Uranus

Neptune

Pluto

Photos Courtesy of NASA

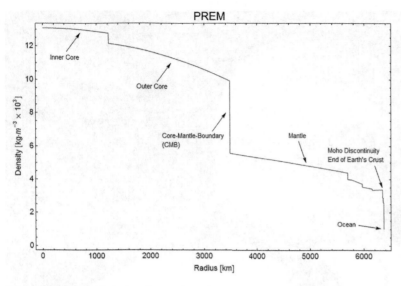

● **FIGURE 23.1** This plot shows the relationship between the thickness and density of Earth's internal layers.

lighter silicate minerals, including abundant quartz. (●**FIGURE 23.1**) shows the density and thickness of Earth's interior layers.

Although the Jovian planets are made mostly of relatively light elements, planetery differntiation concentrated the heaviest of these elements in the center of each planet. This concentration of the heaviest elements in the center of the planets, along with the intense pressures there, combined to form a relatively small, rocky or metal core. Surrounding this rocky core is a much thicker layer of swirling liquid and atmospheric gas.

TABLE 23.1 provides an overview of the eight major planets. Due to the differences in composition, the terrestrial planets are much denser than the Jovian planets. However, the Jovian planets are much larger and more massive than Earth and its neighbors.

TABLE 23.1 Comparison of the Eight Major Planets

Planet	Distance from Sun (millions of kilometers = 1)	Radius (compared to radius of Earth = 1)	Mass (compared to mass of Earth = 1)	Density (compared to density of water = 1)	Composition of Planet	Density of Atmosphere (compared to Earth's atmosphere = 1)	Number of Moons
Terrestrial Planets							
Mercury	58	0.38	0.06	5.4	Rocky with metallic core	One-billionth	0
Venus	108	0.95	0.82	5.2		90	0
Earth	150	1	1	5.5		1	1
Mars	229	0.53	0.11	3.9		0.01	2
Jovian Planets							
Jupiter	778	11.2	318	1.3	Liquid hydrogen surface with liquid metallic mantle and solid core	Dense and turbulent	62
Saturn	1,420	9.4	94	0.7			62
Uranus	2,860	4.0	15	1.3	Hydrogen and helium outer layers with solid core	Similar to Jupiter except that some compounds that are gases on Jupiter are frozen on the outer planets	27
Neptune	4,490	3.9	17	1.7			13

THE TERRESTRIAL PLANETS

As the planets coalesced, the primordial Solar System was crowded with asteroids, meteoroids, comets, and other chunks of rock, gases, and ices. This space debris crashed into the planets, pock-marking their surfaces with millions of craters.

Over geologic time, craters can be obliterated by tectonic events, weathering, and erosion. However, if these processes do not occur, then the craters remain for billions of years. Thus, scientists learn a lot about a planet's history simply by observing crater density. With this background, let us compare the geology and the atmospheres of the four terrestrial planets.

Atmospheres and Climates of the Terrestrial Planets

MERCURY **Mercury**, which is the closest planet to the Sun and therefore initially the hottest, has lost essentially all of its atmosphere. Today it is a rocky sphere with a radius of 2,400 kilometers, less than 0.4 that of Earth's radius. Mercury makes a complete circuit around the Sun faster than any other planet; each Mercurial year is only about 88 Earth days long. Mercury rotates slowly on its axis, so there are only three Mercurial days every two Mercurial years. Because Mercury is so close to the Sun and its days are so long, the temperature on its sunny side reaches 427°C, hot enough to melt lead. In contrast, the temperature on its dark side drops to −175°C, cold enough to freeze methane. The lack of an atmosphere is partly responsible for these extremes of temperature, because there is no wind to carry heat from the hot, sunlit regions to the dark, frigid shadows.

In 1991, radar images of Mercury revealed highly reflective regions at the planet's poles. Data indicated that these regions were composed of ice. Mercury's spin axis is almost perpendicular to its orbital plane around the Sun, so the Sun never rises or sets at the poles but remains low on the horizon throughout the year. Because the Sun is so low in the sky, regions inside meteorite craters are perpetually in the shade. With virtually no atmosphere to transport heat, the shaded regions have remained below the freezing point of water for billions of years.

VENUS Recall from Chapter 1 that **Venus** closely resembles Earth in size, density, and distance from the Sun. As a result, Venus and Earth probably had similar atmospheres early in their histories. However, Venus is closer to the Sun than Earth is, and therefore it was initially hotter. One hypothesis suggests that because of the higher temperature, water never condensed—or if it did, it quickly evaporated again. Because there were no seas for carbon dioxide to dissolve into, most of the carbon dioxide also remained in the atmosphere. Water and carbon dioxide combined to produce a runaway greenhouse effect, and surface temperatures became unbearably hot. The molecular weight of water is less than half that of carbon dioxide. Because water is so light, the hot surface temperature boiled the water into space and the solar wind swept most of the vapor toward the outer reaches of the Solar System. Today, the Venusian atmosphere is 90 times denser than that of Earth. Thus, atmospheric pressure at the surface of Venus is equal to the pressure 1,000 meters beneath the sea on our planet. The Venusian atmosphere is more than 97 percent carbon dioxide, with small amounts of nitrogen, helium, neon, sulfur dioxide, and other gases. Corrosive sulfuric acid aerosols float in a dense cloud layer that perpetually obscures the surface. Due to greenhouse warming, the Venusian surface is hotter than that of Mercury, and almost certainly hot enough to destroy the complex organic molecules necessary for life.

EARTH On Earth, outgassing from the mantle modified the primordial atmosphere. As life evolved, complex interactions among the geosphere, hydrosphere, and biosphere further altered the atmosphere, creating conditions favorable for life. This complex sequence of events is discussed in Chapter 17.

MARS Today the surface of **Mars**, the fourth planet from the Sun, is frigid and dry. The surface temperature averages −56°C and never warms up enough to melt ice. At the poles, the temperature can dip to −120°C, freezing carbon dioxide to form dry ice. The atmosphere at the Martian surface is as thin as Earth's atmosphere 43 kilometers high, which for us is the outer edge of space. Although water ice exists in the Martian polar ice caps and in the soil, there is currently no liquid water on Mars.

However, abundant evidence indicates that the Martian climate was once much warmer and that water flowed across the surface. Photographs from Mariner and Viking spacecraft show eroded crater walls and extinct streambeds and lake beds. One giant canyon, Valles Marineris, is approximately 10 times longer and 6 times wider than Grand Canyon in the American Southwest (●**FIGURE 23.2**). Massive alluvial fans at the mouths of Martian canyons indicate that floods probably raced across the land at speeds up to 270 kilometers per hour.

In January 2004, NASA landed two mobile robots, called **Spirit** and **Opportunity**, on the Martian surface. Each robot was equipped with six wheels and a drive

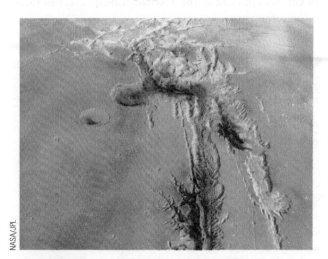

NASA/JPL

● **FIGURE 23.2** The giant Martian canyon, Valles Marineris, was eroded by flowing water and is many times larger than the Grand Canyon in the American Southwest.

Mercury The closest planet to the Sun.

Venus The second planet from the Sun; resembles Earth in size and density.

Mars The fourth planet from the Sun.

Spirit One of two mobile robots that landed on Mars in 2004.

Opportunity One of two mobile robots that landed on Mars in 2004.

system, communications equipment to receive instructions from Earth and transmit data, and solar panels for power. Scientific research was conducted with a variety of sensitive instruments: a camera, a magnifying glass, a grinding wheel to burrow into a rock and expose fresh surfaces, a device to analyze the chemical elements in rock, two instruments to identify minerals by analyzing their visible and non-visible spectral radiation, and a magnet to collect magnetic dust particles. One of the main functions of these rovers was to look for evidence of the existence of water in the Martian past.

The Spirit rover landed on a dry crater and did not detect any evidence of water. However, the rover Opportunity discovered layered sedimentary rocks near its landing site in Meridiani Planum (●**FIGURE 23.3**). Chemical and mineralogical analysis indicated the abundance of iron-rich minerals that usually form in the presence of water. Opportunity also found sulfate salts that had clearly precipitated from a water-rich solution. Several weeks later, Opportunity took spectacular images of ripple marks preserved on the floor of a long-dead lake or ocean.

Meanwhile, Spirit climbed off the lava-covered Gusev plain and found precipitated sulfur salts on the crater rim, indicating that some water had existed here, too. Although water was definitely present in the Martian past, the lakes and oceans were either cold, short-lived, or covered with ice. This conclusion is derived mainly from the observation that carbonate rocks, such as limestone, precipitate from warm, carbon dioxide–rich oceans. It seems probable that carbon dioxide was present in the Martian atmosphere, but there are no carbonates.

The data and information supplied about Mars by the Spirit and Opportunity rovers directly informed scientists about the leading topics to be addressed by the Curiosity Mars rover (●**FIGURE 23.6**), which landed successfully in Gale Crater on August 6, 2012 after being launched from Florida's Cape Canaveral on November 26, 2011. The Curiosity rover is the current centerpiece of the Mars Science Laboratory, a long-term collaboration among the U.S. National Aeronautics and Space Administration (NASA), California Institute of Technology and Jet Propulsion Laboratory. The main scientific goals of the Mars Science Laboratory are to determine whether Mars could ever have supported life, investigate the role that water may have played in supporting life on Mars, and better document the Martian climate and geology. Under these overall goals, the mission has several specific objectives: 1) document the occurrence of any organic compounds or other building blocks of life including carbon, nitrogen, phosphorus, oxygen and sulfur, and identify any signs of life; 2) examine and document the chemistry, mineralogy, and isotopic geochemistry of Martian soils and interpret the processes that have formed them; 3) assess the evolution of the Martian atmosphere over the roughly 4 billion year age of the planet and how water and carbon dioxide currently are distributed and move around the planet; and 4) characterize the broad spectrum of radiation experienced on the Martian surface and by instruments deployed on Mars for long time periods. This last objectives is directly related to preparing for a future crewed mission to Mars, although these will come after the next scheduled launch of a Mars rover mission in 2020.

To address the long list of scientific goals and objectives, the Mars Curiosity rover is equipped with an arsenal of scientific instruments, a powerful computer, and an internal radioisotopic energy source fueled by 4.8 kg (11 lbs.) of Plutonium-238 dioxide supplied by the U.S. Department of Energy (●**FIGURE 23.4**). Heat from the radioactive decay

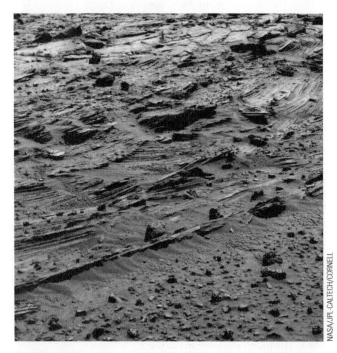

●**FIGURE 23.3** On March 8, 2006, the Opportunity rover photographed these cross-bedded sands on Mars that show sedimentary structures similar to those commonly found on Earth.

●**FIGURE 23.4** This photo shows the glowing-hot heat source for Pathfinder's internal engine – a block of Plutonium-239. Radioactive decay of the Plutonium provides heat that in turn drives Curiosity's Radioisotopic thermoelectric Power Generator.

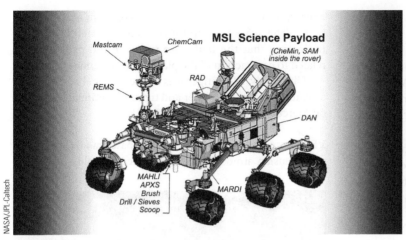

MSL Science Payload
(CheMin, SAM inside the rover)

Mastcam
ChemCam
REMS
RAD
DAN
MAHLI
APXS
Brush
Drill / Sieves
Scoop
MARDI

NASA/JPL-Caltech

Clockwise from upper left:

Mastcam is the Mast Camera instrument
ChemCam is the Chemistry and Camera instrument
RAD is the Radiation Assessment Detector instrument
CheMin is the Chemistry and Mineralogy instrument
SAM is the Sample Analysis at Mars instrument
DAN is the Dynamic Albedo of Neutrons instrument
MARDI is the Mars Descent Imager instrument
MAHLI is the Mars Hand Lens Imager instrument
APXS is the Alpha Particle X-ray Spectrometer instrument
The brush, drill, sieves and scoop are tools on the rover's robotic arm
REMS is the Rover Environmental Monitoring Station

● **FIGURE 23.5** Sketch of the Curiosity Rover with arrows pointing to the location of each instrument shown by abbreviation in the figure and described in the accompanying list.

of the Plutonium-238 drives the internal radioisotopic thermoelectric generator – Curiosity's internal engine. Because the half-life of Plutonium-238 is 87.7 years, this heat source is capable of powering Curiosity far longer than the rover is expected to last.

Curiosity carries nine separate scientific instruments, as well as a drilling system and environmental monitoring station that collectively provide it with the capacity to undertake numerous specific investigations in addition semi-continually monitoring environmental radiation, temperature, and atmospheric pressure. ●**FIGURE 23.5** shows a sketch of the Curiosity rover and onboard instrumentation and ●**FIGURE 23.6** is a selfie the Curiosity's cameras took of the rover in action on Mars.

On August 6, 2017, the community of scientists associated with the Curiosity rover's mission celebrated five years of continuous operation. This celebration is remarkable, because the rover was only expected to last for two years when it was first launched. As of this writing (September 2018, Curiosity has been operational for 2168 Martian sols. (A sol is a Martian day which, because Mars rotates more slowly on its axis than Earth, lasts for about 24 hours and 39 minutes.)

What have we learned so far from the Mars Curiosity rover? Perhaps the most important result from this ongoing mission is its impact on the development of future Earth and

Planetary Scientists: huge crowds gathered to watched the landing of the Curiosity rover on Mars, and results of the mission have been made accessible through the mission's official website (footnote: https://www.nasa.gov/mission_pages/msl/index.html),

As for specific results, the Mars Science Laboratory lists six major results of the mission as of May 8, 2015: 1) Mars could have the right environment to have supported living microbes and found all of the key elemental ingredients; 2) organic carbon was detected in Martian rocks, indicating that the raw ingredients needed to start live existed there at one time; 3) Curiosity observed a ten-fold increase in the concentration of atmospheric methane over a two-month period after observing prior background levels, suggesting that somewhere on Mars methane is being generated; 4) the Martian atmosphere has higher concentrations of heavier isotopes of hydrogen, carbon, and argon, suggesting that the ligher isotopes of these elements were stripped away by the solar wind but that they once existed in what was a much thicker atmosphere in the Martian past; 5) Sedimentary structures indicative of flowing water were observed by Curiosity, indicating that in its past flowing water existed in some places; and 6) between the time it left Earth's atmosphere en route to Mars and 2015, the Curiosity rover experienced more natural radiation than the current lifetime limit allowed by NASA for its astronauts. Thus, any future crewed mission to Mars must come after we develop the technology to reduce this radiation to acceptable levels. This development is likely to require the results from many future uncrewed Mars missions, beginning with the planned launch of the March 2020 Mission which will carry with it an as yet-unnamed rover scheduled to land on the Martian surface in 2021 and undertake a prime mission that is one Martian year (687 days) in length.

NASA/JPL-Caltech

● **FIGURE 23.6** A selfie of the Mars Curiosity on the Red Planet.

Geology and Tectonics of the Terrestrial Planets

MERCURY Little was known about the surface of Mercury before the spring of 1974, when the spacecraft Mariner 10 passed within a few hundred kilometers of the planet. Images relayed to Earth revealed a cratered surface remarkably similar to that of our Moon (●FIGURE 23.7).

Recall that craters formed on all planets and their moons during intense meteorite bombardment early in the history of the Solar System. However, tectonic activity and erosion have erased Earth's early meteorite craters. Yet, Mercury is so close to the Sun that its water and atmosphere boiled off into space, so no wind, rain, or rivers have eroded its surface. Today, 4-billion-year-old craters look as fresh as if they formed yesterday.

For many years, scientists presumed that any water that had been part of Mercury or its early atmosphere had long boiled off. This presumption was shown to be incorrect, however, by NASA's Messenger space craft which was launched in 2004 and took five years to reach Mercury and used close 'flybys' of Earth once and Venus twice and the gravity of these planets reach Mercury on January 14, 2008. Messenger subsequently flew closely by Mercury three times before achieving a stable orbit over three years later on March 18, 2011. For the next four years, Messenger used its instruments to image 100% of Mercury's surface, vastly improving the quantity and quality of information about the planet. Among the most unexpected was the discovery of water-ice at the surface and just below the surface within meteorite impact craters near Mercury's poles. Because the tilt of Mercury on its axis is only about one degree, deep impact craters in the polar regions never experience direct sunlight, and Mercury's on-board instruments were able to document the occurrence of water-ice in these locations (●FIGURE 23.8 and ●FIGURE 23.9).

Another unexpected discovery from the Messenger mission is that, although its surface is covered by meteor craters, it also shows numerous scarps from active thrust faults. The scarps generally crosscut the meteor craters, indicating that the faults are younger and probably are related to contraction of the planet as it continues to cool.

Ultimately, Messenger was limited by fuel and its orbit around Mercury got tighter and tighter. It ultimately crashed on Mercury's surface on April 30, 2015.

VENUS Astronomers use spacecraft-based radar to penetrate the Venusian atmosphere and produce photolike images of

NASA/JPL-Caltech

● **FIGURE 23.8** This image of Mercury's surface from the Messenger Dual Imaging System is color-coded for temperature and shows a wide range from from -223C (purple) to more than 125C (red). The coldest regions in purple mostly occur in meteor impact craters that are located near Mercury's poles and that never receive direct sunlight.

● **FIGURE 23.9** A photomosaic of Mercury's north pole from the Messenger mission. The orbiter not only captured high resolution images of the entire surface of the planet, but it imaged the occurrence of water-ice (yellow) within many of the meteor craters on the planet's poles. Because the tilt of Mercury on its axis is only about one degree, parts or all of the floors of these crater are always in shadow and capable of preserving ice despite Mercury's close proximity to the Sun.

NASA

● **FIGURE 23.7** Mercury has a cratered surface. The craters are remarkably well preserved because there is no erosion or tectonic activity on Mercury. The photograph shows an area 580 km from side to side.

its surface. Gravity studies, also conducted by spacecraft, are used to infer the density of rocks near the Venusian surface. These density measurements provide information about the planet's mineralogy and internal structure. The most spectacular data were obtained by the orbiting Magellan spacecraft, which was launched in May 1989. In October 1994, its mission 99 percent accomplished and federal funding running out, scientists sent Magellan on one final suicide mission into the Venusian atmosphere to provide additional information about its density and composition. During the early history of the Solar System, thick swarms of meteorites bombarded all the planets. However, Magellan's detailed maps show few meteorite craters on Venus. This observation indicates that the Venusian surface was reshaped after the major meteorite bombardments. More detailed analysis shows that most of the landforms on Venus are 300 to 500 million years old.

James Head, Magellan investigator, suggests that a catastrophic series of volcanic eruptions occurred 300 to 500 million years ago, creating volcanic mountains and covering much of the surface with basalt flows (●FIGURE 23.10). According to this model, rising mantle plumes generated magma that repaved the planet in a short time with a rapid series of cataclysmic volcanic eruptions. Head concludes: "The planet's entire crust and lithosphere turned itself over. It certainly makes a strong case that catastrophic events are in the geological records of planets and it ought to make us think about the possibilities of such catastrophic events in Earth's past." An alternative hypothesis contends that the resurfacing occurred more slowly.

Crater abundances indicate that most tectonic activity on Venus stopped after the intense volcanic activity 300 to

500 million years ago. Some evidence indicates that the volcanic activity has ceased permanently because the planet's interior cooled. This cooling may have occurred because radioactive elements floated toward the surface during the volcanic events, removing the mantle heat source. However, other data imply that Venus remains volcanically active and that another repaving event may occur in the future.

Most Earth volcanoes form at tectonic plate boundaries or over mantle plumes, so the discovery of volcanoes on Venus led planetary geologists to look for evidence of plate tectonic activity there. Radar images from spacecraft show that 60 percent of Venus's surface consists of a flat plain. Two large and several smaller mountain chains rise from the plain (●FIGURE 23.11). The tallest mountain is 11 kilometers high—2 kilometers higher than Mount Everest. The images also show large, crustal fractures and deep canyons. If Earth-like horizontal motion of tectonic plates caused these features, then spreading centers and subduction zones should exist on Venus. The images show no features like a mid-oceanic ridge system, transform faults, or other evidence of lithospheric spreading. However, geophysicists have located 10,000 kilometers of trenchlike structures that they believe to be subduction zones.

Despite the apparent existence of subduction zones, the most popular current model suggests that Venusian tectonics have been dominated by mantle plumes. In some regions, the rock has melted and erupted from volcanoes by processes similar to those that formed the Hawaiian Islands. In other regions, the hot, Venusian mantle plumes have lifted the crust to form nonvolcanic mountain ranges. Some geologists have suggested the term **blob tectonics** to describe Venusian tectonics because Venus is dominated by rising and sinking of the mantle and crust. In contrast, tectonic activity on Earth causes significant horizontal movement of its plates.

Mantle plumes on Earth may initiate rifting of the lithosphere and formation of a spreading center. Why have spreading centers not developed over mantle plumes on Venus? Perhaps surface temperature on Venus is so high that the surface rocks are more plastic than those on Earth. Therefore, rock flows plastically rather than fracturing into lithospheric plates. It is also possible that Venus has a thicker lithosphere, which can move vertically but does not fracture and slide horizontally.

EARTH Earth is large enough to have retained considerable internal heat. This heat drives convection within the mantle that produces tectonic motion. In turn, Earth tectonics continuously reshapes the surface of our planet and is partially responsible for the atmosphere that sustains us. This topic was discussed in detail in Unit 2.

MARS The Martian surface consists of old, heavily cratered plains and younger regions that have been altered by tectonic activity (●FIGURE 23.12). Lava flows much like those

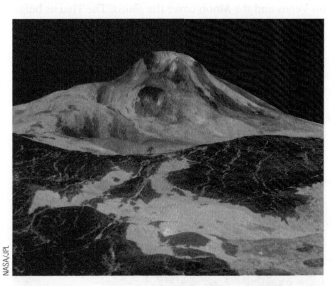

● **FIGURE 23.10** The volcano Maat Mons, on Venus, has produced large lava flows, shown in the foreground. According to one hypothesis, a catastrophic series of volcanic eruptions altered the surface of Venus 300 to 500 million years ago. This image was produced from radar data recorded by the Magellan spacecraft. Simulated color is based on color images supplied by Soviet spacecraft.

NASA/JPL

blob tectonics Tectonic activity dominated by rising and sinking of the mantle and crust, believed to predominate on Venus, as opposed to the horizontal movement of plates associated with tectonic activity on Earth.

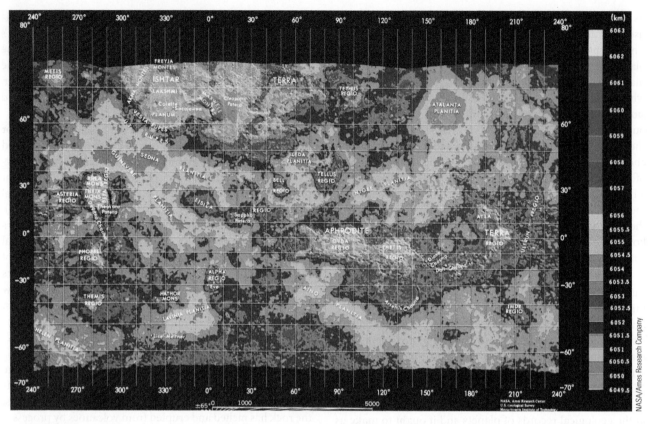

● **FIGURE 23.11** Scientists used radar images from the Pioneer Venus orbiter to produce this map of Venus. The lowland plains are shown in blue, and the highlands are shown in yellow and red-brown. The colors are assigned arbitrarily.

on Venus and the Moon cover the plains. The Tharsis bulge is the largest plain, crowned by Olympus Mons, the largest volcano in the Solar System (●**FIGURE 23.13**). Olympus Mons is nearly 3 times higher than Mount Everest, with a height of 25 kilometers and a diameter of 500 kilometers. Its central crater is so big that Manhattan Island would easily fit inside.

● **FIGURE 23.12** The cratered plain on the left in this photo of the Mars surface is a geologically old surface, whereas the mountains on the lower right are evidence of more recent tectonic activity.

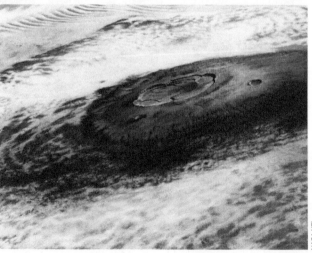

● **FIGURE 23.13** Olympus Mons is the largest volcano on Mars and also the largest in the Solar System.

One hypothesis suggests that the geology of Mars is similar to that of Venus and is dominated by blob tectonics. Supporters of this concept point out that the largest volcanic mountain on Earth, Mauna Kea in the Hawaiian Islands, is limited in size because the mountain is riding on a tectonic plate. Therefore, it will drift away from the underlying hot spot before it can grow much larger. In contrast, because horizontal movement on Mars is nonexistent or very slow, Olympus Mons has remained stationary over its hot spot. As a result, it has grown bigger and bigger.

Another line of evidence for Venusian-type tectonics on Mars is that tremendous parallel cracks split the crust adjacent to the Tharsis bulge (●FIGURE 23.14). If this bulge lay near a tectonic plate boundary, there would be folding or offsetting of the cracks. However, the cracks are neither folded nor offset. Therefore, scientists suggest that a rising mantle plume formed the Tharsis bulge and its volcanoes. The parallel cracks may be the result of stretching as the crust uplifted.

An alternative hypothesis suggests that Earth-like, horizontal tectonics is occurring on Mars. As evidence, researchers have identified what they believe to be strike-slip faults. To support this hypothesis, scientists note that a linear distribution of volcanoes on and near the Tharsis bulge is similar to linear chains of volcanoes along terrestrial subduction zones.

Observations reported in 2005 show very few meteor craters in some lava flows, indicating that eruptions have occurred as recently as a few million years ago (●FIGURE 23.15). Thus, the planet has remained hot and active until relatively recent times.

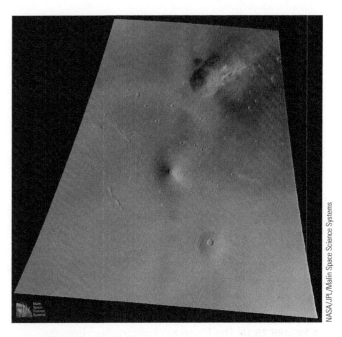

● **FIGURE 23.15** Elysium Mons is one of three large Martian volcanoes that occur on the Elysium Rise. The volcano rises about 12.5 kilometers above the surrounding plain.

THE MOON: OUR NEAREST NEIGHBOR

Most planets have small, orbiting satellites called *moons*. The Earth's Moon is close enough so that we can see some of its surface features with the naked eye. In the early 1600s, Galileo studied the Moon with a telescope and mapped its mountain ranges, craters, and plains. Galileo thought that the plains were oceans and called them seas, or **maria** (Chapter 22). The word *maria* is still used today, although we now know that these regions are dry, barren, flat expanses of volcanic rock (●FIGURE 23.16). Much of the lunar surface is heavily cratered, similar to that of Mercury (●FIGURE 23.17).

A Soviet orbiter took the first close-up photographs of the Moon in 1959. A decade later the United States landed the first of six manned Apollo spacecraft on the lunar surface (●FIGURE 23.18). The Apollo program was designed to answer several questions about the Moon: How did it form? What is its geologic history? Was

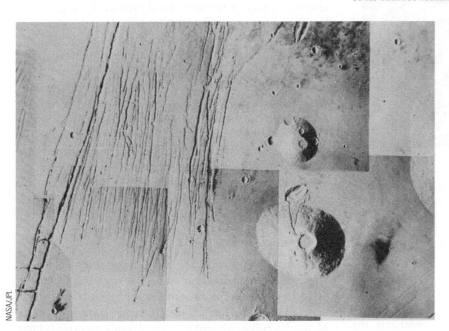

● **FIGURE 23.14** The huge parallel cracks near the Tharsis bulge on Mars are neither folded nor offset. Scientists speculate that the bulge and the cracks were formed by a rising mantle plume. However, the absence of folding or offset movement provides evidence that horizontal tectonics is not active in this region. Two large volcanoes appear at right.

maria Dry, barren, flat expanses of volcanic rock on the Moon, first thought to be seas.

NASA

● **FIGURE 23.16** The Moon as photographed from the Apollo spacecraft from a distance of 18,000 kilometers. Even from this distance, we see cratered regions and the maria, which are smooth lava flows.

NASA

● **FIGURE 23.17** Most of the lunar surface is heavily cratered. In places where smaller craters lie within the larger ones, scientists deduce that the larger craters formed first.

it once hot and molten like Earth? If so, does it still have a molten core, and is it tectonically active?

Formation of the Moon

According to the most popular current hypothesis, the Moon was created when a huge object—the size of Mars or even larger—smashed into Earth shortly after our planet formed. This massive bolide plowed through Earth's mantle, and silica-rich rocks from the mantles of both bodies

NASA

● **FIGURE 23.18** The six Apollo manned Moon landings between 1969 and 1972 answered many scientific questions about the origin, structure, and history of the Moon.

vaporized and created a cloud around Earth. The vaporized rock condensed and aggregated to form the Moon (●**FIGURE 23.19A**).

Perhaps the single most significant discovery of the Apollo program was that much of the Moon's surface consists of igneous rocks. The maria are mainly basalt flows. The highland rocks are predominantly anorthosite, a feldspar-rich igneous rock not common on Earth. Additionally, the rock of both the maria and the highlands has been crushed by meteorite impacts and then welded together by lava. Since igneous rocks form from magma, it is clear that portions of the Moon were once hot and liquid.

How did the Moon become hot enough to melt? Earth was heated initially by energy released by collisions among particles as they collapsed under the influence of gravity. Later, radioactive decay and intense meteorite bombardment heated Earth further. But what about the Moon? Radiometric dating shows that the oldest lunar igneous rocks formed before there was enough time for radioactive decay to have melted a significant amount of lunar rock. Thus, gravitational coalescence and meteorite bombardment must have been the main causes of early melting of the Moon.

History of the Moon

The Moon formed about 4.45 to 4.5 billion years ago, shortly after Earth. So much energy was released during the Moon's rapid accretion that, as the lunar sphere grew, it melted to a depth of a few hundred kilometers, forming a magma ocean. Meteorite bombardment kept the Moon's outermost layer molten. Eventually, meteorite bombardment diminished enough for the Moon's surface to cool. The igneous rocks of the lunar highlands are about 4.4 billion years old, indicating that the highlands were solid by that time.

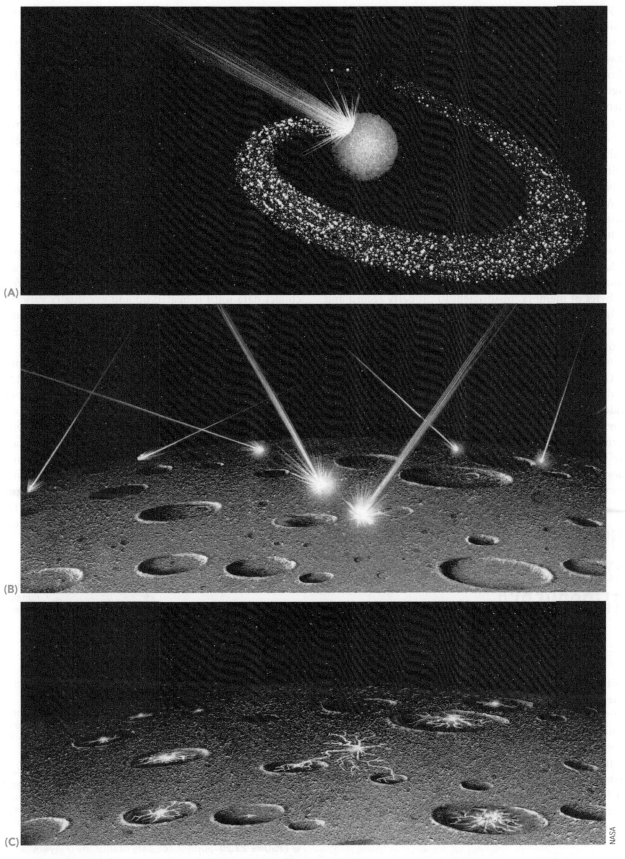

● FIGURE 23.19 (A) According to the most widely held hypothesis, the Moon was formed about 4.5 billion years ago when a Mars-sized object struck Earth, blasting a cloud of vaporized rock into orbit. The vaporized rock rapidly coalesced to form the Moon. (B) During its first 0.6 billion years, intense meteorite bombardment cratered the Moon's surface. (C) The dark, flat maria formed about 3.8 billion years ago as lava flows spread across portions of the surface. Today, all volcanic activity has ceased.

Swarms of meteorites, some as large as Rhode Island, bombarded the Moon again between 4.2 and 3.9 billion years ago (●**FIGURE 23.19B**). In the meantime, radioactive decay was also heating the lunar interior. As a result, by 3.8 billion years ago, most of the Moon's interior was molten and magma erupted onto the lunar surface. The maria formed when lava filled circular meteorite craters (●**FIGURE 23.19C**). This episode of volcanic activity lasted approximately 700 million years. The Moon and Earth shared a similar history until this time, but the Moon is so much smaller that it soon cooled and has remained geologically inactive for the past 3.1 billion years.

Apollo astronauts left seismographs on the lunar surface. Initial analysis of the data indicated that the energy released by moonquakes is only one-billionth to one-trillionth that released by earthquakes on our own planet. However, in 2005, 25 years after the first reports were released, scientists realized that the computers used in the study had only 64 kilobytes of memory, insufficient to properly record and analyze the data. Under reexamination, Yosio Nakamura from the University of Texas detected numerous deep moonquakes, indicating the Moon has a hot, possibly molten, core.

Data from the lunar Prospector spacecraft, released in March 1998, indicated that water may be present in shadowed craters at the lunar poles. In 2009, NASA instruments confirmed the presence of water molecules in the Moon's polar regions, although in relatively small amounts. The water on the Moon could be used to support human colonies or to synthesize hydrogen fuel from a Moon base for exploration of more distant planets.[2]

THE JOVIAN PLANETS: SIZE, COMPOSITIONS, AND ATMOSPHERES

Jupiter

Jupiter is the largest planet in the Solar System, 71,000 kilometers in radius. It is composed mainly of hydrogen and helium, similar to the composition of the Sun. However, Jupiter is not quite massive enough to generate fusion temperatures, so it never became a star.

Jupiter has no hard, solid, rocky crust on which an astronaut could land or walk. Instead, its surface is a vast sea 12,000 kilometers deep, made of cold, liquid molecular hydrogen (H_2) and atomic helium (He). Beneath the hydrogen/helium sea is a layer where temperatures are as

high as 30,000°C and pressures are as great as 100 million times the Earth's atmospheric pressure at sea level (●**FIGURE 23.20**). Under these extreme conditions, hydrogen molecules dissociate to form atoms. Pressure forces the atoms together so tightly that the electrons move freely throughout the packed nuclei, much as electrons travel freely among metal atoms. As a result, the hydrogen conducts electricity and is called **liquid metallic hydrogen**. Flow patterns in this fluid conductor generate a magnetic field 10 times stronger than that of Earth.

Beneath the layer of metallic hydrogen, Jupiter's core is a sphere about 10 to 20 times as massive as Earth's. It is probably composed of metals and rock surrounded by lighter elements such as carbon, nitrogen, and oxygen.

The Galileo spacecraft was launched in October 1989 and rendezvoused with Jupiter in December 1995. Once in orbit around the gas giant, the spacecraft launched a suicide probe to parachute through the outer atmosphere and collect data until it heated up and eventually vaporized. As expected, the atmosphere was primarily hydrogen and helium, with smaller concentrations of ammonia, water, and methane.

More than 300 years ago, two European astronomers reported seeing what came to be called the **Great Red Spot** on the surface of Jupiter. Although its size, shape, and color have changed from year to year, the spot remains intact to this day (●**FIGURE 23.21**). If the Earth's crust were peeled off like an orange rind and laid flat on the Jovian surface, it would fit entirely within the Great Red Spot. Measurements show that the Great Red Spot is a giant hurricane-like storm. Hurricanes on Earth dissipate after a week, yet this storm on Jupiter has existed for centuries. Other wind systems, rotating in linear bands around the planet, have also persisted for centuries (●**FIGURE 23.22**).

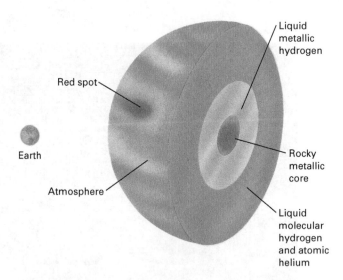

● **FIGURE 23.20** Jupiter consists of four main layers: a turbulent atmosphere, a sea consisting mainly of liquid molecular hydrogen with smaller amounts of atomic helium, a layer of liquid metallic hydrogen, and a rocky metallic core. Earth is drawn to scale on the left.

---

2. Keep in mind that "water on the Moon" does not refer to seas or even puddles. It means molecules of water and hydroxyl that interact with molecules of rock and dust in the top millimeters of the Moon's surface.

● **FIGURE 23.21** Jupiter's Great Red Spot dwarfs the superimposed image of the Earth (to scale) shown on the bottom right. The vivid colors were generated by computer enhancement.

One important mission of the Galileo suicide probe was to measure atmospheric pressures, temperatures, and wind speeds below the visible outer layers. In the outer atmosphere, where the pressure was 0.4 bar, the Galileo probe measured a wind speed of 360 kilometers per hour with a temperature of −140°C. After falling 130 kilometers, the probe reported that the wind speed had increased to 650 kilometers per hour, the pressure was 22 bar, and the temperature was about +150°C. Buffeted by winds, squeezed by intense pressure, and heated beyond the tolerance of its electronics, the spacecraft stopped transmitting. Scientists calculate that 40 minutes later, when the spacecraft had sunk to deeper levels of the Jovian atmosphere, the temperature had increased to 650°C, the pressure increased to 260 bar, and the aluminum shell liquefied. As the probe continued to fall, pressure and temperature rose until the titanium hull melted; streaming metallic droplets flashed into vapor as the probe vanished.

On Earth, winds are driven by the Sun's heat. But Jupiter receives only about 4 percent of the solar energy that Earth receives. Moreover, strong winds rip through the Jovian

Jupiter The largest planet in the Solar System and fifth from the Sun.

liquid metallic hydrogen A form of hydrogen under extreme temperature and pressure, which forces the atoms together so tightly that the electrons move freely throughout the packed nuclei, and as a result the hydrogen conducts electricity.

Great Red Spot A giant hurricane-like storm on the surface of Jupiter that has existed for centuries.

● **FIGURE 23.22** This colorful, turbulent complex cloud system of Jupiter was photographed by the Voyager spacecraft. The sphere on the left, in front of the Great Red Spot, is Io; Europa lies to the right against a white oval.

atmosphere far below the deepest penetration of solar light and heat. For these reasons, scientists deduce that the Jovian weather is not driven by solar heat but by heat from deep within the planet itself. This heat accumulated when Jupiter first formed and gravitational contraction of the planet caused the mass in the middle to become compressed and heated. Although it is considered to have a solid or rocky core, it is not certain whether any radioactive decay taking place there is contributing to Jupiter's internal heat. Regardless, as heat from Jupiter's interior slowly rises, it warms the lower atmosphere, which then transmits heat by convection to higher levels. The weather systems are much more long-lasting than those on Earth because Jupiter's interior heat flux changes only over hundreds to thousands of years.

Saturn

Saturn, the second-largest planet, is similar to Jupiter. It has the lowest density of all the planets, so low that the entire planet would float on water if there were a basin of water large enough to hold it. Such a low density implies that, like Jupiter, it must be composed primarily of hydrogen and helium, with a relatively small core of rock and metal. In fact, in many ways Saturn and Jupiter are alike. For example, Saturn's atmosphere is similar to that of Jupiter. Dense clouds and great storm systems envelop the planet.

Saturn has long been famous for its rings, which are visible with a high quality amateur telescope. In 1981, the space craft Voyager 2 used its onboard photopolarimeter – an instrument for measuring polarized light – to observe

Saturn's rings up close for the first time. Voyager was launched on August 20, 1977, reached Saturn in 1981 en route to Uranus and Neptune which it has since passed on its way out of the Solar System.

Only one spacecraft has visited Saturn since Voyager 2, and that is the Cassini spacecraft which was launched on October 15, 1997 and entered orbit around Saturn on July 1, 2004, Cassini proceeded to undertake nearly 13 years of close observations related to numerous investigations of Saturn and made dozens of orbital loops between the planet and its rings and between several of the rings themselves (●**FIGURE 23.23** and **23.24**). Like Mercury's Messenger, however, Cassini eventually ran too low on fuel to maintain a

● **FIGURE 23.23** The spacecraft Cassini makes one of its final dives into Saturn's upper atmosphere in September, 2017 in this artistic rendering.

● **FIGURE 23.24** An example of an image of Saturn's rings from the Cassini spacecraft. This image is a mosaic acquired by Cassini on April 25, 2007 at a distance of about 725,000 kilometers (450,000 miles). This perspective is from south to north and looks straight through the rings.

stable orbit orbit and over a period of several days slowly plunged into the planet it was sent to study. On August 15, 2017, at 11:55am GMT, Earth received the final signal from Cassini as the space craft hurled downward into Saturn's upper atmosphere. When last heard from, Cassini was operating what was left of its thrusters at 100% capacity to slow it down as it continued to stream real-time observations back to Earth before burning up like a falling star.

Uranus and Neptune

Uranus and **Neptune** are so distant and faint that they were unknown to ancient astronomers. The Voyager II spacecraft, launched in 1977, flew by Jupiter in 1979 and by Saturn in 1981. It encountered Uranus by 1986 and Neptune in 1989. The journey from Earth to Neptune covered 7.1 billion kilometers and took 12 years. The craft passed within 4,800 kilometers of Neptune's cloud tops, only 33 kilometers from the planned path. The strength of the radio signals received from Voyager measured 1 ten-quadrillionth of a watt (11016). It took 38 radio antennas on four continents to absorb enough radio energy to interpret the signals.

Both Uranus and Neptune are enveloped by thick atmospheres composed primarily of hydrogen and helium, with smaller amounts of carbon, nitrogen, and oxygen compounds. Beneath the atmosphere, their outer layers are molecular hydrogen, but neither body is massive enough to generate liquid metallic hydrogen. Their interiors are composed of methane, ammonia, and water, and the cores are probably a mixture of rock and metals. Uranus and Neptune are denser than Jupiter and Saturn, because these outermost giants contain relatively larger, solid cores.

Spacecraft and the Hubble Space Telescope have revealed rapidly changing weather on Uranus and Neptune. On Neptune, winds of at least 1,100 kilometers per hour rip through the atmosphere, clouds rise and fall, and one region is marked by a cyclonic storm system called the Great Dark Spot, similar to Jupiter's Great Red Spot. According to one controversial hypothesis, under the intense pressure near the core, methane decomposes into carbon and hydrogen and the carbon then crystallizes into diamond. Convection currents carry the heat released during the formation of diamond to the planet's surface to power the winds.

Voyager II recorded that the magnetic field of Uranus is tilted 58 degrees from its axis. This was unexpected, as current explanations suggest that the magnetic fields of all planets should be roughly aligned with the spin axis. At first,

scientists thought that Voyager II just happened to pass Uranus during a magnetic field reversal. However, Voyager II later recorded that the magnetic field on Neptune is tilted 50 degrees from its axis. Because the probability of catching two planets during magnetic reversals is extremely low, there must be another explanation. However, at present no satisfactory hypothesis has been developed.

MOONS OF THE JOVIAN PLANETS

The Moons of Jupiter

In 1610, Galileo discovered four tiny specks of light orbiting Jupiter. He reasoned that they must be satellites of the giant planet. By 1999, astronomers had identified 16 moons orbiting Jupiter. By 2009, 62 were known, although at least 52 are relatively small, with irregular orbits. The four discovered by Galileo, referred to as the *Galilean moons*, are the largest and most widely studied: Io, Europa, Ganymede, and Callisto.

IO The innermost moon of Jupiter, **Io**, is about the size of Earth's Moon and is slightly denser. Because it is too small to have retained heat generated during its formation or by radioactive decay, many astronomers expected that it would have a cold, lifeless, cratered, Moon-like surface. However, images beamed to Earth from the Voyager spacecraft showed huge masses of gas and rock erupting to a height of 200 kilometers above the satellite's surface. This was the first evidence of active, extraterrestrial volcanism (●**FIGURE 23.25**). Images from the later Galileo spacecraft showed 100 volcanoes erupting simultaneously, making Io the most active volcanic body in the Solar System.

● **FIGURE 23.25** Voyager I captured an image of a volcanic explosion on Io (shown on the horizon). The eruption is ejecting solid material to an altitude of about 200 kilometers.

Saturn The second-largest planet and sixth from the Sun; marked by its distinctive rings.

Uranus The seventh planet from the Sun; similar to Neptune in size, composition, and atmosphere.

Neptune The eighth planet from the Sun; similar to Uranus in size, composition, and atmosphere.

Io The innermost moon of Jupiter and the most active volcanic body in the Solar System.

Recall that the gravitational field of our Moon causes the rise and fall of ocean tides on Earth. At the same time, Earth's gravity distorts lunar rock. Thus, Earth's gravitation is responsible for deep-focus moonquakes. Jupiter is 300 times more massive than Earth, so its gravitational effects on Io are correspondingly greater. In addition, the three nearby satellites—Europa, Ganymede, and Callisto—are large enough to exert significant gravitational forces on Io, but these forces pull in directions different from that of Jupiter. This combination of oscillating and opposing gravitational forces causes so much rock distortion and frictional heating that volcanic activity is nearly continuous on Io.

Astronomers infer that meteorites bombarded Io and the other moons of Jupiter, as they did all other bodies in the Solar System. Yet, the frequent lava flows on Io have obliterated all ancient landforms, giving it a smooth and nearly crater-free surface.

EUROPA The second-closest of Jupiter's moons, **Europa**, is similar to Earth in that much of its interior is composed of rock and much of its surface is covered with water. One major difference is that, on Europa, the water is frozen into a vast, planetary ice crust. The Galileo spacecraft transmitted images showing a fractured, jumbled, chaotic terrain resembling patterns created by Arctic ice on Earth (●FIGURE 23.26).

FIGURE 23.27 shows a smooth region overlying an older, wrinkled surface. Data suggest that this smooth region is young ice formed when liquid water erupted to the surface and froze. Thus, pools or oceans of liquid water or a water/ice slurry probably lies beneath the surface ice. Astronomers estimate that this surface crust is 10 kilometers thick in many regions. Calculations show that the subterranean oceans are warmed by tidal effects similar to, though weaker than, those that cause Io's volcanism. Scientists speculate that the chemical and physical environment in these subterranean oceans is favorable for life. There is no evidence whatsoever that life actually exists there, but the possibility is tantalizing.

GANYMEDE AND CALLISTO The Galileo spacecraft measured a magnetic field on **Ganymede**, indicating that this moon has a convecting, metallic core. Other measurements imply that the core is surrounded by a silicate mantle covered by a water/ice crust (●FIGURE 23.28). The surface ice is so cold that it is brittle and behaves much like rock. Photographs show two terrains on Ganymede: one is densely cratered, and the other contains fewer craters but many linear grooves. The cratered regions were formed by ancient meteorite storms. The grooved regions probably

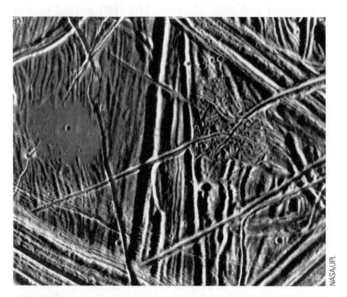

● **FIGURE 23.27** The smooth, circular region in the center-left of this photograph was formed when subsurface water rose to the surface of Europa and froze, covering older wrinkles and fractures in the crust.

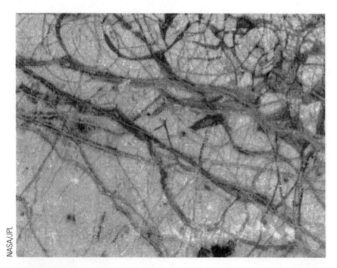

● **FIGURE 23.26** This jumbled terrain on Europa resembles Arctic pack ice as it breaks up in the spring. Scientists estimate that in this region, the ice crust is a few kilometers thick and is floating on subsurface water.

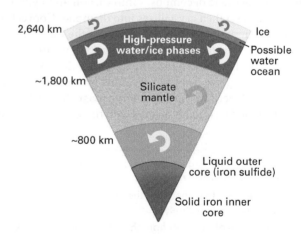

● **FIGURE 23.28** Recent data suggest that Ganymede has a conducting, convecting core; a silicate mantle; and surface layers consisting of ice and water.

NASA/JPL

● **FIGURE 23.29** A close-up of a young terrain on Ganymede shows numerous grooves less than a kilometer wide. One likely explanation is that these grooves were formed by recent tectonic activity.

developed when the crust cracked and water from the warm interior flowed over the surface and froze, much as lava flowed over the surfaces of the terrestrial planets and the Moon. Lateral displacements of the grooves and ridges are indicative of Earth-like, horizontal, plate tectonic activity (●**FIGURE 23.29**).

Callisto, the outermost Galilean moon, is heavily cratered, indicating that its surface is very old. Its craters are shaped differently from those on either Ganymede or the Earth's Moon. Perhaps they have been modified by ice flowing slowly across its surface. Recent measurements indicate

Europa The second-closest of Jupiter's moons; similar to Earth in that much of its interior is composed of rock and much of its surface is covered with water, although the water is frozen into a vast planetary ice crust.

Ganymede A moon of Jupiter marked by a convecting, metallic core and a brittle water/ice crust that behaves much like rock.

Callisto Jupiter's outermost Galilean moon; marked by a heavily cratered surface.

Titan Saturn's largest moon; the only moon in the Solar System with an appreciable atmosphere.

Triton The largest of Neptune's moons; marked by craters filled with ice or frozen methane.

that a subterranean ocean may exist on Callisto, but the jury is still out on a similar feature on Ganymede.

Saturn's Moons

Sixty-two moons orbit Saturn. Saturn's largest moon, **Titan**, is larger than the planet Mercury. Titan is unique because it is the only moon in the Solar System with an appreciable atmosphere. This atmosphere has been retained because Titan is relatively massive and extremely cold. The major constituents are nitrogen, mixed with methane (CH_4) and smaller concentrations of trace gases. The average temperature on the surface of Titan is −178°C, and the atmospheric pressure is 1.5 times greater than that on Earth's surface. These conditions are close to the temperatures and pressures at which methane can exist as a solid, liquid, or vapor.

After a seven-year journey, the Cassini spacecraft reached Titan in December 2004, and released a probe named Huygens that parachuted through the outer clouds and landed on the surface. Images showed a surprisingly Earth-like landscape, with steep-sided hills and features that looked like riverbeds, eroded hillsides, coastlines, and sandbars. The best evidence indicates that these topographic features were formed by wind, tectonic activity, and flowing liquids (●**FIGURE 23.30**). During the extreme cold on Titan, water is permanently locked up as ice, but the temperature and pressure are such that methane in the Titan atmosphere could exist in the liquid, vapor, or solid states. Thus in the past, methane rain has fallen from the clouds and methane rivers flowed across Titan's surface. Liquid methane may remain on the planetary surface today. This situation is analogous to Earth's environment, where water can exist as liquid, gas, or solid and frequently changes among those three states.

Methane, the simplest organic compound, reacts with nitrogen and other materials in Titan's environment to form more complex organic molecules. These organic compounds do not decompose at low temperature, so the satellite's surface is likely to be covered by a tarlike organic goo. It is possible that a similar layer collected on early Earth and later underwent chemical reactions to form life. However, Titan is so cold that life probably has not formed there.

The Moons of Uranus and Neptune

Twenty-seven known moons orbit Uranus. Several of the moons are small and irregularly shaped, indicating that they may be debris from a collision with a smaller planet or moon.

Neptune has at least 13 moons. The largest is **Triton**, which is about 75 percent rock and 25 percent ice. Like many other planets and moons, its surface is covered by impact craters, mountains, and flat, crater-free plains. While

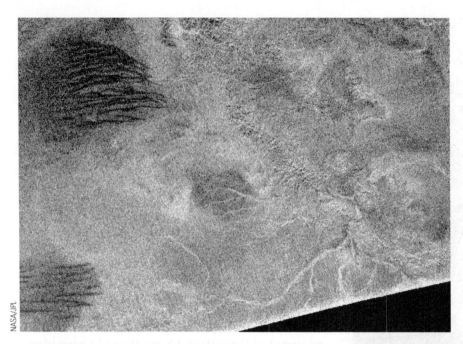

NASA/JPL

● **FIGURE 23.30** Radar images of Titan obtained in February 2005 show a well-developed drainage pattern in the lower-right of the image, and apparent sand dunes. Previous images show features that may have been formed by tectonic processes. Titan appears to have a young and dynamic surface that is modified by volcanism, tectonism, erosion, and impact cratering.

the maria on Earth's Moon are blanketed by lava, those on Triton are filled with ice or frozen methane.

PLANETARY RINGS

Although all the Jovian planets have one or more rings, by far the most spectacular are those of Saturn, which are visible from Earth even through a small telescope. Photographs from space probes show seven major rings, each containing thousands of smaller ringlets (●**FIGURES 23.31** and **23.32**). The entire ring system is only 10 to 25 meters thick, less than the length of a football field. However, the ring system is extremely wide. The innermost ring is only 7,000 kilometers from Saturn's surface, whereas the outer edge of the most distant ring is 432,000 kilometers from the planet, a distance greater than that between Earth and our Moon. Thus, the ring system measures 425,000 kilometers from its inner to its outer edge. A scale model of the ring system with the thickness of a compact disk would be 30 kilometers in diameter.

Saturn's rings are composed of dust, rock, and ice. The particles in the outer rings are only a few ten-thousandths of a centimeter in diameter (about the size of a clay particle), but the innermost rings contain chunks as large as a flying barn. Each piece orbits the planet independently.

Saturn's rings may be fragments of a moon that never coalesced. Alternatively, they may be the remnants of a moon that formed and was then ripped apart by Saturn's gravitational field. If a moon were close enough to its planet, the tidal effects would be greater than the gravitational attraction holding the moon together, and it would break up. Thus, a solid moon cannot exist too close to a planet. Images from Cassini spacecraft show that gravitational forces from Saturn's moons have herded the ring particles into intricate spirals and twists.

Vadim Sadovski/Shutterstock.com

● **FIGURE 23.31** An image of Saturn shows its spectacular ring system.

NASA/JPL

● **FIGURE 23.32** Voyager I took this color-enhanced close-up view of Saturn's rings and ringlets.

PLUTO AND OTHER DWARF PLANETS

In recent years, **Pluto** has been a controversial figure in our Solar System. Although NASA's New Horizons spacecraft reached Pluto in 2015, our highest-resolution photographs are of poor quality compared with those of other planets (●**FIGURE 23.33**). Until 2006, Pluto was considered the ninth planet in our Solar System. However, in 2006, the International Astronomical Union voted to remove Pluto from the list of planets. It has been grouped with similar space objects as a *dwarf planet*—a new classification to be described at the end of this section.

Pluto has three moons: Nix, Hydra, and Charon. Nix and Hydra are very small, but Charon is nearly half of Pluto's diameter. By measuring the orbits of these two bodies, astronomers determined the densities of Pluto and Charon, deducing that each contain about 35 percent ice and 65 percent rock.

Infrared measurements show that Pluto's surface temperature is about −220°C. Spectral analysis of Pluto's bright surface shows that it contains frozen methane. Its atmosphere is extremely thin and composed mainly of carbon monoxide, nitrogen, and some methane.

Pluto's orbit is highly elliptical, at times bringing it closer to the Sun than Neptune is. It may be related to numerous similar icy bodies that are part of the Kuiper Belt. These **Kuiper Belt objects** are residual planetesimals left over from the formation of the Solar System. One Kuiper Belt object, named Eris, is 5 percent larger than Pluto. Three other Kuiper Belt objects found so far are at least half the size of Pluto and may have moons of their own. These bodies, while having enough gravitational force to pull themselves into spherical shapes, cannot be considered planets because they do not meet one of the International Astronomical Union's criteria: they are not large enough to dominate and gravitationally clear their orbital regions of all or most other objects. The International Astronomical Union has classified them as prototypes of new objects called **dwarf planets**.

(A) Ground-based telescope   **(B)** Hubble Space Telescope

● **FIGURE 23.33** (A) Ground-based image of Pluto and its moon Charon and (B) a similar view from the Hubble Space Telescope.

Pluto Once considered to be the ninth planet from the Sun in our Solar System, reclassified in 2006 as a dwarf planet.

Kuiper Belt objects Tiny ice dwarfs, similar to Pluto, orbiting in a disk-shaped region at the outer reaches of the Solar System.

dwarf planets A body (for example, Pluto or Eris) that orbits the Sun, is not a satellite of a planet, and is massive enough to pull itself into a spherical shape, but is not massive enough to clear out other bodies in and near its orbit.

asteroids Small celestial bodies in orbit around the Sun, primarily in the region between Mars and Jupiter.

ASTEROIDS, COMETS, AND METEOROIDS

Asteroids

Astronomers have discovered a wide ring between the orbits of Mars and Jupiter that contains tens of thousands of small orbiting bodies, called **asteroids**. The largest asteroid, Ceres, has a diameter of 930 kilometers. Three others are about half that size, and most are far smaller. The orbit of an asteroid is not permanent like that of a planet. If an asteroid passes near a planet without getting too close, the planet's gravity pulls the asteroid out of its current orbit and deflects it into a new orbit around the Sun. Thus, an asteroid may change its orbit frequently and erratically. Many asteroids orbit near Earth, and some even cross Earth's orbit. If an asteroid passes too close to a planet, it will crash into its surface. As discussed previously, ancient asteroid impacts may have caused mass extinctions on Earth. Statistical analysis shows that there is a 1 percent chance that an asteroid with a diameter greater than 10 kilometers will strike Earth in the next 1,000 years.

One series of images of the asteroid Mathilde show an impact crater larger than the asteroid's mean radius. How could an object sustain such an impact without breaking apart? According to one hypothesis, Mathilde is not solid rock, as Earth is. Instead it is a compressed mass of fractured rock and rubble. Thus, the bolide impact scattered the fractured rock without transmitting force through a solid, brittle body. However, craters, grooves, and surface rocks on the nearby asteroid Eros indicate that it is a solid body.

Comets

Occasionally, a glowing object appears in the sky, travels slowly around the Sun in an elongated elliptical orbit, and then disappears into space (●**FIGURE 23.34**). Such an object is called a **comet**, after the Greek word for "long-haired." Despite their fiery appearance, comets are cold, and their light is reflected sunlight.

Comets originate in the outer reaches of the Solar System, and much of the time they travel through the cold void beyond Pluto's orbit. A comet is composed mainly of water/ice mixed with frozen crystals of methane, ammonia, carbon dioxide, and other compounds. Smaller concentrations of dust particles, composed of silicate rock and metals, are mixed with the lighter ices.

When a comet is millions of kilometers from the Sun, it is a ball without a tail. As the comet approaches the Sun and is heated, some of its surface vaporizes. Solar wind blows some of the lighter particles away from the comet's head to form a long tail. At this time, the comet consists of a dense, solid **nucleus**, a bright outer sheath called a **coma**, and a long **tail** (●**FIGURE 23.35**). Some comet

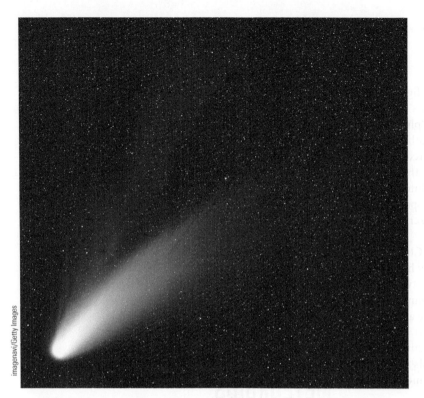

● **FIGURE 23.34** Hale–Bopp was the brightest comet seen from Earth in decades. It was brightest in March and April of 1997.

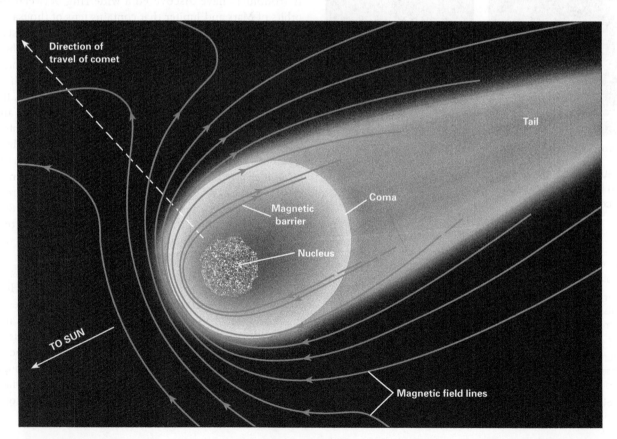

Direction of travel of comet

Tail

Magnetic barrier

Coma

Nucleus

TO SUN

Magnetic field lines

● **FIGURE 23.35** The nucleus of this comet has been enlarged several thousand times to show detail. When the comet interacts with the solar wind, magnetic field lines are generated as shown. Ions produced from gases streaming away from the nucleus are trapped within the field, creating the characteristically shaped tail.

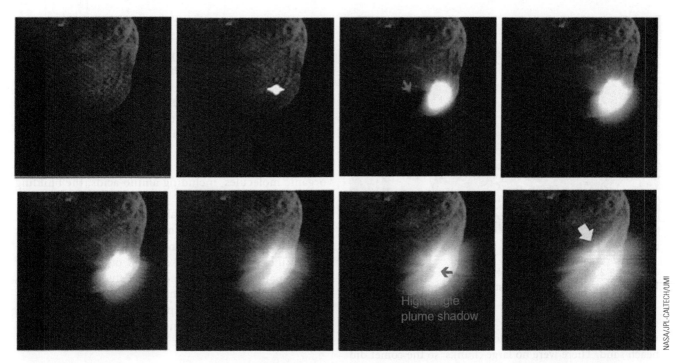

● **FIGURE 23.36** The eight images depict the development of the ejecta plume formed when Deep Impact's probe collided with comet Tempel 1 on July 3, 2005. The red arrows in images 3 and 7 highlight shadows cast by the ejecta. The yellow arrow on image 8 indicates the zone of avoidance in the up range direction. The eight images were spaced 0.84 second apart.

tails are more than 140 million kilometers long, almost as long as the distance from Earth to the Sun. As a comet orbits the Sun, the solar wind constantly blows the tail so that it always extends away from the Sun. By terrestrial standards, a comet tail would represent a good, cold laboratory vacuum—yet viewed from a celestial perspective it looks like a hot, dense, fiery arrow.

Halley's Comet is named after the English astronomer Edmond Halley, who concluded in 1705 that earlier reports of a comet that had approached Earth in 1531, 1607 and 1682 actually were the same comet returning over and over again. Haley predicted in his 1705 publication

'*Synopsis of the Astronomy of Comet*' that the comet would return in the year 1758. Although Halley did not live to see the comet return in 1758, it was named after him when it did so.

Halley's Comet passed so close to Earth in 1910 that its visit was a momentous event. When the comet returned to the inner Solar System in 1986, it was studied by six spacecraft as well as by several ground-based observatories. Its nucleus is a peanut-shaped mass approximately 16-by-8-by-8 kilometers, about the same size and shape as Manhattan Island. The cold, relatively dense coma of Halley's Comet had a radius of about 4,500 kilometers when it passed by.

In one spectacular experiment conducted in 2005, astronomers fired Deep Impact, a 372-kilogram metal probe, into comet Tempel 1, while a mother ship recorded the effects of the impact. The probe triggered two quick flashes, the first when the collision heated the surface to thousands of degrees, and the second, a few milliseconds later, when the now-molten probe penetrated deeper into the nucleus and ejected a layer of volatile material (●**FIGURE 23.36**). Photographs taken just prior to the collision reveal a landscape sculpted by outgassing, melting, and natural impacts—all evidence of a complex geological history. Analysis of the impact ejecta suggests that comet Tempel 1

comet An interplanetary orbiting body composed of loosely bound rock and ice, which forms a bright head and extended fuzzy tail when it approaches the Sun. It appears to be fiery hot, but it is actually a cold object and its "flame" is reflected light.

nucleus The dense, solid core of a comet.

coma The bright outer sheath of a comet, surrounding the nucleus.

tail The long trailing portion of a comet, always pointing away from the Sun, formed when solar winds blow away lighter particles from the comet's head. What appears to be a fiery arrow is actually reflected light from the Sun.

● **FIGURE 23.37** Scientists believe that this meteorite is a fragment from the asteroid Vesta.

was composed of fine dust, more like talcum powder than beach sand. There were no large chunks, so the comet did not have a solid ice crust. In the near-surface interior, the space probe detected organic compounds, which could form the basis for living organisms.

Meteoroids

As tens of thousands of asteroids race through the Solar System in changing paths, many collide and break apart, forming smaller fragments and pieces of dust. A **meteoroid** is an asteroid or a fragment of a comet that orbits through the inner Solar System. If a meteoroid travels too close to Earth's gravitational field, it falls. Friction with the atmosphere heats it until it glows. To our eyes it is a fiery streak in the sky, which we call a **meteor** or, colloquially, a shooting star. Most meteors are barely larger than a grain of sand when they enter the atmosphere and vaporize completely during their descent. Larger ones, however, may reach

Earth's surface. A meteor that strikes Earth's surface is called a **meteorite** (●**FIGURE 23.37**).

Most meteorites are **stony meteorite** and are composed of 90 percent silicate rock and 10 percent iron and nickel. The 90:10 mass ratio of rock to metal is similar to the mass ratio of the mantle to the core in Earth. Therefore, geologists think that meteorites reflect the primordial composition of the Solar System and are windows into our Solar System's past. Most stony meteorites contain small grains about 1 millimeter in diameter called **chondrules**, which contain organic molecules, including amino acids, the building blocks of proteins.

Some meteorites are metallic and consist mainly of iron and nickel, the elements that make up Earth's core, while the remainder are stony-iron, containing roughly equal quantities of silicates and iron-nickel. Some of our knowledge of Earth's mantle and core comes from studying meteorites, which may be similar to the mantles and cores of other planetary bodies. With the exception of rocks returned from the Apollo Moon missions, meteorites are the only physical samples we have from space.

---

meteoroid A small interplanetary body, most often an asteroid or comet fragment, traveling in an irregular orbit through the inner Solar System.

meteor A falling meteoroid that enters Earth's atmosphere and glows as it vaporizes; colloquially called a *shooting star*.

meteorite A meteor that does not completely vaporize and that strikes Earth's surface.

stony meteorite A meteorite with a mass ratio of rock to metal similar to the mass ratio of Earth's mantle to Earth's core, thus reflecting the primordial composition of the Solar System and representing a window into its past. Most meteorites are stony meteorites.

chondrules A small grain about 1 millimeter in diameter embedded in a meteorite, often containing amino acids or other organic molecules.

Key Concepts Review

- Table 23.1 provides an overview of the eight major planets. Due to the differences in composition, the terrestrial planets are much denser than the Jovian planets. However, the Jovian planets are much larger and more massive than the Earth and its neighbors.
- Mercury is the closest planet to the Sun. It has virtually no atmosphere, and it rotates slowly on its axis; therefore, it experiences extremes of temperature.

Mercury's surface is heavily cratered from meteorite bombardment that occurred early in the history of the Solar System. Venus has a hot, dense atmosphere as a result of a runaway greenhouse effect. Its surface shows signs of recent tectonic activity, probably resulting from vertical upwelling, called blob tectonics. Mars is a dry, cold planet with a thin atmosphere, but its surface bears signs of tectonic activity and ancient water erosion. The mobile robots Spirit and Opportunity detected clear signs that water once existed on Mars.

- Earth's Moon probably formed from the debris of a collision between a Mars-sized body and Earth. The Moon was heated by the energy released during condensation of the debris, by radioactive decay, and by meteorite bombardment. Evidence of ancient volcanism exists, but the Moon is cold and inactive today. In 2009, NASA instruments confirmed the presence of water molecules in the Moon's polar regions, although in relatively small amounts.
- Jupiter, Saturn, Uranus, and Neptune are all large planets with low densities. Jupiter and Saturn have dense atmospheres, surfaces of liquid hydrogen, inner zones of liquid metallic hydrogen, and cores of rock and metal. Uranus and Neptune have higher proportions of rock and ice than Jupiter and Saturn do, and their magnetic fields are not in line with their axes of rotation—a phenomenon for which no satisfactory hypothesis has yet been developed.
- The largest moons of Jupiter are Io, which is heated by gravitational forces and exhibits extensive volcanic activity; Europa, which is ice covered; and Ganymede and Callisto, which are large spheres of rock and ice. Titan, the largest moon of Saturn, has an atmosphere rich in nitrogen and methane. Photos of its surface show evidence that landscape was sculpted by wind, flowing liquids, and tectonic activity. Triton, Neptune's largest moon, is about 75 percent rock and 25 percent ice.
- All the Jovian planets have one or more rings, but the most spectacular are those of Saturn, which are visible from Earth using just a small telescope. Saturn's rings are made up of many small particles of dust, rock, and ice. They might have formed from a moon that was fragmented or from rock and ice that never coalesced to form a moon.
- Pluto has a low density and is a small planet composed of ice and rock. It may be related to numerous similar icy bodies known as Kuiper Belt objects.
- Asteroids are small, planet-like bodies in solar orbits between the orbits of Mars and Jupiter. A comet is an object that orbits the Sun in an elongated, elliptical orbit. Comets are composed mainly of water/ice and other frozen volatiles, mixed with smaller concentrations of dust, silicate rock, and metals. When a comet is in the inner part of the Solar System, it consists of a small, dense nucleus composed of ice and rock; an outer sheath or coma composed of gases, water vapor, and dust; and a long tail made up of particles blown outward by the solar wind. A meteorite is a meteoroid (a piece of matter orbiting through the inner Solar System) that falls to Earth. Most meteorites are stony, and some contain organic molecules. About 10 percent of all meteorites are metallic.

Important Terms

asteroids (p. 551)

blob tectonics (p. 539)

Callisto (p. 549)

chondrules (p. 554)

coma (p. 552)

comet (p. 552)

dwarf planets (p. 551)

escape velocity (p. 531)

Europa (p. 548)

Ganymede (p. 548)

Great Red Spot (p. 544)

Io (p. 547)

Jovian planets (p. 531)

Jupiter (p. 544)

Kuiper Belt objects (p. 551)

liquid metallic hydrogen (p. 544)

maria (p. 541)

Mars (p. 535)

Mercury (p. 535)

meteor (p. 554)

meteorite (p. 554)

meteoroid (p. 554)

Neptune (p. 547)

nucleus (p. 552)

Opportunity (p. 535)

Pluto (p. 551)

Saturn (p. 546)

solar wind (p. 531)

Spirit (p. 535)

stony meteorite (p. 554)

tail (p. 552)

terrestrial planets (p. 531)

Titan (p. 549)

Triton (p. 549)

Uranus (p. 547)

Venus (p. 535)

Review Questions

1. List the eight planets in order of distance from the Sun, and distinguish between the terrestrial and the Jovian planets.

2. Give a brief description of Mercury. Include its atmosphere, surface temperature, surface features, and speed of rotation on its axis.

3. Compare and contrast the atmospheres of Mercury, Venus, the Moon, Earth, and Mars.

4. Venus boiled, life evolved on Earth under moderate temperatures, and Mars froze. Discuss the evolution of the atmospheres and climates on these three planets.

(Refer back to Chapter 1 for additional information to answer this question.)

5. Discuss how the information from Spirit and Opportunity creates a picture of the climate history of Mars.

6. Discuss the evidence that the Martian atmosphere was once considerably different from what it is today.

7. Why are there fewer meteorite craters visible on Venus than on Mercury?

8. Explain how Venusian tectonics differs from tectonics on Earth.

NASA/JPL-CALTECH/ESA

24

In the constellation Aquarius, 650 light-years away, a dead star about the size of Earth and called a white dwarf, can be seen at the center of the image as a white dot. It is spewing out massive amounts of hot gas and intense ultraviolet radiation, creating a planetary nebula. In this false-color image, NASA's Hubble and Spitzer space telescopes have imaged the nebula. The colorful gaseous material seen was once part of the central star, but was lost in the death throes of the star as it became a white dwarf.

STARS, SPACE, AND GALAXIES

LEARNING OBJECTIVES

LO1   Discuss the big bang and the formation of our universe.

LO2   Explain the birth of a star from a nebula.

LO3   Describe the structure and composition of the Sun.

LO4   Discuss the location of main sequence stars on the Hertzsprung–Russell diagram.

LO5   Summarize the life and death sequence of stars.

LO6   Describe ways astronomers can demonstrate the existence of black holes.

LO7   Compare at least three different types of galaxies.

LO8   Discuss the structure of our galaxy, the Milky Way.

LO9   Describe how studying quasars is look back in time.

LO10   Explain how dark matter accounts for "missing" mass in the Universe.

INTRODUCTION

Earth is only one planet orbiting one star among roughly 100 billion stars in our Milky Way galaxy. In turn, the Milky Way is only one galaxy of billions in the Universe. Looking beyond the Solar System into galactic and intergalactic space, we must stretch our minds to nearly unimaginable distances, look backward in time to events that occurred billions of years before Earth formed, and attempt to fathom energy sources powerful enough to create an entire Universe.

IN THE BEGINNING: THE BIG BANG

In our search for the origin and history of Earth, we have looked back more than 4.6 billion years to the time when a diffuse cloud of dust and gas coalesced to form the Solar System. But now, as we look into our galaxy and beyond, we ask: How did that cloud of dust and gas form? As our search for answers deepens, we finally ask: How and when did the Universe begin?

Before we explore the origin of the Universe, we must ask an even more fundamental question: Did it begin at all? One possibility is that the Universe has always existed and there was no beginning, no start of time. An alternative hypothesis is that the Universe began at a specific time and has been evolving ever since.

In 1929, Edwin Hubble observed that all galaxies are moving away from each other. By projecting the galactic motion backward in time, he reasoned that they must have started moving outward from a common center, and at the same time. Therefore, scientists calculated that in the beginning the entire Universe was compressed into a single, infinitely dense point. This point was so small that we cannot compare it with anything we know or can even imagine. According to modern theory, this point exploded. But it was no ordinary explosion. It cannot even be compared to a hydrogen bomb or supernova explosion. This explosion, called the **big bang**, instantaneously created the Universe. Matter, energy, and space came into existence with this single event. It was the start of space and time.

Astronomers calculate the timing of the big bang by measuring speeds of galaxies and the distances among them. They then calculate backward in time to determine when they were all joined together in a single point. In 2003, astronomers combined a variety of techniques to determine that the Universe is 13.7 billion years old.

Scientists generally start their discussions of the origin of the Universe when it was a trillionth of a trillionth of a billionth of a second old. How can we reach that far back in time with any degree of certainty? In one series of experiments, scientists study the collision behavior of particles at very high velocities in modern particle accelerators. These results are then compared with observations of deep space. In the more than 80 years since Hubble's pioneering work, the experiments, observations, and calculations have led to disagreements, paradoxes, and unsolved mysteries. Yet, the preponderance of evidence is so persuasive that almost all astronomers agree with the fundamental premise of the big bang theory.

Three lines of evidence and logic support the big bang theory. The first is the expansion of the Universe, discussed previously. To understand the second two—the creation of helium through primordial nucleosynthesis and the discovery of cosmic background radiation—we must study the first 300,000 years in the life of the Universe.

Let us go back to that unimaginably distant time when the Universe was a trillionth of a trillionth of a billionth of a second old. At that time, the Universe was only about as big as a grapefruit and the temperature was about 100 billion degrees on the Kelvin scale (written 100 *billion K*; the degree symbol is not used).[1] But this primordial Universe was expanding rapidly as it was propelled outward by the force of the explosion that formed it. As the Universe expanded, it cooled. During the first second, the Universe cooled to about 10 billion degrees, 1,000 times the temperature in the center of the modern Sun (●**FIGURE 24.1**). At such high temperatures, atoms do not exist. Instead, the Universe consisted of a plasma of radiant energy, electrons, protons, neutrons, and extremely light particles called neutrinos.

Over the next few minutes, the most exciting action came from the behavior of protons and neutrons. Recall that protons are positively charged particles. At large distances (by subatomic standards) two positive protons repel each other, like two north poles of a magnet. However, if protons are hot enough, they collide with sufficient energy to overcome the repulsion. At very close distances, two protons attract and bond together. This bonding is called *fusion*. Hot protons also fuse with hot neutrons.

---

1. Astronomers generally report temperature on the Kelvin scale, which is based on fundamental thermodynamic properties ($0°C = 273$ K). At high temperatures, the 273-degree difference between the two is negligible.

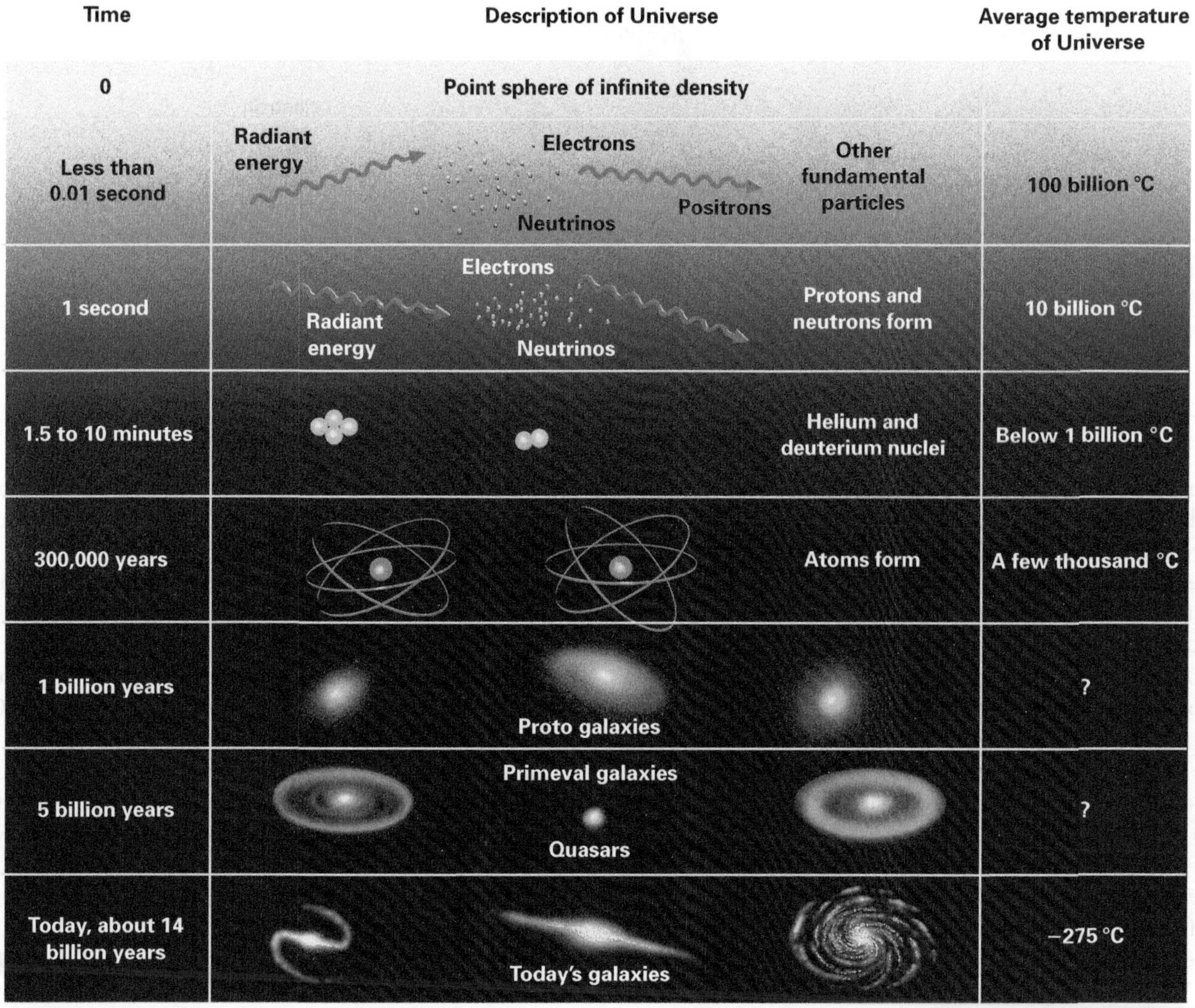

Time	Description of Universe			Average temperature of Universe
0	Point sphere of infinite density			
Less than 0.01 second	Radiant energy	Electrons, Neutrinos, Positrons	Other fundamental particles	100 billion °C
1 second	Radiant energy	Electrons, Neutrinos	Protons and neutrons form	10 billion °C
1.5 to 10 minutes			Helium and deuterium nuclei	Below 1 billion °C
300,000 years			Atoms form	A few thousand °C
1 billion years		Proto galaxies		?
5 billion years		Primeval galaxies, Quasars		?
Today, about 14 billion years		Today's galaxies		−275 °C

● **FIGURE 24.1** A brief pictorial outline of the evolution of the Universe.

A hydrogen nucleus consists of a lone proton. At the right temperature, four protons (hydrogen nuclei) fuse together to form a nucleus of the next-heavier element, helium (●**FIGURE 24.2**). This nuclear fusion is identical to reactions that occur in a star or in a hydrogen bomb.

When the Universe was less than 1.5 minutes old, it was so hot that helium nuclei were blasted apart almost as soon as they formed. After 10 minutes, the Universe was so cool that fusion could no longer occur. Thus, the formation of helium nuclei, called the **primordial nucleosynthesis**, occurred over a time span of 8.5 minutes. Enough fusion had occurred during this time so that 6 percent of the total nuclei in the Universe were helium, while 94 percent remained as hydrogen nuclei. Although primordial

nucleosynthesis created a trace of lithium (the next-heaviest element after helium), it did not produce any heavier elements. There was no carbon, nitrogen, or oxygen, so life could not have evolved, and no silicon or metals, so solid planets could not have evolved. The Universe was still in its infancy.

When scientists calculate the expected ratio of hydrogen to helium in the modern Universe from the conditions of the primordial Universe, they arrive at a number almost exactly in agreement with the observed abundances. This agreement provides the second line of evidence and logic to support the big bang theory. As the Universe continued to expand, it finally cooled to a few thousand degrees by about 300,000 years after the big bang. At this crucial temperature, electrons became attached to the hydrogen and helium nuclei. Thus, the first atoms formed and, in a sense, the modern Universe was born.

Before atoms existed, the Universe was a chaotic plasma that scattered and absorbed photons. As a result, photons could never move far in any direction. The Universe was fog like or opaque. But when electrons combined with nuclei to

big bang An event 10 to 20 billion years ago, thought to mark the beginning of the Universe when all matter exploded from a single infinitely dense point.

primordial nucleosynthesis The formation of helium nuclei that occurred by nuclear fusion during the first 8.5 minutes in the life of the Universe.

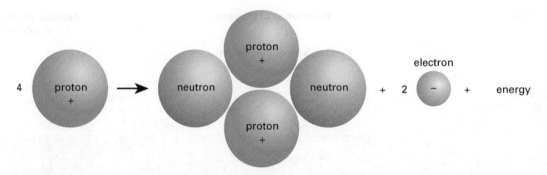

● **FIGURE 24.2** In a series of reactions, four hydrogen nuclei fuse to form helium. Two electrons and tremendous quantities of energy are released.

form hydrogen and helium atoms, conditions changed radically. An atom absorbs light in only a relatively few specific wavelengths. All the rest of the light passes through unhindered. Thus, a hydrogen helium-filled Universe is like a pair of sunglasses. Sunglasses filter out certain wavelengths of light but permit other wavelengths to pass through.

In the 1960s, an astrophysicist at Princeton named Robert Dicke predicted that we should be able to detect the primordial photons that began moving in a straight line as soon as atoms formed. He calculated that continued expansion of the Universe would have cooled the primordial photons to about 2.7 K (2.7°C above absolute zero). In 1964, Arno Penzias and Robert Wilson at Bell Laboratories detected a very faint photon radiation that was 2.7 K and was uniformly distributed throughout space. Thus, experimental observation agreed precisely with Dicke's calculation.

Walk outside at any time of day at any place on Earth and turn your palm upward to the sky. Every second, a million billion low-energy photons will strike your palm. These photons are in the microwave band of the electromagnetic spectrum. The energy is so faint and weak that you could never feel it. Yet, this **cosmic background radiation** began traveling through space when the Universe was only 300,000 years old. Today, the cosmic background radiation is called the echo of the big bang. The prediction and discovery of the cosmic background radiation provides the third convincing line of evidence to support the big bang theory.

THE NONHOMOGENEOUS UNIVERSE

Penzias and Wilson observed that the cosmic background radiation was uniform throughout space. This measurement implied that the Universe was homogeneous in its infancy. However, the modern Universe is clearly not homogeneous. Matter is concentrated into stars, stars are clumped into galaxies, and galaxies are grouped into clusters containing tens of thousands of galaxies. Even

the clusters group into superclusters. Most of the space between the clusters and superclusters contains no galaxies at all (●**FIGURE 24.3**).

The question then arose: How and when did an initially uniform, homogeneous Universe concentrate into stars, galaxies, and clusters? According to one hypothesis, the original, grapefruit-sized Universe had to obey laws of quantum mechanics, the same laws that describe modern atoms. Calculations based on quantum mechanics showed that the earliest Universe contained tiny waves of energy, space, and time. These waves, like sound waves, contained alternating regions of higher and lower densities. If this model were correct, then these bands of varying energy densities would have created tiny temperature differences in the cosmic background radiation.

In the 1980s, physicists calculated that the temperature differences would be about 0.0001°C, far too small to have been detected by Penzias and Wilson's radio telescope. In order to search for these temperature differences, they had to build a much more precise radio telescope. Because the atmosphere interferes with microwave transmissions, they needed to mount the antenna on a satellite.

In 1989, astronomers launched the Cosmic Background Explorer (COBE) satellite that was capable of measuring 0.0001°C temperature differences in the cosmic background radiation. As it slowly scanned the Universe, COBE registered numerous fluctuations in the background temperature. After mapping the temperature of the Universe for three years, COBE scientists reported that cosmic background radiation varies by tiny amounts from one region to another. These data suggested that the primordial Universe was not homogeneous as Penzias and Wilson had inferred. Matter and energy had concentrated into clumps during the earliest infancy of the Universe, perhaps in the first billionth of a second. George Smoot, a researcher on the project and winner of the 2006 Nobel Prize in Physics, called these variations "the imprints of tiny ripples in the fabric of space-time put there by the primeval explosion." In 2003, data from a second spacecraft, called the WMAP satellite, confirmed and refined the COBE results.

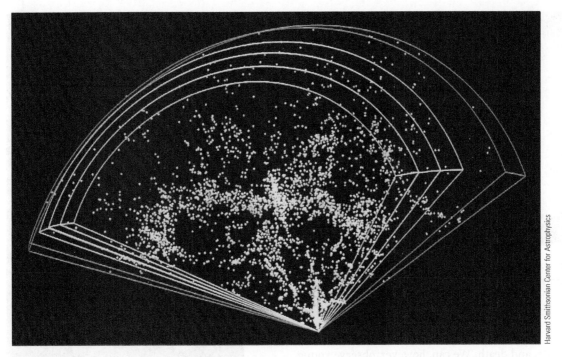

Harvard Smithsonian Center for Astrophysics

● **FIGURE 24.3** A three-dimensional, computer-generated drawing of a portion of the Universe. Notice that matter is unevenly distributed.

THE BIRTH OF A STAR

The cosmic background radiation emanated from the primordial sea of atoms, particles, and energy of the big bang. This radiation is called **first-generation energy**. When you look up into the sky on a clear, moonless night, you see stars sparkling in the blackness. If you owned a powerful telescope, you could peer deeper into space and detect distant galaxies, or even the curious structures called quasars. All of these objects emit what astronomers call **second-generation energy**. They use the term *second-generation* because concentrated matter—stars, galaxies, or quasars—emit this light. To understand our Universe, we must learn how matter in the subtly nonhomogeneous early Universe clumped together to form concentrated bodies that emit light and energy of their own.

---

cosmic background radiation Low-energy, microwave radiation that began traveling through space when the Universe was only 300,000 years old and now pervades all space in the Universe.

first-generation energy Cosmic background radiation that emanated from the primordial sea of atoms, particles, and energy of the big bang.

second-generation energy Radiation emitted by stars, galaxies, quasars, and other forms of concentrated matter.

nebula A cloud of interstellar gas and dust; plural *nebulae*.

Over time, denser regions of the Universe drew matter inward by gravity to form huge clouds of hydrogen and helium; such a cloud of interstellar gas and dust is called a **nebula**. Within each cloud, matter further agglomerated into billions of smaller bodies. As the atoms in a nebula accelerated inward under the force of gravity, they collided rapidly with one another. Thus, the center of each cloud became very dense and hot. Under the intense heat, electrons were stripped away from their atoms, leaving a plasma of positively charged nuclei and negatively charged electrons. If the cloud were originally large enough, the gravitational attraction accelerated the nuclei until they collided with enough energy to fuse. When fusion started, the collapsing portion of the original nebula became a star.

Fusion generates energetic photons and tremendous quantities of heat. If a dense sphere of hydrogen the size of a pinhead were to fuse completely, it would release as much energy as is released by burning several thousand tons of coal. Both the photons and the hot particles generated by fusion accelerate outward against the force of gravity. Thus, two opposing processes occur in a star. Gravity pulls particles inward, but at the same time fusion energy drives them outward (●**FIGURE 24.4**). The balance between these two processes determines the diameter and density of a star of a given mass. At equilibrium, a star with an average mass has a dense core surrounded by a less-dense shell.

A star exists for millions, billions, or tens of billions of years, and we can never observe one long enough to watch

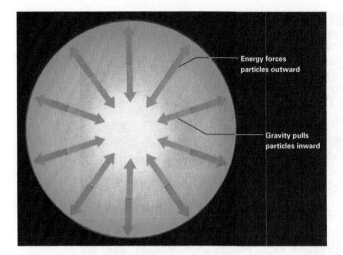

● **FIGURE 24.4** The diameter and density of a star are determined by two opposing forces. Gravity pulls particles inward, while energy from fusion reactions in the core forces particles outward.

(A)

its birth, life, and death. We can, however, observe young, middle-aged, and old stars and thus piece together the story of stellar evolution. It is as if an alien came to Earth for one day to observe the life of a human being. It would take a significant investment of time to observe the birth, growth, and death of a single person, but the alien could observe babies, children, middle-aged people, and old people and thus infer the course of a human life.

One nebula near the belt in the constellation Orion is a nursery for the birth of new stars (●FIGURE 24.5). The stars in one portion of the nebula are at least 12 million years old. Stellar ages decrease progressively toward the southwest until the stars are less than 1 million years old in the region near Orion's sword. This age progression implies that star formation moved steadily from one end of the nebula to the other.

Astronomers are uncertain how the first stars formed in a large nebula. However, once the first stars began to generate energy, they heated the surrounding gas in the nebula. The heated gas expanded so that its pressure and density increased. From the rate of stellar evolution, astronomers calculated that one or more of the initial stars were massive and short-lived. These stars quickly exhausted their supply of fuel and exploded. Hot gases burst into the surrounding nebula, creating a shock wave that increased the pressure and density even more. When the pressure and gas density reached a critical value, gravitational forces between molecules were strong enough to pull the particles together. Matter separated into discrete regions falling toward common centers. Recent observations by the Hubble Space Telescope show that these regions of varying density appear as towering pillars of light and shadow (●FIGURE 24.6). Hundreds of young stars, only 8 million to 300,000 years old, are forming along the edges of the cloud.

Let us look at the structure of our own Sun and then return to study the life cycle of a star.

(B)

● **FIGURE 24.5** (A) The constellation Orion was named after a mythical Greek hunter. The three stars across the center of the constellation show the hunter's belt, and the points of light angling down from the belt show his sword. One of the objects in his belt is an emission nebula. (B) NASA's Spitzer and Hubble space telescopes have captured this view of the Orion nebula, which is the brightest spot in Orion's sword. Young stars are evolving within this stellar nursery.

J. Hester and P. Scowen (ASU) and NASA

● **FIGURE 24.6** The Hubble Space Telescope has provided a close-up view of stellar nurseries near Orion's belt. The pillarlike structures are about three light-years tall and are composed of cold gas and dust. Energy from nearby stars and shock waves from exploding stars have concentrated the gas in the pillars and initiated star formation.

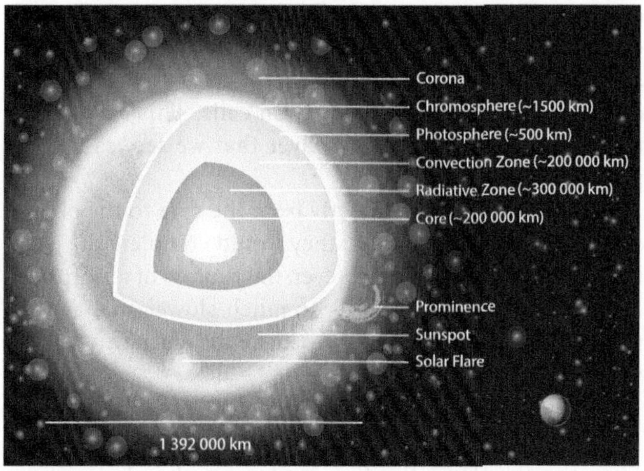

Siberian Art/Shutterstock.com

● **FIGURE 24.7** Structure of the Sun. Fusion occurs in the Sun's core. Energy escapes first by radiation, then by convection. The photosphere is the thin layer visible from Earth. Surface features are shown in more detail in Figures 24.8 through 24.10.

THE SUN

Nobody has ever traveled to the Sun—and nobody ever will. We have never sent spacecraft into the solar atmosphere to collect samples of the torridly hot gases that glow on its exterior. The core will be forever opaque and invisible because any material, any instrument, or any spacecraft would vaporize long before it penetrated into the solar interior. All of our information about the Sun is derived from indirect evidence, yet scientific reasoning is so powerful that we have a reasonably complete and accurate picture of this life-giving orb.

By studying the orbital properties of the planets, scientists can calculate the gravitational force of the Sun and hence its mass, approximately 2×10^{30} kilograms. Although this huge number is almost impossible to comprehend, our Sun is a star of average mass. The Sun's diameter is 1.4 million kilometers (109 Earth diameters).

Recall from Chapter 22 that astronomers can determine the composition of a star by analyzing its emission and absorption spectra. From these studies, we have learned that hydrogen accounts for 92 percent of the Sun's atoms, and helium is second in abundance at 7.8 percent. All the remaining elements make up only 0.2 percent. The Sun's structure is shown in ●**FIGURE 24.7**.

The Sun's Inner Structure

In the late 1800s, scientists calculated that the Sun emitted so much energy that it could not be powered by the burning of conventional fuels. But if the Sun was not a giant fire, what was it? In 1905, Einstein announced his theory of relativity and shortly thereafter other scientists began to unravel the mystery of nuclear reactions. These studies showed that the Sun is powered by hydrogen fusion. Every second, the Sun converts 4 million tons of hydrogen into energy and radiates this energy into space. Yet, the Sun is so massive that there is no detectable change in mass from year to year, or even from century to century.

Our understanding of the solar **core** is derived from calculations of hydrogen fusion reactions and gravitational forces. From these studies, scientists are reasonably certain that the core of the Sun is extremely hot, more than 15 million K, and its density is 150 times that of water. The Sun's core is 140,000 kilometers in diameter and contains enough hydrogen to fuel its fusion reaction for another 5 billion years. When the hydrogen in the core is used up, the Sun will change drastically, as described in sections on "Stars: The Main Sequence" and "The Life and Death of a Star."

Most of the energy generated by fusion in the Sun's core is emitted as radiation of high-energy photons. When a photon flies out of the core, atoms almost immediately absorb it in a broad region surrounding the core called the **radiative zone**. Atoms reemit photons soon after they absorb them, but the gas density in the region is so high that a photon is quickly reabsorbed again. This process of absorption and

core The center of the Sun, where hydrogen fusion takes place.

radiative zone An inner zone of a star surrounding the core where radiation energy is transmitted by absorption and emission.

emission occurs repeatedly until the photon reaches the **convective zone**. The convective zone is much cooler on its surface than in its interior. As a result, hot gas rises, cools, and then sinks in huge convection cells. This convection transports heat to the outer layer that we see, called the **photosphere**.

If photons traveled directly from the core to the photosphere at the speed of light, they would complete the journey in about 4 seconds. However, the process of absorption, emission, and convection is so much slower that energy requires about a million years to make the journey.

The Photosphere

Even though the entire Sun is gaseous, the photosphere is called the solar atmosphere because it is cool and diffuse compared to the core. It is a thin surface veneer, only 400 kilometers thick. The pressure at the middle of the photosphere is about 1/100 that of Earth's atmosphere at sea level, and its average temperature is 5,700 K, about the same as that in Earth's core. Fusion does not occur at this relatively low temperature. Thus, the core heats the photosphere, and the sunlight we see from Earth comes from this thin, glowing atmosphere of hydrogen and helium.

The photosphere has a granular structure, with each grain about 1,000 kilometers across, or about the size of Texas. Convection currents that carry energy to the surface form the granules. If you could watch the Sun's surface at close range, the granules would appear and disappear like bubbles in a pot of boiling water. The bright, yellow granules in ●FIGURE 24.8 are formed by hot, rising gas, and the darker, reddish regions are cooler areas of descending gas.

Large, dark spots called **sunspots** also appear regularly on the Sun's surface. A single sunspot may be small and last for only a few days, or it may be as large as 150,000 kilometers in diameter and remain visible for months. Because a sunspot is 1,000 degrees cooler than the surrounding area, it radiates only about half as much energy and hence appears dark compared to the rest of the photosphere. Since heat normally flows from a hot body to a cooler one, a large, cool region on the Sun's surface should quickly be heated and disappear. Yet, some sunspots persist for long periods.

Astronomers believe that the Sun's magnetic fields restrict mixing of the photosphere and inhibit warming of sunspots. The magnetic fields associated with sunspots also produce flares, by accelerating glowing gases to velocities in excess of 900 kilometers per hour and sending shock waves smashing through the solar atmosphere. The flares emit charged particles that interrupt radio communication and cause *aurorae* (the northern and southern lights) on Earth.

The Sun's Outer Layers

The photosphere is not bounded by a sharp, outer surface. A turbulent, diffuse, gaseous layer called the **chromosphere** lies above the photosphere and is about 2,000 to 3,000 kilometers thick. The chromosphere can be seen with the naked eye only when the Moon blocks out the photosphere during a solar eclipse. Then it appears as a narrow, red fringe at the edge of the blocked-out Sun. Jets of gas called **spicules** shoot upward from the chromosphere, looking like flames from a burning log. An average spicule is about 700 kilometers across and 7,000 kilometers high and lasts about 5 to 15 minutes.

The *corona* is an even more diffuse region that lies beyond the chromosphere. It is normally invisible but, as you learned in Chapter 22, during a full solar eclipse the corona appears as a beautiful halo around the Sun (●FIGURE 24.9). Its density is one-billionth that of the atmosphere at Earth's surface. In a physics laboratory, that would be a good vacuum.

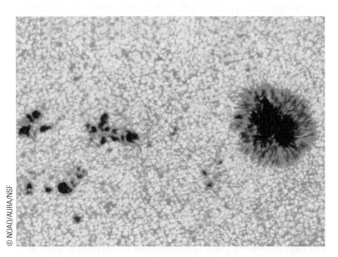

● **FIGURE 24.8** A close-up of the Sun's surface shows granular structures and sunspots. The large black structures are sunspots. Each small bright-yellow dot is a rising column of hot gas about 1,000 kilometers in diameter. The darker, reddish regions between the dots are cooler, descending gases.

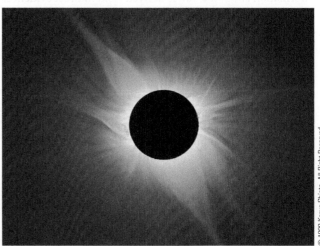

● **FIGURE 24.9** During a solar eclipse the photosphere is blocked by the Moon. The thin red streaks beyond the Moon's outline are portions of the chromosphere, and the large white zone is the corona.

● **FIGURE 24.10** This prominence rises approximately 505,000 kilometers above the photosphere.

NOAO/AURA/NSF

The corona is extremely hot, about 2 million K. How does the photosphere, at 5,700 K, heat the corona to 2 million K? According to one hypothesis, twisting magnetic fields accelerate particles in the corona. When particles are moving quickly, their temperature is high.

One feature found in the corona is a **prominence**. Solar prominences are red, flamelike jets of gas that rise out of the corona and travel as much as 1 million kilometers into space (●**FIGURE 24.10**). Some prominences are held aloft for weeks or months by the Sun's magnetic fields.

---

convective zone The subsurface zone in a star where energy is transmitted primarily by convection.

photosphere The surface of the Sun visible from Earth; also called the solar atmosphere.

sunspots A comparatively cool, dark region on the Sun's surface caused by a magnetic disturbance.

chromosphere A turbulent, diffuse, gaseous layer of the Sun that lies above the photosphere.

spicules A jet of gas at the edge of the Sun, shooting upward from the chromosphere.

prominence A red, flamelike jet of gas rising from the Sun's corona.

apparent brightness or luminosity The luminosity of a star as seen from Earth.

absolute brightness or luminosity The brightness of a star as it would appear if it were a fixed distance from Earth.

light-year The distance traveled by light in one year, approximately 9.5×10^{12} kilograms.

parsec A distance used in astronomy, equal to about 3.26 light-years.

Hertzsprung–Russell or H–R diagram A graph that plots absolute stellar luminosity against temperature.

main sequence A band running across a Hertzsprung–Russell diagram that contains most of the stars, which are fueled by hydrogen fusion.

The high temperature in the corona strips electrons from their atoms, reducing hydrogen and helium to bare nuclei in a sea of electrons. These nuclei and electrons are moving so rapidly that some fly off into space, forming the solar wind that extends outward toward the far reaches of the Solar System.

STARS: THE MAIN SEQUENCE

From our view on Earth, the Sun is the biggest and brightest object in the sky. However, the Sun looks so big and bright only because it is the closest star. When astronomers began to study stars in detail, one of their first efforts was to catalog them by their brightness. The **apparent brightness** (or **apparent luminosity**) of a star is its brightness as seen from Earth. A star can appear luminous either because it is intrinsically bright or because it is close. (Think of a car headlight. The light appears to become brighter as the car approaches. The headlight bulb is not changing in luminosity; the car is just moving closer.) The **absolute brightness** (or **absolute luminosity**) is how bright a star would appear if it were a fixed distance away. Astronomers have chosen a standard distance of 10 parsec, or 32.6 light-years, to calculate absolute luminosities. Light-years and parsecs are common units of distance. One **light-year** is the distance traveled by light in a year, 9.5 trillion (9.5×10^{12}) kilometers. A **parsec** is a distance equal to 3.26 light-years. By convention, astronomers use a scale in which the absolute luminosity of a star is divided by the absolute luminosity of the Sun. Thus, the Sun has a luminosity of 1.0, bright objects have a luminosity greater than 1, and faint objects have a value less than 1.

A second visible property of a star is its color. Different stars have different colors; some are reddish, while others are yellow or blue. The color of a star is a measure of its temperature. Blue is the hottest, yellow is intermediate, and red is the coolest. (The same relationship can be observed in a flame from a welding torch. The hottest, central portion of the flame is blue, and the cooler, outer edge is red.)

Between 1911 and 1913, Ejnar Hertzsprung and Henry Russell plotted the absolute luminosities of stars against their temperatures, as measured by the color. ●**FIGURE 24.11** is a **Hertzsprung–Russell diagram,** or **H–R diagram**. In an H–R diagram, luminosity increases from bottom to top and temperature increases from right to left. Thus hot, big, luminous stars are on the upper-left, and cool, small, less-luminous stars are on the lower-right.

The interesting observation about the H–R diagram is that about 90 percent of all stars fall along a sinuous band from upper-left (bright and hot) to lower-right (dim and cool). This band is called the **main sequence**. Thus, luminosity increases with temperature in main-sequence stars. Our Sun is more or less in the middle of the main sequence.

All main-sequence stars are composed primarily of hydrogen and helium and are fueled by hydrogen fusion.

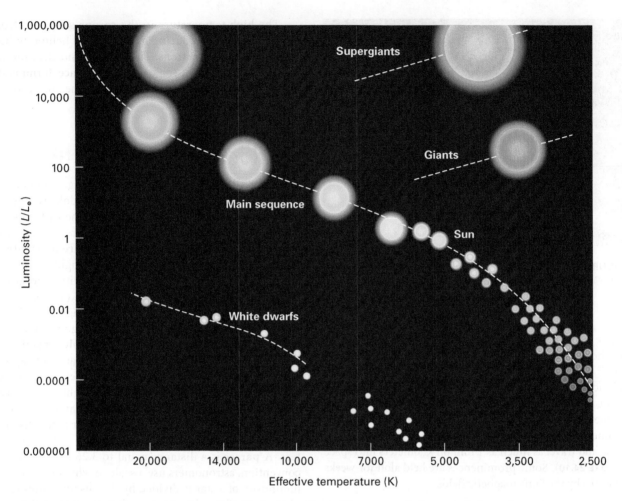

● **FIGURE 24.11** A Hertzsprung–Russell (H–R) diagram. About 90 percent of all stars fall along the main sequence, the sinuous band running from the upper-left to lower-right of the diagram. Our Sun is an average-sized star with a surface temperature of about 6,000 K; it glows yellow.

The major reason for differences in temperature and luminosity among main-sequence stars is that some are more massive than others. Because the force of gravity is stronger in massive stars than in less-massive stars, hydrogen nuclei are packed more tightly and move more rapidly. As a result, fusion is more rapid and intense in massive stars, so they are hotter and more luminous (upper-left of the H–R diagram). The least-massive stars, shown on the lower-right, are cool and less luminous. Notice that our Sun is in the yellow band about halfway along the main sequence.

To explain the stars that do not lie on the main sequence, we must consider the life and death of a star.

THE LIFE AND DEATH OF A STAR

Stars about the Same Mass as Our Sun

In a mature star such as our Sun, hydrogen nuclei fuse to form helium but the helium nuclei do not fuse to form heavier elements. For fusion to occur, two nuclei must collide so energetically that they overcome their nuclear repulsion. Hydrogen nuclei contain one proton, but helium contains two. Therefore, the repulsion between two helium nuclei is much greater than that between two hydrogen nuclei. This stronger repulsion can be overcome if the nuclei are moving very rapidly. Because higher temperature causes nuclei to move more rapidly, helium fusion occurs only when a star becomes much hotter than the temperature at which hydrogen fusion occurs.

The life cycle of a star with a mass similar to our Sun is shown in ●**FIGURE 24.12**. The Sun is now midway through its mature phase as a main-sequence star. It has been shining for about 5 billion years and will continue to shine much as it is today for another 5 billion years (●**FIGURE 24.12A**). During this entire period, the Sun produces energy by hydrogen fusion and remains on the main sequence.

---

red giant A stage in the life of a star when its core is composed of helium that is not undergoing fusion, but a hot shell of hydrogen around the core is fusing at a rapid rate, producing enough energy to cause the star to expand. Its core is hotter, but its expanded surface is cooler and emits red light.

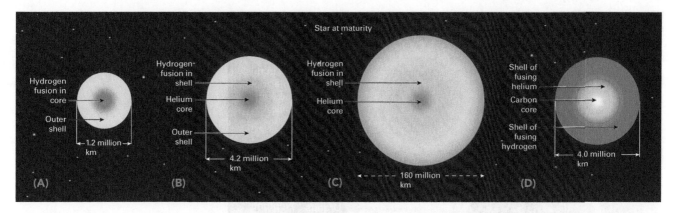

● **FIGURE 24.12** Evolution of a star. (A) At maturity, the energy in a star the size of our Sun is derived from hydrogen fusion in the core. (B) When most of the hydrogen in the core is consumed, the star first contracts, then expands as hydrogen fusion starts in the outer shell. (C) In the red giant phase, gravitational coalescence in the core and hydrogen fusion in the outer shell produce hundreds of times as much energy as was produced when the star was mature. (D) The star contracts again when helium fusion initiates in the core. The four parts in this diagram cannot be drawn to scale as the star's diameter varies from 1.2 million kilometers to 160 million kilometers. Also, in all cases the core has been drawn larger than scale to show detail.

After about 10 billion years, the outer shell of a star such as our Sun still contains large quantities of hydrogen, but most of the hydrogen in the core has fused to helium (●**FIGURE 24.12B**). The star's behavior now changes drastically. Because the hydrogen in the core is nearly used up, hydrogen fusion slows down, less nuclear energy is produced, and the core cools. As it cools, the outward pressure of particles and energy decreases. Then the core starts to contract under the force of gravity. This gravitational contraction causes the core to grow hotter. It seems a paradox that when the nuclear reactions decrease, the core becomes hotter, but that is what happens.

As the core heats up, the rising temperature initiates hydrogen fusion in the outer shell. The star is now heated by both the gravitational coalescence in the core and the

hydrogen fusion in the outer shell. As a result, the star releases hundreds of times as much energy as it did when it was mature. This intense energy output now causes the outer parts of the star to expand and become brighter. The star has become a **red giant** (●**FIGURE 24.12C**). A red giant is hundreds of times larger than an ordinary star. Its core is hotter, but its surface is so large that heat escapes and the surface cools. This cool surface emits red light (recall that red wavelengths have the lowest energy of the visible spectrum). These sudden changes in energy production and diameter move the star off the main sequence (●**FIGURE 24.13**). Thus, a red giant is brighter and cooler than a main-sequence star of the same mass.

Five billion years from now, the hydrogen in our Sun's core will be exhausted and the Sun will expand into

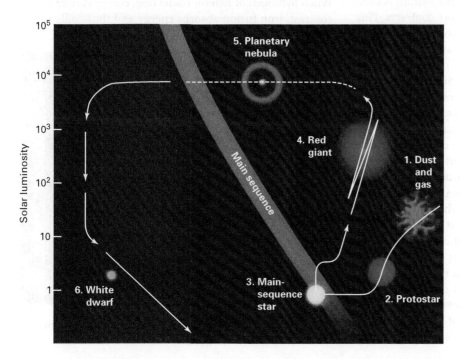

● **FIGURE 24.13** A star the mass of our Sun passes through six major stages in its life cycle: (1) After the original nebula condenses, (2) the protostar glows from the heat of gravitational coalescence; (3) the star enters the main sequence when hydrogen fusion starts in the core; (4) after hydrogen fusion ends in the core, the star leaves the main sequence and passes through the red giant stage; and (5) finally it explodes to produce a planetary nebula, and then (6) the remaining core glows as a white dwarf.

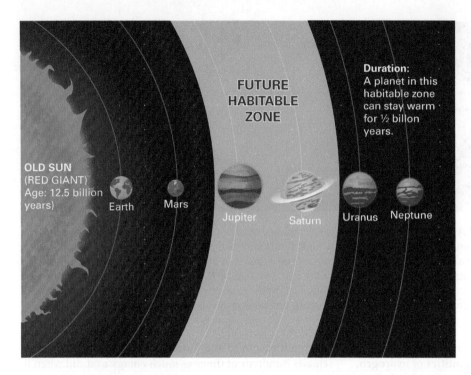

FUTURE HABITABLE ZONE

Duration: A planet in this habitable zone can stay warm for ½ billon years.

OLD SUN (RED GIANT) Age: 12.5 billion years)

Earth   Mars

Jupiter   Saturn   Uranus   Neptune

● **FIGURE 24.14** The Sun will engulf Earth when it expands to its red giant stage.

a red giant. It will engulf Mercury, Venus, and Earth (●**FIGURE 24.14**). Perhaps the heat will blow much of Jupiter's atmosphere away, exposing a rocky surface.

The core of a red giant condenses under the influence of gravity and gets hotter until its temperature reaches 100 million K. At this temperature, helium nuclei begin to fuse to form carbon nuclei. When helium fusion starts, radiant energy pushes outward once again, and the core expands. The star cools, its outer layers contract, and it enters a second stable phase.

Gradually, as more helium fuses to carbon, the carbon accumulates in the core just as helium did during the earlier life of the star (●**FIGURE 24.12D**). When the helium is used up, fusion ceases again and the carbon core contracts. This gravitational contraction causes the core to heat up again.

What happens next depends on the star's initial mass. Astronomers express the mass of a star relative to that of the Sun: 1 **solar mass** is the mass of the Sun. In a star with a mass about the same as our Sun, contraction of the carbon core is not intense enough to raise its temperature sufficiently to initiate fusion of the carbon nuclei. However, gravitational contraction of the carbon core does release enough energy to blow a shell of gas out into space. This shell is called a **planetary nebula** (●**FIGURE 24.15**). Meanwhile, the material remaining in the star contracts until atoms are squeezed so tightly together that only the pressure exerted by the electrons prevents further compression. A dying star as massive as our Sun will eventually shrink until its diameter is approximately that of Earth. Such a shrunken star no longer produces energy, and it glows solely from its residual heat produced during past eras. The star has become a **white dwarf** (●**FIGURE 24.16A**). It will continue to cool slowly over tens of billions of years, but it will never change diameter again. No further nuclear

reactions will occur. Its gravitational force is not strong enough to overcome the strength of the electrons, so it will never contract further.

Stars with a Large Mass

Some stars do not die so gently (●**FIGURE 24.16B**). If the star is larger than 1.44 solar masses, a white dwarf does not form. Instead, as helium fusion ends, gravitational contraction produces enough heat to fuse carbon. Renewed fusion produces increasingly heavier elements in a stepwise process until iron forms. Iron is different from the lighter elements. When hydrogen or helium nuclei fuse, energy is released. In contrast, iron fusion absorbs energy and thus cools a star.

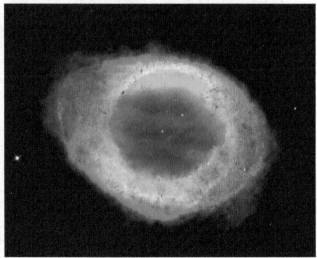

● **FIGURE 24.15** The Ring Nebula is a sphere of gas and dust expelled as a dying star exploded.

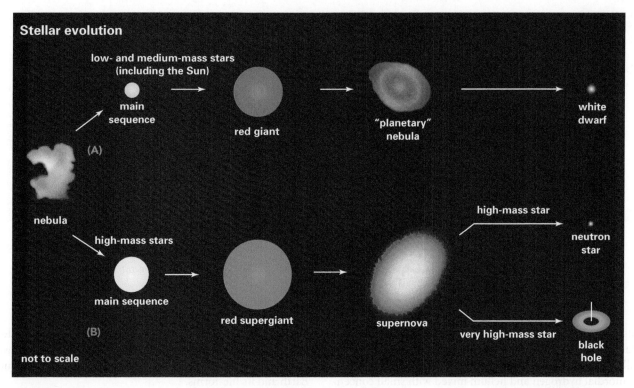

● FIGURE 24.16 (A) A star with a mass about the same as our Sun passes through its life cycle until it becomes a white dwarf. (B) A more massive star follows a different route, ending as a neutron star or a black hole, depending on its initial mass.

When this happens, the thermal pressure that forced the stellar gases outward diminishes and the star collapses under the influence of gravity. This collapse releases large amounts of heat. Within a few seconds—a fantastically short time in the life of a star—the star's temperature reaches trillions of degrees and the star explodes to create a **supernova**. For a brief period, a supernova shines as brightly as hundreds of billions of normal stars and may even emit as much energy as an entire galaxy. To observers on Earth, it appears as though a new, brilliant star suddenly materialized in the sky, only to become dim and disappear to the naked eye within a few months. What happens after that is covered in section on "Neutron Stars, Pulsars, and Black Holes."

On February 24, 1987, an astronomer named Ian Shelton was carrying out research unrelated to supernovas. When he developed one of his photographic plates, he saw a bright star where previously there was only a dim one (●**FIGURE 24.17**). He walked outside, looked into the sky, and saw the star with his naked eye. This was the first supernova explosion visible to the naked eye since 1604, five years before the invention of the telescope.

A supernova explosion is violent enough to send shock waves racing through the atmosphere of the star, fragmenting atomic nuclei and shooting subatomic particles in all directions. Many of the nuclear particles collide with sufficient energy both to fuse and to split apart. These processes form all the known elements heavier than iron. Thus, in studying the evolution of stars, scientists learned how the heavy elements originated.

First- and Second-Generation Stars

Hydrogen and helium were the first elements to form when our Universe was 300,000 years old. Originally, all the stars in the Universe were composed entirely of these two lightest elements. These old, first-generation stars are called **population II stars**, and a few still exist.

Within the cores of population II stars, hydrogen fused to helium, changing the ratio of these two elements in the Universe. As stellar evolution continued, helium fused to carbon, and carbon fused to heavier elements, up to iron. Elements heavier than iron formed during supernova explosions of massive population II stars.

solar mass A standard unit for expressing the mass of a star relative to the mass of our Sun; 1 solar mass equals the mass of the Sun.

planetary nebula A nebula created when a star about the size of our Sun explodes and blows a shell of gas out into space.

white dwarf A stage in the life of a star when fusion has halted and the star glows solely from the residual heat produced during past eras. White dwarfs are very small stars.

supernova An exploding star that is releasing massive amounts of energy.

population II star An old star, with lower concentration of heavy elements than a population I star.

(A)   (B)

© Australian Astronomical Observatory

● **FIGURE 24.17** During the supernova explosion of 1987, a relatively insignificant star (A) (see arrow) became the brightest object in that portion of the sky (B).

As described earlier, when a star dies it blasts gas and dust into space to form a new nebula. Eventually, these nebulae condense once again into new, second-generation stars called **population I stars**. Population I stars begin life with primordial hydrogen and helium mixed with small concentrations of heavy elements that were inherited from population II stars and from supernova explosions. The Sun is a population I star that condensed from a nebula containing all the natural elements.

Thus, our Solar System was born from the debris of dying stars. Solar systems containing Earth-like planets and

living organisms could never form around population II stars because these stars do not have all the necessary elements. As you can see, in studying the life cycle of stars, we also learn about the origins of the elements that make up Earth and its life-forms.

NEUTRON STARS, PULSARS, AND BLACK HOLES

Neutron Stars and Pulsars

In a supernova explosion, most of the matter in a star is blasted into a nebula, but a substantial fraction remains behind, compressed into a tight sphere. In the 1930s, scientists developed a hypothesis to explain what happens within this sphere. If it is between 2 and 3 solar masses, the gravitational force is so intense that the star cannot resist further compression the way a white dwarf does. Instead, the electrons and protons are squeezed together to form neutrons:

Electrons + Protons = Neutrons

The neutrons then resist further compression and remain tightly packed. This ball of compressed neutrons is called a **neutron star**. A neutron star is extremely dense—approximately 10^{13} kg/cm^3 (●**FIGURE 24.18**). If the entire Earth were as dense, it would fit inside a football stadium.

The first neutron star was discovered by accident in 1967, 30 years after scientists postulated their existence. Jocelyn Bell Burnell, a graduate student at the time, was studying radio emissions from distant galaxies. In one part of the sky, she detected a radio signal that switched on and off with a frequency of about one pulse every

One tablespoonful of normal star material weighs about as much as a pencil.

One tablespoonful of white dwarf material weighs about as much as an elephant.

Everest

One tablespoonful of neutron star material weighs about as much as Mount Everest.

● **FIGURE 24.18** A normal star has a relatively low density. A white dwarf is more dense, and a neutron star has even higher density.

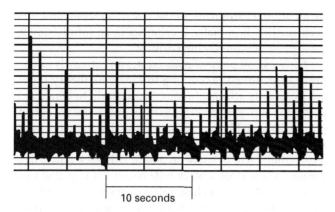

● FIGURE 24.19 Pulsar signals appear as sharp spikes on the recording of a radio telescope.

---

population I star A relatively young star formed from material ejected by an older, dying star; composed mainly of primordial hydrogen and helium, with a small percentage of heavier elements. Our Sun is a population I star.

neutron star A small, extremely dense star, created from remnants of the supernova explosion of a large star and composed almost entirely of compressed neutrons.

pulsar A neutron star that emits a pulsating radio signal.

1.33 seconds (**●FIGURE 24.19**). Many radio signals arrive at Earth from outer space, but the emissions Burnell heard were unusual because they were sharp, regular, and spaced a little more than a second apart. If such a signal were fed into the speaker of a conventional radio, you would hear a "click, click, click," evenly spaced, with one click every 1.33 seconds.

At first, astronomers considered the possibility that they might be a signal from intelligent life, so they called the signals LGM, for "little green men." But when Burnell found a similar pulsating source in a different region of the sky, scientists ruled out the possibility that two life-forms in different parts of the Universe would send similar signals. Once it was established that the signals did not originate from intelligent beings, their sources were called **pulsars**. But naming the source did not explain it.

The first step toward identifying pulsars was to estimate their sizes. Not all parts of an object in space are equidistant from Earth (**●FIGURE 24.20**). If a large sphere emits a sharp burst of energy from its entire surface, some of the photons start their journey closer to Earth than others and therefore arrive sooner. A person on Earth listening to the radio noise hears not a sharp click but a more prolonged "cliiiiiick," because it takes a while for all the radio waves to arrive.

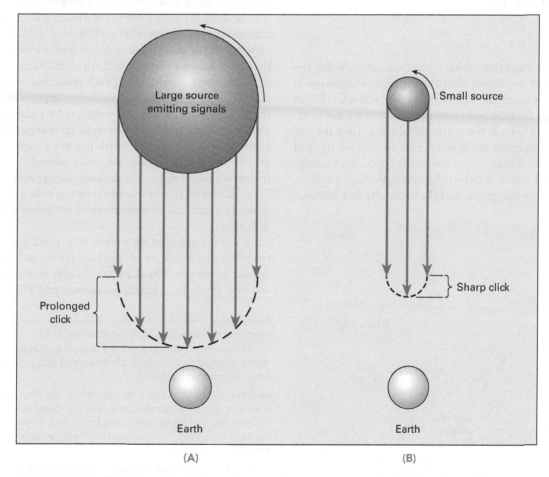

● FIGURE 24.20 (A) A sharp signal from a larger sphere arrives over a longer time interval than (B) a sharp signal from a smaller sphere.

Alternatively, a signal from a small source is much sharper. Pulsar signals are sharp, indicating that the source must be unusually small for an energetic object in space—30 kilometers in diameter or less. The smallest star previously recorded was a white dwarf 16,000 kilometers in diameter. Scientists then reasoned that the pulsar detected by Burnell might be the long-searched-for neutron star.

Astronomers suggested that the radio signals are emitted by an electromagnetic storm on the surface of a neutron star. Thus, a pulsar is a neutron star that emits intermittent, but regular, radio signals. According to this hypothesis, as the star rotates, so does the storm center. A receiver on Earth detects one click per revolution of the star, just as a lookout on a ship sees a lighthouse beam flash periodically as the beacon rotates (●FIGURE 24.21). If a pulse is received once every 1.33 seconds, the pulsar must rotate that rapidly. White dwarfs can again be ruled out since they are too big to rotate so rapidly. But neutron stars are small enough to rotate that fast.

Astronomers searched for an example to test the hypothesis. They focused a radio telescope on the Crab Nebula, where a supernova explosion occurred during the Middle Ages. A pulsar signal was found exactly where the supernova had occurred about 950 years ago—precisely where a neutron star should be.

Black Holes

If a star with more than about 5 solar masses explodes, the core remnant remaining after the supernova explosion is thought to be more massive than 2 to 3 solar masses. When a sphere this massive contracts, the neutrons are not sufficiently strong to resist the gravitational force. Then the star shrinks to a diameter much smaller than a neutron star and becomes a **black hole**. Such a collapse is impossible to imagine in earthly terms. A tremendous mass, perhaps a trillion, trillion, trillion kilograms, shrinks to the size of a pinhead,

then continues to shrink to the size of an atom, and then even smaller. Eventually, it collapses to an infinitesimally small point of infinitely high density.

Such a small point of mass creates an extremely intense gravitational field. According to Einstein's theory of relativity, gravity affects photons. Thus, starlight bends as it passes the Sun. If an object is massive and dense enough, its gravitational field becomes so intense that light and other radiant energy cannot escape. So, just as you cannot throw a ball from Earth to space because it falls back down, light cannot escape from a black hole because it is pulled back downward. Because no light can escape such an object, it is invisible; hence the name *black hole*. If you were to shine a flashlight beam, a radar beam, or any kind of radiation at a black hole, the energy would be absorbed. The beam could never be reflected back to your eyes; therefore, you would never see it again. It would be as though the beam just vanished into space. Similarly, if a spaceship flew too close to a black hole, it would be sucked in. No engine could possibly be powerful enough to accelerate the rocket back out, for no object can travel faster than the speed of light.

The search for a black hole is therefore even more difficult than the search for a neutron star. How do you find an object that is invisible and can neither emit nor reflect any energy? In short, how do you find a hole in space? Although it is theoretically impossible to see a black hole, astronomers can observe the effects of its gravitational field. Many stars exist in pairs or small clusters. If two stars are close together, they orbit around each other. Even if one becomes a black hole, the two still orbit about each other, but one is visible and the other invisible. The visible one appears to be orbiting an imaginary partner. Astronomers have studied several stars that seem to orbit in this unusual manner. In several cases, the invisible member of the pair has a mass equal to or greater than 3 solar masses. Because a normal star of 3 solar masses would be visible, the invisible partner may be a black hole. However, the simple observation that a star moves around an invisible companion does not prove that a black hole exists.

If a star were orbiting a black hole, great masses of gas from the star would be sucked into the black hole, to disappear forever (●FIGURE 24.22). As this matter started to fall into the hole, it would accelerate, just as a meteorite

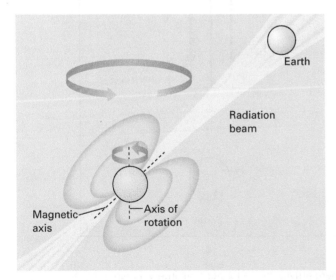

● **FIGURE 24.21** The radiation beam of a pulsar is detected only when it sweeps across Earth.

Earth

Radiation beam

Magnetic axis

Axis of rotation

---

black hole An infinitesimally small region of space, created after the supernova explosion of a huge star, that contains matter packed so densely that light cannot escape from its intense gravitational field.

gamma ray burst A burst of high-energy radiation emanating from space. Longer bursts come from the death throes of massive stars, while shorter bursts form when a neutron star is sucked into a black hole or when two neutron stars collide to form a black hole.

galaxy A large volume of space containing many billions of stars, held together by mutual gravitational attraction.

elliptical galaxy A galaxy with an oval shape and no spiraling arms; can be either giant or dwarf.

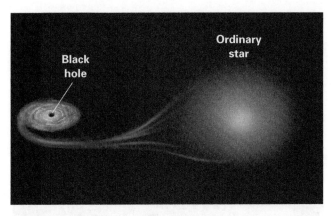

● **FIGURE 24.22** If a black hole and an ordinary star are in orbit around one another, gases from the star will eventually be sucked into the black hole.

accelerates as it falls toward Earth. The gravitational field of a black hole is so intense that particles drawn into it would collide against each other with enough energy to emit X-rays. Thus, just as a falling meteorite glows white-hot as it enters Earth's atmosphere, anything tumbling into a black hole would glow even more energetically; that is, it would emit X-rays. These X-rays might then be detected here on Earth. But, you may ask, if light cannot escape from a black hole, how can the X-rays escape? The answer is that the X-rays are produced and escape from just outside the edge of the black hole. Thus, matter being sucked into a black hole sends off one final message before being pulled into the void from which no message can ever be sent.

An important experiment, therefore, was to focus an X-ray telescope on portions of the sky where a star appeared to orbit an invisible partner. Such telescopes must be located aboard space satellites because X-rays do not penetrate Earth's atmosphere. In the 1980s, an orbiting X-ray telescope detected X-ray sources adjacent to stars that appeared to orbit an unseen partner. More recent observations by ground-based observatories and the Hubble Space Telescope convince most astronomers that black holes exist.

Gamma Ray Bursts

If you could "see" electromagnetic radiation 10 million times more energetic than visible light, the Universe would appear to be a much more erratic, violent environment than we

normally perceive. On average, once a day a flash of high-frequency radiation, called a **gamma ray burst**, spews out of the heavens from a random direction, with a brief pulse containing more power than we receive from the Sun. Gamma ray bursts have a duration ranging from a few milliseconds to slightly more than two seconds. In 2002, astronomers determined that the longer bursts emanated from the death-throe explosion of a very massive star. But the origin of the short bursts proved more elusive. Imagine standing on the rim of a football stadium and trying to locate the source, intensity, and fine frequency structure of a camera flash somewhere in the crowd. In 2004, NASA scientists launched the Swift space-based observatory, equipped with an X-Ray Telescope (XRT), an UltraViolet/Optical Telescope (UVOT), and a Burst Alert Telescope (BAT), all aligned and coordinated for instant observations over a wide range of frequencies. From these studies scientists determined that short, gamma ray bursts form when a neutron star is sucked into a black hole (●**FIGURE 24.23**) or when two neutron stars collide to form a black hole.

GALAXIES

In the late 1700s, a French astronomer, Charles Messier, was studying comets. He recorded more than 100 fuzzy objects in the sky that clearly were not stars. When these objects were studied in the 1850s with more powerful telescopes, many were observed to have spiral structures like pinwheels. They were not comets—but what were they, and how far away were they? These questions were not answered until 1924, when Edwin Hubble determined that they were farther away than even the most distant known stars. In order for us to see them at all, they must be much more luminous than a star. Hubble concluded that each object is a **galaxy**, a large volume of space composed of billions of stars held together by gravity. Today we recognize that galaxies and clusters of galaxies form the basic structure of the Universe.

A galaxy with an oval shape and no spiraling arms is called an **elliptical galaxy**. About one-third of the galaxies in our region of the Universe are elliptical. Giant elliptical galaxies contain up to 1013 solar masses and may have a diameter larger than our Milky Way galaxy (which has a spiral shape, discussed next). However, most ellipticals are dwarf and contain only a few million solar masses. Recent

Dana Berry/NASA

● **FIGURE 24.23** Short gamma ray bursts can be formed when a neutron star is sucked into a massive black hole.

Source: "In a Flash NASA Helps Solve 35-year-old Cosmic Mystery," National Aeronautics and Space Administration, October 5, 2005, online at http://www.nasa.gov/mission_pages/bursts/short_burst_oct5.html

observations indicate that dwarf ellipticals may be the most common type of galaxy, far from our immediate galactic neighborhood.

Almost two-thirds of the bright galaxies in our region of the Universe are **spiral galaxies**. The stars in a spiral are arranged in a thin disk, with arms radiating outward from a spherical center or nucleus (●**FIGURE 24.24**). The stars in the outer arms rotate around the nucleus like a giant pinwheel. A typical spiral galaxy contains about 100 billion stars. Nearly half of the spirals are **barred spiral galaxies**, having a straight bar of stars, gas, and dust extending across the nucleus (●**FIGURE 24.25**). A few percent of all galaxies are lens-shaped or irregular and show no obvious pattern.

Normally, we think of galaxies as visible objects composed of collections of stars. Galaxy Virgo H121 is a swirling cloud of hydrogen-rich gas, more massive than many visible galaxies, yet it contains no stars and emits no visible light, only radio-wave radiation. Astronomers hypothesize that *dark galaxies* like Virgo H121 have such a low density of matter that gravitational forces haven't been strong enough to initiate star formation.

Galactic Motion

Recall from Chapter 22 that astronomers measure the relative velocities of distant objects by measuring the frequency

● **FIGURE 24.25** A barred spiral galaxy, NGC 1365, displays a conspicuous bar running through its nucleus.

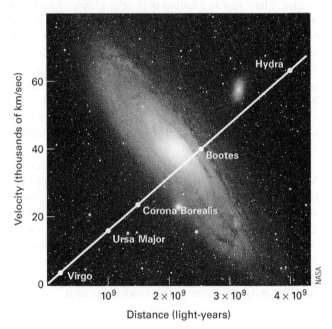

● **FIGURE 24.26** Hubble's law states that the velocity of a galaxy is directly proportional to its distance from Earth.

● **FIGURE 24.24** The dense nucleus of the spiral galaxy NGC 2997 is surrounded by spiraling arms composed of billions of stars.

of light they emit. In 1929, five years after he had described galaxies, Hubble noted that frequency of emitted light from almost every galaxy is shifted toward the red end of the spectrum. Hubble interpreted this red shift to mean that all the galaxies are flying away from us and from each other: the Universe is expanding. Moreover, he observed that the most-distant galaxies are moving outward at the greatest speeds, whereas the closer ones are receding more slowly. This relationship is known as **Hubble's law** (●**FIGURE 24.26**). Using Hubble's law, the distance from Earth to a galaxy can be calculated by measuring the galaxy's red shift.

THE MILKY WAY

Our Sun lies in a barred spiral galaxy called the **Milky Way** (●FIGURE 24.27). The Milky Way's galactic disk is 2,000 light-years thick, and recent studies indicate that it is 200,000 light-years in diameter. The Milky Way contains about 400 billion stars. It has also swallowed up numerous independent dwarf galaxies that now rotate in the outer-spiral arms. Because the disk is thin, like a CD or DVD, an observer on Earth sees relatively few stars perpendicular to its plane. Thus, most of the night sky contains a diffuse scattering of stars with large expanses of black space between them. However, if you look into the plane of the disk, you see a dense band of stars from horizon to horizon (●FIGURE 24.28). This band is commonly called the Milky Way, although astronomers use the term to describe the entire galaxy. The galactic disk rotates about its center once every 200 million years, so in the 4.6-billion-year history of Earth, we have completed about 23 rotations.

A spherical **galactic halo**—a cloud of dust and gas—surrounds the Milky Way's galactic disk. This halo is so large that even though it is extremely diffuse, it contains as much as 90 percent of the mass of the galaxy. Many dim and relatively old stars exist within this halo. Some are concentrated in groups of 10,000 to 1 million stars. Each group is called a **globular cluster**. The galactic halo and globular clusters are probably remnants of one or more protogalaxies that condensed to form the Milky Way. The spherical structure of the halo suggests that the entire galaxy was once spherical.

---

spiral galaxies A galaxy characterized by arms that radiate out from the center like a pinwheel.

barred spiral galaxies A spiral galaxy with a straight bar of stars, gas, and dust extending across the nucleus.

Hubble's law A law that states that the velocity of a galaxy is proportional to its distance from Earth. Thus, the most distant galaxies are traveling at the highest velocities.

Milky Way The barred spiral galaxy in which our Sun and Solar System are located.

galactic halo A spherical cloud of dust and gas that surrounds the galactic disk of a spiral galaxy such as the Milky Way.

globular cluster A concentration of older stars that shared a common origin and are held together by gravity, lying within the galactic halo of a spiral galaxy such as the Milky Way.

The Nucleus of the Milky Way

Photographs of other spiral galaxies show that the galactic nucleus shines much more brightly than the disk. The concentration of stars in the galactic nucleus is perhaps 1 million times greater than the concentration in the outer disk. If you could visit a planet orbiting one of these stars, you would never experience night, for the stars would light the planet from all directions. However, stable solar systems could not exist in this region because gravitational forces among nearby stars would rip planets from their orbits.

It is impossible to see into the nucleus of our own galaxy because visible light does not penetrate through the interstellar nebulae that lie in the way. Looking into the nucleus is like trying to see a ship on a foggy day. However, just as sailors use radar to penetrate the fog, astronomers study the center of the galaxy by analyzing radio, infrared, and X-ray emissions that travel through the clouds. These studies provide a picture of the nucleus of the Milky Way.

A cloud of dust and gas orbits the Milky Way's galactic nucleus. Infrared measurements show that this cloud is so hot that it must be heated by an energy source with 10 to 80 million times the output of our Sun. The cloud is ring-shaped, with a hole in its center, like a doughnut. Relatively recently, 10,000 to 100,000 years ago, a giant explosion blew out the center of a larger cloud and created the hole. X-ray and gamma ray emissions tell us that matter is now accelerating inward, back into the center at a rate of 1 solar mass every 1,000 years (●FIGURE 24.29).

By measuring the orbits of stars close to the galactic center, astronomers can measure the gravitational force of the galactic center, and hence its mass. Calculations show that an unseen object, 3.6 million times as massive as our Sun, lies at the heart of our galaxy. This object can only be a black hole. According to the most widely accepted hypothesis, early in the history of the galaxy, one huge star or many large ones formed in the galactic center. These stars were large enough to pass through the main sequence quickly and then collapse to become the black hole that now forms the center of our galaxy.

Galactic Nebulae

Many large nebulae exist within our galaxy. Even though a nebula looks spectacular when viewed through a powerful

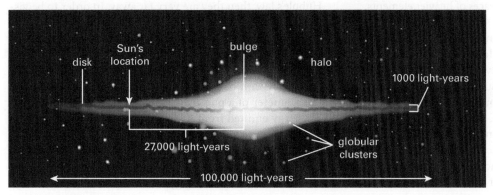

● **FIGURE 24.27** An artist's drawing of the Milky Way galaxy shows the galactic disk, which is 200,000 light-years in diameter and 2,000 light-years thick. The distance from the Sun to Earth is only 8 light-minutes.

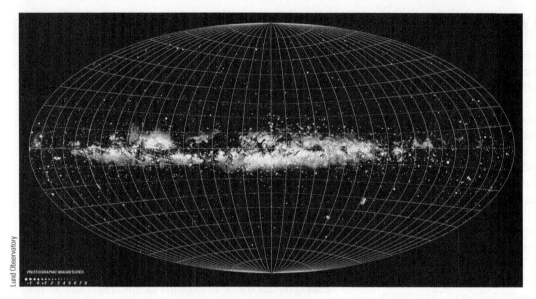

Lund Observatory

PHOTOGRAPHIC MAGNITUDES
–1 0 +1 2 3 4 5 6 7 8

● **FIGURE 24.28** The Milky Way appears as a solid band of light. However, this light is produced by billions of stars in the plane of the galactic disk.

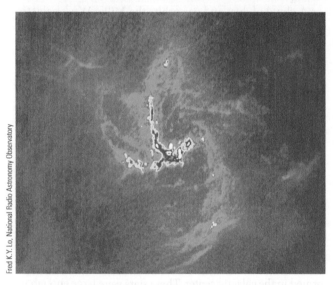

Fred K.Y. Lo, National Radio Astronomy Observatory

● **FIGURE 24.29** The bright region in the center of this image is a small but powerful radio source from the center of the galaxy. Gases falling toward the center produce the red, orange, and blue streamers.

telescope, the densest nebulae are only 10^{-13} as dense as the Earth's atmosphere at sea level, and less dense ones are only 10^{-18} as dense as our atmosphere! A nebula consists mainly of hydrogen and helium, with traces of heavier elements.

As explained in section on "The Birth of a Star," a huge system of nebulae lies in the constellation Orion. The entire structure is 100 light-years in length from north to south, and even though it is extremely diffuse, it contains a mass equivalent to 200,000 of our Suns. ●**FIGURE 24.30** shows a close-up of the Horsehead Nebula, which appears near the easternmost star of Orion's belt. The brightly colored region surrounding the horsehead is an **emission nebula**, a nebula so hot that its atoms and molecules glow like a neon light. A dark region (shaped like a horse's head in Figure 24.30) occupies the middle of one of the red emission nebulae.

This dark region is an **absorption nebula**, where the dust and gases are frigidly cold, about 13 K (−260°C), and absorb light from the emission nebula behind it. At these extremely low temperatures, most elements condense to form molecules. Although the two lightest elements, hydrogen and helium, constitute more than 99 percent of the cloud's total mass, the Orion absorption nebula contains all or most of the natural elements. Astronomers have detected 100 different molecules in the Orion absorption nebula. Most are simple—such as hydrogen, water, carbon monoxide, and ammonia—but several larger, organic molecules have also been detected. This nebula is the birthplace of stars.

QUASARS

In the 1960s, astronomers were studying many objects that look like stars but emit extremely large amounts of energy as radio waves. These objects were perplexing because normal stars, such as our Sun, emit mostly visible and ultraviolet light. During the following decade, astronomers discovered that the spectra of many of these objects coincide with that of hydrogen, except that they have a very large red shift. Such objects are now called **quasar** (●**FIGURE 24.31**).

Recall that a large red shift indicates that an object is moving away from Earth at a high speed. If quasars obey Hubble's law, then they are very far away. In order to be visible when they are so far away, quasars must emit tremendous amounts of energy—10 to 100 times more than an entire galaxy!

Quasars emit erratic bursts of energy. Recall from our discussion of pulsars that astronomers can estimate the diameter of a pulsar by the sharpness of a short burst of energy emitted from it. Similar techniques indicate that some quasars are several light-months in diameter. By comparison, the distance from Earth to the nearest star is 4 light-years. Thus, a quasar is much smaller than a galaxy but emits much more energy. Furthermore, it emits energy

● **FIGURE 24.30** The Horsehead Nebula appears close to the bright star on Orion's belt shown in the left-center of this photograph. The bright white lights in the photograph are other stars. The reddish glows are emission nebulae, and the horsehead-shaped dark area is an absorption nebula in front of an emission nebula.

over a wide range of wavelengths. Most quasars are very distant from Earth. One common hypothesis for their origin is that a massive black hole, perhaps of 1 billion solar masses, lies at the center of a quasar and that the quasar emits energy as gas and dust accelerate into the hole.

Looking Backward into Time

Many quasars are 8 to 12 billion light-years away. If an object is 10 billion light-years away, the light we see today started its journey 10 billion years ago. Therefore, we see what was happening long ago but not what is happening today. If the object blew up and disappeared 9 billion years ago, we will not know about it for another billion years! Thus, when we look at close objects, we see what happened recently, but when we look at distant objects, we see what happened in

emission nebula A hot nebula that glows because it is emitting light energy.

absorption nebula A cold, dark nebula that absorbs light.

quasar An object, less than 1 light-year in diameter and very distant from Earth, that emits an extremely large amount of energy as radio waves.

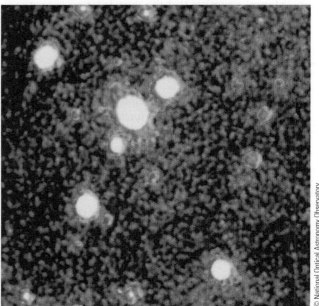

● **FIGURE 24.31** Quasar 3C 275.1 is the brightest object near the center of this photograph. An elliptical gas cloud surrounds the nucleus. This quasar is about 2 billion light-years from Earth. The smaller objects are normal galaxies.

the distant past. One goal of building more powerful telescopes is to study more distant objects and therefore probe further back in time.

The oldest quasars must have formed when the Universe was young. Moreover, quasars contain heavy elements, and heavy elements form only in the supernova explosions of dying stars. Therefore, quasars are second-generation structures; they formed after earlier stars were born, evolved, exploded, and died. Hence, stars must have formed and passed through their life cycle less than 2 billion years after the Universe formed.

DARK MATTER

The Milky Way rotates slowly, like a giant pinwheel. Using laws of motion developed by Kepler and Newton in the 17th century, astronomers have attempted to calculate the mass of our galaxy from their knowledge of its rotational speed. However, these calculations produced a giant anomaly. The Milky Way is much more massive than we can account for if we add up all the known matter within it.

Satellite studies of the nonhomogeneity of galaxies (section on "The Nonhomogeneous Universe") allow scientists to calculate the mass and gravitational force of the primordial Universe. These studies indicate that matter we can see comprises only 4 percent of the mass of the Universe! The other 96 percent is invisible and is called **dark matter**. What is it? Not many years ago, many astrophysicists speculated that dark matter might be composed of planetlike objects that do not radiate energy (and are therefore invisible) and black holes that cannot radiate energy. However, in 2003, results from the WMAP satellite convinced most scientists that the story is far more exotic, as explained next.

THE END OF THE UNIVERSE

From the first instant of the big bang, the Universe has been affected by two opposing forces. Matter is expanding outward as a result of the big bang, but at the same time gravity is pulling all matter back inward toward a common center. If the gravitational force of the Universe is sufficient, all the galaxies will eventually slow down, reverse direction, and fall back to the center, forming another point of infinite density. This hypothesis is called a **closed Universe** (●**FIGURE 24.32A**). Some scientists have speculated that if the Universe is closed, it will collapse and then explode again to form a new Universe. In turn, the new Universe will expand and then collapse, creating a continuous chain of Universes. However, no mathematical model explains how another big bang would occur. Instead, current models predict that in a closed system, the Universe would collapse into a mammoth black hole.

Another possibility is that the gravitational force of the Universe is not sufficient to stop the expansion and that the galaxies will continue to fly apart forever. Within each galaxy, stars will eventually consume all their nuclear fuel and stop producing energy. As the stars fade and cool, the galaxies will continue to separate into the cold void. This scenario is called the **open Universe** (●**FIGURE 24.32B**).

In the mid- to late 1990s, astronomers were attempting to measure the deceleration of the Universe. If they could learn how fast it was slowing down, then they reasoned that they could calculate whether the Universe is open or closed. Astronomers made these measurements by studying the speed (red shift) of distant supernovae. The result was totally unexpected. The expansion of the Universe is not slowing down at all. Instead, it is accelerating!

The concept of an **accelerating Universe** is totally baffling. Objects accelerate only when they are acted on by a force. In this case, the force must be some sort of repulsion that acts against gravity and pushes the galaxies apart.

By 2003, WMAP data had convinced most astrophysicists that dark matter has two components. **Dark mass**, which makes up 23 percent of the Universe, is an as-yet-undetected particle that exerts a conventional gravitational attraction. Another 73 percent of the Universe is **dark energy**—the poorly understood repulsive force that is causing the modern Universe to accelerate. As explained earlier, only 4 percent of the Universe is "ordinary" matter that makes stars shine, rivers flow, and people walk through grassy meadows on a spring day.

WHY ARE WE SO LUCKY?

Planets and living organisms are composed of midmass and heavy elements. But, as we have learned, primordial population II stars were originally formed from the two lightest elements: hydrogen and helium. Thus, in order for life to have evolved, population II stars had to pass through their life cycles and produce carbon and other midmass elements.

---

dark matter Mysterious invisible matter that may comprise up to 96 percent of the mass of the Universe.

closed Universe A model in which the gravitational force of the Universe is so great that all the galaxies will eventually slow down, reverse direction, and fall back to the center, forming another point of infinite density.

open Universe A model in which the gravitational force of the Universe is not sufficient to stop the current expansion and the galaxies will continue to fly apart forever.

accelerating Universe A model in which an unidentified force is overpowering gravity and causing the expansion of the Universe to accelerate.

dark mass One component of dark matter, comprising a collection of as-yet-undetected particles, large enough to make up 23 percent of the Universe, that exert conventional gravitational attraction.

dark energy One component of dark matter, comprising the poorly understood repulsive force that causes the modern Universe to accelerate.

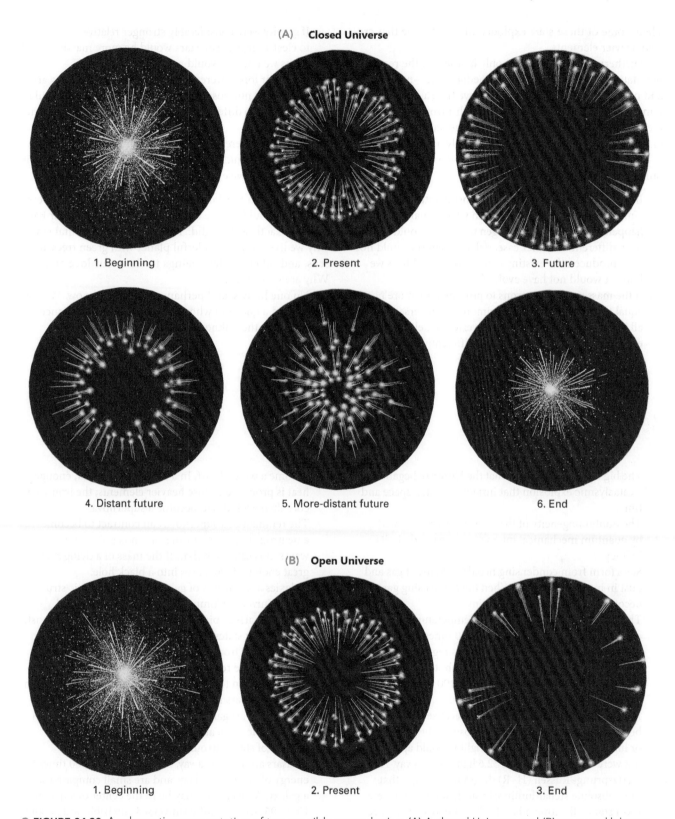

(A)   Closed Universe

1. Beginning

2. Present

3. Future

4. Distant future

5. More-distant future

6. End

(B)   Open Universe

1. Beginning

2. Present

3. End

● **FIGURE 24.32** A schematic representation of two possible cosmologies: (A) A closed Universe and (B) an open Universe. Recent data imply that the Universe may not only be open but also accelerating.

Then, some of these stars exploded in supernovae that created heavier elements.

Furthermore, life can exist only if atoms in the resultant population I stars fuse at an intermediate rate—not too fast and not too slow—so solar systems have time and the proper temperature for planets to evolve with favorable environments. These conditions are met only if the starting conditions of the big bang fit within an extraordinarily narrow window.

The conditions for life in the Universe would be possible only under the following conditions:

- If the initial temperature and expansion rate of the Universe were greater or less than what actually happened, the ratio of hydrogen to helium would have been different. In either case, stellar evolution would not have produced the existing conditions—and life as we know it would not have evolved.
- If the mass ratio of neutrons to protons were more nearly equal than they were in the primordial Universe, almost all the hydrogen would have converted to helium. Stellar evolution would have proceeded differently—and life as we know it would not have evolved.

- If gravity were considerably stronger relative to electromagnetism, stars would be more massive—and life as we know it would not have evolved.
- If nuclear forces were stronger or weaker than they are, primordial nucleosynthesis and stellar evolution would have been so different—and life as we know it would not have evolved.
- If electrons were more massive than they are, complex chemistry would not be possible—and life as we know it would not have evolved.

There is no fundamental reason why all the universal constants should agree exactly so that our Sun and our Earth formed exactly as they did. There is no fundamental reason why we live on this wonderful planet with green trees, azure seas, and other sentient beings to share our love and joys. Why are we so lucky?

No one knows, and perhaps no one will know. We have formulated a question whose answer transcends science and ventures into the unknowable.

Key Concepts Review

- The big bang theory states that the Universe began with a cataclysmic explosion that instantly created space and time.
- The nonhomogeneity of the Universe was generated by quantum mechanical interactions very early in its infancy.
- Stars form from condensing nebulae, clouds of gas and dust in interstellar space. When the condensing gases become hot enough, fusion begins.
- The central core of the Sun has a temperature of about 15 million K and a density 150 times that of water. Hydrogen fusion occurs in this region. The visible surface of the Sun, called the photosphere, is only 5,700 K and has a pressure of 1/100 of Earth's atmosphere at sea level.
- A star can appear luminous either because it really is bright or because it is close. The absolute brightness or luminosity of a star is how bright it would appear if it were a fixed distance of 32.6 light-years away. A Hertzsprung–Russell (H–R) diagram is a graph that plots absolute stellar luminosity against temperature. Most stars fall within a band called the main sequence.
- When the hydrogen in the core of a mature star is exhausted and hydrogen fusion ends, gravity compresses the core and the temperature rises. Fusion then begins in the outer shell, and the star expands to become a red giant. After helium fusion in an average-mass star ends, the star releases a planetary nebula and then shrinks to

become a white dwarf. In a more massive star, enough heat is produced to fuse heavier elements; the iron-rich core then explodes to become a supernova.
- The remnant of a supernova can contract to become a neutron star. If a neutron star emits pulses of radio waves, it is called a pulsar. If the mass of a dying star is great enough, it collapses into a black hole.
- Galaxies and clusters of galaxies form the basic structure of the Universe. Commonly, galaxies are elliptical, spiral, or barred spiral, although other shapes also exist. Hubble's law states that the most distant galaxies are moving away from Earth at the greatest speeds, while closer ones are receding more slowly.
- Our Sun lies in the Milky Way, a barred spiral galaxy. In addition to stars in the main disk, the galaxy contains captured dwarf galaxies, diffuse clouds of gas and dust between the stars, and a galactic halo and globular clusters of stars surrounding the disk.
- Quasars are very far away, emit as much as 100 times the energy of an entire galaxy, and are small compared with a galaxy. A black hole may lie in the center of a quasar.
- Up to 96 percent of the Universe is invisible and is called dark matter.
- In trying to determine whether the Universe is open or closed, astronomers learned that the expansion of the Universe is accelerating.
- Life could have formed under only a very narrow range of fundamental physical constants.

Important Terms

absolute brightness (or absolute luminosity) (p. 565)

absorption nebula (p. 576)

accelerating Universe (p. 578)

apparent brightness (or apparent luminosity) (p. 565)

barred spiral galaxies (p. 574)

big bang (p. 558)

black hole (p. 572)

chromosphere (p. 564)

closed Universe (p. 578)

convective zone (p. 564)

core (p. 563)

cosmic background radiation (p. 560)

dark energy (p. 578)

dark mass (p. 578)

dark matter (p. 578)

elliptical galaxy (p. 573)

emission nebula (p. 576)

first-generation energy (p. 561)

galactic halo (p. 575)

galaxy (p. 573)

gamma ray burst (p. 573)

globular cluster (p. 575)

Hertzsprung–Russell or H–R diagram (p. 565)

Hubble's law (p. 574)

light-year (p. 565)

main sequence (p. 565)

Milky Way (p. 575)

nebula (p. 561)

neutron star (p. 570)

open Universe (p. 578)

parsec (p. 565)

photosphere (p. 564)

planetary nebula (p. 568)

population I stars (p. 570)

population II stars (p. 569)

primordial nucleosynthesis (p. 559)

prominence (p. 565)

pulsars (p. 571)

quasar (p. 576)

radiative zone (p. 563)

red giant (p. 567)

second-generation energy (p. 561)

solar mass (p. 568)

spicules (p. 564)

spiral galaxies (p. 574)

sunspots (p. 564)

supernova (p. 569)

white dwarf (p. 568)

Review Questions

1. Outline the three main forms of evidence to support the big bang theory.

2. Describe the Universe when it was a trillionth of a trillionth of a billionth of a second old.

APPENDIX A

	English Unit	Conversion Factor	Metric Unit	Conversion Factor	English Unit
Length	Inches (in)	2.54	Centimeters (cm)	0.39	Inches (in)
	Feet (ft)	0.305	Meters (m)	3.28	Feet (ft)
	Miles (mi)	1.61	Kilometers (km)	0.62	Miles (mi)
Area	Square inches (in^2)	6.45	Square centimeters (cm^2)	0.16	Square inches (in^2)
	Square feet (ft^2)	0.093	Square meters (m^2)	10.8	Square feet (ft^2)
	Square miles (mi^2)	2.59	Square kilometers (km^2)	0.39	Square miles (mi^2)
Volume	Cubic inches (in^3)	16.4	Cubic centimeters (cm^3)	0.061	Cubic inches (in^3)
	Cubic feet (ft^3)	0.028	Cubic meters (m^3)	35.3	Cubic feet (ft^3)
	Cubic miles (mi^3)	4.17	Cubic kilometers (km^3)	0.24	Cubic miles (mi^3)
Weight	Ounces (oz)	28.3	Grams (g)	0.035	Ounces (oz)
	Pounds (lb)	0.45	Kilograms (kg)	2.20	Pounds (lbs)
	Short tons (st)	0.91	Metric tons (mt)	1.10	Short tons (st)
Temperature	Degrees Fahrenheit (°F)	$-32° \times 0.56$	Degrees centigrade (Celsius) (°C)	$\times 1.80 + 32°$	Degrees Fahrenheit (°F)

Examples:

10 inches = 25.4 centimeters; 10 centimeters = 3.9 inches

100 square feet = 9.3 square meters; 100 square meters = 1080 square feet

50°F = 10.1°C; 50°C = 122°F

APPENDIX B

Earth Science Mineral Identification

To identify most minerals in hand sample, geologists use physical properties such as color, luster, crystal form, hardness, cleavage, specific gravity, and several others. The mineral identification table (B1) is divided into two parts: The first part lists minerals with a metallic luster, and the second part lists minerals with a nonmetallic luster. After determining luster, ascertain hardness and note that each part of the table is arranged with minerals with increasing hardness. Thus, if you have a nonmetallic mineral with a hardness of 6, it must be augite, hornblende, plagioclase, or one of the two potassium feldspars (orthoclase or microcline). If this hypothetical mineral is dark green or black, it must be augite or hornblende, and if it has two cleavage planes intersecting at nearly right angles, it is augite.

TABLE B1 Mineral Identification Tables

Metallic Luster

Mineral	Chemical Composition	Color	Hardness Specific Gravity	Other Features	Comments
Graphite	C	Black	1–2 2.09–2.33	Greasy feel; writes on paper; 1 direction of cleavage	Used for pencil 'lead'; mostly in metamorphic rocks
Galena	PbS	Lead gray	2½ 7.6	Cubic crystals; 3 cleavage directions at right angles	The ore of lead; mostly in hydrothermal rocks
Chalcopyrite	$CuFeS_2$	Brassy yellow	3½–4 4.1–4.3	Usually massive; greenish black streak; iridescent tarnish	The most common copper mineral; mostly in hydrothermal rocks
Magnetite	Fe_3O_4	Black	5½–6½ 5.2	Strong magnetism	An ore of iron; an accessory mineral in many rocks
Hematite	Fe_2O_3	Red brown	6 4.8–5.3	Usually granular or massive; reddish brown streak	Important iron ore; an accessory mineral in many rocks
Pyrite	FeS_2	Brassy yellow	6½ 5.0	Cubic and octahedral crystals	Found in some igneous and hydrothermal rocks and in sedimentary rocks associated with coal

Nonmetallic Luster

Mineral	Chemical Composition	Color	Hardness Specific Gravity	Other Features	Comments
Talc	$Mg_3Si_4O_{10}(OH)_2$	White, green	1 2.82	1 cleavage direction; usually in compact masses; soapy feel	Formed by the alteration of magnesium silicates; mostly in metamorphic rocks
Clay minerals	Varies	Gray, buff, white	1–2 2.5–2.9	Earthy luster; particles too small to observe properties	Found in soils, mudrocks, slate, phyllite
Chlorite	$(Mg,Fe)_6(Si,Al)_4O_{10}(OH)_8$	Green	2 2.6–3.4	1 cleavage direction; occurs in scaly masses	Common in low-grade metamorphic rocks such as slate
Gypsum	$CaSO_4{\cdot}2H_2O$	Colorless, white	2 2.32	Elongated or fibrous crystals; vitreous, silky, or earthy luster	The most common sulfate mineral; found mostly in evaporite deposits

Muscovite (Mica)	KAl$_2$(AlSi$_3$O$_{10}$)(OH)$_2$	Colorless	2–2½ 2.7–2.9	1 cleavage direction; cleaves into thin sheets	Common in felsic igneous rocks, metamorphic rocks, and some sedimentary rocks
Biotite (Mica)	K(Mg,Fe)$_3$ AlSi$_3$O$_{10}$(OH)$_2$	Black, brown	2½ 2.9–3.4	1 cleavage direction; cleaves into thin sheets	Occurs in both felsic and mafic igneous rocks, in metamorphic rocks, and in some sedimentary rocks
Calcite	CaCO$_3$	Colorless, white	3 2.71	3 cleavage directions at oblique angles; cleaves into rhombs; reacts with dilute HCl	The most common carbonate mineral; main component of limestone and marble
Anhydrite	CaSO$_4$	White, gray	3½ 2.9–3.0	Crystals with 2 cleavage directions; usually in granular masses	Found in limestones, evaporite deposits, and the cap rock of salt domes
Halite	NaCl	Colorless, white	2–2½ 2.2	3 cleavage directions at right angles; cleaves into cubes; cubic crystals; salty taste	Occurs in evaporite deposits
Dolomite	CaMg(CO$_3$)$_2$	White, yellow, gray, pink	3½–4 2.85	3 cleavage directions at oblique angles; reacts with dilute hydrochloric acid when powdered	The main constituent of dolostone; also found associated with calcite in limestone and marble
Fluorite	CaF$_2$	Colorless, purple, green, brown	4 3.18	4 cleavage directions; cubic and octahedral crystals	Occurs mostly in hydrothermal rocks and in limestone and dolostone
Augite	Ca(Mg,Fe,Al)(Al,Si)$_2$O$_6$	Black, dark green	6 3.25–3.55	Short 8-sided crystals; 2 cleavage directions; cleavages nearly at right angles	The most common pyroxene mineral; found mostly in mafic igneous rocks
Hornblende	(Ca,Na)$_2$(Mg,Fe,Al)$_5$ (Si,Al)$_8$O$_{22}$(OH)$_2$	Green, black	6 3.0–3.4	Elongated, 6-sided crystals; 2 cleavage directions intersecting at 56° and 124°	A common rock-forming amphibole mineral in igneous and metamorphic rocks
Plagioclase feldspar	Varies from CaAl$_2$Si$_2$O$_8$ to NaAlSi$_3$O$_8$	White, gray	6 2.56	2 cleavage directions at right angles	Common in igneous rocks and a variety of metamorphic rocks; also in some arkosic sandstones
Microcline	KAlSi$_3$O$_8$	White, pink, green	6 2.56	2 cleavage directions at right angles	Common in felsic igneous rocks, some metamorphic rocks, and arkoses
Orthoclase	KAlSi$_3$O$_8$	White, pink	6 2.56	2 cleavage directions at right angles	
Olivine	(Fe,Mg)$_2$SiO$_4$	Olive green	6½ 3.3–3.6	Small mineral grains in granular masses; conchoidal fracture	Common in mafic igneous rocks
Quartz	SiO$_2$	Colorless, white, gray, pink, purple, black	7 2.67	6-sided crystals; no cleavage; conchoidal fracture	A common rock-forming mineral in all rock groups and hydrothermal rocks; also occurs in varieties known as chert, flint, agate, and chalcedony
Garnet	(Fe,Ca,Mg,Mn)$_3$ (Fe,Al,Cr)$_2$Si$_3$O$_{12}$	Dark red, green	7–7½ 4.32	12-sided crystals common; uneven fracture	Found mostly in gneiss and schist
Zircon	Zr$_2$SiO$_4$	Brown, pink	7½ 3.9–4.7	4-sided, elongated crystals	Most common as an accessory in granitic rocks
Topaz	Al$_2$SiO$_4$(F, OH)$_2$	Colorless, white, yellow, blue	8 3.5–3.6	High specific gravity; 1 cleavage direction	Found in pegmatites, granites, and hydrothermal rocks
Corundum	Al$_2$O$_3$	Gray, blue, pink, brown	9 4.0	6-sided crystals and high hardness are distinctive	An accessory mineral in some igneous and metamorphic rocks

660-kilometer discontinuity A boundary in the mantle, at a depth of about 660 kilometers, where seismic wave velocities increase because pressure is great enough that the minerals in the mantle recrystallize to form denser minerals.

A horizon The layer of soil below the O horizon, composed of a mixture of humus, sand, silt, and clay; combines with the O horizon to form topsoil.

aa Lava that has a jagged, rubbly, broken surface.

abrasion A mechanical weathering process that consists of the grinding and rounding of rock and mineral surfaces by friction and impact.

absolute age The age of a geologic event based on the absolute number of years ago it occurred.

absolute brightness or luminosity The brightness of a star as it would appear if it were a fixed distance from Earth.

absolute humidity The mass of water vapor in a given volume of air, expressed in grams per cubic meter (g/m^3).

absorption nebula A cold, dark nebula that absorbs light.

absorption of radiation The process that occurs when energy is absorbed: the energy of a photon is converted to electrical, chemical, vibrational, or heat energy, and the photon disappears.

absorption spectrum A spectrum of radiation wavelengths that are absorbed when light passes through a substance, such as a star; absorbed wavelengths appear as dark lines crossing a full-color spectrum.

abyssal plains The flat, level, largely featureless parts of the ocean floor between the Mid-Oceanic Ridge and the continental rise.

accelerating Universe A model in which an unidentified force is overpowering gravity and causing the expansion of the Universe to accelerate.

accreted terranes A mappable, fault-bounded landmass that originates as an island arc or a microcontinent that is later added onto a continent.

accuracy The degree to which a measurement reflects the measure's 'true' value; the degree of 'correctness' of a quantity.

acid precipitation, also called acid rain Rain, snow, fog, or mist that has become acidic after reacting with air pollutants.

active continental margin A continental margin that occurs at a convergent or transform plate boundary.

adiabatic temperature changes Temperature changes caused by compression or expansion of gas without gain or loss of heat.

advection fog Fog that forms when warm, moist air from the sea blows onto cooler land, where the air cools and water vapor condenses at ground level.

aerosol In pollution terminology, a particle or particulate that is suspended in air.

ages The shortest period of geologic time. Epochs are divided into ages.

air mass A large body of air that has approximately the same temperature and humidity at any given altitude throughout.

air-fall tuff A tuff formed during an eruption by fallout of ash from the atmosphere.

albedo The reflectivity of a surface; surfaces that reflect more light have a higher albedo.

alfisol A soil formed in semiarid to humid climates, typically under hardwood cover. Characterized by accumulation of clay in the B horizon and relatively high fertility, making it productive for agriculture.

alluvial fan A fan-shaped accumulation of sediment created where a steep mountain stream rapidly slows down as it reaches a relatively flat plain.

alluvium Sediment deposited by moving water.

altostratus High-altitude stratus clouds.

anaerobic A condition in which oxygen is not present; anaerobic bacteria can survive in the absence of oxygen.

Andean margin A continental margin characterized by subduction of an oceanic lithospheric plate beneath a continental plate; also called a *continental arc*.

andisol Young soil developed on volcanic parent material and containing abundant unweathered volcanic glass and other volcanic debris, resulting in a high cation exchange capacity and high fertility.

angle of repose The maximum slope or steepness at which loose material remains stable. If the slope becomes steeper than the angle of repose, the material slides.

angular unconformity An unconformity in which younger sediment or sedimentary rocks rest on the eroded surface of tilted or folded older rocks.

anion An ion that has a negative charge.

Antarctic Circumpolar Current A strong west-to-east ocean current that circulates clockwise around Antarctica and prevents warmer water from getting close to the shores of the continent.

anticline A fold in rock that arches upward; the oldest rocks are in the middle.

anticyclone A high-pressure region with its accompanying system of outwardly directed rotating winds that develop where descending air spreads over Earth's surface.

apparent brightness or luminosity The luminosity of a star as seen from Earth.

aquifer A body of rock that can yield economically significant quantities of groundwater; should be both porous and permeable.

aquitards A body of sediment or rock that has low porosity and permeability and that inhibits the flow of groundwater.

Archean Eon A division of geologic time 3.8 to 2.5 billion years ago. The oldest-known rocks formed at the beginning of, or just prior to, the start of the Archean Eon.

architectural element A top-down hierarchical classification for describing the 3-D spatial arrangement of sedimentary deposits at scales ranging from an entire sedimentary basin to individual grains. The overall geometry of each sedimentary lithology, all surfaces of erosion, and all sedimentary structures are examples of architectural elements.

arête A sharp narrow ridge of rock between adjacent valleys or between two cirques, created when two alpine glaciers moved along opposite sides of the mountain ridge and eroded both sides.

aridisol A soil formed in arid or semiarid environments and characterized by very low organic content, water deficiency, and precipitation of salts in the B horizon.

artesian well A well drilled into a confined aquifer, in which the water rises without pumping and in some cases flows to the surface.

artificial levee A wall built along the banks of a stream to prevent rising floodwater from spilling out of the channel onto the floodplain.

ash-flow tuff A volcanic rock formed when a pyroclastic flow solidifies.

aspect The orientation of a slope with respect to the Sun; the direction toward which the slope faces.

asteroids Small celestial bodies in orbit around the Sun, primarily in the region between Mars and Jupiter.

asthenosphere The portion of the upper mantle just beneath the lithosphere, extending from a depth of about 100 kilometers to about 350 kilometers below the surface of Earth and consisting of weak, plastic rock where magma may form.

atmosphere The gaseous layer above the Earth's surface, mostly nitrogen and oxygen, with smaller amounts of argon, carbon dioxide, and other gases. The atmosphere is held to Earth by gravity and thins rapidly with altitude.

atmospheric pressure, often called barometric pressure The pressure of the atmosphere at any given location and time.

atoll A circular coral reef that forms a ring of islands around a central lagoon and that is bounded on the outside by the deep water of the open sea. Usually forms on top of a subsided seamount.

avulsion The process by which a stream channel is abandoned and the water and sediment diverted down a new channel.

axial surface An imaginary surface that connects all the points of maximum curvature in a fold.

B horizon The soil layer just below the A horizon, containing less organic matter and where ions and clays leached from the A and E horizon accumulate; also called *subsoil*.

backshore The uppermost zone of a beach, consisting typically of a dry sandy surface that slopes gently landward but that is washed over by waves during large storms.

bajada A broad, gently sloping depositional surface formed by the merging of alluvial fans from closely spaced canyons and extending outward into a desert valley.

banded iron formation A marine chemical sedimentary rock formed mostly between 2.7 and 1.9 billion years ago and consisting of centimeter-scale interbeds of iron oxide and chert. Formed as atmospheric oxygen alternately went through periods of accumulation as a waste product from photosynthesizing cyanobacteria and periods of withdrawal during widespread precipitation of iron oxide minerals.

banded iron formations Iron-rich sedimentary rocks composed of alternating iron-rich and silica-rich layers; source of most of the world's supply of iron.

banks The rising slopes bordering the sides of a stream channel.

bar Unit of measurement for atmospheric pressure. One bar is roughly equal to atmospheric pressure at sea level.

barchan dune A crescent-shaped dune, highest in the center, with the tips pointing downwind; typically forms in rocky deserts where there is a general shortage of sand.

barometer A device used to measure barometric pressure.

barred spiral galaxies A spiral galaxy with a straight bar of stars, gas, and dust extending across the nucleus.

barrier island A long, narrow, low-lying island that extends parallel to the shoreline.

basal slip Movement of a glacier in which the entire mass slides over bedrock.

base level The deepest level to which a stream can erode its bed. The ultimate base level is usually sea level.

basement rock The older igneous and metamorphic rock that lies beneath the thin layer of sedimentary rocks and soil covering much of Earth's surface; forms the "basement" of the crust.

basin A bowl-shaped synclinal structure, commonly filled with sediment.

batholith A large pluton, exposed across more than 100 square kilometers of Earth's surface.

bauxite A gray, yellow, or reddish-brown rock, composed of a mixture of aluminum oxides and hydroxides, that formed as a residual deposit; the principle source of aluminum.

baymouth bar A spit that extends partially or completely across the entrance to a bay.

beach drift The gradual movement of sediment (usually sand) along a beach, parallel to the shoreline, when waves strike the beach obliquely but return to the sea directly.

beach Any strip of shoreline washed by waves and tides.

bed load The total mass of a stream's sediment load that is transported along the bottom or in intermittent contact with the bottom of the streambed.

bed The floor of a stream channel.

bedding Layering that develops as sediments are deposited; also called *stratification*.

bedrock The solid rock that lies beneath soil or unconsolidated sediments; it can be igneous, metamorphic, or sedimentary.

Benioff zone A three-dimensional zone of earthquake foci within and along the upper portion of a subducting plate; formed by release of strain as the subducting plate scrapes past the overriding plate.

big bang An event 10 to 20 billion years ago, thought to mark the beginning of the Universe when all matter exploded from a single infinitely dense point.

biodegradable pollutants Pollutants that decay naturally in a reasonable amount of time by being consumed or destroyed by organisms that live in soil or water.

biogenic structures Any physical trace left in the sedimentary record by a fossil organism; includes tracks, trails, burrows, and root casts.

biome A community of plants growing in a large geographic area characterized by a particular climate.

biomineralize The process by which living organisms produce minerals.

bioremediation The use of microorganisms to decompose an environmental contaminant.

biosphere The zone of Earth comprising all forms of life in the sea, on land, and in the air.

bitumen A thick, sticky, oil-like substance that permeates tar sands and can be converted to crude oil.

black hole An infinitesimally small region of space, created after the supernova explosion of a huge star, that contains matter packed so densely that light cannot escape from its intense gravitational field.

black smokers A jet of black water spouting from a fracture or vent in the seafloor, commonly near a mid-oceanic ridge. The black color is caused by precipitation of fine-grained metal sulfide minerals as the hydrothermal solutions cool on contact with seawater.

blob tectonics Tectonic activity dominated by rising and sinking of the mantle and crust, believed to predominate on Venus, as opposed to the horizontal movement of plates associated with tectonic activity on Earth.

blue shift A shift toward shorter (blue) wavelengths, observed in the spectrum of a distant galaxy or other object that is moving toward Earth; caused by the Doppler effect.

body waves Seismic waves that travel through the interior of Earth, carrying energy from the earthquake's focus to the surface.

bolide A large piece of space debris, such as an asteroid, that crashes into a planet.

braided streams A stream that flows in many shallow, interconnecting channels that are usually separated by emergent sediment bars; formed because the stream's capacity has been exceeded by its sediment supply.

breaks To collapse or crash, as when a wave approaches a beach and the front of the wave rises over its base, growing steeper until it collapses forward.

butte A flat-topped mountain, smaller and more towerlike than a mesa, characterized by steep cliff faces.

C horizon The lowest soil layer, consisting of weathered bedrock.

calcrete A hardpan that forms in the B soil horizon in arid and semiarid regions when calcium carbonate precipitates and cements the soil particles together.

caldera A large circular depression created by the collapse of the magma chamber after an explosive volcanic eruption.

Callisto Jupiter's outermost Galilean moon; marked by a heavily cratered surface.

capacity factor A measure of the actual to total potential output of an energy source over a period of time.

capacity The maximum quantity of sediment that a stream can transport at any one time.

capillary action The process by which water is pulled upward through the soil due to the natural attraction of water molecules to soil particles and the cohesion of water.

carbonate platforms Extensive accumulations of limestone, such as the Florida Keys and the Bahamas, formed in warm regions on an isolated continental shelf where terrigenous clastic sediment does not muddy the water and reef-building organisms thrive.

carbonate rocks Bioclastic sedimentary rocks composed of the carbonate minerals (minerals based on the CO_3^{-2} anion).

carbonates Minerals containing the anionic group carbonate, CO_3^{2-}; an example is calcite ($CaCO_3$).

catastrophism The principle stating that infrequent catastrophic geologic events alter the course of Earth history, in contrast to the principle of gradualism.

cation exchange capacity (CEC) Ability of a soil to release cations, typically by exchanging basic cations K+, Na+, Ca++, or Mg++ for H+ with plant rootlets.

cation A positively charged ion.

cavern An underground cavity or series of chambers created when groundwater dissolves large volumes of rock, usually limestone; also called a *cave*.

celestial spheres The hypothetical series of concentric transparent spheres surrounding Earth in Aristotle's model of the Universe. Aristotle postulated that the Sun, Moon, planets, and stars are embedded in the spheres.

cenote A deep, usually clear lake formed when groundwater is exposed at the surface through collapse of caves in limestone bedrock.

Cenozoic Era The latest of the three Phanerozoic eras, 65 million years ago to the present.

chalk A very fine-grained limestone made up of the remains of tiny marine microorganisms.

channel characteristics Features describing the shape and roughness of a stream channel.

chemical remediation Treatment of a contaminated area by injecting it with a chemical compound that reacts with the pollutant to produce harmless products.

chemical weathering The decomposition of rock when it chemically reacts with air, water, or other agents in the environment, altering its chemical composition and mineral content.

chemoautotroph An organism, typically a bacterium, that derives its metabolic energy by oxidizing inorganic compounds.

chemosynthesis A process in which bacteria produce energy from hydrogen sulfide or other inorganic compounds and thus are not dependent on photosynthesis.

chlorofluorocarbons (CFCs) Organic compounds containing chlorine and fluorine, which rise into the upper atmosphere and destroy the ozone layer there.

chondrules A small grain about 1 millimeter in diameter embedded in a meteorite, often containing amino acids or other organic molecules.

chromosphere A turbulent, diffuse, gaseous layer of the Sun that lies above the photosphere.

cinder cone A small volcano, typically less than 300 meters high, made up of loose pyroclastic fragments blasted out of a central vent; usually active for only a short time.

cinders Glassy, pyroclastic volcanic fragments 4 to 32 millimeters in size.

cirque A steep-walled, spoon-shaped depression eroded into a mountain peak by a glacier.

cirrus Wispy, high-altitude clouds composed of ice crystals.

Clean Water Act A federal law passed in 1972 mandating the cleaning of the nation's rivers, lakes, and wetlands and forbidding the discharge of pollutants into waterways.

cleavage The tendency of some minerals to break along flat surfaces, which are planes of weak bonds in the crystal. When a mineral has well-developed cleavage, sheet after sheet can be peeled from the crystal, like peeling layers from an onion.

climate The characteristic weather of a region, averaged over several decades; refers to yearly cycles of temperature, wind, rainfall, and so on, and not to daily variations.

closed Universe A model in which the gravitational force of the Universe is so great that all the galaxies will eventually slow down, reverse direction, and fall back to the center, forming another point of infinite density.

cloud condensation nuclei Very small particles suspended in the atmosphere on which water vapor condenses.

coal bed methane Methane that is chemically bonded to coal. The methane can be recovered by removing the groundwater from a coal bed, which decreases the pressure and allows the methane to separate from the coal as a gas.

cold front A front that forms when moving cold air collides with stationary or slower-moving warm air. The dense cold air distorts into a blunt wedge and pushes under the warmer air, creating a narrow band of violent weather commonly accompanied by cumulus and cumulonimbus clouds.

column A cave deposit formed when a stalactite and a stalagmite meet and grow together.

columnar joints Regularly spaced cracks that commonly develop in lava flows, grow downward starting from the surface, and typically form five- or six-sided columns.

coma The bright outer sheath of a comet, surrounding the nucleus.

comet An interplanetary orbiting body composed of loosely bound rock and ice, which forms a bright head and extended fuzzy tail when it approaches the Sun. It appears to be fiery hot, but it is actually a cold object and its "flame" is reflected light.

compaction Increased packing together of sedimentary grains, usually resulting from the weight of overlying sediment; causes a decrease in porosity and contributes to lithification.

competence A measure of the largest particles that a stream can transport.

composite cone A steep-sided volcano formed by an alternating series of lava flows and pyroclastic eruptions and marked by repeated eruption.

conduction The transport of heat by direct collision among atoms or molecules.

cone of depression A cone-shaped depression in the water table, created when water is pumped out of a well more rapidly than it can flow through the aquifer to the well. If water continues to be pumped from the well at such a rate, the water table will drop.

confined aquifer An inclined aquifer sandwiched between layers of impermeable rock; typically, the water in the lower part of the aquifer is under pressure from the weight of water above.

conformable A term describing sedimentary layers that were deposited continuously, without detectable interruption.

conglomerate A clastic sedimentary rock that consists of lithified gravel.

constellations A group of stars that seem to form a pattern when viewed from Earth. Many of the patterns have been named by astronomers.

consumption Any process that uses water and then returns it to Earth far from its source.

continental drift The theory proposed by Alfred Wegener that Earth's continents were once joined together and later split and drifted apart. The continental drift theory has been replaced by the more complete plate tectonics theory.

continental rifting The process by which a continent is pulled apart at a divergent plate boundary.

continental rise An apron of sediment at the foot of the continental slope that merges with the deep seafloor.

continental shelf A shallow, very gently sloping portion of the seafloor that extends from the shoreline to ~200 meters water depth at the top of the continental slope.

continental slope The relatively steep (averaging 3 degrees but varying between 1 degree and 10 degrees) submarine slope between the continental shelf and the continental rise.

convection The upward and downward flow of fluid material in response to density changes produced by heating and cooling. Convection occurs slowly in Earth's mantle and much more quickly in the oceans and the atmosphere.

convective zone The subsurface zone in a star where energy is transmitted primarily by convection.

conventional petroleum reservoir A porous, permeable sedimentary rock that is saturated with trapped oil.

convergent boundary A plate boundary where two tectonic plates move toward each other and collide.

core The center of the Sun, where hydrogen fusion takes place.

core The dense, metallic, innermost region of Earth's geosphere, consisting mainly of iron and nickel. The outer core is molten, but the inner core is solid.

Coriolis effect A deflection of air or water currents caused by the rotation of Earth.

corona The outer atmosphere of the Sun, normally invisible but appearing as a halo around a black Moon during a solar eclipse.

correlation The process of establishing the age relationship of rocks or geologic features from different locations on Earth; can be done by comparing sedimentary characteristics of the layers or the types of fossils found in them. There are two types of correlation: *time correlation* (age equivalence) and *lithologic correlation* (physical continuity of the rock unit).

cosmic background radiation Low-energy, microwave radiation that began traveling through space when the Universe was only 300,000 years old and now pervades all space in the Universe.

country rock The older rock already in an area, cut into by a younger igneous intrusion or mineral deposit.

crater A bowl-like depression at the summit of a volcano, created by volcanic activity.

creep A form of mass wasting in which loose material moves very slowly downslope, usually at a rate of only about 1 centimeter per year and usually on land with vegetation. Trees on a creeping block tilt downhill and grow to have a trunk shaped like a pistol butt.

crescent moon A lunar phase in which the Moon appears as a thin crescent, when either waxing or waning.

crest The highest part of a wave.

crevasse A fracture or crack in the brittle upper 40 meters of a glacier, formed when the glacier flows over uneven bedrock.

cross-bedding A sedimentary structure in which wind or water deposits sets of beds that are inclined to the main sedimentary layering.

crust The outermost layer of Earth's geosphere, ranging from 4 to 75 kilometers thick and composed of relative low-density silicate rocks.

crystal face The flat surface that develops if a crystal grows freely in an uncrowded environment. Under perfect conditions, the crystal that forms will be symmetrical.

crystal habit The characteristic shape of an individual crystal, and the manner in which aggregates of crystals grow.

crystal settling A process in which the crystals that solidify first from a cooling magma settle to the bottom of the magma chamber because the minerals are more dense than magma; the ultimate result is a layered body of igneous rock, each layer containing different minerals.

crystal A solid element or compound whose atoms are arranged in a regular, periodically repeated pattern.

crystalline structure The orderly, repetitive arrangement of atoms in a crystal.

cumulonimbus Towering storm clouds that form in columns and produce intense rain, thunder, lightning, and sometimes hail.

cumulus Fluffy white clouds with flat bottoms and billowy tops.

current A continuous flow of water in a particular direction.

cyanobacteria Blue-green algae that were among the earliest photosynthetic life-forms on Earth.

cycle A sequential process or phenomenon that returns to its beginning and then repeats itself over and over.

cyclone A low-pressure region with its accompanying system of inwardly directed rotating winds. In common, nonscientific usage the term often refers to a variety of different violent storms including hurricanes and tornados.

dark energy One component of dark matter, comprising the poorly understood repulsive force that causes the modern Universe to accelerate.

dark mass One component of dark matter, comprising a collection of as-yet-undetected particles, large enough to make up 23 percent of the Universe, that exert conventional gravitational attraction.

dark matter Mysterious invisible matter that may comprise up to 96 percent of the mass of the Universe.

deep-sea currents Vertical and horizontal flow of water below a depth of 400 meters in the oceans, caused mainly by gravity-driven differences in water density.

deflation Erosion by wind.

delta A fan-shaped accumulation of sediment formed where a stream enters a lake or ocean; includes a nearly flat delta top that is partly onshore and partly offshore, a more steeply dipping delta front located offshore, and a muddy prodelta located further offshore at the base of the delta front.

delta front The more steeply sloping, usually submerged, outer edge of a delta beyond the delta top.

delta top The upper surface of a delta, including the parts above and below water.

desert pavement A continuous cover of closely packed gravel- or cobble-sized clasts left behind when wind erodes smaller particles such as silt and sand.

desert Any region that receives less than 25 centimeters (10 inches) of rain per year and consequently supports little or no vegetation.

desertification A process by which semiarid land is converted to desert by human mismanagement or by climate change.

dew point The temperature at which the relative humidity of air reaches 100 percent and the air becomes saturated.

dew Moisture that is condensed onto objects from the atmosphere, usually during the night, when the ground and leaf surfaces become cooler than the surrounding air.

dike A sheetlike igneous rock, cutting through layers of country rock, that forms when magma is injected into a fracture.

discharge The volume of water flowing downstream over a specified period of time, usually measured in units of cubic meters per second (m^3/sec) or cubic feet per second (cfs).

disconformity A type of unconformity in which the sedimentary layers above and below the unconformity are parallel.

disseminated ore deposit A large, low-grade hydrothermal deposit in which metal-bearing minerals are widely scattered throughout a rock body; not as concentrated as a hydrothermal vein.

dissolution A chemical weathering process in which mineral or rock dissolves, forming a solution.

dissolved load The total mass of ions dissolved in and carried by a stream at any one time; the ions are derived from chemical weathering.

distributary channels Channels that split from the main stream feeding a delta or alluvial fan and spread out across its surface, depositing sediment in the process.

divergent boundary A plate boundary where tectonic plates move apart from each other and new lithosphere is continuously forming; also called a *spreading center* or a *rift zone*.

diversion system A pipe, canal, or other infrastructure that transports water from its natural place and path in the hydrologic cycle to a new place and path to serve human needs.

doldrums A vast low-pressure region of Earth near the equator with hot, humid air and where local squalls and rainstorms are common but steady winds are rare.

dome A circular or elliptical anticlinal structure resembling an inverted cereal bowl.

Doppler effect The observed change in frequency of light or sound that occurs when the source of the wave is moving either toward or away from the observer.

downcutting Downward erosion by a stream into its bed, usually by cutting a V-shaped valley along a relatively straight path.

downhill creep The gradual downhill movement, under the force of gravity, of soil and loose rock material on a slope. Facilitated by the freeze-thaw cycle, in which soil particles move orthogonal to the slope surface during freezing but directly downward during thawing.

drainage basin The region that is drained by a single stream.

drainage divide A topographic high separating drainage basins.

drift Any rock or sediment transported and deposited by a glacier or by glacial meltwater.

drumlins Elongate hills, usually occurring in clusters, formed when a glacier flows over and reshapes a mound of till or stratified drift.

dry adiabatic lapse rate The rate at which dry air cools adiabatically as it rises—10°C for every 1,000 meters above sea level. Rising air cools at the dry adiabatic lapse rate until it cools to its dew point and condensation begins.

dune A mound or ridge of wind-deposited sand.

dwarf planets A body (for example, Pluto or Eris) that orbits the Sun, is not a satellite of a planet, and is massive enough to pull itself into a spherical shape, but is not massive enough to clear out other bodies in and near its orbit.

E horizon Soil horizon in which organic acids derived from overlying O and A horizons leach soluble cations and translocate them downward along with clays.

earthflow A viscous flow of fine-grained sediment or fine-grained sedimentary rock that is saturated with water and moves downslope as a result of gravity; usually slow moving, typically less than one to several meters per day.

earthquake A sudden motion or trembling of Earth caused by the abrupt release of slowly accumulated elastic energy in rocks.

eccentricity A term referring to the elliptical shape of Earth's orbit around the Sun; the more elliptical the orbit, the more eccentric it is said to be.

echo sounder An instrument that emits timed sound waves that reflect off the seafloor, return, and are recorded; the data are used to measure water depth and define the topography of the seafloor.

ecosystem A complex community of individual organisms interacting with each other and with their physical environment and functioning as an ecological unit in nature.

Ekman transport The natural process by which surface water moved by wind drags the layer of ocean water below it, which in turn drags the layer below it, and so forth, to a depth that depends on the wind strength. Only the surface layer responds directly to the wind. The layers below the surface respond both to the directional movement of the water layer directly above and to the Coriolis effect. Deeper layers are deflected more by the Coriolis effect than shallower levels.

El Niño An episodic weather pattern occurring every 3 to 7 years in which the trade winds slacken in the Pacific Ocean and warm water accumulates off the coast of South America and causes unusual rains and heavy snowfall in the Andes.

electromagnetic radiation Radiation consisting of an oscillating electric and magnetic field, including radio waves, infrared, visible light, ultraviolet, x-rays, and gamma rays.

electromagnetic spectrum The entire range of electromagnetic radiation from very-long-wavelength (low frequency) radiation to very-short-wavelength (high frequency) radiation.

element A substance that cannot be broken down into other substances by ordinary chemical means.

elliptical galaxy A galaxy with an oval shape and no spiraling arms; can be either giant or dwarf.

eluviation The removal and downward movement of dissolved ions and clays from the O, A, and E horizons by infiltrating water.

emergent coastline A coastline that was previously underwater but has become exposed to air, because either the land has risen or sea level has fallen.

emission nebula A hot nebula that glows because it is emitting light energy.

emission of radiation The process that occurs when energy, in the form of a photon, is emitted as an electron falls out of an excited state, with the equivalent loss of energy from the emitting substance.

emission spectrum A spectrum created when absorbed radiation is reemitted from its source.

end moraine A ridge of till that forms at the end, or terminus, of a glacier that is neither advancing nor retreating and whose terminus has remained in the same place for years.

energy resources Geologic resources—including petroleum, coal, natural gas, and nuclear fuels—used for heat, light, work, and communication.

entisol A very young soil typically lacking horizons and formed on unconsolidated parent material. All soils not classified with a different order are classified as entisols, so much diversity exists within this order.

eons The largest unit of geologic time. The most recent eon, the Phanerozoic Eon, is further subdivided into eras.

epicenter The point on Earth's surface directly above the initial rupture point (focus) of an earthquake.

epochs A geologic time unit longer than an age and shorter than a period.

equatorial upwelling Oceanic upwelling in which surface currents flowing westward on both sides of the equator are deflected poleward and are replaced by upward flow of deeper, nutrient-rich waters.

equilibrium altitude line, or ELA The boundary between the zone of accumulation and the zone of ablation.

equinoxes Either of two times during the year—on or about March 21 and September 22—when the Sun shines directly overhead at the equator and every portion of Earth receives 12 hours of daylight and 12 hours of darkness.

eras A geologic time unit. Eons are divided into eras and, in turn, eras are subdivided into periods.

erosion The removal of weathered rocks that occurs when water, wind, ice, or gravity transports the material to a new location.

erratics Boulders, usually different from bedrock in the immediate vicinity, that were transported to their present location by a glacier.

escape velocity The speed that an object must attain to escape the gravitational field of a planet or other object in space.

esker A long, snakelike ridge formed as the channel deposit of a stream that flowed within or beneath a melting glacier.

Europa The second-closest of Jupiter's moons; similar to Earth in that much of its interior is composed of rock and much of its surface is covered with water, although the water is frozen into a vast planetary ice crust.

eustatic sea level change A global sea level change caused by three different processes: the growth or melting of glaciers, changes in water temperature, and changes in the volume of the Mid-Oceanic Ridge.

eutrophic lake A relatively shallow lake characterized by abundant nutrients, thus sustaining multiple living organisms.

evaporation fog Fog that forms when air is cooled by evaporation from a body of water, commonly a lake or river and typically in late fall or early winter when the air is cool but the water is still warm. The water evaporates, but the vapor cools and condenses to fog.

evolution The biological theory that life-forms have changed in their physical and genetic characteristics over time.

excited state A state of physical energy higher than the lowest energy level (or *ground state*) of an electron in an atom or molecule.

exfoliation A weathering process resulting in fracture when concentric plates or shells split away from a main rock mass like the layers of an onion; frequently explained as a form of pressure-release fracturing, but many geologists consider it could result from hydrolysis-expansion.

exosphere The outermost layer of the in which gas molecules are so diffuse they do not collide. Merges outward into interstellar space.

extrusive igneous rock Igneous rock formed from material that has erupted through the crust onto the surface of Earth; usually finely crystalline. Also called *volcanic rock*.

eyepiece The lens, closest to the eye in a refracting telescope, that magnifies the image.

facing direction The direction, up or down, corresponding to the original stratigraphic up-direction of layered sedimentary rocks in which younger layers overlie older layers. In overturned and recumbent folds, one limb is facing down.

fall Form of rapid mass wasting in which unconsolidated material falls freely or bounces down steep slopes or cliffs.

fault creep A continuous, slow movement of solid rock along a fault, resulting from a constant stress acting over a long time. Creeping faults do not usually have large earthquakes.

fault zone An area of numerous, closely spaced faults.

fault A fracture in rock along which one side has moved relative to the other side. Compare with *joint*.

feedback mechanism A reaction whereby a small initial perturbation in one component affects a different component of Earth's systems, which amplifies the original effect, which perturbs the system even more, which leads to an even greater effect, and so on.

First Great Oxidation Event The sudden increase in Earth's atmospheric oxygen concentration from trace amounts to appreciable quantities that occurred approximately 2.4 billion years ago, probably because of a combination of biological and geochemical processes.

first-generation energy Cosmic background radiation that emanated from the primordial sea of atoms, particles, and energy of the big bang.

fissures Breaks, cracks, or fractures in rocks.

fjords A deep, narrow, glacially carved valley on a high-latitude seacoast that was later flooded by encroaching seas as the glaciers melted.

flash flood A rapid, intense, local flood of short duration, usually following a rainstorm.

flood basalt Basaltic lava that erupts gently and in great volume from vents or fissures at Earth's surface, to cover large areas of land and form lava plateaus.

flood A relatively high stream discharge that overtops the stream banks, covering land that is not usually underwater.

floodplain That portion of a river valley adjacent to the channel; it is built upward by sediment deposited during floods and is covered by water during a flood.

flow Form of rapid mass wasting in which loose soil or sediment moves downslope as a slurry-like fluid, not as a consolidated mass; may occur slowly (less than 1 centimeter per year for some earthflows) or rapidly (several meters per second for some mudflows and debris flows).

fly ash Noncombustible minerals that escape into the atmosphere when coal burns, eventually settling as gritty dust.

focus The initial rupture point of an earthquake, typically located below Earth's surface.

fold A bend in rock.

foliation The layering in metamorphic rocks resulting from regional dynamothermal metamorphism.

footwall A term to describe the lower side of an inclined fault or vein (i.e., the rock one would walk on).

forearc basin A sedimentary basin between the oceanic trench and the magmatic arc, either in an island arc or at an Andean margin.

foreland basin A sedimentary basin formed by downward flexing of continental crust by the weight of thrust plates in a compressional orogen.

foreshocks Small earthquakes that precede a large quake by a few seconds to a few weeks.

foreshore or intertidal zone The part of a beach that lies between the high-tide and low-tide lines and is exposed to the air at low tide but covered by water at high tide.

fossil fuels Energy resources including petroleum, coal, and natural gas, which formed from the partially decayed remains of plants and animals; they are nonrenewable and unrecyclable.

fossils The imprint, remains, or any other trace of a plant or animal preserved in rock.

fracture The manner in which minerals break, other than along planes of cleavage.

frequency The number of complete wave cycles, from crest to crest, that pass by any point in a second.

front In meteorology, the boundary between a warmer air mass and a cooler one.

frontal wedging A process by which a moving mass of cool, dense air encounters a mass of warm, less-dense air; the cool, denser air slides under the warm air mass, forcing the warm air upward to create a weather front.

frost wedging A mechanical weathering process in which water freezes in a crack in rock, and the resulting expansion wedges the rock apart.

frost Ice crystals formed directly from vapor when the dew point is below freezing.

full moon The lunar phase when the Moon appears round and fully illuminated, because it is on the opposite side of Earth from the Sun and its entire sunlit area is visible.

Gaia The term (Greek for "Earth") used by James Lovelock to refer to our planet, which he likened to a living creature due to the interconnectivity of all of Earth's systems.

galactic halo A spherical cloud of dust and gas that surrounds the galactic disk of a spiral galaxy such as the Milky Way.

galaxy A large volume of space containing many billions of stars, held together by mutual gravitational attraction.

gamma ray burst A burst of high-energy radiation emanating from space. Longer bursts come from the death throes of massive stars, while shorter bursts form when a neutron star is sucked into a black hole or when two neutron stars collide to form a black hole.

Ganymede A moon of Jupiter marked by a convecting, metallic core and a brittle water/ice crust that behaves much like rock.

gelisol A high-latitude soil formed over permafrost that is no deeper than two meters. Characterized by an organic-rich A horizon that usually extends to the permafrost boundary.

gem A mineral that is prized for its rarity and beauty rather than for industrial use.

geocentric A model that places Earth at the center of the Universe.

geologic structure Any feature produced by rock deformation, such as a fold or a fault; also refers to the combination of all such features in an area or region.

geologic time scale A chronological arrangement of geologic time subdivided into eons, eras, periods, epochs, and ages.

geosphere The solid Earth, consisting of the entire planet from the center of the core to the outer crust.

geothermal energy Energy extracted from Earth's heat.

geyser A type of hot spring that periodically erupts with violent jets of hot water and steam; eruptions occur when groundwater recharging a geyser's subsurface plumbing system becomes superheated and forces a small volume of water out of the geyser's vent. The release of this water lowers pressure at deeper levels, causing water there also to flash to steam and initiating the main eruption.

gibbous moon A bright moon, either waxing or waning, with only a sliver of dark visible.

glacial striations Parallel grooves and scratches in bedrock that form as rocks are dragged along at the base of a glacier.

glaciation A time when alpine glaciers descend into lowland valleys and continental glaciers grow over high-latitude continents; usually used in reference to Pleistocene Glaciation.

glacier A massive, long-lasting accumulation of compacted snow and ice that forms on land and moves downslope or spreads outward under its own weight.

glaze An ice coating that forms when rain falls on subfreezing surfaces.

globular cluster A concentration of older stars that shared a common origin and are held together by gravity, lying within the galactic halo of a spiral galaxy such as the Milky Way.

Gondwana The southern part of Pangea, consisting of what is now South America, Africa, Antarctica, India, and Australia.

graben A wedge-shaped block of rock that has dropped downward between two normal faults, forming a valley.

graded stream A stream with a smooth, concave-upward profile, in equilibrium with its sediment supply; it transports all the sediment supplied to it, with neither erosion nor deposition in the streambed.

gradient The steepness or vertical drop of a stream over a specific distance.

gradualism A principle stating that geological change occurs as a consequence of slow or gradual accumulation of small events, such as the slow erosion of mountains by wind and rain. More recently, scientists studying biological evolution use the term to describe a theory of evolution that proposes that species change gradually in small increments.

Great Red Spot A giant hurricane-like storm on the surface of Jupiter that has existed for centuries.

greenhouse effect An increase in the temperature of the planet's surface caused when infrared-absorbing gases in the atmosphere trap energy from the Sun.

groins A narrow barrier or wall built on a beach, perpendicular to the shoreline, to trap sand transported by currents and waves.

ground moraine A moraine formed when a glacier recedes steadily and deposits till in a relatively thin layer over a broad area.

groundwater Subsurface water contained in the soil and bedrock of the upper few kilometers of the geosphere, comprising about 0.63 percent of all water in the hydrosphere.

Gulf Stream One of many oceanic surface currents. It begins near the Gulf of Mexico and travels northward up the Atlantic coast, grows wider and slower as it moves to the northeast, and bathes the coastlines of western Europe and Scandinavia with exceptionally warm water.

guyot A flat-topped seamount, formed when the top of a sinking island, usually of volcanic origin, is eroded by wave energy.

gyres A circular or elliptical current in either water or air.

Hadean Eon The earliest time in Earth's history, ranging from 4.6 billion years ago to 3.8 billion years ago.

hail Large ice globules varying from 5 millimeters to a record 14 centimeters in diameter that fall from cumulonimbus clouds.

half-life The time it takes for half of the atoms of a radioactive isotope in a sample to decay.

halons Compounds containing bromine and chlorine, which rise into the upper atmosphere and destroy the ozone layer there.

hanging valley A small glacial valley lying high above the floor of the main valley.

hanging wall A term to describe the upper side of an inclined fault or vein (i.e., the rock hanging above one's head).

hardness The resistance of a mineral to scratching, controlled by the bond strength between its atoms.

hardpan General term for a soil layer that is relatively impervious to water and impenetrable to plant roots. Commonly forms from precipitation of salts in a soil B horizon by either downward or upward translocation.

heat A measure of the total energy in a sample.

heliocentric A model that places the Sun at the center of the Solar System.

Hertzsprung–Russell or H–R diagram A graph that plots absolute stellar luminosity against temperature.

histosol A very organic-rich soil, typically formed in a poorly drained area where stagnant water inhibits organic decay. Typically composed of thick O and A horizons. Can be mined as peat.

horn A sharp, pyramid-shaped rock summit where three or more cirques intersect near the summit.

horse latitudes A calm, high-pressure region of Earth lying at about 30 degrees north and south latitudes, in which generally dry conditions prevail and steady winds are rare.

horst The block of rock between two grabens, which has moved relatively upward along normal faults as the grabens have settled downward.

hot spot The hot upper mantle rock located within a plume and associated with a volcanic center that forms on the overlying lithosphere.

hot springs A spring formed where hot groundwater flows to the surface.

Hubble's law A law that states that the velocity of a galaxy is proportional to its distance from Earth. Thus, the most distant galaxies are traveling at the highest velocities.

humid continental climate A midlatitude climate characterized by hot summers, cold winters, and precipitation throughout the year.

humid subtropical climate A midlatitude climate characterized by hot, humid summers but cooler winters and rainfall that falls throughout the year.

humidity The amount of water vapor in the air.

humus The dark, organic component of soil consisting of litter that has decomposed enough so that the origin of the individual pieces cannot be determined.

hurricane A tropical storm occurring in North America or the Caribbean whose wind exceeds 120 kilometers per hour; called a *typhoon* in the western Pacific and a *cyclone* in the Indian Ocean.

hydraulic fracturing The process of fracturing an unconventional reservoir—usually an organic-rich shale—by forcing large volumes of pressurized fluid into it.

hydrologic cycle The continuous circulation of water among the hydrosphere, the atmosphere, the biosphere, and the geosphere; also called the *water cycle*.

hydrolysis A chemical weathering process in which a mineral reacts with water to form a new mineral that has water as part of its crystal structure.

hydrosphere All of Earth's water, which circulates among oceans, continents, glaciers, and atmosphere.

hydrothermal processes Geologic processes in which hot water or steam dissolves metals and minerals from rocks or magma; the solutions then seep through cracks before cooling, to create ore deposits.

hydrothermal vein deposit A rich, sheetlike mineral deposit that forms when economically valuable minerals precipitate from hot water solutions along a fault or other fracture.

ice sheet, or continental glacier A glacier that covers an area of 50,000 square kilometers or more and spreads outward in all directions under its own weight.

ice shelf A thick mass of ice that floats on the ocean surface but is connected to and fed from a glacier on land.

icebergs A large chunk of ice that breaks from a glacier into a body of water.

icefall A section of a glacier consisting of numerous crevasses and towering ice pinnacles.

Igneous rock Rock that forms when magma cools and crystallizes.

inceptisol A young soil exhibiting weak horizons and developed in subhumid to humid environments. Typically retains abundant unweathered material.

index fossils A fossil that dates the layers where it is found because it came from an organism that is abundantly preserved in rocks, was widespread geographically, and existed as a species or genus for only a relatively short time.

industrial minerals Rocks or minerals that have economic value, exclusive of metal ores, fuels, and gems.

inertia The tendency of an object to resist a change in motion.

interglacial periods A relatively warm, ice-free time separating glaciations.

intermittent streams A stream that does not maintain some flow all the time and may be dry for long periods.

intrusive igneous rock A rock formed when magma solidifies within Earth's crust without erupting to the surface; usually medium to coarsely crystalline. Also called *plutonic rock*.

Io The innermost moon of Jupiter and the most active volcanic body in the Solar System.

ion An atom with an electrical charge, either positive or negative.

island arc A gently curving chain of volcanic islands in the ocean formed by the convergence of two plates, each bearing ocean crust, and the resulting subduction of one plate beneath the other.

isobars Lines on a weather map connecting points of equal air pressure.

isostasy The concept that the lithosphere floats on the asthenosphere as an iceberg floats on water.

isotherms Lines on a weather map connecting areas with the same average temperature.

isotopes Atoms of the same element that have different numbers of neutrons.

jet streams Narrow bands of high-altitude, fast-moving wind.

joint A fracture along which the rock on either side of the break does not move. Compare with *fault*.

Jovian planets The outer planets—Jupiter, Saturn, Uranus, and Neptune—all massive, with relatively small rocky or metal cores surrounded by swirling liquid and gaseous atmospheres.

Jupiter The largest planet in the Solar System and fifth from the Sun.

kame and kettle topography Rolling hill topography associated with glacial kames and kettles.

kame A small mound or ridge of stratified drift deposited by a stream that flows on top of, within, or beneath a glacier.

karst topography A landscape that forms over limestone or other soluble rock and is characterized by abundant sinkholes, disappearing streams, and caverns.

kerogen The waxy, solid organic material in oil shales that yields oil when the shale is heated; the precursor of liquid petroleum.

kettle lake A lake that forms in a depression created by a receding glacier, filled with the water from the melting glacier.

kettles A small depression formed by a block of stagnant ice; many fill with water to become a kettle lake.

key bed A thin, widespread, easily recognized sedimentary layer that can be used for correlation because it was deposited rapidly and simultaneously over a wide area.

Koeppen Climate Classification A climate classification system describing Earth's principal climate zones, used by climatologists throughout the world.

Kuiper Belt objects Tiny ice dwarfs, similar to Pluto, orbiting in a disk-shaped region at the outer reaches of the Solar System.

lagoon A sheltered body of water separated from the sea by a reef or barrier island.

lake A large, inland body of standing water that occupies a depression in the land surface.

landslide A general term for mass wasting (the downslope movement of rock and regolith under the influence of gravity) and the landforms it creates.

latent heat Stored heat; the energy released or absorbed when a substance changes from one state to another, by melting, freezing, vaporization, condensation, or sublimation.

lateral erosion The action of a low-gradient stream as it cuts into and erodes its outer bank while simultaneously depositing sediment onto its inner bank; results in slow lateral migration of the channel and, through time, formation of wide, flat alluvial valleys.

lateral moraine A ridgelike moraine that forms from sediment on or adjacent to the sides of a mountain glacier.

Laurasia The northern part of Pangea, consisting of what is now North America and Eurasia.

lava plateau A broad plateau covering thousands of square kilometers, formed by the accumulation of many individual lava flows that occur over a short period of geologic time.

lava Fluid magma that flows onto Earth's surface from a volcano or fissure. Also, the rock formed by solidification of the same fluid magma.

leaching The chemical dissolution of ions from the O and A soil horizons and their removal, usually downward into the B horizon where they accumulate.

light pollution The nighttime glow of city lights that competes with the light of distant stars and reduces the vision of a telescope.

light-year The distance traveled by light in one year, approximately 9.5×10^{12} kilograms.

limbs The sides of a fold in rock.

liquefaction A geological process in which a soil or sediment loses its shear strength during an earthquake and becomes a fluid.

liquid metallic hydrogen A form of hydrogen under extreme temperature and pressure, which forces the atoms together so tightly that the electrons move freely throughout the packed nuclei, and as a result the hydrogen conducts electricity.

lithification The process by which loose sediment is converted to solid rock.

lithosphere The cool, rigid, outer part of Earth, which includes the crust and the uppermost mantle, is about 100 kilometers thick and makes up Earth's tectonic plates.

litter Leaves, twigs, and other plant or animal materials that have fallen to the surface of the soil but have not decomposed.

loam The most fertile soil, a mixture especially rich in sand and silt with generous amounts of organic matter.

loess A homogenous, porous deposit of windblown glacial silt, typically unlayered, that forms vertical bluffs and cliffs. Fertile, agriculturally productive soils—usually mollisols—commonly form on loess.

loess An accumulation of windblown silt derived from glacial erosion.

loess Deposits of windblown glacial silt.

longitudinal dunes A long, symmetrical dune that forms as a result of two different wind directions with comparable magnitude.

longshore current A current generated when waves strike a shore at an angle, producing flow parallel and close to the coast. Some longshore currents are capable of transporting sand for hundreds of kilometers along the coastline.

lunar eclipse A phenomenon that occurs when Earth lies directly between the Sun and the Moon, causing Earth's shadow to fall on the Moon and darken it.

luster The quality and intensity of light reflected from the surface of a mineral.

magma Molten rock generated from melting of any rock in the subsurface; cools to form igneous rock.

magmatic processes Geologic processes that form ore deposits as liquid magma solidifies into igneous rock.

magnetic reversal A change in Earth's magnetic field in which the north magnetic pole becomes the south magnetic pole and vice versa; has occurred on average every 500,000 years over the past 65 million years.

magnetometer An instrument that measures the strength and, in some cases, the direction of a magnetic field.

main sequence A band running across a Hertzsprung–Russell diagram that contains most of the stars, which are fueled by hydrogen fusion.

manganese nodules A potato-shaped rock found on the ocean floor and rich in manganese and other metals precipitated from seawater through biomineralization.

mantle plume A relatively small rising column of mantle rock that is hotter than surrounding rock. As pressure decreases in a rising plume, the rock partially melts, forming magma.

mantle The rocky, mostly solid layer of Earth's geosphere lying beneath the crust and above the core. The mantle extends from the base of the crust to a depth of about 2,900 kilometers.

maria Dry, barren, flat expanses of volcanic rock on the Moon, first thought to be seas.

marine west coast climate A midlatitude climate characterized by relatively mild but wet winters and cool summers, with little temperature difference between seasons.

Mars The fourth planet from the Sun.

mass extinction A sudden, catastrophic event during which a significant percentage of all life-forms on Earth become extinct.

mass wasting The downslope movement of earth material, primarily caused by gravity. *(See also landslide.)*

meanders A series of twisting curves or loops in the course of a stream.

medial moraine A moraine formed in or on the middle of a glacier by the merging of lateral moraines as two glaciers flow together.

Mediterranean climate A midlatitude climate characterized by dry summers, rainy winters, and moderate temperatures.

Mercalli scale A scale of earthquake intensity that expresses the strength of an earthquake based on its destructive power and its effects on buildings and people; does not accurately measure the energy released by the quake.

Mercury The closest planet to the Sun.

mesa A flat-topped mountain, shaped like a table, that is smaller than a plateau and larger than a butte.

mesopause The ceiling of the mesosphere; the boundary between the mesosphere and the thermosphere.

mesosphere The layer of air that lies above the stratopause, extending upward from about 55 kilometers to about 80 kilometers above Earth's surface.

Mesozoic Era The part of geologic time roughly from 251 to 65 million years ago. Dinosaurs rose to prominence and became extinct during this era.

metamorphic grade The intensity of metamorphism that formed a rock; the maximum temperature and pressure attained during metamorphism.

metamorphic halo The zone surrounding an intrusive igneous body in which the country rock has been metamorphosed by heat and hydrothermal fluids from the cooling magma.

metamorphic rock A rock formed when igneous, sedimentary, or other metamorphic rocks recrystallize in response to elevated temperature, increased pressure, chemical change, and/or deformation.

metamorphism The process by which rocks change texture and mineral content in response to variations in temperature, pressure, chemical conditions, and/or deformation.

meteor A falling meteoroid that enters Earth's atmosphere and glows as it vaporizes; colloquially called a *shooting star*.

meteorite A meteor that does not completely vaporize and that strikes Earth's surface.

meteoroid A small interplanetary body, most often an asteroid or comet fragment, traveling in an irregular orbit through the inner Solar System.

microplastic debris Small pieces of plastic, typically a few millimeters or centimeters long, that result from accumulation and concentration of plastic garbage in parts of the ocean. The debris results from wave and wind energy that breaks plastic litter down into smaller and smaller pieces. The effects of microplastics in the ocean ecosystem are presently unknown.

microwave radar A satellite-based instrument that measures the travel time of microwave pulses that reflect off of the sea surface. The processed data allow for the detection of subtle swells and depressions on the sea surface, which are controlled by seafloor topography and can be used to map the seafloor.

Mid-Atlantic Ridge The portion of the Mid-Oceanic Ridge system that lies in the middle of the Atlantic Ocean, halfway between North America and South America to the west, and Europe and Africa to the east.

Mid-Oceanic Ridge system The undersea mountain chain that forms at the boundary between divergent tectonic plates within oceanic crust. It circles the planet like the seam on a baseball, forming Earth's longest mountain chain.

Milky Way The barred spiral galaxy in which our Sun and Solar System are located.

mineral reserves A term to describe the known supply of ore in the ground; can be used on a local, national, or global scale.

mineral resources Economically valuable geological materials including both metal ore and nonmetallic minerals.

mineral A naturally occurring inorganic solid with a definite chemical composition and a crystalline structure.

Mohorovičić discontinuity The boundary between the crust and the mantle, identified by a change in the velocity of seismic waves; also called the *Moho*.

Mohs hardness scale A scale, based on a series of 10 fairly common minerals and numbered 1 to 10 (from softest to hardest), used to measure and express the hardness of minerals.

mollisol Grassland soil characterized by rich A horizon, high cation exchange capacity, and B horizon rich in base cation salts; very fertile soil.

moment magnitude An earthquake scale in which the surface area of fault movement, the offset produced on the fault, and a measure of rock strength are multiplied together; moment magnitude scale closely reflects the total amount of energy released.

monsoon A seasonal wind and weather system caused by uneven heating and cooling of land and adjacent sea, generally blowing from the sea to the land in the summer when the continents are warmer than the ocean, and from land to sea in winter when the ocean is warmer than the land.

moraine A mound or ridge of till deposited directly by glacial ice.

mud cracks Irregular polygonal downward-tapering fractures that develop when mud dries; may be preserved when the mud is lithified.

mudflow A form of rapid mass wasting that involves the downslope movement, usually on unvegetated land, of fine-grained soil particles mixed with water; can be slow moving, as slow as 1 meter per year, or as fast as a speeding car.

mudstone A clastic sedimentary rock that consists of clay- and silt-sized particles.

native elements Minerals that consist of only one element and thus the element occurs in the native state (not chemically bonded to other elements). Only about 20 elements occur in the native state as solids.

natural gas A mixture of naturally occurring light hydrocarbons composed mainly of methane, CH_4, that is used for home heating and cooking and to fuel large electric generation plants.

natural selection The process by which the hereditary traits of individuals within a population change so as to provide a competitive advantage. Those individuals with the competitive advantage tend to survive (i.e., are naturally selected) and produce more offspring with the same key trait. Over many generations the entire population evolves to possess the trait.

neap tides The relatively small tides that occur when the Moon is 90 degrees out of alignment with the Sun and Earth.

nebula A cloud of interstellar gas and dust; plural *nebulae*.

Neptune The eighth planet from the Sun; similar to Uranus in size, composition, and atmosphere.

neutron star A small, extremely dense star, created from remnants of the supernova explosion of a large star and composed almost entirely of compressed neutrons.

new moon The lunar phase during which the Moon is dark when viewed from Earth, because it has moved to a position between Earth and the Sun and its sunlit side faces away from us.

nimbostratus Stratus clouds from which rain or snow falls.

nonbiodegradable pollutants Pollutants that do not decay naturally in a reasonable amount of time, including some industrial compounds, toxic inorganic compounds, and nontoxic sediment that muddies streams and habitats.

nonconformity A type of unconformity in which layered sedimentary rocks lie on an erosion surface cut into igneous or metamorphic rocks.

nonmetallic mineral resources Economically useful rocks or minerals that are not metals; examples include salt, building stone, sand, and gravel.

nonpoint source pollution Pollution that is generated over a broad area, such as fertilizers and pesticides spread over agricultural fields.

normal fault A fault in which the hanging wall has moved downward relative to the footwall.

normal lapse rate The vertical temperature structure of the atmosphere; in other words, the rate at which air that is neither rising nor falling cools with elevation.

normal magnetic polarity A magnetic orientation the same as that of Earth's current magnetic field.

nuclear fuels Radioactive isotopes, such as those of uranium, used to generate electricity in nuclear reactors.

nucleus The dense, solid core of a comet.

O horizon The uppermost layer of soil, named for its organic component; the combined O and A horizons are called *topsoil*.

objective lens The lens, farthest from the eyepiece in a refracting telescope, that collects light from a distant object.

occluded front A front that forms when a faster-moving cold air mass traps a warm air mass against a second mass of cold air. Precipitation occurs along both frontal boundaries, resulting in a large zone of inclement weather.

ocean acidification A decrease in pH of the world's oceans resulting from higher atmospheric carbon dioxide concentrations that, in turn, cause more carbon dioxide to dissolve in ocean water. Once dissolved, the carbon dioxide forms carbonic acid, decreasing the pH.

oceanic island A submarine mountain (seamount) that rises above sea level.

oceanic trench A long, narrow, steep-sided depression of the seafloor formed where a subducting oceanic plate sinks into the mantle, causing the seafloor to bend downward like a flexed diving board.

oligotrophic lake A deep lake characterized by nearly pure water but with low concentrations of nutrients, thus sustaining relatively few living organisms.

open Universe A model in which the gravitational force of the Universe is not sufficient to stop the current expansion and the galaxies will continue to fly apart forever.

Opportunity One of two mobile robots that landed on Mars in 2004.

ore minerals Minerals from which metals or other elements can be profitably recovered.

organic activity A mechanical weathering process in which a crack in a rock is expanded by tree or plant roots growing there.

orogen The belt of rock that is deformed in an orogeny.

orogeny The process of mountain building; all tectonic processes associated with mountain building.

orographic lifting Lifting of air that occurs when air flows over a mountain.

outgassing The release of volatiles from Earth's mantle and crust during volcanic eruptions at the surface.

outwash plain A broad, gently-sloping surface formed when outwash spreads onto a wide valley or plain beyond a glacier.

outwash streams A stream that emerges from below the snout of a glacier and carries glacial sediment further downslope.

outwash Sediment deposited by streams flowing from the terminus of a melting glacier.

oxbow lake A crescent-shaped lake created where a meander loop is cut off from a stream and the ends of the meander become plugged with sediment.

oxidation A chemical weathering process in which a mineral decomposes when it reacts with oxygen.

oxisol A soil formed in a hot, humid climate and characterized by intensive leaching of soluble cations from the A horizon, little ability to retain nutrients, and very poor fertility. Very insoluble iron and aluminum oxides are concentrated.

ozone hole An unusually low ozone concentration in the stratosphere that is centered roughly over Antarctica.

P waves Body waves that travel faster than other seismic waves and are the first or "primary" waves to reach an observer; formed by alternating compression and expansion of rock parallel to the direction of wave travel.

pahoehoe Lava with a smooth, billowy, or ropy surface.

Paleozoic Era The part of geologic time occurring during a period from 541 to 251 million years ago. During this era invertebrates, fishes, amphibians, reptiles, ferns, and conebearing trees were dominant.

Pangea The supercontinent that existed when all Earth's continents were joined together, about 300 million to 200 million years ago, first identified and named by Alfred Wegener.

parabolic dune A crescent-shaped dune with tips pointing into the wind; forms in moist semidesert regions and along seacoasts where sparse vegetation is present to anchor the tips of the dune.

parallax The apparent change in position of an object due to the change in position of the observer.

parent rock Any original rock before it is changed by weathering, metamorphism, or other geological processes.

parsec A distance used in astronomy, equal to about 3.26 light-years.

partial melting The process in which a silicate rock only partly melts as it is heated, forming magma that is more silica rich than the original rock.

particle or particulate In pollution terminology, any small piece of solid matter larger than a molecule, such as dust or soot.

passive continental margin A margin that occurs where continental and oceanic crust are firmly joined together and where little tectonic activity occurs. Not a plate boundary.

paternoster lakes A series of lakes in a glacial valley, strung out like beads and connected by short streams and waterfalls.

peat A loose, unconsolidated, brownish mass of partially decayed plant matter; a precursor to coal.

pediment A broad, gently sloping erosional surface that forms along the front of desert mountains uphill from a bajada, usually covered by a patchy veneer of gravel only a few meters thick.

pelagic sediment Muddy ocean sediment made up of the skeletons of tiny marine organisms.

penumbra A wide band outside of the umbra, where only a portion of the Sun is hidden from view during a solar eclipse.

percentage base saturation The proportion of a soil's cation exchange capacity that is saturated by basic cations K+, Na+, Ca++, or Mg++.

perched aquifer A local aquifer formed where a layer of impermeable rock or sediment exists above the regional water table and creates a locally saturated zone.

perennial streams A stream that maintains some flow even during dry seasons.

periods A geologic time unit longer than an epoch and shorter than an era.

permeability The ability of a solid material such as a rock to transmit water or another fluid through its pore network; depends on the size, shape, and interconnectedness of the pores within the material.

persistent bioaccumulative toxic chemicals (PBTs) Nonbiodegradable toxins that are released into and accumulate in the environment.

petroleum source rock The shale or other sedimentary rock from which oil or natural gas originates.

petroleum A complex liquid mixture of hydrocarbons, formed from decayed plant and animal matter, that can be extracted from sedimentary strata and refined to produce propane, gasoline, and other fuels. Also called *crude oil* or simply *oil*.

pH scale A logarithmic scale that measures the acidity of a solution. A pH of 7 is neutral; numbers lower than 7 represent acidic solutions, and numbers higher than 7 represent basic ones.

Phanerozoic Eon The most recent 541 million years of geologic time, including the present, represented by rocks that contain evident and abundant fossils.

photons A particle of light; the smallest particle or packet of electromagnetic energy.

photosphere The surface of the Sun visible from Earth; also called the solar atmosphere.

photosynthesis The process by which chlorophyll-bearing plant cells convert carbon dioxide and water to organic sugars, using sunlight as an energy source; oxygen is released in the process.

mechanical weathering *or* **physical weathering** The disintegration of rock into smaller pieces by physical processes without altering the chemical composition of the rock.

phytoplankton Plankton that conduct photosynthesis like land-based plants and that are the base of the food chain for marine animals.

pillow basalt Molten basaltic lava that solidified under water, forming spheroidal lumps of basalt that resemble a stack of pillows.

placer deposit A surface mineral deposit formed along stream beds, beneath waterfalls, or on beaches when water currents slow down and deposit high-density minerals.

planetary nebula A nebula created when a star about the size of our Sun explodes and blows a shell of gas out into space.

planets A celestial body that revolves in a fixed orbit around a star.

plankton Small marine organisms that live mostly within a few meters of the sea surface, where sunlight is available, and that conduct most of the photosynthesis and nutrient consumption in the ocean and form the base of the marine food web.

plastic deformation Deformation that occurs without fracture after a rock's elastic limit is reached, and it continues to deform, like putty, while still solid. The rock keeps its new shape and does not store the energy used to deform it; thus, earthquakes do not occur when rocks deform plastically.

plastic flow Movement of a glacier in which the ice flows as a viscous fluid.

plate boundary A fracture or boundary that separates two tectonic plates.

plate tectonics A theory of global tectonics stating that the lithosphere is segmented into several plates that move about relative to one another by floating on and sliding over the plastic asthenosphere. Seismic and tectonic activity occur mainly at the plate boundaries.

plateau A large, elevated area of relatively flat land.

playa lake A temporary desert lake that dries up during the dry season.

playa The dry desert lake bed of a playa lake.

Pleistocene Glaciation The most recent series of glaciations occurring during the Pleistocene Epoch and beginning around 2 million years ago. Most climate models indicate that Earth is still in the Pleistocene Glaciation.

plume of contamination The slow-growing three-dimensional zone within an aquifer and/or body of surface water affected by a dispersing pollutant.

Pluto Once considered to be the ninth planet from the Sun in our Solar System, reclassified in 2006 as a dwarf planet.

pluton A body of intrusive igneous rock.

pluvial lakes A lake formed in a topographic basin as the result of a moist climate during a glacial interval.

point bar A deposit of sediment in the slower water on the inside of a meander.

point source pollution Pollution that arises from a specific site such as a septic tank or a factory.

polar easterlies Persistent polar surface winds in the Northern Hemisphere that flow from east to west.

polar front The low-pressure boundary at about 60 degrees latitude formed by warm air rising at the convergence of the polar easterlies and prevailing westerlies.

polar jet stream A jet stream that flows along the polar front.

pollution The reduction of the quality of a resource by the introduction of impurities.

population I star A relatively young star formed from material ejected by an older, dying star; composed mainly of primordial hydrogen and helium, with a small percentage of heavier elements. Our Sun is a population I star.

population II star An old star, with lower concentration of heavy elements than a population I star.

pore space The empty space between particles of rock, sediment, or soil.

porosity The proportional volume of pores or open space within a material; indicates the maximum possible volume of fluid that could be held within the material.

Precambrian A term referring to all of geologic time before the Paleozoic Era, encompassing approximately the first 4 billion years, or roughly 90 percent, of Earth's history. Also refers to all rocks formed during that time.

precession The circling or wobbling of Earth's axis as the planet travels in its orbit, like that of a wobbling top.

precipitation A chemical reaction that produces a solid salt, called a *precipitate*, from a liquid solution.

precision The degree to which multiple measurements are close to each other, irrespective of their accuracy.

pressure gradient A measure of the change in air pressure over distance, used to determine wind speed.

pressure-release fracturing A mechanical weathering process in which tectonic forces lift deeply buried rocks upward and then erosion removes overlying rock and sediment—the net result of which is to remove the pressure from overlying material, causing the rock to expand and fracture.

pressure-release melting Melting caused by a decrease in pressure, expansion of rock volume, and melting. Usually occurs in the asthenosphere.

prevailing westerlies The winds that blow steadily toward the poles from the southwest in the Northern Hemisphere and from the northwest in the Southern Hemisphere, between 30 and 60 degrees north and south latitudes.

primordial nucleosynthesis The formation of helium nuclei that occurred by nuclear fusion during the first 8.5 minutes in the life of the Universe.

principle of crosscutting relationships The obvious principle that a rock or feature must first exist before anything can happen to it; thus, if a dike of basalt cuts across a layer of sandstone, the basalt is younger.

principle of faunal succession The principle that species succeeded one another through time in a definite order, so the relative ages of fossiliferous sedimentary rocks can be determined by their fossil content.

principle of included fragments The principle that a rock unit must first exist before pieces of it can be broken off and incorporated into another rock unit.

principle of original horizontality The principle that most sediment is deposited as nearly horizontal beds, and therefore most sedimentary rocks started out with nearly horizontal layering.

principle of superposition The principle that in undisturbed layers of sediment or sedimentary rock, the age becomes progressively younger from bottom to top; younger layers always accumulate on top of older layers.

prodelta The fine-grained, outermost edge of a delta, located offshore beyond the delta front.

prominence A red, flamelike jet of gas rising from the Sun's corona.

Proterozoic Eon The portion of geologic time occurring during a period from 2.5 billion to 541 million years ago.

pulsar A neutron star that emits a pulsating radio signal.

pump jack The above-ground portion of a reciprocating piston pump in an oil well.

pyroclastic flow An extremely destructive incandescent mixture of volcanic ash, larger pyroclastic particles, minor lava, and hot gas that forms from collapse of an eruptive column and flows rapidly along Earth's surface.

pyroclastic rock Rock made up of liquid magma and solid rock fragments that were ejected explosively from a volcanic vent.

quasar An object, less than 1 light-year in diameter and very distant from Earth, that emits an extremely large amount of energy as radio waves.

radiation fog Fog that occurs when Earth's surface and the air near the surface cool by radiation during the night, and water vapor in the air condenses because it cools below its dew point.

radiative zone An inner zone of a star surrounding the core where radiation energy is transmitted by absorption and emission.

radiometric dating The process of measuring the absolute age of rocks, minerals, and fossils by measuring the concentrations of radioactive isotopes and their decay products.

rain-shadow desert A desert formed on the downwind side of a mountain range.

recessional moraine A moraine that forms at the new terminus of a glacier as the glacier stabilizes temporarily during retreat.

recharge To replenish an aquifer by the addition of water.

red giant A stage in the life of a star when its core is composed of helium that is not undergoing fusion, but a hot shell of hydrogen around the core is fusing at a rapid rate, producing enough energy to cause the star to expand. Its core is hotter, but its expanded surface is cooler and emits red light.

red shift A shift toward longer (red) wavelengths, observed in the spectrum of a distant galaxy or other object that is moving away from Earth; caused by the Doppler effect.

reef A wave-resistant ridge or mound built by corals or other marine organisms.

reflecting telescopes A type of telescope that uses a mirror or mirrors to collect and focus an image, which is then reflected to the eyepiece by another mirror.

refracting telescope A type of telescope that uses two lenses: one to collect light from a distant object and another to magnify the image.

refraction The bending of a wave that occurs when it approaches the shore at an angle; the end of the wave in shallow water slows down, while the end in deeper water continues at a faster speed.

regolith The thin layer of loose, unconsolidated, weathered material that overlies bedrock. Some earth scientists and engineers use the terms *regolith* and *soil* interchangeably; soil scientists identify soil as only the upper layers of regolith.

relative age An approach to determining the timing of geologic events based on the order in which they occurred, rather than on the absolute number of years ago in which they occurred.

relative humidity The ratio of the amount of water vapor in a given volume of air divided by the maximum amount of water vapor that can be held by that air at a given temperature, expressed as a percentage.

remediation The treatment of a contaminated area, such as an aquifer, to remove or decompose a pollutant.

residual ore deposits A mineral deposit formed from relatively insoluble ions left in the soil near Earth's surface after most of the soluble ions were dissolved and removed by abundant water.

residual soils A soil formed from the weathering of bedrock below.

resolution The capability to which an instrument can detect or measure detail, such as the capability of an optical telescope to define individual craters on the Moon.

retrograde motion When viewed from Earth, the apparent motions of the planets in which they temporarily move backward (westward) with respect to the stars, before resuming their original eastward motion.

reverse fault A fault in which the hanging wall has moved up relative to the footwall.

reversed magnetic polarity Magnetic orientations in rock that are opposite to the current orientation of Earth's magnetic field.

revolution Orbiting around a central point. A satellite revolves around Earth, and Earth revolves around the Sun.

reworking The process by which sediment is deposited, then re-eroded and transported further.

Richter scale A scale of earthquake magnitude that expresses the amount of energy released; calculated from the amplitude of the largest body wave on a standardized seismograph, although not a precise measure of earthquake energy.

rift valley An elongate depression that develops at a divergent plate boundary. Examples include continental rift valleys and the rift valley along the center of the Mid-Oceanic Ridge system.

Ripple marks Small, semiparallel ridges and troughs formed mostly in sand by wind, water currents, or waves; often preserved when the sediment is lithified.

rock cycle The sequence of events in which rocks are formed, destroyed, altered, and reformed by geological processes.

rock dredge An open-mouthed steel net dragged along the seafloor behind a research ship for the purpose of sampling rocks and sediment from the ocean floor.

rock record The rocks that currently exist on Earth and contain the record of its history.

rock-forming minerals The nine minerals or mineral groups that are most abundant in the Earth's crust and that combine to make most rocks. They are olivine, pyroxene, amphibole, mica, the clay minerals, quartz, feldspar, calcite, and dolomite.

rockslide A subcategory of slide mass wasting in which a segment of bedrock slides downslope along a fracture and the rock breaks into fragments and tumbles down the hillside; also called a *rock avalanche*.

rotation Turning or spinning on an axis. Tops and planets rotate on their axes.

runoff Surface water that flows to the oceans in streams and rivers.

S waves Seismic body waves that travel slower than P waves and are the "secondary" waves to reach an observer; sometimes called *shear waves* due to the shearing motion in rock caused as the waves vibrate perpendicular to the direction they travel.

salinity The total quantity of dissolved salts in seawater, expressed as a percentage.

salinization A process whereby salts accumulate in soil that is irrigated heavily, lowering soil fertility.

salt cracking A chemical weathering process in which salts that are dissolved in water in the pores of rock crystallize, exerting an outward pressure on pore walls and pushing the mineral grains apart.

saltation carpet When a very strong wind blows across dry sand, so many grains saltate at once that a sand-air fluid is formed, usually a few centimeters or decimeters high. This sand-air fluid is dominated by grain-to-grain collisions and is unaffected by turbulence from the wind.

saltation The asymmetric jumping movement of sedimentary particles that are ejected off the bed through impact by another particle and are carried downstream by wind or water for some distance before falling back to the bed surface. Most sand grains in desert environments move via saltation by wind.

saltwater intrusion A condition along the coasts of oceans and inland salt water bodies in which excessive pumping of fresh groundwater causes salty groundwater to invade an aquifer.

San Andreas Fault zone A zone of strike-slip faults extending from Cape Mendicino in northern California to the northern Gulf of California in Mexico; fault zone forms the transform boundary between the Pacific Plate and the North American Plate and is the source of many earthquakes.

saturation The maximum amount of water vapor that air can hold.

Saturn The second-largest planet and sixth from the Sun; marked by its distinctive rings.

scarp A break in the land surface caused by an earthquake.

scavenging The process by which hydrothermal fluids sweep through large volumes of country rock and dissolve low concentrations of metals, concentrating them elsewhere as an ore deposit.

sea arch An arch created when a short cave is eroded all the way through a narrow headland.

sea stack A pillar of rock left when a sea arch collapses or when the inshore portion of a headland erodes faster than the tip.

seafloor drilling A process in which drill rigs mounted on offshore platforms or on research vessels cut cylindrical cores from both the sediment and rock of the seafloor. Following their extraction, the cores are brought to the surface for study.

seafloor spreading The hypothesis that segments of oceanic crust are separating at the Mid-Oceanic Ridge.

seamount A submarine mountain, usually of volcanic origin, that rises 1 kilometer or more above the surrounding seafloor.

Second Great Oxidation Event The process, occurring about 600 million years ago, when the oxygen concentration in Earth's atmosphere increased abruptly a second time in response to interactions among chemical, physical, and biological processes.

secondary and tertiary recovery techniques Methods of extracting oil or natural gas by artificially augmenting the reservoir energy or fluid composition, as by injection of water, pressurized gas, solvents, or other fluids.

second-generation energy Radiation emitted by stars, galaxies, quasars, and other forms of concentrated matter.

sediment gravity flows Underwater mixtures of sediment and water that flow downslope.

sediment Solid rock or mineral fragments that are transported and deposited by wind, water, gravity, or ice; that are weathered by natural forces, precipitated by chemical reactions, or secreted by organisms; and that accumulate in loose, unconsolidated layers.

Sedimentary rock Rock formed when sediment becomes compacted and cemented through the process of lithification.

sedimentary structures Any feature of sedimentary rock formed by physical processes during or shortly after deposition; examples include stratification, cross-bedding, ripple marks, and tool marks.

sediment-gravity flow Sediment gravity flows occurs when gravity acts on the sediment particles and moves the fluid; this is in contrast to rivers where the fluid moves the particles.

seismic reflection profiler A device that emits a high-energy timed seismic signal that penetrates into and reflects from layers of sediment and rock beneath the seafloor; the data are used to construct an image of rock and sediment layers along with other geologic structures below the seafloor.

seismic waves An elastic wave that travels through rock, produced by an earthquake or explosion.

seismogram The physical or digital record of earthquake waves as measured by a seismograph.

seismograph An instrument that records seismic waves.

seismology The study of earthquakes and the nature of Earth's interior based on evidence from seismic waves.

sequence stratigraphy The study of sedimentary rock relationships using a chronostratigraphic framework in which genetically-related stratigraphic units that are separated by surfaces of erosion or non-deposition are recognized based on their lithology, fossil content, sedimentary structures, stratal geometry and other physical properties.

sequence A succession of layered sediments or sedimentary rocks that are genetically related and deposited during one full cycle of change in available accommodation space, such as one full cycle of sea level change.

shield volcano A large, gently sloping volcanic mountain formed by successive flows of basaltic magma.

shoreface The seaward-sloping seafloor surface extending from the mean low tide line to the mean fair-weather wave base. Includes that part of a beach in which incoming waves reach the bottom, steepen, and break, releasing much mechanical energy.

silicate tetrahedron The fundamental building block of all silicate minerals, a pyramid-shaped structure consisting of one silicon atom bonded to four oxygen atoms to form the anionic group silicate, $[SiO_4]^{4-}$.

silicates Minerals whose crystal structure contains silicate tetrahedra.

sill A sheetlike igneous rock, parallel to the grain or layering of country rock, that forms when magma is injected between layers.

sinkhole A depression on Earth's surface caused by the collapse of a cavern roof or by the dissolution of surface rocks, usually limestone.

slaty cleavage A metamorphic foliation producing a parallel fracture pattern that cuts across the original sedimentary bedding.

sleet Small spheres of ice that develop when raindrops form in a warm cloud and freeze as they fall through a layer of cold air at lower elevation.

slide Form of rapid mass wasting in which the rock or soil initially moves as a consolidated unit along a fracture surface.

slip face The steep leeward side of a dune, typically at the angle of repose for loose sand, so that the sand flows or slips down the face, where it is deposited.

slip The distance that rocks on opposite sides of a fault have moved relative to each other.

slump A type of slide in which blocks of material slide downslope as a consolidated unit over an upward-concave, curved fracture in rock or regolith; trees on the slumping blocks tilt uphill. The uphill portion of the slump usually consists of several tilted slide blocks, whereas the toe of the slump usually consists of rumpled, folded sediment.

smog Visible, brownish air pollution formed through chemical reactions that involve incompletely combusted automobile gasoline, nitrogen dioxide, oxygen, ozone, and sunlight.

soil horizon A layer of soil that is distinguishable from other layers because of differences in appearance and in physical and chemical properties.

soil order The highest hierarchical classification of soils by the National Resource Conservation Service. Twelve soil orders are recognized.

soil series The lowest hierarchical classification of soils by the National Resource Conservation Service. Over 20,000 soil series are recognized, including 50 designated "state soils."

soils The upper layers of regolith that support plant growth. Some earth scientists and engineers use the terms *soil* and *regolith* interchangeably.

solar cell A device that produces electricity directly from sunlight; also sometimes called a *photovoltaic (PV) cell*.

solar eclipse A phenomenon that occurs when the Moon passes directly between Earth and the Sun; the Moon casts its shadow on Earth, thus blocking the Sun's light.

solar mass A standard unit for expressing the mass of a star relative to the mass of our Sun; 1 solar mass equals the mass of the Sun.

solar wind A stream of electrons and positive ions radiating outward from the Sun at high speed.

solstice Either of two times per year when the Sun is furthest from the equator and shines directly overhead at either 23.5 degrees north latitude (on or about June 21) or 23.5 degrees south latitude (on or about December 22). The solstice on or about June 21 marks the longest day of the year in the Northern Hemisphere and the shortest day in the Southern Hemisphere. The solstice on or about December 22 marks the longest day in the Southern Hemisphere and the shortest day in the Northern Hemisphere.

specific gravity The weight of a substance relative to the weight of an equal volume of water.

specific heat The amount of energy required to raise the temperature of 1 gram of a substance by 1°C.

spectrum An ordered array of colors; a pattern of wavelengths into which a beam of light or other electromagnetic radiation is separated.

spheroidal weathering The combined mechanical and chemical weathering of fractured crystalline bedrock into spheroidally shaped boulders; caused by the faster weathering rate of sharp bedrock corners (where at least three faces of rock can be attacked by weathering), over edges and the faster weathering of edges over faces.

spicules A jet of gas at the edge of the Sun, shooting upward from the chromosphere.

spiral galaxies A galaxy characterized by arms that radiate out from the center like a pinwheel.

Spirit One of two mobile robots that landed on Mars in 2004.

spit A small ridge of sand or gravel extending from a beach into a body of water.

spodosol Sandy, acidic soil developed in moist, temperate environments, commonly in coniferous or mixed coniferousdeciduous forests. Leaching has translocated base cations downward, resulting in a well-developed E horizon.

spring tides The very high and very low tides that occur when the Sun, Moon, and Earth are aligned and their gravitational fields combine to create a strong tidal bulge.

spring A seep or flow of groundwater onto the surface; commonly occurs where the water table intersects the land surface.

stable air A parcel of warm, dry air that does not rise rapidly, does not ascend to high elevations, and does not lead to cloud formation and precipitation.

stagnant ice Glacial ice that has broken away from the front of a terrestrial glacier and is no longer connected to the glacier's ice-delivery system.

stalactite An icicle-like dripstone of precipitated calcite that hangs from the ceiling of a cavern.

stalagmite A cone-shaped deposit of calcite formed over time as drops of water fall to the same spot on the floor of a cavern, release carbon dioxide gas upon impact, and precipitate the mineral as a result of the increase in pH.

stationary front A front at the boundary between two stationary air masses of different temperatures.

steppes A vast, semiarid, grass-covered plain found in dry climates such as those in southeast Europe, central Asia, and parts of central North America.

stock A pluton exposed over less than 100 square kilometers of Earth's surface; similar to a batholith, but smaller.

stony meteorite A meteorite with a mass ratio of rock to metal similar to the mass ratio of Earth's mantle to Earth's core, thus reflecting the primordial composition of the Solar System and representing a window into its past. Most meteorites are stony meteorites.

storm surge Abnormally high coastal waters and flooding, created by a combination of strong onshore winds and the low atmospheric pressure of a storm that raises the sea surface by several meters.

storm tracks Paths repeatedly followed by storms.

stratified drift Glacial drift that was first carried by a glacier and then transported and deposited in layers by a stream.

stratocumulus Low, sheetlike clouds with some vertical structure.

stratopause The ceiling of the stratosphere; the boundary between the stratosphere and the mesosphere.

stratosphere The layer of air above the tropopause, extending upward to about 55 kilometers.

stratovolcano A steep-sided volcano formed by an alternating series of lava flows and pyroclastic deposits and marked by repeated eruption.

stratus Horizontally layered clouds that spread out into a broad sheet, usually creating dark, overcast skies.

streak The color of the fine powder of a mineral, usually obtained by rubbing the mineral on an unglazed porcelain streak plate.

stream A moving body of water confined in a channel and flowing downslope; a river is a large stream fed by smaller ones.

stress The force exerted against an object, usually measured as force per unit area or pressure.

strike-slip fault A vertical fault across which rocks on opposite sides move horizontally.

subarctic The northernmost edge of the humid continental climate zone, lying just south of the arctic.

subduction complex The structurally complicated mass of rock consisting of deformed seafloor sediment and fragments of basaltic oceanic crust and upper mantle material that is scraped from the upper layers of the subducting slab in a subduction zone and added to the overriding plate.

subduction zone A long, narrow region at a convergent boundary where a lithospheric plate is sinking into the mantle during subduction; also referred to as *subduction boundary*.

subduction The process in which two lithospheric plates of different densities converge and the denser one sinks into the mantle beneath the other.

submarine canyons A deep, V-shaped, steep-walled trough eroded into a continental slope and, in some cases, outer shelf. Funnels sediment from the continental shelf across the slope to the continental rise.

submarine fan A large, fan-shaped accumulation of sediment deposited on the deep seafloor, usually within and beyond the mouth of submarine canyons.

submarine hydrothermal ore deposits Ore deposits that form when hot seawater dissolves metals from seafloor rocks and then, as it rises through the upper layers of oceanic crust, cools and precipitates the metals.

submergent coastline A coastline that was previously above sea level but has been drowned, because either the land has sunk or sea level has risen.

subsidence The irreversible sinking or settling of Earth's surface.

subtropical jet stream A jet stream that flows between the trade winds and the westerlies.

sulfides Minerals containing the anion sulfide, S^{1-}. An example is pyrite (FeS_2).

Sumatra-Andaman earthquake An earthquake of moment-magnitude 9.2, the third largest ever recorded; occurred on December 26, 2004, off the Sumatran coast and produced a devastating tsunami that contributed to the deaths of an estimated 286,000 people.

sunspots A comparatively cool, dark region on the Sun's surface caused by a magnetic disturbance.

supercontinent A continent, such as Alfred Wegener's Pangea, consisting of all or most of Earth's continental crust joined together to form a single, large landmass. At least three supercontinents are thought to have existed during the past 2 billion years, and each broke apart after a few hundred million years.

supercooling A condition in which water droplets in air do not freeze even when the air cools below the freezing point.

superheated water Water that has a temperature above the boiling point but that does not boil because it is under pressure.

super-heating Heating of a substance above a phase-transition (gas to liquid or liquid to solid) without the transition occurring. For example, high pressure can keep solid rock from melting even though it is above its melting temperature.

supernova An exploding star that is releasing massive amounts of energy.

supersaturation A condition in which the relative humidity of the air exceeds 100 percent.

surf The chaotic, turbulent waves breaking along the shore.

surface currents Horizontal flow of water in the upper 400 meters of the oceans, caused by wind blowing over the sea surface.

surface mine A hole excavated into Earth's surface for the purpose of recovering mineral or fuel resources.

surface waves Seismic waves that radiate from the earthquake's epicenter and travel along the surface of Earth or along a boundary between layers within Earth.

suspended load The total mass of a stream's sediment load that is carried within the flow by turbulence and is free from contact with the streambed.

syncline A fold in rock that arches downward, and whose center contains the youngest rocks.

system Any combination of interacting components that form a complex whole.

taiga A biome of conifers that can survive the extreme winters of the subarctic.

tail The long trailing portion of a comet, always pointing away from the Sun, formed when solar winds blow away lighter particles from the comet's head. What appears to be a fiery arrow is actually reflected light from the Sun.

talus An accumulation of loose, angular rocks at the base of a cliff, created as rocks broke off the cliff as a result of frost wedging.

tar sands Sand deposits saturated with heavy oil and an oil-like substance called bitumen.

tarn A small lake at the base of a cirque.

tectonic accretion Tectonic accretion is the process of material such as island arcs and sediment being added to tectonic plates at subduction zones.

tectonic plates The segments of Earth's outermost, cool, rigid shell, comprising the lithosphere. Tectonic plates float on the weak, plastic rock of the asthenosphere beneath.

temperate rainforests A rainforest that grows in marine west coast climates where rainfall is more than 100 centimeters per year and is relatively constant throughout the year, particularly along the northwest coast of North America from Oregon to Alaska.

temperature A measure of heat in a substance, proportional to the average speed of atoms and molecules in a sample.

terminal moraine An end moraine that forms when a glacier is at its greatest advance before beginning to retreat.

terminus The end, or foot, of a glacier.

terrestrial planets The four Earth-like planets closest to the Sun—Mercury, Venus, Earth, and Mars—which are composed primarily of nonvolatile metals and silicate rocks.

terrigenous sediment Sediment composed of sand, silt, and clay eroded from the continents.

texture The size, shape, and arrangement of mineral grains, or crystals, in a rock.

thermal expansion and contraction A mechanical weathering process that fractures rock when temperature changes rapidly, causing the surface of the rock to heat or cool faster, and thereby to expand or contract faster, than the rock's interior.

thermocline The boundary between the upper warm layers and deeper cool layers of water in a lake.

thermohaline circulation The force behind deep-sea currents, created by differences in water temperature and salinity and therefore differences in water density.

thermosphere An extremely high and diffuse region of the atmosphere lying above the mesosphere, from about 80 kilometers upward.

three-cell model A model of global wind patterns that depicts three convection cells in each hemisphere, bordered by alternating bands of high and low pressure.

threshold effect A reaction whereby the environment initially changes slowly (or not at all) in response to a small perturbation, but after the threshold is crossed, an additional small perturbation causes rapid change.

thrust fault A type of reverse fault that is nearly horizontal, with a dip of 45 degrees or less over most of its extent.

tidal bulge A bulge that forms on the surface of Earth's oceans facing directly toward and away from the moon. The bulge facing the moon forms due to gravitational attraction of the ocean water by the moon, whereas the bulge facing away from the moon forms to compensate for the displacement of mass in the opposite bulge.

tidal current A current, channeled by a bay with a narrow entrance or by closely spaced islands, caused by the rise and fall of the tides.

tidal range The vertical distance between low and high tide at a particular point on Earth's oceans.

tides The cyclic rise and fall of ocean water caused by the gravitational force of the Moon and, to a lesser extent, of the Sun.

till Glacial drift that was deposited directly by glacial ice.

tillite Till that was deposited by glaciers so long ago that it became lithified into solid rock.

tilt The angle of Earth's axis with respect to a line perpendicular to the plane of its orbit around the Sun. Earth's axis is tilted by about 23.5 degrees.

Titan Saturn's largest moon; the only moon in the Solar System with an appreciable atmosphere.

Tōhoku earthquake An earthquake of moment-magnitude 9.0, the fourth largest ever recorded, that occurred on March 11, 2011, off the Japanese island of Honshu; produced a gigantic tsunami that contributed to the deaths of about 17,600 people and led directly to the third-worst nuclear accident in history.

topsoil The fertile, dark-colored surface soil; the combined O and A soil horizons.

tornado A small, intense, short-lived, funnel-shaped storm that protrudes from the base of a cumulonimbus cloud.

trade winds The winds that blow steadily toward the equator from the northeast in the Northern Hemisphere and from the southeast in the Southern Hemisphere, between 5 and 30 degrees north and south latitudes.

transform boundary A plate boundary where two tectonic plates slide horizontally past one another.

transform faults A strike-slip fault between two offset segments of a mid-ocean ridge or along a strike-slip plate boundary.

transformation The change of soil constituents from one form to another, such as the hydrolysis of feldspar to clay.

translocation The vertical, usually downward, movement of physical or chemical soil constituents from one horizon to another.

transpiration Direct evaporation of water into the atmosphere from the leaf surfaces of plants.

transported soil A soil formed by the weathering of regolith that is transported from somewhere else and deposited.

transverse dunes A relatively long, straight dune with a gently sloping windward side and a steep lee face that is perpendicular to the prevailing wind; forms where sand is plentiful and evenly dispersed.

travel-time curve A graph in which first arrival times of P and S earthquake waves are plotted against distance from epicenter; the separation distance between the resulting two curves determines the distance between the seismograph location and the epicenter.

tributaries A stream that feeds water into another stream or river.

Triton The largest of Neptune's moons; marked by craters filled with ice or frozen methane.

tropic of Cancer The latitude 23.5 degrees north of the equator. On or about June 21, the summer solstice in the Northern Hemisphere, sunlight strikes Earth from directly overhead at noon at this latitude.

tropic of Capricorn The latitude 23.5 degrees south of the equator. On or about December 22, the summer solstice in the Southern Hemisphere, sunlight strikes Earth from directly overhead at noon at this latitude.

tropical cyclone or tropical storm A broad, circular storm with intense low pressure that forms over warm oceans.

tropical monsoon climate A tropical climate with distinct wet and dry seasons, characterized by high rainfall during the wet months and a relatively short dry season.

tropical rainforests A forest growing in a tropical climate with abundant, year-round rainfall; characterized by tall trees that branch only near the top, creating a dense canopy of leaves that block out most of the light, and with soggy ground and water dripping everywhere.

tropical savanna climate A tropical climate with distinct wet and dry seasons but with relatively low annual rainfall, characterized by large areas of grassland supporting herds of grazing animals.

tropopause The top of the troposphere; the boundary between the troposphere and the stratosphere.

troposphere The layer of air that lies closest to Earth's surface and extends upward to about 17 kilometers.

trough The lowest part of a wave.

tsunami A large, destructive sea wave produced by an undersea earthquake or volcano; sometimes erroneously called a *tidal wave*, although it has nothing to do with astronomical tides.

tundra A biome dominated by low-lying grasses, mosses, flowers, and a few small bushes that grow in the Arctic and high alpine regions.

turbidity currents A highly turbulent mixture of sediment and water that flows rapidly downslope in a subaqueous setting. Capable of causing substantial subaqueous erosion.

turnover A process, occurring in fall and spring in temperate climates, in which a lake's surface water changes temperature in response to seasonal weather changes and convection mixes the water to equalize temperature throughout the lake.

typhoon A tropical storm occurring in the western Pacific Ocean whose wind exceeds 120 kilometers per hour; called a hurricane in North America and the Caribbean and a cyclone in the Indian Ocean.

ultisol Strongly weathered soil formed in semihumid or humid environments. Intense weathering has removed most base cations, resulting in low fertility. Includes red clay soils of SE United States.

umbra The narrow band of the Moon's shadow where the Sun is completely blocked out during a solar eclipse.

unconformity A physical surface between layers of sediment or sedimentary rock that represents an interruption in the deposition of sediment, including an interruption that involves partial erosion of the older layer. Also the physical surface between eroded igneous and overlying sedimentary layers deposited following erosion. Types of unconformities include disconformity, angular unconformity, and nonconformity.

unconventional reservoirs A sedimentary rock that is capable of producing oil with the application of special techniques, such as hydraulic fracturing.

underground mine A mine consisting of subterranean passages that commonly follow ore veins or coal seams.

underthrusting The process by which one continent subducts beneath the other during a continent–continent collision.

uniformitarianism A principle stating that the geologic processes operating today also operated in the past.

unstable air A parcel of warm, moist air that rises rapidly, ascends to high elevations, and leads to formation of towering clouds and heavy rainfall.

upslope fog Fog that forms when air cools as it rises along a land surface.

upwelling A rising ocean current that transports cold water and nutrients from the depths to the surface.

Uranus The seventh planet from the Sun; similar to Neptune in size, composition, and atmosphere.

urban heat island effect Warmer temperatures in a city compared to the surrounding countryside, caused by factors in the urban environment itself.

U-shaped valley A glacially eroded valley with a broad, characteristic U-shaped cross section.

vent An opening in a volcano, typically in the crater, through which lava and rock fragments erupt.

Venus The second planet from the Sun; resembles Earth in size and density.

vertisol Clay-rich, dark-colored soil characterized by periodic desiccation and deep cracking. Nearly impermeable and sticky when wet.

vesicles Holes in lava rock that formed when the lava solidified before bubbles of gas or water could escape.

volcanic ash The smallest pyroclastic particles, less than 2 millimeters in diameter.

volcano A hill or mountain formed from lava and rock fragments ejected from a volcanic vent.

waning Becoming smaller; the 14 to 15 days after a full moon and before a new moon when the visible portion of the Moon decreases every day.

warm front A front that forms when moving warm air collides with a stationary or slower-moving cold air mass. The moving warm air rises over the denser cold air, cools adiabatically, and the cooling generates clouds and precipitation.

wash A streambed that is dry for most of the year.

water scarcity A situation in which the aggregate demand for water by all users, including the environment, exceeds the available supply.

water table The top surface of the zone of saturation; the water table separates this zone from the zone of aeration above.

wave base The depth below which surface wave energy does not reach. Roughly as deep as half a single wavelength and limited by wind-driven surface waves to about 200 meters on the outer continental shelf.

wave height The vertical distance from the crest to the trough of a wave.

wave orbitals The circular motion of water that occurs as a surface wave passes by. Wave orbitals decrease in diameter with depth, ultimately disappearing altogether at wave base.

wave-cut cliff A cliff created when waves erode the headland of a rocky coastline into a steep profile.

wave-cut platform The horizontal or gently sloping platform left when a wave-cut cliff is eroded back.

wavelength The distance between successive wave crests (or troughs).

waxing Becoming full; the 14 to 15 days after a new moon and before a full moon when the visible portion of the Moon increases every day.

weather The state of the atmosphere at a given place and time as characterized by temperature, wind, cloudiness, humidity, and precipitation.

weathering The decomposition and disintegration of rocks and minerals at Earth's surface by chemical and physical processes.

welded tuff An ash-flow tuff that compacts from the weight of overlying tuff deposits and fuses together because of the residual heat from the pyroclastic flow.

well A hole dug or drilled into Earth, generally for the production of water, petroleum, natural gas, brine, or sulfur, or for exploration.

wet adiabatic lapse rate The rate at which rising moist air cools adiabatically after it has reached its dew point and condensation has begun—varies depending on moisture content from 5°C to 9°C for every 1,000 meters it rises.

wetlands Regions that are water soaked or flooded for all or part of the year; includes *swamps, bogs, marshes, sloughs, mudflats,* and *floodplains.*

white dwarf A stage in the life of a star when fusion has halted and the star glows solely from the residual heat produced during past eras. White dwarfs are very small stars.

wind Horizontal airflow caused by pressure differences resulting from unequal heating of Earth's atmosphere. Winds near Earth's surface always flow from a region of high pressure toward a low-pressure region.

withdrawal Any process that uses water and then returns it to Earth locally.

zone of ablation The lower-altitude part of an alpine glacier where more snow melts in summer than accumulates in winter, and where the melting snow leaves behind a surface of old, hard glacial ice.

zone of accumulation The higher-altitude part of an alpine glacier where snow accumulates from year to year and forms glacial ice.

zone of aeration A subsurface zone located between the ground surface and the water table and in which the pores are mostly filled with air; also called the *unsaturated zone.*

zone of saturation A subsurface zone below the water table in which all porosity within soil and bedrock is filled with water.

zooplankton Tiny marine animals that live mostly within a few meters of the sea surface and feed on phytoplankton.

INDEX